Jatropha, Challenges for a New Energy Crop

Nicolas Carels • Mulpuri Sujatha • Bir Bahadur
Editors

Jatropha, Challenges for a New Energy Crop

Volume 1: Farming, Economics and Biofuel

Springer

Editors
Nicolas Carels
Functional Genomics and Bioinformatics
Fiocruz/IOC, Manguinhos
Rio de Janeiro, Brazil

Mulpuri Sujatha
Principal Scientist
Crop Improvement Section
Directorate of Oilseeds Research
Hyderabad, India

Bir Bahadur
Kakatiya University
Warangal, India

ISBN 978-1-4614-4805-1 ISBN 978-1-4614-4806-8 (eBook)
DOI 10.1007/978-1-4614-4806-8
Springer New York Heidelberg Dordrecht London

Library of Congress Control Number: 2012950584

Printed on acid-free paper

Springer is part of Springer Science+Business Media (www.springer.com)

Foreword

In our world, where energy demand is increasing and access to fossil fuels becomes uncertain and expensive, a growing interest in renewable energy is developing. In particular, financial efforts and scientific investments have been made to find substitutes to replace gasoline in transportation. Plant sugars, oils, and, more generally, plant biomass are proposed as possible options to produce liquid bioenergy.

In this context, the 1980s saw the extensive cultivation of crops and wild species with high carbohydrate or fat content for biofuels. With the food price increase of 2007, partly attributed to the cultivation of crop plants for biofuels, renewed interest has been paid to exploit nonedible plants for their biomass, high sugar, or oil content.

Among a dozen of wild tropical plants, *Jatropha curcas* L. was designated as the best candidate, perhaps because of its pantropical distribution (a plant known to have more than 30 common names in Asia, Africa, and America), and certainly because of its already intense culture in Asia (India, China, Thailand, etc.) and Africa (Mali, Burkina Faso, Senegal, etc.). Thanks to its seed oil easily converted to liquid biofuel, which meets the American and European standards; its press cake used as fertilizer; and its ability to grow in extreme conditions (on marginal soils without irrigation), *Jatropha* was dubbed "the green gold of the desert" and taken by a lot of investors or project developers to tackle the challenges of energy supply and greenhouse gas (GHG) emission reduction. Such enthusiasm quickly replaced by negative feedback goes against all the advantages previously put forward for growing *Jatropha*.

Less than a century ago, nobody would have thought that this plant would be the source of so much interest and controversy. Originated from America, this multipurpose plant was introduced in the seventeenth century via the Cape Verde in African and Asian continents where it is generally grown as a living fence. Described by Linneus, its scientific name indicates a pharmacological potential (jatros = doctor, trophe = food) that is confirmed by the numerous traditional medicinal uses. The first intensive cultivation has been done in the early twentieth century on the Cape Verde islands, but it was intended to produce, from seed oil, soap and candles for Southern Europe. Its potential as a biofuel source was revealed later, triggering the cultivation in many tropical countries of this undomesticated plant. However, the returns from these experiences have never been collected.

The main objective of this work is to provide the reader with the most recent scientific studies on *Jatropha curcas*. This book gives an update on the economic importance of *Jatropha* in the parts of the world where its culture is well developed, especially in India, China, Brazil, and the Sahel region of Africa. In response to many questions, the second chapter is devoted to the physiology of the plant and presents its ability to grow and produce under extreme conditions of culture or in the presence of microbial inoculants. Finally, the last part presents all the advantages provided by the cultivation of *Jatropha*, from its by-product production, its use for rehabilitation of degraded lands to its potential as an energy crop.

Montpellier, France Claudine Campa

Preface

Jatropha, Challenges for a New Energy Crop—Volume 1 aims to report on the state of the art of scientific investigations that were made during the past 10 years on the new crop *Jatropha curcas*. The progresses obtained on the knowledge of this abstemious, semi-wild species are already impressive and were mainly achieved in just a decade (2001–2011). This knowledge extends from basic Jatropha physiology and biological reproduction to the basic agronomic practices and systems for its productive management, but also the complete set of biotechnological tools, such as in vitro culture, genetic transformation, genomic sequence, genetic map, and markers assisted selection that is necessary for Jatropha's selective breeding. These scientific and technological achievements pave the way for future technological management and domestication of Jatropha as an industrial oilseed crop able to contribute to the feeding of the transport system.

This first volume of a two-book series that forms the first comprehensive compilation by global experts appeared necessary to us in view of the importance that Jatropha demonstrated worldwide by its large-scale cultivation and emerging value for energy business as a biofuel. This reality contrasts with the difficult access to objective information scattered among science media eventually written in different languages. We thought it was necessary to gather the information scattered worldwide in a sort of summary or general agreement of what is known on Jatropha at the moment. The form of a compilation was also necessary because the knowledge on Jatropha is shared over the tropical belt by different teams, in different politico-economic realities, and with different technological and scientific backgrounds. A compilation was the best way to faithfully transmit the point of view of these experts with as few biases as possible. We believe and hope that this compilation will be a valuable source of inspiration for next-generation scientists investigating this new crop, for technologists invested in improving its profitability, as well as for decision makers and policy implementers involved in politics, economics, environment, or social management that are thinking and acting for the development of a world based on sustainability.

The book is presented in five units comprising 30 chapters covering the main aspects of the worldwide economic importance of Jatropha as well as its physiology, farming, oil processing, by-products, biodiesel, and biofuel combustion. It aims to give a kind of comprehensive picture on the whole *productive chain* of

Jatropha and is supposed to help the reader to make a mental representation concerning the potential of Jatropha as a crop. By contrast, volume two is dedicated to Jatropha as a *biological system* with the purpose of understanding what can be improved in Jatropha and how this can be achieved.

We wish to express our gratitude to all the contributors from all over the world for readily accepting our invitations for not only sharing their knowledge but for admirably integrating their expertise on scattered information from diverse fields in composing the chapters and enduring editorial suggestions to finally produce this venture that we hope to be a success. We greatly appreciate their commitment.

We also acknowledge the support received from many colleagues in the preparation of the manuscripts as well as to our family members and relatives for bearing with us, our commitment to the book.

We thank Ms Hannah Smith, Associate Editor at Springer Science Publishers, and her team for their help and excellent cooperation for bringing out the book in an excellent and readable getup.

Finally, we apologize for any mistakes, omissions, or failures that may subsist in this work.

Rio de Janeiro, Brazil — Nicolas Carels
Hyderabad, India — Mulpuri Sujatha
Warangal, India — Bir Bahadur

Author Biography

Nicolas Carels obtained his graduate degree in agronomy in Belgium and did a Ph.D. in plant pathology at *Faculté des Sciences Agronomique de Gembloux* (FSAGx, Gembloux) prior to working as a scientist on the elaboration of the first genetic map of sugarbeet at the end of the 1980s (ICIseed-SES, Belgium). He then moved to Paris at *Institut Jacques Monod* (IJM, CNRS, France) where he did a Ph.D. on the genome organization in plants. He continued his work on genomics in Italy at *Stazione Zoologica 'Anton Dohrn'* (SZN, Naples) and Spain at the *Centro de Astrobiología* of *Instituto Nacional de Técnica Aeroespacial* (INTA-CAB, Madrid, Torrejon de Ardoz) prior to moving to Brazil (Bahia, Ilhéus, UESC) where he contributed to the application of bioinformatics and genomics toward the improvement of cacao and rubber tree for resistance to fungal diseases. He took Jatropha at its beginning when it was declared a strategic crop for the Brazilian economy by President Lula. His investigations covered the measure of the genome size by flow cytometry and the application of reverse genetics to detect QTLs for oil production with the purpose of breeding Jatropha for this trait. He also published an extensive review (ABR) on Jatropha and more recently an overview on bioenergies (InTech) with special concern for climate change mitigation and biodiversity preservation. He is now a federal officer of Fiocruz (Rio de Janeiro, Brazil) and is interested in the exploration of genomics, bioinformatics, and natural products for human health benefit.

Mulpuri Sujatha graduated in plant sciences at the University of Hyderabad (UoH), India. She has a Ph.D. in genetics from Osmania University (OU), Hyderabad, and has worked on intergeneric and interspecific affinities between Ricinus and Jatropha. She has made significant contributions for the genetic improvement of oilseed crops through genetics, tissue culture, and biotechnological tools. Sujatha's important achievements include development of male sterility systems in safflower, sunflower, and niger, and reliable and efficient tissue culture and transformation protocols for sunflower, castor, niger, safflower, and Jatropha. The genetic transformation protocols developed are being used for the development of insect-resistant transgenics through deployment of suitable Cry genes in castor, development of transgenic male sterility and fertility restoration system in safflower and development of transgenics for resistance to necrosis disease in sunflower. Her experience in molecular markers resulted in mapping of downy mildew resistance gene (*Pl13*) in sunflower besides the development of appropriate molecular markers for distinguishing toxic and non-toxic accessions of *Jatropha curcas*.

Bir Bahadur graduated from Nizam College with a postgraduate degree from University College, Osmania University, Hyderabad, India. He obtained his Ph.D. in plant genetics from Osmania University and was closely associated with Prof. J.B.S. Haldane, FRS, renowned geneticist of the twentieth century. Prof. Haldane advised, guided, and encouraged Bahadur to study heterostyly and incompatibility in Indian plant species, a subject first studied by Charles Darwin in England about 160 years ago. He has made significant contributions in several areas of plant biology especially in incompatibility, mutagenesis, morphogenesis, tissue culture, organism asymmetry, and application of SEM in plant sciences. He has published over 250 research papers, which were well received and quoted in national and international journals, including a number of theses and several publications on Jatropha and Castor. He served Osmania and Kakatiya Universities as lecturer, reader, and professor, and has also served as chairman, head of department, dean of the Faculty of Science, Kakatiya University, Warangal. He has taught genetics, biotechnology, and reproduction of plants for over 40 years and accumulated research experience in these areas for about 50 years. He was a post-doctoral fellow at the Institute of Genetics of Hungarian Academy (Budapest); recipient of the Royal Society Bursary, London; and Honorary Research Fellow at the Birmingham University (UK). He has been invited speaker of over 100 conferences including Max Plank Institute, Koln (Germany); Institute of Genetics (Budapest), Birmingham (UK); University of Texas, Houston (USA); Missouri University, St Louis (USA); Sabrao conference, Szukoba, Tokyo (Japan); Indian Science Congress; etc. He has authored/edited eight books and was editor-in-chief of both proceedings of Andhra Pradesh Akademi of Sciences (Hyderabad, India) and Journal of Palynology (Lucknow, India). He is on the editorial boards of several journals in India. He is recipient of Best Teacher Award by AP state government and Prof. Vishwamber Puri Gold Medal from the Indian Botanical Society for his original contributions in various aspects of plant sciences. He is fellow of over dozen professional bodies in India and abroad including the Linnean Society, London; Institute of Biology and Chartered Biologist, London; and New York Academy of Sciences. He has been recently awarded the Bharath Jyoti Award for his sustained academic and research career at New Delhi. Presently he is on the Board of Directors of Sribiotech, Hyderabad, India and Emeritus Professor, Shandan Post-Graduate College, Osmania University, Hyderabad.

Contents

Part III Farming

Part IV Byproducts

Contributors

Avinash Kumar Agarwal Department of Mechanical Engineering, Engine Research Laboratory, Indian Institute of Technology, Kanpur, India

Carlos Juliano Brant Albuquerque Fazenda Experimental de Uberlândia, Uberlândia, Minas Gerais, Brazil

Alexandre Alonso Alves Embrapa Agroenergia, Brasilia, DF, Brazil

Rafael Costa Amaral Departamento de Ciências Exatas e Tecnológicas, Universidade Estadual de Santa Cruz Rod, Ilhéus, Bahia, Brazil

K. Anitha National Bureau of Plant Genetic Resources Regional Station, Hyderabad, India

S. Anitha Department of Biotechnology, Sri Krishna Devaraya University, Anantapur, India

A. Aravinda Department of Biotechnology, Sri Krishna Devaraya University, Anantapur, India

N.R. Banapurmath Department of Mechanical Engineering, B.V.B. College of Engineering and Technology, Hubli, India

V.A. Bapat Emeritus Scientist, CSIR Department of Botany / Biotechnology, Shivaji University, Kolhapur, India

Y. H. Basavarajappa Department of Mechanical Engineering, Tontadarya College of Engineering, Gadag, India

Roger Bayala Ecole Nationale Supérieure d'Agriculture (ENSA), Université de Thiès, Thiès, Sénégal

Klaus Becker Institute for Animal Production in the Tropics and Subtropics, (480b), University of Hohenheim, Stuttgart, Germany

J.K. Bhavanasi Nandan Biomatrix Limited, Aparna Crest, Hyderabad, India

Sébastien Bonnet The Joint Graduate School of Energy and Environment, King Mongkut's University of Technology Thonburi, Bangkok, Thailand

Debashish Bora Division of Medicinal Aromatic & Economic Plants, North East Institute of Science & Technology (CSIR), Jorhat, Assam, India

Nicolas Carels Laboratório de Genômica Funcional e Bioinformática, Fundação Oswaldo Cruz (FIOCRUZ), Instituto Oswaldo Cruz (IOC), Rio de Janerio, RJ, Brazil

P.P. Chakrabarti Centre for Lipid Research, Indian Institute of Chemical Technology, Hyderabad, India

Girish Chander Resilient Dryland Systems, International Crops Research Institute for the Semi Arid Tropics (ICRISAT), Hyderabad, India

Supattra Charoenpakdee Biology Program, Faculty of Science and Technology, Pibulsongkram Rajabhat University, Phitsanulok, Thailand

Alok Chaube Jabalpur Engineering College, Jabalpur, India

D.R. Chaudhary Discipline of Wasteland Research, Central Salt & Marine Chemicals Research Institute (Council of Scientific and Industrial Research, New Delhi), Bhavnagar, Gujarat, India

Jitendra Chikara Discipline of Wasteland Research, Central Salt & Marine Chemicals Research Institute (Council of Scientific and Industrial Research, New Delhi), Bhavnagar, Gujarat, India

Rosenira Serpa da Cruz Departamento de Ciências Exatas e Tecnológicas, Universidade Estadual de Santa Cruz Rod, Ilhéus, Bahia, Brazil

Bernard Dell Sustainable Ecosystems Research Institute, Murdoch University, Murdoch, WA, Australia

N. Devanna Department of Chemistry, Jawaharlal Nehru Technological University, Hyderabad, AP, India

Rakshit K. Devappa Institute for Animal Production in the Tropics and Subtropics, (480b), University of Hohenheim, Stuttgart, Germany

Prathibha Devi Department of Botany, Molecular Genetics & Biotechnology Laboratory, Osmania University, Hyderabad, India

Atul Dhar Engine Research Laboratory, Department of Mechanical Engineering, Indian Institute of Technology Kanpur, Kanpur, India

Djiby Dia Bureau d'Analyses Macro-Économiques, Institut Sénégalais de Recherches agricoles, Dakar, Sénégal

Mário Sérgio Carvalho Dias Fazenda Experimental de Pouso Alegre, Pouso Alegre, Minas Gerais, Brazil

Ibrahima Diédhiou Ecole Nationale Supérieure d'Agriculture (ENSA), Université de Thiès, Thiès, Sénégal

Papa Madiallacké Diédhiou UFR des Sciences Agronomiques d'Aquaculture et de Technologies Alimentaires Université Gaston Berger de Saint-Louis (UGB), Saint-Louis, Sénégal

Marcos Antônio Drumond Embrapa Semiárido, Petrolina, PE, Brazil

Sérgio Luiz Ferreira-Silva Departamento de Bioquímica e Biologia Molecular, Laboratório de Metabolismo de Plantas, Universidade Federal do Ceará, Fortaleza, Ceará, Brazil

José Faustino Souza de Carvalho Filho Departamento de Ciências Exatas e Tecnológicas, Universidade Estadual de Santa Cruz Rod, Ilhéus, Bahia, Brazil

George Francis Live Energies GmbH, Stuttgart, Germany

JATROPOWER Bio-trading Pvt Ltd, Coimbatore, India

Shabbir H. Gheewala The Joint Graduate School of Energy and Environment, King Mongkut's University of Technology Thonburi, Bangkok, Thailand

Center for Energy Technology and Environment, Ministry of Education, Bangkok, Thailand

Arup Ghosh Discipline of Wasteland Research, Central Salt & Marine Chemicals Research Institute, (Council of Scientific and Industrial Research, New Delhi), Bhavnagar, Gujarat, India

N.M. Girish Department of Mechanical Engineering, B.V.B. College of Engineering and Technology, Hubli, India

Animesh Gogoi Division of Medicinal Aromatic & Economic Plants, North East Institute of Science & Technology (CSIR), Jorhat, Assam, India

Nívio Poubel Gonçalves Fazenda Experimental do Gorutuba, Nova Porteirinha, Minas Gerais, Brazil

Adrita Goswami Division of Medicinal Aromatic & Economic Plants, North East Institute of Science & Technology (CSIR), Jorhat, Assam, India

Jorge Martinez Herrera Centro De Desarrollo De Productos Bioticos-IPN, Yautepec, Morelos, Mexico

S. S. Hiremath Department of Mechanical Engineering, Basaveshwar Engineering College, Bagalkot, India

R.S. Hosmath Department of Mechanical Engineering, B.V.B. College of Engineering and Technology, Hubli, India

Shashi Kumar Jain School of Energy and Environment Management, RGPV, Bhopal, India

Jamaluddin Emeritus Scientist CSIR, Department of Biological Science, R.D. University, Jabalpur, India

Raphael Muzondiwa Jingura Chinhoyi University of Technology, Chinhoyi, Zimbabwe

T. Sudhakar Johnson Natural Remedies, Bangalore, India

Asha A. Juwarkar Eco-Restoration Division, National Environment Engineering Research Institute (NEERI), Nagpur, India

Mohameth Kâne Ecole Nationale Supérieure d'Agriculture (ENSA), Université de Thiès, Thiès, Sénégal

K.R. Karanam Nandan Biomatrix Limited, Aparna Crest, Hyderabad, India

B. Karunakar Department of Biochemistry, National P.G. College, Nandyal, India

S.V. Khandal Department of Mechanical Engineering, KLEIT, Chikodi, India

G.P. Kumar Plant Nutrition Division, Defence Food Research Laboratory, Mysore, Karnataka, India

G. Raja Krishna Kumar Plant Metabolic Engineering Group, Reliance Life Sciences Pvt. Ltd., Dhirubhai Ambani Life Sciences Center, Navi Mumbai, India

Sunil Kumar Rajiv Gandhi Proudyogiki Vishwavidyalaya, Bhopal, India

Bruno Galvêas Laviola Embrapa Agroenergia, Brasilia, DF, Brazil

Keat Teong Lee School of Chemical Engineering, Universiti Sains Malaysia, Engineering Campus, Nibong Tebal, Pulau Pinang, Malaysia

Steven Lim School of Chemical Engineering, Universiti Sains Malaysia, Engineering Campus, Nibong Tebal, Pulau Pinang, Malaysia

Ivon Pinheiro Lobo Departamento de Ciências Exatas e Tecnológicas, Universidade Estadual de Santa Cruz Rod, Ilhéus, Bahia, Brazil

Saisamorn Lumyong Faculty of Science, Department of Biology, Chiang Mai University, Chiang Mai, Thailand

Harinder P. S. Makkar Livestock Production Systems Branch, Animal Production and Health Division, Rome, Italy

Cristian Jimenez Martinez Escuela Nacional de Ciencias Biologicas-IPN, Dpto Alimentos, DF, Mexico

Banna Mbaye Ecole Nationale Supérieure d'Agriculture (ENSA), Université de Thiès, Thiès, Sénégal

Danielle de Lourdes Batista Morais Fazenda Experimental do Gorutuba, Nova Porteirinha, Minas Gerais, Brazil

Kalpataru Dutta Mudoi Division of Medicinal Aromatic & Economic Plants, North East Institute of Science & Technology (CSIR), Jorhat, Assam, India

Khadidiatou Ndir Ecole Nationale Supérieure d'Agriculture (ENSA), Université de Thiès, Thiès, Sénégal

Bassiaka Ouattara Centre d'Etudes Régional pour l'Amélioration de l'Adaptation à la Sécheresse (CERAAS), Thiès, Sénégal

Rodrigo Meirelles de Azevedo Pimentel Avenida Santa Cruz, n°500, Bairro Santa Cruz, Minas Gerais, Brazil

R. B. N. Prasad Centre for Lipid Research, Indian Institute of Chemical Technology, Hyderabad, India

D. Muralidhara Rao Department of Biotechnology, Sri Krishna Devaraya University, Anantapur, India

José Carlos Fialho de Resende Fazenda Experimental de Montes Claros, Montes Claros, Minas Gerais, Brazil

Rodrigo Barros Rocha Embrapa Rondônia, Porto Velho, RO, Brazil

Siddhartha Proteem Saikia Division of Medicinal Aromatic & Economic Plants, North East Institute of Science & Technology (CSIR), Jorhat, Assam, India

Sujatha Samala University College of Pharmaceutical Sciences, Kakatiya University, Warangal, India

Evandro Nascimento Silva Departamento de Bioquímica e Biologia Molecular, Laboratório de Metabolismo de Plantas, Universidade Federal do Ceará, Fortaleza, Ceará, Brazil

José Tadeu Alves da Silva Fazenda Experimental de Montes Claros, Montes Claros, Minas Gerais, Brazil

Joaquim Albenísio Gomes Silveira Departamento de Bioquímica e Biologia Molecular, Laboratório de Metabolismo de Plantas, Universidade Federal do Ceará, Fortaleza, Ceará, Brazil

Fúlvio Rodriguez Simão Fazenda Experimental do Gorutuba, Nova Porteirinha, Minas Gerais, Brazil

Felipe Oliveira Souza Departamento de Ciências Exatas e Tecnológicas, Universidade Estadual de Santa Cruz Rod., Ilhéus, Bahia, Brazil

P.G. Tewari Department of Mechanical Engineering, B.V.B. College of Engineering and Technology, Hubli, India

A.V. Tumbal Department of Mechanical Engineering , B.V.B. College of Engineering and Technology, Hubli, India

K.S. Varaprasad Directorate of Oilseeds Research, Hyderabad, India

Ciddi Veeresham University College of Pharmaceutical Sciences, Kakatiya University, Warangal, India

Norma Guemes Vera Universidad Autonoma de Hidalgo, Hidalgo , Mexico

Ricardo Almeida Viégas Departamento de Engenharia Florestal, Universidade Federal de Campina Grande, Centro de Saúde e Tecnologia Rural, Patos, Paraíba, Brazil

Idrissa Wade Ecole Nationale Supérieure d'Agriculture (ENSA), Université de Thiès, Thiès, Sénégal

Zanxin Wang School of Development Studies, Yunnan University, Kunming, Yunnan province, China

Suhas P. Wani Resilient Dryland Systems, International Crops Research Institute for the Semi Arid Tropics (ICRISAT), Hyderabad, India

S.K. Yadav Eco-Restoration Division, National Environment Engineering Research Institute (NEERI), Nagpur, India

V.S. Yaliwal Department of Mechanical Engineering, S.D.M. College of Engineering and Technology, Dharwad, India

Part I
Worldwide Importance of Jatropha

Chapter 1
The Birth of a New Energy Crop

Nicolas Carels

Concerning the Socio-Economic Context of Biofuel Investigation

To live and enjoy life is surely a surprising experience in all times, but the particular moment where this book is being planned and written is certainly a very interesting and crucial period in the history of humanity. The main challenge for humanity is the auto-control of its worldwide population. Some experts argue that there will be no problem to feed nine billion people when we are already seven billion on earth now. This optimism sounds obsolete when one thinks objectively that the problem will be posed exactly in the same terms at the moment where the human population will effectively reach nine billions. Actually, resources will be far too less than they are now at seven billion and so what? Chinese people already understood that the size of human population is thc rcal challenge and even if it still with a growing population, Chinese government took a long time ago the courageous initiative of an intensive family planning program that, among other things, is allowing only one child per couple without payment of deterrent taxes (Orleans 1975; Nathans 2010). Given the difficulties associated to the acceptance of this planning program, the Chinese population continued to grow, but at a much slower rate and it is expected to reach about 1.5 billion around 2030, before starting to decrease (Attané 2002). As an expected consequence, new problems are emerging, such as accelerated population aging (Hvistendahl 2010) or widening gap between rich and poor, but other positive effects were obtained, such as to bring the consciousness of everyone on the importance of the issues related to demography through education for all (http://www.china.org.cn/e-white/familypanning/). Education for all had the consequence to boost the Chinese economy despite the aging of the population. In turn, a more profitable economy should give a return to

N. Carels (✉)
Laboratório de Genômica Funcional e Bioinformática, Fundação Oswaldo Cruz (FIOCRUZ), Instituto Oswaldo Cruz (IOC), Rio de Janerio, RJ, Brazil
e-mail: nicolas.carels@gmail.com

N. Carels et al. (eds.), *Jatropha, Challenges for a New Energy Crop: Volume 1: Farming, Economics and Biofuel*, DOI 10.1007/978-1-4614-4806-8_1,

alleviate better life to the old people. The classical notion that it is the young people that should pay for the old ones is now obsolete. Actually, young people suffer unemployment because of substitution by machines, thus it is these machines that should be expected to pay for old people and not young ones. Again, there is some bias in the system just because basic economic rules are not respected. At this point, one would ask whether humanity should pay the price to go through wars or starvation to regulate its population density because the decision of making the simple accounting of how much square meters of earth planet is allowed for humans failed to be taken. Absence of human population size regulation by itself would have the mathematical consequence to be enforced in a way following the predator–prey equation (Lonngren and Bai 2008) just because earth's geometry will obviously not change.

Another disastrous consequence of an ever growing human population is the ever disappearing biodiversity, the so called ***sixth extinction*** (Wake and Vredenburg 2008). Biodiversity is an important integral component of the earth as we know it; it is a capital that has been accumulating during the past three billion years and that has been historically erroneously evaluated to ***zero*** by economists' procedures. Besides its invaluable environmental services associated, biodiversity offers a capital of tools that manage material transformation at very low energetic levels. These tools are the living cells and their enzymes enclosed in complex multifunctional organisms. These organisms are, thus, information systems that produce material transformations in various timescales. Genomics has recently shown that the most complex organisms, the higher eukaryotes, all host 15.000–25.000 units of information (genes) whatever, plant, insect, mammals or other higher eukaryotes. Thus, the similarity of quantitative information contrasts with the variety of shape/form and functions. We can draw from this observation that the capital given to us by nature is more in the state of the information network of an organism as an instantaneous picture of live processes than in its genes themselves. Through biodiversity, nature offers to us a book of variations on the way to coordinate the expression of genes to accomplish given functions and, this, on the basis of a content of information common to all life forms. In a sense, life is a quantum computing activity based on carbon chemistry that human technology is aiming to reproduce in the mineral paradigm with electronic components. When a species is extinct, this invaluable capital is destroyed for ever because humans do not have time in evolutionary scale, three billions years, to reconstruct it. Actually, we are just starting to investigate how these networks work when they are precisely entering a phase of extinction (Wake and Vredenburg 2008). For example, we can ask what would be medicine, and more generally biotechnology, today without ***polymerase chain reaction*** (PCR). PCR is a legacy from the biodiversity of Yellowstone National Park (USA), but did Yellowstone receive a fee for this gift to humanity? Certainly not! The tendency to consider gifts from nature for free is another general bias of economy in total contradiction with the theory itself and it can be seen as a human distortion in its mathematical context. Recently, this fact has been recognized and taxes on pollution have been started to be implemented in a number of countries. Thus, it is of central importance to keep the nature's capital alive and in its diversity at least to warrant our technological future and this without considering the environmental services that biodiversity offers to humans; services that are absolutely necessary to a healthy human living condition. It is unacceptable for humanity to lose the nature's capital for facts of ignorance, negligence or convenience. According to the concept of Gaia (Lovelock and Margulis 1974), earth is a living dynamic system and we cannot look at as it would not exist; this would be synonym of human extinction in a difficult to predict time delay (Lonngren and Bai 2008). Climate stabilization is one example of service from complex natural systems. Biodiversity has numerous stabilizing effects on the environment and climate (Odling-Smee 2007; Tomkiewicz 2006), ecosystems (Yue et al. 2011) and (consequently) human populations (Simões et al. 2010). Fortunately, even if slowly, the economical value of feral nature (Costanza et al. 1997) is starting to be recognized at economic level and *The **Economics of Ecosystems and Biodiversity*** (TEEB) has been created (http://www.teebweb.org/) with this objective.

World population regulation is possible as shown by China, but it is, at least, necessary to take the decision of its planning and management. The price of the disregard shown by the few later human generations around this issue will have to be paid by the future generations to come and there is no way to escape from this fact. Thus, the best strategy would be to reduce/stop the rate of population growth before entering overpopulation and suffering its deleterious consequences. China and India know the price to pay for people over crowding, but other countries unfortunately seem to have difficulties in learning their lesson. The politics that is followed by most if not virtually all countries is to adapt food resources to the needs of an ***ever growing population*** by technology adjustments. This strategy only has a limited future. Any normal person considering this question will definitely agree on the fact that it will not work for ever. The concept of ***infinite resource availability*** on which worldwide economies are based should never have existed and has been the major failure of economic theories (Nadeau 2006). It is on the basis of this concept that population size is growing worldwide. The concept of infinite resource availability does not hold good if it is not included in that of ***sustainability***, which limits the extent of the first to what is realistic, a concept that former economists from the period of ***industrial revolution*** intentionally forget to include in their calculations. This logic has been possible as far as resources were far from their ***peak production,*** but when resources are passing their peak production, a recession is to be expected if alternative resources are not found to maintain the *services* that their provide. This dynamics applies to the case of energy.

Considering energy, the ***infinitely available resource*** is not more crude oil since the one available at low price is probably reaching its peak, but rather the solar one. Experts believe that peak of world oil production should not occur before at least 30–40 years from now. The OPEC nations are currently operating at near full capacity. According to the expected solar life, solar energy will remain "*infinite*" during a large number of human generations. However, solar energy is still expensive and the transition to its exploration can only be partial in its present state of the art. Thus, the joining of a myriad of renewable energies, according to regional opportunities, will be necessary to the successful substitution of fossil and nuclear fuels to feed the energy grid (see Carels 2011 for a review). The model from de Vries et al. (2007) showed the following: (i) electricity from solar energy is typically available from Northern Africa, South Africa, the Middle East, India, and Australia; (ii) wind is concentrated in temperate zones such as Chile, Scandinavia, Canada, and the USA; (iii) biomass can be produced on vast tracts of abandoned agricultural land typically found in the USA, Europe, the Former Soviet Union (FSU), Brazil, China and on grasslands and savannas in other locations. In many areas of India, China, Central America, South Africa and equatorial Africa, these energy sources are found in areas where there is already a large demand for electricity (or there will be such demand in the near future). A combination of electricity from wind, biomass and/or solar sources (Eugenia Corria et al. 2006) may yield economies-of-scale in transport and storage systems. Regions with high ratios of solar-wind-biomass that potentially meet the current demand for electricity include Canada (mainly wind), African regions (solar-photovoltaic and wind), the FSU (wind and biomass), the Middle East (solar-photovoltaic) and Oceania (all sources). In other regions (such as Southeast Asia and Japan), the solar-wind-biomass supply is significantly lower than the demand for electricity. Ratios of around one are found in Europe and South Asia. The potentials just described depend on many parameters, and their achievement will depend on future land-use policies (de Vries et al. 2007; Miles and Kapos 2008).

Why Biofuels?

Numerous low emission scenarios have demonstrated that the goals of the Kyoto Protocol cannot be achieved without providing a large role for biofuels by 2050 in the global energy economy (Vertès et al. 2006). Among the reasons why biofuels are appropriate for such a transition, one may distinguish: (i) their simplicity; (ii) their production via well-known agricultural technologies; (iii) their potential for mitigation of climate warming without complete restructuring of the current working energy system; (iv) the use of existing engines for their transportation (even considering the conventional turbofan used in aviation) (Kleiner 2007; Rothengatter 2010); (v) their potential to facilitate worldwide mobilization around a common set of regulations; (vi) their potential as a directly available energy source with good public acceptance; (vii) their more uniform distribution than the distributions of fossil fuel and nuclear resources; and (viii) their potential to create benefits in rural areas, including employment creation.

Biodiesel from palm oil and bioethanol from sugarcane are currently the two leaders of plant bioenergy production per hectare. Palm and sugarcane are being grown in increasing amounts; however, the continuous increase in their production is not sustainable and will not resolve the enormously increasing demands for energy. Oil palm yields ~5,000 $L\,ha^{-1}$. In Brazil, the best bioethanol yields from sugarcane are 7,500 $L\,ha^{-1}$. Most of the energy needed for growing sugarcane and converting it to ethanol is gained from burning its wastes (i.e., *bagasse*). For every unit of fossil energy that is consumed by producing sugarcane ethanol, ~8 units of energy are recovered (Bourne 2007). Frequent droughts in many Asian countries have made difficult for them to replicate Brazil's success with sugarcane, which needs an abundant water supply. Thailand and Indonesia are tapping the potential with palm oil.

The general feeling is that first-generation biofuels are already reaching saturation because of the limited availability of arable lands. However, Brazil has additional lands available for sugarcane and *Jatropha curcas* L. (hereafter referred to as Jatropha) cultivation, whereas India is promoting Jatropha on its extensive wastelands. The development of these fuels has already been a success because they have demonstrated that engine running on ethanol or biodiesel is feasible without significant technological modification and can, at least, be used to power public transport.

Why *Jatropha curcas?*

Most traditional biofuels (such as ethanol from corn, wheat, or sugar beets and biodiesel from oilseeds) are produced from classic agricultural food crops that require high-quality agricultural land for growth. The biofuel economy will grow rapidly during the twenty-first century (Demirbas 2008). Currently, approximately 84% of the world biodiesel production is met by rapeseed oil. The remaining portions are from sunflower oil (13%), palm oil (1%), soybean oil and others (2%) (Gui et al. 2008). More than 95% of biodiesel is still produced from edible oils. In Brazil, ~85% of B5 (5% biodiesel in fossil diesel) is sustained by soybean oil. To overcome this undesirable situation, biodiesel is increasingly being produced from non-edible oils and waste cooking oil (WCO). Non-edible oils offer the advantage that they theoretically do not compete with edible oils on the food market.

Because of increased land use for biofuel production, biofuel crops are now competing with food crops (Odling-Smee 2007) and they are expected to have substantial effects on the economy. In addition, the use of food crop for biofuel production promote deforestation and creates a "*carbon debt*" by releasing 17–420 times more CO_2 than the annual greenhouse gas (GHG) reductions that these biofuels would provide by displacing fossil fuels. In contrast, biofuels from waste biomass or from biomass grown on marginal lands planted with perennial species incur little or no carbon debt (Fargione et al. 2008; Searchinger et al. 2008).

Considering biofuels, Jatropha offers several advantages compared to other oilseeds: (i) Jatropha can thrive on wasteland, where poor populations are generally found. Thus, it is expected to have a positive social effect by attracting government's investment, stabilizing the population in rural areas, and providing the population with incomes and energy; (ii) the feasibility of its biodiesel has been proved (see Carels 2009 for a review); (iii) it can be burnt as neat oil or biodiesel in conventional ***compression ignition*** (CI) engines (see Carels 2009 for a review); (iv) its environmental impact is lower than that of palm oil as long as no natural ecosystems are removed for its implantation (Darussalam 2007; Laurance 2007; Malhi et al. 2008; Stone 2007; Venter et al. 2008). On the contrary, Jatropha is used for waste land reclaiming (Francis et al. 2005; Pandey et al. 2012); and (v) its life cycle justifies its exploitation (Pandey et al. 2011) even if it is still to be improved.

Regional Contributions and Development Worldwide

Scientific investigations on Jatropha started in the 1980s. In contrast to Brazil, where it stalled soon after its beginning and only recovered in 2006, it has been continuing in India. India, is certainly the country that most contributed to Jatropha's promotion worldwide. As summarized in the article title: "the little shrub that could – maybe" of Fairless (2007), the symbolic of a semi-wild species able to thrive in marginal conditions and give oil suitable for biodiesel production, has certainly drawn the attention of the scientific community on Jatropha. From 2008, the scientific community did a huge work and drawn Jatropha to the light by successively mastering its interspecific hybridization (Reddy et al. 1987; Sujatha and Prabakaran 2003; Sujatha 2006; Basha and Sujatha 2009; Parthiban et al. 2009; Popluechai et al. 2009; Karanam and Bhavanasi 2010), performing the transesterification of its oil (see Carels 2009 for a review), characterizing its biodiesel combustion in CI engines (see Carels 2009 for a review), determining it genome size (Carvalho et al. 2008), setting the bases for its DNA marker assisted selection (Sudheer-Pamidiamarri et al. 2009), sequencing the transcriptome of it seeds (Costa et al. 2010; Gomes et al. 2010; King et al. 2011; Natarajan and Parani 2011), sequencing and annotating its genome (Sato et al. 2011) and finally setting up a preliminary genetic maps (Wang et al. 2011); a set of victories that justified the preparation of this book.

Scientific community worldwide had different contributions to the development of Jatropha research mostly according to the technological profile of its regional specificities. Africa is developing an approach where Jatropha is considered on its own and managed with minimal care. Jatropha is taken for what it can release without significant investment, due to its ability to thrive in minimal conditions, and for its ability to thrive in semi-arid climate, which makes it a good candidate to struggle against desertification. As pointed above, India under the pressure of its large dependency on foreign importation of crude oil, has given a strong attention to the biological parameters of Jatropha and to the setting of first insides in its marker-assisted breeding for its rational improvement. Coming later, China also strongly invested in research due to its unsustainable long term dependency on coal for energy and its limited amount of lands suitable for food crops. Through Kazusa DNA Research Institute, Japan offered the genome sequence to the world. Due to its tradition as a world center of high technology, Singapore is investing in biotechnology to shape Jatropha through genetic engineering. Brazil, with its tradition for biofuel production looks for an integrated approach that would optimize the whole picture for the benefit of the whole chain. Mexico, "*late but not least*" is now giving strong attention to the description of genetic diversity of Jatropha in the country since it appears to be the diversification center of Jatropha and is preparing itself to release certified genotypes worldwide.

In Brazil, Jatropha cultivation (20,000 ha) has only been allowed by law in 2009, which explains that the significant interest for this new crop is only relatively recent. In China, the arable land area per capita is lower than the world average. As a result, most edible oils need to be imported. Since Jatropha is not too demanding in terms of soil quality, it is a good option for Southwest China (Yunnan, Sichuan and Guizhou provinces), which is the most suitable area for its cultivation, but with only a moderate potential since it is limited to ~100,000 ha of marginal lands not suitable for food crops (Wu et al. 2009). Such limited potential makes it difficult to expect significant contribution from biodiesel in this country (Yang et al. 2009); more is to be expected from biomass. In other Asian countries with high dependency (>80%) from foreign fuel, such as Thailand, Cambodia, Vietnam, Myanmar, Laos, Indonesia, and India, eroded land areas from deforested soil are also available for Jatropha cultivation. For over 7 years, the Indonesian government is supporting various national and international agencies as well as research institutes for the investigation of Jatropha with a target of biodiesel production covering 10% of the national fuel consumption. At moment, about 1.5 million hectares are effectively planted (Legowo 2007; Silitonga et al. 2011). Since India imports 70% of its fuel consumption (111 Mio t), any renewable energy is most welcome. Because India is a net importer of edible oil, it emphasizes non-edible oils from plants such as, Jatropha, karanja, neem, mahua, simarouba, etc. Jatropha and karanja are the two important leaders of the Indian plant list for biodiesel production and India has 33 million ha of degraded land available for Jatropha reclaiming because of improper use and population pressure over several years (Francis et al. 2005; Kumar and Sharma 2008; Misra and Murthy 2011). Pilot experiments for Jatropha implementation are also being carried out throughout Africa and this has been reviewed by Henning (2005), but government incentives are still few (Jumbe et al. 2009).

Challenges for Jatropha as a Successful Biofuel Crop

Even if a huge work has been done that allows scientists to have a comprehensive picture of Jatropha as a whole, it remains that this species is a semi-wild species in the process of domestication. There is no reason to complain about such status when one realizes that in ~4 years, the scientific community succeeded in going through the most part of a journey that took 5,000 years for a species such as wheat for being domesticated. In other words, if Jatropha is still far from an industrial crop, it is however rather sure that due to its amazing technological potentialities it went over a point of no return. The new crop Jatropha is now entering in the domain of scientific integration towards complete domestication for profitable biodiesel production. It is to be expected that within a short time scale the architecture and functionalities of Jatropha will be deeply affected by selective breeding and genetic engineering. Among the traits that should be affected in a feedback loop process along the whole chain of biodiesel production are: (i) short-term (few years) or long-term (perennial) system of exploitation; (ii) extensive (with resting period in semi-arid condition) or intensive system (ferti-irrigation) of exploitation; (iii) fertilization; (iv) plant architecture (semi-dwarf or dwarf); (v) discontinuous, but synchronous flowering and fruiting; (vi) dioecious plants to maximize out-crossing and yield/production per plant; (vii) increased seed production over 4 t ha^{-1}; (viii) plant tolerance to pests and diseases (especially to those of the rooting system); (ix) mechanical harvesting; (x) fatty acid composition; (xi) genetic engineering to facilitate the biofuel processing (triglyceride conversion into free fatty acids?, enzymatic conversion); (xii) low cost alkyl ester conversion; (xiii) biofuel composition for optimized combustion in CI engines; (xix) microeconomic modeling of suitable agro-systems of Jatropha growing and exploitation; (xx) life cycle modeling; (xxi) plant growth and fructification modeling according to climate conditions of culture.

Acknowledgements N. Carels is grateful to *Coordenação de Aperfeiçoamento de Pessoal de Nível Superior* (CAPES) and *Fundação Oswaldo Cruz* (FIOCRUZ) for providing a research fellowship from the *Centro de Desenvolvimento Tecnológico em Saúde* (CDTS).

References

Attané I (2002) China's family planning policy: an overview of its past and future. Stud Fam Plann 33(1):103–113

Basha SD, Sujatha M (2009) Genetic analysis of Jatropha species and interspecific hybrids of *Jatropha curcas* using nuclear and organelle specific markers. Euphytica 168:197–214

Bourne JK Jr (2007) Biofuels: green dreams. Natl Geogr Mag 41–59 Available from http://ngm.nationalgeographic.com/2007/10/biofuels/biofuels-text.html [accessed August 10, 2012]

Carels N (2009) *Jatropha curcas*: a review. In: Kader JC, Delseny M (eds) Advances in botanical research. Elsevier, Amsterdam, The Netherlands, pp 39–86

Carels N (2011) The challenge of bioenergies: an overview. In: dos Santos Bernardes MA (ed) Biofuel's engineering process technology, 1st ed. InTech, Rijeka, pp 23–64

Carvalho CR, Clarindo WR, Praça MM, Araújo FS, Carels N (2008) Genome size, base composition and karyotype of *Jatropha curcas* L., an important biofuel plant. Plant Sci 174:613–617

Costa GGL, Cardoso KC, Del Bem LEV, Lima AC, Cunha MAS et al (2010) Transcriptome analysis of the oil-rich seed of the bioenergy crop *Jatropha curcas* L. BMC Genomics 11:462

Costanza R, d'Arge R, de Groot R, Farber S, Grasso M, Hannon B et al (1997) The value of the world's ecosystem services and natural capital. Nature 387:253–260

Darussalam B (2007) Last-Gasp effort to save Borneo's tropical rainforests. Science 317:192

de Vries BJM, van Vuuren DP, Hoogwijk MM (2007) Renewable energy sources: their global potential for the first-half of the 21st century at a global level: an integrated approach. Energy Policy 35:2590–2610

Demirbas A (2008) Biofuels sources, biofuel policy, biofuel economy and global biofuel projections. Energy Convers Manag 49:2106–2116

Eugenia Corria M, Melian Cobas V, Silva Lora E (2006) Perspectives of Stirling engines use for distributed generation in Brazil. Energy Policy 34:3402–3408

Fairless D (2007) The little shrub that could – maybe. Nature 449:652–655

Fargione J, Hill J, Tilman D, Polasky S, Hawthorne P (2008) Land clearing and the biofuel carbon debt. Science 319:1235–1238

Francis G, Edinger R, Becker K (2005) A concept for simultaneous wasteland reclamation, fuel production, and socio-economic development in degraded areas in India: need, potential and perspectives of Jatropha plantations. Nat Resour Forum 29:12–24

Gomes KA, Almeida TC, Gesteira AS, Lôbo IP, Guimarães ACR, de Miranda AB et al (2010) ESTs from seeds to assist the selective breeding of *Jatropha curcas* L. for oil and active compounds. Genomics Insights 3:29–56

Gui MM, Lee KT, Bhatia S (2008) Feasibility of edible oil vs. non-edible oil vs. waste edible oil as biodiesel feedstock. Energy 33:1646–1653

Henning RK (2005) Assessment of the impact of the dissemination of "the Jatropha System" on the ecology of the rural area and the social and economic situation of the rural population (target group) in selected countries in Africa. Available from http://www.underutilized-species.org/Documents/PUBLICATIONS/jatropha_curcas_africa.pdf [Accessed August 10, 2012]

Hvistendahl M (2010) Has China outgrown the one-child policy? Science 329(5998):1458–1461

Jumbe CBL, Msiska FBM, Madjera M (2009) Biofuels development in Sub-Saharan Africa: are the policies conducive? Energy Policy 37(11):4980–4986

Karanam KR, Bhavanasi J (2010) Jatropha interspecific hybrid. US patent US 2010/0287820 A1 1–6

King AJ, Li Y, Graham IA (2011) Profiling the developing *Jatropha curcas* L. seed transcriptome by pyrosequencing. Bioenerg Res 4:211–221

Kleiner K (2007) Civil aviation faces green challenge. Nature 448:120–121

Kumar A, Sharma S (2008) An evaluation of multipurpose oil seed crop for industrial uses (*Jatropha curcas* L.): a review. Ind Crops Prod 28:1–10

Laurance WF (2007) Switch to corn promotes amazon deforestation. Science 318:1721

Legowo EH (2007) Development of alternative energy in Indonesia. 5th Asian petroleum technology symposium, Jakarta, 2007

Lonngren KE, Bai E-W (2008) On the global warming problem due to carbon dioxide. Energy Policy 36:1567–1568

Lovelock JE, Margulis L (1974) Atmospheric homeostasis by and for the biosphere- The Gaia hypothesis. Tellus 26:2–10

Malhi Y, Roberts JT, Betts RA, Killeen TJ, Li W, Nobre CA (2008) Climate change, deforestation, and the fate of the Amazon. Science 319:169–172

Miles L, Kapos V (2008) Reducing greenhouse gas emissions from deforestation and forest degradation: global land-use implications. Science 320:1454–1455

Misra RD, Murthy MS (2011) Jatropha – The future fuel of India. Renew Sustain Energy Rev 15(2):1350–1359

Nadeau R (2006) The environment endgame: mainstream economics, ecological disaster, and human survival. Rutgers University Press, London, p 214

Natarajan P, Parani M (2011) *De novo* assembly and transcriptome analysis of five major tissues of *Jatropha curcas* L. using GS FLX titanium platform of 454 pyrosequencing. BMC Genomics 12:191. doi:10.1186/1471-2164-12-191

Nathans J (2010) China's plan flawed but courageous. Science 330:1625

Odling-Smee L (2007) Biofuels bandwagon hits a rut. Nature 446:483

Orleans L (1975) China's experience on population control: the elusive model. World Dev 3(7–8):497–525

Pandey KK, Pragya N, Sahoo PK (2011) Life cycle assessment of small-scale high-input Jatropha biodiesel production in India. Appl Energy 88(12):4831–4839

Pandey VC, Singh K, Singh JS, Kumar A, Singh B, Singh RP (2012) *Jatropha curcas*: a potential biofuel plant for sustainable environmental development. Renew Sustain Energy Rev 16(5):2870–2883

Parthiban KT, Kumar RS, Thiyagarajan P, Subbulakshmi V, Vennila S, Rao MG (2009) Hybrid progenies in Jatropha – a new development. Curr Sci 96:815–823

Popluechai S, Breviario D, Mulpuri M, Makkar HPS, Raorane M, Reddy AR et al (2009) Narrow genetic and apparent phenetic diversity in *Jatropha curcas*: initial success with generating low phorbol ester interspecific hybrids. hdl:10101/npre.2009.2782.1

Reddy KRK, Swamy NR, Bahadur B (1987) Cross incompatibility between *Ricinus* and *Jatropha*. Plant Cell Incompat Newslett USA 17:60–65

Rothengatter W (2010) Climate change and the contribution of transport: basic facts and the role of aviation. Transp Res Part D 15:5–13

Sato S, Hirakawa H, Isobe S, Fukai E, Watanabe A, Kato M et al (2011) Sequence analysis of the genome of an oil-bearing tree, *Jatropha curcas* L. DNA Res 18:65–76

Searchinger T, Heimlich R, Houghton RA, Dong F, Elobei A, Fabiosa J et al (2008) Use of U.S. croplands for biofuels increases greenhouse gases through emissions from land-use change. Science 319:1238–1240

Silitonga AS, Atabani AE, Mahlia TMI, Masjuki HH, Badruddin IA, Mekhilef S (2011) A review on prospect of *Jatropha curcas* for biodiesel in Indonesia. Renew Sustain Energy Rev 15(8):3733–3756

Simões AF, Kligerman DC, La Rovere EL, Maroun MR, Barata M, Obermaie M (2010) Enhancing adaptive capacity to climate change: the case of smallholder farmers in the Brazilian semi-arid region. Environ Sci Policy 13:801–808

Stone R (2007) Can palm oil plantations come clean? Science 317:1491

Sudheer-Pamidiamarri DVN, Pandya N, Reddy MP, Radhakrishnan T (2009) Comparative study of interspecific genetic divergence and phylogenic analysis of genus *Jatropha* by RAPD and AFLP. Mol Biol Rep 36:901–907

Sujatha M (2006) Genetic improvement of *Jatropha curcas* L. possibilities and prospects. Indian J Agrofor 8(2):58–65

Sujatha M, Prabakaran AJ (2003) New ornamental *Jatropha* through interspecific hybridization. Genet Resour Crop Evol 50:75–82

Tomkiewicz M (2006) Global warming: science, money and self-preservation. C R Chimie 9:172–179

Venter O, Meijaard E, Wilson K (2008) Strategies and alliances needed to protect forest from palm-oil industry. Nature 451:16

Vertès AA, Inui M, Yukawa H (2006) Implementing biofuels on a global scale. Nat Biotechnol 24:761–764

Wake DB, Vredenburg VT (2008) Are we in the midst of the sixth mass extinction? A view from the world of amphibians. Proc Natl Acad Sci USA 105:11466–11473

Wang CM, Liu P, Yi C, Gu K, Sun F, Li L et al (2011) A first generation microsatellite- and SNP-based linkage map of *Jatropha*. PLoS One 6(8):e23632

Wu W, Huang J, Deng X (2009) Potential land for plantation of *Jatropha curcas* as feed- stocks for biodiesel in China. Science China-Earth Sciences 39(12):1–8

Wu WG, Huang JK, Deng XZ (2010) Potential land for plantation of *Jatropha curcas* as feedstocks for biodiesel in China. Sci China 53(1):1–8

Yang H, Zhou Y, Liu J (2009) Land and water requirements of biofuel and implications for food supply and the environment in China. Energy Policy 37(5):1876–1885

Yue TX, Fan ZM, Chen CF, Sun XF, Li BL (2011) Surface modeling of global terrestrial ecosystems under three climate change scenarios. Ecol Model 222:2342–2361

Chapter 2
Importance of *Jatropha curcas* for Indian Economy

Sunil Kumar, Alok Chaube, and Shashi Kumar Jain

Introduction

Energy is a vital commodity as it is commonly recognized that access to energy is closely linked with economic development (DFID 2005). India is targeting gross domestic product (GDP) growth rate of 8–9% in coming years. It is likely to have a significant consumption of energy resources in future for meeting the targeted growth rate and fulfilling the energy needs of its increasing population. It lacks sufficient domestic resources to meet its growing crude oil requirements. In India, it has been projected that there is a need to increase the primary energy supply by at least three to four times from their 2003–04 levels by 2031–32 in order to maintain 8% growth rate. Maximum contribution of renewable energy in an optimistic scenario will be around 5–6% by 2031–32 with an import dependence on crude oil expected to be in the range of 90% in the next decade making energy security a concern (Planning Commission 2006). On account of high targeted economic growth rate and with over 15% of the world's population, India is likely to have a significant consumption of energy resources. India ranks fifth in terms of primary energy consumption and accounts for less than 4% of the global commercial energy demand as it is still in developing phase. With GDP of $4.469 trillion (WEO 2011), India is the third largest economy in purchasing parity terms. However because of its large population, per capita energy consumption is extremely low. On 2011 census basis,

S. Kumar (✉)
Rajiv Gandhi Proudyogiki Vishwavidyalaya, Airport Bypass Road Gandhi Nagar, Bhopal 462 036, India
e-mail: sunilrewa@yahoo.com

A. Chaube
Jabalpur Engineering College, Gokulpur Ranjhi, Jabalpur 482 011, India

S.K. Jain
School of Energy and Environment Management, RGPV, Airport Bypass Road, Gandhi Nagar, Bhopal 462 036, India

N. Carels et al. (eds.), *Jatropha, Challenges for a New Energy Crop: Volume 1: Farming, Economics and Biofuel*, DOI 10.1007/978-1-4614-4806-8_2,

the per capita energy consumption is estimated to be 451 *kg oil equivalent* (kgoe) whereas for developed countries it is around 5,000 kgoe. Despite a global slowdown in economy, India's energy demand has shown an increasing trend. In terms of end-use, increase in energy demand in the transport sector is expected to be particularly high.

Crude oil is the largest consumed fossil fuel after coal in India. Worldwide known crude oil reserves are estimated to be depleted in less than 50 years at the present rate of consumption (Sheehan et al. 1998). Many countries lacking crude oil resources are facing foreign exchange crisis and high inflation rate mainly due to import of crude oil (Demirbas 2005). For oil deficit countries, renewable energy is to be looked as possible resource for meeting energy demand challenges and for fulfilling the targeted growth.

Dependence on imported crude oil, environmental issues and employment in rural areas are reasons for replacement of fossil fuels by biofuels (Senthilkjumar and Gunasekaran 2005). Biofuels can be viable substitutes in transport sector for petroleum products, which account for one third of India's total imports. Researchers have put forward that the need for wide spread usage of renewable energy to meet the energy demand in India may compete with the use of land and water resources, which could have an adverse impact on food security (Rajagopal et al. 2007). Self-sustainable energy sources are likely to hold the key to economic development of India in future. India cannot depend on certain group of countries to meet its ever growing energy needs, but it is mandatory to seriously implement bioenergy development programs as a part of environmental sustainability in the form of clean development mechanism (CDM) (Kumar et al. 2010).

Brundtland Report introduced environmental protection by defining sustainable development as "development that meets the needs of the present generation without compromising the ability of future generations to meet their own needs" (WCED 1987). Global initiatives related to mitigating climate change, like Kyoto Protocol and Asia Pacific Partnership on Clean Development and Climate Change to which India is also a party, could give a thrust to biofuel based energy technologies, which are still in the infancy stage in India.

The *clean development mechanism* (CDM), established by the Kyoto Protocol, promotes the industrialized nations to provide resources to developing countries in order to support their sustainable development, while at the same time reducing the global green house gases (GHGs).

High speed diesel (HSD) is the largest consumed petro-product in India on account of better mileage, power and lower price rate compared to petrol (gasoline) (Kumar et al. 2008a). Among various alternatives to diesel, Planning Commission of India has identified *Jatropha curcas* L. (Jatropha), a non-edible oilseed tree whose oil can be easily converted into biodiesel with properties very similar to diesel (Planning Commission 2003). Jatropha is a drought-resistant, perennial plant living up to 50 years and has the capability to grow on marginal soils. It requires little irrigation and grows in all types of soils, thus making Jatropha a more sustainable choice than other vegetable oils (NOVOD 2007, 2008; Altenburg et al. 2009; Reubens et al. 2011). Jatropha biodiesel can be used for decentralized micro-grid electricity generation at village or taluka (suburb) level and as a replacement for diesel fuel in irrigation pump sets, diesel generators and also as an alternative to kerosene. Thus, Jatropha biodiesel has to be seen not in isolation, but as part of a total management system of environmental and energy resources.

Prevailing Energy Scenario in Transport Sector of India

India is emerging as one of the fastest growing economies in the world with a GDP growth ranging between 6 and 8% consistently for the past few years and this trend is expected to continue in the future. Because energy is the key driver of this growth, its easy and cheap availability is of prime importance to sustain the continuous economic growth of the India. The available future projections show that the energy demand is expected to be more than three to four times of the current level in the next 25 years. With this pace of growth, India is expected to face formidable challenges in meeting out its energy needs in a sustainable manner at reasonable cost. Critical analysis of the current energy scenario as well as future plans suggests that India still has a long way to go in ensuring energy security to its people. Some of the following summarized statistics clearly helps in establishing this fact:

- Data from 2012 (April 1st) indicate that India has total reserves (proved and inferred) of 1,201 *million metric tons* (MMT) of crude oil and 1,437 billion cubic meters of natural gas.
- Net petro-products import was 121,897 MMT during the year 2010–11.
- Crude oil production during 2010–11 at 37.68 MMT is almost stagnant.
- The consumption of petroleum products during 2010–11 was 141.78 MMT (including sales through private imports) among which diesel accounted for 59.99 MMT.
- In the year 2010–11, the increase in diesel consumption in absolute terms was nearly three times larger than the increase in petrol consumption.

National Autofuel Policy envisages technologies for producing biofuels from renewable sources by providing *research and development* (R&D) as well as other support. It seeks to ensure sustainable, safe, affordable and uninterrupted supplies of auto fuels of high quality to support social and economic development. One of the key factors for meeting this policy objective is to diversify the sources and reduce dependence on any single source of supply (National Autofuel Policy 2003). This policy, lends support and encouragement for biofuels.

Rationale Behind Usage of Bioenergy

Mass utilization of diesel in India imposes a threat to meeting the future energy needs, if the unexpected volatilities in the price of petroleum persists in future and government of India enforces oil marketing companies to sell diesel at uncapped

Table 2.1 Categorization of waste-land in India (Planning Commission 2003)

Type of land	Area (Mha)
Forest area	3
Boundary plantations	3
Agro forestry	2
Cultivable fallow lands	2.4
Waste lands under integrated watershed development	2
Strip land like: railway, road & canal	1
Total	13.4

Table 2.2 Some key socio-economic indicators of India (TERI India)

	1990s	2020s	2050s
Population projections (million)	846	1354	1888
GDP projections (Rs Crore)	886,933	5,094,093	14,298,068
Demand for food grains (million tons)	176	295	450
Required food grain productivity (tons/ha)	1.7	2.3	3.5
Demand for electricity by urban households (mtoe)	3.9	23.4	69.4

price (Kumar et al. 2008b). Its demand is expected to raise up to 100 MMT with an assumption of 6% per annum growth rate on a very conservative basis by 2020. With an approximate import dependency of 90%, energy security favours the adoption of 20% diesel blending with Jatropha biodiesel (B20). Table 2.1 shows that in India 13.4 Mha of waste land can be used for Jatropha plantations (Planning Commission 2003).

The key macro-indicators of India: population, GDP, energy requirement and food grain demand are shown in Table 2.2. The data clearly shows that with the growth of population, the per capita food intake and energy requirement also increased since the 90s. Additional population increase may result in competition between food and biofuels. Further studies are required to estimate the possible usage of waste land in sustainable harmony with flora, fauna and environment.

Social education is needed for the production and usage of proper biofuel crops according to local constraints. Concerns have been raised that cultivators may willingly opt for alternative crops with higher rate of return on biofuel production in the short term. Farming is an energy intensive activity requiring power for irrigation, ploughing and processing. The *Indian government* (GoI) initiative to promote the plantations of Jatropha saplings under National Rural Employment Guarantee Scheme (NREGS) is a sincere move towards the integration of Jatropha to energy production. In India, researchers have observed that Jatropha biodiesel and its blends with diesel can be used in existing diesel engines without any modifications (Banapurmath et al. 2008; Sahoo and Das 2009; Sahoo et al. 2009; Kumar et al. 2012a). In the longer run, the economic sustainability of Jatropha biodiesel will definitely prove to be the best bet for India as far as the economic viability of biodiesel with respect to diesel is concerned (Kumar et al. 2008c).

Table 2.3 Employment generation potential from Jatropha biodiesel production in Man Days/hectare (Adholeya and Dadhich 2008)

Stage	First year[a]	Second year
Nursery	68	0
Plantation	122	29
Post-harvesting	0	56
Oil extraction	0	15
Transesterification	0	14
Total	190	114

[a]The values in the table are given per hectare

With rapid increase in population in Asia, arable land area is decreasing and it is already only 0.1 hectare per person, on average, in several densely populated countries, which means that it cannot be used for biofuel plantations. Establishing biofuel plantations like Jatropha on degraded soils can be a win-win strategy provided that these soils are adequately restored and specific problems (e.g., nutrient and water imbalance, loss of top soil, shallow rooting depth, drought stress, salinization, compaction, crusting) alleviated.

Considering households, the average energy requirements per capita is 20% less in rural areas compared to urban areas (Pachauri 2004). It is obvious that a significant rise in energy consumption is expected from the improvement of living condition standards and population increase (Parikh and Lior 2009). Rural bioenergy is still the predominant form of energy used by people in less developed countries such as India. Thus, meeting income generation and irrigation management through renewable sources provide a large potential for sustainable development.

In rural areas, particularly in remote locations, the distribution of energy generated from fossil fuels can be difficult and expensive. Renewable energy can facilitate economic and social development in communities if projects of sustainable development are intelligently designed and carefully executed with local inputs and cooperation. In poor areas, the renewable energy projects would absorb a significant part of participants' small incomes. Investigations in this direction have been based on the following concepts namely: renewable energy sources can be replenished in a short period of time and it is clean, i.e., it produces lower or negligible levels of greenhouse gases and other pollutants when compared with the conventional energy sources they replace (Demirbas and Demirbas 2007).

One of the major synergic effects on the economic return of a state investment in biodiesel would be the availability of facilities for power generation in close proximity to the area of biomass production. Such structural investments can result in manifold increase in employment opportunities. As observed from Table 2.3 it is evident that 190 man days of employment in the first year and 114 man days in the second year per hectare for poor people living in rural areas may prove to be a potential source of income generation.

Considering the average man days for first and second year from Table 2.3 for 150 days employability in a year, Jatropha cultivation on 13.4 Mha of wasteland will result in 300 days/year employment for roughly 6.5 million people through social schemes of GoI, such as National Rural Employment Guarantee Schemes

(NREGS) for people of rural areas. Because of the uneven distribution of wealth and the large population size, India is passing through social unrest in many parts of the country leading to large scale violence in many forms. The creation and development of such local opportunities in poor rural areas would also help in relaxing social unrest due to poverty.

CDM is a potential tool for climate change mitigation that allows the compensation of an excess in an industrial country by a mitigation project to be conducted in a developing country. The fees of CDM could be paid back to the stake holders to invest in local power generation powered by Jatropha biodiesel or its blend with fossil diesel. Actually, direct use of Jatropha oil, pure or in blends with fossil diesel, would be preferable to biodiesel since it would save a huge quantity of energy by avoiding the transesterification process that is needed to transform Jatropha's oil into biodiesel. Under the CDM system, industrialized countries are committed to contribute by financing, technology transfer and other supports necessary to the success of target projects. The increased flow of these resources to developing countries is intended in principle to support their sustainable development, while at the same time reducing the global GHGs emissions since it is becoming practically impossible to achieve this commitment in the developed countries (Akorede et al. 2010). Energy production using Jatropha biodiesel could be a viable project that Annex I parties (which have ratified the Kyoto Protocol) could embark upon to assist most developing countries since it is glaringly evident that convulsive electric power supply has always been the bane of development in these countries. However, recent studies have revealed that CDM has not been able to achieve its purpose as mandated by Kyoto Protocol since its inception in 1997 due to certain factors such as delay in ratification and absence of binding targets for developing nations (Olsen 2007). On the other hand, it is widely believed that widespread adoption of localized power generation systems based on eco-friendly fuels can play a key role in creating a clean, reliable energy with substantial benefits including environmental ones.

Shifting to Sustainable Renewable Energy Regime in India

Fossil fuels will continue to play a dominant role in the energy scenario in India in the next few decades. However, conventional or fossil fuel resources are limited, non-renewable, polluting and, therefore, need to be used prudently. On the other hand, renewable energy resources are indigenous, non-polluting and virtually inexhaustible. India is endowed with abundant renewable energy resources. Therefore, their use should be encouraged in every possible way. India's energy security would remain vulnerable until the partial or total replacement of petroleum based fuels by alternative fuels developed from indigenously produced renewable feedstocks. Considering biofuels, the country has a ray of possible alternatives to warrant energy security. Biofuels are environment friendly fuels and their utilization would address global concerns about the control of carbon emissions. The transportation sector has been identified as a major polluting sector. Use of biofuels has, therefore, become

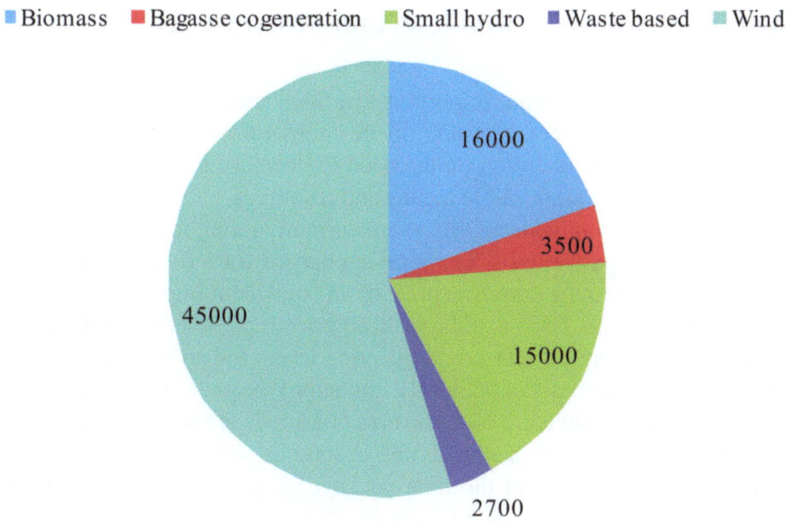

Fig. 2.1 India's technical potential of renewable power in MW (ABPS 2009)

compelling in view of the tightening automotive vehicle emission standards to curb air pollution (National Biofuel Policy 2008).

Figure 2.1 shows the huge potential capacity for renewable energy in India. Biofuels have great potential in remote areas with non-uniform topography where it is not economically sustainable to build power plants or to transport fuels. It is critical to note that India met its 50% of edible oil consumption requirements through imports to warrant a total domestic consumption of around 16.91 MMT in 2010–11 (ADM Investor Services 2011). Even with this large import dependence for edible oil, it hardly reaches an average consumption rate around 14 kg per capita, which is much lower compared to that of developed countries (44–48 kg per capita). Due to heavy dependency on import of edible oil, the production of biodiesel from edible oil as feedstock in India is neither feasible nor affordable. Therefore, it is necessary to explore non-edible oilseeds, such as Jatropha for sustainable biodiesel production as indicated by the policy statement of the Government of India (GoI) (Planning Commission 2003).

India has a total geographical area of 328 Mha out of which around 142 Mha are used for agriculture. By 2030 Indian population is expected to rise up to 1.5 billion from around 1.21 billion presently. To feed this large population it will require around 185 Mha of agricultural land even with very conservative estimates and with the assumption of constant land productivity. The promotion of biofuels at the cost of food products would have catastrophic consequences on the Indian social equity and peace. Thus, the promotion of biodiesel needs to incorporate measures and regulations to ensure the role of the agricultural industry and the first and foremost is to achieve food security for its population. National Biofuel Policy needs to focus on the increased use of waste land to promote environment friendly biofuels.

Economics of Jatropha Biodiesel and Fossil Diesel

Cost vulnerability of petroleum imports is a serious issue for the policy makers as drastic fluctuations in prices of crude oil in international market during last few years has drawn serious attention on the need to decrease the nation's dependency on fossil fuels by using all possible means and resources.

The *oil marketing companies* (OMCs) are currently rating their products from the refineries on import parity basis, i.e., a price charged for a domestically produced good that is set equal to the domestic price of an equivalent imported good. The difference between the cost price and the realized price represents the under-recoveries of OMCs. At prevailing current crude oil price level (Indian basket rules at US $ 122.95 per barrel on the basis of 06.3.2012 quotation) the under recoveries may go up to 140,000 (Exchange rate Rs./$ 50.03) crores (1 crore is ten million) for 2011–12.

Value of crude oil import, is almost 30% of total import value in India. The overall share of transport sector in the total package of petroleum, oil and lubricants (POL) demand is estimated to be 37% (Sethi 2009).

According to statistics, petro-diesel consumption was around 59.99 MMT in 2010–11 and its consumption is continuously growing in India (PPAC 2012). This requires sufficient quantity of Jatropha biodiesel for blending in required proportion at an affordable price to meet the targets of National Biodiesel Mission Phase-II and National Biofuel Policy. GoI has failed in providing the 10% ethanol blended petrol planed under its earlier directive. Important aspect of this failure seems to be non-congruence of opinion between Indian OMCs and ethanol producers related to pricing of ethanol. Availability of sufficient quantity of Jatropha biodiesel for blending with petro-diesel at an affordable price holds the key to successful adoption of bioenergy program in India. One of most important aspect to ensure timely implementation of National Biofuel Policy that should enter in application by 2017 is the economic parity of Jatropha biodiesel with petro-diesel. Table 2.4 below indicates that the expected sale prices of Jatropha biodiesel in India may be around Rs 46–50 per liter. The variation in the cost may depend on the procurement cost of Jatropha seeds.

This indicative price of locally generated biodiesel is about Rs 46.45 per liter without any taxation, is just comparable to *retail selling price* (RSP) of petro-diesel in India (Rs 40.91 per liter at Delhi as on 09-03-2012), though this prices may vary

Table 2.4 Tentative cost of biodiesel production in Indian Rupees (Rs)

Cost component	Rate (Rs/kg)	Quantity (kg)	Cost (Rs)
Seed	16	3.28	52.48
Cost of collection and oil extraction	2.36	1.05	2.478
Less cake produced	1	2.23	−2.23
Transesterification cost	6.67	1	6.67
Less cost of glycerol produced	50	0.095	−4.75
Cost of biodiesel per kg	—	—	54.65
Cost of biodiesel per liter (specific gravity of 0.85 at 15 ^{0}C)	—		46.45

Table 2.5 Central excise and customs Tariff—updated from 2011 June 25 (PPAC 2011)

	Customs			Central excise	
Particulars	Basic customs duty	Additional customs duty (CVD)	Additional customs duty	Basic cenvat duty	Additional excise duty
HSD	2.5%	NIL	Rs 2.00/liter	Nil	Rs 2.00/l
HSD (branded)	—	—	—	Rs 3.75/l	Rs 2.00/l

Table 2.6 Sales tax rate on diesel in Indian states w.e.f. 2011 Dec 01 (PPAC 2011)

State	Rate (%)
Chhattisgarh	25.00
Gujarat	24.63
Madhya Pradesh	24.23
Maharashtra	24.00
Kerala	22.83
Andhra Pradesh	22.25
Tamil Nadu	21.43
Uttaranchal	19.24
West Bengal	18.97
Goa	18.00
Delhi	12.24

from state to state and from producer to producer. Capped and subsidized pricing leads to under recoveries for OMCs. Present petro-diesel costing results in about Rs 10.94 under recovery for OMCs at US $ 120.99 per barrel crude oil pricing for Indian basket (Fortnight, 2012 Feb 16–29). The tax component includes Rs 7.59 comprising customs duty as Rs 1.07 per liter, (specific excise duty as Rs 2.06 per liter and value added tax (VAT) as Rs 4.46 per liter (PPAC 2012). At same rate of taxation, biodiesel costs around Rs 54 per liter. Hence, if under-recovery is also added in present petro-diesel pricing, actual selling price of petro-diesel comes about Rs 52–53 per liter, which is well comparable with computed price of biodiesel.

As clearly evident in Table 2.5, the GoI has levied basic customs duty at the rate of 2.5% at cost and freight price plus import charges along with additional customs duty of Rs 2.00/liter. Excise duty at the rate of Rs. 2.00/l is also applicable in overall price buildup of diesel. Table 2.6 shows state sales tax (VAT) charged by certain states in India. Data clearly reveals that overall taxes/duties are in order of 40–45% of the total selling price.

Petrol is being sold at market-based pricing and diesel is still being sold at capped pricing, this issue may itself be a serious hindrance in production of Jatropha biodiesel and its economics in India. Considering the fact that Indian economy is heavily dependent on crude oil import bill, the National Policy on Biofuels was formulated by GoI to achieve 20% blending of biofuel in both petrol and petro-diesel by 2017. The policy has suggested removal of all the central taxes levied on biodiesel and accorded 'declared goods' status, which will ensure a uniform 4% VAT on the products across the country. Moreover, the policy envisages having certification following the norms of *Bureau of Industrial Standards* (BIS) for blending of biofuel in both petrol and diesel.

Economics of Blended Diesel with Jatropha Biodiesel

Even after the declaration of Biodiesel Purchase policy by the government (MoPNG 2005), there is a lot of confusion among farmers, industrial companies, NGOs, etc. concerning a possible higher economic returns from Jatropha plantations.

OMCs have the largest network for selling of liquid fuels and they would be the natural choice for providing the marketing support for biodiesel. The BIS specifications, BIS 15607:2005 for pure biodiesel (B100) is already available considering all the important properties of biodiesel.

R&D efforts carried out in the transportation sector have already established that 20% biodiesel blended with petro-diesel requires no modification in engine specifications when used as fuel.

Fig. 2.2 reveals that government of India is receiving huge amount of revenue from petroleum sector in the form of royalty, oil development cess, custom and excise duties and sales tax. Moreover the taxes on petrol are more than that on diesel at present. Taxation policy of GoI is itself biased towards diesel. Thus, the GoI may feel that loss in revenue may be one of the reasons to delay introduction of biodiesel blended diesel in Indian market. One of most important aspect is that introduction of biodiesel will indeed help to reduce the import bill burden. It is sure that government is going to lose around 20% of revenue on account of selling the diesel blended with 20% Jatropha biodiesel when Jatropha biodiesel pricing is pegged with petro-diesel selling price under zero duty regime. GoI in its Biofuel policy unveiled in September 2008 has decided to set 2017 as deadline to sell B20 blend in the country. But the presented data itself tells the reasons for the delay in the announcement of that policy.

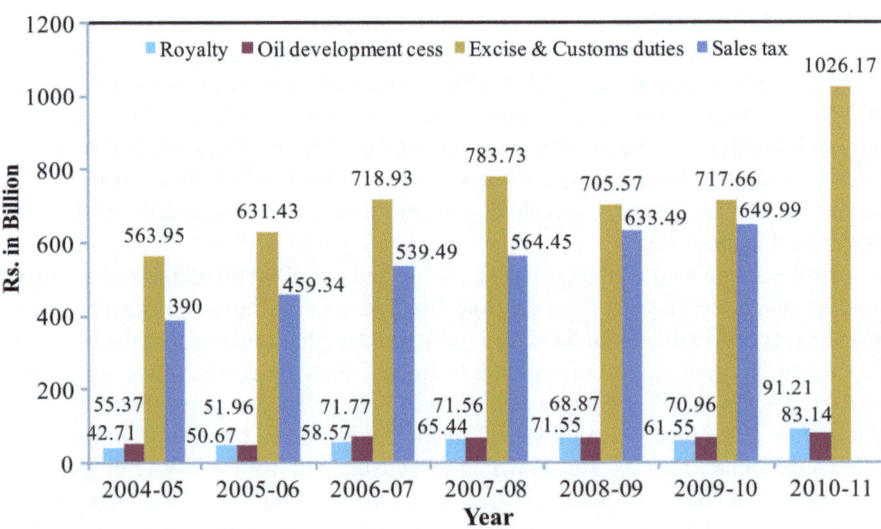

Fig. 2.2 Contribution of oil sector to centre/state government resources (MoPNG 2011)

Impact of Petroleum on Indian Economy

Petroleum is the largest consumer of foreign exchange reserves in India as all the payment for the yearly petroleum bill is made in US dollars. Table 2.7 presents substantial yearly increase in import of crude oil.

The fact that crude oil prices rose to US $ 147 per barrel in July 2008 and that crude oil is currently being traded in the range of US $ 95–115 per barrel (Fig. 2.3), shows the volatile nature of the crude oil pricing in the international market. Moreover the import of crude oil is made by payment in US dollars. In July 2008 it was priced at approximately Rs. 40 but now it is priced at around Rs 52–53. On account of these factors OMCs in India are reporting huge under-recoveries. Impact on inflation is also visible due to rise in price of Indian crude basket. It is clearly evident that economic health of a developing country like India is severely influenced by price of crude oil and its products.

Cost volatility of petroleum imports is a serious issue for the policy makers as drastic fluctuations in prices of crude petroleum oil in international market during the last 2 years has drawn serious attention on meeting nation's energy needs by using all possible sources.

Table. 2.7 Import bill for crude oil in India in crores (PPAC 2011)

Year	2003–04	2004–05	2005–06	2006–07	2007–08	2008–09	2009–10	2010–11
Crude oil	83,528	117,003	171,702	219,029	272,699	348,304	375,277	455,276

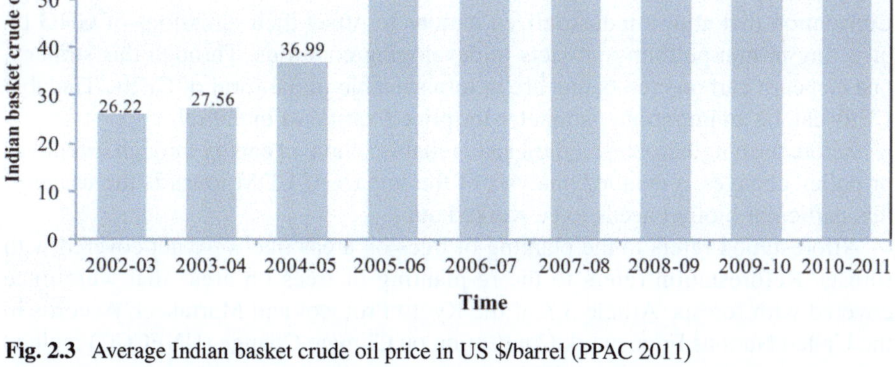

Fig. 2.3 Average Indian basket crude oil price in US $/barrel (PPAC 2011)

Table 2.8 Sector-wise consumption of diesel in India (Sethi 2009)

Sector	Consumption (%)
Agriculture	13.8
Power generation (Gen sets)	9.7
Passenger vehicles	18.9
Commercial vehicles	50.4
Industry	7.2

Table 2.9 Biodiesel emissions compared to petro-diesel (Planning Commission 2003)

Emissions	B100 (%)	B20 (%)
Total unburned hydrocarbons	–93	–30
Carbon monoxide	–50	–20
Particulate matter	–30	–22
NOx	13	2
Carbon dioxide	–80	—

Diesel consumption stands at almost forty percent of total petroleum products consumed and its sector wise consumption is shown in the Table 2.8 below. Diesel consumption in India as per the available statistics stands at 59.99 MMT in 2010–11(PPAC 2012).

At the rate of 8% increase, the diesel consumption may reach a figure of 95.6 MMT by the year 2016–17. Available data states total number of vehicles on road will be around 372.7 millions by 2035, which will require 371 million ton of oil equivalent (mtoe) of fuel to power India (ADB 2006).

Impact of Biodiesel on CDM

GHGs like CO_2 are reduced when using biodiesel or diesel blended with biodiesel as shown in Table 2.9. The CDM could be used to support investment in Jatropha-based fuel production. The CDM is a provision of the UN Climate Change Convention that allows industrialized nations to offset their emissions of GHG by investing in non-polluting projects in developing countries. Through this strategy, one can earn carbon credits that are in turn saleable in the form of CERs. Tradable CERs can be an important parameter for price fixation of biodiesel.

Before pushing Jatropha biodiesel as a reliable source of energy through incentives or policy changes, a detailed analysis of the impact of CDM towards the usage of this particular biofuel needs to be worked out.

Afforestation refers to the planting of trees on areas that were not covered with forests. Reforestation refers to the re-planting of trees on areas that were once covered with forests. Article 3.3 of the Kyoto Protocol and Marrakech Accords in the United Nations Framework Convention on Climate Change (UNFCCC) include

Table 2.10 CO_2 sequestration by Jatropha plantations

Year	First	Second	Third	Fourth	Fifth
Growth of tree (%)	20	40	60	80	100
25 kg CO_2 sequestration/plant at maturity level	5	10	15	20	25
Total CO_2 sequestered in kg for 1,600 plants/hectare	8,000	16,000	24,000	32,000	40,000
Total CO_2 sequestered in ton for 1,600 plants/hectare	8	16	24	32	40
Amount accrued from CER[a] in US$ at $10.47/CER	83.76	167.52	251.28	335.04	418.8

[a]price of CER on NCDEX Mumbai, India on 15th Dec 2011

afforestation and reforestation as possible measures to reduce carbon dioxide level in the atmosphere. Afforestation and reforestation (AR) are defined by the UNFCCC as direct human-induced conversion of non-forested land to forested land through planting, seeding and/or the human-induced promotion of natural seed sources. Afforestation and reforestation differ only in that the activity will be afforestation if the land on which it takes place had not been forested for at least 50 years, whereas reforestation refers to land that did not contain forest before 1990 for reforestation project activities. The first phase of CDM till the year 2012 is open to reforestation and afforestation projects in developing countries like India. The main features of a possible CDM project state that it should sequester up to 8,000 tons of carbon annually and low-income communities should implement the project.

Jatropha plantations projects are entitled for trade of CER for CO_2 sequestration. Assuming a density of 1,600 plants per hectare, Jatropha will produce, 5 years after plantation, about 200 kg of biomass, including roots with a dry matter content of about 25 % , i.e., a biomass of 80 tons dry matter per hectare. About half of that weight is CO_2, i.e., 40 tons (Jatropha plant has a very light wood with a density ranging from 0.33 to 0.37) (Hooda and Rawat 2006). The trade of certified emission certificates pays around US $10.47 (spot traded price of Rs 562 per CER on NCDEX India as on Dec 15th, 2011, at exchange rate of Rs. 53.68 per US $) per ton of CO_2 sequestration at current market price, which is about US $ 679.2 per hectare as shown in Table 2.10 below.

Considering 13.4 million hectares of waste land, one may calculate that US $ 5.61 billion (13.4 million * 418.8) may be earned by virtue of CERs in the context of CDM due to Jatropha plantations and biodiesel production, which is bound to improve the economic viability of Jatropha biodiesel in India.

On one hand, the combustion details of petro-diesel ($C_{16}H_{34}$), assuming a net calorific value of 43.38 GJ/t, are modeled by equation (2.1) (Peterson and Hustrulid 1998):

$$C_{16}H_{34} + 24.5O_2 + 92.12N_2 \rightarrow 16\,CO_2 + 17H_2O + 92.12N_2 \quad (2.1)$$

On the other hand, the combustion of biodiesel ($C_{18}H_{34}O_2$) can be represented by the equation (2.2) (Rajesh et al. 2008):

$$C_{18}H_{34}O_2 + 25.5O_2 + 3.76N_2 \rightarrow 18\,CO_2 + 17H_2O + 95.88N_2 \quad (2.2)$$

As shown above in (2.1), 3.11 kg of CO_2 is produced for each kg of diesel fuel used, whereas as per (2.2), 2.81 kg of CO_2 is produced for each kg of biodiesel fuel used. Hence, the same amount of CO_2 would be produced for an equivalent energy content of biodiesel as shown in (2.2). Available data indicates that Jatropha biodiesel has a calorific value of 38,000 kJ/kg while diesel has 43,380 kJ/kg. This means that 14.15% more Jatropha biodiesel (mass basis) is necessary to produce the same energy as diesel, whereas the theoretical carbon balance shows that 11.07% more Jatropha biodiesel would produce the same amount of CO_2. The difference is mainly due to the oxygen content associated with biodiesel. Thus, the combustion of biodiesel and petro-diesel emits about the same amount of CO_2 for similar energy output. Reported *life cycle assessment* (LCA) analysis of Jatropha biodiesel shows that CO_2 emissions are reduced by 80–85% as compared to petro-diesel due to its renewable nature.

Net energy ratio (NER) defined as ratio of energy output to energy input can be taken as an indicator of energy efficiency of any fuel. For petro-diesel, reported value of NER is 0.8328 (Sheehan et al. 1998). Whereas for Jatropha biodiesel under rainfed conditions and without allocating primary process energy to co-products (like: glycerine and seed cake) the reported NER is 1.2 (Kumar et al. 2012b). NER may be improved further either with higher seed yield or by reducing process energy requirement for conversion of Jatropha oil into biodiesel. Transesterification process requires 20% of total process energy. Use of Jatropha oil in existing diesel engines requires preheating due to its very high viscosity.

Conclusions

The greatest impediment to overall development in India is the positive rate of urbanization and population growth. It is necessary to define the fraction of farmland, waste land or barren land that could be used for the production of biodiesel in a sustainable manner without conflicting with food security and environmental issues. Biofuel development can provide supplementary incomes for rural people including the eradication of gender disparity. Proper credit mechanism needs to be developed to help small farmers to cultivate and sell crop profitably.

Renewable energy generation in rural areas may be a viable option of climate change mitigation in India under CDM. Governmental policies may need to be adjusted for subsidizing renewable power for the poor rural areas, the so called *rural poor* (http://www.fao.org/worldfoodsummit/english/fsheets/ifad.pdf).

Development of localized biodiesel supply chain may help in providing valuable energy to fulfill the daily needs of rural people. Worldwide extensive research is going on to develop diesel engine that can run efficiently on biodiesel or even pure oil. Design and development of efficient chulhas (cook-stove) for cooking using Jatropha biodiesel may be a turning point in improving the living standards of the rural poor. Thus, after analyzing all the aspects it can be concluded that, production and utilization of Jatropha biodiesel as prospective fuel needs top most priority in India to achieve sustainable energy regime.

It is desired that world agriculture needs to keep up with continuous growth of world population and, moreover, to develop further in order to reduce the number of undernourished people and to promote health and welfare. However, in the use of these advancements the precautionary principle is imperative in order that food safety, as well as environment protection and biodiversity should be ensured. Otherwise, unwise application of technological tools may further deteriorate human health and environmental quality and compromise future development of human societies.

It is worth mentioning that the undernourishment and the food security problems in the world are critical; thus, their relation to the production of biofuels must be studied. Countries with a better climatic and land potential for the development of biofuels have significant possibilities of developing their agricultural regions, which can improve the population's life condition substantially, by rising their income.

The role of governments to elaborate regulatory marks regarding the use and distribution of land is of utmost importance, given that one of the possible disadvantages of biofuel programs may be the concentration of land ownership. Such resource concentration may generate more poverty, monoculture, forest destruction and ultimately aggravate the environmental impacts. The environment destruction itself may worsen the socio-economic network and generate additional poverty entering a vicious circle.

The major constraint in adopting the Jatropha biodiesel is to provide required marginal lands for its production, developing a supply chain and ensuring its price competitiveness with petro-diesel. Another major area of concern is to restrict the expansion of biodiesel production to land normally dedicated to food crops in order to avoid unnecessary food security threats in India. Thus a major challenge is to devise a public policy to promote the cultivation of Jatropha only on marginal and waste lands, but with incentive to facilitate its integration to the socio-economic system. These incentives may count with the creation of seed collection centers, warranty on oil prices and price parity with any other possible alternative energy source so that its production can be made on a competitive and sustainable basis. Present Jatropha biodiesel economics show that:

(a) The local price of biodiesel, around Rs 46.45 per liter without any taxation, is just comparable to that of petroleum diesel in India (Rs 40.91 per liter at Delhi). Present petro-diesel cost is around Rs 10.94 under recovery for oil marketing companies at 120.99 $ per barrel of crude oil for Indian basket.
(b) Present petro-diesel pricing at Delhi includes Rs 7.59 of taxes. At a same rate of taxation, biodiesel costs around Rs 54 per liter. Hence, if under-recovery is also added in present petro-diesel pricing, actual selling price of petro-diesel comes at about Rs 53 per liter, which is well comparable with biodiesel.
(c) For 13.4 Mha of waste land US $ 5.61 billion (13.4 million * 418.8) may be earned by virtue of CERs in the context of CDM due to Jatropha plantations and biodiesel production. CDM is bound to improve the economic viability of Jatropha biodiesel in India.

(d) 13.4 Mha Jatropha cultivation may result in 300 days/year employment for roughly 6.5 million people.
(e) LCA analysis of Jatropha biodiesel shows that CO_2 emissions are reduced by 80–85% as compared to petro-diesel due to its renewable nature.

Considering the importance of energy as the backbone for economic development of a nation like India, it is of paramount importance to look for a sustainable alternative to petro-diesel. Current volatilities in crude price have already worsened the overall economic situation of major OMCs. The need of the hour is to study and implement the biodiesel program, which is already successfully operational in countries like Germany. Economics are strongly in favour of Jatropha biodiesel to be a sustainable alternative to diesel.

References

ABPS (2009) Report on conceptual framework for renewable energy certificate mechanism for India, submitted to Ministry of New and Renewable Energy, Government of India. Available from http://mnre.gov.in/pdf/MNRE_REC_Report.pdf

Adholeya A, Dadhich PK (2008) Production and technology of bio-diesel: seeding a change. The Energy and Resources Institute (TERI), New Delhi, pp 1–9

ADM Investor Services (2011) Market outlook for Europe, Russia and India. Available from http://www.admisi.com/assets/49/Outlook-Europe-Russia-India-20110414.pdf

Akorede MF, Hizam H, Pouresmaeil E (2010) Distributed energy resources and benefits to the environment. Renew Sust Energ Rev 14:724–734

Altenburg T, Dietz H, Hahl M, Nikolidakis N, Rosendahl C, Seelige K (2009) Biodiesel in India. Value chain organisation and policy options for rural development. Bonn, German Development Institute. Available from http://www.die-gdi.de/CMS-Homepage/openwebcms3.nsf/(ynDK_contentByKey)/ANES-7PKDWS/$FILE/Studies%2043.2009.pdf

Asian Development Bank (ADB) (2006) Energy efficiency and climate change considerations for onroad transport in Asia, Version 9a. Available from http://www.adb.org/Documents/Papers/Energy-Efficiency-Transport/CCTS.pdf

Banapurmath NR, Tewari PG, Hosmath RS (2008) Performance and emission characteristics of a DI compression ignition engine operated on Honge, Jatropha and sesame oil methyl esters. Renew Energy 33:1982–1988

Demirbas A (2005) Biodiesel production from vegetable oils by super critical methanol. J Sci Ind Res 64:858–865

Demirbas AH, Demirbas I (2007) Importance of rural bioenergy for developing countries. Energy Convers Manage 48:2386–2398

Department for International Development (DFID) (2005) Energy as a key variable in eradicating extreme poverty and hunger: a gender and energy perspective on empirical evidence on MDG #1. Available at http://www.dfid.gov.uk/r4d/PDF/Outputs/Energy/R8346_mdg_goal1.pdf

Hooda N, Rawat VRS (2006) Role of bio-energy plantations for carbon-di-oxide mitigation with special reference to India. Mitig Adopt Strateg Glob Change 11:445–467

Kumar S, Chaube A, Jain SK (2008a) Jatropha biodiesel a promising C.I. engine alternate fuel in India. Indian J Appl Life Sci 4:1–5

Kumar S, Chaube A, Jain SK (2008b) Jatropha biodiesel: a prominent renewable biofuel in India. Indian J Appl Life Sci 4:14–19

Kumar S, Chaube A, Jain SK (2008c) Economic sustainability of Jatropha biodiesel in India. J Environ Res Dev 3:292–300

Kumar S, Chaube A, Jain SK (2010) Issues pertaining to substitution of diesel by Jatropha biodiesel in India. J Environ Res Dev 4:877–884

Kumar S, Chaube A, Jain SK (2012a) Experimental evaluation of C.I. engine performance using diesel blended with Jatropha biodiesel. Intl J Energ Environ 3:471–484

Kumar S, Singh J, Nanoti SM, Garg MO (2012) A comprehensive life cycle assessment (LCA) of Jatropha biodiesel production in India. Bioresour Technol. Article in press

Ministry of Petroleum and Natural Gas (MoPNG) (2005) Bio-diesel purchase policy, Government of India. Available from www.petroleum.nic.in/Bio-Diesel.pdf

Ministry of petroleum and natural gas (MoPNG) (2011) Basic statistics on Indian petroleum & natural gas, Government of India. Available from http://www.petroleum.nic.in/petstat.pdf

National Autofuel Policy (2003) Available from www.petroleum.nic.in/autopolicy.pdf. Accessed 2003

National Oilseeds & Vegetable Oils Development (NOVOD) Board (2007) Jatropha. An alternate source for biodiesel. Available from http://www.novodboard.com/Jatropha-english.pdf

National Oilseeds & Vegetable Oils Development (NOVOD) Board (2008) 3rd R&D report on tree borne oilseeds 2007–2008. Available from http://www.novodboard.com/3rd%20R&D-Report.pdf

National Policy on Biofuels (2008) Ministry of new & renewable energy, Government of India. Available from www.mnre.gov.in/policy/biofuel-policy.pdf

Olsen KH (2007) The clean development mechanism's contribution to sustainable development: a review of the literature. Clim Chang 84:59–73

Pachauri S (2004) Elasticities of electricity demand in urban Indian households. Energy Policy 32:1723–1735

Parikh J, Lior N (2009) Energy and its sustainable development for India, Editorial Introduction and commentary to the special issue of energy the international journal. Energy 34:923–927

Peterson CL, Hustrulid T (1998) Carbon cycle for rapeseed oil biodiesel fuels. Biomass Bioenergy 14:91–101

Petroleum Planning and Analysis Cell (PPAC) (2011) Ministry of petroleum and natural gas, Government of India. Available from http://www.ppac.org. Accessed 9 Mar 2012

Planning Commission (2003) Report of the committee on development of bioFuel, Government of India. Available at http://planningcommission.nic.in/reports/genrep/cmtt_bio.pdf. Accessed 16 Apr 2003

Planning Commission (2006) Report of the expert committee on integrated energy policy, Government of India. Available at http://planningcommission.nic.in/reports/genrep/rep_intengy.pdf. Accessed Aug 2006

Rajagopal D, Sexton SE, Roland-Holst D, Zilberman D (2007) Challenge of biofuel: filling the tank without emptying the stomach? Environ Res Lett 2:1–9. doi:10.1088/1748-9326/2/4/044004

Rajesh S, Raghavan V, Shet USP, Sundararajan T (2008) Analysis of quasi-steady combustion of Jatropha bio-diesel. Intl Commun Heat Mass Transf 35:1079–1083

Reubens B, Achten WMJ, Maes WH, Danjon F, Aerts R, Poesen J et al (2011) More than biofuel? *Jatropha curcas* root system symmetry and potential for soil erosion control. J Arid Environ 75:201–205

Sahoo PK, Das LM (2009) Combustion analysis of Jatropha, Karanja and Polanga based biodiesel as fuel in a diesel engine. Fuel 88:994–999

Sahoo PK, Das LM, Babu MKGP, Arora P, Singh VP, Kumar NR, Varyani TS (2009) Comparative evaluation of performance and emission characteristics of jatropha, karanja and polanga based biodiesel as fuel in a tractor engine. Fuel 88:1698–1707

Senthilkjumar V, Gunasekaran P (2005) Bioethanol production from cellulosic substrates: engineered bacteria and process integration challenges. J Sci Ind Res 64:845–853

Sethi V (2009) IEA seminar on global oil market outlook & stability, New Delhi, India. Available at http://www.iea.org/work/2009/India_oil/ppac.pdf

Sheehan J, Cambreco V, Duffield J, Garboski M, Shapouri H (1998) Life cycle inventory of biodiesel and petroleum diesel for use in an urban bus. Final report by US Department of Agriculture and US Department of Energy. Available at http://www.nrel.gov/docs/legosti/fy98/24089.pdf

The Energy and Resources Institute India (TERI). Socio-economic scenarios for climate change impacts in India. Keysheet 3. Available at https://www.decc.gov.uk/assets/decc/what%20we%20do/global%20climate%20change%20and%20energy/tackling%20climate%20change/intl_strategy/dev_countries/india/india-climate-3-socio-econ.pdf

Urban F, Benders RMJ, Henri CM (2009) Energy for rural India. Appl Energy 86:S47–S57

World Commission on Environment and Development (WCED) (1987) Our common future. Oxford University Press, New York

World Economic Outlook (WEO) (2011) International Monetary Fund. Available at http://www.imf.org/external/pubs/ft/weo/2011/02/weodata/weoreport.aspx

Chapter 3
Status of Bioenergy Research and Jatropha in India: A Review

Prathibha Devi

Introduction

Escalating world fuel prices, growing demand for energy, concerns about environmental effects of fossil fuel use and global warming are the key issues facing mankind today. The increasing interest in renewable energy sources or bioenergy around the world has focused on biofuels, which are of biological and renewable origin, like bioethanol, biodiesel, and biomass for energy. Bioenergy appears to offer hope for addressing these concerns while also providing new opportunities for farmers in developing and developed countries alike. Bioenergy is already making a substantial contribution to meeting this global energy demand. It could sustainably contribute between a quarter and a third of global primary energy supply in 2050. It is the only renewable source that can replace fossil fuels in all energy markets—in the production of heat, electricity, and fuels for transport. India is positioned to maximize the gains from large availability of solar, wind, biomass and hydro-energy sources. India, with rapidly growing economy, is increasingly looking to biofuels to meet its energy needs. India is currently the world's fourth largest producer of ethanol and is expanding its biodiesel industry. India also has long term projects using biogas and has started new initiatives to use bioenergy to supply electricity, heat and light to isolated rural populations (Gonsalves 2006).

A specific plant or substance used for bioenergy is called a feedstock. Feedstocks are usually converted into a more easily usable form, usually a liquid fuel. Forestry, agricultural and municipal residues, and wastes are the main feedstocks for generation of electricity and heat from biomass. In addition, a very small share of sugar, grain, and vegetable oil crops are used as feedstocks for the production of liquid biofuels. Today, biomass supplies some 50 EJI globally, which represents 10% of

P. Devi (✉)
Molecular Genetics & Biotechnology Laboratory, Department of Botany,
Osmania University, Hyderabad 500 007, India
e-mail: prathi56@yahoo.com

N. Carels et al. (eds.), *Jatropha, Challenges for a New Energy Crop: Volume 1: Farming, Economics and Biofuel*, DOI 10.1007/978-1-4614-4806-8_3,

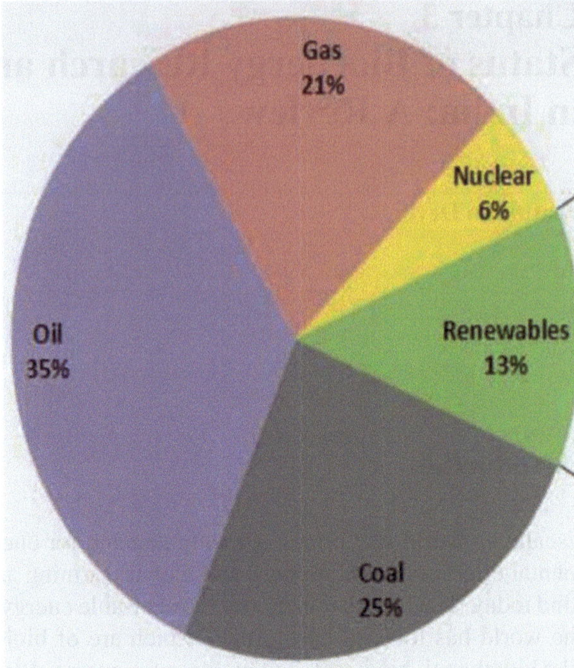

Fig. 3.1 Share of bioenergy in the world primary energy mix (Source: IEA 2006; IPCC 2007. 11 EJ=1018 Joules (J)=1015 kilojoules (kJ)=24 million tonnes of oil equivalent (Mtoe))

global annual primary energy consumption and is mostly traditional biomass used for cooking and heating (Fig. 3.1).

Bioenergy development in the Asia-Pacific region is faced with many challenges. Fuelwood and charcoal remain the dominant energy sources in the region (Heruela 2010). Biofuels can contribute to climate change mitigation by substituting fossil fuels. The Asia-Pacific region led by China and India had been experiencing the fastest economic growth in the world. Economic growth in many countries is leading to consumption patterns that approach those of industrialized countries. The rapid growth is contributing significantly to increasing environmental pressure, with serious consequences for the environment, human health and long-term prosperity.

Across the developing world, several crops are now being examined for their suitability for production of biofuels. Many bioenergy routes are being used to convert a range of raw biomass feedstocks into a final energy product. The productivity of biomass feedstocks needs to be increased through improved agricultural practices and biotechnological methods (Naresh 2008; Leela et al. 2009, 2011; Naresh et al. 2012). Developing countries, especially the tropical and sub-tropicals have a comparative advantage for bioenergy production because of greater availability of land, favorable climatic conditions for agriculture and lower labour costs. However, they face socio-economic and environmental implications affecting the potential to benefit from the increased local and global demand of bioenergy. The interrelationship between land uses and the competing needs of energy and food security is a key issue in bioenergy that should be considered. In addition, the effects of large-scale bioenergy production on global commodity prices are a significant trade concern. Bioenergy production may also entail harmful environmental effects such as deforestation and loss of biodiversity. Regulation is required to reduce the negative impacts of large-scale production, as well as to ensure the use of technologies with the best cost-effective ratios and rates of energy conversion. The stages of bioenergy development from research to commercialization have already been done and are supported by the government policies and regulation commitments. These include the know-how for the construction of new bioenergy plants, development of plantations for production of raw material and marketing policies. However, rural self-sufficiency energy programmes need to be developed for biofuel in order to open job opportunities and eliminate poverty at isolated or remote villages by empowering the society to fulfill their energy needs.

Biofuels have the potential of helping to power the transportation sector that is responsible for around 30% of current energy usage. There has been a lot of debate on the production of biofuels from food crops since they may endanger food security by displacing food resources from their original target. In addition, food crop are not necessary the most productive concerning biofuels. Therefore, production of biofuels must be cost effective, environmentally safe and sustainable. Biofuels need to be integrated within a broader framework of investment in rural infrastructure and human capital.

Extensive research has shown that *Jatropha curcas* is the most favourite biodiesel oilseed as it offers several advantages, such as its ability to grow on marginal lands with low inputs, easy propagation, low gestation period, uselessness for grazing animal like cattle or sheep, pest resistance, high seed yield and oil content, and production of protein rich manure. Large-scale plantation schemes should therefore be promoted to generate employment and income and make biodiesel affordable to the poor. Several types of biofuels are presented below followed by a discussion on the latest status of biodiesel-yielding plants including Jatropha and the future outlook and prospects of biodiesel.

Types of Biofuels

Liquid Biofuels

Liquid biofuels are used to supplement or replace traditional petroleum-based transportation fuels and can be used in existing vehicles with little or no modification to engines and fueling systems. They can also be used for heating and electricity production. The two most common kinds of liquid biofuels are ethanol and biodiesel, but other liquid fuels are being developed.

Bioethanol

Ethanol as biofuel has been known since early 1900 with the Ford motor of Model T running on corn ethanol. Ethanol is currently produced in large quantities by fermenting the sugar or starch portions of agricultural raw materials. The feedstocks used for ethanol production include sugar cane in Brazil, maize in North America, sugar beets in France, etc. The top three ethanol producers are Brazil, the US and China. Because producing ethanol from sugar and starch competes with food production, new methods are being developed to produce ethanol from cellulose, which makes up the bulk of all plants and trees. Cellulosic ethanol is referred to as a second-generation biofuel. Jerusalem artichoke (*Helianthus tuberosus* L.) is one of the most suitable materials for ethanol production as it contains nearly 20% of carbohydrates, of which 70–90% is inulin. Researchers at The Institute for Sustainable and Renewable Resources, in Danville, Virginia, are studying the potential use of Jerusalem artichoke as a feedstock for producing ethanol as an alternative for farming communities affected by declining tobacco and textile markets (Christiansen 2009). Ethanol production from Jerusalem artichoke by *Zymomonas mobilis* was reported by Tatcha et al. (2007).

Biodiesel

Biodiesel refers to any diesel-equivalent biofuel made from renewable biological materials, such as vegetable oils or animal fats consisting of long-chain saturated hydrocarbons (methyl or ethyl esters of long chain fatty acids). It is produced by chemically processing oils (transesterification) obtained from both edible oil plants, such as rapeseed, soybean and the fruits of oil palms and non-edible oil plants like Jatropha and Pongamia. Waste cooking oil can also be converted to biodiesel (IEA Bioenergy)

Other Liquid Biofuels

- *Biobutanol*: Butanol (called "biobutanol" if derived from biomass) is an alcohol similar to ethanol. Butanol is currently more expensive to produce than ethanol.
- *Pure Plant Oil (PPO)*: PPO (also known as SVO, Straight Vegetable Oil) is a diesel type fuel that occurs naturally in plant oils such as rapeseed, Jatropha, etc. Waste cooking oil can also be converted to PPO which is used in its pure form or mixed with diesel/biodiesel.
- *Biokerosene*: Kerosene is used to power jet engines. However, biokerosene is not produced for large-scale aviation because aviation fuels need to meet special requirements. A synthetic biokerosene is produced using biomass feedstocks.
- *Pyrolysis oil*: Also known as "bio-oil", is a liquid biofuel created through pyrolysis.
- *Biomass-to-liquids (BTL)*: Chemical processes that transform biomass into liquid fuels.
- *Biogasoline*: It is under development by Shell and Virent Energy Systems, biogasoline is a synthetic form of gasoline made from cellulosic feedstocks.

Gas Biofuels

Gas biofuels include Biogas or Biomethane, Biopropane and Synthetic natural gas (SNG).

Solid Biofuels

Solid biofuels include wood, manure or charcoal burned as fuel as well as more recent innovations like high-density clean burning pellets. Solid biomass can be burned for heat or to produce electricity either by itself or as part of a co-firing power plant.

Wood: Wood can be utilized for bioenergy in the following forms, such as, firewood, wood charcoal (charcoal), wood-fired biomass boilers, wood gasification (wood gas)—especially waste wood, wood pellets and wood residues (waste wood).

Charcoal: Char or biomass-derived black carbon, is a form of charcoal produced from biomass and commonly used as a fertilizer or soil amendment. Char may also have the potential to sequester large amounts of carbon in the soil. Much of the interest in char has been stimulated by research on *terra preta*, "dark earth" in the Amazon.

Biomass pellets: Biomass can be pressed into pellets. Due to their low moisture content, regular shape and high density, pellets can be burned very efficiently and are often used for heating or electricity generation.

Biomass

Biomass is the most important renewable energy source used in the world today, but the utilization of bioenergy is very low and needs to increase (Mayfield and Foster 2008). Large scale utilization of biomass for energy is still limited to a few countries, the largest user being United States which generates more than 70% of its energy needs. Energy, economic, and environmental benefits can be derived from the use of biomass for bioenergy and bio-based products.

On a world-wide basis, roughly 80% of the total energy supply comes from fossil fuel sources, 13% from renewable energy sources, and 7% from nuclear power. In the United States, 86% of the total energy consumed is obtained from fossil fuels, 6% from renewable energy sources, and 8% from nuclear power. With China, Brazil and India in the lead, the world is going through a fast industrialization process, which is significantly increasing the global demand for fossil fuel energy. In this context, biomass emerges as an attractive modern energy source provided it can be economically justified. Many countries have realized the need to improve the efficiency of energy generation, distribution, consumption, to harness local resources as a way to increase the security of the energy supply, reverse fossil fuel dependency, and to improve trade balance. The formation of biofuel markets is likely to benefit developing countries, which, in general, have favorable conditions for growing biomass. Further, biofuels produced from energy crops can make an important contribution to reduction of greenhouse gas emissions (Gutterson and Zhang 2009).

Biodiesel

The concept of using vegetable oil as a fuel dates back to 1895 when Dr. Rudolf Diesel developed the first diesel engine to run on vegetable oil. While there are numerous interpretations being applied to the term biodiesel, the term biodiesel usually refers to an ester, or an oxygenate, made from the oil and methanol (in other words, the name "biodiesel" can be applied to any transesterified vegetable oil that makes it suitable for use as a diesel fuel). Recent advances in oil extraction and transesterification could reduce the cost of making biodiesel from plant oils. Biodiesel can be used in diesel engines either as a stand alone in pure form (B100) or blended with petro diesel. Much of the world uses a system known as the "B" factor to state the amount of biodiesel in any fuel mix. For example, fuel containing 20% biodiesel is labeled B20, while pure biodiesel is referred to as B100.

Biodiesel can be derived from the triglycerides (fats) of either plants or animals, though a very large percentage of biodiesel. The plant-based bio-diesel is sometimes known as botadiesel, and the animal-based biodiesel as zoodiesel (Source:
http://www.oilgae.com/energy/sou/ae/re/be/bd/def/def.html
http://www.bdpedia.com/biodiesel/research/research.html
http://www.castoroil.in/reference/plant_oils/uses/fuel/bio_fuels%20may%2009%202006.html).

Zoodiesel: Biodiesel from animal fat is much less prevalent than biodiesel from plants, but a good amount of research is going on in this area. A few studies suggest that biodiesel from animal fats could cost significantly less (about 20%) than that from plant oils because animal fat is cheaper than fats from plant oils.

Botadiesel: As mentioned earlier, plant oils form the feedstock for a very large percentage of the biodiesel in use today. While the most popular plant oils exploited for biodiesel are from sunflower, soybean, Jatropha, corn, canola, safflower and rapeseed oil, experiments are ongoing with many more plants to check if their oils could be suitable candidates for biodiesel. More information on the various plant oils that can be used for biodiesel is given in a later section.

Biodiesel from Algae

Biodiesel from algae has been widely discussed among experts in the petroleum industry and conservationists who are looking for a more reliable and safer source of energy that is both renewable and easy to attain. One of the key reasons for consideration of algae as feedstock for oil is their yields. The DOE (Department of Energy, Govt of USA) has reported that algae yields 30 times more energy per acre than land crops such as soybeans, and some estimate even higher yields up to 56,781.1 litres per acre. Besides keeping the earth clean and free from pollution, the algal biodiesel fuels help to utilize a resource that is available in abundance just waiting to be harnessed and exploited. Once the algae are grown and harvested, there are different ways of extracting the oil. Regardless of the extraction method, the resulting product is a vegetable oil called 'green crude', similar to crude oil, which is further transformed into biodiesel through a 'transesterification' process which is similar to that used for the seed-oil of Jatropha or Pongamia.

Algae represent the third generation feedstock for biodiesel, with much higher yields than second generation crops like Jatropha and Pongamia. Algae yields could reach as much as 50 tons of biodiesel per ha per year against 2 tons for competing feedstocks, such as Jatropha. On lipid content alone (in terms of % of lipid content by dry weight), algae score over all other oil crops in terms of the yield of biomass. Oil yields per unit area from algae can be further increased, and it is one of the most researched topics currently. There could be some challenges in converting algae oil into biodiesel using the transesterification process owing to the high FFA of algae oil. Some recent publications in the area of biodiesel from algae include Cheng et al. (2009). In this study, *Chlorella protothecoides* utilized hydrolysate of Jerusalem artichoke tuber as carbon source and accumulated lipid *in vivo*, with lipid content as high as 44% by dry mass, and a carbon source to lipid conversion ratio of about 25% in a 4-day scale cultivation.

Why Is Jatropha so Exciting from a Renewable Energy Standpoint?

In India, the Center for Jatropha Promotion and Biodiesel has issued reports identifying up to 13 additional (in addition to Jatropha) non-food oilseeds and trees that can be utilized for biofuel production and carbon emissions control. The Center has identified Camelina (*Camelina sativa*), Flax (*Linum usitatissimum*), Jojoba (*Simmondsia chinensis*), Jerusalem artichoke (*Helianthus tuberosus*), Kenaf (*Hibiscus cannabinus*), Karanja (Pongamia: *Pongamia pinnata*), Kokum (*Garcinia indica*), Moringa (*Moringa oleifera*), Mahua (*Madhuca indica*), Neem (*Azadirachta indica*), Castor bean (*Ricinus communis*), Simarouba (*Simarouba glauca*) and Tumba (*Citrullus colocynthis*) as the most promising. Among these, Jojoba is native to the semiarid regions of southern Arizona, southern California and northwestern Mexico. Rajasthan's state government has allotted 110 ha of wastelands including 70 ha for Jojoba plantation. Pongamia is one of the few nitrogen fixing trees to produce seeds containing 30–42% oil. Mahua is one of the forest tree with a large potential of non-edible oil production of about 60 million tons per annum in India. The oil extracted from the castor bean already has a growing international market, assured by more than 700 uses, ranging from medicines and cosmetics to biodiesel, plastics and lubricants. Each ha of castor bean planted in arid and semi-arid regions produces 350–900 kg of oil per ha. Although seed oil from corn (*Zea mays*) has been in use, activity in the field of bioenergy has been accelerating in the last few years with oil from palm, soybean, rape and from many other oilseeds (apart from those mentioned above), such as *Shorea robusta, Mesua ferra* and *Mallotus philippines*, with areas under cultivation rapidly increasing (Naresh 2008; Leela 2010; Naresh et al. 2012). Unlike the first generation biodiesel and biofuel crops, such as soybean or corn, *J. curcas* is a non-food crop used as a feedstock for second-generation biofuel and presents one of the most exciting possibilities as a future solution to energy problems, especially that of transportation fuel. *J. nana* has also been presented as a promising biofuel source (Bhagat and Kulkarni 2009).

Jatropha may thrive in non-agricultural and marginal lands not suitable for food crops and can use waste water for its growth. Fertilizer and pesticide requirements as well as crop management costs, are relatively lower for Jatropha than for many other energy crops. Jatropha is a perennial and it yields oil for over 30 years. Jatropha is being cultivated in almost all the continents and a large number of countries for oil and byproducts (Abreu 2008; Achten et al. 2008). India, Malaysia, African nations and other Asian countries can greatly benefit from cultivation of biodiesel crops to meet their ever growing diesel requirements. Jatropha is an "energy species" that has to be domesticated as a "tree crop" (Heller 1996; Henning 2008; Makkar and Becker 1997; NOVOD 2007; Openshaw 2000; Wiesenhütter 2003).

Studies that would focus as much on yield as on performance of biodiesel are being undertaken around the world. The main challenge of Jatropha promotion in rural areas would come from the communities for whom the scope of petrocrop adoption would need to be attractively and profitably packaged along with a demystified plantation and processing technology. Site-specific cultivation packages and agroforestry models for Jatropha need to be developed and mass mobilization/awareness campaigns designed and implemented to institutionalize the process and to achieve the desired scale of Jatropha plantation. However, many of the actual investments and policy decisions on developing Jatropha as an oil crop have been made without the backing of sufficient scientific knowledge. Realizing the true potential of *Jatropha* requires separating facts from the claims and half-truths.

The analyses of *Jatropha* seeds revealed that the percentage of crude protein, crude fat and moisture were 24.6%, 47.2% and 5.5%, respectively (Akintayo 2004). The seeds can be transported without deterioration and at low cost due to their high specific weight. Jatropha seeds contain 30–40% oil that can be easily expressed for processing (transesterification) and refinement to produce biodiesel (Akintayo 2004; Gubitz et al. 1999; Mahanta et al. 2008; Parawira 2010). The byproducts of biodiesel processing are nitrogen-rich press cake and glycerol, which are said to have good commercial value as fertilizer and as a base for soap and cosmetics, respectively. Makkar et al. (1998) found that crude protein was 56% in Cape Verde, 61% in Nicaragua, 56% in Nigeria and 64% in non-toxic Mexican *J. curcas* varieties.

A Jatropha tree absorbs around 8 kg of CO_2 every year. About 2,500 trees can be planted in a ha, thus, resulting in 20 tons of CO_2 sequestration per year for the lifetime of 40–50 years. Moreover, each ha produces an average of 3,785.4 litres of biodiesel per year and 3,500 kg of biomass. The use of biodiesel results in the reduction of 3.2 kg CO_2/litre produced by diesel. At 78% efficiency, biodiesel will reduce 2.5 kg of CO_2/litre or 9.2 tons of CO_2 for every ha of plantation. The biomass produced after the oil extraction will further result in carbon reduction based on the amount of electricity generated from it.

Cultivation, Economic Importance and Other Aspects of Jatropha

The essential practices that will enhance existing Jatropha cropping systems, and bring the crop to an ideal model for industrial cultivation include:

- Selection of elite lines of Jatropha for higher productivity per acre and their molecular fingerprinting (Basha and Sujatha 2007; Ranade et al. 2008; Leela et al. 2009).
- Developing cost-effective commercial nurseries
- Healthy practices to reduce incidence of diseases
- Value addition to Jatropha projects with use of co-products.

(Source: European Biodiesel Board, Malaysian Palm Oil Board, National Biodiesel Board, USA http://www.osti.gov/bridge/servlets/purl/924080-y8ATDg/924080.PDF).

The three main methods of Jatropha cultivation are nursery raising, direct seeding in the main field and vegetative propagation. The best and recommended method for Jatropha cultivation is nursery method. Micropropagation is catching up in a big way (Sujatha et al. 2005; Naresh 2008; Dhillon et al. 2009; Leela 2010; Leela et al. 2011; Naresh et al. 2012). It starts bearing from the third year, but is commercially viable from the fifth year. Optimal use of water is the most important criteria for growing Jatropha. Minimum rainfall required to produce fruits is 600 mm/ha/year.

The seed yield of Jatropha varies under different agro-climatic conditions and planting density. The average seed yield per ha from the fifth year for Jatropha under irrigated farming is about 5 tons. Cultivation costs of Jatropha are estimated to be in the range of US$ 994/year per ha, including seed or sapling cost, land preparation and sowing, which however, varies based on the region. The total cost for extracting and refining Jatropha oil through expeller press process is US$ 0.37 per gallon and US$ 0.5 through solvent extraction process. Based on the data from various surveys reported by the Comprehensive Jatropha Report (2011), the average seed prices ranged from US$ 0.05 to 0.25 in Africa, US$ 0.2 to 0.4 in Latin America, and US$ 0.2 to 0.4 in Asia.

The overall Jatropha oil manufacturing process is: Sowing—>Cultivation—>Harvest—>Seed Dehulling and Cleaning—>Oil Extraction—>Oil Filtration and Purification—>Oil Refining. Jatropha meal is the extract left behind after the extraction of oil from Jatropha seeds. Currently, it can be sold on the market as an organic fertilizer and for producing biogas, among others. The toxins in the Jatropha meal can be denatured or inactivated by heat and the detoxified Jatropha meal can be used as animal feed. The efficiency and profitability of the Jatropha biofuel industry increases with the commercial usage of by-products like Jatropha cake and glycerol generated during the process of oil extraction.

In addition to its emerging use as a biofuel, Jatropha oil could be a good source of industrial enzymes, biopolymers and biopharmaceuticals. Similar to other sources of biomass, different parts of the Jatropha plant (and not just its seeds) can be used to produce a wide range of fuels, using a range of processes—chemical, thermo chemical and biochemical processes.

Jatropha Plantations in India

The National Mission on Jatropha Biodiesel

The Government of India started its BioFuel mission in April 2003, on development of BIO-FUEL, under the auspices of the Planning Commission of India, and it presented its report that recommends a major multi-dimensional programme to replace 20% of India's diesel consumption. The National Planning Commission has integrated the Ministries of Petroleum, Rural Development, Poverty Alleviation and the Environmental Ministry and others. It has announced the BioFuel Policy on September 11th, 2008. The Union Cabinet in its meeting gave its approval for the National Policy on Biofuel prepared by the Ministry of New and Renewable Energy as well as for setting up an empowered National Biofuel Coordination Committee, headed by the Prime Minister of India and a Biofuel Steering Committee headed by Cabinet Secretary. One of the objectives was to blend petro-diesel with a planned 13 million tons of bio-diesel by 2013 (>l000 times compared to the present world Jatropha cultivation and production) produced mainly from non-edible Jatropha oil and, on a smaller part, from Pongamia. To plant 11 million ha of Jatropha, the program is to become a "National Mission" and has to mobilize a large number of stakeholders including individuals, communities, entrepreneurs, oil companies, business, industry, the financial sector as well as Government and most of its institutions.

The National Mission on Biodiesel, is therefore proposed in two phases as below:

1. Phase I consisting of a Demonstration Project was aimed to be implemented by the year 2006–07 with an investment of Rs. 1,500 crore ($300 million) on 400,000 ha.
2. As a follow up of the Demonstration Project, Phase II consisted of a self sustaining expansion of the programme that began in the year 2007 and led to produce the biodiesel required in 2011–12.

About 200 districts in the states of Andhra Pradesh, Bihar, Chhattisgarh, Jharkhand, Gujarat, Goa, Himachal Pradesh, Haryana, Karnataka, Kerala, Madhya Pradesh, Maharashtra, Orissa, Punjab, Rajasthan, Tamil Nadu have been identified on the basis of availability of wasteland, rural poverty ratio, *below poverty line* (BPL), census and agro-climatic conditions suitable for Jatropha cultivation (Figs. 3.2 and 3.3). Each district is treated as a block and under each block 15,000 ha of Jatropha

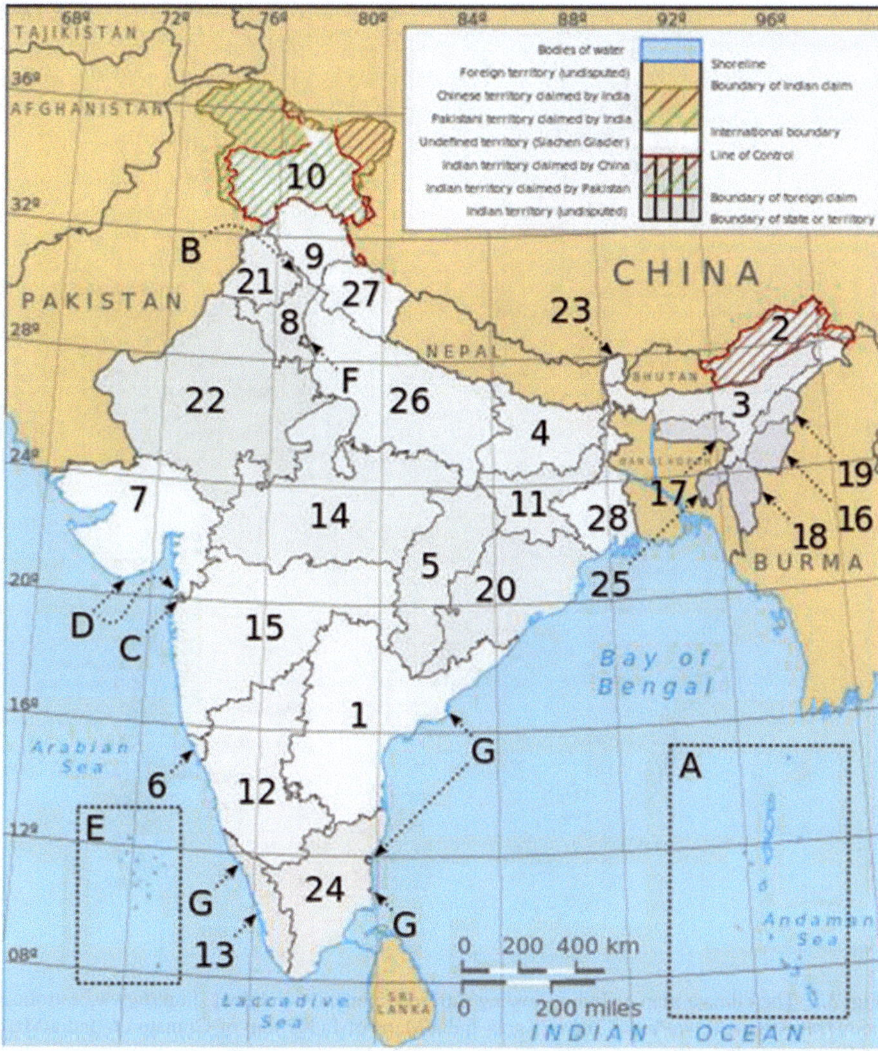

Fig. 3.2 A clickable map of India showing 28 states and seven union territories of India (WIKIPEDIA).

States

Andhra Pradesh
Arunachal Pradesh
Assam
Bihar
Chhattisgarh
Goa
Gujarat
Haryana
Himachal Pradesh
Jammu and Kashmir
Jharkhand
Karnataka
Kerala
Madhya Pradesh
Maharashtra
Manipur
Meghalaya
Mizoram
Nagaland
Orissa
Punjab
Rajasthan
Sikkim
Tamil Nadu
Tripura
Uttar Pradesh
Uttarakhand
West Bengal

Union territories

Andaman and Nicobar Islands
Chandigarh
Dadra and Nagar Haveli
Daman and Diu
Lakshadweep
National Capital Territory of Delhi
Pondicherry

Fig. 3.3 The climate map of India showing different climatic conditions. http://www.besttofind.com/Travel/Asia-Travel/India-Travel/Know-India/Maps-of-India/Map-of-Climate-of--India/Map-of-Climate-of--India.htm.

plantation will be undertaken through below poverty line (BPL) farmers. It is proposed to provide green coverage to about three million ha of wasteland through plantation of Jatropha in the identified districts over a period of 3 years. The details of plantations being developed in India are given in Table 3.1.

Biodiesel is now being produced in India for use in stationary engines and in trains, trucks and tractors. Biodiesel is rapidly replacing both kerosene (which was used illegally and inefficiently) and diesel as a more efficient, cheap, and clean alternative for large engines. A lot of planning is going on to cultivate non-edible oilseed trees such as Pongamia on barren land to use their oil for biodiesel production. The Railways of India have been very supportive of the program and various state governments recommend planting these plants in unused areas through aid from the government sectors (Table 3.1).

Table 3.1 Details of Jatropha research and plantations in India

Institute, University or Company	Particulars of research and plantations
The Government of India started National BioFuel mission in April 2003 and announced BioFuel Policy on 11th September 2008 with a view to complete it by 2013	One of the objectives is to blend petro-diesel with a planned 13 Million t of biodiesel by 2013. The aims included the following: • To plant 11 Million ha of Jatropha, to mobilize a large number of stakeholders including individuals, communities, entrepreneurs, oil companies, business, industry, the financial sector as well as Government and most of its institutions • About 200 districts in some potential states have been identified on the basis of availability of wasteland, rural poverty ratio, below poverty line (BPL), census and agro-climatic conditions suitable for Jatropha cultivation • It is proposed to provide green coverage to about three Million ha of wasteland through plantation of Jatropha in the identified districts over a period of 3 years The National Mission on Biodiesel, was proposed in two phases as below: • Phase I consisted in a Demonstration Project to be implemented by the year 2006–07 with an investment of Rs. 1,500 crore ($300 million) on 400,000 ha • Phase II consisted in a self sustaining expansion of the programme that began in 2007 and led to produce the biodiesel required in 2011–12
Department of Biotechnology (DBT), Ministry of Science and Technology, Government of India (2006–2012)	The DBT has planted five lakh plants, and brought 200 ha of area under plantation. Further, it has initiated two Jatropha missions, *viz.* the Biofuel Mission and the Jatropha Mini Mission to select good germplasm, develop quality planting material, and standardize agro-techniques
Council of Scientific and Industrial Research (CSIR), Government of India (2005–2015)	The Council of Scientific and Industrial Research (CSIR) has initiated a network programme for genetic improvement in association with the industry under its prestigious New Millennium Technology Leadership Initiative (NMTLI) programme
National Bureau of Plant Genetic Resources, Regional Station, Ranchi, India (2003–2012)	Several projects are in progress including the ongoing CSIR funded project on genetic improvement of Jatropha
Other Indian National Scientific laboratories engaged in the field (2004–2012)	The Indian institutes carrying out researches on different aspects of Jatropha including the genetic improvement and crop management practices are: Central Salts and Marine Chemicals Research Institute (CSMCRI), Karnataka Watershed Development Agency (KAWAD), Social Uplift Through Rural Action (SUTRA), and Central Research Institute for Dryland Agriculture (CRIDA)

(continued)

Table 3.1 (continued)

Institute, University or Company	Particulars of research and plantations
Indian Ministry of Environment and Forests and the Forest Departments of several State Governments (2004–2012)	They are working on thousands of ha of Jatropha plantations in India as part of a Clean Development Mechanism (CDM) project
Indian Railways, Government of India (2004–2012)	They planned to plant Jatropha along 25,000 kilometers route along the railway track and to replace 10% of their total petro-diesel consumption with Jatropha oil. The project was started on a pilot scale
	Further, the Indian Railways has started to use the Jatropha oil (blended with diesel fuel in various ratios) to power its diesel engines with great success
	Currently the diesel locomotives that run from Thanjavur to Nagore (south India) section and Tiruchirapalli to Lalgudi, Dindigul and Karur (south India) sections of India run on a blend of Jatropha and diesel oil
Indian oil marketing (Government owned) companies (2004–2012)	The three National oil marketing companies—Indian Oil Corporation (IOC), Hindustan Petroleum Corporation Limited (HPCL) and Bharat Petroleum Corporation Limited (BPCL)—are planning to take up cultivation of Jatropha across more than 180,000 acres in two states of India
	• IOC and HPCL are forming a joint venture with the Chhattisgarh government to take up large-scale Jatropha farming across 74,100 acres and 37,000 acres, respectively. IOC is also planning to enter UP as well as some other states
	• BPCL has formed a company called Bharat Renewable Energy in association with Hyderabad-based Nandan Biomatrix—an R&D company—and Shapoorji Pallonji Company, for producing biodiesel from Jatropha in Uttar Pradesh across 70,000 acres. The company will be investing Rs 2,200 crore in the next 7 years to produce one million tons of biodiesel from Jatropha plantations
	• Individual farmers are ready to plant Jatropha in the waste lands and want buy-back agreement like contract farming with good price (but the price offered presently is generally very low). Bundelkhand area (of Uttar Pradesh and Madhya Pradesh states) has lot of waste land (600,000–700,000 ha), which is best suited for Jatropha and on the other hand Districts of Eastern Uttar Pradesh) have saline and waste lands, which are also suitable for plantation

(continued)

Table 3.1 (continued)

Institute, University or Company	Particulars of research and plantations
Several State Governments including those of Andhra Pradesh, Tamil Nadu, Haryana, Uttaranchal and Rajasthan. (2004–2012)	They are implementing developmental schemes on *Jatropha curcas* and programmes to cultivate thousands of ha. Some of them are: • Andhra Pradesh: Andhra Pradesh government under its Rain Shadow Area Development (RSAD) has allocated about 15 lakh ha of land for growing Jatropha and Karanja for producing oil for biodiesel. Andhra Pradesh has entered into a formal agreement with Reliance Industries for Jatropha planting. The company has selected 80 ha of land at Kakinada to grow Jatropha for high quality bio-diesel fuel. • Rajasthan: The Centre for Jatropha Promotion & Biodiesel has proposed to bring 2.2 lakh ha under Jatropha. Announced an aggressive plan to produce two million tons of biodiesel a year from Jatropha seeds starting in 2012 • Maharashtra: In September 2007, the HPCL joined hands with the Maharashtra State Farming Corporation Ltd (MSFCL) for a Jatropha seed-based bio-diesel venture. As part of the project, a Jatropha plant would be grown on 200 acres in Nasik and Aurangabad. In November 2005, the Maharashtra Government aimed to cultivate Jatropha on several hundreds of ha in the state, with half the land going to the public sector and the other half to the private sector. On July 1 2006, Pune Municipal Corporation took the lead among Indian cities in using biodiesel from Jatropha in over 100 public buses • Karnataka: Farmers in semi-arid regions of Karnataka are planting Jatropha as it is well suited to those conditions • Haryana: The Planning Board is growing Jatropha on 20,000 ha. 19 districts have gone for Jatropha plantation. In 12 districts, 820 acres of land has been brought under Jatropha cultivation with the involvement of 146 g panchayats • Chhattisgarh: Planning to bring at least one million ha of land under Jatropha cultivation by 2012 • Uttaranchal: The Uttaranchal Biofuels Board has brought two lakh ha under Jatropha plantation by 2012. Here, plantation in 29 *panchayats* covering an area of 350 ha has been done • Kerala: Kerala is planning a massive Jatropha planting campaign
Several Agricultural Universities of India (2005–15)	Jatropha plantation and genetic improvement projects are in progress

(continued)

Table 3.1 (continued)

Institute, University or Company	Particulars of research and plantations
Several Central and State Universities of India (2004–2012)	Jatropha improvement projects including the project funded by UGC at Osmania University in Andhra Pradesh in which, a total of 48 accessions have been analyzed through molecular methods and several of the accessions have been raised in the plantation (Naresh 2008; Leela et al. 2009, 2011; Naresh et al. 2012)
Private Companies: Reliance Company, Centre for Jatropha Promotion & Biodiesel (CJP), Tree Oils Company, Nandan Biomatrix, SG Biofuels, etc. (2006–2020)	• Sources indicate that Reliance Industries is also in talks with the Chhattisgarh government for a similar venture. RIL's arm, Reliance Life Sciences, has taken up a 50-acre Jatropha plant pilot project in Kakinada, the landfall point of the gas from its blocks in the Krishna-Godavari basin (Andhra Pradesh) • Taken up Jatropha plantation in an area of few thousand ha in Andhra Pradesh/Rajasthan/Maharashtra/Gujarat states. CJP is a private company comprised of farmers, scientists, botanists and economists with Jatropha plantation of around 42,000 ha • Tree Oils, Nandan Biomatrix and SG Biofuels companies have several acres with Jatropha plantation in A.P. state and venturing into seed oil production plants. Nandan Biomatrix is venturing into industrial scale Jatropha projects established in the states of Uttar Pradesh, Orissa, Gujarat, Karnataka, Andhra Pradesh and Madhya Pradesh
Non-governmental organizations engaged in the field (2002–2012)	Several NGOs have brought more than 14,000 ha under Jatropha cover with a target of 100,000 ha. They are: Rural Community Assistance Corporation (RCAC), women self-help groups, Child Development and Rehabilitation Centre (CDRC), Society for Rural Initiatives for Promotion of Herbals (SRIPHL)
Daimler Chrysler and Hohenheim University (2004–2009)	They have successfully conducted a research project in two different climatic zones of India of 20 ha of Jatropha trees planted on wastelands

Biodiesel blends are being used to run state transport corporation buses in Karnataka state. The University of Agriculture Sciences at Bangalore has identified many elite lines of Jatropha. Large-scale plantations have been initiated in North-East India and Jharkhand by D1 Williamson Magor Bio Fuel Limited, a joint venture between D1 Oils of U.K. and Williamson Magor of India. The hilly areas of the North-East are ideal for growing hardy, low-maintenance plants. Indian Oil Corporation has tied up with Indian Railways to introduce biodiesel crops over one million square kilometers. Also, Jharkhand and Madhya Pradesh have tied up with Indian Oil to cultivate large tracts of land with Jatropha. In order to organize the industry, the BioDiesel Society of India has been formed to encourage energy plantations for increasing feedstock supplies.

Jatropha Research

Conventional agriculture uses vegetative cuttings and seed cultivation of *J. curcas*. There are several problems of seed cultivation like high heterogeneity in seeds, difficulty in maintaining uniformity in planting materials, seasonal practices and control of seed borne diseases which may be transmitted to seedlings. Further, the quality in terms of oil content cannot be ensured because plants are heterozygous. However, a breeding programme with targeted pollination could lead to elite varieties (http://www.lib-pdf.com/ppt/coconut-oil-process-production.html). Whereas, vegetative propagation through cuttings may produce about 100–500 clones from one mother plant, tissue culture through micropropagation will yield thousands of clones. Hence, a good protocol of micropropagation needs to be developed to propagate elite varieties of Jatropha (http://ec.europa.eu/research/agriculture/agenda_jatropha.htm).

Initial Jatropha cultivation should start with genetically superior strains in terms of its oil content and oil quality. However, long term strategy of production of biofuel will centre on biotechnological approaches, which only can assure high oil content and quality. These would require large amount of quality planting materials to meet the demand in the future http://www.lib-pdf.com/ppt/coconut-oil-process-production.html

To meet the huge demand for supply of quality Jatropha planting material, efficient propagation (clonal and micropropagation) methods are being improvised (Sujatha and Mukta 1996; Rajore and Batra 2005; Sujatha et al. 2005; Nannapat et al. 2006; Datta et al. 2007; Naresh 2008; Leela 2010; Leela et al. 2011; Naresh et al. 2012). Micropropagation using nodal explants resulted in minimum genetic variation (Pierik 1991).

Biodiesel Business Worldwide

The Jatropha industry is in its very early stages, covering a global area estimated at 900,000 ha. More than 85% of Jatropha plantings are in Asia, chiefly Myanmar, India, China and Indonesia. Africa accounts for around 12% or approximately 120,000 ha, mostly in Madagascar and Zambia, and also in Tanzania and Mozambique. Latin America has approximately 20,000 ha of Jatropha, mostly in Brazil. The Jatropha plantations are projected to grow to 12.8 million ha by 2015. By then, Indonesia is expected to be the largest producer in Asia with 5.2 million ha, Ghana and Madagascar together will have the largest area in Africa with 1.1 million ha, and Brazil is projected to be the largest producer in Latin America with 1.3 million ha (Gexsi 2008).

Biodiesel Business in India

While the total global biodiesel production was 7.2 million tons in 2006 it has gone up to 32 million tons in 2012, India's biodiesel production is less than 10,000 tons a year. India, with the world's second largest population and rapidly growing economy, is looking towards biofuels to meet its energy needs. India is currently the world's fourth largest producer of ethanol. India also has long term projects using biogas as well as a variety of initiatives to use bioenergy to supply electricity, heat and light to isolated rural populations (An Assessment of the Biofuels Industry in India, UNCTAD 2011 web report).

After the announcement of the Indian Biofuel Policy in 2008, around 150 companies, led by state-owned oil marketing companies and Reliance Industries (RIL), the country's largest private company, evinced interest in production of biodiesel. The move came in response to the government's plan to introduce a 20% blending of the non-fossil fuel with diesel. Apart from the data presented on Indian private companies in Table 3.1 (Reliance Company, Centre for Jatropha Promotion & Biodiesel (CJP), Tree Oils Company, Nandan Biomatrix, SG Biofuels, etc.), the biodiesel producing companies in India include Mumbai-based Bharat Renewable Energy and the government-owned Hindustan Petroleum who have already planted more than a million acres of Jatropha to provide a million metric tons of biodiesel by 2015. Naturol Bioenergy, in Hyderabad (in Andhra Pradesh), has a 300 tons facility and has been shipping biodiesel to the EU and United States since 2008 because India can't blend the alternative fuel into gasoline and diesel yet. Two other Indian biodiesel companies have followed Naturol's lead *viz.*, Cleancities Biodiesel, which began commercial production in 2008 and started shipping biodiesel to United States. Emami Group, a third company in Kolkata, has established a biodiesel plant in the port city of Haldia (West Bengal) and started biodiesel production from waste cooking oil and palm oil.

Other recent biodiesel companies are (An Assessment of the Biofuels Industry in India, UNCTAD 2011 web report): the Jatropha mission of Uttar Pradesh (a joint venture of BPCL), Nandan Biomatrix, Shapoorji Pallonji, and Shirke Biohealthcare (supported by Uttar Pradesh Government) that produces 10 tons of biodiesel daily and supplies to Pune Municipal Transport. They have set up a 60,000 ha plantation in Papua New Guinea and a 300,000 ha project in Indonesia. Jain Irrigation (Jalgaon in Maharashtra) has 42 ha under Jatropha and is setting up a 150,000 tons per day plant in Chhattisgarh with one million ha to be brought under Jatropha. Nova Bio Fuels (Panipat in Haryana) is planning a 30 tons per day plant. Reliance Energy has a pilot project of 80 ha in Jamnagar (Gujarat). Bharat Petroleum has invested US$ 9.4 million in Andhra Pradesh and D1 Oils UK has invested Rs. 6 crores in Tamil Nadu (Fig. 3.2).

Future Prospects

Declining fossil fuel reserves and growing concern about carbon dioxide emissions that are driving climate change have focused world attention on the need to reduce dependence on fossil fuels. In turn, this has increased interest in promoting bioenergy, including biofuels, as a renewable energy source. Bioenergy could sustainably contribute between a quarter and a third of global primary energy supply by 2050. It is the only renewable source that can replace fossil fuels in all energy markets—in the production of heat, electricity, and fuels for transport (Bauen et al. 2009).

Liquid biofuels can be used in transportation with only a few changes to the existing distribution infrastructure and have become an extremely important form of bioenergy. Producing liquid biofuels from food crops using conventional technology is also being pursued for rural development. However, the use of biofuels is creating a lot of uncertainty and will continue to do so until it can be shown that biofuels can be competitive, environmental friendly, renewable, sustainable and do not endanger food security.

India is highly dependent on fuel imports to satisfy its energy needs. Under this situation, Jatropha has emerged as a possible solution for India's biodiesel demands of the future due to its suitability to Indian climate and its multiple advantages including the ability to grow on marginal lands, low cost of production and high potential for employment generation. The yield of Jatropha is also influenced by the planting material and management practices. Usually, the average earnings from Jatropha cultivation range from Rs. 10,000–30,000 per acre. The increasing support from the government for Jatropha cultivation (in India) and the financial institutions coming forward to finance farming activities, the market value of Jatropha is expected to increase in the coming decade. Jatropha cultivation and biofuel production around the globe is all set to come of age in the years ahead.

References

Abreu F (2008) Alternative by-products from Jatropha. In: International consultation on pro-poor Jatropha development. IFAD, Rome. Available at http://www.ifad.org/events/jatropha/2008

Achten WMJ, Verchot L, Franken YJ, Mathijs E, Singh VP, Aerts R et al (2008) Jatropha bio-diesel production and use. Biomass Bioenergy 32:1063–1084

Akintayo ET (2004) Characteristics and composition of *Parkia biglobbossa* and *Jatropha curcas* oils and cakes. Bioresour Technol 92:307–310

Basha SD, Sujatha M (2007) Inter and intra population variability of *Jatropha curcas* characterized by RAPD and ISSR markers and development of population specific SCAR markers. Euphytica 156:375–386

Bauen A, Göran B, Martin J, Marc L, François V (2009) Bioenergy—a sustainable and reliable energy source. A review of status and prospects, IEA Bioenergy Netherlands Report

Bhagat BR, Kulkarni DK (2009) *Jatropha nana* Dalz. & Gibs.: a plant for future energy. In: The Jatropha system an integrated rural approach. The Jatropha journal, an online journal- http://www.jatropha.de/Journal/index.htm

Cheng Y, Wenguang Z, Chunfang G, Kenneth L, Yang G, Qingyu W (2009) Biodiesel production from Jerusalem artichoke (*Helianthus tuberosus* L.) tuber by heterotrophic microalgae *Chlorella protothecoides*. J Chem Tech Biotechnol 84:777–781

Christiansen RC (2009) Belgian ethanol plant begins production. In: Ethanol producer magazine: web exclusive posted 2 Jan 2009, at 10:20 a.m. CST. http://www.ethanolproducer.com/articles/5212/belgian-ethanol-plant-begins-production/

Comprehensive Jatropha Report 2011, Bottlenecks, cultivation trends. 220 pages last updated Feb 2011 http://www.biozio.com/ref/report/jat/jatropha_biodiesel.html

Datta MM, Mukherjee P, Ghosh B, Jha BT (2007) *In vitro* clonal propagation of biodiesel plant (*Jatropha curcas* L.). Curr Sci 93:1438–1442

Dhillon RS, Hooda MS, Pundeer JS, Ahlawat KS, Kumari S (2009) Development of efficient techniques for clonal multiplication of *Jatropha curcas* L., a potential biodiesel plant. Curr Sci 96:25

GEXSI—The Global Exchange for Social Investment 2008 Global market study on Jatropha. Final report. Prepared for the World Wide Fund for Nature (WWF). London/Berlin, 2008

Gonsalves JB (2006) An assessment of biofuels industry in India. In: United nations conference on trade and development lucas Assunção, climate change programme and biotrade initiative, UNCTAD

Gubitz GM, Mittelbach M, Trabi M (1999) Exploitation of the tropical oil seed plant *Jatropha curcas* L. Bioresour Technol 67:73–82

Gutterson N, Zhang J (2009) Important issues and current status of bioenergy crop policy for advanced biofuels. Biofuels Bioprod Bioref 3:441–447

Heller J (1996) Physic nut. *Jatropha curcas* L. Promoting the conservation and use of underutilised and neglected crops, vol 1. Institute of Plant Genetics and Crop Plant Research/International Plant Genetic Resources Institute, Gatersleben/Rome

Henning RK (2008) Identification, selection and multiplication of high yielding Jatropha curcas L plants and economic key points for viable Jatropha oil production costs. Paper presented to: international consultation on pro-poor Jatropha development Rome, IFAD

Heruela CS (2010) Biomass energy status, opportunities and challenges in the Asia-pacific region.,IEA, Paris http://www.fftc.agnet.org/library/eb/617/

Leela T, Naresh B, Reddy MS, Madhusudhan NCH, Prathibha C (2011) Morphological, physico-chemical and micropropagation studies in *Jatropha curcas* L. and RAPD analysis of the regenerants. Appl Energy 88:2071–2079

Leela T, Suhas PW, Seetha K, Naresh B, Sreedevi TK, Hoisington DA et al (2009) AFLP-based molecular characterization of elite germplasm collection of *Jatropha curcas* L., a biofuel plant. Plant Sci 4:505–513

Leela T (2010) Molecular characterization, physicochemical analysis and micropropagation of *Jatropha curcas* L. Ph. D. thesis Osmania University, Hyderabad, India

Mahanta N, Gupta A, Khare SK (2008) Production of protease and lipase by solvent tolerant *Pseudomonas aeruginosa* PseA in solid state fermentation using *Jatropha curcas* seed cake as substrate. Bioresour Technol 99:1729–1735

Makkar HPS, Aderibigbe AO, Becker K (1998) Comparative evaluation of non-toxic and toxic varieties of *Jatropha curcas* for chemical composition, digestibility, protein degradability and toxic factors. Food Chem 62:207–215

Makkar HPS, Becker K (1997) Potential of *Jatropha curcas* seed meal as a protein supplement to livestock feed, constraints to its utilisation and possible strategies to overcome constraints. In: Gübitz GM, Mittelbach M, Trabi M (eds) Biofuels and industrial products from *Jatropha curcas*. Symposium "Jatropha 97", Managua

Mayfield C, Foster D (2008) Understanding bioenergy resources www.forestencyclopedia.net/p/p1140

Nannapat T, Chockpisit T, Aree T (2006) In vitro induction of shoots and roots from Jatropha curcas L. explants. J Hort Sci Biotech 83:106–112

Naresh B, Srikanth Reddy M, Vijayalakshmi P, Veena R, Prathibha D (2012) Physico-chemical screening of accessions of *Jatropha curcas* for biodiesel production. Biomass Bioenergy 40:155–161

Naresh B (2008) Physicochemical characterization and molecular analysis of *Jatropha curcas* L. Ph.D. thesis, Hyderabad, India

NOVOD (2007) *Jatropha*—An alternate source for biodiesel. New Delhi: National Oil Seeds and Vegetable Oils Development Board. Available at http://www.novodboard.com

Openshaw K (2000) A review of *Jatropha curcas*: an oil plant of unfulfilled promise. Biomass Bioenergy 19:1–15

Parawira W (2010) Biodiesel production from *Jatropha curcas*: a review. Sci Res Essays 5:1796–1808

Pierik RLM (1991) Commercial micropropagation in Western Europe and Israel. In: Debergh PC, Zimmerman RH (eds) Micropropagation technology and applications, Kluwer, The Netherlands, 155–65

Rajore S, Batra A (2005) Efficient plant regeneration via shoot tip explants in *Jatropha curcas* L. J Plant Biochem Biotech 14:73–75

Ranade SA, Srivastava AP, Rana TS, Srivastava J, Tuli R (2008) Easy assessment of diversity in *Jatropha curcas* L. plants using two single-primer amplification reaction (SPAR). Biomass Bioenergy 32:533–540

Sujatha M, Makkar HPS, Becker K (2005) Shoot bud proliferation from axillary nodes and leaf sections of non-toxic *Jatropha curcas* L. Plant Growth Regul 47:83–90

Sujatha M, Mukta N (1996) Morphogenesis and plant regeneration from tissue cultures of *Jatropha curcas* L. Plant Cell Tiss Organ Cult 44:135–141

Tatcha O, Thanonkeo P, Thanonkeo S, Yamada M (2007) Ethanol production from Jerusalem artichoke by *Zymomonas mobilis* in batch fermentation. KMITL Sci Tech J 7:55–60

Wiesenhütter J (2003) Use of the physic nut (*Jatropha curcas* L.) to combat desertification and reduce poverty. CCD Project Document. GTZ, Bonn

Chapter 4
Socio-Economy, Agro-Ecological Zones, Agronomic Practices and Farming System of *Jatropha curcas* L. in Sub-Saharan Africa

Raphael Muzondiwa Jingura

Introduction

Sub-Saharan Africa (SSA) is a geographical region that covers the area of the African continent that lies south of the Sahara. This region is made up of 49 countries of which 32 have some activities related to cultivation of *Jatropha curcas* L. (hereafter referred to as Jatropha). The GEXSI report (2008a) stated that there were 119,000 ha under Jatropha production in SSA and the acreage is expected to reach two million ha by 2015. The major drivers for Jatropha production in Africa are four-fold:

(a) Protection of the environment and mitigation of climate change
(b) Means to ensure sustainable energy supply
(c) Desire to reduce dependency on external energy sources (energy security)
(d) Promotion of rural development

Commercial production of Jatropha in SSA is a relatively new industry that is still burgeoning and expected to witness both spatial and temporal transformation. There is strong political support for Jatropha production in SSA. This is a catalyst for both public and private investment in the Jatropha industry. By 2008, five countries in the region (Senegal, Nigeria, Mali, Ethiopia and Zimbabwe) had policies promoting Jatropha production (GEXSI 2008a). Other countries were expected to follow suit.

Traditionally, farming systems in sub-Saharan Africa are characterised by large-scale and smallholder farming. The enterprise mix is dominated by mixed crop/livestock systems with special focus on food and fibre production. Mixed farming systems involve complementary interactions of crops and livestock and are very important in SSA (Powell and Williams 1995). The entry of energy crops such as Jatropha into these farming systems adds a third dimension to the dual food and

R.M. Jingura (✉)
Chinhoyi University of Technology, Post Bag 7724, Chinhoyi, Zimbabwe
e-mail: rjingura@cut.ac.zw

N. Carels et al. (eds.), *Jatropha, Challenges for a New Energy Crop: Volume 1: Farming, Economics and Biofuel*, DOI 10.1007/978-1-4614-4806-8_4,

fibre production approach. Jatropha production needs to be appropriately situated in this new context of food-fibre-fuel production. Based on projections that Jatropha will be a major energy crop, SSA as a region needs to carve a competitive niche in this industry.

In this paper, we explore the suitability of the SSA region for Jatropha production, mainly from agro-ecological and socio-economic perspectives. We further look at agronomic practices in vogue and highlight attendant problems. The focus is on situating Jatropha within the agro-ecological context of SSA, description of agronomic practices in the region and analysis of attendant problems. We conclude by looking at the new dynamics being introduced into smallholder farming systems as a result of cultivation of Jatropha.

J. curcas in Africa

Jatropha curcas L. originates from Central America and is generally believed to have been distributed by Portuguese seafarers to Africa via the Cape Verde Islands (Henning 2003a). According to history, Cape Verde, Madagascar and Guinea exported Jatropha seeds to France during the first half of the twentieth century until the trade disappeared in the 1970s (Henning 2004). Cape Verde alone also used to export 35,000 t per year of Jatropha seed to Portugal during the same period (Brittaine and Lutaladio 2010). Today Jatropha is grown in many countries in SSA and this region is now one of the major Jatropha producing zones in the world.

Jatropha has a long history in Africa. It was originally planted as a hedge plant in many countries and was also used for artisanal soap production and medicinal purposes. In 2008, the total length of Jatropha hedges in Africa was estimated at 75,000 km yielding potentially 60,000 t of seeds per year (Muok and Källbäck 2008). Today Jatropha is grown as an energy crop in SSA with significant public and private investments in its production. However, there are regional disparities in Jatropha production across the five regions of Africa. There is minimal production in Northern Africa mainly due to its aridity, although there are several projects in Egypt that use sewage water for irrigation (GEXSI 2008b). Jatropha activities in SSA are summarised in Table 4.1.

Agro-Ecological Zones of SSA

Adaptation of energy crops to climatic conditions appears to be a low priority or even irrelevant as a factor of importance to governments in SSA when considering biofuels programmes (Batidzirai et al. 2006). Climate adaptation is an important factor which should be taken into consideration when developing biofuels programmes. It is well known that crop productivity depends on physical environmental conditions, which influence the availability of basic resources such as light, heat

Table 4.1 Jatropha activities in Sub-Saharan Africa

Sub-region	Major countries	Description of activities
Western Africa	Mali, Cape Verde, Ghana, Nigeria, Cameroon	Mali and Cape Verde have long tradition of cultivation. Large –scale projects in Ghana, Nigeria and Cameroon
Eastern Africa	Ethiopia, Uganda, Kenya, Tanzania	Tanzania and Ethiopia have large projects. Activities are growing in the other countries
Central Africa	Democratic Republic of Congo (DRC), Central Africa Republic	DRC has very favourable climate for Jatropha cultivation. Jatropha is not a major crop in this region
Southern Africa	Madagascar, Mozambique, Zambia	Most countries in the region (apart from Botswana, Angola, and South Africa) have ambitious Jatropha projects

Source: GEXSI (2008b)

Table 4.2 Summary of climatic and physical conditions suitable for cultivation of Jatropha

Characteristic	Tolerance parameters	Optimum conditions	References
Annual mean temperature	18–40°C	26–27°C	Gour (2006), Trabucco et al. (2010)
Average minimum temperature	8–9°C	—	Trabucco et al. (2010)
Average maximum temperature	35–45°C	—	Trabucco et al. (2010)
Annual precipitation (mm)	250–1,500	900–1500	Benge (2006), Trabucco et al. (2010)
Frost tolerance	Sensitive to frost and very low temperatures (<7°C)	—	Gour (2006)
Altitude (m)	0–1,500	< 500	Singh et al. (2006), Muok and Källbäck (2008)
Average Slope	≤ 15°	—	Wu et al. (2009)
Thornthwaite humidity index	100–33.3	—	Wu et al. (2009)

and water. Before we describe the *agro-ecological zones* (AEZ) of SSA, it is important to define climatic conditions that are considered to be optimum for Jatropha production from an ecological perspective.

Several studies have examined the correlation between Jatropha production and agro-climatic conditions, but there are no universally accepted criteria to guide Jatropha plantations (Wu et al. 2009). However, there is a consensus in literature that temperature, annual precipitation, soil quality and land slope are the key factors (Wu et al. 2009). A summary of climatic and physical conditions suitable for cultivation of Jatropha is given in Table 4.2.

The area in SSA suitable for Jatropha cultivation, known as the 'Jatropha belt' is delimited by the Tropics of Cancer and Capricorn (Basili and Fontini 2010). Universally, the 'Jatropha belt' is defined to be stretching 30°N and 35°S (Jongschaap

Fig. 4.1 Areas of Sub-Saharan Africa suitable for Jatropha cultivation (Source: Parsons 2005)

et al. 2007). Areas in SSA that have suitable agro-climatic conditions for cultivation of Jatropha are shown in Fig. 4.1. The dark areas represent prime Jatropha growing regions and cover 10.8 million square kilometres or one billion ha (Parsons 2005). Average annual rainfall in these regions exceeds 800 mm and the minimum temperature of the coldest month is greater than 2°C (Parsons 2005). The light green areas have average annual rainfall above 300 mm and minimum temperature of the coldest month greater than 2°C (Parsons 2005). These areas cover 5.8 million square kilometres or 580 million ha and are also viable regions for growing Jatropha (Parsons 2005).

When describing SSA in terms of *agro ecological zones* (AEZ), temperature and moisture are the main factors that determine the types of crops that can be grown in a specific area. Temperature zones in SSA are either tropical or sub-tropical. Tropics are areas where all months have monthly average temperature, corrected to sea level, above 18°C and sub-tropics below 18°C (Food and Agricultural Organisation—FAO 1994). The moisture zones are demarcated in terms of the length of the growing period (LGP). The LPG is the period in days during the year when soil moisture from rainfall is greater than half potential evapotranspiration (FAO 1994). In terms of moisture zones, SSA can be divided into arid (<70 days LGP), semi-arid (70–180 days LGP), humid (180–270 days LGP) and sub-humid (>270 days LGP). FAO (1994) classified countries in SSA into four AEZ as shown in Table 4.3. The information in Table 4.3 is useful to map areas (Fig. 4.1) in SSA that are suitable for cultivation of Jatropha

Table 4.3 Classification of Sub-Saharan African countries by agro-ecological zones

Agro-ecological zone	Region	Countries
Warm arid & semi-arid tropics (AEZ 1)	West Africa	Cape Verde, Chad, The Gambia, Mali, Mauritania, Niger, Senegal and parts of Benin, Burkina Faso and Nigeria
	East Africa	Djibouti, Somalia, Sudan and parts of Ethiopia, Kenya, Tanzania and Uganda
	Southern Africa	Botswana, Namibia, Swaziland and parts of Angola, Madagascar, Malawi, Mozambique, Zambia and Zimbabwe
Warm humid tropics (AEZ 2)	West Africa	Guinea, Guinea Bissau, Togo, and parts of Benin, Burkina Faso and Nigeria
	East Africa	Parts of Ethiopia, Tanzania and Uganda
	Southern Africa	Comoros, and parts of Angola, Madagascar, Malawi, Mozambique, Zambia and Zimbabwe
Warm humid tropics (AEZ 3)	West Africa	Cameroon, Central African Republic Côte d'Ivore, Equatorial Guinea, Gabon, Ghana Sierra Leone and parts of Nigeria
	Central Africa	Democratic Republic of Congo, Central African Republic
	Southern Africa	Madagascar and Mauritius
Cool tropics (AEZ 4)	—	Burundi, Lesotho, Rwanda and parts of Angola, Ethiopia, Kenya, Madagascar and Tanzania

Source: FAO (1994)

from an ecological perspective. Soil characteristics are also important and Jatropha just like any other crop benefits from good soils as shall be explained later. Soil properties have an important effect on productivity of Jatropha (Openshaw 2000).

Agronomic Practices in Jatropha Cultivation

In addition to agro-ecological conditions, viability of Jatropha production also depends on other factors such as quality of planting material, agronomic practices and crop management. Optimisation of agronomic practices is a prerequisite for high crop performance. Data on agronomic practices in the cultivation of Jatropha in SSA are limited, discontinuous and often not supported by scientific evidence. There is very little site-specific information on best practices in Jatropha production. The agronomic practices in vogue are mainly based on generalisations or extrapolations from areas outside SSA.

Jatropha Production Systems in SSA

There are several production systems for Jatropha that apply throughout SSA.

Production Systems

Jatropha production systems in SSA can be divided into five categories as follows:

(a) Plantations
(b) Plantation plus out-grower schemes
(c) Out-grower schemes
(d) Smallholder production
(e) Hedges

Plantations are large schemes that can be under public or private ownership and are usually larger than 5 ha. Some of these plantations are on wastelands or marginal lands. In the model of *plantation plus out-grower*, the out-growers are linked to a commercial plantation, which provides support to the farmers in the form of planting material, inputs and agronomic advice (Brittaine and Lutaladio 2010). *Out-growers* are mainly smallholder farmers who are contractually associated with a central organisation. In the pure *out-grower* model the farmers are not contractually linked to a commercial plantation, but to a central organisation. For example, in Zimbabwe the National Oil Company of Zimbabwe has contracted smallholder farmers to grow Jatropha and in turn the farmers sell the seeds to the company. The pure *small holder production* model involves smallholder farmers who have no contractual obligations with anyone. These farmers grow Jatropha on their own will and choose their buyers.

Types of Plantations

There are basically three types of plantations of Jatropha that are found in SSA (Gesellschaft für Technische Zusammenarbeit—GTZ 2009). These are:

(a) Monoculture plantations
(b) Intercropped plantations
(c) Livestock protection hedges

Monoculture plantations follow the traditional pure plantations practice, which is common for most commercial crops grown on a large-scale. Jatropha grown in monoculture plantations is amenable to the usual challenges faced with monoculture, such as disease and pest problems. *Intercropped plantations* incorporate other crops, such as maize, beans, elephant grass, lablab and others. Plant spacing for Jatropha is generally 2×2 m, 2.5×2.5 m or 3×3 m and this provides an intra-row space that can be occupied by other crops. Generally, *intercropping* is feasible in the first 2 or so years of growth when the plant has not yet fully grown up. *Hedges* are usually found around crop fields or vegetable gardens where they provide protection against animals.

Kenyan experience has shown that *monoculture* and *intercrop plantations* are not profitable and the recommendation has been to promote Jatropha as a hedge crop (GTZ 2009). Wahl et al. (2009) in Tanzania also concluded that Jatropha plan-

Table 4.4 Former use of land under Jatropha production in Africa

Former land use	Proportion of projects (%)
Non-wasteland	30
Agriculture (non-food crops)	58
Secondary forests	7
Agriculture (food crops)	5
Primary forests	0

Source: GEXSI (2008a)

tations were not viable at that time and *intercropping* with annual crops was only cost-effective in the first 2 years of growth. The subject of viability shall be discussed later in this chapter. The GEXSI report (2008a) showed that 70% of Jatropha schemes on a global basis is performed under *intercropping*.

Land Types Used for Jatropha Cultivation

One of the contentious issues in Jatropha production in SSA has been the type of land that is used for cultivation of the crop. Use of land suitable or being used for production of food crops is largely not recommended for reasons of food security. It would appear that in some countries in SSA Jatropha production has encroached onto fertile lands used for food production. In Mozambique and Tanzania, the farmers were reported to be growing Jatropha on fertile soils replacing food crops (Justica Ambiental and União Nacional de Campeneses—JA and UNAC 2009; Wahl et al. 2009). In terms of land use and sustainability issues around Jatropha production systems in Africa, data from GEXSI (2008a) showed that Jatropha production in the region had not led to reduction in food production or destruction of forests (Table 4.4). Information given in Table 4.4 shows that use of land reserved for production of food crops is minimal. However, these data are not conclusive and there is a need for further elaboration of this issue across all countries in SSA.

Crop Establishment and Management Practices

There are generally three broad methods used in establishment of Jatropha plantations. These are vegetative propagation, direct seeding and transplantation of seedlings from nurseries. Each of these methods has its own merits, demerits, cost and labour implications. All the methods are used as shall be explained later in this section. Direct seeding is the easiest and cheapest method. Other important agronomic practices include planting station preparation, fertilisation, canopy management, watering regimes, disease and pest management, weed control and harvesting. There are no standards or best practices in place that take into account site-specific conditions.

Hedges

The traditional use of Jatropha in SSA has been as a live fence to protect food and cash crops against marauding animals. Hedges are established both vegetatively and generatively. There is not much documented information on the general agronomic practices for hedges. Inter- and intra-row spacing varies and post-establishment management has no specific standard. In Tanzania, spacing for hedges varies from 5 to 50 cm depending on the type of livestock species being protected against, for example, goats or chickens (Wahl et al. 2009). In Zimbabwe and southern Africa the recommended spacing for hedgerows for soil conservation is 15–25 cm between trees in one row or two rows separated by 15–25 cm (FACT 2006). This gives about 4,000–6,700 trees per kilometre. The attractiveness of hedges is that they have low inputs and maintenance requirements and almost zero opportunity costs.

Plantations

The transformation of Jatropha from a traditional hedge plant to an energy crop is a recent development in SSA. There were over 97 Jatropha projects in Africa covering 119,000 ha in 2008 (GEXSI 2008a); 85% were established from seedlings transplanted from nurseries, 45% by direct seeding and 40% from cuttings (GEXSI 2008a). A country-specific example is Kenya where, 45% of the plantations in the country were established by direct seeding, 31% from seedlings and 19% from cuttings (GTZ 2009).

Generally, sources of planting materials are hedges or they are provided by agents promoting Jatropha cultivation in different countries. Spacing distances vary, but generally are 2×2 m, 2.5×2.5 m or 3×3 m giving rise to varying plant populations ranging from 1,111 to 2,500 plants per ha (Wahl et al. 2009). The general recommendation is to increase spacing on marginal lands to reduce intra-specific competition for water and nutrients.

Management practices of crop post-establishment vary considerably. The most critical practices include weed and pest control, fertiliser application, canopy management (pruning) and watering regimes. There is not much data on crop management practices in different parts of SSA as a region. What is apparent is that some notable plant management takes place, despite the notion that Jatropha requires little attention as a crop. Varying irrigation regimes are used, mainly in large-scale plantations. There is limited data available on water needs, water productivity and water-use efficiency of Jatropha. Experience in Mozambique has shown that Jatropha requires a lot of care in the first 18 months of growth. Farmers in Mozambique provide 5–7 litres of water per day and per plant to supplement rainfall in this early phase of growth (JA and UNAC 2009). This makes reliance on unpredictable rainfall problematic during early growth of Jatropha. Thus, irrigation is quite common in Jatropha production systems in SSA. Data from Kenya indicate that at least 40% of Jatropha farmers practiced irrigation (GTZ 2009).

Other practices such as pruning and fertilizer application are carried out as well. Jatropha is a plant bearing terminal flowers and pruning is critical to increase the number of branches and terminals capable of producing fruits. There is a strong correlation between the number of branches and the amount of fruit produced (JA and UNAC 2009). Ideally each plant is supposed to have close to 100 branches, but data from Mozambique show that after 2 to 3 years of growth Jatropha plants were observed to have 18–30 branches (JA and UNAC 2009). This indicates lack of appropriate canopy management in order to optimize both the number of branches and tree height for harvesting purposes.

Use of fertilisers, especially organic types is quite common in SSA. In Kenya, 50% of farmers use organic fertilizers in Jatropha production (GTZ 2009). In Tanzania, no farmers were observed to apply fertilizers or any other inputs (Brittaine and Lutaladio 2010). Jatropha seems to respond to both organic and inorganic fertiliser application. For example, in India treatment of Jatropha plantations with 3 $t\,ha^{-1}$ of Jatropha seedcake increased seed yield by 120% and 93% at different plant densities of 833 (4×3 m) and 1,667 (3×2 m) plants ha^{-1}, respectively (Ghosh et al. 2007).

Pest and Disease Management

One of the claimed attributes of Jatropha is its reported tolerance to pests and diseases. However, information originating from SSA seems to suggest that Jatropha, like any other plant, is vulnerable to several diseases and pests. Contrary to the belief that toxicity and insecticidal properties naturally protect Jatropha from pests and diseases, several pests and diseases have been reported and some of these are summarised in Table 4.5.

Table 4.5 provides information on the susceptibility of Jatropha to pests and diseases. Pests and diseases can cause economic damage and there is a need for proper prevention and their management. What is missing in most countries are appropriate integrated pest and disease management regimes. Thus, as breeding programmes for Jatropha unfold, resistance to pests and diseases is a trait that needs consideration.

Table 4.5 Pests and diseases of Jatropha reported in SSA

Country	Pest/Disease	Reference
Mozambique	Cercospora leaf spot, collar rot, root rot	JA and UNAC (2009)
Kenya	Golden flea beetle (*Aphtona* sp.), leaf spotting, powdery mildew (*Oidium* sp.)	GTZ (2009)
Zimbabwe	Stem borer (*Ostrinia furnacalis* or *Xyleborus* sp.), golden flea beetle, fungus of *Cercospora* species (frogeye)	FACT (2006)
Tanzania	Scutellarid bug (*Scutellera nobilis*), golden flea beetle, stem borer, powdery mildew	Wahl et al. (2009)

Seed Yield

The most important economic trait in commercial production of Jatropha is the seed yield. Very little empirical data has been collected regarding actual seed yields of Jatropha in SSA. The information that is available is highly variable, not standardised and difficult to verify. However, what is clear is that harvesting of seeds is labour intensive due to the heterogeneous or asynchronous fruiting of the plants. This is problematic in smallholder farming situations where family labour is overstretched in most instances.

Benchmark figures for seed yield indicate that with optimum rainfall of 900–1,200 mm yields can be up to 5 $t\,ha^{-1}$ (Maes et al. 2009). Other ranges of seed yield available include 0.4–12 $t\,ha^{-1}$ (Openshaw 2000), 0.1–8 $t\,ha^{-1}$ (Heller 1996) and a theoretical optimum of 7.8 $t\,ha^{-1}$ for mature plantations after 3 to 4 years of growth (Jongschaap et al. 2007). These figures are used here to serve as a reference point to the yield data emanating in the SSA region. Seed yields that have been reported include 1.65 $t\,ha^{-1}$ in Tanzania (Brittaine and Lutaladio 2010), less than 1 kg per tree in Mozambique (JA and UNAC 2009), 0.63 $t\,ha^{-1}$ in Mali (FACT 2006) and up to 0.86 kg per tree in Kenya (GTZ 2009). Reported yields are accompanied by little or no information on variables such as genetic provenance, age of plantations, propagation method used, canopy management regime, rainfall, tree densities, soil types and soil fertility management (Brittaine and Lutaladio 2010). This makes very difficult to relate the yields to any parameters.

Seed yield given per tree should be treated with caution. Henning (2008) showed that there is too much variation among individual trees in terms of seed yield. The results showed that annual yield variation of 19 trees ranged from 0 to 850 g dry seed per tree. Thus, according to Achten et al. (2008) expectations that Jatropha will yield up to 12 $t\,ha^{-1}$ of dry seeds result from illegitimate extrapolation from individual plants. The yield dynamics of Jatropha are mainly caused by variation due to cross-pollination (Paramathma et al. 2006). Also, there is little data available on seed yield from mature stands of Jatropha and most seed yield data available appear to be from young plants. Age of plantation and seed yield are positively correlated. One of the reasons generally given for the failure of some Jatropha projects in SSA is unviable seed yields. Clearly, the prospects for improving seed yield are based on improving the Jatropha germplasm that is available for commercial production.

Jatropha in Smallholder Farming Systems

One of the major drivers for promotion of Jatropha cultivation is to enhance rural development and benefit smallholder farmers through introduction of a new enterprise. The integration of Jatropha into smallholder farming systems has had no major conflict with other agricultural activities (Henning 2003b). Smallholder farmers are increasingly playing a vital role in the cultivation of Jatropha. In Africa, about two thirds of Jatropha projects in 2008 integrated smallholder farmers either as out-growers or standalone farmers (GEXSI 2008a). In Tanzania, for example, more than 10,000 smallholder farmers have established Jatropha plantations (Wahl et al. 2009).

Role of Jatropha in Smallholder Farming Systems

There have been several attempts to introduce Jatropha into smallholder farming systems in SSA, not just as a hedge crop, but also as a commercial multi-purpose plant. Most notable projects include the *United Nations Industrial Development Organisation/United Nations Development Programme* (UNIDO/UNDP) Multifunctional Energy Platforms (MFP), which was carried out in rural areas in countries such as Mali, Tanzania, Senegal, Ghana, Guinea and Côte d'Ivoire. The scope of MFPs includes use of Jatropha to combat rural energy poverty through, firstly, cultivation of Jatropha, secondly, installation of a power plant with a generator running on pure Jatropha oil and thirdly, the local power would supply electricity to the community (United Nations Department of Economics and Social Affairs 2007).

The other project is the 'Jatropha System', a GTZ project in the early to mid 1990s which had four main aspects of rural development. These were: promotion of women through artisanal soap production from Jatropha oil; poverty reduction by protecting crops from marauding animals, selling seeds and oil; erosion control by planting hedges as windbreaks; and supply of energy (Henning 2004). Over and above these projects, Jatropha cultivation has grown to become integrated into smallholder farming systems as a result of massive political support and futuristic projections. The role of Jatropha in these systems includes social, ecological and economic issues.

The ecological issues will be explored further. One of the major concerns in SSA is the escalation of desertification due to the fragility of soils in this region. There are large areas with degraded soils. Jatropha is being used in the rehabilitation of degraded soils, erosion control and soil improvement. When Jatropha is planted on degraded soils, its taproot extracts nutrients from deep soil and nutrients are also returned to the soil through leaf fall, debris and other organic remains (Brittaine and Lutaladio 2010). Jatropha controls soil erosion as its taproot anchors the plant in the ground and the profusion of lateral and adventitious roots near the surface binds the soil (Brittaine and Lutaladio 2010). However, rooting system is influenced by the propagation method as vegetatively propagated plants do not develop a taproot. Jatropha in hedge rows reduces wind erosion by reducing wind speed. The plant improves water infiltration when planted in lines to form contour bunds.

The soil protection and improvement function of Jatropha is a very fundamental issue in the smallholder sector. Land degradation and fragile soils are among the most serious environmental problems afflicting smallholder agriculture in SSA. There is empirical evidence that Jatropha has the potential to contribute to erosion control and soil quality improvement. Work done by Ogunwale et al. (2008) in India on a degraded entisol showed the following results: 11% average increase in *mean weight diameter*

Table 4.6 Roles of Jatropha in smallholder farming systems

Domain	Role and impact
Social	Economic empowerment of women through soap production, impact on gender issues with regard to ownership of hedges and plantations and labour divisions, alternative energy options impact on health, women and children welfare
Ecological	Protection of fragile ecosystems through soil quality improvement, impact on biodiversity, whether it is competing with other crops or has any allelopathic effects, impact on water, forests and land use
Economic	Income opportunities, opportunities arising from the value chain (production, transportation and processing), income from *clean development mechanism* (CDM) and carbon credits
Environmental	Balance of greenhouse gas emissions, erosion control

(MWD) of the soil (the higher the MWD score is, the more "water-stable" is the soil); 2% increase in soil macro-aggregate turnover; and 6–30% improvement in macro-aggregate stability. Much more work still need to be done and these early results highlight the potential of Jatropha to contribute to soil improvement in the smallholder sector. The roles of Jatropha in smallholder farming systems are summarised in Table 4.6.

Experiences with Jatropha Cultivation in the Smallholder Sector

As indicated earlier, farming systems in the smallholder sector exist in ecologically, socially and economically fragile environments. The introduction of Jatropha as a new energy crop in this environment has introduced new dynamics in smallholder agriculture. Firstly, it has brought about a new option to diversify income streams in rural economies. What is worth noting is that the attraction of Jatropha has been based on its reported numerous advantages. These include high yields on marginal soils, degraded soil reclamation, drought tolerance, pest and disease tolerance, low nutrient requirements and low maintenance requirements, amongst others. The reality is that these are not always achievable in totality or in combination. It is worthwhile to consider the experiences in SSA with regard to the performance of Jatropha.

Low Seed Yield

Most plantations established in the late 1980s to the 1990s were abandoned due mainly to low productivity and/or high labour costs (Jongschaap et al. 2007). Reasons given for not recommending Jatropha cultivation, for example by JA and UNAC (2009) in Mozambique and Wahl et al. (2009) in Tanzania, are centred on low seed yields and poor viability. As shown earlier, Jatropha productivity has been low and acclaimed seed yield levels have not been attained. Seed yields have been highly variable and often less than 1 t per ha. There is no convincing data available for seed yields above 2 or 3 t per ha. Plants with seed yields above 2 kg per tree are ideal. This translates to about 4.5 $t\,ha^{-1}$.

Generally Jatropha in the smallholder sector is subjected to harsh climatic conditions, marginal soils and poor agronomic practices that do not allow high yields. Smallholder farmers who ventured into Jatropha cultivation expecting a crop that would thrive on marginal soils and with low maintenance requirements (JA and UNAC 2009) have been disappointed. The reality check has shown that Jatropha, just like other crops, requires the following:

(a) Good quality germplasm
(b) Good soils
(c) Supply of additional nutrients
(d) Reliable rainfall or irrigation
(e) Good crop management practices

Given the fact that farm sizes in the smallholder sector range between 2 and 5 ha and that land is a limiting factor, there is a need to optimise production per unit area since low yields are not attractive for Jatropha cultivation. Perhaps it is reasonable to acknowledge that the low seed yields are consistent with a plant still under domestication and will improve in course of time.

Lack of Best Practices

Crop productivity depends on the effects of environmental conditions that can be mitigated or amplified by agronomic practices (Trabucco et al. 2010). The range of agronomic practices includes selection of planting material, propagation methods, watering regimes, fertilizer use, canopy management, pest and disease control. These practices mitigate specific environmental constraints. Best practices are required in order to establish efficient production systems. From an agronomic or silvicultural perspective, development of production packages is a fundamental issue and these packages need to be in place to guide farmers.

Unlike with other commercial agricultural crops, which have planting materials certified according to performance characteristics, there is currently no certified planting material for Jatropha in most parts of SSA. The planting material consists of seeds, seedlings and stem cuttings, which have not been improved to supply elite planting materials. Jatropha has longevity of up to 50 years. This makes it less flexible than annual crops and without plant improvement programs plantations will be based on inferior planting materials (Jingura 2011).

However, knowledge on propagation methods of Jatropha is well established and most farmers are aware of the options available to them. What is not clear is standardisation of planting time so as to coincide with optimum conditions for rapid and successful establishment in specific areas. This is important; given that rates of plant survivability, particularly during establishment, determines the success of plantations.

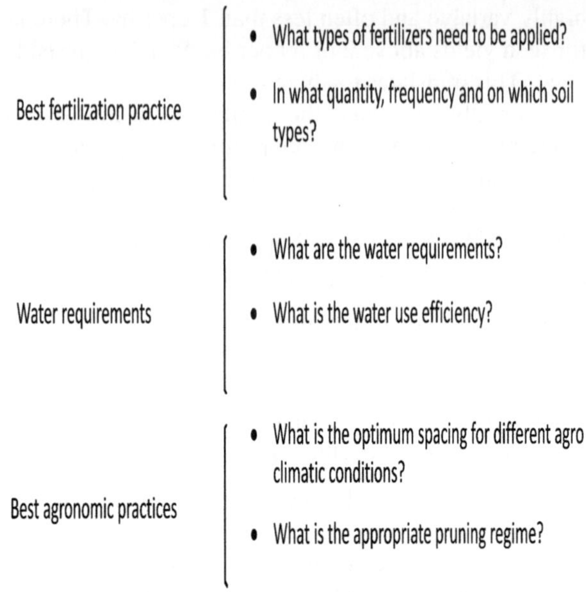

Fig. 4.2 Some important questions regarding Jatropha cultivation

Knowledge gaps in agronomic practices also exist in terms of water requirements for Jatropha as well as growth and yield responses to inputs. Experiences in SSA have shown that Jatropha requires additional water, even in areas with more than 1,000 mm of annual rainfall, mainly due to the uneven distribution of rainfall in a given rainy season.

Jatropha has been largely promoted in SSA as a crop suitable for production on waste or marginal lands with little or no requirement for additional nutrients. Yield of Jatropha on poor quality land has been as low as 0.2 $t\,ha^{-1}$. Such low yields are uneconomical and to realize biological yield potential of above 5 $t\,ha^{-1}$, judicious use of fertilizers will be required. There is insufficient data on the response of Jatropha to fertilizers under different growing conditions to make specific recommendations for optimum crop nutrition (FACT 2006). Both inorganic and organic fertilizers have been used in Jatropha production. Application rates for Jatropha are based on judicious extrapolation from other related crops such as castor beans. Appropriate fertilizer application regimes need to be established for Jatropha and site-specific trials are also needed. Figure 4.2 highlights the issues concerning best practices for Jatropha cultivation.

Viability Problems

Without providing a detailed economic analysis it is important to indicate how Jatropha has fared as a commercial crop. Early Jatropha projects in 1980s were based on use of Jatropha oil as an energy source and for soap production. These early attempts, for example in Mali and Zambia, were found to be inviable as the price of the oil was not competitive (Benge 2006). This led to a lull in Jatropha cultivation until the resurgence in the new millennium due to soaring oil prices and the search for alternative fuels.

Studies in Tanzania (Wahl et al. 2009) and Kenya (GTZ 2009) showed that Jatropha cultivation was not a viable enterprise under the conditions that prevailed at that time. Reasons offered for poor viability included the following:

(a) High requirement for labour
(b) High opportunity cost when grown on fertile land
(c) Low producer prices
(d) Low seed yields

Evidence from Zimbabwe also shows that the Jatropha project that was initiated by the Plant Oil Producers Association in 1992 was abandoned after it was realised that the profit margins were not as big as originally expected (Henning 2003b). The indication is that commercial viability of Jatropha cultivation needs to be improved. The true potential of Jatropha as a cash crop depends on the successful development of the agronomic practices needed to domesticate a semi-wild plant as well as the creation of a market that ensures reasonable prices (GTZ 2009).

Conclusion

The place of Jatropha as an energy crop in SSA is well established despite the current challenges with crop cultivation. The suitability of Jatropha to agro-ecological conditions in this region is not in doubt. The need to optimise the production of Jatropha as an energy crop is an important imperative. The supply of elite planting material and development of appropriate agro-techniques remain the most critical issues in the endeavours to transform Jatropha into a viable crop.

What is required is to develop thorough understanding of the crop in order to develop improved plant types that are more responsive to input supply and biologically more efficient in terms of seed yield and quality norms. This is not without precedence in agriculture. Early claims about Jatropha seem not to have stood the test of time and the plant cannot be relegated to wasteland status. Jatropha requires suitable agronomic and crop management practices befitting a commercial crop. Jatropha production has to be attractive to farmers in terms of economic viability in comparison with other crops.

Acknowledgements The author thanks Mr. Reckson Kamusoko for assisting in gathering the information necessary to this report. I express my gratitude to my colleagues at Chinhoyi University of Technology for critical review of the manuscript. I am also indebted to Nicolas Carels for his encouragement and assistance in many ways.

References

Achten WMJ, Verchot L, Franken YJ, Mathijs E, Singh VP, Aerts R et al (2008) Jatropha biodiesel production and use. Biomass Bioenergy 32(12):1063–1084

FAO—Food and Agricultural Organisation (1994) Agroecological zones framework and database for the review of CGIAR priorities and strategies. Available from: http://www.fao.org/wairdocs/tac/x5756e/x5756e0j.htm. Accessed 22 Oct 2011

Basili M, Fontini F (2010) Biofuel from *Jatropha curcas*: environmental sustainability and option value. Available from http://www.unisi.it/d12/20100519144204853.biofuel pdf. Accessed 15 Sept 2011

Batidzirai B, Faaj APC, Smeets E (2006) Biomass and bioenergy supply from Mozambique. Energy Sustain Dev 10(1):54–81

Benge M (2006) Assessment of the potential of *Jatropha curcas* (biodiesel tree), for energy production and other uses in developing countries. Available from http://www.echotech.org/mambo/index.php?option¼com_docman& task¼doc_view&gid¼179&Itemid¼68. Accessed 10 Oct 2011

Brittaine R, Lutaladio N (2010) *Jatropha*: A smallholder bioenergy crop—the potential for pro-poor development, vol 8, Integrated Crop Management. FAO, Rome

FACT (2006) Jatropha handbook. First Draft. Available from http://www.fact-fuels.org/. Accessed 23 Oct 2011

Gesellschaft für Technische Zusammenarbeit. Jatropha reality check (2009) A field assessment of the agronomic and economic viability of *Jatropha* and other oilseed crops in Kenya. Available from http://www. www.worldagroforestry.org/downloads/publications/PDFs/B16599.PDF. Accessed 10 Oct 2011

GEXSI (2008a) Global market study on Jatropha. Final report. Available from http://www.slideshare.net/gizosmit/gexsi-global-jatropha-study-abstract-presentation. Accessed 10 Aug 2011

GEXSI (2008b) Global market study on Jatropha. Project inventory: Africa. Available from http://www.jatropha-allianc.org/fileadmin/documents/GEXSI_Jatropha-Project-Inventory_Africa.pdf. Accessed 12 Aug 2011

Ghosh A, Patolia JS, Chaudhary DR, Chikara J, Rao SN, Kumar D et al (2007) Response of Jatropha curcas under different spacing to Jatropha de-oiled cake. Expert seminar on *Jatropha curcas* L. Agronomy and genetics, FACT Foundation, Wageningen, 26–28 March 2007

Gour VK (2006) Production practices including post harvest management of *Jatropha curcas*. In: Singh B, Swaminathan R, Ponraj V (eds) Proceedings of the conference "Biodiesel towards energy independence – focus on Jatropha" New Delhi. pp 223–51

Heller J (1996) Physic nut. *Jatropha curcas* L. Promoting the conservation and use of underutilised and neglected crops, vol 1. Institute of Plant Genetics and Crop and Plant Research/International Plant Genetics Resource Institute, Gatersleben/Rome

Henning RK (2003a) The Jatropha booklet. A guide to the Jatropha system and its dissemination in Africa. Bagani GbR, Weissensberg

Henning RL (2003b) *Jatropha curcas* L. in Africa: assessment of the impact of the dissemination of the Jatropha system on the ecology of the rural area and the social and economic situation of the rural population (target group) in selected countries in Africa. Available from http://www.underutilized-species.org/Documents/PUBLICATIONS/jatropha_curcas_africa_pdf. Accessed 10 Aug 2011

Henning RK (2004) The Jatropha system—an integrated approach of rural development. Available from http://www.jatropha.de. Accessed 20 Sept 2011

Henning RK (2008) Identification, selection and multiplication of high yielding *Jatropha curcas* L. plants and economic key points for viable Jatropha oil production costs. International consultation on pro-poor Jatropha development, Rome, IFAD. Available from http://www.ifad.org/events/jatropha/. Accessed 20 Jun 2011

Jingura RM (2011) Technical options for optimization of production of Jatropha as a biofuel feedstock in arid and semi-arid areas of Zimbabwe. Biomass Bioenergy 35:2127–2132

Jongschaap REE, Corré WJ, Bindraban PS, Brandenburg WA (2007) Claims and facts on *Jatropha curcas* L. Global *Jatropha curcas* evaluation, breeding and propagation programme. Plant Research International B.V., Wageningen. U.R. report 158. Available from http://www.fact-fuels.org/media_en/Claims_ and_ Facts_ on Jatropha_WUR. Accessed 10 Nov 2011

Justica Ambiental (JA), União Nacional de Campeneses (UNAC) (2009) Jatropha! A socio-economic pitfall for Mozambique. Available from http://www.swissaid.ch/global/PDF/entwicklungspolitik/agrotreibstofffe/Report_Jatropha_JA_and_UNAC.pdf. Accessed 15 Sept 2011

Maes WH, Trabucco A, Achten WMJ, Muys B (2009) Climatic growing conditions of *Jatropha curcas* L. Biomass Bioenergy 33:1481–1485

Muok B, Källbäck L (2008) Feasibility study of *Jatropha curcas* as a biofuels feedstock in Kenya. African Centre for Technology Studies. Available from http://www.acts.or.ke. Accessed 20 Sept 2011

Ogunwale JO, Chaudhary DR, Ghosh A, Daudu CK, Chikara J, Patolia JS (2008) Contribution of *Jatropha curcas* to soil quality improvement in a degraded Indian entisol. Acta Agric Scand B Plant Soil Sci 58:245–251

Openshaw K (2000) A review of *Jatropha curcas*: an oil plant of unfulfilled promise. Biomass Bioenergy 19:1–5

Paramathma M, Reeja S, Parthiban KT, Malarvizhi D (2006) Development of inter-specific hybrids in Jatropha. In: Singh B, Swaminathan R, Ponraj V (eds) Proceedings of the conference "Biodiesel towards energy independence—focus on Jatropha". New Delhi. pp 136–142

Parsons K (2005) Jatropha in Africa—fighting the desert and creating wealth. Available from http://www.na-wa-ro.eu/de/download/jatropha/156-jatropha-in-africa/download.html. Accessed 13 June 2011

Powell JM, Williams TO (1995) An overview of mixed farming systems in sub-Saharan Africa. In: Powell JM, Fernandez-Rivera S, Williams TO, Renard C (eds) Proceedings of conference on "Livestock and sustainable nutrient cycling in mixed farming systems of sub-Saharan Africa". Addis Ababa. pp 21–26

Singh L, Bargali SS, Swamy SL (2006) Production practices and post-harvest management in Jatropha. In: Singh B, Swaminathan R, Ponraj V (eds) Proceedings of the conference "Biodiesel towards energy independence - focus on Jatropha" New Delhi. pp 252–267

Trabucco A, Achten WMJ, Bowe C, Aerts R, van Orshoven J, Norgrove L et al (2010) Global mapping of *Jatropha curcas* yield based on response of fitness to present and future climate. GCB Bioenergy 2(3):139–151

United Nations Department of Economics and Social Affairs (2007) Small-scale production and use of liquid biofuels in sub-Saharan Africa: perspectives for sustainable development. UNDESA, New York: Background paper no. 2

Wahl N, Jammandas R, Baur H, Munster C, Iiyama M (2009) Economic viability of *Jatropha curcas* L. plantations in Northern Tanzania—Assessing farmers' prospects via cost-benefit analysis. ICRAF, Nairobi. Working paper no. 97. World Agroforestry Center

Wu WG, Huang JK, Deng XZ (2009) Potential land for plantation of *Jatropha curcas* as feedstock for biodiesel in China. Science China Earth Sci. doi:10.1007/s11430-009-0204-y

Chapter 5
The Importance of Jatropha for Brazil

Bruno Galvêas Laviola, Alexandre Alonso Alves, Rodrigo Barros Rocha, and Marcos Antônio Drumond

Introduction

In this chapter, we first present the Brazilian potential for biofuel production and the National Program for Production and Use of Biodiesel. We then review the raw materials currently used for biodiesel production in the country and their limitations. At the same time we present the characteristics that make some alternative oilseed species interesting as sources of vegetable oil for biodiesel production, highlighting why physic nut (*Jatropha curcas* L.) has been seen as one of the best alternatives, perhaps the best one. We then effectively demonstrate the importance of physic nut for Brazil describing the dense research program funded by the Brazilian Government, which is currently underway to try to solve most of the challenges (also reviewed) in making this crop effectively a biofuel crop.

Brazilian Potential for Biofuels Production

Fossil fuels currently supply most of the world's energy, even though it represents a finite resource (Vermerris 2008). Aiming at reducing the fossil fuels dependency, alternative sources of energy have been pursued in the last few years. In view of the urgent need to develop new technologies that may enable the widespread use of environment friendly forms of energy, biofuels in general and biodiesel in particular

B.G. Laviola (✉) • A.A. Alves
Embrapa Agroenergia, PqEB, Brasilia, DF 70770-901, Brazil
e-mail: bruno.laviola@embrapa.br

R.B. Rocha
Embrapa Rondônia, BR 364, Km 5,5, Porto Velho, RO 76815-800, Brazil

M.A. Drumond
Embrapa Semiárido, BR 428, Km 152, Petrolina, PE 56302-970, Brazil

N. Carels et al. (eds.), *Jatropha, Challenges for a New Energy Crop: Volume 1: Farming, Economics and Biofuel*, DOI 10.1007/978-1-4614-4806-8_5,

(La Rovere et al. 2011) are receiving considerable attention throughout the world and especially in Brazil (Laviola et al. 2012b; Laviola and Dias 2008; Laviola et al. 2010; Rosado et al. 2010).

In this context, it must be emphasized that Brazil is a country with a huge potential for production not only of biofuels, but also of other derivatives from vegetable oil. Such enormous capacity can potentially supply part of the national and international market demand on biofuels. The country possesses a privileged location in the tropical region of the world with high incidence of solar energy and adequate rainfall regime in most regions. Moreover, Brazil possesses a large amount of available land making possible the adequate planning of the land use in a sustainable basis, i.e., without compromising biome composition and structure. Current estimates indicate that in Brazil there are near 90 million hectares of available lands for agriculture expansion, without considering the 210 million hectares of degraded pastures (BAP 2006), from which some level of technology can be recovered and used for the production of food and biofuels. Nevertheless, there also exists in the country more than 200 oilseed species with different potentialities and different adaptation to the various natural occurring environmental conditions (Beltrão 2005). Some of such species can be incorporated in production systems and used for biofuel production and/or for other ends with higher aggregated value. The Brazil´s challenge is, then, to take maximum advantage of its biodiversity, of its regional potentialities to concomitantly enhance biofuel production and obtain greater social gains from the biodiesel production. This can be undoubtedly done by applying technology (agricultural, processing, etc.) not only on traditional crops, but also on undomesticated (to be explored) oilseed species (Laviola and Alves 2011). To achieve such goal, the Brazilian Federal Government has recently launched the National Program for Production and Use of Biodiesel (or in Portuguese: Programa Nacional de Produção e Uso de Biodiesel—PNPB).

The PNPB and Raw Materials Currently Used for Biodiesel Production

In 2005, the Brazilian Federal Government launched the National Program for Production and Use of Biodiesel grounded in the law nº 11.097 from January 13, 2005. The PNBP was elaborated to equate fundamental questions to the country, such as (i) reduction of foreign petroleum dependency, (ii) generation of jobs and income, (iii) social inclusion and reduction of pollutant emission and (iv) regional disparities in terms of development, by addressing social, strategic, economic and environmental aspects of the biodiesel production chain. It is determined that research should seek the substitution of part of the fossil diesel with biodiesel. It also established the need to diversify the sources of oilseeds used for biodiesel production, since today, near 87% of the biodiesel produced in the country is produced from soybean oil (animal fat represents the second main source of fatty acids for biodiesel production, followed by cotton). All other oilseed species if grouped

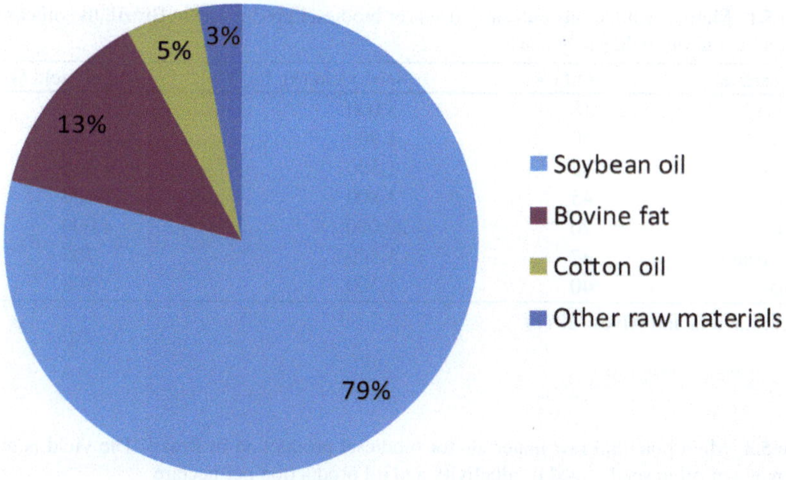

Fig. 5.1 Relative contributions of raw materials to the biodiesel matrix in Brazil (Source: Brazilian National Agency of Petroleum, Natural Gas and Biofuels –March 2011)

represent less than 1% of the biodiesel production chain (Fig. 5.1). The challenges and strategies to be met by the Brazilian biodiesel program involve technical-scientific bottlenecks in the raw material production, industrial processing and integration with existing productive chains.

Aiming at gradual substitution of petroleum diesel by biodiesel, the PNPB established as its initial goal to fix 2% (B2) as the minimum percentage of biodiesel addition to the petroleum diesel in any part of the national territory by 2008 and as 5% (B5) by 2013. However, with the rapid expansion of soybean production and industrial processing units, the government decided to anticipate its goals and officially introduced B5 by 2010. Nowadays, the capacity of biodiesel production is 2 times higher than the compulsory demand for biofuel, being then possible to immediately regulate B10 when the plant oil offer will be sufficient to meet the demand. However, to progress in the regulation of new mixtures, other questions need to be weighted, such as the raw material availability and diversification.

Generally two distinct groups of species are recognized: (i) the traditional raw materials, for which technological aspects of its production have already been addressed and (ii) the potential alternative raw materials, for which research is still required. Oil extracted from species like canola (oilseed rape; *Brassica napus* L.), soybeans (*Glycine max* L.), castor beans (*Ricinus communis* L.) and sunflower (*Helianthus annuus* L.) have been traditionally used for biodiesel production (Vermerris 2008; Yuan et al. 2008). Current estimates indicate that 600 kg ha^{-1} is the actual oil productivity of these traditional species (Table 5.1). In Brazil, besides being an oilseed species with low energetic productivity, soybean is the only crop among those species considered to be traditional raw materials for biodiesel production that totally meets the basic parameters of a program with the dimensions of the PNPB (Laviola and Alves 2011). These parameters belong to the (i) technological

Table 5.1 Main raw materials currently used for biodiesel production in Brazil, the oil content in seeds, seed and oil yield per hectare

Raw material	Oil (%)	Seed yield (kg ha^{-1})	Oil yield (kg ha^{-1})
Soybean	18	3,000	540
Cotton	20	1,900	360
Sunflower	42	1,500	630
Peanut	45	1,800	800
Oil palm	20	20,000	4,000
Castor beans	47	1,500	705
Canola	40	1,300	500

Source: Laviola and Alves (2011)

Table 5.2 Main potential raw materials for biodiesel production in Brazil. The yield is given in percentage of oil in seeds, seed productivity and oil production per hectare

Raw material	Oil (%)	Seed or fruit yield (kg ha^{-1})	Oil yield (kg ha^{-1})
Physic nut	35	4,500	1,500
Macaw palm	20	20,000	4,000
Inajá	20	17,500	3,500
Tucumã	20	12,000	2,400
Babaçu[a]	5	10,000	500

[a]Mainly used for energy cogeneration. (Source: Laviola and Alves 2011)

domain—Brazil is one of worldwide leaders in soybean research and development (what allowed for example its cultivation with minimum dependency on nitrogen fertilization); (ii) production scale—today less than 20% of the whole soybean production is sufficient to supply the current demand of the PNPB (other oilseed species such as cotton, sunflower and castor beans are not productive enough to support B2); and (iii) logistics—soybean is the only species among those considered to be traditional raw materials for biodiesel production that can be grown and produced in all regions of the country and with a whole distribution system to drain the production already installed. Even though all these characteristics contribute to make soybean the main raw material in the context of the PNPB nowadays, it is important to highlight that the continuous development of other oilseed species with higher energetic productivity is necessary, especially when considering questions related to diversification and regionalization of raw materials. Actually, the energy segment is too much important to assume the risk of its dependency on only one or few crop(s) in the long term. The importance of raw material diversification can be easily understood from the larger potential productivity of many alternative crops as presented in Tables 5.1 and 5.2. Moreover, it is necessary to remember that as much as 40–60% of biodiesel production costs are related to feedstock costs (BAP 2006). Thus, increasing the average oil productivity should prove to be an effective approach to increase the competitiveness of biodiesel when compared to gasoline and ethanol.

In this last context, the use of non-edible species is highly desirable as it offers the advantage of not competing with food crops for area and resources. However, such competition between food and fuel crops is not expected in the short term in Brazil, because Brazil is a country with continental dimensions and millions of hectares of land still available for agriculture expansion. But this does not necessarily apply to other countries, where food agriculture occupies almost the entire agricultural available lands. In such countries, the cultivation of non-edible species will almost certainly compete with food crops for land and resources. It is noteworthy that research should aim at improving the production system and also at incorporating new un-domesticated energetic species to the list of biofuel crops, e.g., physic nut and some tropical palms such as Macaw palm also known as Macaúba or *Acrocomia aculeata* (Jacq.) Lodd. ex Mart. (Moura et al. 2010; Rosado et al. 2010).

Alternative Feedstocks for Biodiesel Production

The PNPB established that research should seek new sources of raw materials with higher energetic productivity than soybean in order to enhance oil yield up to 4,000–6,000 kg ha^{-1}. The PNPB also indicate that research should aim at identifying new oil species so as to maximize biodiesel production according to region potentialities and biodiesel requirement.

Because Brazil holds ~20% worldwide biodiversity, a number of alternative oilseed species are available for biodiesel production. A number of species can also be used as source of biomass for cogeneration process. For example, Macaw palm is an alternative for the expansion of biofuel production in the Brazilian savannah, since its estimated oil yield is extremely high (Table 5.2). If one considers that the Macaw palm produces pretty much the same yield as oil palm and that oil palm has been improved via traditional breeding over decades, it is evident that the Macaw palm is an excellent choice as an alternative oilseed species. The macaw palm can be used in sustainable production systems (i.e., by extractivism or exploring the native individuals in their natural environment). Another possibility is to produce the Macaw palm in consortium with other crops species, which warrant farmers with incomes even in the first years after palm plantation. Other highly productive palms, such as Inajá (*Maximiliana maripa* (Aublet) Drude) and Tucumã (*Astrocaryum aculeatum* G. Meyer) (Table 5.2), grow spontaneously in the Northern region of Brazil and can potentially be explored in a variety of sustainable forms. By contrast, Babaçu palm (*Orbignya phalerata*, Mart.) is well adapted to Northeast Brazil and has been over the years used to produce a highly calorific charcoal. Recently its potential for biodiesel production has also been demonstrated even though its oil yield is not comparable to other potential palms (Table 5.2). Along with these palms, physic nut has been considered one of the best alternative species for sustainable oil production due to its large geographic extension covering almost every region of the country and other characteristics which are reviewed.

Why Physic Nut?

As mentioned above, physic nut has been considered as one of the best alternative oilseed species. There are a number of reasons for that, such as its (i) hardiness, (ii) easy propagation, (iii) drought tolerance, (iv) high oil content, (v) short life cycle, (vi) rapid growth, (vii) adaptation to a wide range of agro-climatic conditions, (viii) bushy/shrubby nature, and (ix) multiple uses of different plant parts (Achten et al. 2010). In addition, physic nut can be grown on wastelands or degraded lands provided pH, fertility and water deficit correction; consequently, it does not compete for land with food crops.

A single physic nut plant from non-elite material can currently produce as high as 2.5 kg of seeds (Drumond et al. 2009), with oil content varying between 30% and 40% (Ginwal et al. 2004; Rao et al. 2008; Sunil et al. 2008). In such a scenario, a single plant can produce 0.75–1.0 kg of oil. Since 1,250 plants can be cultivated in one hectare with 4×2 m spacing, physic nut has nowadays the potential to produce between 930 and 1,250 kg of oil per ha. Such estimate however will change radically as soon as its selective breeding will be undertaken. Other desirable characteristics are high oil quality (Figure 5.2), low cost of oil conversion into biodiesel either by chemical (Berchmans and Hirata 2008) or biological transesterification (Modi et al. 2007) and satisfaction of technical requirements of Brazilian, US and European standards for biodiesel.

The popular claims concerning drought tolerance, low nutrient requirement, pest and disease resistance increased the expectations on this species to unreachable levels (Achten et al. 2010) even though most of these claims are yet to be supported by scientific evidence (Achten et al. 2010). The truth is that, besides the advantageous characteristics mentioned above, physic nut cultivation is challenging as it is a quasi-undomesticated species requiring substantial investment in research and development (R&D) (Fig. 5.2). We describe the main challenges that need to be addressed in order to make physic nut a viable biofuel crop in the near future.

Challenges in Making Physic Nut a Viable Biofuel Crop

Physic nut can be considered as a quasi-undomesticated species because neither elite cultivars nor defined production systems are yet available (Carels 2009). In Brazil, current commercial plantations are in the initial phase of implantation and less than 4 years old (Laviola et al. 2010). In order to boost this installation processes, basic knowledge on seed production techniques is paramount as well as information on propagation systems, planting density, pruning systems, mineral nutrition and fertilization and perhaps most important correct management practices (Fig. 5.2). Along with such limitations, one has also to considerer the regulation on cultivar and pesticides or fungicides registration, which is eventually not available.

Physic nut is also susceptible to a wide range of diseases and pests. One of the most important diseases that currently affect physic nut plantations in Brazil is powdery mildew caused by the fungus *Oidium heveae* B.A. Steinm (Ramakris and Radhakri 1963). Mildew occurs on seedlings and on trees in the field. The main sign of the infection is the presence of a whitish mycelium over the leaves and shoots

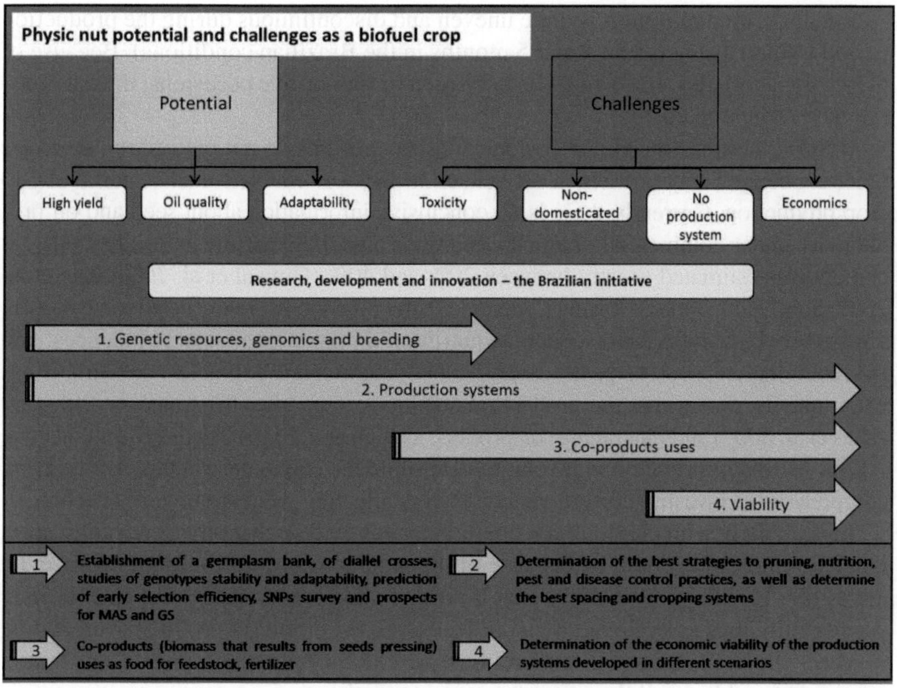

Fig. 5.2 Physic nut potential and challenges. Clear boxes indicate the physic nut characteristics that makes it a potential biofuel crop as well as those that currently hamper its wide adoption as a biofuel crop. Arrows indicate the main research fields and their description (*bottom*) that are included in the Embrapa's network for physic nut research (Source: Durães et al. 2011)

(Ramakris and Radhakri 1963). Even though the infection can be mitigated with the application of fungicides, the use of resistant plants is seen as a best strategy in the mid to long-term. However, virtually nothing is known concerning genetics and molecular aspects of the physic nut-*Oidium heveae* interaction and dedicated studies are necessary to obtain such information.

Another important aspect of physic nut is its toxicity. The most problematic toxic components are probably a number of phorbol esters found in high concentrations mainly in the seeds (Devappa et al. 2010b, 2011; Goel et al. 2007). As phorbol esters are compounds known to cause severe biological effects, including inflammation and tumor promotion (Devappa et al. 2010b), removal of such compounds is imperative in order to make use of the press cake, perhaps the physic nut co-product with the highest potential as animal feed. Physic nut seeds also contain a trypsin inhibitor, lectins and phytates that are antinutritional factors limiting the potential use of the press cake for animal feeding (Devappa et al. 2011; Goel et al. 2007). However, both trypsin inhibitor and lectins can be completely denatured and destroyed by exposure to elevated temperature as occurs before and/or during the oil extraction process technology (Devappa et al. 2011). The addition of synthetic phytases might mitigate phytate problem. Alternatively, press cake can be returned to soil and used as a crop fertilizer (Devappa et al. 2010a).

Finally, fruit maturation is quite uneven and discontinuous during the production period (which last between 4 and 5 months in the Brazilian conditions). Because of these characteristics, fruits must be harvested in four or five occasions, thus increasing the production cost.

Besides these technical and biological aspects of physic nut cultivation, there are other challenges, most of which are related to lack of reliable statistics pertaining to crop production. Currently there is no conclusive information about seed and oil productivity under commercial planting conditions also. In experimental evaluations, oil content was estimated to vary between 20% and 40% (Ginwal et al. 2004; Rao et al. 2008; Sunil et al. 2008). However, seed yield and oil content are expected to be heavily affected by (i) abiotic factors, such as nutrition, drought, management practices, and (ii) biotic factors, such as pests, diseases, age. With regard to this last concern, it has been initially shown that the production stability occurs after the fourth or fifth year (Dias et al. 2007) and that plants can produce as high as 2.5 kg of seeds (Drumond et al. 2009). As this information is pivotal to determine the economic viability of the crop, comprehensive investigations of genetic factors affecting yield are urgently needed.

In agreement with previous considerations, it is evident that physic nut cultivation is not as viable as it could be because of the innumerous questions that must yet be answered. However, we believe that a vibrant and focused research program may successfully make physic nut a viable feedstock for biodiesel production in couple of years. Below we review the Embrapa research program on physic nut that is supposed to address all the concerns mentioned above.

The Embrapa Research Initiative

The Embrapa research initiative in physic nut was built on the idea that an integrated program is necessary for making physic nut a viable biofuel crop in a short to mid-term period. Fourteen areas were identified as priority, i.e., selective breeding, seeds and seedlings production, pruning, spacing, fertilization, irrigation, weeds, pests, diseases, harvest, post-harvest, biodiesel, co-products, transversal studies and zoning. An action plan named BRJATOPHA (Research, Development and Innovation in physic nut for Biodiesel Production) was elaborated encompassing more than 130 research actions and 22 institutions (Fig. 5.3) and granted by FINEP (a Brazilian agency for the promotion and financing of innovation through scientific and applied research in private and public sectors).

This integrated program effectively began in 2008 and was just evaluated through an inquiry covering 39 research fields that encompasses all the 14 major areas discussed above and identified as priority. Researchers were asked upon the current stage of development in each field based on their experience according to a zero to five scale: (0) Without knowledge, i.e., researcher does not have enough knowledge to express his opinion, (1) No information, i.e., there is no information or research results available, (2) Common sense, i.e., there is no information available, but practices from other crops may be adopted, (3) Low knowledge, i.e., initial research

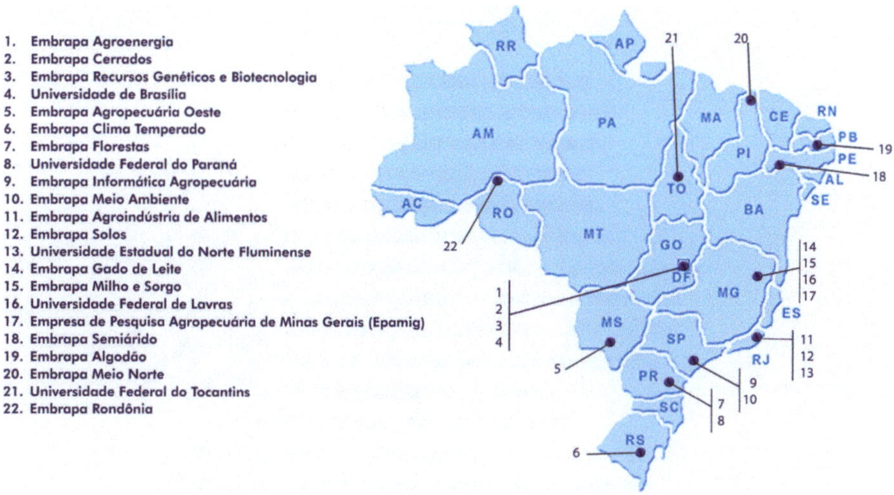

Fig. 5.3 The Embrapa's network for physic nut research involves 22 institutions distributed across the country and more than 130 research programs

results does not support recommendations for commercial applications, (4) Medium knowledge, i.e., partial research results support recommendations for commercial applications with restrictions, (5) Elevated knowledge, i.e., conclusive results supporting recommendations for commercial applications.

The result of this inquiry (Fig. 5.4) indicated that, in the opinion of the researchers, there are some fields, e.g., seeds technology in which research has had tremendous advances since 2008 and in which some technology transfer actions can now be implemented. It also indicated that some research fields still require attention in order to reach a stage where technology transfer actions can begin, e.g., selective breeding and cultivar development.

Genetic Resources, Genomics and Selective Breeding

Despite the lack of some basic and technical information, *J. curcas* is being disseminated in various Brazilian regions and recent analyses indicate that more than 40,000 ha of physic nut are already planted (Rosado et al. 2010) even though no elite (improved) cultivars are available (Carels 2009). As a result, much of the research effort is concentrated on building resources that may allow the genetic improvement of the species (Achten et al. 2010; Divakara et al. 2010; Rosado et al. 2010). In the perspective of future breeding efforts, Embrapa assembled and characterized a germplasm bank (Fig. 5.5) with about 200 accessions (Laviola et al. 2010; Rosado et al. 2010). This germplasm bank represents most of the genetic variability of the species in Brazil (Rosado et al. 2010). The initial characterization

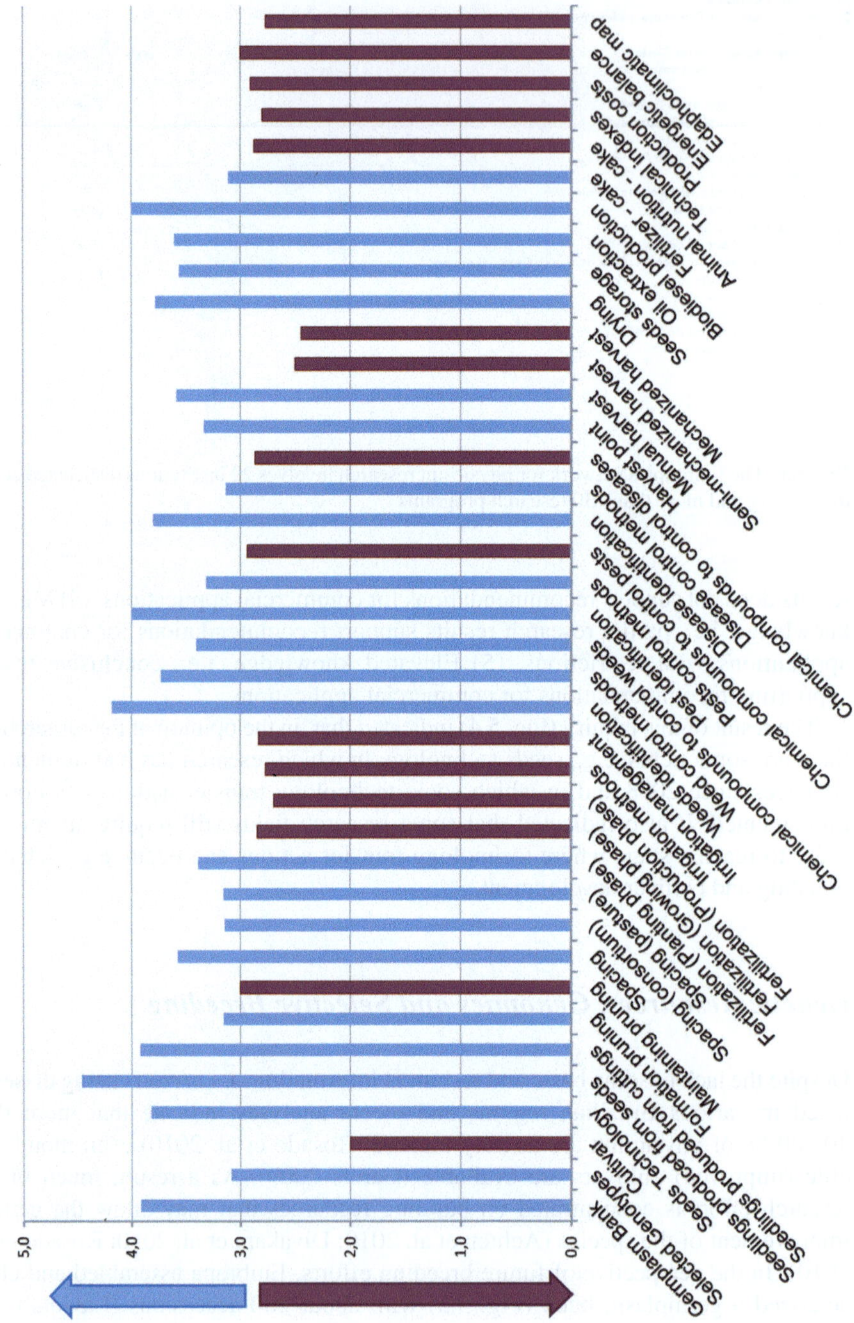
0,0
1,0
2,0
3,0
4,0
5,0
Germplasm Bank
Selected Genotypes
Seeds Technology
Cultivar
Seedlings produced from seeds
Seedlings produced from cuttings
Formation pruning
Maintaining pruning
Spacing pruning
Spacing
Spacing (Consortium)
Spacing (pasture)
Fertilization (Planting phase)
Fertilization (Growing phase)
Fertilization (Producing phase)
Irrigation methods
Weeds managements
Weed identification
Weed control methods
Chemical compounds to control weeds
Pest identification
Pests control methods
Chemical compounds to control pests
Disease identification
Diseases control methods
Chemical compounds to control diseases
Harvest point
Manual harvest
Semi-mechanized harvest
Mechanized harvest
Seeds storage
Drying
Oil extraction
Biodiesel production
Fertilizer production
Animal nutrition - cake
Technical - cake
Production indexes
Energetic costs
Edaphoclimatic map

Fig. 5.4 Current knowledge in different physic nut research fields (*x axis*) and their respective knowledge score (*y axis*). The scores are obtained according to the following rating: (*0*) Without knowledge, i.e., researcher does not have enough knowledge to express his opinion, (*1*) No information, i.e., there is no information or research result available, (*2*) Judgment, i.e., there is no information, but practices from other crops may be adopted, (*3*) Low knowledge, i.e., initial research results (does not support recommendations of commercial applications), (*4*) Medium knowledge, i.e., partial research results (supports recommendations of commercial applications with restrictions), (*5*) Elevated knowledge, i.e., conclusive results (support recommendations of commercial applications). *Blue bars* indicate a specific field for which a considerable amount of information has already been gathered. By contrast, *red bars* indicate a specific field for which considerable amount of research still needs to be done. The blue arrow indicates that there is an urgent need of more refined research and of technology transfer actions. The *red arrow* indicates that there is an urgent need to scale up research in order to get enough information for the fields marked in *red*

Fig. 5.5 Physic nut germplasm bank established by Embrapa in Planaltina (Distrito Federal, Brazil). The germplasm bank contains near 230 accessions and it is now being enriched with seeds collected in the center of origin of the species (i.e., Mexico)

of such bank, using phenotypic (Laviola et al. 2010) and molecular data (Rosado et al. 2010), indicated that although limited, there is enough genetic variability available for breeding purposes (Laviola et al. 2012b). Sources of variability for *Oidium* spp. resistance, seed toxicity and rate of male/female flowers, for example, have been characterized in this bank.

It seems that although limited, the genetic variability is sufficient to ensure that elevated genetic gains can be obtained with selective breeding for superior individuals and families. It is important to note that genetic diversity estimation based on neutral molecular markers not always mimic the genetic variability for the traits of interest, since markers can be located in genes not directly involved in the genetic control of the trait, but also in non-coding regions. Therefore, we do believe that the accessions represented in the Brazilian germplasm constitute a population that can be used as a starting material for breeding purposes. Of course, since the variability is limited, the addition of accessions collected at the centers of origin and diversification of the species is highly desirable and necessary to ensure that genetic gains can still be obtained in the long term.

Two main strategies are considered for increasing physic nut seed yield: the plant management and the selection of superior genotypes (Openshaw 2000). Pallet and Sale (2006) observed an additive relationship between these factors due to the expression of superior genotypes in favorable environmental conditions. Based on these considerations, Embrapa initiated a breeding program (Table 5.3) with the purpose to select materials of physic nut with high seed yield and/or oil production by recurrent selection. Genetic improvements for grain yield will rely on intra-population recurrent selection, manipulation of genetic variance and vegetative

Table 5.3 Embrapa's program of physic nut improvement

Activity	Year 1		Year 2		Year 3		Year 4		Year 5		Year 6		Year 7	
	1° S	2° S	1° S	2° S	1° S	2° S	1° S	2° S	1° S	2° S	1° S	2° S	1° S	2° S
Collection	■													
AGB[a] implantation		■												
AGB enrichment					■		■		■		■			
Agronomic characterization			■	■	■	■	■	■	■	■	■	■		
Genotypic characterization				■		■		■		■		■		
Crosses					■		■		■		■		■	
Mass and recurrent selection					■		■		■		■		■	
Evaluation of GxE interaction						■		■		■		■		■
Elite cultivars and clones														■

[a]"AGB" is for active germplasm bank. "S" is for semester

Fig. 5.6 Evaluation of a cloning methodology by mini-cutting (**a**) of physic nut to propagate elite genotypes (**b**, **c**, **d**) for the Embrapa's breeding program

propagation of superior genotypes. Asexual propagation allows breeders to maximize the genetic variation that is transmitted to the next generation. Vegetative propagation allows breeders to shorten the time required to release improved materials to the market by selecting superior genotypes in heterogeneous populations. Since physic nut is an allogamous species, the use of clones may also be an alternative to generate more uniform plantations. In this context, the use of mini-cuttings has been considered promising (Fig. 5.6). Field validations are still needed in order to evaluate the development of the seedlings produced by this technology.

In terms of breeding for seed production, it must be stressed that several traits, such as seed size, seed oil content, seed yield and plant branching seems to determine the final oil production in physic nut (Heller 1996). However, refined evaluations of these oil yield components showed that more than 90% of the oil yield variability is related to the seed yield variability, indicating that biomass production is the major determinant in oil production (Spinelli et al. 2010). In this context, it should be mentioned that despite some optimistic prognosis of seed yields over 3 $t\,ha^{-1}$, only small seed yields (1 $t\,ha^{-1}$ or less) have been reported in different environmental conditions (Costa et al. 2010; Ginwal et al. 2004; Raiger et al. 2011), indicating that much can potentially be gained with breeding efforts. The genetic basis of physiological stress induced by water deficit or pests and diseases affecting the production of physic nut are being investigated.

The intra-population recurrent selection is expected to increase the frequency of favorable alleles for traits of interest, including seed yield. An initial study of the breeding potential of the Brazilian material suggested that three consecutive measurements are necessary to predict accurately and efficiently the true breeding value of an individual. Genetic correlations among the numerous traits of interest were also obtained providing starting points for indirect selection of the best individuals and to better design management practices (Laviola et al. 2012b).

However, as no elite cultivars are currently available and because breeding approaches based on recurrent selection for elite genotypes are only in their initial stages, rapid development and release of improved physic nut cultivars is highly desirable. It is well known that the productivity is generally a polygenic trait, based on quantitative gene expression, with low heritability and, thus, strongly affected by environment factors. In Physic nut however, genetic studies were carried out to assess the variability and inheritance of this trait. These studies showed that both individual heritability and heritability of progenies are elevated (~70%), providing evidence that selection for genotypes with higher yield is possible (Laviola et al. 2010, 2012a). Different strategies have been used for the early selection of superior genotypes and Drumond et al. (2010) selected genotypes in irrigated conditions with a yield of 2.12 kg plant^{-1} 12 months after planting. With regard to other environments, Laviola et al. (2010) observed that seed yield ranged from 0 to 0.18 kg plant^{-1} 12 months after planting in the semi-arid conditions of the Brazilian savannah. Rocha et al. (2011) observed a genetic progress of 99% when selective breeding according to mass selection was applied to the ten more productive genotypes, which corresponds to an estimated grain yield of 1.23 kg plant^{-1} or 2.25 t ha^{-1}. Another study conducted in Brazil (Laviola et al. 2012b) showed that (i) early selection of physic nut is possible, (ii) it can promote high genetic gains on the initial stages of the breeding program, and (iii) genotypes with increased productivity can be potentially selected and released as improved cultivars. Moreover, it showed that genetic improvement of physic nut can be performed by early recurrent selection as an alternative to the slower, but more accurate recurrent selection with three consecutive measurements (Laviola et al. 2012b).

In addition to breeding methods, such as (i) assessment of the variation in wild sources, or germplasm collections, (ii) selection of superior or elite genotypes, (iii) recurrent selection of naturally breeding populations, the physic nut improvement should also benefit from the application of genomic breeding strategies, e.g., the use of molecular markers (Gomes et al. 2010; Johnson et al. 2011; Liu et al. 2011; Mastan et al. 2012) in marker assisted selection (MAS) (Grattapaglia and Kirst 2008) and genomic selection (GS) (Grattapaglia and Resende 2011; Resende et al. 2012a, 2012b). The rationale is that the physic nut small genome size (~400 Mb) is a favorable feature for the use of such tools (Carvalho et al. 2008). In depth studies are needed to test the efficiency in integrating genomic breeding tools in physic nut improvement. By adopting the strategy to sequence and align cDNA from various physic nut individuals, a large SNP (Single Nucleotide Polymorphism) discovery initiative is currently underway in Brazil (Gomes et al. 2010; Silva-Junior et al. 2011). Initial results indicated that thousands of SNPs were discovered and additional validation of such SNPs could prove to be efficient to build a chip, with hundreds to thousands genotypable SNPs. DArT

Fig. 5.7 Occurrence of *Oidium* in physic nut fruits (**a**), "Green leafhopper" (*Empoasca kraemeri*) causing severe defoliation in a physic nut plantation (Mato Grosso do Sul, Brazil) (**b**)

Fig. 5.8 Comparative growing vigor in non-toxic genotypes (**a** and **b**) and toxic genotype (**c**), 12 months after planting. The toxic accessions are much more vigorous than the non-toxic genotypes

(Diversity Array Technology) markers are also being developed in collaboration with Diversity Arrays Technology Pty Ltd. Such resources (SNPs and DArTs chip) could potentially be used to assist breeding efforts, either by MAS strategies (MARS—marker assisted recurrent selection or GS—genomic selection) in the near future. Powdery mildew resistance, pest resistance (Fig. 5.7) and non-toxicity (Fig. 5.8) are among the traits that could be the initial targets for MAS, and productivity as well as oil content among the traits that are to be targeted by GS. It is interesting to note that the development of commercial non-toxic cultivars could contribute to aggregate value to the press cake that results of the oil extraction, also contributing to reduction of the environmental passive (nowadays it is necessary to detoxify the press cake). However, since genotypes that do not produce phorbol esters are less vigorous and produce less oil than the toxic cultivars (Fig. 5.8), it is probable that MAS will have to be implemented in backcrossing schemes. It is also being considered to cross non-toxic genotypes with high yielding individuals followed by selection of high yielding and vigorous non-toxic plants.

Since Synthetic Genomics Inc. (SGI), Asiatic Centre for Genome Technology (ACGT) and Kazusa DNA Research Institute have completed the physic nut genome project recently (Sato et al. 2010), the annotation of 95% of the physic nut genes are available freely by online access (http://www.kazusa.or.jp/jatropha/). The genome information by Kazusa is expected to boost the discovery of new molecular markers. As tissue culture protocols are available for regeneration of physic nut genotypes (Jha et al. 2007), genetic transformation and gene transfer, can perhaps be an alternative to modify and, or introduce genes for important traits, especially for those for which variability is rather limited.

Production Systems

One of the most popular claims about physic nut is that it tolerates low technological level in its cropping conditions. However, in such conditions the fact is that its productivity is low, making it economically unviable (Laviola and Dias 2008) Although physic nut is indeed a species considered rustic and adaptable to different environmental conditions, it is now clear that physic nut needs cultivation technologies (fertilizer application, pest and disease control, management practices, etc.) to sustain economic levels of oil production.

In order to establish a production system, an important issue that we are considering is the plant phenology. All management practices depend upon this information. Although the phenological phases are not rigid, the duration of each phase is likely to be influenced by environmental conditions that occur in a specific region or year. Considering this, we have attempted to study the plant´s phenology in one of the main Brazilian vegetation, the savannah (also known as cerrado). The results of such effort indicate that initial establishment of the plants is probably one of the most important phases as any stress that occurs in this phase can compromise the plants architecture and hence, the plants yield. It is now clear that, following the second year, the phenological phases occur in repeated cycles. During the raining period, with the rise of the temperature and humidity, the plant's vegetative growth reaches its maximum, with the formation of new shoots and inflorescences. In the Brazilian conditions, fruiting occurs on average 3 months after the start of the growing season, mostly in the terminal region of the shoots recently grown. In that respect, it is noteworthy that the production of new inflorescences is highly dependent on continuous vegetative growth. This indicates that in this period it is important to manage the plants properly, i.e., by avoiding the occurrence of biotic and abiotic stresses so as to allow the plant to express all its potential. At the end of the growing season, with the fall of temperature and humidity, the deciduous stage begins. In this period physic nut loses its leaves and remains in rest conditions until the beginning of the next vegetative growth phase.

As outlined above, the phenological data support perhaps all decisions on management practices to be performed. Aiming at providing the plants an ideal environment for the vegetative growth stadium, the control of invading seeds, pests and diseases is imperative. According to Dias et al. (2007), a number of pests and diseases from other plant species may affect physic nut. Currently there are no specific pesticides and fungicides recommended for the culture and because of that situation,

Fig. 5.9 Pruning in physic nut. The formation pruning is to be performed 45 days after planting (**a–c**). The maintenance pruning is carried out when plants are 2 m high and is seen as a way to standardize the plant`s architecture and production (**d**)

an integrated management approach is been pursued. Such an approach considers the adoption of cultural, chemical and biological strategy concurrently to control the undesirable biotic interactions. In that context, the development of cultivars with enhanced resistance/tolerance to such stresses is seen as one of the best approaches. Initial results are encouraging in that respect, since resistant genotypes for many of these stresses were already identified in the Brazilian germplasm bank (unpublished data).

While trying to develop a production system devoted to physic nut cultivation, it is important to determine the plant's responses to mineral nutrition. Initial results gathered among many Brazilian culture conditions indicate that physic nut responds well to fertilizer applications and a correct balance of nutrients is required to warrant plants with satisfactory fruit yield (Laviola and Dias 2008). Thus, annual fertilization is necessary to supply enough amounts of nutrients to support the plant's vegetative growth and reproductive development.

Another issue that has been a matter of study in the context of the Embrapa research initiative on physic nut is the effect of pruning. Two types of pruning are being tested, i.e., formation pruning and production pruning (Dias et al. 2007) (Fig. 5.9). The objective of the formation pruning is to standardize the plants (in

Fig. 5.10 Mechanical harvesting systems in physic nut from front (**a**) and lateral (**b**) (photos taken in Bioauto Farm, Nova Mutum, Mato Grosso, Brazil)

terms of number of basal branches and height) and to favor the basal shoot ramification by breaking the apical dormancy (the apical meristem is actually removed). On the other hand, the objective of production pruning is to renew the productive ramifications as well as to reduce the plant size, which facilitates cultural practices and fruit harvesting. Even though the effects of these techniques are still being evaluated, it seems to be very promising as it contributes to standardize fruit production (unpublished data).

Considering fruit production, the most important aspect that is currently being studied is the maturation process. In general, maturation is quite uneven during the production period (which last between 4 and 5 months in the Brazilian conditions) and progressive with a mix of immature and mature fruits on the same raceme. Because of these characteristics, fruits must be harvested in four or five occasions, burdening the production cost (discussed below). Two approaches are currently being investigated to understand the determinants of maturation process in physic nut. The first is to gather genetic individuals that display fruit maturation concentrated in discontinuous and short periods (as few as possible). Such approach depends on the existence of genetic variability for this trait in the germplasm bank and on its inheritance. The second approach relies on the application of an exogenous chemical that will interfere with the plant physiology and allow the standardization of fruit maturation. Both approaches are in initial stages and no conclusive results are yet available.

Nowadays physic nut fruits are harvested manually. Mechanized (Fig. 5.10) and semi-mechanized harvest systems are currently being developed in Brazil by taking advantage on the knowledge gathered with other perennial cultures, such as coffee, for example. Correct dimensioning of such systems however depends on the size of the genetic material to be harvested and on the plants density, which ultimately depends on planting spacing. These issues are still being studied and correct dimensioning of the harvest systems will be possible. These mechanized and semi-mechanized systems are expected to improve the harvest efficiency besides lowering significantly the costs of physic nut production.

By-Products

The cake that results from pressing the physic nut seeds may potentially constitute an excellent organic fertilizer since it is rich in nitrogen, phosphorous and potassium (Achten et al. 2008). Since the press cake is also rich in protein (43–63% depending on the method used for oil extraction), it could theoretically be used along with the fruits shell as a high protein supplement in ruminant feeds, however, such use is currently impossible given the high toxicity level of the press cake. As previously mentioned, a number of phorbol esters occur in high concentrations in the seeds. Removal of phorbol esters during processing is achievable (Rakshit et al. 2008), however the possible presence of toxic degradation products from phorbol esters after treatment cannot be ruled out (Achten et al. 2010). Aiming at solving such a problem, we are currently focusing on two different, yet complementary approaches to completely remove phorbol esters from physic nut seeds.

The first approach relies on the identification of phorbol ester free genotypes in our germplasm bank. The current data indicate that as toxicity, in terms of phorbol esters presence seems to be expressed qualitatively (Sujatha et al. 2005) and may be regulated by only one or a few genes (see Sato et al. 2010; Gomes et al. 2010, for a discussion). It has also been suggested that the trait is subjected to maternal inheritance, i.e., the phenotype of the maternal parent is passed on to the progeny. Such mode of inheritance was suggested earlier by Sujatha et al. (2005), however, it is still to be proven in the next couple of years. In that respect, non-toxic genotypes are common in the Mexican germplasm (Mastan et al. 2012) and we are now confirming the genetic basis of such trait in our genetic material through the generation of large segregating populations and genetic mapping. The intention is to map the locus (loci) responsible for (non) toxicity with molecular markers that are currently in development to assist selective breeding for non-toxic genotypes. If the non-toxicity is indeed inherited from the mother to the progeny, this can have a profound effect on breeding programs worldwide, as the main traits subjected to improvement generally does not follow such mode of inheritance. The incorporation of trait subjected to maternal inheritance in the breeding programs will force breeders to consider this information for cross designs.

The second approach relies on the application of the following treatments: (i) thermoplastic extrusion process associated with chemical additives, (ii) solvent washing and (iii) biotransformation (by means of fungus, such as yeast). The main focus here is to detoxify the press cake, either through the removal of phorbol esters or modification of phorbol ester molecules so that the new form loses their toxic activity. The absence of phorbol esters guarantee that no soil (Gressel 2008), animal or human contamination occurs. *Phorbol zero* cultivars would add tremendous value to the culture especially for small farmers (the majority of Brazilian physic nut growers), since it would allow the press cake recycling through animal feed.

Perspectives of Physic Nut Cultivation According to Different Scenarios

The determination of the economic viability of physic nut cultivation is an important issue. Therefore, it is necessary to consider that physic nut cultivation can occur in different scenarios: (i) large production systems with high technological level, (ii) small production systems with mid-technological level and (iii) small production systems with low technological level associated or not with livestock or/and other intercrops, such as oleaginous species, grasses or even food crops.

Considering the first scenario, it is clear that improved genotypes are necessary. Such improved genotypes must have large seed and oil productivity, since physic nut is the main product. The genotypes are also expected to be highly responsive to improved environmental conditions, since fertilization and irrigation are likely to be the default situation in this scenario. Genotypes highly tolerant to biotic and abiotic stresses, such as the ability to grow in marginal areas are the ones that most likely will support large physic nut plantations (in order to avoid food crop displacement). Such improved genotypes are expected to be commercialized as mixed populations of elite vegetative clones to maximize the genotypic potential of the individuals whose features are optimized to growing conditions. Mechanization of both cultural practices, such as (i) weeds, pest and disease control and (ii) fruit crops will be a *sine qua non* condition, since the costs of manual operation of these services is too expensive to make the business viable. All these operations are undoubtedly bound to increase costs. However, with the use of elite individuals, fertilization, irrigation and mechanized practices, the marginal return of physic nut planting is expected to be maximized and the gains to be a function of the venture scale. Moreover, the commercialization of co-products, such as press cake should generate extra profit when non-toxic genotypes will be available and contribute to increasing the viability of physic nut plantation.

It can be argued that growing physic nut on an industrial scale with high inputs (irrigation, fertilizers, and pesticides) would lead to the failure of many principles of sustainability of biofuel production and, thus, expose this new crop to negative pressures from environmentalists that could claim other oilseed species eventually better matching their considerations. However, this concern can be ruled out, since PNBP stimulates planting of a diverse array of oilseed species and encourages the diversification of oilseed materials in order to fit local conditions to the most possible extent. In addition, Brazil fortunately has a huge amount of land available for agriculture and in our opinion physic nut can be grown together with other oilseed species without competing risks between food and fuel crops, since oilseed species are thought to be regulated on a regional basis.

Keeping the second and the third scenarios in view, it is clear that genotypes will have to be improved considering that adequate management practices will not always be available. In this way, genotypes with robust features are desirable as such genotypes maintain their productivity potential even if the environmental conditions are not at the optimum. In these last scenarios, the press cake will probably assume another level of importance, since it can be used to feed livestock created in the same farms. In order to keep the implantation costs as low as possible, improved genotypes will probably be commercialized by means of their seeds. Enhanced seed and oil productivity will also be important, but another important issue could be to keep the wild type of continuous fruit maturation. This characteristic is important for small farmers, since it minimizes the period without production and fruit harvesting can rely on family labour, reducing the impact of its costs. A last issue that needs to be considered is that, in such cases, profit can be both direct and indirect. Direct profit comes from de-commercialization of seed, oil or press cake while indirect profit comes from the use of the press cake as an organic fertilizer or as feed for livestock. Indirect profit can also be obtained by the valorization of intercrops. In that respect, the intercrop production may indeed be an important factor to make small production systems viable. In all cases, economic studies are underway to determine the exact conditions needed to make physic nut cultivation a viable business.

Conclusions

The report presents the importance of physic nut for the Brazilian economy considering its potentials, challenges and the concerns covered by the research that is underway in the country. We intentionally avoid focusing on the economic aspects of physic nut cultivation because we believe that it can only be considered after the availability of improved cultivars and suitable production systems for the different Brazilian regions. As discussed above, some more years will be necessary to develop such improved cultivars and to determine suitable production systems even with worldwide initiative.

However, our cautious, but yet optimistic vision is that the Brazilian research network may successfully contribute to make physic nut a viable feedstock for biodiesel production in the near future. The increasing demand for biodiesel and the potential demand for jet fuel (biokerosene) in the near future is a guarantee that the biofuel market will grow continuously in the upcoming years and also provides stimulus to scale up physic nut research.

Acknowledgments We acknowledge the financial support from the Brazilian Ministry of Science, Technology and Innovation (MCTI) to BRJATROPHA as well as the students, research collaborators and breeders for their constant scientific inputs and useful discussions.

References

Achten WMJ, Nielsen LR, Aerts R, Lengkeek AG, Kjær ED, Trabucco A et al (2010) Towards domestication of *Jatropha curcas*. Biofuels 1:91–107

Achten WMJ, Verchot L, Franken YJ, Mathijs E, Singh VP, Aerts R et al (2008) Jatropha biodiesel production and use. Biomass Bioenergy 32:1063–1084

BAP (2006) Brazilian agroenergy plan. Embrapa, Brasilia

Beltrão NEM (2005) Agronegócio das oleaginosas no Brasil. Informe Agropecuário 26:14–17, Portuguese

Berchmans HJ, Hirata S (2008) Biodiesel production from crude *Jatropha curcas* L. seed oil with a high content of free fatty acids. Bioresour Technol 99:1716–1721

Carels N (2009) *Jatropha curcas*: a review. Adv Bot Res 50:39–86

Carvalho CR, Clarindo WR, Praça MM, Araújo FS, Carels N (2008) Genome size, base composition and karyotype of *Jatropha curcas* L., an important biofuel plant. Plant Sci 174:613–617

Costa AAA, Junior NP, Aranda DAG (2010) The situation of biofuels in Brazil: new generation technologies. Renew Sustain Energy Rev 14:3041–3049

Devappa RK, Makkar HPS, Becker K (2010a) Biodegradation of *Jatropha curcas* phorbol esters in soil. J Sci Food Agric 90:2090–2097

Devappa RK, Makkar HPS, Becker K (2010b) Jatropha toxicity. A review. J Toxicol Environ Heal B 13:476–507

Devappa RK, Makkar HPS, Becker K (2011) Jatropha Diterpenes: a review. J Am Oil Chem Soc 88:301–322

Dias LAS, Leme LP, Laviola BG, Pallini Filho A, Pereira OL, Carvalho M et al (2007) Cultivo de pinhão manso (*Jatropha curcas* L.) para produção de óleo combustível. Universidade Federal de Viçosa, Viçosa, Portuguese

Divakara BN, Upadhyaya HD, Wani SP, Laxmipathi Gowda CL (2010) Biology and genetic improvement of *Jatropha curcas* L. A review. Appl Energy 87:732–742

Drumond MA, Santos CAF, Oliveira VR, Anjos JB, Evangelista MRV (2009) Desempenho agronômico de genótipos de pinhão manso irrigado no Semiárido pernambucano aos 12 e 24 meses de idade. I Congresso Brasileiro de Pesquisa em Pinhão Manso. F&B Comunicação e eventos, Brasília. Portuguese

Drumond MA, Santos CAF, Oliveira VR, Martins JC, Anjos JB, Evangelista MRV (2010) Agronomic performance of different genotypes of physic nut in the semi-arid zone of Pernambuco state. Ciência Rural 40:44–57

Duraes FOM, Laviola BG, Alves AA (2011) Potential and challenges in making physic nut (*Jatropha curcas* L.) a viable biofuel crop: the Brazilian perspective. Perspectives in Agriculture, Veterinary Science, Nutrition and Natural Resources 6:043

Ginwal HS, Rawat PS, Srivastava RL (2004) Seed source variation in growth performance and oil yield of *Jatropha curcas* Linn. in central India. Silvae Genet 53:186–192

Gomes KA, Almeida T, Gesteira A, Lobo IP, Guimarães ACR, de Miranda AB et al (2010) ESTs from seeds to assist the selective breeding of *Jatropha curcas* L. for oil and active compounds. Genomics Insights 3:29–56

Goel G, Makkar HPS, Francis G, Becker K (2007) Phorbol esters: structure, biological activity, and toxicity in animals. Intl J Toxicol 26:279–288

Grattapaglia D, Kirst M (2008) *Eucalyptus* applied genomics: from gene sequences to breeding tools. New Phytol 179:911–929

Grattapaglia D, Resende MDV (2011) Genomic selection in forest tree breeding. Tree Genet Genomes 7:241–255

Gressel J (2008) Transgenics are imperative for biofuel crops. Plant Sci 174:246–263

Heller J (1996) Physic nut—*Jatropha curcas* L. Promoting the conservation and use of underutilized and neglected crops. International Institute of Plant Genetic and Crop Plant Research & Plant Genetic Resource Institute, Gatersleben/Rome

Jha TM, Priyanka M, Datta MM (2007) Somatic embryogenesis in *Jatropha curcas* L. an important biofuel plant. Plant Biotechnol Rep 1:135–140

Johnson TS, Eswaran N, Sujatha M (2011) Molecular approaches to improvement of *Jatropha curcas* Linn. as a sustainable energy crop. Plant Cell Rep 30:1573–1591

La Rovere EL, Pereira AS, Simoes AF (2011) Biofuels and sustainable energy development in Brazil. World Dev 39:1026–1036

Laviola BG, Alves AA (2011) Matérias primas potenciais para a produção de biodiesel: situação atual e perspectivas. In: Lana RP, Guimarães C, Lima GS, Veloso CM, Patino HO (eds) Simpósio Brasileiro de Agropecuária Sustentável: O uso de Tecnologias Limpas e Agroenergia. Arka editora, Viçosa. Portuguese

Laviola BG, Alves AA, Gurgel FL, Rosado TB, Rocha RB, Albrecht JC (2012a) Estimates of genetic parameters for physic nut traits based in the germplasm two years evaluation. Ciência Rural 42:429–435

Laviola BG, Alves AA, Gurgel FL, Rosado TB, Costa RD, Rocha RB (2012b) Estimate of genetic parameters and predicted gains with early selection of physic nut families. Ciência e Agrotecnologia 36:163–170
Laviola BG, Dias LAD (2008) Nutrient concentration in *Jatropha curcas* L. leaves and fruits and estimated extraction at harvest. Rev Bras Ciênc Solo 32:1969–1975
Laviola BG, Rosado TB, Bhering LL, Kobayashi AK, Resende MDV (2010) Genetic parameters and variability in physic nut accessions during early developmental stages. Pesquisa Agropecuária Brasileira 45:1117–1123
Liu P, Wang CM, Li L, Sun F, Liu P, Yue GH (2011) Mapping QTLs for oil traits and eQTLs for oleosin genes in jatropha. BMC Plant Biol 11:132. doi:10.1186/1471-2229-11-132
Mastan SG, Sudheer PDVN, Rahman H, Reddy MP, Chikara J (2012) Development of SCAR marker specific to non-toxic *Jatropha curcas* L. and designing a novel multiplexing PCR along with nrDNA ITS primers to circumvent the false negative detection. Mol Biotechnol 50:57–61
Modi MK, Reddy JRC, Rao BVSK, Prasad RBN (2007) Lipase-mediated conversion of vegetable oils into biodiesel using ethyl acetate as acyl acceptor. Bioresour Technol 98:1260–1264
Moura EF, Ventrella MC, Motoike SY (2010) Anatomy, histochemistry and ultrastructure of seed and somatic embryo of *Acrocomia aculeata* (Arecaceae). Sci Agric 67:399–407
Openshaw K (2000) A review of *Jatropha curcas*: an oil plant of unfulfilled promise. Biomass Bioenergy 19:1–15
Pallet RN, Sale G (2006) The relative contributions of tree improvement and cultural practice toward productivity gains in Eucalyptus pulpwood stands. For Ecol Manage 193:33–43
Raiger HL, Dua RP, Sharma SK, Phogat BS (2011) Performance and stability of jatropha (*Jatropha curcas*) for seed yield and its components. Indian J Agric Sci 81:125–128
Rakshit KD, Darukeshwara J, Raj KR, Narasimhamurthy K, Saibaba P, Bhagya S et al (2008) Toxicity studies of detoxified Jatropha meal (*Jatropha curcas*) in rats. Food Chem Toxicol 46:3621–3625
Ramakris TS, Radhakri PN (1963) *Jatropha curcas* L.—a collateral host for *Oidium heveae*, stein. Curr Sci 32:428
Rao GR, Korwar GR, Shanker AK, Ramakrishna YS (2008) Genetic associations, variability and diversity in seed characters, growth, reproductive phenology and yield in *Jatropha curcas* L. accessions. Trees Struct Funct 22:697–709
Resende MDV, Resende MFR, Sansaloni CP, Petroli CD, Missiaggia AA, Aguiar AM et al (2012a) Genomic selection for growth and wood quality in Eucalyptus: capturing the missing heritability and accelerating breeding for complex traits in forest trees. New Phytol 194:116–128
Resende MFR, Munoz P, Acosta JJ, Peter GF, Davis JM, Grattapaglia D et al (2012b) Accelerating the domestication of trees using genomic selection: accuracy of prediction models across ages and environments. New Phytol 193:617–624
Rocha RB, Ramalho AR, Laviola BG, Santos AR, Spinelli VM, Holanda ZF (2011) Genetic progress in grain yield of the physic nut (*Jatropha curcas* L.). Biomass Bioenergy, in press
Rosado TB, Laviola BG, Faria DA, Pappas MR, Bhering LL, Quirino B et al (2010) Molecular markers reveal limited genetic diversity in a large germplasm collection of the biofuel crop *Jatropha curcas* L. in Brazil. Crop Sci 50:2372–2382
Sato S, Hirakawa H, Isobe S, Fukai E, Watanabe A, Kato M et al (2010) Sequence analysis of the genome of an oil-bearing tree. *Jatropha curcas* L. DNA Res 1–12. doi:10.1093/dnares/dsq030
Silva-Junior O, Rosado T, Laviola B, Pappas M, Pappas G, Grattapaglia D (2011) Genome-wide SNP discovery from a pooled sample of accessions of the biofuel plant *Jatropha curcas* based on whole-transcriptome Illumina resequencing. BMC Proc 5:P57. doi:10.1186/1753-6561-5-S7-P57
Spinelli VM, Rocha RB, Ramalho AR, Marcolan AL, Vieira Júnior JR, Fernandes CF et al (2010) Primary and secondary yield components of the oil in physic nut (*Jatropha curcas* L.). Cienc Rural 40:1752–1758
Sujatha M, Makkar HPS, Becker K (2005) Shoot bud proliferation from axillary nodes and leaf sections of non-toxic *Jatropha curcas* L. Plant Growth Regul 47:83–90
Sunil N, Varaprasad KS, Sivaraj N, Kumar TS, Abraham B, Prasad RBN (2008) Assessing *Jatropha curcas* L. germplasm *in situ*—A case study. Biomass Bioenergy 32:198–202
Vermerris W (2008) Genetic improvement of bioenergy crops. Springer, New York
Yuan JS, Tiller KH, Al-Ahmad H, Stewart NR, Stewart CN (2008) Plants to power: bioenergy to fuel the future. Trends Plant Sci 13:421–429

Chapter 6
Producing Jatropha Biodiesel in China: Policies, Performance and Challenges

Zanxin Wang

Introduction

Along with the rapid economic growth, China's consumption of energy is continuously increasing. Since 1998, the consumption of energy in China has been increasing at a rate of 5%, which is three times higher than the world average. China is now the second largest importer of oil in the world. In 2005, China imported 168 million tons of crude oil, which accounts for 42.9% of the total amount of domestic consumption. As projected, China will reach the peak of energy consumption, that is, 4.18 billion tons of standard coal equivalents, in 2020 (EPASEA-NIC 2008). To meet its demand for energy, China highly depends on importing energy. According to the International Energy Agency, China's import of petroleum will account for about 77% of its total demand by 2020, and 84% by 2030 (IEA 2005). Thus, it is a hard task to meet the demand for diesel in China in the long term (Li 2001). In the next two decades, China will face a severe energy supply challenge.

Given abundant coal reserves compared to the insufficient supply of oil and gas, China has mainly relied on coal for its energy in the past decades. China's energy structure is dominated by coal, which accounted for 68.8% of the total energy consumption in 2006 (EPASEA-NIC 2008). Along with the consumption of fossil energy, air pollution and the emission of *carbon dioxide* (CO_2) have also attracted the concerns of Chinese government and academia. As pollution from the burning of coal becomes more and more, it is an urgent task to search for environment-friendly alternatives to fossil fuel. Moreover, along with the worldwide concerns on global warming, China is facing international pressures for the reduction of green-

Z. Wang (✉)
School of Development Studies, Yunnan University, No. 2, Northern Cuihu Road, Kunming city, Kunming, Yunnan province 650 091, China
e-mail: wzxkm@hotmail.com

The Beijer Institute of Ecological Economics, The Royal Swedish Academy of Sciences, Box 50005 SE-10405, Lilla Frescativägen, Stockholm, Sweden

N. Carels et al. (eds.), *Jatropha, Challenges for a New Energy Crop: Volume 1: Farming, Economics and Biofuel*, DOI 10.1007/978-1-4614-4806-8_6,

house gas (GHG) emission. The emission of CO_2 equivalent is projected to be 5.2 billion tons and 5.6 tons in 2015 and 2020, respectively (EPASEA-NIC 2008).

The increases in crude oil prices and the concern for environmental protection gave an impetus to the search of renewable alternative sources of oil (Shay 1993; Runge and Senauer 2007; Hazell and Pachauri 2006). Among the renewable energy to be developed, liquid biofuels, including bioethanol and biodiesel, is a priority for development because it can be used in existing engines. Moreover, the production and use of biofuel has the potential of reducing dependence on petroleum, improving environmental quality, lowering the emission of greenhouse gas (GHG), promoting rural development, and providing farmers job opportunities (Krawczyk 1996; FAO 2008; Wiesenthal et al. 2009). As in many other countries, the production of biofuel in China is seen as an alternative to reduce the pressure induced by the increasing oil scarcity and to achieve sustainable development.

In 2005, China became the world's third largest biofuel producer with an output of one million tons, behind Brazil and the US (USDA 2006). In 2006, the National Development and Reform Commission (NDRC) of China set the threshold of 15% biofuel as the target to be meet by the fuel transportation system by 2020 (USDA 2006). The State Forest Administration (SFA) of China set the target of 13 million hectares to be planted with oilseed species suitable for biodiesel production by 2010. The expectation of the Ministry of Science and Technology (MOST) is to reach the production of 1.5-2 million tons of biodiesel by 2010 and 12 million tons by 2020 (GTZ 2006).

The production of first generation biofuels whose feedstock is agricultural crops will have a negative effect on food security if produced in large quantities (Runge and Senauer 2007; Huang et al. 2009). In 2007, food prices on the international market increased by 15.6% and those in China by 10.8% on a year to year basis (NBSC 2008). Using woody plants produced from non-agricultural land as feedstock to produce second generation biofuel will not affect food security and can be more conducive to respect the environment than first generation biofuels. That is, sustainable biofuel production will be better achieved with a shift from the production of first generation biofuel to that of second generation biofuel including lignocellulosic ethanol from biomass crops and biodiesel from oilseed tree species (FAO 2008; Flavin 2008). The rise in food prices has thus triggered the production priority from first generation to second generation biofuel.

While the second generation bioethanol is facing a technological challenge, second generation biodiesel is attracting more and more attention throughout the world due to its technological easiness of production. Second generation biodiesel is a promising fuel, but its production is challenged by high cost and limited availability of oil resources. The availability and sustainability of sufficient supplies of less expensive feedstock will be a crucial determinant in delivering a competitive biodiesel to commercial fueling stations. One way of reducing biodiesel production costs is to use less expensive fatty acid containing feedstocks, such as inedible oils, animal fats, waste food oil and byproducts of vegetable oil refining (Veljkovic et al. 2006). Fortunately, biodiesel can be produced from inedible vegetable oils, mostly from seed-bearing trees and shrubs. There are 151 families, 697 genera and 1,554 species of oilseed plants of which 30 species have the potential of becoming feedstock for biodiesel production. However, only a few of these species can be cultivated as biodiesel feedstock plantations on waste land. So far, the Ministry of Science and Technology of China is targeting at producing biodiesel from *J. curcas* (JCL), *Pistacia chinensis* Bunge, *Xanthoceras sorbifolia* Bunge, and *Cornus wilsoniana* Wangerin.

JCL receives the largest attention because it does not compete with food uses, it has excellent oil characteristics and it is distributed all over tropical and subtropical climates across the developing world (Heller 1996; Openshaw 2000). In China, the diesel oil consumption was 67.15 million tons in 2000, 85.3 million tons in 2005 and will be 133 million tons in 2015 (Li 2001). Thus, JCL biodiesel has a large market potential.

Since China has the largest population in the world and the lowest farmland *per capita* (less than 0.1 hectare), it is thus impossible to use farmland to produce biofuel on a large scale (Wang et al. 2005). However, China has a lot of mountains and marginal lands which are suitable for the plantation of arborous and shrub oilseed plants. The cultivation of these plants can not only provide raw material for the biofuel industry, but also promote the restructuring of rural industry and increase farmers' income. Moreover, the cultivation of these plants on marginal or waste land in a sustainable way can also contribute to ecological rehabilitation. Being such an oilseed, JCL is considered as a priority feedstock species in China, as in many other countries.

Biofuel Policies in China

The development of renewable energy has become an important component of China's energy development and sustainable development strategies. The Chinese government has adopted many policies and measures to develop renewable energy and address the issue of climate change. The promotion of biofuel was listed in the "National plan for scientific and technological industrialization programs" in May 2003. Accordingly, the Ministry of Science and Technology launched the "10th 5-year plan for biofuel technology development" in 2004. The agricultural and forest biomass program, supported by a national special fund, was started in 2005. According to its "Middle and Long Term Development Plan of Renewable Energy," China set an ambitious target of an annual production of 10 million tons of biofuel by 2020 (NDRC 2008). The "law for renewable energy" was also approved in 2005 and came into force starting from January 1, 2006. In 2007, the first national standard for biodiesel was designed and put into practice. In China's "11th five-year plan framework for national economic and social development", it was highlighted that more efforts should be invested in the development of renewable energy, especially by improving the production capacity of power stations running on biomass, bioethanol and biodiesel. These policies and measures laid a solid foundation for biodiesel as a substitute to fossil diesel to enter fuel market and provided a legal basis for providing subsidized loan and tax discount to companies involved in biodiesel production.

In 2007, concerns over food security led China's central government to control the use of grain-based feedstock for biofuel production and to reorient the country's bioenergy plans toward perennial crops to be grown on marginal land. That is, neither biofuel should be derived from feedstock competing with crops from foodlands nor should they compete with human and livestock access to food. In addition, biofuels should not be harmful to the environment (MOA 2007). China has the largest human population, accounting for 22% of the world population, but it has only 120 million hectares farmland, i.e., 7% of the total world area available for farming. Thus, the limited farmland will be used to grow food crops so as to feed the large population. The expansion of grain-derived biofuel is prohibited in order to alleviate the pressure on food security, while non-crop feedstock is encouraged, including sweet sorghum, cassava, sweet potato and other non-grain crops. As a non-grain crop, JCL has emerged as a high potential biodiesel feedstock because of its adaptability to diverse growing conditions. Provincial governments in Southwest China have drafted plans to increase the area of JCL planting by over one million hectares in the next decade (Weyerhaeuser et al. 2007).

To promote the development of low-carbon economy is another justification for the production of biofuel. Chinese government has announced many related policies since 2006, including the "National Assessment Report on Climate Change", "China's National Climate Change Program", "Decision on the strengthening of energy saving work", and "The comprehensive work plan for energy saving and emission reduction". All of these policies have laid out a development plan for the energy system, and will contribute a lot to the control and mitigation of GHG emissions in China. In particular, in November 2009, Chinese government announced its target of reducing the emission of CO_2 by 40–45% per unit of GDP in 2020 over the level of 2005. To meet this target, biofuel, especially second generation biofuel, is considered as an important alternative for fossil fuel (Wang et al. 2010; Liu et al. 2010).

To encourage the development of biofuels industry, a set of incentive policies was implemented in 2007; they include (i) mandatory mixing of 10% bioethanol in gasoline in nine provinces to secure the biofuel market; (ii) waiving the 5% consumption tax on bioethanol and refunding the 17% value added tax; and (iii) direct subsidies to biofuel plants of about $ 200 per ton (Qiu et al. 2008). These policies resulted in a rapid increase in the output of biofuel in China. In 2007, China is among the largest bioethanol producers in the world with an output of 1.3 million tons of bioethanol.

Subsidies to biofuel feedstock producers are indispensable to promote the integration of non-grain biofuel in the present Chinese energy market. In 2007, the Ministry of Finance of China released a policy to encourage the use of non-food feedstock to produce biofuels by providing subsidies and other forms of financial support to people involved in the production of biofuels (China Daily 2007). Specifically, farmers involved in feedstock production for biofuel manufacture started to receive a subsidy of 3,000 yuan (1 yuan = $ 0.16; April 2012) for each hectare of forest and 2,700 yuan for each hectare of non-grain agricultural crops. As far as JCL plantation is concerned, the producers are to be subsidized by a rate of 3,000 yuan for each hectare of JCL plantation.

JCL and Its Development in China

JCL in China

JCL is a small tree or shrub belonging to the family of Euphorbiacae. JCL has great potential for the production of biodiesel due to its adaptability to the environment, especially to drought, but also to soil of poor quality. JCL was introduced into China 300 years ago; its growth in the wild is found in the tropical or subtropical area, such as the dry and hot valleys of Guangdong, Guangxi, Hainan, Fujian and Taiwan, especially Yunnan, Guizhou and Sichuan (Wu and Zong 2007). In Yunnan Province, JCL can typically grow at an altitudinal range of 600–1,400 m above sea level (Zeng 2006). In Yunnan Province, JCL mainly grows between 600 and 1,400 m and can survive in plains, hills and valley slopes between 700 and 1,600 m (Yuan et al. 2007). In Guizhou Province, JCL is mainly distributed in the southern and southwest parts, mainly in dry and hot valleys in the basins of Nanpan River, Beipan River and Hongshui River at an altitude range of 275–800 m (Chen et al. 2006). In Sichuan Province, JCL is mainly distributed in the dry-hot valleys below 1,600 m (Xu et al. 2008).

According to preliminary statistics, the natural growth of JCL in Yunnan, Guizhou, and Sichuan provinces is about 33,000 hectares (Wu and Zong 2007). Over the past several years, the area of JCL plantation has been gradually increased as more effort has been made in the development of biodiesel. It is reported that the area of JCL plantation was 680,000 hectares in Yunnan, more than 17,300 hectares in Sichuan in 2007, and 15,300 hectares in Guangxi by the end of July 2008. Despite its strong ecological adaptation, existing JCLs are mainly distributed on the sides of villages and homestead, banks of ditches and channels, alluvion of rivers, valleys and the lower end of slopes. For example, the distribution of existing JCL resources in Yunnan Province is shown in Table 6.1.

Before cultivation for biodiesel production, JCL has been used (i) for fencing or hedging along rivers and road sides and its seed oil was mainly used to produce lubricant and lacquer (Zheng et al. 2008); (ii) as raw material for green manure in Nanpan River, Beipan River of Guizhou and Panzhihua of Sichuan; (iii) for soap fabrication in Qiaojia County of Yunnan Province; and (iv) recently to prevent soil erosion in Binchuan County of Yunnan Province and Panzhihua, in Liangshan prefectures of Sichuan Province (Xiang et al. 2008).

Table 6.1 The distribution of existing JCL resources in Yunnan Province

Area	Forests	Clusters on four sides[a]	Scattered trees	Total
Ha	7162.01	8066.76	2823.87	18052.65
%	39.67	44.69	15.64	100

[a]*Four sides* refer to sides of homesteads, villages, roads and ditches; *scattered plots* refer to unused small patches of lands which are sporadically distributed

Table 6.2 Seed yields of JCL grown at different sites

Code of samples	Coordinates of sample sites	Elevation (m)	Age (yr)	Height (m)	Crown width	Number of fruits
1	23°05′37″N; 103°10′47.3″E	183	10	5.3	7.5	300
2	23°36′55.5′N; 102°27′22. 2″E	1150	10	3.3	4.0	450
3	23°34′21.9″N; 103′52′2.22″E	1476	10	2.8	3.0	300
4	23°0′6.72″N; 103°40′3. 84″E	1220	15	4.8	3.5	600
5	22°42′7.3″N; 102°57′6.1″E	774	10	5.1	3.7	500
6	23°18′34″N; 1 02°39′6″E	258	5	4.4	4.8	600
7	23°21′54.4″N; 102°25′60.8″E	858	8	4.5	3.5	300
8	23°17′46. 5″N; 102°36′43.3″E	482	15	4.2	4.5	500
9	23°0′47″N; 102°13′14.3″E	1116	7	5.1	4.4	600
10	23°47′22.8″N; 103°16′33.1″E	1092	20	5.0	4.5	1,000

Source: He et al. (2007)

Yield of JCL seeds varies greatly with eco-climatic conditions. In their literature survey of JCL seed yields in India, Mali, Nicaragua, Paraguay, Thailand and Cape Verde, Achten et al. (2008) showed that the annual yield of JCL seeds was 0.3–5 tons per hectare with a few reports of 6.7 tons per hectare in India and 8 tons per hectare in Mali. At a conference on the production of biodiesel from JCL held in Wageningen in the Netherlands in March 2007, participating experts agreed that a yield of 4–5 tons per hectare is feasible. However, the upper limit of seed yield (5 tons per hectare) has never been confirmed by JCL growers. As soon as the first year of plantation, JCL can bear fruit in small quantities and the fruit yield can reach 4–5 kg per tree 4 or 5 years later if the plantation is well-managed. However, if trees are planted on barren hills and fed by rain, the yield per tree is not expected to overcome 1–1.25 kg (Kumar et al. 2003).

In China, He at al. (2007) found that the seed yield is between 300 and 600 fruits per tree with an average of 200 fruits per tree and a maximum of 1,500 fruits per tree according to a survey of 10 JCL plots in Honghe Prefecture of Yunnan whose trees were between 5 and 20 years old (Table 6.2). Because there are usually three seeds in each fruit and the weight of a single seed is 0.5–0.75 g (Zheng et al. 2008), the seed yield is 0.54–1.08 kg per tree with an upper limit of 2.70 kg per tree. In Guizhou province, Chen et al. (2006) found that seed yield of JCL was between 2 and 4 kg per tree older than 5 years. After studying the population of JCL in Sichuan, Xu et al. (2008) found that the seed yield of JCL in the wild was 750–2,250 kg ha^{-1}. In summary, the seed yield of JCL is 0.6–4.5 tons per hectare at the present levels of technology and management in China.

In each fruit, there are usually three seeds and occasionally, only two seeds. The weight of the seeds is about 55–65% of that of the fruit (Table 6.2). The JCL seeds are black and toxic. The dry JCL fruit is mainly composed of water (4.4–4.7%), protein (17.8–28.9%), fat (52.9–57.4%), and cellulose (3.7–4.3%) (Chen et al. 2006).

The characteristics of JCL seeds from different regions of Yunnan are shown in Table 6.3. According to the nine samples taken from different parts of Yunnan, the weight per 100 seeds is about 48.2–72.3 g, of which the kernel accounts for more than 60%. The oil rate ranges from 32.2 to 40.2% of seeds and 50.0–61.3% of kernels. These oil yields were obtained using the mechanical extraction method, which is widely used in Yunnan. However, larger oil rate can be obtained from seeds by chemical extraction.

Development of JCL Biodiesel in China

China started the research and development of JCL biodiesel two decades ago. During the period of 1986–1990, the project of "Demonstration project on the development and application of wild plant oil as alternative fuel" was initiated in Sichuan province. This research was conducted in locations where JCL is found in the wild and considered factors such as cultivation technique, extraction of JCL oil and biodiesel fabrication. Since then, further researches have been done on the distribution, growth, seed yield, oil characteristics, provenance test, and techniques for JCL plantation in the three southwest provinces. In particular, some elite JCL varieties screened according to their sites of origin, adaptability, seed quality and seed oil content were used to establish mother plantations. Among the most important projects that were implemented later on, one can note "the study and applied techniques of oilseed plants" by the Chinese Academy of Science and "the joint project of poverty alleviation through the development of green energy in minority ethnic areas" by the Chinese government and the UNDP. These investigations boosted knowledge accumulation on production techniques, germplasm collection, physiological analysis, biochemical characterizations and genetic diversity in JCL.

The establishment of JCL plantation on a massive scale started mainly in southwest China, including Yunnan, Sichuan and Guizhou provinces. Mountains are the major terrain in these provinces, especially in Yunnan where 94% of the land area is upland or mountains. The government's strategy for JCL plantation has focused on barren lands, which is a specific term used in Chinese agriculture and forest management, referring to lands that are not being used for obvious productive purposes, including some lands that are in fact used for grazing livestock. The provinces just quoted have large portion of land that can be used for JCL plantation, since much of the upland or mountains are barren land. The barren

Table 6.3 The characteristics of JCL seeds in different sites of Southwest China

Site	Yuanmo	Liuku	Binchuan	Panzhihua	Mangshi	Daluo	Yongren	Shuangbai	Yuanyang
Weight per 100 seeds (g)	52.5	71.0	67.8	55.2	59.3	72.3	50.9	68.9	68.3
Weight per 100 kernels (g)	33.5	44.2	46.0	33.2		47.2	31.3	45.3	41.2
Oil percentage in seeds (%)	35.4	32.2	35.7	36.8	37.6	33.5	34.3	40.2	37.0
Oil percentage in kernels (%)	55.5	51.7	52.6	62.8	56.7	51.3	55.7	61.2	61.3

Sources: Zhang et al. (2001), Wang and Long (2009) and author's surveys land is mainly owned by the government or village collectivities enjoying the right to use the land contracted to individual households. Therefore, a significant portion of the land area planned for JCL plantation is likely to be collectively owned, but under contract of individual households.

In 2006, the Ministry of Finance, the National Development and Reform Commission, the Ministry of Agriculture, the State Taxation Administration and the State Forestry Administration jointly enacted the "Statute on supporting financial as well as tax policies for the development of bioenergy and biochemical engineering". Since then, the three provinces have made plans for the establishment of 1.667 million hectares of JCL plantations on waste or marginal lands within 10–15 years, including 0.667 million hectares in Yunnan, 0.6 million hectares in Sichuan, and 0.4 million hectares in Guizhou. According to the State Forestry Administration, the total areas of JCL plantation in the three provinces was 0.15 million hectares, which accounts for more than 95% of China's total JCL plantations by the end of 2008. During the same year, the national development and reform committee of China approved the participation of some small and giant companies, such as Sino-Pec and Petro China to pilot projects for the production of JCL biodiesel.

The Production Process of JCL Biodiesel

The JCL biodiesel production chain can be divided into four stages: (i) production of JCL seeds through cultivation; (ii) extraction and processing of JCL oil; (iii) distribution and retailing of final products; and (iv) consumption of JCL biodiesel. Here we define *JCL biodiesel* either as the refined JCL oil or the JCL methyl esters (JME), both blended with fossil diesel. Of course, in the case of JCL oil considered as an end product, no conversion or transesterification reaction is required.

The Production of JCL Seeds

Despite many literatures on the establishment and management of JCL plantations, the basic process of JCL seed production can be summarized as shown in Fig. 6.1.

JCL trees can be propagated by seedling, cutting and micro-propagation. Since the survival rate of cutting plantation is low and the cost of micro-propagation is much higher than that of seedling or cutting, JCL plantation is mainly established with seedlings.

Land preparation mainly comprises of terracing, digging holes (40×40×40 cm), applying green manures and soil erosion control. However, not all sites need every of the activities depending on their conditions. Irrigation is mainly applied in the first year after plantation is established. In the dry season, the plantation will be watered twice a week. Also, for some areas with a level of annual rainfall larger than 2,500 mm, irrigation is not necessary.

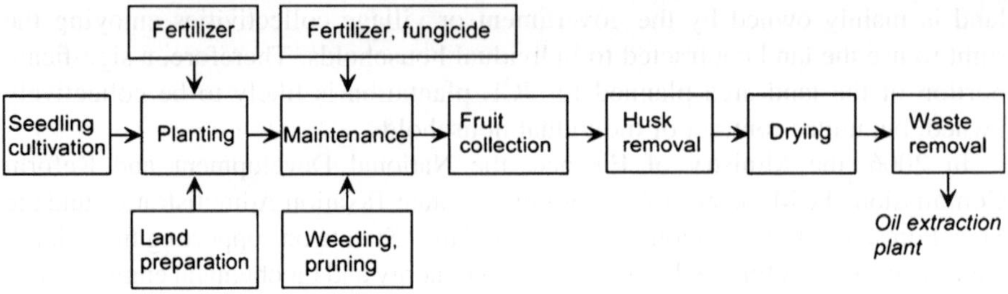

Fig. 6.1 The process of JCL seed production

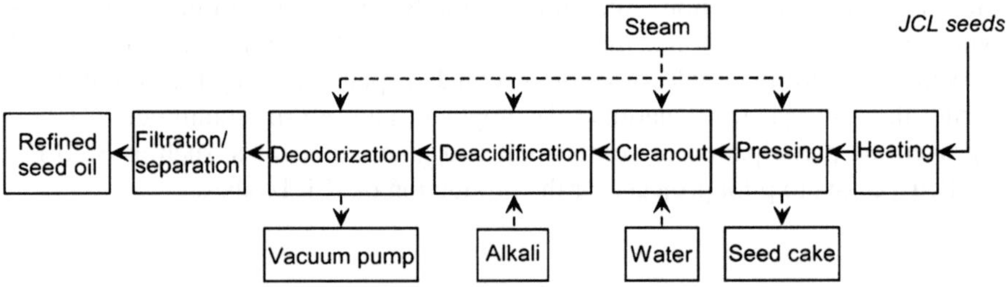

Fig. 6.2 The process of oil extraction with mechanical expeller

The physico-chemical conditions of the JCL plantation are commonly corrected with chemical fertilizers, green manures and JCL seed cakes. In the course of site preparation, green manure is put into the holes. Chemical fertilizers are applied after seedlings are transplanted for the first 3 years. Each 3 years, about 100 g of urea and 250 g of NPK are applied to each tree in the first part of the year, followed by top dressing with 100 g of urea for each tree. From the fourth year to the 30th year of plantation, JCL seed cake and green manure are used instead of chemical fertilizers. According to Openshaw (2000), one kg of seed cake is equivalent to 0.15 kg of NPK fertilizer (N:P:K=4:2:1).

To control weeds and fungi, the application of herbicides and fungicides is necessary. The main herbicide used is glyphosate; it is applied using simple sprayer. In dry season, the control of weeds by manual cutting is also applied. JCL is a disease-tolerant plant and can be used to produce insecticides. The damage from pest attack is not significant in China. However, carbendazim should be applied to control diseases caused by fungi.

Extraction and Conversion of JCL Oil

Under this industrial activity, the JCL oil is extracted and converted into biodiesel. The JCL oil is mainly extracted using two methods: chemical method using solvent extraction with n-hexane and the mechanical method using an engine driven-expeller. Because of its simple technology, the latter is widely used (Fig. 6.2). The oil is then converted into biodiesel through transesterification by methyl esterification.

The ripe fruits are plucked from JCL trees, sun dried and then de-husked. To prepare seeds for mechanical extraction, they should be heated at sun for several hours or roasted for 10 min so as to reduce the oil viscosity. If chemical extraction is chosen, the shelling of seeds can increase the yield of oil.

The characteristics of JCL oil are shown in Table 6.4. JCL oil is not suitable for direct use in engines because of its viscosity. The high viscosity of JCL oil may result in incomplete fuel combustion, formation of carbon deposits in engines and reduction of engine life span. Significant reduction in viscosity can be achieved by dilution of vegetable oil with diesel in varying proportions (Pramanik 2003). Among blends, the blends containing up to 30% (v/v) JCL oil have viscosity values close to that of fossil diesel and up to 50% diesel can be substituted by JCL oil without any major operational difficulties of compression ignition (CI) engine. Forson et al. (2004) showed that the 97.4% diesel/2.6% JCL fuel blend produced the maximum values of brake power and brake-thermal efficiency as well as the minimum values of specific fuel consumption and, thus, JCL can be used as an ignition-accelerator additive for diesel fuel.

Many scholars have analyzed emissions by mixtures of JCL biodiesel and diesel. As compared with fossil diesel, the combustion of B10, a blend of 10% of JCL biodiesel and 90% of diesel, reduces the emissions of *nitrogen oxide* (NO_x), *hydrocarbon* (HC), *carbon monoxide* (CO) and *particulate matter* (PM) by 5.3%, 9.1%, 15.9%, and 8.9%, respectively, but the deposited carbon is slightly higher (Hu et al. 2010). These results are consistent with those of Prasad et al. (2000) concerning the emission of NO_x, but are in contradiction with those of Kumar et al. (2003) regarding the emission of CO. The emission of *aldehydes*, *sulphur dioxide* (SO_2) and CO_2 by fossil diesel has also been reported to decrease upon increasing contribution of JCL biodiesel to blends (Lou et al. 2010).

Although significant reduction in viscosity can be achieved by dilution of JCL oil with diesel in varying proportions, the oil quality will be improved and there will be less long-term problems if JCL oil is converted into biodiesel. The process of oil conversion into biodiesel is as shown in Fig. 6.3.

JCL biodiesel satisfies the international standards of diesel fuel for CI engines (Table 6.5). Compared to diesel, JME have higher cetane number and flash point, a similar calorific value and lower sulfur content. Therefore, JCL biodiesel has an overall performance close to that of diesel and thus can be used as a substitute for diesel (Chen et al. 2006).

Environmental and Economic Performances of JCL Biodiesel

Only the mechanical method has been investigated here because it is far more widely used than the chemical method. Since JCL oil can be used in engines after it is blended with fossil diesel or converted into JME through transesterification reaction, JCL biodiesel can be JME or a mixture of JCL oil and diesel. Here we analyze the environmental, economic and energy (3E) performance of JCL seed oil while

Table 6.4 Characteristics of JCL oil in China

Calorific value (MJ/kg)	Flash point (°C)	Cetane value	Cloud point (°C)	Specific gravity (g/cm3)	Water content (%)	Acid number (mg KOH/g)	Iodine number (mg iodine/g)	Saponification number (mg/g)
39.6 ~ 41.8	110 ~ 240	51.0	2	0.91 ~ 0.92	0.24	7.62–16.82	100.85	213.14

Source: Chen et al. (2006), She et al. (2005), Liu et al. (2007), Zhang et al. (2001)

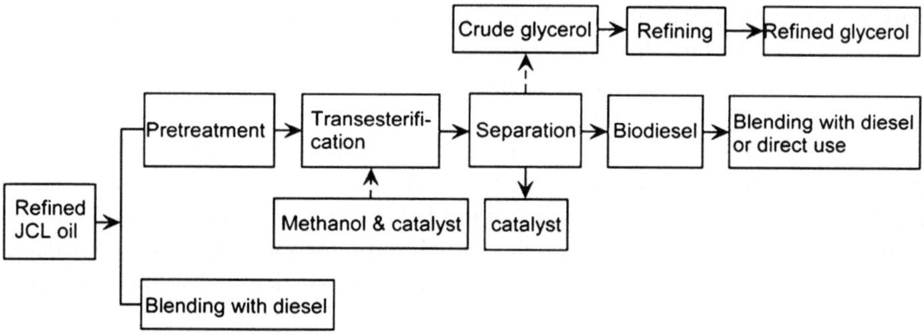

Fig. 6.3 The process of JCL biodiesel production

Table 6.5 Physical and chemical properties of JCL biodiesel and diesel

Characteristics	JCL seed oil	JCL methyl ester	Standard method	German DIN51606 standards	Diesel
Gravity, 20°C, g/cm³	0.875	0.8784	SH/T0604	0.875 ~ 0.900 (15°C)	0.84 ~ 0.85
Kinematic viscosity, mm²/S	33.49 (40°C)	7.320 (20°C)	GB/T265	315 ~ 510 (40°C)	112 ~ 315 (40°C)
Calorific value, MJ/kg	37.55	40	GB/T384	≥32	42.6 ~ 45.0
Flash point, °C	124	>170	GB/T261	≥110	80
Cloud point, °C	−9	+1	GB/T3535	–	−14
Carbon residual (%)	–	<0.05	GB/T17 144	≤0.05	–
H_2SO_4 residual (%)	–	<0.005	GB/T2433	–	–
Sulfur content, (mg/l)	24.3	32 (0.0036%)	SH/T253 -92	≤0.01	1.0 ~ 1.2
Cetane No.	51	56.1	–	≥49	47.8

Source: Chen et al. (2006), Luo et al. (2011)

that of JME is available in Wang et al. (2011). As mentioned earlier, the lifecycle of JCL biodiesel consists of four stages: JCL cultivation, oil extraction and refining, biodiesel production, and use in engines.

The analysis includes the assessments of 3E performance associated with direct inputs and construction of facilities, such as manufacturing machines, housing, etc. as well as with manual labor, such as planting, pruning, fruit collection and drying, weeding, etc. The calculations are based on one hectare of JCL plantation for 30 years.

Indices used to assess the 3E performance are financial net present value (NPV), CO_2 equivalent balance (CEB) and net energy ratio (NER). The NPV is the difference between the present value of the flow of revenues and the present value of the flow of costs. The CEB is the difference between the total sequestrated CO_2, reduced CO_2 emissions and total emitted CO_2 equivalent. The NER is the ratio of total energy outputs to the total energy inputs.

The allocation of costs to co-products is based on their monetary values on the market. The credits for allocating environmental and energy burdens are based on displacement effects. Although parts of the JCL tree can be exploited for a number of uses, such as medicines, insecticides, fuel, fertilizer, etc., this study considered seed cake as a fertilizer and the parts involved in fuel stock as substitutes to coal.

Carbon Balance at Production Level

When JCL oil is directly blended with fossil diesel, its lifecycle carbon balance is shown in Table 6.6. It shows that the production and use of JCL oil has a positive carbon balance (Wang and Lu 2011). The effect of carbon emission reduction by the diesel substitution with JCL oil tends to become more consistent when seed yield is increased. The major contributors to carbon balance are JCL biomass stock and JCL oil. The combustion of fruit husks can also reduce carbon emission when they are used to replace coal as a fuel.

The major GHG emissions come from the application of fertilizers together with the transportation of seeds and fertilizers. The JCL plantation will be fertilized with chemical fertilizers, green manure and JCL seed cakes. In the course of site preparation, green manure is put into the holes. Chemical fertilizers are applied after seedlings are transplanted throughout the first three years. In each of the three years, about 100 g of urea and 250 g of NPK will be applied to each tree in the first part of the year, followed by top dressing with 100 g of urea for each tree. From the fourth year to thirtieth year of plantation, JCL seed cake and green manure will be used instead of chemical fertilizers. According to Openshaw (2000), one kg of seed cake is equivalent to 0.15 kg of NPK fertilizer (N:P:K=4:2:1). Although much direct energy is used in the oil extraction stage, the GHG emission associated to this processing step is relatively small.

Economic Performance of JLC Oil Production

The Cost of Seed Production

The production of JCL seeds starts from seedling cultivation to seeds delivered to oil extraction plants, involving activities including transplanting, site preparation, tending, seed collection, etc. To estimate the revenue of seeds, we used the present price of JCL seed that is 2 yuan/kg.

Taking a seed yield of 1,485 kg/ha (around 500 pieces of fruits per tree) as an example, the total cost at the seed production stage is 73,609 yuan/ha and its components are

Table 6.6 The life cycle CO_2 equivalent of JCL oil for a period of 30 years

Annual seed yields, (kg/ha)	297	891	1485	2079	2673	3267	3861
Emission from applied fertilizers (kg/ha)	–	–	–	6719.65	–	–	–
Emission from the transportation of seeds and fertilizers (kg/ha)	235.76	604.01	972.25	1340.49	1708.73	2076.98	2445.22
Sequestration by JCL plantation (ton/ha)	12.39	37.16	61.93	86.70	111.48	136.25	161.02
Reduction by coal substitution with JCL fruit husks (kg/ha)	64.39	193.18	321.97	450.75	579.54	708.33	837.11
Emission from oil extraction and refining (kg/ha)	34.45	103.36	172.27	241.18	310.09	379.00	447.91
Reduction by substituting fertilizers with seed cake (ton/ha)	1.23	3.69	6.16	8.62	11.08	13.55	16.01
Emission from *distribution* kg/ha	2.86	8.58	14.3	20.02	25.73	31.45	37.17
Reduction by substituting diesel with JCL oil (ton/ha)	7.42	22.26	37.11	51.95	66.79	81.64	96.48
Life cycle CO_2 eq. balance (ton/ha)	14.11	55.88	97.64	139.41	181.17	222.94	264.70

Source: Wang and Lu (2011)

as shown in Table 6.7. Obviously, the main costs are associated with fruit drying and husk removal, fruit collection, fertilization, fungicide treatments and weeding.

The seed yield of JCL varies with the local conditions. The FNPV of seed production ($FNPV_s$) and the breakeven price of seeds would be different as seed yield changes. Assuming that the JCL tree has an annual yield of 100, 300, 500, 700, 900, 1,100, 1,300 or 1,500 fruits per tree, which are corresponding to an annual seed yield of 297, 891, 1485, 2,079, 2,673, 3,267, 3,861 or 4,455 kg/ha, respectively, the relationship between the $FNPV_s$ and yield is as shown in Fig. 6.4. It reveals that the $FNPV_s$ tends to increase as the seed yield is increased. When the price of seeds is 2 yuan/kg, the $FNPV_s$ will be positive only if the seed yield is higher than 3 tons/ha. Obviously, the yield of seeds is one of the important determinants of the financial feasibility of biodiesel production.

Table 6.7 The cost components of JCL seed production

Cost component	Description	Percentage
1. Seedling cultivation	Given land renting, seed production, seed treatment, nursery treatment, fertilization and tending, the cost of each seedling is 0.11 Yuan. At an assumed survival rate of 85%, 1950 seedlings are needed for each hectare of land	0.59
2. Land renting	750 Yuan/ha for a period of 30 years	2.06
3. Site preparation & transplanting	Contracted to reforestation company at the cost of 1,800 Yuan/ha	4.95
4. Fertilization	Including fertilizer and associated labor cost for a period of the first 3 years. Fermented seed cake is used as fertilizer instead of chemical fertilizers in the fourth year and forwards	18.74
5. Weeding	Including: a manual weeding and one herbicide treatment each year during the first 5 years	6.87
6. Spraying fungicide	Including: annual application, labor and fungicide/pesticide costs	11.05
7. Pruning	Twice a year. Costs of labor and instruments	1.84
8. Fruit collection	By hand: labor cost only	20.48
9. Drying and husk removal	By hand: labor cost only	30.72
10. Transportation Cost	Including seeds and fertilizer transportation costs: C = 200 Yuan + 1.5 Yuan/km*mileage for a 5 tonnage truck	2.69
Total	Including costs incurred from seedling cultivation to seed transportation until the oil plant	100

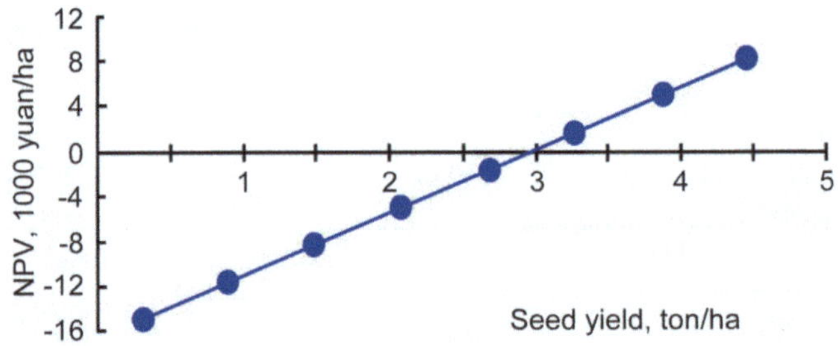

Fig. 6.4 The relationship between seeds yield and the $FNPV_s$

Cost of Mechanical Extraction of JCL Oil

When a mechanical extraction method is used, the stage of oil extraction starts from the heating of seeds to the refined JCL oil. When the yield of seeds is 1,485 kg/ha, the total cost of the processing of JCL seeds is 2,321.91 yuan/ton and its components are shown in Table 6.8. It shows that the major cost is from seed purchase (seed price is 2 yuan/kg), which accounts for 86.13% of the total cost while the sum of all other costs accounts for about 14.87% only.

Table 6.8 Components of the mechanical extraction and refining costs per ton of JCL seeds

Cost components	Description	Percentage (%)
1. Cost of machinery for each ton of JCL seeds	Calculated with the straight-line average service life method with a salvage value of 5% (Jin et al. 2009). The machinery includes mechanical expeller and the equipment for oil refining	2.06
2. Coal	0.1 ton × 450 Yuan/ton	1.94
3. Electricity	17.5 kilowatt × 1 Yuan/kilowatt	0.70
4. Labor	3 workers, 50 Yuan/(worker × day)	6.01
5. Workshop renting	20,000 Yuan/Year, 240 working days/year	2.43
6. Seed cost	Seed price is 2 Yuan/kg	86.13
7. NaOH for deacidification	The present local price of NaOH is 3,700 Yuan per ton. The acidic value of crude oil is about 12.35 mg KOH/g	0.47
8. Transportation of oil	The distance between the plant and oil distribution is assumed to be 10 km, on average. The cost is 100 Yuan for a distance of 10 ~ 30 km	0.26
Total		100

As shown in Table 6.3, the crude oil content ranges from 32.2% to 40.2% in seeds when oil is extracted using an engine driven-expeller. According to researchers' recent survey at an oil extraction plant of Er-Kang Science and Technology Co. Ltd. in Yunnan province, the yield of refined oil is a little lower at ~30.4%. Based on the specific gravity of JCL oil and total costs (Table 6.8), the cost of producing 1 liter JCL oil is calculated to be 6.99 yuan/liter when the seed price is 2 yuan/kg.

Financial Analysis of JCL Oil Final Product

When both the stages of seed production and oil extraction are considered together or produced by a single producer and JCL oil is used as an end-product, the financial feasibility associated with different seed yields is shown in Table 6.9. Since most oil plants are not far away from refueling station and the shared transportation cost for each unit of JCL oil is negligible, the cost associated with oil distribution is neglected.

Table 6.9 shows that the major cost is incurred at the seed production stage and the financial net present value tends to decrease as the seed yield increases. The reason is that the cost increases with the JCL oil production when the cost is higher than the revenue of producing an extra unit of oil. In short, the more JCL oil is produced, the more the producer will lose. As expected, both the total cost and revenue increase with seed yield. However, the breakeven price of JCL oil decreases as the seed yield of JCL plantation increases. Moreover, when seed yield increases, the percentage of the costs associated with seed production tends to decrease even if the total cost increases.

Table 6.9 Financial analysis of JCL oil final product

Annual seed yield, kg/ha	297	891	1485	2079	2673	3267	3861
Total cost, 10^4 Yuan/ha	2.16	3.14	4.11	5.08	6.06	7.03	8.01
Cost of seed production (%)	95.58	90.87	88.38	86.85	85.82	85.06	84.50
Cost of oil extraction (%)	4.42	9.13	11.62	13.15	14.18	14.94	15.50
Total revenue, 10^4Yuan/ha	0.55	1.64	2.73	3.82	4.92	6.01	7.10
FNPV, 10^4Yuan/ha	−1.62	−1.50	−1.38	−1.26	−1.14	−1.03	-0.91
Output of JCL oil, 10^4 l/ha	0.28	0.80	1.33	1.86	2.40	2.93	3.46
Breakeven price of JCL oil, Yuan/l	23.32	11.28	8.87	7.83	7.26	6.90	6.64

Subsidies Required for JCL Oil Production

Subsidies are required for both JCL seed producer and oil processors if a given level of biodiesel production is targeted. At the stage of seed production, subsidies can be provided on the basis of seed weight or the area of JCL plantation. Assuming that both the seed producers and processors will receive a margin of 10%, the two kinds of subsidy rates are shown in Table 6.10. It shows that subsidies are required if the seed yield is lower than 3 tons/ha and the amount to be paid tends to decrease as the seed yield increases. Obviously, the current rate of subsidy, 3,000 yuan/ha of biofuel feedstock is sufficient to stimulate the production of JCL seeds by companies at a wide range of geo-climatic conditions.

At the stage of oil extraction, subsidies can be provided based on the volume of JCL oil. As previously calculated, the break-even prices of oil extracted with mechanical methods is 6.99 yuan/l, respectively. Based on the present local price of diesel, 5.89 yuan/l, the subsidy is found to be 1.88 yuan/l, when a margin of 10% is assumed. Obviously, the increase of diesel price might provide incentive for investors to produce JCL oil. For better environment and energy security, it is desirable for government to provide producers subsidy as long as there is an economic justification.

When the whole production chain is considered, subsidy is required even if the seed yield is as high as 3.86 tons per ha. A single producer who produces both JCL seeds and oil will lose even if it receives subsidy for seed production. That is, subsidy is also required for oil extraction and refining. However, there is no specific incentive policy for oil extraction and refining in China yet, which is probably the reason why the JCL biodiesel industry grows only very slowly and few JCL oil processing plants are established.

Energy Efficiency of Production and Use of JCL Oil

According to the average calorific values and specific gravities of diesel and JCL oil, it was calculated that 1 liter of diesel has a calorific value equivalent to 0.994 liters of JCL oil. Thus, we consider in the following that, in terms of energy content,

Table 6.10 Subsidies required for the production of JCL oil

Annual seed yield, (kg/ha)		297	891	1485	2079	2673	3267	3861
Required subsidy	For seeds production (yuan/kg)	5.4	1.4	0.6	0.3	0.1	0	0
	For oil extraction and refining (yuan l^{-1})	1.88	1.88	1.88	1.88	1.88	1.88	1.88
	For the whole production chain (yuan l^{-1})	19.76	6.52	3.87	2.72	2.10	1.70	1.41

1 liter of JCL oil is equivalent to 1 liter of diesel. The energy efficiency is an important indicator to evaluate the eco-performance of a renewable energy resource. It was assessed with the following indicators:

$$EE_1 = \frac{HV_{biodiesel}}{HV_{energy\text{-}input}} \tag{1}$$

$$EE_2 = \frac{HV_{biodiesel} + HV_{co\text{-}products}}{HV_{energy\text{-}input}} \tag{2}$$

$$EE_3 = \frac{HV_{biodiesel}}{HV_{direct\ energy\text{-}input}} \tag{3}$$

where EE_1 is the ratio of the *heat value* (HV) of biodiesel per unit to the energy used to produce that unit of biodiesel; $HV_{biodiesel}$ is the heat value of biodiesel; $HV_{energy\text{-}input}$ is the heat value of the energy necessary to produce one unit of biodiesel; EE_2 is the ratio of the heat value of biodiesel and co-products and that of the input energy; EE_3 is the ratio of the heat value of biodiesel and that of direct energy inputs.

The major energy use for the production of JCL biodiesel is from the application of herbicides, insecticides, chemical fertilizers as well as the extraction and refining of oil, which are common for the end products of Jatropha oil and JME.

The energy attributed to chemicals is mainly an indirect energy input, i.e., the energy inputs for the production of these chemicals. For example, the energy used for the production of glyphosate and carbendazim are 454 MJ/kg and 397 MJ/kg (Tzilivakis et al. 2005), respectively. Because of a low level of mechanization, the energy inputs in JCL management is 2.28 MJ/h (Kallivroussis 2002); it is mainly due to fruit collection and drying, husk removal, planting and tending. The direct energy input is mainly used in the extraction and refining of JCL oil, as well as for the transportation of seeds, fertilizers and end products. When JME is the end product, additional energy inputs are required for the oil transesterification.

The energy consumption for buildings (office and workshops) as well as the machineries for oil extraction and further processing was estimated according to the energy coefficients in Yáñez Angarita et al. (2009). The energy content of the produced biodiesel was estimated according to the thermal performance and emission of JCL oil

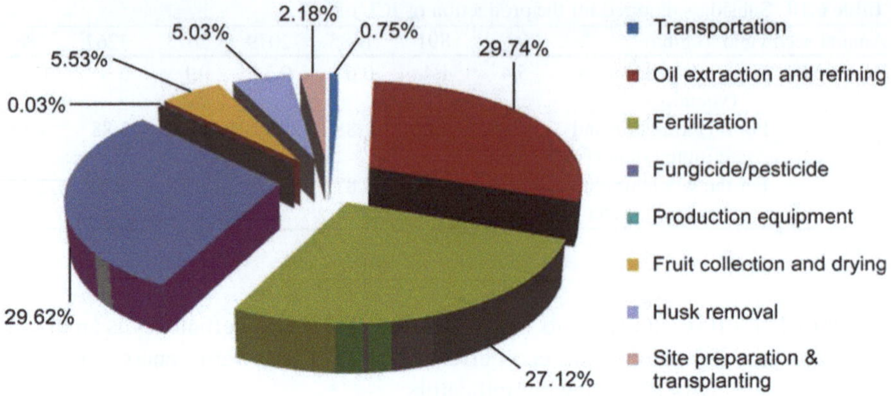

Fig. 6.5 The composition of the lifecycle energy input of JCL oil

Table 6.11 The energy efficiency of production of JCL oil

Annual seed yield (kg/ha)	297	891	1485	2079	2673	3267	3861
EE_1	0.44	1.09	1.57	1.92	2.20	2.42	2.60
EE_2	1.86	4.66	6.68	8.19	9.37	10.32	11.09
EE_3	10.65	11.12	11.22	11.26	11.29	11.30	11.31

given in Agarwal and Agarwal (2007). Limited by data, the energy consumption of machinery is only composed of the portions in the production of cast steel, structural steel and stainless steel used in the machinery sets while those in the processes of fabrication were not included.

When refined JCL oil was the end product, its energy inputs were counted starting from its processing site to its distribution site. For example, when the seed yield is 1,485 kg/hm^2 and oil is extracted with a mechanical method, the composition of energy input is shown in Fig. 6.5. The major energy input is for oil extraction and refining, fungicide or pesticide application and fertilization. The energy output includes the chemical energy in the JCL oil, which is 37.83 ~ 39.20 MJ/l (Augustus 2002; Sahoo 2009; Baitiang et al. 2008) and the biomass energy in twigs, woods and leaves.

The energy efficiencies for different levels of seed yield are shown in Table 6.11. The results show that both E_2 and E_3 are greater than one, which means that JCL is energy-efficient for these parameters, but E_1 depends on seed yield and the method used for oil extraction. When mechanical extraction method is applied, E_1 will only be greater than one when the seed yield is greater than 1,200 kg/ha. Because of higher energy consumption for chemical extraction, the energy efficiency will not be greater than one even if the level of seed yield is as high as 3,861 kg/ha (data not shown).

Therefore, since E_2 is greater than one for all levels of seed yields, it is concluded that the production and use of JCL oil will be energy-efficient if all energy outputs including twigs, woods and leaves are fully used. If only the direct energy input is considered the energy efficiency is higher, as shown by EE_3.

The production of either JCL oil or JME is not financially sustainable as it can be concluded from the comparison of JME lifecycle and 3E performances (Wang et al. 2011). Transesterification reaction incurs additional costs. However, as a co-product of JME, glycerin shares 8.1% of the total costs. As a result, the FNPV of JME is only slightly higher than that of JCL oil. Similarly to the case where JCL oil is used as an end product, the lifecycle carbon balance of JME is also positive. Intuitively, more GHG will be emitted for the production of JME because more inputs are needed. However, as a coproduct of JME, glycerol adds a credit to the lifecycle carbon balance of JME, and consequently the lifecycle carbon balance of JME is higher than that of JCL oil. For both Jatropha oil and JME, the E1, E2 and E3 are greater than one. However, because an additional process of transesterificaiton reaction is required, the energy efficiency of JME is slightly lower than that of JCL oil.

Challenges and Suggestions

The promotion of JCL biodiesel is directly driven by the scarcity of oil and by the social pressure for the reduction of GHG emission. The development of rural areas and the rehabilitation of degraded ecological system are other indirect drivers. The Chinese policy framework has provided a good opportunity for the production of JCL biodiesel. However, despite this opportunity, the production of JCL biodiesel is still facing many challenges, which are to be dealt with before JCL biodiesel is produced on a massive scale.

Challenges

Biodiesel is theoretically an environment-friendly alternative to fossil fuel and has huge potential to mitigate the future energy needs of the country and to impart economic prosperity in the poor and backward areas of the country. However, the production of biodiesel is still an emerging industry. The commercialization of JCL biodiesel in China is fairly recent since commercial seedling production began in 2005. Although China has set an ambitious target at establishing biodiesel bases and achieving energy independence, there is still lack of sufficient scientific and technological information for the production of JCL biodiesel. For example, what is the optimal combination of inputs so as to minimize the production cost and how can the JCL seed yield be improved? Although the research on JCL cultivation in China can date back to 1970s with much effort carried out with germplasm development and tissue culture, little attention has been given to the technology for the industrial production of its biodiesel, especially concerning the valorization of its by-products. Before rapidly scaling up JCL plantation acreage, more scientific and technological studies are required so as to avoid unnecessary costs and to lay a foundation for the growth of biodiesel industry responding to society needs.

The ecological impacts of the production system of JCL biodiesel also deserve concern. Being an exotic species in most growing areas, the impact on biodiversity due to a change of land use towards JCL large scale cultivation is expected to be negative. Impact on biodiversity will be especially negative if natural systems, such as forests, are to be cleared. Heavy machinery and high fertilizer application are also the drivers towards a negative impact. However, JCL can generate environmental benefits too. Besides its potential contribution to the reduction of GHG emission, JCL plantations may generate economic value by facilitating the ecological restoration of barren lands. Although no literature on the impacts of a change in land use due to JCL plantation is available so far, JCL is strongly believed to be able to control and prevent soil erosion and it has ever been practically used for erosion control in southwest China. Because the production of JCL biodiesel is at present not financially sustainable, the study of ecological impacts as an economic input associated with biodiesel production can also provide more information on the economic justification of JCL. To design polices effective for the promotion of JCL biodiesel requires information from both systematic and site-specific studies.

In China, the development of biodiesel industry is still in its infancy. While much effort has been taken towards the establishment of JCL plantations, the mechanism for its industrialization only received very few concerns. There is a lack of institutional management to coordinate seed producers, oil processors, oil distributors, and supporting agencies. For example, farmers are only involved in providing labors and renting land to companies involved in seed production and are not voluntary for JCL planting and exploration because the production of JCL seeds is not cost effective. Companies are motivated to establish JCL plantation because of government's subsidies. While subsidies are provided in terms of area of JCL plantation rather than seed yield, the incentive function of subsidy cannot be guaranteed and the distribution of subsidy among producers should be clarified according to the differences of production stages. Moreover, there is also a lack of national recommendations for standards of biodiesel quality.

The basic structure of many current institutions reflects views of land, natural resources and people, which are fundamentally different from those proposed under ecosystem management with its themes of holism, dynamism, complexity and uncertainty (Kessler 1994). The difference is partly a result of the complexity of the subject. It is also because we have failed to recognize the linkages between the ways people relate to nature and the character of our institutions. The biodiesel production system is a social ecological system with human production activities interacting with the ecosystem, but the exact requirements for institutional change are unclear. In order to manage the social-ecological system of JCL biodiesel in a sustainable way, a better understanding of the relationships between institutions and the production system is required.

The biggest challenge for the industrialization of JCL biodiesel is the lack of sufficient land for JCL plantation so as to meet the policy target for biodiesel production. For example, the consumption of diesel in Yunnan province was 5.56 million tons in 2010. Assuming the constant diesel consumption in future, about 0.83 million tons of biodiesel is required in order to meet the B15 target. However, the land areas suitable for JCL plantation is estimated to be 0.675 million hectares (Liu et al. 2010). According to the current yield of JCL seeds, the maximum annual output of biodiesel would be 0.62 million tons if all suitable lands are used. Obviously, the potential biodiesel production will not allow the meeting of the policy target. Moreover, as compared to other provinces in China, Yunnan has the largest land areas suitable for JCL plantation and its consumption of diesel is relatively small. That is, the potential of B15 in other provinces is lower than that in Yunnan province. Therefore, the promotion of biodiesel alone cannot achieve the policy target and other alternatives should also be developed.

Governments across the world have approved legislative instruments to foster the biofuel industry according to one or several different targets, such as oil dependency reduction (e.g., EU, USA, China), setting up domestic market (e.g., Brazil), reducing carbon emission (e.g., EU, USA) or/and contributing to rural development (e.g., China, India). External incentives are applied to achieve the biofuel targets set by national administrations. Biofuels are being developed in a very complex, dynamic and diverse context. In assessing performance of biofuel policies, there is a need for sustainable development framework using criteria based on the potential social, economic and environmental impacts (Amiguna et al. 2011). However, many policies are being recommended and developed by focusing on a single or a few specified targets while the sustainability of the production system is not well addressed. For example, there is no incentive policy for oil extraction and refining yet. Moreover, biofuel policies have distributional implications for consumers and producers, farm and nonfarm sectors, global trade in food and biofuels, and the price of land and other scarce resources (Khanna et al. 2010). As a type of biofuel, JCL biodiesel is facing the same policy problems in its development course.

Suggestion

To invest in JCL biodiesel requires an economic justification in a framework of social benefit cost analysis. As the biodiesel production is not financially sustainable yet, synergic services other than biodiesel production should be investigated for JLC with the purpose to increase its attractiveness. Such synergic services are related to the concept of *natural capital*, which consists of both non-renewable resources (e.g., minerals and soil reserves) and renewable resources (e.g., plants, animals and water). The natural capital supports the production of goods and services on which society depends (Chapin et al. 2009). Natural capital and its derived goods and services are the preconditions or the basis for the economic development. Actually, it is impossible for humans to create human-made capital without the support from natural capital (Daly 1990). The sustainable production of biodiesel can thus be considered as an investment in natural capital. Gains in natural capital can be made when biofuel development is integrated with ecological restoration because the costs for ecological restoration can be reduced and the ecological service increased. Ecological restoration using JCL can lead to an increase in natural capital in terms of soil improvements and water conservation; biodiesel-driven restoration of degraded ecosystems can also increase terrestrial carbon sequestration (Zhuang et al. 2010).

The environmental and energy lifecycle performance of JCL biodiesel can be improved by (i) increasing the oil extraction efficiency; (ii) selective breeding for high yielding varieties concerning seed quantity and weight as well as oil quality and quantity; (iii) creating specific machinery for de-husking and fruit collection in order to reduce the cost of seed production; (iv) value addition from co-products of biodiesel production. For example, the leaves can be used to feed silkworm, the pruned twigs can be used as a source of medicinal compounds and the raw material for pesticide production; (v) optimization of ecosystem management and green manure that would reduce the energy consumption associated to pesticides, herbicides, fertilizers, oil extraction and refining; (vi) by applying low energy input herbicides, such as atrazine and cyanazine or by manual weeding and by using fruit husks as an alternative to coal (Wang et al. 2011).

Social parameters of JCL biodiesel development including the psychological factors (cognitive and cultural) that shape farmer's values and beliefs must be integrated to policy strategy (English 2008), since the understanding of biodiesel economics is important, but not sufficient. Integrated package of policy measures also needs to enforce and implement the concepts of sustainable development and natural capital. According to such recommendations, a set of sustainability index and indicators should be designed to guide policy makers and production managers. From the mix of policy tools available to support the promotion of the JCL biodiesel industry, it is necessary to identify the most effective and sustainable ones. Alternatively, new tools should be created if those currently available are insufficient.

Before agreements on institutional arrangement for JCL biodiesel production can be set, additional institutional analyses of JCL production system are needed concerning (i) the extent to which existing laws, policies, and regulations may constrain or promote the development and implementation of JCL production policies, programs, and practices; (ii) the design of institutional mechanisms for a national approach of the economical-ecological management of biodiesel production; (iii) the adoption of an economical-ecological management approach as a management philosophy eventually requiring internal organizational changes and new arrangements among resource management agencies; and (iv) the development of methodological approaches for the investigation of institutional questions concerning the objective related to the sustainable management of JCL biodiesel production. The answer to these institutional questions will play an important role in achieving the sustainable production of JCL biodiesel. For example, an institutional decision supporting the integration of JCL biodiesel production with the domestic and/or international carbon market may provide producers with extra revenue from selling carbon credits.

Acknowledgement This research report is based on investigations funded by the National Natural Science Foundation of China (71063024) and the Economy and Environment Program for Southeast Asia (EEPSEA, No. 003591-180).

References

Achten WMJ, Verchot L, Franken YJ, Mathijs E, Singh VP, Aerts R et al (2008) Jatropha bio-diesel production and use. Biomass Bioenergy 32:1063–1084

Agarwal D, Agarwal AK (2007) Performance and emissions characteristics of JCL oil (preheated and blends) in a direct injection compression ignition engine. Appl Therm Eng 27:2314–2323

Amiguna B, Musangob JK, Stafford W (2011) Biofuels and sustainability in Africa. Renew Sustain Energy Rev 15:1360–1372

Anon (2007) Biofuel makers to get subsidy. China Daily 27 (No. 8620).

Augustus GDPS, Jayabalan M, Seiler GJ (2002) Evaluation and bioinduction of energy components of *Jatropha curcas*. Biomass Bioenergy 23:161–164

Baitiang T, Suwannakit K, Duangmukpanao T, Sukjamsri C, Topaiboul S, Chollacoop N et al (2008) Effects of biodiesel and JCL oil on performance, black smoke and durability of single-cylinder diesel engine. J Met Mat Miner 18(2):181–185

Chapin FS, Kofinas IIIGP, Folke C (2009) Principles of ecosystem stewardship: resilience-based natural resource management in a changing world. Springer, NY

Chen B, Deng B, Yu J, Huang H (2006) The survey study of *Jatropha curcas* L. in Guizhou Province. For By-prod Spec China 6:55–57

Daly H (1990) Toward some operational principles of sustainable development. Ecol Econ 2:1–6

IEA (International Energy Agency) (2005) World energy outlook 2005. Paris, International Energy Agency

English M (2008) Socioeconomic considerations with biofuels production. The proceedings of China–US workshop on bioenergy consequences for global environmental change. Beijing, China pp 42–44

FAO (2008) Bioenergy, food security and sustainability towards an international framework. HLC/08/INF/3. Rome

Flavin C (2008) Time to move to a second generation of biofuels. Worldwatch Institute, Washington, DC

Forson FK, Oduro EK, Hammond-Donkoh E (2004) Performance of *Jatropha curcas* L. oil blends in a diesel engine. Renew Energy 29(7):1135–1145

GTZ (German Technical Cooperation). 2006. Liquid Bio-fuel for Transportation: Chinese Potential and Implications for Sustainable Agriculture and Energy in the 21st Century. Beijing: GTZ

Hazell P, Pachauri RK (2006) Bioenergy and agriculture: promises and challenges. International Food Policy Research Institute 2020, Washington, DC. Focus 14

He L, Lang N, Ma H, Zheng K, Peng M (2007) Analysis on chemical components in Seeds of *Jatropha curcas* from Honghe prefecture. J West China For Sci 36(4):69–74

Heller J (1996) Physic Nut. *Jatropha curcas* L. Promoting the conservation and use of underutilized and neglected crops. Institute of Plant Genetics and Crop Plant Research/International Plant Genetic Resources Institute, Gatersleben/Rome

Hu Z, Tan P, Lou D (2010) Experimental research of diesel taxis fueled by jatropha biodiesel blends. J Tongji Univ (Nat Sci) 38(6):898–902

Huang J, Qiu H, Michiel K, Erika M, van Veen W (2009) Impacts of bioethanol development on China's regional agricultural development. China Econ Quar 8(2):727–742

Jin X, Wang H, Liu J (2009) Financial analysis. Press of Renmin University of China, Beijing

Kallivroussis L, Natsis A, Papadakis G (2002) The energy balance of sunflower production for biodiesel in Greece. Biosyst Eng 81(3):347–354

Kessler WB (1994) Significant barriers to further progress of ecosystem management. Paper presented at the institutional problem analysis workshop, Stevenson, 20–22 Oct 1994. On file at the Water Resources Research Center, University of Arizona, Tucson

Khanna M, Scheffran J, Zilberman D (2010) Handbook of bioenergy economics and policy. Natural Resource Management and Policy 33, Springer, Washington. p 439

Krawczyk T (1996) Biodiesel—Alternative fuel makes inroads but hurdles remain. Inform 7:801–829

Kumar MS, Ramesh A, Nagalingam B (2003) An experimental comparison of methods to use methanol and JCL oil in a compression ignition engine. Biomass Bioenergy 25:309–318

Liu, Y. Lu, H. Liang, B.; and Chen, P(2007) Pre-esterification of *Jatropha curcas* L. seed oil for biodiesel production. China oils and fats. 32 (7):43–46

Li Z (2001) Analysis of the demand and supply diesel in Chinese market. Intl Petroleum Econ 9(4):5–6

Liu W, Lu D, Zhang L, Wang L, Zhao J, Li S et al (2010) The framework and science basis for China's low-carbon economic development. Shangwu Press, Beijing

Lou D, Shi J, Hu Z, Li B (2010) Research on unregulated emissions in diesel engine fueled with biodiesel from *Jatropha curcas* oil. Chinese Intern Combust Engine Eng 31(5):69–73

Luo F, Guo J, Wang Z, Liang Y (2011) Injection and performance of diesel engine fueled with *Jatropha curcas* oil. J Jiangsu Univ (Natural Science edition) 32:287–291

MOA (Ministry of Agriculture) (2007) Development planning of China's bioenergy industry (2007–2016), China's Ministry of Agriculture Beijing, China

NBSC (National Bureau of Statistics of China) (2008) China statistical yearbook. China Statistical, Beijing

NDRC (2008) Annual compilation of costs and benefits of China's agricultural production. China Statistical, Beijing

Openshaw K (2000) A review of *Jatropha curcas*: an oil plant of unfulfilled promise. Biomass Bioenergy 19:1–15

EPASEA-NIC (Environmental Planning Academy of State Environmental Administration and National Information Center) (2008) Analyzing and predicting China's environmental economic status and trends in 2008–2020. China Environmental Science, Beijing

Pramanik K (2003) Properties and use of *Jatropha curcas* oil and diesel fuel blends in compression ignition engine. Renew Energy 28:239–248

Prasad CMV, Krishna MVSM, Reddy CP, Mohan KR (2000) Performance evaluation of non-edible vegetable oils as substitute fuels in low heat rejection diesel engines. Proceedings of the Institution of Mechanical Engineers Part D. J Automobile Eng 214:181–187

Qiu H, Huang J, Keyzer M, Van Veen W (2008) Policy options for China's bio-ethanol development and the implications for its agricultural economy. China World Econ 16:112–124

Runge, C., and B. Senauer (2007). How Biofuels Could Starve the Poor. Foreign Affairs. http://www.foreignaffairs.com/articles/62609/c-ford-runge-and-benjamin-senauer/how-biofuelscould-starve-the-poor; 2007 [15.06.09]

Sahoo PK, Das LM (2009) Process optimization for biodiesel production from Jatropha, Karanja and Polanga oils. Fuel 88:1588–1594

Shay EG (1993) Diesel fuel from vegetable oils: status and opportunities. Biomass Bioenergy 4:227–242

She Z, Liu D, Tan P (2005) Methylation of *Jatropha curcas* L. seed oil. China Oils Fats 30(9):34–36

Tzilivakis J, Warner DJ, May M, Lewis KA, Jaggard K (2005) An assessment of the energy inputs and greenhouse gas emissions in sugar beet production in the UK. Agric Syst 85:101–119

USDA (United States Department of Agriculture) (2006) Bio-fuels: an alternative future for agriculture. GAIN report number: CH6049

Veljkovic VB, Lakicevic SH, Stamenkovic OS, Todorovic ZB, Lazic KL (2006) Biodiesel production from tobacco (Nicotiana tabacum L.) seed oil with a high content of free fatty acids. Fuel 85:2671–2675

Wang T (2005) A survey of the woody plant resources for biomass fuel in China. Sci Technol Rev 5:12–14

Wang Y, Liu Y, Li X, Hu F, Cai C, Yang D et al (2010) China sustainable development strategy report 2010: green development and innovation. Science, Beijing

Wang Z, Calderon M, Ying L (2011) Lifecycle assessment of the economic, environmental and energy performance of *Jatropha curcas* L. biodiesel in China. Biomass Bioenergy 35:2893–2902

Wang Z, Lu Y. (2011) Lifecyle environmental, economic and energy performance of the seed oil of *Jatropha curcas* L. as biodiesel. Res Environ Yangtze Basin 20(1):61–67

Weyerhaeuser H, Tennigkeit T, Su Y, Kahrl F (2007) Biofuels in China: an analysis of the opportunities and challenges of *Jatropha curcas* in Southwest China. ICRAF Working Paper Number 53. Available from https://jatropha.uni-hohenheim.de/fileadmin/einrichtungen/jatropha/Biofuels_in_China-An_Analysis_of_the_Opportunities_and_Challenges_of_*Jatropha_curcas*_in_Southwest.pdf

Wiesenthal T, Leduc G, Christidis P, Schade B, Pelkmans L, Govaerts L et al (2009) Biofuel support policies in Europe: lessons learnt for the long way ahead. Renew Sustain Energy Rev 13(4):789–800

Wu Z, Zong W (2007) Prospect for producing bioenergy from *Jatropha curcas* L. Sci News 14:15

Xiang Z, Luo Q, Hu M, Xiang Y (2008) The geographical provenances and distribution of *Jatropha curcas* L. in China. China For Sci Technol 22(6):13–19

Xu J, Fei S, He Y, Cai X, Chen X, Lei C (2008) The quantitative characteristics and regeneration of *Jatropha curcas* populations in Sichuan Province. J Sichuan For Sci Technol 29(1):1–6

Yáñez AEE, Lora EES, da Costa RE, Torres EA (2009) The energy balance in the palm oil-derived methyl ester (PME) life cycle for the cases in Brazil and Colombia. Renew Energy 34(21):2905–2913

Yuan L, Zhao Q, Kang P, Yang L, Zhao J, Gou P et al (2007) Investigation of geographical distribution and evaluation of *Jatropha curcas* in Yunnan province. Southwest China J Agric Sci 20(6):1283–1286

Zeng J (2006) A energy plant with a promising potential for development—JCL. Yunnan For 27(2):21

Zhang W, Song H, Wei X, Liu Z (2001) The adaptability of *Jatropha curcas* in Yuanmo County. Agric Technol 21(1):22–25

Zheng K, Lang N, Zhang R, Peng M, Guo W (2008) The distribution and growth status of *Jatropha curcas* L. in Honghe. J For Sci Western China 36:101–104

Zhuang J, Gentry RW, Yu GR, Sayler GS, Bickham JW (2010) Bioenergy sustainability in China: potential and Impacts. Environ Manage 46:525–530

Part II
Physiology

Chapter 7
Physiological Mechanisms Involved with Salt and Drought Tolerance in *Jatropha curcas* Plants

Joaquim Albenísio Gomes Silveira, Evandro Nascimento Silva, Sérgio Luiz Ferreira-Silva, and Ricardo Almeida Viégas

Introduction

Jatropha curcas L., more commonly known as the physic or purge nut, is an oilseed plant species that is native to tropical America and now thrives in many parts of the tropics and semiarid regions in Africa, Asia and South America. This species grows in marginal areas frequently subjected to dry and hot conditions and associated with poor soils, where most other crops are not able to survive (Francis et al. 2005; Tang et al. 2011). Recently, it has received special attention because of its high oil content and quality of seeds. Therefore, *J. curcas* is potentially an important crop for biofuel production and a universally acceptable source of energy (Kumar et al. 2008).

Although the importance of *J. curcas* plants as a bioenergy source is recognized, the key physiological processes involved in drought and salt tolerance are poorly understood. An improved understanding of its stress physiology is essential for the adoption of strategies to make its productivity competitive in marginal areas subjected to adverse environmental conditions, such as drought, salinity and low soil fertility combined with high temperature and radiation. These factors are widespread in tropical regions, especially in semiarid areas, which are located mainly in Asia, Africa and Brazil, where *J. curcas* is becoming widely cultivated.

Plant tolerance to abiotic stress has been defined as the ability to maintain adequate rates of growth and metabolism under adverse conditions (Munns and Tester 2008). The genetic and physiological mechanisms that control plant tolerance are

J.A.G. Silveira (✉) • E.N. Silva • S.L. Ferreira-Silva
Departamento de Bioquímica e Biologia Molecular, Laboratório de Metabolismo de Plantas, Universidade Federal do Ceará, CP 6004CEP60455-970, Fortaleza, Ceará, Brazil
e-mail: silveira@ufc.br

R.A. Viégas
Departamento de Engenharia Florestal, Universidade Federal de Campina Grande, Centro de Saúde e Tecnologia Rural, CP 64 CEP 58700000, Patos, Paraíba, Brazil

N. Carels et al. (eds.), *Jatropha, Challenges for a New Energy Crop: Volume 1: Farming, Economics and Biofuel*, DOI 10.1007/978-1-4614-4806-8_7,

complex because tolerance to abiotic stress is a quantitative trait that depends strongly on environmental factors. In addition, because plants exhibit several complex processes during plant stress, such as metabolic plasticity, it is difficult to determine the basic processes that account for tolerance to a specific abiotic stress factor (Flowers 2004). Plants display several common physiological mechanisms to cope with salinity and drought, but some adaptive mechanisms are stress dependent.

The most important adaptive processes that are common to drought and salt tolerance are (i) osmotic adjustment, (ii) water, photosynthesis and gas exchange control, (iii) oxidative protection, (iv) molecular signaling and (v) regulation of gene expression. The most important specific mechanisms involved with salt tolerance are (i) ionic homeostasis (especially Na^+/K^+ homeostasis), (ii) nutritional balance, (iii) ionic toxicity and ion-specific signaling and (iv) gene expression involved with cellular protection. However, protection against drought might also involve specific mechanisms, such as drought-triggered signaling and drought-specific gene expression.

Plant responses to salinity and drought differ widely among species and cultivars; they also depend on the stress characteristic exposed to the plants. For instance, the type of water deficit frequently interferes with the plant response. Among other factors, water deficit can be induced through (i) a quick or slow soil dehydration; (ii) the type of substrate; (iii) the presence of artificial osmotic agents, such as PEG and mannitol; and (iv) the stress intensity. Similarly, salt response depends on salt composition, concentration, exposure time and plant age. Several studies involving the physiological aspects of *J. curcas* response to drought and salt stress have previously been reported by our group (Silva et al. 2009a, b, 2010a, b, c, 2011; Rodrigues et al. 2012) and several other authors (Eswaran et al. 2010; Tang et al. 2011; Díaz-López et al. 2012; Gimeno et al. 2012).

In this review, we will present current results regarding the most important processes related to drought and salt tolerance in *J. curcas,* including osmotic adjustment, photosynthesis and oxidative protection. Overall, the data reported suggest that *J. curcas* has both biochemical and physiological characteristics that confer drought tolerance and relative salt sensitivity during its initial growth phase.

Physiological Responses of *J. curcas* to Salinity

General Mechanisms Displayed by Glycophytes in Response to Salinity

Salinity is characterized by an excess of soluble salts in the root medium that reaches a concentration higher than 40 mM or an electric conductivity above 4 dS m^{-1} in the soil solution (Munns and Tester 2008). The salt excess initially induces an osmotic effect, especially at concentrations higher than 50 mM, in most glycophytes. Practically all crop plants are glycophytes, i.e., plants growing in non-saline soils or non-halophytes. In these species, the salt excess causes a rapid decrease in the plant growth rate, primarily in the meristematic tissues, which causes a significant reduction in plant size. This effect is due to, at least in part, a strong increase in stomatal closure accompanied by a decrease in photosynthesis. In general, one might distinguish two types of glycophyte species based on the mechanisms involved in root uptake and saline ion transport to the shoots.

The *salt excluder* species (with an inefficient salt exclusion mechanism) show low selectivity for saline ions in the plasmalemma of the roots and xylem parenchyma cells. These species rapidly accumulate huge amounts of saline ions in root that reach, in the short-term (days), shoot parts, including old and young leaves (Munns and Tester 2008). If these species do not have an efficient partitioning mechanism for salt storage in the vacuoles of the oldest tissues, the ion concentration will rapidly reach toxic levels in the cytosol of the young leaves. In general, these types of glycophytes are relatively more susceptible to saline toxicity. Thus, salt-sensitive species do not manage ion exclusion mechanisms efficiently. Paradoxically, the most salt tolerant plant species, the halophytes, are *includers*; they accumulate extremely high amounts of salts in the leaves under high soil salinity (Silveira et al. 2009).

Some glycophyte *salt includers* might regulate high levels of salts by physiological mechanisms, such as high protoplasmatic resistance, mobilization of saline ions via phloem (high recycling rate) and maintaining high K^+/Na^+ ratios in the cytosol (Flowers 2004). In contrast, the *salt excluders* exhibit contrasting mechanisms to cope with salt excess. These species display a great selectivity in the membranes that allows them to stop cell ion uptake above a given threshold of cytoplasmic ion concentration, which suppresses the salt influx and transport rate to the shoots. The main physiological characteristic presented by these species is a coordinate balance between the rates of salt uptake and net photosynthesis to avoid salt accumulation. However, this strategy results in slow plant growth throughout the development period.

In general, *salt excluders* might also have more active Na^+/H^+-antiporters (transporter proteins) in their tonoplast (NHX system) and root plasmalemma (Voigt et al. 2009). The tonoplast antiporters can pump Na^+ ions from the cytosol into the vacuole, avoiding or minimizing their toxic effects in the cytosol. Plasmalemma antiporters (SOS1) might pump Na^+ outside the cytosol, i.e., in the apoplast; however, this hypothesis is still under debate because the toxic ions released back could reach the plasmalemma of nearby cells (Yeo 1998). However, experimental evidences indicate that SOS1 protein is more active in cells of the root epidermis, which would facilitate their diffusion into the soil solution, thus avoiding root toxicity (Shi et al. 2002).

In fact, the division of *includers* or *excluders* is rather conceptual, and plant responses to salt stress can vary widely, even within the same species. However, there is no simple definition or unique strategy used by a given species or even a cultivar because plant behavior depends on stress circumstances, especially with regard to plant-environment interaction and plant phenotypic plasticity. Environmental factors, such as (i) the vapor pressure deficit between the plant and atmosphere, (ii) temperature and (iii) radiation intensity, can completely change the plant response to salinity by exacerbating or minimizing the salt stress effects (Ferreira-Silva et al. 2011). Substrate types and properties, such as soil humidity and soil texture, can also strongly influence the plant response and strategy under salt stress.

J. curcas is an interesting model for evaluating the physiological response to salinity. When cultivated in hydroponic medium, young plants exhibit extremely high affinity for Na^+, with high rates of influx and transport to the leaves. *J. curcas* has a reduced affinity for Cl^- ions compared with Na^+ (Rodrigues et al. 2012). In addition, this species also presents a high affinity for K^+, especially when salinity is low (Silva et al. 2009a; Rodrigues et al. 2012). In hydroponic medium, the ions are almost completely available to root uptake; thus, it is an artificial condition compared with the soil. This condition could be roughly comparable to sandy soil irrigated with saline water. However, when *J. curcas* was grown in lysimeters on sandy-loamy soil and irrigated with saline water for 2 years, the Na^+ and Cl^- contents in leaves were lower than those in the hydroponics medium (Nery et al. 2009; Veras et al. 2010). Thus, *J. curcas* may be considered a *salt includer* species.

Growth, Accumulation of Saline Ions and Salt Sensitivity in J. curcas

Similar to other glycophytes, the growth of *J. curcas* plants is drastically decreased by salinity, especially during the initial growth stage (Silva et al. 2009a). Under hydroponic conditions, salt stress similarly affects the root and leaf dry matter accumulation. After a relative short-term exposure (1 week under NaCl 100 mM) associated with high temperature and high vapor pressure deficit, *J. curcas* plants accumulate huge amounts of saline ions in all organs and tissues (data not published). After a week of salt exposure, the external NaCl concentration sufficient to reduce total dry matter by 50% was 49 mM. The seedlings rapidly accumulated Na^+ in the leaves, reaching concentrations approximately four-times that supplied to the root medium (Silva et al. 2009a). Interestingly, the individual leaf area of the salt-stressed plants was generally not significantly affected.

The quick triggering of necrotic symptoms in the roots and leaves of young *J. curcas* plants was observed upon transfer to a hydroponic medium containing mild concentrations of Na^+ and Cl^- (~50 mM). In the leaves, these necrotic symptoms began as small yellow spots that progressively extended to major necrotic areas (Fig. 7.1). The signals of cellular death and apoptosis due to salt toxicity became evident as the brown areas enlarged progressively (Silva et al. 2011). Under severe salt stress, the necrotic areas progressively increased until death and the fall of the leaf.

The high affinity of *J. curcas* roots and shoots for saline ions, especially Na^+, could limit the expansion of the species in semiarid regions where the irrigation is frequently performed with saline water and where weather and soil conditions are favorable to secondary salinization. Additional studies conducted in Brazil in field conditions or in lysimeters containing high soil amounts (~200 kg) have shown contradictory results concerning the salt sensitivity of *J. curcas* (Fig. 7.2). These studies conducted with adult plants indicate that it is moderately tolerant (Veras et al. 2010). The characterization of salt tolerance under field conditions is difficult to establish because several environmental factors that change over time can

Fig. 7.1 Leaf symptoms of salt toxicity in young *J. curcas* plants grown on hydroponic media. (**a**) without NaCl, (**b**) with 50 mM NaCl for 1 week (osmotic effects) and (**c**) with 50 mM NaCl for 2 weeks (ionic effects) (Source: Silva et al. 2011)

Fig. 7.2 Symptoms of salt toxicity in adult *J. curcas* plants grown in big lysimeters in absence of salt stress (**a**) or irrigated continuously at field capacity with saline water (2.4 dS m^{-1}), i.e., ~24 mM NaCl (**b**). The irrigation was conducted for 2 years, and subsequently, the soil solution (leaching water) reached approximately 70 dS m^{-1} (Courtesy: AEC Sousa, UFCG, Brazil)

affect the plant response. For this reason, salt resistance in adult plants is frequently evaluated in lysimeters under greenhouse conditions, as shown in Fig. 7.3. However, this type of experiment is artificial also and does not mimic well the field conditions. For instance, the salts do not undergo periodic soil leaching during the rainy season.

Salt tolerance occurs mostly in the initial growth phase, in contrast to drought resistance, which must be evaluated, especially in the reproductive stage, when plants are more sensitive (Dasgan et al. 2002; Garrity and O'Toole 1994). Because the initial growth phase is the most important to salt tolerance, one might conclude that *J. curcas* has several physiological characteristics associated with salt sensitivity, such as (i) increased affinity for Na^+ and Cl^-, (ii) reduced efficiency for the storage of saline ions in old leaf tissues, (iii) strong antagonism between Na^+ and K^+, (iv) reduced ability to exclude saline ions from the roots and (v) increased ion transport rates from the roots to the leaves (Silva et al. 2009a; 2011; Rodrigues et al. 2012). Thus, selective breeding for salt tolerant cultivars is required to guarantee the large-scale cultivation of *J. curcas* in semiarid regions.

Fig. 7.3 General aspects of a long-term experiment for the evaluation of salt and drought tolerance of adult *J. curcas* plants cultivated in lysimeters of ~200 kg of soil (courtesy: AEC Sousa, UFCG, Brazil)

Na^+ Toxicity and K^+ Protection

It has been reported that elevated levels of K^+ in the external medium exert beneficial effects on plants exposed to high levels of Na^+, thus restricting salt toxicity (Voigt et al. 2009; Rodrigues et al. 2012). Moreover, the maintenance of a high cytosolic K^+/Na^+ ratio is a key feature of salt tolerance (Apse and Blumwald 2007). Under optimal conditions, the cytosolic K^+ content is estimated to be approximately 150 mM and is associated with a negligible Na^+ level (Carden et al. 2003). However, under conditions of salt stress, the excessive Na^+ accumulation in the cytosol is associated with K^+ efflux and the cytosolic K^+/Na^+ ratio falls dramatically below a critical level (Shabala and Cuin 2007). This Na^+-induced K^+ loss from the cells is the result of an induced membrane depolarization induced by NaCl leading to the activation of depolarization-activated outward-rectifying K^+ channels (Cuin et al. 2008).

Fig. 7.4 A high K^+ concentration (10 mM) mitigates the toxic effects induced by Na^+ (50 mM). (**a**): Leaf symptoms of Na^+ toxicity (black arrow) on control treatment (K_0Na_1), where the *J. curcas* plantlets are supplemented with Na^+ (50 mM). (**b**): comparison of control (K_0Na_1) and treated (K_1Na_1) plantlets. (**c**): Leaf symptoms of Na^+ toxicity (*black arrow*) on a plantlet (K_1Na_1) treated with a supplement of K^+ (10 mM) and Na^+ (50 mM). The photos were taken 1 week after treatment (source: Rodrigues et al. 2012)

The competitive interaction between K^+ and Na^+ involves root transport systems, which involve transporter proteins and ion channels (Apse and Blumwald 2007). The K^+ transport systems are relatively well known, but those involved with Na^+ uptake are less understood (Britto et al. 2010). Experimental evidence has shown that Na^+ influx might be mediated through several transport systems, especially via high-affinity K^+ transport channels (Dreyer and Blatt 2009; Szczerba et al. 2009). At high concentrations in the root medium, Na^+ can also be absorbed using low-affinity systems via non-selective cation channels (Voigt et al. 2009; Britto et al. 2010).

The data obtained by our group support the idea that the leaf K^+/Na^+ ratios ranging from 1.0 to 2.0 were strongly favorable for photosynthesis and plant growth in *J. curcas* grown under 50 mM NaCl (Rodrigues et al. 2012). The data reinforce the importance of the K^+/Na^+ ratio as an indicator of ionic homeostasis in presence of toxic NaCl concentration. However, the physiological mechanisms that regulate this balance at the whole plant level have been insufficiently studied to date. Because the leaf tissue might simultaneously present high concentrations of both Na^+ and K^+ without showing any sign of Na^+ toxicity, it is probable that in some plant glycophyte species high amounts of Na^+ are stored in the vacuoles, as in halophytes, but in lesser concentrations (Whang and Showalter 2004; Silveira et al. 2009; Britto et al. 2010).

Recently, we demonstrated that Na^+ and K^+ present a strong and reciprocal interaction at the whole plant level in *J. curcas* young plants (Rodrigues et al. 2012). Increased K^+ concentration (10 mM) in the root medium mitigates the toxic effects caused by 50 mM NaCl, whereas in the absence of K^+, the leaves showed typical symptoms of NaCl toxicity (Fig. 7.4). Taken together, our results show that a supply of high concentrations of K^+ might promote a favorable ion homeostasis. High K^+ affinity and selectivity over Na^+ are associated with high rates of K^+ uptake by the roots and xylem flux and transport toward shoots; both features are essential to promote proper K^+/Na^+ ratios in the plant tissues and to avoid the toxic effects induced by high Na^+ concentration.

Table 7.1 Concentrations of Na^+, Cl^-, K^+ and the K^+/Na^+ ratio in the leaves of young *J. curcas* plants cultivated with and without NaCl 100 mM for 7 and 14 days and after recovery for 3 days. The concentrations of the ions are expressed in mmol (kg DM)$^{-1}$ (Source: Silva et al. 2011)

	NaCl	Days after treatment		
Parameter	(mM)	7 days	14 days	Recovery
Na^+	0	133 Bb	215 Ab	215 Ab
	100	603 Ca	1721 Aa	1458 Ba
Cl^-	0	102 Bb	147 Ab	139 Ab
	100	326 Ca	1498 Aa	1240 Ba
K^+	0	716 Ba	1071 Aa	1123 Aa
	100	581 Ab	423 Cb	496 Bb
K^+/Na^+ ratios	0	5.4 Aa	5.0 Aa	5.2 Aa
	100	0.96 Ab	0.24 Bb	0.34 Bb

Values represented by the same upper case letters, between time of treatment and same lower case letters for each time of treatment are different (Tukey, $p<0.05$)

Photosynthesis and Salt Sensitivity

Photosynthesis and plant growth are among the primary processes most affected by salinity (Chaves et al. 2008). Water stress and salinity can affect photosynthesis directly or indirectly by (i) decreasing CO_2 availability due to diffusion limitations (stomatal closure), (ii) affecting photosynthetic metabolism or (iii) affecting the energetic conversion by the photochemical system apparatus (Souza et al. 2004). In parallel to the impairment of photosynthesis, salinity induces strong alterations of leaf water balance and osmotic homeostasis. It is widely accepted that short-term exposure (days) to salinity induces osmotic effects and that long-term exposure can cause ionic damage to the plant cells (Munns and Tester 2008).

Recently, we observed that the osmotic effects induced by salt excess for 7 days and the toxic ionic effects induced after 14 days of exposure to salt excess (NaCl 100 mM) were responsible for necrotic symptoms and K^+/Na^+ ratio imbalance in the leaves of *J. curcas* (Silva et al. 2011). Indeed, after 7 days of treatment, the leaves showed minor senescence symptoms associated with K^+/Na^+ ratios near 1.0, but after 14 days, the leaves showed large necrotic symptoms and the K^+/Na^+ ratios were ~5.0. The necrotic symptoms persisted even after NaCl removal from the nutritive medium (Silva et al. 2011). Table 7.1 shows ion concentrations at both osmotic and ionic phases.

The reduction of photosynthesis in *J. curcas* during the osmotic phase was due to stomatal closure, i.e., the reduced carboxylation induced by the limitation of CO_2 availability, while both the stomatal and biochemical limitations were observed after 14 days of treatment (ionic phase). As expected, the negative effect caused by salt stress on the leaf gas exchange progressively increased with time as the salt excess increased. In addition, photosynthesis did not recover after salt removal, suggesting that irreversible damage to chloroplasts occurred after 14 days of treatment.

Significant decreases in the intercellular CO_2 concentration (C_I) and carboxylation instantaneous efficiency (P_N/C_I) after 14 days of salt exposure also indicate that salt stress affected photosynthesis by metabolic limitation. The reduction of P_N/C_I ratio is likely associated with a decrease of Rubisco carboxylase activity, which correlates with the accumulation of Na^+ and Cl^- in the leaf tissues. Thus, the reduction of photosynthesis in *J. curcas* may, at least in part, be a direct effect of Na^+ and Cl^- ions, as observed in sorghum (Netondo et al. 2004) and orange (López-Climent et al. 2008) plants.

Regarding the photochemical effects after 7 days of salt treatment in *J. curcas*, the maximum quantum efficiency of PSII photochemistry (F_V/F_M), which represents the maximum efficiency at which light is absorbed by PSII, is used for the reduction of QA and was not affected by salinity; the actual quantum yield of primary photochemistry or PSII operating efficiency, which represents the efficiency at which light is absorbed by PSII, is used for QA reduction. At a given photosynthetically active photon flux density (PPFD), this parameter provides an estimate of the quantum yield of the linear electron flux through PSII (ΔF/Fm'), which decreased significantly after 14 days of exposure to NaCl. Photochemical quenching (qP) was not affected by salt stress, while the non-photochemical quenching (NPQ) increased significantly in plants subjected to salt stress.

The increase of NPQ was already significant after 7 days of treatment and progressively increased after 14 days of salt stress (Silva et al. 2011). The increase in non-photochemical quenching indicates an efficiency energy excess dissipation mechanism at the photosystem II level. Thus, these data revealed reasonable photochemical activity even under ionic stress, which was not compatible with the photosynthetic rates. Altogether, the results of the photosynthesis of young plants of *J. curcas* exposed to excess NaCl provided evidence that this species has adequate stomatal control to restrict CO_2 assimilation and the effective mechanisms of dissipation of energy excess at the photosystem II level, especially by the liberation of heat via the non-photochemical process. However, when the concentration of saline ions reaches very high concentrations, the photosynthetic machinery collapses and irreversible damage occurs.

The accumulation of toxic ions (Na^+ and Cl^-) is accompanied by a decrease in the K^+ concentration and the photosynthetic damage reinforces, once more, that young plants of *J. curcas* are relatively sensitive to excess of NaCl. The absence of any photosynthetic recovery three days after salt removal (recovery phase) also indicates that the high Na^+ and Cl^- concentrations in the leaves might cause acute damage to the photochemical system and gas exchange under conditions of acute stress caused by NaCl 100 mM in *J. curcas*. Thus, salt-induced ionic toxicity is able to induce irreversible damage to the photosystems and to Calvin cycle machinery.

Oxidative Protection Under Excess Salinity

Plants have developed several defensive antioxidative mechanisms against the excessive production of reactive oxygen species (ROS) generated by excess salinity (Miller et al. 2010) or other abiotic and biotic stresses. This defense system involves several enzymes, such as *superoxide dismutase* (SOD), *ascorbate peroxidase* (APX) and *catalase* (CAT), together with the main non-enzymatic *antioxidants ascorbate* (AsA) and *glutathione* (GSH). SOD and APX represent the first line of defense against ROS produced in the chloroplasts, while CAT represents the primary scavenger of H_2O_2 generated by photorespiration.

It has been shown that when *J. curcas* is submitted to heat shock (6 h at 42°C) under excess salinity, both stressful conditions strongly increase electrolyte leakage and lipid peroxidation. The increased electrolyte leakage showed a positive correlation with the lipid peroxidation, suggesting that plasmalemma injuries were, at least partly, a consequence of oxidative damage. Numerous investigations have demonstrated that injuries caused by abiotic stresses in the plant cells are triggered in part by oxidative stress (Asada 1999).

Total SOD, CAT and APX activities were differently regulated by one or combined stresses, such as salinity and heat, for example, in *J. curcas*. APX and CAT activities were strongly stimulated by salinity, while SOD activity was only slightly increased by this factor (data not published). This response is likely related to the specific regulation of gene expression (Cavalcanti et al. 2007). It is well known that the SOD-APX-CAT system protects the photosynthetic machinery from oxidative damages in plants exposed to salt stress. Thus, despite the increase in the activity of antioxidant enzymes, excess salinity causes oxidative damage, as indicated by lipid peroxidation and membrane damage.

The question of whether *J. curcas* has an efficient antioxidant system under excess salinity is still controversial. Our results are difficult to interpret because the stress-induced up-regulation of antioxidant enzymes under *in vitro* conditions does not necessarily confer an effective protection against salt-induced oxidative damage under *in vivo* conditions (Cavalcanti et al. 2004). In fact, we observed that excess salinity causes a reduction of plant growth, photosynthesis and increases lipid peroxidation and membrane damage. The threshold between oxidative changes favorable to plant defense and oxidative damage triggered by salinity is still unknown (Foyer et al. 2009).

Osmotic Adjustment in J. curcas *Under Salinity*

The most important physiological process for salt and drought tolerance in plants is turgor maintenance via effective osmotic adjustments, which can preserve essential metabolic processes and contribute to a sustained growth under adverse conditions (Martínez et al. 2005). The osmotic adjustment in halophytes and glycophytes under excess salinity is accomplished mainly through the accumulation of saline ions, especially in the leaves (Silveira et al. 2009). However, to avoid ionic toxicity and metabolic disturbance in the cytosol, a great part of these ions should be stored in vacuoles.

Under excess salinity, the non-toxic cytosol osmotic homeostasis is obtained mostly through the accumulation of K^+ ions and organic solutes. The major osmo-solutes found in most halophytes and some glycophytes are *glycinebetaine* (GB) and proline (Flowers and Colmer 2008). These N compounds can also act as protective solutes against membrane damage and oxidative stress (Silveira et al. 2003). Unfortunately, few glycophytes display an effective osmotic adjustment under excess salinity because the saline ions are frequently the major solutes that account for the decrease of osmotic potential. Osmotic adjustment through the accumulation of saline ions is poorly effective; in *J. curcas,* the osmotic adjustment is limited because saline ions quickly reach toxic levels and the plants start to suffer damage from long-term exposure to excess salinity (Silva et al. 2009b). This mechanism occurs despite that species to accumulate constitutively prominent amounts of GB in their roots and leaves.

Our group was perhaps the first to study osmotic adjustment in young plants of *J. curcas* subjected to different levels of salinity. We concluded that the saline ions (Na^+ followed by Cl^-) are quantitatively the most important for the osmotic adjustment in leaves (Silva et al. 2009b). The K^+ ion is also important, but Na^+ excess induces a decrease in the K^+ concentration, especially in the leaves. Nitrate has a minor quantitative importance. Among the organic solutes, sugars, amino acids and GB are the most important for osmotic adjustment, but their synthesis is not up-regulated by salinity.

GB has a high endogenous concentration, but proline has no contribution to osmotic adjustment because its concentration is extremely low. Our results suggest that the selection of *J. curcas* genotypes with high GB concentration in the roots and leaves would be a good strategy to improve crop performance under excess salinity. In addition to contributing to osmotic adjustment, this osmo-solute is extremely important to protect the cells against various types of stresses (Silveira et al. 2009). The next strategy is to study GB synthesis in meristematic regions of the shoot and root in *J. curcas* because osmotic adjustment and cell protection in these tissues is essential for maintaining growth and cell expansion under excess salinity.

Mechanisms Involved in the Drought Tolerance of *J. curcas*

General Aspects

Drought is by far the most important environmental stress in agriculture, and several efforts have been made to improve crop productivity under water-limiting conditions. While natural selection has favored mechanisms for adaptation and survival, breeding activity has directed selection toward increasing the economic yield of the cultivated species. Since their domestication, the evolution of crops has been driven by the selection of desired traits recognized at the phenotypic level (Cattivelli et al. 2008). However, the complexity of drought tolerance mechanisms explains the slow progress of selective breeding for yield in drought environments. In recent years, crop physiology and genomics have led to new insights in drought tolerance, providing breeders with new knowledge and tools for plant improvement (Tuberosa and Salvi 2006).

It has been suggested that *J. curcas* may thrive in areas with semiarid climates and marginal soil conditions that are not suitable to most crops (Maes et al. 2009a, b; Tang et al. 2011). Indeed, we showed that this species displayed efficient physiological and biochemical responses to drought involving water relations, photosynthesis (gas exchange and chlorophyll fluorescence), osmotic adjustment and oxidative protection. Although our results have contributed to the understanding of key physiological mechanisms, they have some limitations because these studies were performed with young plants under controlled and greenhouse conditions. However, some other investigations performed under field conditions have indicated that *J. curcas* effectively displays biochemical and physiological characteristics associated with drought tolerance (Pompelli et al. 2010).

Further studies on adult plants under field conditions are necessary to determine the full implications of the drought-tolerance displayed by *J. curcas*. Such information is necessary to weigh existing knowledge and will be essential to the selective breeding of this new crop to develop drought tolerant genotypes that are able to function under stressful conditions. Although the isolated effects of different abiotic stresses on plant metabolism have been extensively studied, relatively little is known about the combined impact of drought with others stresses commonly found under field conditions. For example, recent studies with *J. curcas* revealed that the plant response to the combination of drought and heat is different from that obtained for drought or heat stresses applied individually (Silva et al. 2010a).

Water Preservation in Leaves as an Efficient Mechanism of Drought Resistance in **J. curcas**

One of the main responses of *J. curcas* to low water availability is its ability to maintain a good leaf hydration status. In a recent study, we observed that the decrease in leaf Ψ_w of drought-stressed plants was associated with an adequate leaf hydration status, as indicated by the high values of relative water content, which was similar to that of the non-stressed control plants (Fig. 7.5) (Silva et al. 2010a). Maes et al. (2009a) worked with 4-week-old *J. curcas* and observed similar results. This strategy appears to be common in species from semiarid regions, such as the cowpea, when subjected to drought (Souza et al. 2004) and salinity stress (Cavalcanti et al. 2004).

Moreover, the preservation of water in the leaves of *J. curcas* under severe drought stress is associated with absence of any injury symptoms, such as leaf drying or chlorosis. However, adult plants subjected to long-term water deficit, such as 2 years with 25% field capacity under greenhouse in big lysimeters exhibited typical symptoms, such as growth reduction and leaf chlorosis (Fig. 7.6). In fact, water stress, as indicated by Ψw reduction, is influenced by leaf age and capacity for osmotic adjustment (Bajji et al. 2001).

Experimental evidence suggests that the drought tolerance of *J. curcas* is associated with efficient water preservation in the leaves, such as efficient stomatal

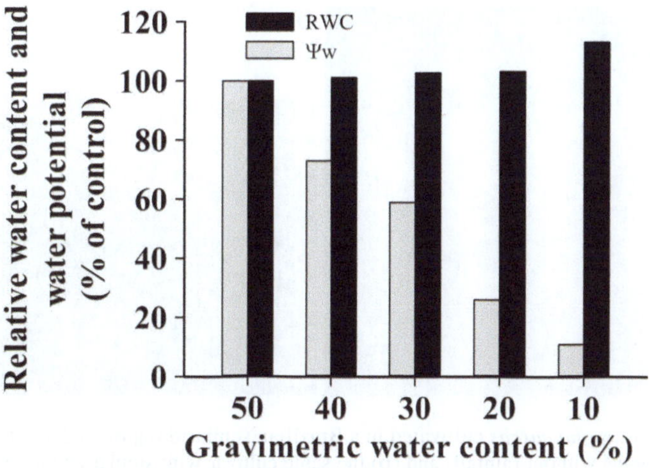

Fig. 7.5 Changes in the leaf relative water content (RWC) and water potential (Ψw) in *J. curcas* plants exposed to different water deficits for 10 days. The absolute values of the control (100%) were RWC = 74% and Ψw = −0.55 MPa (unpublished data)

Fig. 7.6 Visual aspects of *J. curcas* grown under controlled conditions. (**a**) Well-watered, (**b**) under water stress corresponding to 25% of field capacity for 2 years (Courtesy: AEC Sousa, UFCG, Brazil)

control, coordinate photosynthetic regulation and photodamage protection in photosystem II (Silva et al. 2010a; Díaz-López et al., 2012). In addition, *J. curcas* is a succulent species that can store significant amounts of water in the stem and branches and supply it back to leaves and fruits during dry seasons (Maes et al. 2009a).

Fig. 7.7 (**a**) *Jatropha curcas* cultivated in a Brazilian semiarid region under extreme dry conditions (8 months without rainfall) and (**b**) the same cultivar with similar age grown in the same region under irrigation ("evergreen") (Courtesy: Fazenda Tamanduá, Paraiba, Brazil)

Moreover, as a deciduous plant, *Jatropha* presents an avoidance strategy through the fall of leaves when cultivated under extreme dry conditions, as has been observed frequently in the semiarid regions of Brazil (Fig. 7.7a).

Some observations under field conditions have provided clue that the intensity of the leaf drop appears to be associated with that of the of drought level. Indeed, when physic nut plants are cultivated in areas with regular rainfall, only a part of its leaves drop due to senescence of the oldest leaves. In contrast, under irrigation conditions *Jatropha* is maintained as "evergreen plants" as shown in Fig. 7.7b; it is important to note that both plants shown in Fig. 7.7 were of the same age. However, the "evergreen plants" exhibited low rates of flowering, fruiting and seed yield (R.A. Viégas—personal communication).

Role of Inorganic and Organic Solutes in Osmotic Adjustment

The physiological and developmental mechanisms, which allow a species to tolerate prolonged periods of water deficit, can involve numerous attributes. One means of increasing drought tolerance is through the accumulation of osmotically active solutes, so that turgor and turgor-dependent processes might be maintained. The osmotic adjustment associated with partial stomatal opening allows the maintenance of CO_2 assimilation, water uptake, cell enlargement and plant growth despite water stress (Alves and Setter 2004). In this context, the *osmotic adjustment* (OA) has been considered as an essential mechanism for drought tolerance and it has drawn much attention during the last years (Hessine et al. 2009).

Our findings showed that in *J. curcas*, the strategy of water loss restriction through the leaf is associated with an effective osmotic adjustment mechanism (Silva et al. 2010b).

Table 7.2 Relative contribution of the inorganic solutes in the osmotic adjustment of the leaves and roots of *J. curcas* plants subjected to different ratios of vermiculite and gravimetric water content (mass of water per mass of dry soil) for 10 days (mean of 4 replicates) (Source: Silva et al. 2010b)

	Na^+		Cl^-		K^+		NO_3^-	
	Osmolality (%)							
Gravimetric water content, %	leaf	root	leaf	root	leaf	root	leaf	root
50	15.7a	17.2a	13.7b	9.5b	32.7a	30.1a	5.1a	5.0a
40	17.9a	16.0a	13.2b	12.0a	30.4a	32.9a	5.0a	4.3b
30	16.4a	18.7a	15.4a	9.9b	23.9b	29.0b	4.6a	3.8b
20	16.4a	16.3a	13.9b	8.3c	23.3b	28.0b	4.8a	3.9b
10	13.3b	16.7a	16.4a	6.6d	21.3b	25.6c	5.3a	4.1b

The same letters are not significantly different at $P=0.05$ by Tukey's test

The OA capacity is due to the net increase of the relative contribution of some inorganic and organic solutes to the osmotic potential. Our recent data revealed that inorganic solutes, especially Na^+ Cl^- and K^+, are involved in the OA in the leaf and root tissues of both stressed and non-stressed *J. curcas* plants (Silva et al. 2010b). We observed that the Na^+ concentration did not change significantly in the leaves and roots of plants under drought stress. By contrast, the Cl^- concentration in the stressed plants increased in the leaves and slightly decreased in the roots (Table 7.2).

The Cl^- and Na^+ ions showed a significant contribution to OA in the leaves and roots of non-stressed and stressed plants. In addition, the K^+ concentration in these two organs was high, and its relative contribution to the osmotic potential in *J. curcas* plants under water stress was higher than the other inorganic ions (Table 7.2). The K^+ ion is quite soluble and plays a key osmo-regulatory role in the turgor regulation of guard cells (Shabala and Cuin 2007). Moreover, K^+ is involved in charge balancing in the cytoplasm and the enzymatic activation mechanisms, which are essential for cell function.

In spite of its minor quantitative contribution to OA, the leaf concentration of NO_3^- increases under drought stress (Table 7.3). The increase of NO_3^- levels in the leaves likely originates from the inhibition of reductase activity, which has been observed in many species, even under mild stress (Kameli and Losel 1995). Although the NO_3^- contribution to the OA is not quantitatively comparable to other solutes, its relative participation in the regulation of the osmotic potential can be important in the organs of stressed and non-stressed plants.

Organic solutes, such as *total soluble sugars* (TSS), *total free amino acids* (TFAA) and GB, are also significantly involved in the OA of leaves and roots in *J. curcas* (Table 7.3). The leaf TFAA contents increased only under more acute drought stress. However, the leaf and root TSS contents increased in all water stress treatments. Sucrose and reducing sugars were the unique solutes that presented significant increases in the net concentrations in response to increases in the intensity of the water deficit. Our data support the hypothesis proposed by Iannucci et al. (2002) that soluble sugars are the organic solutes that most contribute to OA in the leaves and roots of higher plants under water stress conditions.

Table 7.3 Relative contribution of organic solutes to the osmotic adjustment of the leaves and roots of *J. curcas* plants subjected to different ratios of vermiculite gravimetric water content for 10 days (mean of four replicates) (source: Silva et al. 2010b)

	TFAA		TSS		GB		Proline	
	% of osmolality							
Gravimetric water content (%)	leaf	root	leaf	root	leaf	root	leaf	root
50	8.7b	10.1a	7.7d	10.6c	7.4b	8.6a	0.1b	0.1b
40	7.7b	7.7b	9.8c	9.7c	7.1b	9.2a	0.1b	0.1b
30	7.9b	9.7a	16.2b	12.4b	6.6b	8.7a	0.1b	0.1b
20	10.2a	8.1b	18.2a	19.6a	8.4a	8.3b	0.1b	0.1b
10	9.5a	9.1a	20.7a	21.4a	8.8a	10.0a	0.2a	0.2a

The same letters are not significantly different at P=0.05 by Tukey's test

Among several organic solutes studied, the proline content did not change in the roots under water stress, but its accumulation increased in the leaves of plants under the same conditions, but without insignificant contribution to the osmotic potential. By contrast, the leaf and root GB concentrations were high in the well-watered and drought-stressed plants. In addition, if the GB accumulation is exclusively confined to the cytosol (approx. 10% of the cell volume), a 10-fold increased concentration could be expected (Silveira et al. 2009). In this case, GB in concert with soluble sugars and K^+ would contribute mostly to the OA in the cytoplasm of the leaf and root cells of *J. curcas* plants. There is evidence that water stress-induced GB synthesis is an adaptive response because it might function as a non-toxic osmolyte or an osmoprotectant primarily in the cytoplasm, but also in organelles, such as chloroplasts (Bajji et al. 2001). Overall, these observations suggest that the increased GB concentration is a mechanism for adaptation to drought in *J. curcas*, and this GB concentration contributes to OA.

Photosynthesis in J. curcas *Under Drought Conditions*

Among the effects caused by water stress in plants, photosynthesis is one of the most sensitive biological processes (Chaves et al. 2008; Guerfel et al. 2009). The inhibitory effects of drought on photosynthesis might be associated with low CO_2 availability caused by (i) low stomatal and mesophyll conductance (Flexas et al. 2004, 2007) and/or (ii) impairments in photochemistry and carbon metabolism (Lawlor and Cornic 2002; Peeva and Cornic 2009). Stomatal closure is an early response to drought and an efficient way to reduce transpiration in water-limiting environments. Biochemical limitations also play an important role in photosynthesis regulation under prolonged periods of drought stress (Yin et al. 2005; Flexas et al. 2006; Chaves et al. 2008).

In *J. curcas*, we demonstrated that increased levels of water deficit cause a progressive reduction in net photosynthesis (P_N) and stomatal conductance (g_S) (Fig. 7.8), but this effect might be more of an acclimation and/or tolerance response

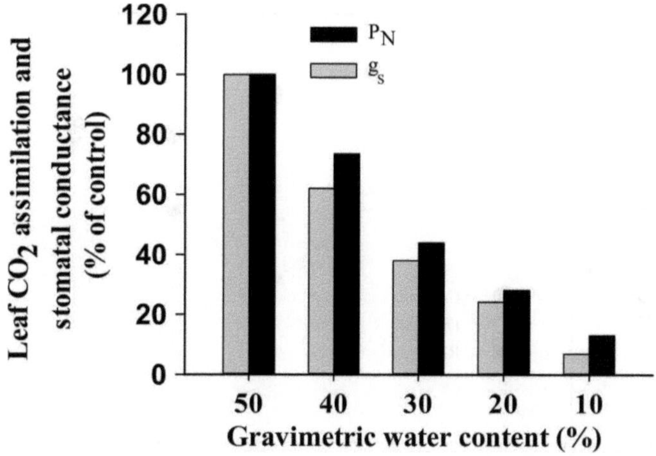

Fig. 7.8 Relative variation in CO_2 assimilation rate (P_N) and stomatal conductance (g_S) in *J. curcas* plants exposed to water-deficit treatments for 10 days. The absolute values of the control (100%) were $P_N = 9.98$ μmol m^{-2} s^{-1} and $g_S = 0.30$ mol m^{-2} s^{-1} (mean of four replicates) (Source: Silva et al., in press)

to low water availability than a harmful effect imposed by water stress. This hypothesis is supported by the fact that when *J. curcas* plants were subjected to a drought for 10 days (no water supply), a complete recovery in the photosynthesis occurred from the tenth day after re-watering (data not shown). These results showed that drought significantly decreased leaf CO_2 assimilation and stomatal conductance; however, at the end of the recovery period, these changes were fully reversible (Fig. 7.9). It is well reported in the literature that a decrease in P_N through stomatal limitation associated with a reduction in transpiration to reduce water loss is a common characteristic in plants typically adapted to semi-arid environments (Souza et al. 2004).

Based on chlorophyll fluorescence parameters, we also demonstrated that water deficit induces changes in photochemical activity and photosynthetic pigment content (Silva et al. 2010a). Under increasing levels of water deficit, we observed reductions in the (i) quantum yield of PSII ($\Delta F'/F_M'$), (ii) apparent electron transport rate (ETR'_S) and (iii) chlorophyll content. In fact, the changes in $\Delta F'/F_M'$ and ETR'_S are potentially associated with damage caused by primary electron acceptors of PSII (plastoquinone) due to reduced quinone accumulation leading to the deactivation of the electron transport chain in thylakoid membranes (Drodzova et al. 2004; Chagas et al. 2008).

The reduction in the chlorophyll content can be considered as an additional strategy of photo-protection because the reduction of light absorption decreases energetic pressure at the PSII level (Chaves et al. 2008). Moreover, increased non-photochemical quenching (NPQ) and the relative excess of light energy (EXC) observed in response to drought, respectively indicate the (i) existence of a non-radiative energy dissipation mechanism using thermal processes and (ii) over-exci-

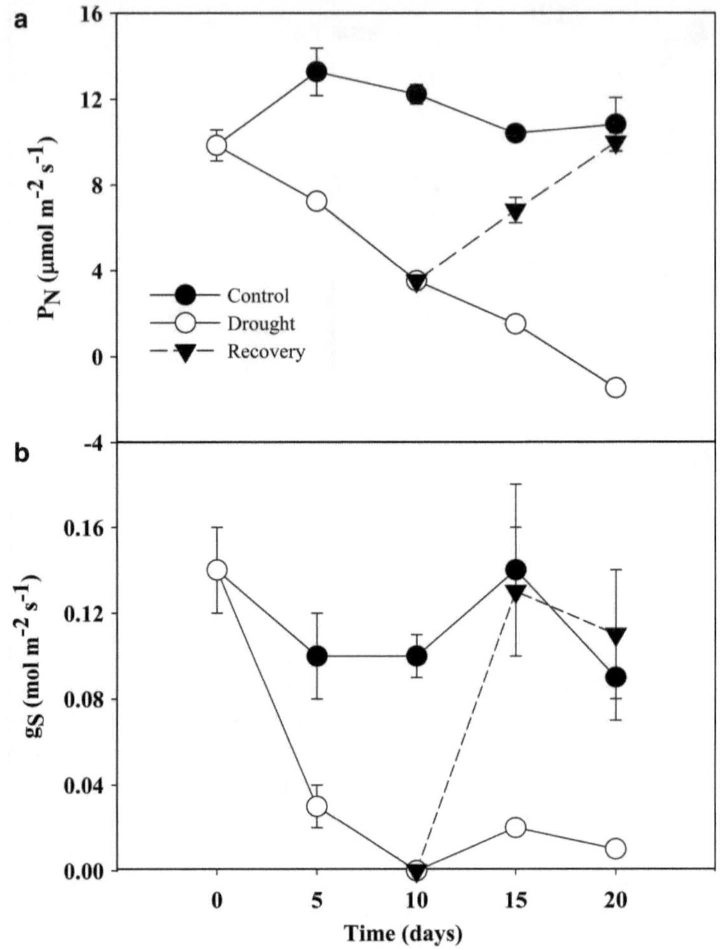

Fig. 7.9 Leaf CO_2 assimilation rate (**a**) and stomatal conductance (**b**) in *J. curcas* plants exposed to drought for 20 days and after a period of recovery (mean of four replicates) (unpublished data)

tation of the PSII complex through light energy, which must be deactivated by dissipation processes to avoid photochemical damage. These photoprotective mechanisms help to maintain the high oxidative state of the primary electron acceptors of PSII, lowering the probability of photo-damage and photo-oxidative stress in the chloroplast components (Chagas et al. 2008). Figure 7.10 shows the effects of drought on photochemical activity and chlorophyll content.

Water stress also induces a progressive increase of photorespiration in *J. curcas* plants (Fig. 7.11). The increase of photorespiration has been considered as a protective mechanism, consuming photochemical products and mitigating damage at the PSII level (Chaves et al. 2008; Foyer et al. 2009). Although a decrease in photochemical activity has been observed in *J. curcas* plants under water deficit, we argue

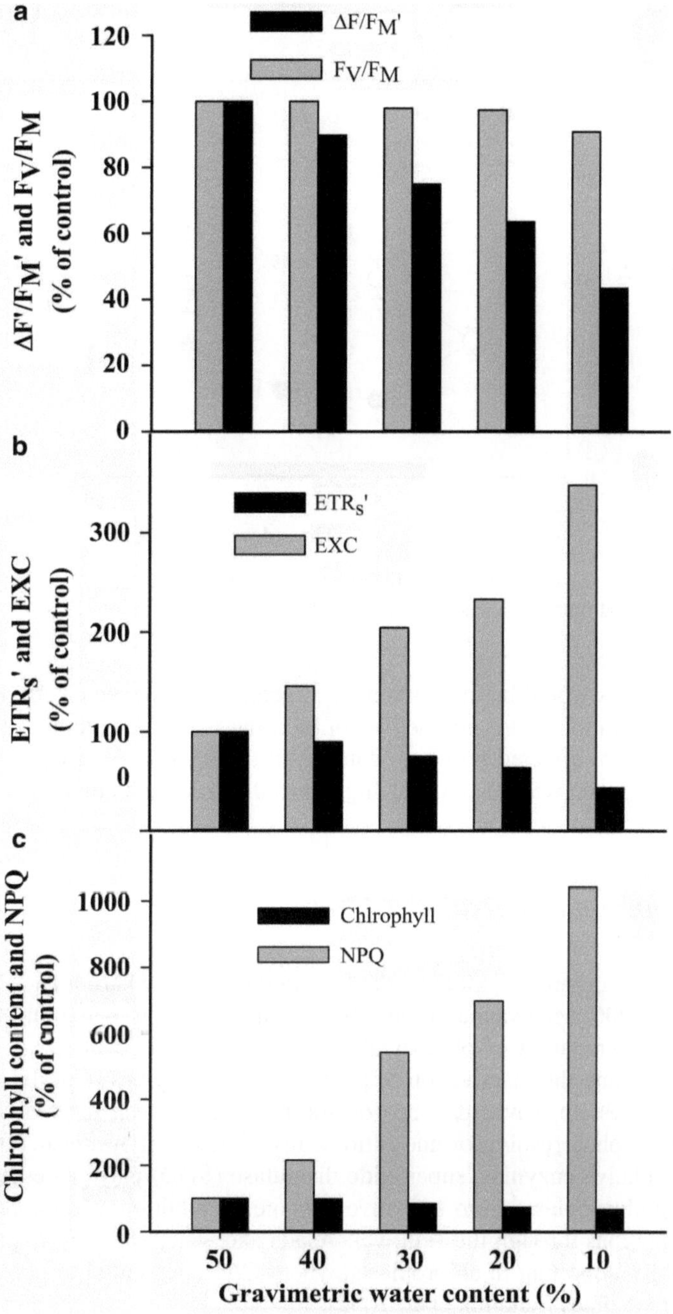

Fig. 7.10 Effects of drought on the photochemical activity and chlorophyll content of *J. curcas*. (**a**): Relative variation in maximal (F_V/F_M) and actual ($\Delta F'/F_M'$) quantum yield of PSII. (**b**): Apparent electron transport rate (ETR_S') and relative energy excess (EXC). (**c**): Chlorophyll content and non-photochemical quenching (NPQ) in plants exposed to water-deficit treatments for 10 days. The absolute values of the control (100%) were $\Delta F'/F_M'=0.663$; $F_V/F_M=0.774$; $ETR_S'=106.3$ μmol m^{-2} s^{-1}; EXC=0.18; total chlorophyll content=0.88 mg (g FW)$^{-1}$ and NPQ=0.247 (mean of four replicates) (unpublished data)

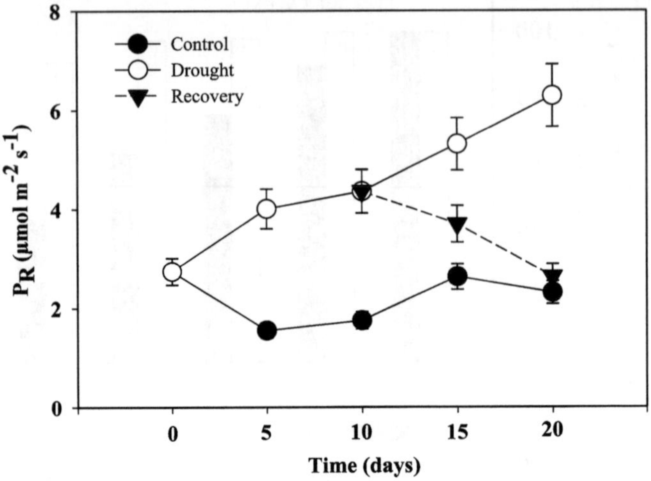

Fig. 7.11 Photorespiration in *J. curcas* plants exposed to drought for 20 days and after a period of recovery (mean of four replicates) (unpublished data)

that leaf tissues would be more strongly affected if alternative electron sinks, such as photorespiration, were non-active. In fact, the maximal quantum yield of PSII (F_V/F_M) was not affected by photo damage in the PSII reaction centers caused by water deficit, even with plants showing reduced chlorophyll content.

Protection Against Oxidative Stress

Water stress can induce oxidative damage in the leaves (Sharma and Dubey 2005; Guo et al. 2006; Manivannan et al. 2007), causing the peroxidation of membrane lipids, the degradation of photosynthetic pigments (Cavalcanti et al. 2004; Silva et al. 2010a) and the inactivation of photosynthetic enzymes (Guerfel et al. 2009). Oxidative stress in leaves depends on the balance between the reactions of CO_2 assimilation, photorespiration and antioxidative defenses (Foyer et al. 2009). Among the antioxidative enzymes, superoxide dismutase (SOD) plays an essential role in protecting chloroplasts from oxidative damage. In addition, *ascorbate peroxidase* (APX) functions through the Halliwell-Asada and water-water cycle pathways and is especially important in chloroplasts, where CAT is essential to photorespiration in the peroxisomes (Palatnik et al. 2002).

In many plant species, multiple photo-protective and antioxidant mechanisms have evolved to withstand the oxidative stress induced by drought. In this context, our results demonstrated that *J. curcas* is able to overcome the accumulation of superoxide and hydrogen peroxide in chloroplasts by increased SOD and APX activities when subjected to water stress. However, the decrease in CAT activity induced by drought should be a limiting factor for H_2O_2 scavenging and oxidative protection in this species, especially under acute water deficit (data not published).

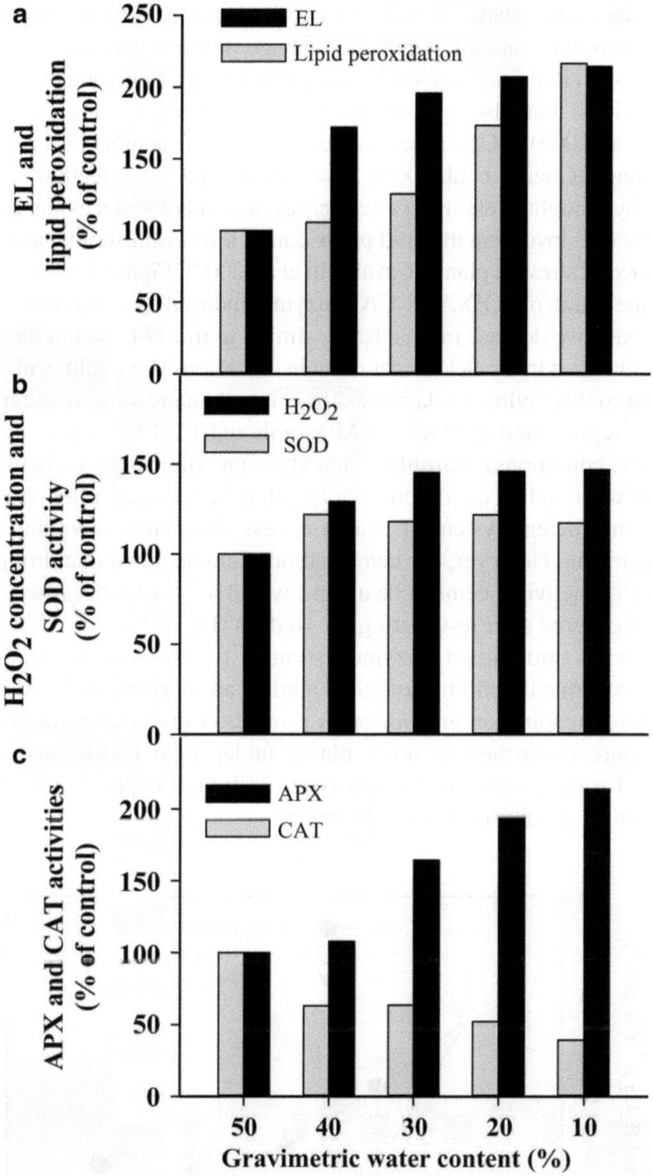

Fig. 7.12 Oxidative damage and response of the SOD-APX-CAT complex induced by water deficit in *J. curcas*. (**a**): Relative variation in electrolyte leakage (EL) and lipid peroxidation. (**b**): H_2O_2 concentration and SOD activity. (**c**): APX and CAT activities. *J. curcas* plants were exposed to water deficit for 10 days (unpublished data)

The occurrence of oxidative damage through increases in H_2O_2 has also been shown through lipid peroxidation and membrane damage in the leaf tissues. In fact, several studies have shown that lipid peroxidation is a natural metabolic process under normal aerobic conditions, but also one of the most investigated consequences of ROS (Cavalcanti et al. 2007). ROS causes the peroxidation of membrane lipids, leading to oxidative damage (Guerfel et al. 2009). However, despite the induction of strong lipid peroxidation by drought stress in *J. curcas* leaves, no experimental evidence to date has shown suitable data involving the lipid peroxidation levels (thresholds) and physiological performance of stressed plants (Cavalcanti et al. 2007). Figure 7.12 shows oxidative damage and the action of SOD-APX-CAT enzymes induced by water deficit in *J. curcas* plants. This oxidative defense mechanism is similar to that obtained in the adult plants of *J. curcas* cultivated in the field under drought conditions (Pompelli et al. 2010).

In addition to displaying a relatively efficient enzymatic system under conditions of water stress represented by SOD and APX (although CAT activity is strongly inhibited under these conditions), *Jatropha* plants show an efficient photochemical system. Indeed, under water deficit conditions, photosystem II does not suffer photoinhibition and triggers an efficient system of energy excess dissipation involving non-photochemical quenching. However, our investigations demonstrated that drought-induced reduction in CAT activity seems to be a non-reversible process because CAT activity was slightly recovered after re-watering for 10 days (Fig. 7.13).

As CAT is an antioxidant enzyme essential to oxidative protection, further molecular, biochemical and physiological studies are needed to elucidate its effective role in photorespiration and *Jatropha* protection under drought conditions. If these results are confirmed in adult plants under field conditions, the genetic improvement for the production of genotypes with high expression and activity of catalases could be an important genetic target.

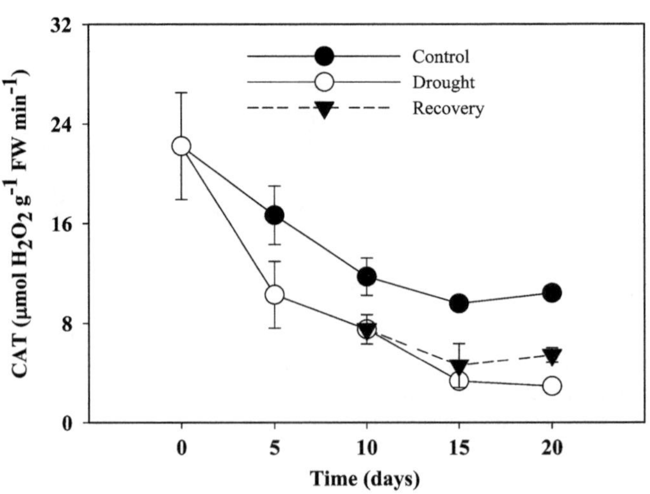

Fig. 7.13 Catalase activity in *J. curcas*. Plants were exposed to drought for 20 days, and the catalase activity was scored after a period of recovery (mean of four replicates) (unpublished data)

Comparative Responses Involving Water and Salt Stresses

Regarding water relations, both water and salt stresses reduced the water potential, but even in these conditions, a good degree of leaf hydration was maintained. Plants submitted to salt and drought excesses showed significant reductions of gas exchange in their leaves due to stomatal closure and biochemical alterations, as evidenced by the disturbance in the electron transport chain. However, some mechanisms of excess energy dissipation were activated under these environmental conditions.

Concerning the ability of *J. curcas* to overcome salt and water stresses by OA, we found that under mild to high salinity (50–100 mM NaCl), the largest relative contribution to osmotic potential was from Na^+, Cl^-, amino acids and GB. In contrast, under increasing water deficit, the levels of K^+, soluble sugar and GB showed the largest relative contribution to OA. Interestingly, under both stressful conditions, the high level of GB in the leaf and root tissues of *J. curcas* seems to be constitutive; i.e., its high concentration is independent of the adverse environmental conditions.

Although increased membrane damage was observed, as indicated by the increase in the lipid peroxidation and electrolyte leakage in *Jatropha* leaves exposed to both salt and drought stresses, the ROS-scavenging system, represented by SOD-APX-CAT enzymes, was relatively efficient to mitigate oxidative damage. Interestingly, in contrast to drought, salinity stimulated both CAT and APX activities. As previously mentioned, the question of H_2O_2 scavenging produced during photorespiration under drought conditions requires further elucidation in the *Jatropha* leaves (unpublished data).

Conclusion and Perspectives

Salinity

The determination of the degree of stress tolerance of a crop is difficult, especially in a woody plant like *J. curcas*. For example, there are several approaches and criteria for the classification of salt tolerance of a species or cultivar to salinity. In general, salt tolerance is reported as the electrical conductivity of the soil corresponding to a 50% reduction of biomass or yield. A mixture of salts like NaCl, $MgCl_2$, $CaCl_2$ and Na_2SO_4, corresponding to the native composition of soil salts should be utilized in salinity experiments than just NaCl because they are better to mimic the real conditions and consequently the plant responses to salinity. *J. curcas* is considered moderately salt sensitive because an increase of salt concentration corresponding to 47 mM or approximately 4.7 dS m^{-1} cause a 50% reduction of total dry matter in seedlings. In addition, *J. curcas* exhibits a strong affinity for Na+ion (a salt includer characteristic) and this saline ion effectively competes with K^+. Unfortunately, studies to determine the degree of resistance of this species under field conditions at the productive phase or even in the mature vegetative stage have not been reported.

In tropical semi-arid regions of India, Africa and Brazil, where the cultivation of *J. curcas* is increasing, problems of primary salinity and secondary salinization caused by irrigation can be critical. In these regions, the expansion of this new crop will inevitably aggravate the problem of soil salinization as a consequence of irrigation. To avoid future problems, preventive care, such as drainage, water and soil management, will have to be taken. However, it is important to encourage selective breeding for tolerance to salt stress if *J. curcas* participates in the energy matrix in the near future. Unfortunately, such politics are difficult to implement in those countries and perhaps worldwide. It seems clear that this process should begin with the classical genetic approach. Molecular marker-assisted selection should boost this process; however, the selective process will only be possible if the potential, limits and parameters of this new crop are known. Thus, further investigations in molecular genetics will be essential to effectively characterize the genes involved in salt resistance.

Jatropha producing countries will also have to promote research by multidisciplinary and decentralized groups in (i) genetics and breeding, (ii) biochemistry and physiology and (iii) omics (i.e., genomics, transcriptomics, proteomics and metabolomics). These multidisciplinary studies might elucidate mechanisms of tolerance to salinity. Therefore, there are still significant challenges ahead until we can describe the mechanisms of tolerance to salinity and obtain cultivars of *J. curcas* tolerant to this factor. Unfortunately, this tolerance to salinity has not even been reached in key crops. Thus, specifically in the case of *J. curcas*, it is essential to select salt-resistant cultivars and/or lines and explain the genetic, biochemical and physiological foundations of mechanisms involved with productivity and stress tolerance to exchange them with productive cultivars. The data obtained in our studies suggest that it would be interesting to obtain cultivars with fewer *salt includer* mechanism through the increased expression of genes involved with increased exclusion and/or decreased transport of Na^+; high ion storage within the vacuoles of the roots, stems and leaves; high recycling among leaves and roots by phloem and high K^+/Na^+ ratio in leaves.

Drought Tolerance

The geographical distribution of *J. curcas* strongly suggests that this species is drought tolerant. However, the features of the physiological parameters of drought tolerance were based, until now, on a narrow genetic basis that was not well characterized. Furthermore, most investigations have concentrated on individuals grown in the greenhouse, primarily at the vegetative stage. In contrast to salinity, when the increased sensitivity of plants occurs at the vegetative phase, several experiments have provided evidence that increased crop sensitivity to water stress occurs mainly at the reproductive stage.

Few observations under field conditions in the semiarid regions of Brazil have provided evidence that *J. curcas* display complex ecophysiological mechanisms associated with development and productivity in response to water availability.

Under non-irrigated conditions, this species maintains its leaves with high photosynthesis and adequate metabolic activity for perhaps 4 months. When the dry period progresses, the leaves undergo progressive metabolic activity loss (e.g., decrease in the levels of chlorophyll). In general, this period coincides with a large increase in the incidence of pests and diseases. It is unclear if this process is associated with a reduction in plant defense mechanisms associated with water restriction and/or if it is controlled by hormones as a part of the physiology of the entire plant.

In fact, the possibility of the sustainable production of *Jatropha* without irrigation in dry and hot regions, including semiarid regions, is controversial, partly because the physiology of this species is not yet sufficiently known and the plant breeding has presented little progress in countries with dry weather conditions. However, the development of "evergreen crops" with irrigation also requires the development of physiological studies and genotypes that respond adequately to water investment under adverse conditions of high temperature, high light and poor soils.

Some specific recommendations can be made to improve the performance of *J. curcas* under dry conditions, i.e., lower branching by pruning or selective breeding for small plants and increasing the depth and branching of roots because the root system of this species is not well developed. Concerning the most important genes and processes that could be selected, the up-regulation of catalase expression and activity under drought conditions associated with a reduction in photorespiration could be a good option. It would be also important to promote studies on drought stress in the presence of factors involved in other abiotic stresses, such as high temperature and low air humidity.

Because osmotic adjustment is a crucial slow process for drought tolerance and *J. curcas* produces high endogenous levels of glycinebetaine, it is also important to promote the breeding of cultivars that produce high concentrations of this solute. The same line could be followed to increase the concentrations of sucrose, fructose and glucose, as we found that these chemicals improve drought tolerance. The accumulation of these solutes in the meristematic tissues is essential, especially in the roots, which support continued growth under adverse conditions.

Acknowledgements The authors would like to thank the Fundação Cearense de Apoio ao Desenvolvimento Científico e Tecnológico (FUNCAP) and Fundação Cearence de Pesquisa e Cultura (FCPC) for financial support (Project 2155/Programa Núcleos de Excelência, PRONEX) and the Fazenda Tamanduá for supplying the *Jatropha curcas* seeds.

References

Alves AAC, Setter TL (2004) Abscisic acid accumulation and osmotic adjustment in cassava under water deficit. Environ Exp Bot 51:259–271

Apse MP, Blumwald B (2007) Na^+ transport in plants. FEBS Lett 58:2247–2254

Asada K (1999) The water-water cycle in chloroplasts: Scavenging of active oxygens and dissipation of excess photons. Annu Rev Plant Physiol Plant Mol Biol 50:601–639

Bajji M, Lutts S, Kinet J-M (2001) Water deficit effects on solute contribution to osmotic adjustment as a function of leaf ageing in three durum wheat (*Triticum durum* Desf.) cultivars performing differently in arid conditions. Plant Sci 160:669–681
Britto DT, Ebrahimi-Ardebili S, Hamam AM, Coskun D, Kronzucker HJ (2010) ^{42}K analysis of sodium-induced potassium efflux in barley: mechanism and relevance to salt tolerance. New Phytol 186:373–384
Carden DE, Walker DJ, Flowers TJ, Miller AJ (2003) Single-cell measurements of the contributions of cytosolic Na^+ and K^+ to salt tolerance. Plant Physiol 131:676–683
Cattivelli L, Rizza F, Badeck FW, Mazzucotelli E, Mastrangelo AM, Francia E et al (2008) Drought tolerance improvement in crop plants: an integrated view from breeding to genomics. Field Crops Res 105:1–14
Cavalcanti FR, Oliveira JTA, Martins-Miranda AS, Viegas RA, Silveira JAG (2004) Superoxide dismutase, catalase and peroxidase activities do not confer protection against oxidative damage in salt-stressed cowpea leaves. New Phytol 163:563–571
Cavalcanti FR, Lima JPMS, Ferreira-Silva SL, Viégas RA, Silveira JAG (2007) Roots and leaves display contrasting oxidative response during salt stress and recovery in cowpea. J Plant Physiol 164:591–600
Chagas RM, Silveira JAG, Ribeiro RV, Vitorello VA, Carrer H (2008) Photochemical damage and comparative performance of superoxide dismutase and ascorbate peroxidase in sugarcane leaves exposure to paraquat-induced oxidative stress. Pest Biochem Physiol 90:181–188
Chaves MM, Flexas J, Pinheiro C (2008) Photosynthesis under drought and salt stress: regulation mechanisms from whole plant to cell. Ann Bot 125:1–10
Cuin TA, Betts SA, Chalmandrier R, Shabala S (2008) A root's ability to retain K^+ correlates with salt tolerance in wheat. J Exp Bot 59:2697–2706
Dasgan HY, Aktas H, Abak K, Cakmak I (2002) Determination of screening techniques to salinity tolerance in tomatoes and investigation of genotype responses. Plant Sci 163:695–703
Díaz-López L, Gimeno V, Simón I, Martínez V, Rodríguez-Ortega WM, García-Sanchez F (2012) *Jatropha curcas* seedlings show a water conservation strategy under drought conditions based on decreasing leaf growth and stomatal conductance. Agric Water Manag 105:48–56. doi:10.1016/j.agwat.2012.01.001
Dreyer I, Blatt MR (2009) What makes a gate? The ins and outs of Kv-like K^+ channels in plants. Trends Plant Sci 14:383–390
Drodzova IS, Pustovoitova TN, Dzhibladze TG, Barabanshchikova NS, Zhdanova NE, Maevskaya SN et al (2004) Endogenous control of photosynthetic activity during progressive drought: influence of final products of photosynthesis. Rus J Plant Physiol 51:668–675
Eswaran N, Parameswaran S, Sathram B, Anantharaman B, Kumar GRK, Tangirala SJ (2010) Yeast functional screen to identify genetic determinants capable of conferring abiotic stress tolerance in *Jatropha curcas*. BMC Biotechnol 10:23
Ferreira-Silva SL, Voigt EL, Silva EN, Maia JM, Fontenele AV, Silveira JAG (2011) High temperature positively modulates oxidative protection in salt stressed cashew plants. Environ Exp Bot 74:162–170
Flexas J, Bota J, Loreto F, Cornic G, Sharkey TD (2004) Diffusive and metabolic limitations to photosynthesis under drought and salinity in C3 plants. Plant Biol 6:269–279
Flexas J, Ribas-Carbó M, Bota J, Galmés J, Henkle M, Martinez-Canellas S et al (2006) Decreased Rubisco activity during water stress is not induced by decreased relative water content but related to condition of low stomatal conductance and chloroplast CO_2 concentration. New Phytol 172:73–82
Flexas J, Diaz-Espejo A, Galmés J, Kaldenhoff R, Medrano H, Ribas-Carbo M (2007) Rapid variations of mesophyll conductance in response to changes in CO_2 concentration around leaves. Plant Cell Environ 30:1284–1298
Flowers TJ (2004) Improving crop salt tolerance. J Exp Bot 55:307–319
Flowers TJ, Colmer TD (2008) Salinity tolerance in halophytes. New Phytol 179:945–963
Foyer CH, Bloom AJ, Queval G, Noctor G (2009) Photorespiratory metabolism: genes, mutants, energetics and redox signaling. Annu Rev Plant Biol 60:455–484

Francis G, Edinger R, Becker K (2005) A concept for simultaneous wasteland reclamation, fuel production, and socioeconomic development in degraded areas in India. Need, potential and perspectives of Jatropha plantations. Nat Res Forum 29:12–24

Garrity DP, O'Toole JC (1994) Screening rice for drought resistance at the reproductive phase. Field Crops Res 39:99–110

Gimeno V, Syvertsen JP, Simón I, Nieves M, Díaz-López L, Martínez V, García-Sánchez F (2012) Physiological and morphological responses to flooding with fresh or saline water in *Jatropha curcas*. Environ Exp Bot 78:47–55

Guerfel M, Ouni Y, Boujnah D, Zarrouk M (2009) Photosynthesis parameters and activities of enzymes of oxidative stress in two young 'Chemlali' and 'Chetoui' olive trees under water deficit. Photosynthetica 47:340–346

Guo YP, Zhou HF, Zhang LC (2006) Photosynthetic characteristics and protective mechanisms against photooxidation during high temperature stress in two citrus species. Sci Hort 108:260–267

Hessine K, Martínez JP, Gandour M, Albouchi A, Soltani A, Abdelly C (2009) Effect of water stress on growth, osmotic adjustment, cell wall elasticity and water-use efficiency in *Spartina alterniflora*. Environ Exp Bot 67:312–319

Iannucci A, Russo M, Arena L, Di Fonzo N, Martiniello M (2002) Water deficit effects on osmotic adjustment and solute accumulation in leaves of annual clovers. Eur J Agron 16:111–122

Kameli A, Losel DM (1995) Contribution of carbohydrates and solutes to osmotic adjustment in wheat leaves under water stress. J Plant Physiol 145:363–366

Kumar N, Sudheer DVN, Pamidimarri MK, Boricha G, Muppala PR (2008) Effects of NaCl on growth, ion accumulation, protein, proline contents and antioxidant enzymes activity in callus cultures of *Jatropha curcas*. Biologia 63:378–382

Lawlor DW, Cornic G (2002) Photosynthetic carbon assimilation and associated metabolism in relation to water deficits in higher plants. Plant Cell Environ 25:275–294

López-Climent MF, Arbona V, Pérez-Clemente RM, Gómez-Cadenas A (2008) Relationship between salt tolerance and photosynthetic machinery performance in citrus. Environ Exp Bot 62:176–184

Maes WH, Trabucco A, Achten WMJ, Muys B (2009a) Climatic growing conditions of *Jatropha curcas* L. Biomass Bioenergy 33:1481–1485

Maes WH, Achten WMJ, Reubens B, Raes D, Samson R, Muys B (2009b) Plant–water relationships and growth strategies of *Jatropha curcas* L. seedlings under different levels of drought stress. J Arid Environ 73:877–884

Manivannan P, Jaleel AC, Kishorekumar A, Sankar B, Somasundaram R, Sridharan R et al (2007) Changes in antioxidant metabolism of *Vigna unguiculata* (L.) Walp. By propiconazole under water deficit stress. Colloids Surf B Biointerfaces 57:69–74

Martinez JP, Kinet JM, Bajji M, Lutts S (2005) NaCl alleviates polyethylene glycolinduced water stress in the halophyte species *Atriplex halimus* L. J Exp Bot 56:2421–2431

Miller G, Suzuki N, Ciftci-Yilmaz S, Mittler R (2010) Reactive oxygen species homeostasis and signalling during drought and salinity stresses. Plant Cell Environ 33:453–467

Munns R, Tester M (2008) Mechanisms of salinity tolerance. Annu Rev Plant Biol 59:651–681

Nery AR, Rodrigues LN, Silva MBR, Fernandes PD, Chaves LHG, Neto JD, Ghey HR (2009) Growth of Jatropha irrigated with saline water in greenhouse. Rev Bras Eng Agric Ambient 13:551–558

Netondo GW, Onyango JC, Beck E (2004) Sorghum and salinity: II. Gas exchange and chlorophyll fluorescence of sorghum under salt stress. Crop Sci 44:806–811

Palatnik JF, Valle EM, Federico ML, Gómez LD, Melchiorre MN, Paleo AD et al (2002) Status of antioxidant metabolites and enzymes in a catalase-deficient mutant of barley (*Hordeum vulgare* L.). Plant Sci 162:363–371

Peeva V, Cornic G (2009) Leaf photosynthesis of *Haberlea rhodopensis* before and during drought. Environ Exp Bot 65:310–318

Pompelli MF, Barata-Luís R, Vitorino HS, Gonçalves ER, Rolim EV, Santos MG et al (2010) Photosynthesis, photoprotection and antioxidant activity of purging nut under drought deficit and recovery. Biomass Bioenergy 34:1207–1215

Rodrigues CRF, Silva EN, Dutra ATB, Viégas RA, Silveira JAG (2012) Transport and partitioning of K^+ alleviates toxic effects of Na^+ ions in *Jatropha curcas* young plants. Rev Bras Ciên Solo. in press

Shabala S, Cuin TA (2007) Potassium transport and plant salt tolerance. Physiol Plant 133:651–669

Sharma P, Dubey RS (2005) Drought induces oxidative stress and enhances the activities of antioxidant enzymes in growing rice seedlings. Plant Growth Reg 46:209–221

Shi H, Quintero FJ, Pardo JM, Zhu JK (2002) The putative plasma membrane Na^+/H^+ antiporter SOS1 controls long-distance Na^+ transport in plants. Plant Cell 14:465–477

Silva EN, Silveira JAG, Fernandes CRR, Dutra ATB, Aragão RM (2009a) Ion uptake and growth of *Jatropha* under different salinity levels. Rev Ciên Agron 40:240–246

Silva EN, Silveira JAG, Rodrigues CRF, Lima CS, Viégas RA (2009b) Contribution of organic and inorganic solutes to osmotic adjustment of physic nut under salinity. Pesq Agric Bras 44:437–445

Silva EN, Ferreira-Silva SL, Fontenele AV, Viégas RA, Silveira JAG (2010a) Photosynthetic changes and protective mechanisms against oxidative damage subjected to isolated and combined drought and heat stresses in *Jatropha curcas* plants. J Plant Physiol 167:1157–1164

Silva EN, Ferreira-Silva SL, Viégas RA, Silveira JAG (2010b) The role of organic and inorganic solutes in the osmotic adjustment of drought-stressed *Jatropha curcas* plants. Environ Exp Bot 69:279–285

Silva EN, Ribeiro RV, Ferreira-Silva SL, Viégas RA, Silveira JAG (2010c) Comparative effects of salinity and water stress on photosynthesis, water relations and growth of Jatropha curcas plants. J Arid Environ 74:1130–1137

Silva EN, Ribeiro RV, Ferreira-Silva SL, Viégas RA, Silveira JAG (2011) Salt stress induced damages on the photosynthesis of physic nut young plants. Sci Agric 68:62–68

Silveira JAG, Viegas RA, Rocha IMA, Moreira ACDM, Moreira RA, Oliveira JTA (2003) Proline accumulation and glutamine synthetase activity are increased by salt-induced proteolysis in cashew leaves. J Plant Physiol 160:115–123

Silveira JAG, Araújo SAM, Lima JPMS, Viégas RA (2009) Roots and leaves display contrasting osmotic adjustment mechanisms in response to NaCl-salinity in *Atriplex numularia*. Environ Exp Bot 66:1–8

Souza RP, Machado EC, Silva JAB, Lagoa AMMA, Silveira JA (2004) Photosynthetic gas exchange, chlorophyll fluorescence and some associated metabolic changes in cowpea (*Vigna unguiculata*) during water stress and recovery. Environ Exp Bot 51:45–56

Szczerba MW, Britto DT, Kronzucker HJ (2009) K^+ transport in plants: physiology and molecular biology. J Plant Physiol 166:447–466

Tang M, Liu X, Deng H, Shen S (2011) Overexpression of *JcDREB*, a putative AP2/EREBP domain containing transcription factor gene in woody biodiesel plant *Jatropha curcas*, enhances salt and freezing tolerance in transgenic *Arabidopsis thaliana*. Plant Sci 181:623–631

Tuberosa R, Salvi S (2006) Genomics-based approaches to improve drought tolerance of crops. Trends Plant Sci 11:405–412

Veras RP, Laime EMO, Fernandes PD, Soares FAL, Freire EA (2010) Plant height, stem diameter and production of *Jatropha* irrigated under different salinity levels. Rev Bras Eng Agric Ambient 15:582–587

Voigt EL, Caitano RF, Maia JM, Ferreira-Silva SL, Macêdo CEC, Silveira JAG (2009) Involvement of cation channels and NH_4^+-sensitive K^+ transporters in Na^+ uptake by cowpea roots under salinity. Bio Plant 53:764–768

Whang LW, Showalter AM (2004) Cloning and salt-induced ABA independent expression of choline mono-oxygenase in *Atriplex prostata*. Physiol Plant 120:405–412

Yeo A (1998) Molecular biology of salt tolerance in the context of whole-plant physiology. J Exp Bot 49:915–929

Yin C, Peng Y, Zang R, Zhu Y, Li C (2005) Adaptive responses of *Populus kangdingensis* to drought stress. Physiol Plant 123:445–451

Chapter 8
Role of Microbial Inoculants on Growth and Development of *Jatropha curcas* L.

Jamaluddin

Introduction

Jatropha curcas L. the biofuel plant has attracted the attention of agriculturist and forester. It has the potential to become the world's ecofriendly key energy crops since the biodiesel prepared form its oil has proper combustion and low emission. Large scale trials are being made in nursery and fields, but no significant information have been documented on the role of microbial inoculants and their interaction with inorganic fertilizers. Soils of tropical regions are mainly low in available nutrients and moisture. The symbiotic association between *arbuscular mycorrhizae fungi* (AMF) and plant roots is widespread in nature (Safir 1994). Positive effects of AMF inoculation on agricultural cropping systems have been investigated and include improved uptake of phosphorus and other nutrients besides, increased stress tolerance and beneficial plant growth regulation. Plant roots release a wide range of compounds, which are involved in complex communication processes in the rhizosphere. Phosphorus solubilizing bacteria are known to play a major role in the solubilization of unavailable forms of soil phosphorus and uptake in both native and applied forms and because of this potential it has been extensively used for enhancing plant productivity. Therefore, looking at the enormous potential and use of microbes the following studies were undertaken.

1. Effect of root exudates on mycorrhization of AMF.
2. Effect of rhizospheric mycoflora on growth and development of *J. curcas* in the field.

Jamaluddin (✉)
Department of Biological Science, R.D. University, Jabalpur 482 001, India
e-mail: jamaluddin_125@hotmail.com

N. Carels et al. (eds.), *Jatropha, Challenges for a New Energy Crop: Volume 1: Farming, Economics and Biofuel*, DOI 10.1007/978-1-4614-4806-8_8,

3. Effect of plant-microbe interaction on growth and development of *J. curcas* by application of microbes in nursery trials.
4. Studies on phosphatase enzyme activity, total leaf phosphorus, proteins (root and leaf) of different microflora.

Effect of Root Exudates on AMF Colonization

Surface sterilized seeds of *J. curcas* were germinated and grown in sterilized mixture of soil, sand and peat in equal proportion for 21 days at 25°C with 12–12 h (day-night) photoperiod. Then, roots from uprooted plants were placed in 150 ml sterilized substrate for 24 h to collect root exudates. The filtered exudates were stored at −20°C until use.

Seedlings of *J. curcas* were inoculated with spores of AMF, i.e., *Glomus mosseae, Gigaspora sp.* and *Acaulospora sp.* trapped in maize roots. After 6 week growth, seedlings were separated in 10 replicates and watered with 100 ml of root exudates to each plant. The controls of this experiment were (1) seedlings inoculated with AMF alone and (2) seedling watered with root exudates alone. Height weeks after treatment, height, collar diameter, total leaf phosphorus and total protein were estimated as described by Jackson (1973) and Lowry et al. (1952). The soil phosphatase enzyme activity was measured by spectrophotometry at 420 nm as described by Tabatbai and Bremmer (1969) and Singh and Jamaluddin (2008).

Determination of Root Colonization

To visualize roots colonized with AMF, roots were cut into 0.5 cm fragments and boiled in 10% KOH at 90°C for half an hour, rinsed and acidified with 1% HCl. Following acidification, roots were stained with 0.05% trypan blue in lactophenol, boiled for 5 min in a water bath and destained with lactic acid and glycerol (1:1). Squashed roots were observed under microscope and root colonization (%) was recorded following the formula:

$$\text{Root colonization}(\%) = \frac{\text{Number of root segments colonized}}{\text{Total number of roots examined}} \times 100$$

The relative root colonization (%) by AMF in *J. curcas* treated with root exudate was 80% compared to 75% and 70% for the control, i.e., plants treated with AMF or root exudates alone (Table 8.2). Relative colonization in *J. curcas* was still low after 6 weeks, but increased after a longer incubation. Significant variation in phosphatase enzyme activity and more of alkaline phosphatase secreted by AMF than acidic phosphatase. Maximum activity was recorded with plant root exudates

Table 8.1 Treatment details of microbes and VAM

No.	Treatments	Per plant
1.	*Pseudomonas fluorescens*	100 ml
2.	*Azospirillum* species	100 ml
3.	Single Super Phosphate	10 g
4.	VAM	100 g
5.	Mixed Treatment (*Pseudomonas* + *Azospirillum* + VAM)	as above
6.	Control	—

followed by AMF and root exudates. Amongst AMF population the *Glomus mosseae, Acaulospora* sp. and *Gigaspora* sp. were observed to be the prominent species. Increased AMF colonization may be due to the flavonoids present in root exudates, which play an important role at the beginning of a signal exchange between the host plant and AMF. The higher level of alkaline phosphatase activity indicates the formation of vacuoles due to the AMF, which appears with the development of hyphae and becomes intense in mature arbuscular and intercellular hyphae (Singh and Jamaluddin 2008). By contrary, Ho and Zak (1979) reported maximum acid phosphatase activity in roots colonized by AMF. Singh and Jamaluddin (2008) also supported the above notion that there is a direct correlation between root fungi and acid phosphatase activity. Enzyme activities varied considerably in the treatments studied. The growth measured in terms of plant height and collar diameter also varied significantly amongst all the treatments studied. A positive correlation was found between growth and enzyme activity, total phosphorus content and protein content amongst all the treatments. Maximum leaf protein was recorded in plants treated with root exudates followed by those treated by AMF or root exudates alone. Any factor influencing the soil microorganisms will indirectly affect soil enzymes activities (Mc Claugherty and Linkins 1990). AMF interact with a wide range of other soil microorganisms in the root, rhizosphere, mycorhizosphere and in the bulk soil. These interactions with specific functional capabilities may influence plant growth (Linderman 1988; Fitter and Garbye 1994; Liang 1994). From the current work it can be concluded that root exudates positively affect the AMF colonization in *J. curcas* and therefore can be used to enhance the colonization which in turn will improve the growth and development of *J. curcas*.

Detection of AM Fungi Using PCR

DNA extraction: Colonized and non colonized root samples (8–10 mg) were crushed by boiling each sample in 800 ml of extraction buffer (1 M Tris–HCl, pH 8.5). The sample was then centrifuged at 12,000 rpm for 10 min before processing for PCR. PCR reaction mix contained 50 pmol of each primer (VANS1 and NS2), 200 μM dNTPs, 20 mM Tris–HCl 100 mM KCl, 0.02% gelatin and 1.25 U *Taq*

Table 8.2 Effect of root exudates on AMF colonization in *J. curcas*

Treatment	Height (cm)	Collar diameter (cm)	Enzyme activity µg/PNP/g soil/h)		Phosphorus (%)	Total leaf protein (µg/ml)	Root colonization (%)
			Alkaline	Acidic			
Root exudates	30±0.002	2.5±0.02	1.5±0.02	0.85±0.001	1.75±0.01	12.6±0.02	80±0.015
AMF+water	25±0.12	2.3±0.01	0.8±0.12	0.64±0.04	1.25±0.002	9.5±0.01	76.4±0.001
Root exudates	22.5±0.005	1.5±0.01	0.5±0.04	0.52±0.002	0.94±0.001	8.2±0.05	70.3±0.004

Values given in the table are mean ± SEm of all the ten replicates, Incubation period; 3 months; Amount of root exudates used in in vivo studies = 30 ml/plant

DNA polymerase. Forty cycles of amplification was carried out in a thermocycler with denaturation at 90°C for 60 s, annealing at 50°C for 45 s, extension at 72°C for 60 s and final extension at 72°C for 10 min. Amplified products were analyzed by electrophoresis on 3% agarose gel and visualized by UV transilluminator after being stained with ethidium bromide.

A rDNA amplification product of approximately 550 bp was detected in colonized roots and supposed to be from the AMF *Glomus intraradices*. This AMF was detected after 1 month of infection. This product was never detected on control plant roots. However, non specific products were amplified as well in roots from the treated plants including those from the control. Thus, the primers we used in this study are not taxon specific since they amplified non-mycorrhizal DNA as well.

Effect of Different Bioinoculants on Growth of J. curcas *and the Status of Fungal Species in Rhizosphere*

Isolation, Preservation and Identification of Fungi

The soil samples collected from the field were analyzed for numbers and kinds of fungi by a standardized soil dilution plate technique (Waksman 1922; Warcup 1952). The preservation of pure cultures of fungal isolates was done on potato dextrose agar slants in test tubes as stock cultures in the refrigerator at 5–7°C. The identification was done by characterizing each fungus in pure culture and making reference with relevant literature.

The species profile of mycoflora showed significant diversity among treated plants and untreated plants. Fungi belonging to 12 different genera were found, namely *Aspergillus, Emrecilla, Mucor, Rhizopus, Curvularia pallescens, Colletotrichum, Trichoderma viride, Fusarium oxysporum, Biospora, Sclerotium, Cladosporium*. Sterile mycelium could be detected from AMF and mixed inoculants. However, these were not observed in all treatments. The relative abundance of these genera varied among treatments. However, the isolated mycoflora have a significant impact on plant growth as shown by data of Table 8.3.

Growth Enzymatic Activity and Nutrient Status of Plants

The growth level reached by treated and untreated plants was measured after one year in terms of height and collar diameter. The phosphatase activity of all bioinoculants was also measured as described by Tabatbai and Bremner (1969). The nutrient status of the plants was assessed in terms of leaf phosphorus content (P%), which was estimated by the vandomolybdate phosphoric yellow color method as reported by Jackson (1973). The total protein in leaf and root was estimated as described by Lowry et al. (1952) and by SDS-PAGE. Data were analysed statistically using ANOVA and differences

Table 8.3 Effect of different bioinoculants and mycoflora on growth and nutrient status of *J. curcas* in field

Treatments	Height (cm)	Collar diameter (cm)	Leaf phosphorus (%)	Root protein (μg/ml)	Leaf protein (μg/ml)	Phosphatase activity (μg PNP/g soil/h)		Fungi recorded
						Alkaline	Acidic	
Pseudomonas fluorescens (PSB)	83.9	8.5	0.09	7.0	6.2	1.0	1.8	*Penicillium sp., Sclerotium, Fusarium, Mucor, Aspergillus flavus, Emerciella, Cladosporium sp.*
Azospirillum sp.	86.4	8.2	0.08	7.5	7.3	1.6	1.35	*Aspergillus sp., Penicillium, Curvularia, Aspergillus niger*
Inorganic phosphate	110.6	10.1	1.06	4.2	5.6	2.1	2.0	*Aspergillus sp., Penicillium, Colletotrichum gloeosporioides, Emericella*
AMF	90.5	9.8	0.2	7.8	7.5	2.8	2.5	*Aspergillus sp., Penicillium, Trichoderma viride, Fusarium, Mucor, Mycelia sterilia*
Mixed culture	96.7	10.7	0.25	6.0	5.9	1.4	1.5	*Aspergillus sp., Penicillium, Trichoderma viride, Mucor, Rhizopus, Mycelia sterilia*
Control	78.5	7.5	0.48	5.5	5.2	0.5	1.2	*Biospora sp., Aspergillus sp., Trichoderma sp., Curvularia sp.*
LSD	13.8	3.5	0.25	3.23	1.83	0.75	0.09	

Values given in the table are mean ± SEm of ten replicates; Amount of inoculum used in in vivo studies: *Azospirillum* and *Pseudomonas* culture: 100 ml/plant; AMF: 1000 g/plant; organic phosphate: 10 g/plant between treatments were decided significant or not by reference to $p<0.05$. AMF were maintained on trap plant (maize or *Panicum maximum*).

Mycoflora exhibited significant impact on the growth and development of *J. curcas*. Remarkable differences of plant growth, measured in terms of plant height and collar diameter, were recorded amongst treatments and controls. Plants fertilized with inorganic phosphate attained a maximum height up to 150 cm and collar diameter up to 10 cm and were followed by those inoculated by AMF, mix culture, *Azospirillum* and *phosphate solubilizing bacteria* (PSB). Significant variations of alkaline and acidic phosphatase activities among inoculants were also observed and the alkaline phosphatase activity was larger than the acidic one. The maximum activity was obtained with inorganic phosphate followed by AMF and mixed culture. PSB also showed significant enzymatic activity. In terms of plant nutrient status, the maximum phosphorus content was obtained with plants fertilized with inorganic phosphate followed by those inoculated with mixed treatment and AMF. However, inoculants failed to induce a qualitative variation in their root and leaf protein content and the same situation has been also detected on acrylamide gel, i.e., all the inoculants shared a common banding pattern, but the band intensities varied from one treatment to another.

Effect of Plant-Microbe Interaction on the Growth of *J. curcas*

Thirty days after germination, *J. curcas* seedlings were treated with 30 ml of spore suspension of *Trichoderma* sp., *Aspergillus* sp. and *Fusarium* sp. and 30 g of soil containing AMF inoculum separately and in combination as well. Plant height and collar diameter was recorded after one month of treatment. Nutrient analysis, root staining and enzymatic studies were also performed as described above.

Interaction of AMF with non-mycorrhizal fungi like *Fusarium* sp., *Trichoderma* sp., and *Aspergillus* sp. exhibited significant impact on the growth and development of *J. curcas*. Maximum effect was recorded from mixed inoculum, which showed significant results in all the parameters investigated (Table 8.4). Isolates secreted more of alkaline phosphates than acidic phosphatases which reflects the preference of AMF for alkalinity. Non mycorrhizal fungi are also playing significant role in supplying the required phosphates to plants. The phosphorus solubilizing property of these fungal agents has also been reported by many other workers (Linderman 1988; Fitter and Garbye 1994; Liang 1994). It is now clear that the microbial interactions with specific plant functions stimulate plant growth.

Field Trial

Two plots were laid out in a random block design with six treatments and ten replicates per treatment (Table 8.1). Five treatments contained AMF (100 g/plant), *Pseudomonas* sp. (100 ml/plant), *Azospirillum* (100 ml/plant), superphosphate (10 g/plant), mixed culture (containing AMF, *Pseudomonas* sp. and *Azospirillum* sp.)

Table 8.4 Effect of plant microbe interactions on growth and nutrient status of *J. curcas* in nursery after three months

Treatments	Height (cm)	Collar diameter (cm)	Leaf Phosphorus (%)	Leaf Protein (μg/ml)	Phosphatase activity (μg PNP/g soil/h)	
					Alkaline	Acidic
Trichoderma sp. (T)	14.7	8.5	0.09	4.2	1.0	1.2
Aspergillus sp. (A)	16.4	8.2	0.08	5.3	1.6	1.5
Fusarium sp. (F)	13.6	10.1	0.35	4.8	2.1	2.0
AMF	21.5	9.8	0.2	6.2	2.8	2.5
AMF+T	18.8	5.6	0.5	5.9	3.2	1.7
AMF+A	22.6	4.7	0.52	6.8	2.3	1.23
AMF+F	24.3	5.9	0.64	6.4	1.8	1.6
Mixed culture	29.7	10.7	0.85	7.9	1.4	1.8
Control	12.5	7.5	0.48	5.2	0.5	1.2
LSD	2.35	1.3	0.2	0.6	0.42	0.33

Table 8.5 Effect of different microbial inoculants on growth and development of *J. curcas*

No.	Treatment	For the first year		For the second year	
		Height (cm)	Collar diameter (cm)	Height (cm)	Collar diameter (cm)
1	*Pseudomonas fluorescens*	75.3	6.8	208.9	18.9
2	*Azospirillum* sp.	77.0	6.25	205.0	19.0
3	Single Super Phosphate	90.1	7.1	211.4	19.5
4	VAM	84.7	6.85	199.7	19.7
5	Mixed Treatment (*Pseudomonas* + *Azospirillum* + VAM)	82.2	7.07	210.8	19.8
6	Control	75.6	5.5	175.3	18.3

Uninoculated plants served as control. A ten-fold diluted culture of bacteria was used. Growth and development of the plant was recorded in terms of collar diameter after 6 months and 2 years of treatment. The height of the plants was also observed.

In the experiment laid out in August 2008 in the field according to treatments of Table 8.5. It is evident that inorganic phosphate (Single Super Phosphate) was suitable to promote *J. curcas* growth in height and collar diameter. Comparatively, the mixed treatment of bioinoculants (*Pseudomonas* + *Azospirillum* + AMF) was also found suitable to increase the level of growth and development of *J. curcas* plants. The results indicated that the differences of height and collar diameter were only negligible compared to those obtained with inorganic phosphate fertilization. The other microbial treatments were also effective in increasing the growth and collar diameter of plants during the two years of evaluation. The untreated plants (control) showed lower growth and collar diameter when compared with treated plants. Thus, the treatment with microbial mixtures was effective for increasing the growth and development of *J. curcas* plants in the field. Another experiment was carried out during August 2009 by using three microbial inoculants including *Pseudomonas*

Table 8.6 Effect of different microbial inoculants on growth and development of *J. curcas*

No.	Treatment	Height (cm)	Collar diameter (cm)
1.	*Pseudomonas fluorescens*	115.3	9.8
2.	*Azospirillum* sp.	105.7	8.63
4.	AMF	122.8	11.3
5.	Mix Microbes (*Pseudomonas* + *Azospirillum* + VAM)	125.0	10.8
6.	Control	88.2	7.1

fluorescens, Azospirillum and AMF. They were given individually and also as a mixture and the control was without microbial inoculants. After 18 months of treatment the height and collar diameter of plants were recorded. It was observed that the bioinoculants gave significant result in increasing the height and collar diameter of *J. curcas* plants. Comparatively, control (untreated) plants showed a lower level of growth. The best treatment observed was mixed inoculants followed by AMF treatment (Table 8.6).

References

Fitter AH, Garbye J (1994) Interaction between mycorrhizal fungi and other soil organisms. Plant Soil 159:123–132

Ho I, Zak B (1979) Acid phosphatase activity of six ectomycorrhizal fungi. Can J Bot 79:1203–1205

Jackson ML (1973) Soil Chemical Analysis. Prentice Hall of India, New Delhi

Liang SF (1994) Function and application of dissolving phosphorus microbial fertilizer. Soil Fertil 2:46–48

Linderman RG (1988) Mycorrhizal interactions with the rhizosphere microflora: the mycorrhizosphere effect. Phytopathology 78:366–371

Lowry OH, Rosebrough NJ, Farr AL (1952) Protein measuring with folin phenol regent. J Biol Chem 193:265–275

Mc Claugherty CA, Linkins AE (1990) Temperature response of enzymes in two forest soils. Soil Biol Biochem 22:29–33

Safir GR (1994) Involvement of cropping systems plants produced compounds and inoculum production in the functioning of VAM fungi. In: Pfilger FL, Lindermann RG (eds) Mycorrhiza and plant health, 9th ed. APS Press, St. Paul

Singh AK, Jamaluddin (2008) Phosphatase activity in the rhizosphere of medicinal plants inoculated with arbuscular mycorrhizal fungi. Mycorrhiza News;19:11–12

Tabatbai MA, Bremner JM (1969) Use of P-nitrophenol phosphate for assay of soil phosphatase activity. Soil Biol Biochem 1:301–307

Waksman SA (1922) A method of counting the number of fungi in the soil. J Bacteriol 7:339–341

Warcup JH (1952) Soil plate method for isolation of fungi from soil. Nature 166:117–118

Part III
Farming

Chapter 9
Cultivation Technology for *Jatropha curcas*

K.R. Karanam and J.K. Bhavanasi

Introduction

Development of cultivation technology is towards maximizing the yields per unit area with the best utilization of natural resources *viz;* land (soil), water (rainfall), sunlight, etc. The limiting resources except sunlight are applied artificially in order to reap good harvests.

Crop management is the most crucial parameter for increasing the productivity of *Jatropha curcas* (hereafter referred to as Jatropha). It includes the components of crop management practices, sustaining productivity, integrating resources for increased use efficiency at field scale.

Jatropha is recommended to be grown on marginal soils, which are not fit for any other crop cultivation. However, sufficient bush clearing, land shaping, land preparation, alignment and staking, pit digging, pit mixture preparation, basal and top dressing for fertilizers application are vital in the first 2 years to ensure initial crop establishment and sustainable yields for a period of 30 years. The research findings on crop management at Nandan Biomatrix Limited (NBL) and their recommendations are mentioned here after:

Agronomy Trials

- Standardization of plant density
- Mapping of soils suitable for cultivation
- Bush clearing and land shaping

K.R. Karanam • J.K. Bhavanasi (✉)
Nandan Biomatrix Limited, Aparna Crest, H.No 8-2-120/112/88&89,
Road No 2, Banjara Hills, Hyderabad 500 034, India
e-mail: jk@nandancleantec.com

N. Carels et al. (eds.), *Jatropha, Challenges for a New Energy Crop: Volume 1: Farming, Economics and Biofuel*, DOI 10.1007/978-1-4614-4806-8_9,

Table 9.1 Plant density per unit area based on annual precipitation

No.	Annual rainfall (mm)	Spacing (m)	Plant density/ha
1	< 500	2×1.5	3,333
2	501–1,000	2×2	2,500
3	1,001–1,800	2×2.5	2,000
4	1,801–2,500	3×2	1,600
5	2,501–3,500	3×3	1,111

Table 9.2 Soil amendments and number of applications (MT=100 kg)

Initial soil PH	Soil amendments	Qty/ha bring to PH 7	No. of years
5.0	Lime	20 MT	2 (10MTx2)
5.5	Lime	16 MT	2(8MTx2)
6.0	Lime	11 MT	2(5.5MTx2)
6.5	—	—	—
7.0	—	—	—
7.5	—	—	—
8. 0	Gypsum	2.5 MT	1
8.5	Gypsum	5.0 MT	2(2.5MTx2)
9.0	Gypsum	5.0 MT	2(2.5MTx2)

- Land preparation, alignment and staking
- Pit preparation
- Pit mixture preparation and refilling of pits
- Transplantation and watering
- Fertilization
- Weeding and inter culture
- Integrated nutrient management
- Water management and water footprint
- Integrated pest management
- Harvesting and yield
- Pruning and canopy architecture

Standardization of plant density: The plant density recommended for low precipitation regions is high with closed spacing while under high rainfall zones, wider spacing with low plant density is recommended (Table 9.1).

Suitability of Jatropha in various soil reactions: Jatropha is recommended for marginal and sub-marginal lands. Based on the initial soil pH, different amendments and application number are to be optimized (Table 9.2).

Rainfall: Jatropha requires a minimum 400 mm rainfall per year. The spread of rainfall is also very important. During the long dry spells, one monthly life saving irrigation is important in its initial 2 years of plants establishment as it is a perennial crop that is expected to stand for 30–35 years. Hence, crop planting is recommended

Table 9.3 Seed yield per plant (Kg) of 3-year old crop under rainfed conditions

Block	Rainfed	Rainfed + life saving irrigation	Variation in yield (%)
A	1.1	1.4	21.4
B	1.4	1.8	21.3
C	1.3	1.6	19.0
D	1.2	1.7	27.5
E	1.5	1.8	16.7
Average	**1.3**	**1.7**	**20.6**

Table 9.4 Spacing for gradient terraces

Field slope (%)	V.I (m)	H.I (m)
0.5	0.45	140
1	0.75	75
2	0.90	45
3	1.05	35
4	1.20	30
5	1.35	27
6	1.5	25
7	1.65	23.7
8	1.80	22.5

in the beginning of the rainy season of the year. Once the crop is established, irrigation is less important, but the yields are dependent upon the spread of rainfall. At NBL, seed yield (Kg plant^{-1}) of third year crop under rainfed conditions is presented in Table 9.3.

Bush clearing and land shaping: It is recommended to cut the bushes and dig the entire root zone. A minimum shaping of land as per slope or gradient is required to avoid soil runoff, to control soil erosion, to improve water retention and to improve ground water recharge (Table 9.4).

Land preparation, alignment and staking: We plough the land with a mould board plough in two directions. Harrowing should be done across the slope and pit marking has to be given in straight lines by maintaining a uniform distance of 2 m between rows and of 2 m as well between plants.

Pit preparation: Since land is already ploughed, making another 6 in. to obtain a 50 cm x 50 cm pit with pick and spade is possible. Pits have to be dug at a distance of 50 cm one of the other. While preparing the pit, the first 25 cm of soil from the pit should be put on one side of the hole and the last 25 cm of soil, on the other side (Fig. 9.1).

Pit mixture preparation and pit refilling: The pit mixture consists of 2 kg farmyard manure (FYM) or 1 kg vermicompost, 6 g bio-inoculants (1 g from each packet supplied by NBL), Urea at 60 g, single super phosphate at 20 g and muriate of potash at 15 g to be mixed thoroughly with the lower portion of 25 cm soil taken out from the pit. While refilling the pits, the upper portion of the soil removed from the

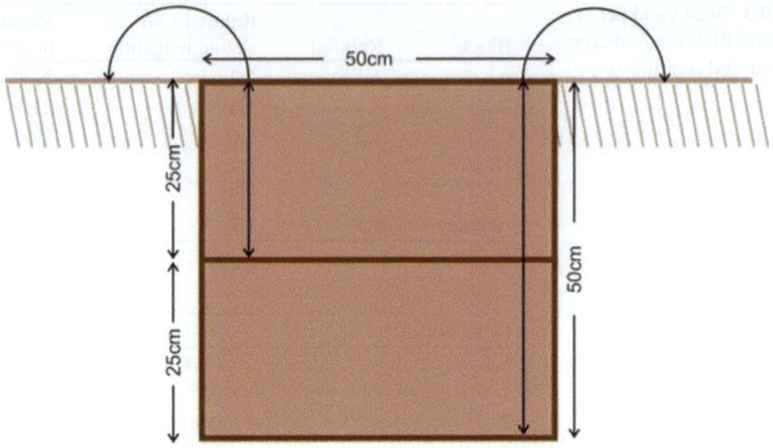

Fig. 9.1 Pit preparation for *J. curcas*

pit should be filled at the bottom of the pit and the lower portion removed from the pit should be mixed thoroughly with compost, bio-inoculants and fertilizers to be filled at the top.

Jatropha succeed in taking up in marginal soils lacking in available nutrients. However, certain nutrients are fixed in the soils in unavailable forms to Jatropha. Addition of bio-inoculants consisting of micro organisms allow the solubilization of phosphorous and potash and make the nutrients available to the plants. The bio-inoculants kit consists of 6 micro organisms and includes (1) vesicular arbuscular mycorrhiza (VAM), (2) phosphorous solubilizing bacteria, (3) *Azotobacter* sp., (4) *Trichoderma viride*, (5) potassium solubilizing bacteria, and (6) *Metarhizium* sp.

VAM: They are capable of extending the nutrient scavenging capacity of the plant on a purely physical basis and also releases compounds, which dissolve the elements under insoluble form such as phosphorus and zinc, thus, improving the uptake of these nutrients.

PSB: *Phosphate solubilizing bacteria* (PSB) increase phosphorus availability to the plant; phosphorus is a major nutrient for plants inducing vigorous growth and also contributing to their disease resistance. Phosphorous helps in root formation as well as plant growth and fruting. The plant by itself utilizes only 10–15% of the phosphate applied. The other 85–90% remain in insoluble form in the soil. PSB is a highly efficient phosphate solubilizing microorganism or a mix of such microorganism that grows and secretes organic dissolving unavailable phosphates into soluble forms and making them available to plants. Thus, the availability of residual phosphates in the soil is improved and external application can be optimized.

Azotobacter: **It contains a highly** efficient nitrogen fixing bacteria of the *Azotobacter* genus, which produces a variety of growth promoting substances like indoleacetic acid

(IAA), gibberellins (GA), vitamin B and antifungal substances. Another important characteristic of the bacteria from this genus is associated to their excretion of ammonia in the rhizosphere in the presence of root exudates. It fixes N_2 at about 15–20 kg/ha under ideal soil conditions and thereby reduces the requirement of nitrogen fertilizers by 10–20%.

***Trichoderma*:** It is an agent used for the biocontrol of soil diseases. It has been used successfully against pathogenic fungi. *T. virides* strains solubilize phosphates and micronutrients. The application *T. virides* strains to plants increases the number of deep roots, thereby increasing the plant's ability to resist drought.

KSB:—*Potassium solubilizing bacteria* (KSB) are used to improve plant growth by stimulating their potassium uptake. *Bacillus mucilaginous* and *Bacillus edaphicus* are examples of microorganisms that are used as biofertilizers. KSB solubilizes potassium rock through production and secretion of organic acids. Potassium has many functions in plant growth, such as smoothing the progress of cell division and growth, increase disease resistance and drought tolerance, regulate the opening and closing of stomata required for osmotic regulation. Potassium is essential for photosynthesis and act as a key to activate the enzymes from carbohydrate and amino acid metabolisms. Furthermore, potassium aids in the transport of assimilates during plant ontogeny and one of its most important influences is to improve oil content in plants. KSB is made of aerobic bacteria, which play an important role in maintaining soil structure by their contribution in the formation and stabilization of soil aggregates.

***Metarhizium anisopliae*:** It is a fungus that naturally grows in soils throughout the world and causes disease in various insects by acting as a parasite; it is known to infect over 200 insect species including termites. It is currently being used as a biological insecticide to control a number of pests, such as grasshoppers, termites, thrips, etc. The fungus does not appear to infect humans or other animals and is considered safe as an insecticide.

Transplantation and watering: The plantlets have to be transplanted at a depth of 4–6″ by grouting the entire root system into the soil (no root zone exposed above the ground) depending on the size of the root system. For initial crop stand, watering has to be done to maintain soil moisture at field capacity, once in 10–15 days depending on the texture and structure of the soil. This watering is required at least for the first three months to ensure the initial crop stand.

Integrated nutrient management (fertilization): The basal dose of nutrient application should be done only during pit preparation as detailed above. During the second year, organic manure at 2.5 ton ha^{-1} needs to be applied. From third year onwards, nutrients from Jatropha leaf fall (kg ha^{-1}) was found sufficient.

Analysis of plant parts for nutrient status and recommendations: Plants from the first to the fifth year of growth were uprooted and digested to know their micro and macro nutrient contents for the calculation of fertilizer requirement.

During winter, Jatropha goes to deciduous stage and leaf senescence happens every year adding 3–4 kg of dry leaf per plant to the soil (Fig. 9.2). Thus, artificial

Fig. 9.2 Deciduous stage showing dry leaf fall

Table 9.5 Nutrient availability from dry leaves

Nutrient	% of dry leaf wt.	Available nutrient (g $plant^{-1}$)	Available nutrient (Kg ha^{-1})
Nitrogen	3.64	127.4	318.5
Phosphorous	0.25	8.75	22
Potash	2.32	81.2	200
Sulphur	0.17	5.95	15
Calcium	2.67	93.45	234
Magnesium	1.05	36.75	92

fertilization is calculated as the amount of nutrients needed to complement that coming from the natural fertilization due to leaf litter falling on the soil each year (Table 9.5).

1. Average No. of leaves per plant (3 years and above): 2000
2. Average weight of dry leave: 1.75 g
3. Dry leaf addition to soil every year: 3.5 Kg $plant^{-1}$
4. Plant density: 2,500/ ha
5. Total leaf fall every year on a dry wt. basis: 8.75 tons ha^{-1}

In addition to the enrichment of soil with nutrients from leaf litter, NBL recommends application of 300 kg of Jatropha deoiled cake per hectare, which adds 96, 36 and 54 kg ha^{-1} of N, P and K, respectively at 3.2, 1.2 and 1.8% availability.

Water management and water footprint: With Indian food supply chain being irrigation dependent, the rising cereal demand is likely to degrade the water resources in the country. A study conducted by USAID reports that cultivation of Jatropha (a non-food biodiesel feedstock) on marginal lands would consume less water

Table 9.6 Comparison of sustainability of Jatropha versus other non-sustainable feed stocks

Feedstock	Water requirement, low to high (mm year^{-1})	Biofuel yield per unit of water (kg mm^{-1})	Growing season/time to full maturity	Growing season/ time to full maturity
Sugarcane	1,500	2,500	1.65	10–12 months
Sorghum	450	650	0.82	4–5 months
Sweet sorghum	450	650	3.45	4–5 months
Oil palm	1,800	2,500	2.33	10–12 years
Jatropha	150	300	2.67	3–4 years
Coconut oil	600	1,200	5.2	5–10 years
Maize/Corn	500	800	0.69	4–5 months
Rapeseed	350	450	0.83	120–150 days
Sugar beet	550	750	11.34	5–6 months
Wheat	450	650	1.09	4–5 months
Soybeans	450	700	0.35	100–150 days
Sunflower	600	750	0.32	100–120 days
Groundnut	400	500	1.13	100–120 days

compared to other conventional biofuel feed stocks (Table 9.6). The study also estimates that Jatropha yields 2.67 kg of biodiesel per mm of water consumed compared to 2.33 kg per mm of water consumed by most demanded feed stocks like Palm Oil.

Weeding and inter: After plant establishment, weeding and interculture cleaning is necessary to avoid crop weed competition. Weeding and interculture cleaning can be done mechanically by ploughing the land in two directions. Weeds within the vicinity of the plants may be removed manually at least 3–4 times in a year (first 2 years) depending on weed density.

Integrated crop management: The other aspects of integrated nutrient management include proper canopy architecture, increasing the harvest index, standardizing the leaf area index, the cropping system and irrigation requirements of the crop.

Pruning and canopy architecture: Pruning is done every year after the harvest in the month of March. This process increases the number of productive branches, avoids summer fruiting and maintains a proper canopy architecture with a height lower than 2 m. Figure 9.3 shows the quantity of pruned biomass that is obtained from each hectare of plantation.

Integrated pest management: Jatropha is considered to be tolerant to many pests and diseases, but this was true as long as it was wild. Once the crop is being domesticated, a wide array of insect pests as well as fungi, bacteria and viruse diseases were observed along with physiological disorders. NBL observed several pests and diseases in its trials whose main are:

1. Scutellarid bugs
2. Mealy bug

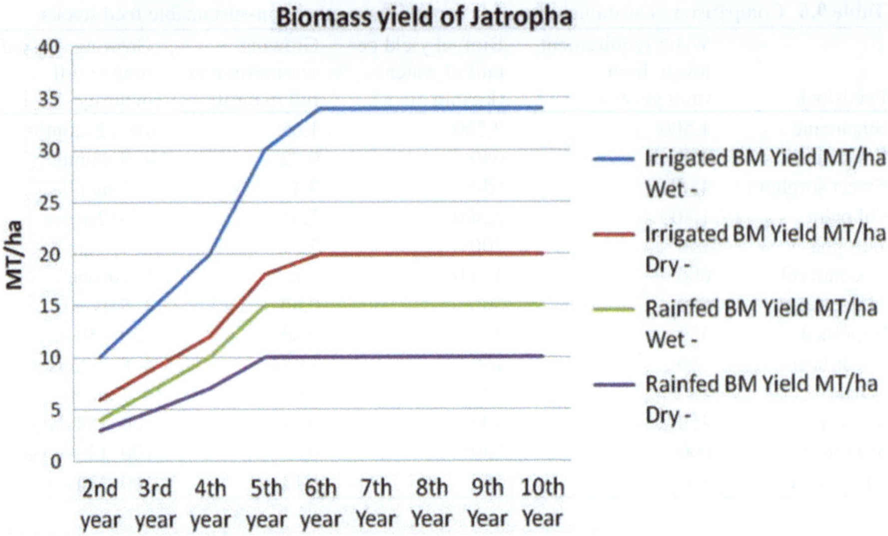

Fig. 9.3 Biomass (BM) yield (MT=100 kg) of *J. curcas*

Table 9.7 Pest incidence and yield loss in *J. curcas*

Insect/Disease	Yield (Kg plant^{-1})	Yield variation (%)
Control	1.4	—
Scutellarid bugs	1.17	16.42
Mealy bug	0.89	36.42
Leaf and inflorescence webber	1.05	25.0
Yellow mite	0.9	35.71
Powdery mildew	1.13	19.28

3. Leaf and inflorescence Webber
4. Yellow Mite
5. Termites
6. Powdery Mildew
7. Mosaic virus

NBL has developed effective control measures to combat these pests, which can otherwise cause damage and reduce yields by 20–30% (Table 9.7).

Propagation Techniques

Seed raised seedlings: NBL has developed propagation techniques of seed raised seedlings based on vegetative propagation to guarantee the homogeneous multiplication of its elite plant material (Fig. 9.4).

Fig. 9.4 Propagation of elite material through seeds and vegetative cuttings

Productivity of Jatropha

The productivity of seeds and oil are the most critical factors that govern the Jatropha's sustainability in the farmers' fields. There have been reports of fruit yield in Jatropha varying from a humble half a kilogram per plant to more than 12 kg per plant.

One to one and half kg per plant is a routine productivity that has been achieved by several institutes and companies in the country. Several projects of research and development conducted under DBT and NOVOD Board (India) have reported this productivity. Apart from these results, several institutes have also reported plants with productivity of 2 kg per plant.

NBL has selected superior plants with productivity of 2 kg per plant and hybridized these plants in their inter-specific and intra-specific hybridization program. This has given them a value of 5.2 kg in the NANDAN-1 hybrid as the highest average yield.

Nandan-1, Nandan-2 and Nandan-3 are intra-specific hybrids developed by NBL, which are tested under both rainfed and irrigated conditions. Five-year-old plantations are available and yields under both the conditions are presented in Fig. 9.5.

Cost of Plantation

Several factors control the costs of plantation such as soil type, irrigation, facility, rainfall amount and regularity, etc. NBL has calculated the number of man days required for various operations of Jatropha cultivation (Table 9.8).

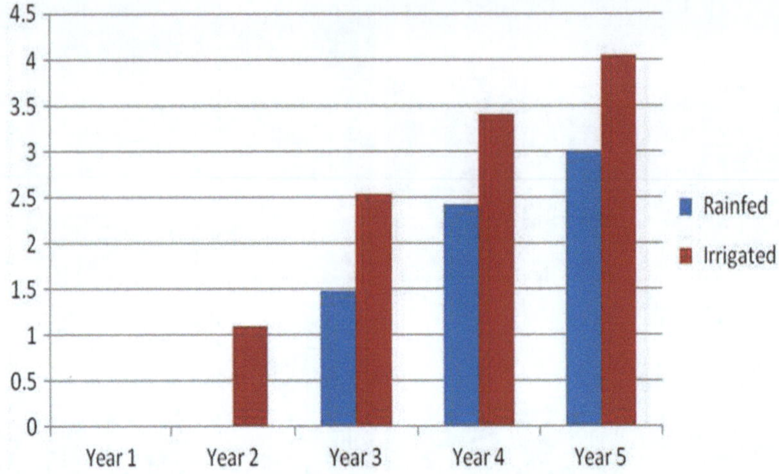

Fig. 9.5 Seed yield (kg plant^{-1}) of Nandan hybrids under rainfed and irrigated conditions

Table 9.8 Number of man days (MD) for different operations of *J. curcas* cultivation

No.	Activity/ha	Man days
1	Bush clearing & land shaping	10 MD
2	Alignment and staking	5 MD
3	Digging of pits of 50 cm × 50 cm × 50 cm @ 50 pits per MD	50 MD
4	Fertilizer application	2 MD
5	Mixing FYM, insecticides, fertilizers & refilling pits @100 pits per MD	25 MD
6	Planting & replanting cost of 100 plants per MD and 5 MD, respectively.	25 MD
7	Weeding & soil working × 2 times per year	10 MD.
8	Harvesting of fruits/seeds @ 2 MD per 100 Kg of seed from third year onwards	
	Total	**127 MD**

Chapter 10
Jatropha Pests and Diseases: An Overview

K. Anitha and K.S. Varaprasad

Introduction

Jatropha curcas L., a non-edible oilseed, drought tolerant, is gaining popularity commercially as a biodiesel plant in India and many other developing countries. It is mostly grown as a medicinal plant or hedge crop and is being advocated for the development of waste lands and dry lands throughout the world. The crop has been expected to be less prone to damage due to different categories of pests and diseases as wild varieties were used as live fence in dry lands. Another reason is that the seed oil is reported to have insecticidal, molluscicidal, nematicidal and fungicidal properties. However, reports of pest outbreaks on *J. curcas* have started appearing with planting *J. curcas* as a regular monocrop in both marginal and arable lands using high yielding and high fertilizer responsive cultivars. Serious problems of economic significance have been reported in *J. curcas* plantations due to attack of fungi, bacteria, viruses, insects and other pests. Cryoconservation did not reduce the incidence of fungi on *J. curcas* seeds though the physiological quality was preserved (Goldfarb et al. 2010). Adverse impact on the crop economics is expected from pest and disease attacks since it could seriously affect production costs.

In view of the considerations above, the global status of insect pests and diseases on *J.curcas* and some species of the *Jatropha* genus has been reviewed and compiled. Below, we report on common and scientific names of pests and diseases along with photos of their corresponding symptoms wherever available. We also report on the management options given in the literature for each of the important pests and diseases.

K. Anitha (✉)
National Bureau of Plant Genetic Resources Regional Station,
Rajendranagar, Hyderabad, 500 030, India
e-mail: aniyagnya@yahoo.com

K.S. Varaprasad
Directorate of Oilseeds Research,
Rajendranagar, Hyderabad, 500 030, India

N. Carels et al. (eds.), *Jatropha, Challenges for a New Energy Crop: Volume 1: Farming, Economics and Biofuel*, DOI 10.1007/978-1-4614-4806-8_10,

Fungal Diseases

Root Rot

Different fungal pathogens, *viz.*, *Fusarium moniliforme, Lasiodiplodia theobromae, Phytophthora* sp., *Rhizoctonia bataticola* or *Macrophomina phaseolina, Sclerotium rolfsii, Clitocybe tabescens*, etc., are known to cause root rot or collar rot in *J. curcas*.

Fusarium moniliforme [*Gibberella fujikuroi*] causes root rot under waterlogging conditions as observed for the first time during routine surveys in Bawal, Haryana, India, causing 20–25% mortality of *J. curcas* (Sharma et al. 2001).

Symptoms: Yellowing of leaves at 45–50 days after planting, followed by defoliation, drying or wilting of the plant from the tip downwards, and finally death. The presence of sooty roots is also observed. The disease is more pronounced during winter months.

Lasiodiplodia theobromae root rot and collar rot was reported for the first time on *J. curcas* grown in the state of Tamil Nadu, India with heavy losses in 2007 (Latha et al. 2009). Pathogenicity was confirmed by artificial inoculations on 1-year-old plants. In China, the pathogen caused gummosis of *J. podagrica* (Fu et al. 2007). The most damaging root and collar rot disease caused by *L. theobromae* was observed on adult plants of *J. curcas* in the states of Minas Gerais, São Paulo (Pereira et al. 2009) and Ceara (Freire and Mosca 2009) in Brazil. Seed health tests undertaken on 213 germplasm accessions of *J. curcas* from Andhra Pradesh, Chattisgarh, Uttaranchal, Kerala and Tamil Nadu states in India revealed that *L. theobromae* was the most prevalent seed borne pathogen, occurring in 83 accessions from all sources (Anitha et al. 2005).

Symptoms: Yellowing, drooping and shedding of leaves, wilting, blackening and decaying of the collar region and rotted roots, followed by death of the plants, especially those undergoing drought stress. Black pycnidia are seen on the collar region and black necrotic patches are observed when the bark is removed from the collar region. A blackening of the vascular region is consistently observed on collar and root regions of ill plants. The disease can also lead to the death of adult plants.

***Phytophthora* sp.** was reported to cause root rot in Ghent, Kenya (http://betterglobe-forestry.com/research/pests-and-diseases.html). Large plantations of *J. curcas* in Chiapas, Mexico had been destroyed by *Phytophthora* sp. in 2009, causing collapse and death of young plants (CABI UK Centre Annual Report 2009). Interspecific hybridization between *J. curcas* and *J. integerrima* was undertaken with the objective of combining desirable traits such as resistance to root rot, tolerance to frost and high oil content (Dhillon et al. 2009a). The hybrids were successfully obtained only when *J. curcas* was used as seed (female) parent. Dhillon et al. (2009b) also found that the cleft grafting of *J. curcas* scions on *J. gossypiifolia* rootstock let to resistant individuals to root rot.

Rhizoctonia bataticola *or* Macrophomina phaseolina

The collar rot disease is more common in the drip irrigated fields, excessively irrigated fields or fields with high water table/poor drainage. The plant tries to recover by producing new roots in the rotted area. Secondary root growth is stimulated by covering the rotted area with soil (http://jatropha.pro). Twenty to thirthy-days-old *J. curcas* plants showing root rot symptoms (10–13% severity) were observed in Bawal, Haryana, India, from April to May 2005 (Sharma and Kumar 2009) and found that *R. bataticola* was responsible for the disease. *R. bataticola* caused basal stem rot of seedlings in nurseries raised in Mexico (CABI UK Centre Annual Report 2009). The epidemiological factors that affected the root rot development were studied (Kumar and Sharma 2010) and found that inoculum density of 20 $g\,kg^{-1}$ soil resulted in 42% *pre-emergence mortality* (PEM) and 23% *post-emergence mortality* (POEM). Inoculum at a depth of 5 cm induced 42% and 31% PEM and POEM, respectively. Irrigation at 20 days interval showed maximum PEM (42%) and POEM (35%). *R. bataticola* or *M. phaseolina* is known to cause collar rot of young seedlings (Heller 1996; Daey Ouwens et al. 2007; Sharma and Sarraf 2007).

Symptoms**:** Small light-brown lesions develop on hypocotyl/collar region, which later turn dark brown and girdle the entire collar region (Fig. 10.1). The infection moves downwards rapidly. The tissue inside the bark shows black coloration.

Seed and Seedling Rots (Sclerotium rolfsii)

Seed and seedling rot of *J. curcas* caused by *S. rolfsii* was observed in a nursery raised in Karnataka state, India (Hegde et al. 2009).

Fig. 10.1 Collar rot (*Macrophomina phaseolina*) affected *J. curcas* plant (Courtesy: http://jatropha.pro)

Symptoms: Seedlings show complete wilting. When such seedlings are uprooted, white mycelium is seen at the basal portion with sclerotial bodies on rotting seeds or seedlings. Hegde et al. (2009) isolated *S. rolfsii* from such wilted plants and could link fungal infection to pathogenicity since seed failed to germinate, seeds rotted or germinated only after 40 days leading discolored seedlings to completely wilt within 4 days.

Armillaria tabescens *Root Rot*

Armillaria tabescens (***Clitocybe tabescens***) was reported to be responsible for Armillaria root rot of *J. curcas* in USA (Agriculture Hand book 165, 1960) and Sudan (Dalia Amin 2001). *A. tabescens* is on the disease list that justifies quarantine in India (Plant Quarantine Order 2003).

Management: There are no cost-effective fungicidal sprays to control *Fusarium* root rot, according to Michigan State University. Therefore, prevention is the common method for controlling root rot. It is indicated to avoid overly moist soil conditions. *M. phaseolina* root rot can be effectively managed by soil drenching with fungicides like carbendazim at 0.1%, hexaconazole at 0.1%, mancozeb at 0.2% or carboxin + thiram at 0.1%. Carbendazim application increased the plant vigour significantly (Hegde and Chavhan 2009/2010). However, there is no registered chemical product for this pathogen on *J. curcas* in Brazil.

The collar rot disease of *J. curcas* can be controlled by using 1% Bordeaux solution (http://www.k4rd.org/jatropha.htm) (TERI 2007). The disease due to *R. bataticola* or *M. phaseolina* can be controlled with 0.2% copper oxychloride (COC) or 1% Bordeaux drenching (Sharma and Sarraf 2007).

Plant Wilt (Fusarium oxysporum)

The pathogen attacks the plant in the seed bed or in the field. The disease spreads from one plant to another, sometimes through flood and open channel irrigation (http://jatropha.pro/wilt.html).

Symptoms: Both young and fully grown plants collapse in a very short time (starting from the bottom) as they do not get water due to the blockage of vascular system by the fungus (Fig. 10.2). When the seedling is attacked, the leaves become pale green, wither, dried (Fig. 10.3) and fall down leaving only the younger leaves from upper plant floors (Fact Foundation 2010).

Management: Several cultural strategies can be followed to control infectious plant wilt. These include healthy seedling selection, use of resistant varieties, avoiding flooding around the plant, improving soil drainage conditions, avoid the creation of a hard crust when using a machine for planting, ridge planting, burn the trash or infected plant materials and avoid inter cropping with plants from the

Fig. 10.2 Wilt (*Fusarium sp.*) affected *J. curcas* seedling (Courtesy: http://jatropha.pro)

Fig. 10.3 Vascular discolouration due to *Fusarium oxysporum* (Courtesy: www.jatropha.pro)

Solanaceae family, which are alternate hosts of *Fusarium* spp. The seed treatment with carbendazim, soil fumigation or drenching with benlate (50 g/100 l) is effective in combating the problem although expensive. The application of *Trichoderma viride* as a biocontrol agent at 250 g ha^{-1} at planting has also given good results.

Leaf Spot (Cercospora jatrophae-curcas)

The incidence *of C. jatrophae-cur*cas responsible for leaf spot symptoms on *J. curcas* has been reported in India (Kar and Das 1988; Sharma and Sarraf 2007). *C. ricinella* was also found responsible for leaf spots on *J. curcas* in Indonesia (http://www.waterlandasiabio.com). *Cercospora* sp. has been isolated from leaf spots observed on *J. curcas* at the University of Florida (http://strawberry.ifas.ufl.edu/DiagnosticLab). The disease is severe during the rainy season and spreads primarily by conidial propagation through rain splash. Temperatures ranging between 24°C and 26°C and above 60% relative humidity (RH) are favorable for the disease development.

Symptoms: Initially, small brownish necrotic irregular spots with a yellowish halo appear on the leaf surface. As the disease advances, these spots enlarge, coalesce and leave characteristic shot hole on the middle. The damage caused by this pathogen on *J. curcas* can be very heavy. Sometimes, black or brown round spots surrounded by a pale green ring are found on both surfaces of leaves. When the spot enlarges, it becomes gray and is surrounded by a brown crown that finally becomes irregular (http://www.fact-foundation.com/). The conidia are hyaline, tapering at one end and 70×3 μm with 2–7 septa.

Management: This disease normally does not hamper fruit production by *J. curcas*. In case of severe incidence, application of 0.2% carbendazim (180–200 litre per acre), mancozeb or any fungicide is helpful. It is recommended to use varieties that are resistant to the pathogen. The systemic resistance of *J. curcas* to pathogens such as *Cercospora* sp. and viruses was increased by soaking seeds in leaf extracts of *Clerodendrum aculeatum* (Debnath and Verma 2008). A minimal care is to adopt the recommended cultural practices that is the removal and burning of infected leaves around plant bases as well as the avoidance of overhead irrigation.

Alternaria *spp.*

Leaf spot disease caused by *A. alternata* (Fr.) Keissler was observed in the *J. curcas* plantation at Marathwada Agricultural University, Parbhani, Maharashtra in 2006 (Hudge and Datar 2010) and in Jabalpur, Madhya Pradesh (Shukla and Jamaluddin 2010), India. Leaf spots caused by *Alternaria* spp. were recorded in India (http://www.rrljorhat.com/) and Kenya (http://betterglobeforestry.com/). The disease incidence was maximum in the month of August and subsequently decreased up to November. High humidity and temperatures in the range of 16–20°C favour the disease development and the attack may reach 70% affecting the fruit yield and oil content (http://www.waterlandasiabio.com/).

***Symptoms*:** Initially yellow spots appear on the leaves, later they turn to brown. If the infection is massive, the plant becomes stunted, and may also die. If the attack occurs at the onset of flower initiation, the buds may die; if it occurs at the end of flowering, the flowers may open, but capsules may not form. If the attack is light, the flowers may dry. The fungus develops very quickly on the capsules when humidity is high and fruits become black. Premature drop of the green blighted fruits also occurs in severe cases. The disease may appear throughout the year, but intensive attack occurs in the rainy season. The disease may spread through seeds, internally or externally. It may also tumble young seedling in the seed bed. Premature leaf fall

Fig. 10.4 Angular leaf spot stage of anthracnose (*Colletotrichum* spp.) symptoms on *J. curcas* leaf (Courtesy: Dra. Daisy de la C. Perez Brito, Yucatan, Mexico)

due to *Alternaria* sp. infection on *J. curcas* was reported in Zimbabwe (http://www.jatropha.de/sudan/jatropha-project).

In vitro effect of temperature and humidity against *A. alternata* leaf spot of *J. curcas* in India was studied by Hudge and Datar (2009). These authors found that the optimum temperature for its growth was 25°C, followed by 30°C and 35°C. Mycelial growth was found to be positively correlated with the incubation period. Maximum sporulation was observed at 25°C, followed by 30°C and 20°C, while sporulation was poor at low temperatures, i.e., 15°C and 10°C. The largest mycelial growth (90 mm) occurred at 100% relative humidity. The optimum relative humidity for growth and sporulation of the fungus lies between 84.5% and 100%.

Management: Seed treatment with contact fungicide is useful to prevent the first phase of its development.

Anthracnose (Colletotrichum *spp.*)

The leaf spot disease caused by *C. gloeosporioides* was recorded on *J. curcas* plants in Sudan (Dalia Amin 2001) and Brazil (Freire and Mosca 2009). Brown angular lesions with yellow halo (Fig.10.4) followed by enlarged necrotic lesions (Fig.10.5) on leaves are the most common symptoms. Crown canker, apical death of seedlings and foliar necrosis were observed on *J. curcas* in fields located in the Yucatan Peninsula in Mexico in 2008 affecting 25% of the production (Torres-Calzada et al. 2011).

C. jatrophae-curcas was found responsible for leaf necrosis in Mexico in 2009 (CABI UK Centre Annual Report 2009). Periodic surveys of different places (Chitrakoot, Seoni, Narsinghpur and Bilaspur) in Madhya Pradesh, India, and in the neighbourhood of *J. curcas* revealed that *J. curcas* plants were infected with *C. dematium* (Gupta et al. 2007; Shukla and Jamaluddin 2010). Naik (1993) reported the incidence and pathogenicity of *C. gloeosporioides* [*Glomerella cingulata*] on

Fig. 10.5 Advanced necrotic lesions (*Colletotrichum* spp.) on *J. curcas* leaf

J. glandulifera in India for the first time. Seed health tests undertaken on 213 germplasm accessions of *J. curcas* from different states in India revealed the presence of *C. acutatum* and *C. graminicola* (Anitha et al. 2005). Stephen and Mark (2010) reported the incidence of anthracnose and leaf spot due to *Colletotrichum* sp. on *J. integerrima* in Lee county, USA by the University of Florida.

Management: Infected leaves should be removed when the plant is dry. Leaves fallen around the base of the plant should be raked up and disposed of. Overhead irrigation should be avoided as far as possible and water should be directed at soil level. Watve et al. (2009) studied the management of *C. gloeosporioides* using biocontrol agents, plant extracts and fungicides and found that *Trichoderma harzianum* (in *in vivo* assay), neem leaf extract (in *in vitro* assay) and carbendazim followed by propiconazole were effective against the disease.

Powdery Mildew (Oidium *spp.*; Erysiphe euphorbiae; E. jatrophe)

Powdery mildew damages leaves and flowers of *J. curcas* in Zimbabwe (http://www.jatropha.de/sudan/jatropha-project) and Kenya (http://betterglobeforestry.com/). *Oidium* sp. was observed on *J. gossypiifolia* in Cuba (Bocourt et al. 2005). Reddy and Reddy (1980) reported *Erysiphe euphorbiae* on *J. gossypiifolia* L. from India. Later Braun (1987) reported *E. jatrophe* Doidge on *J. zeyheri* Sond from

Fig. 10.6 *J. curcas* seedling infected with powdery mildew (Courtesy: http://jatropha.pro)

South Africa. Occurrence of *E. jatrophe* on *J. curcas* plants was reporded in Manipur, India (Sharma et al. 2010).

Powdery mildew occurred in the nursery damaging leaves and stems in Zhenkang County, China in July of 2008 (Li et al. 2009). Infection characteristics of powdery mildew on *J. curcas* were studied in China and the results showed that plantations of *J. curcas* were easily infected by several waves of powdery mildew over the duration of the experiment. The period of August to October corresponded to the peak incidence of the disease given the favorable temperatures and humidity. After November, the disease development decreased, but never stopped (Li et al. 2010).

***Symptoms*:** The white powdery mycelium of the fungus is seen mainly on younger leaves and terminal shoots (Fig. 10.6). Powdery mildew damages leaves and flowers of *J. curcas* more often in areas with low average temperatures and relatively high humidity.

Management: Products based on dithiocarbamate (Zineb, Dithane, Manzate, etc.) may be used to fight the disease (http://jatropha.pro/).

Pestalotiopsis Blight (Pestalotiopsis *spp.*)

J. curcas is reported to be a host of *Pestalotiopsis guepinii* (Mordue 1971). Leaf spot caused by *P. versicolor* (Phillips 1975) and *P. paraguarensis* (Singh 1983) on *J. curcas* was reported earlier. *P. mangiferae* was isolated from affected plants in Allahabad, Uttar Pradesh and it was confirmed that in the absence of bark injury, the pathogen failed to enter the plant system to cause the disease (Pandey et al. 2006).

The incidence and severity of *Pestalotiopsis* blight was recorded in 2006 in all plantations and hedges of *J. curcas* in two agro-climatic zones in Karnataka with very high levels of disease severity index (DSI) (Chavhan et al. 2010). *Pestalotiopsis* sp. is also known to cause leaf spot on *J. curcas* (Sharma and Sarraf 2007).

***Symptoms*:** Stems of affected plants exhibit multiple injuries at the collar region and above. The exposed inner tissue is dark brown to black. Small to extended cankers are formed at these wounds, giving the area a charcoal black appearance (Pandey et al. 2006).

Management: Two sprays of 0.1% mancozeb at 30-day intervals proved to be highly effective in decreasing the disease in the field at very low disease severity index (DSI of 1.17) compared to a DSI of 1.85 in the control (Chavhan et al. 2010).

Fungicidal screening revealed that carbendazim completely inhibited the *P. mangiferae* colony growth even at the lowest concentration (0.1%). Dithane M-45 [mancozeb] was effective only at 0.3%, while captan proved to be ineffective even at 0.3% (Pandey et al. 2006).

Botrytis Decay (Botrytis ricini)

***Symptoms*:** The initial symptoms appear as black spots on the flowers. The disease is a very serious problem in the rainy season especially when it coincides with the capsule forming phase. The disease spreads with rains during the nights when the temperature is cool. The infected flowers start to rot and end up covered. Pathogen spreads from the flower to the fruit capsule easily (http://www.waterlandasiabio.com/rd/).

Rust (Phakopsora jatrophicola; P. arthuriana)

There are two rust species recorded on *J. curcas*—*Phakopsora jatrophicola* and *P. arthuriana*. Although *P. pachyrhizi* has quite a broad host range within the legumes, it has not been recorded on *J. curcas* (http://biofuelexperts.ning.com/).

Incidence of *P. jatrophicola* on *J. curcas* was noticed in Sudan (Dalia Amin 2001). Survey conducted in 2009 revealed that rust caused by *Phakospora* spp. was the most widely spread and damaging disease in Mexico (CABI UK Centre Annual Report 2009), infecting plants of all ages. *J. curcas* infected by rust (*P. jatrophicola*) with 25–30% disease severity was also recorded in Singapore (AVA 2010).

New isolates of *P. jatrophicola* from Central and South America have been tested on six different Australian varieties of *J. gossypiifolia*, which is a declared noxious weed in Australia (Seier et al. 2009). All varieties tested, including *J. curcas*, were susceptible to the rust species. Hence, it was decided to collect and evaluate additional strains of *P. jatrophicola* with an increased host specificity for *J. gossypiifolia*.

Symptoms**:** The disease looks like rust spots with symptoms as small, bright orange, yellow, or brown pustules on the lower surface of the leaf. When touching it with the finger, it leaves a colored spot of spores. Leaf necrosis and defoliation of seedlings are the major symptoms. The pathogen attacks many plants from *Euphorbiaceae*.

Management: Spraying or dusting using sulfur powder and mancozeb fungicide may reduce the intensity of the attack. In anticipation of new resistant varieties, some basic cares such as maximum air circulation, clean up of plant debris especially around plants that have been infected, avoidance of overhead irrigation and water irrigation only during the day to allow plant drying before night are efficient means of inoculum control.

Viral Diseases

Jatropha Mosaic Virus Disease

The *Jatropha mosaic virus* (JMV) was first reported on *J. gossypiifolia* from Puerto Rico and identified as begomovirus (Brown et al. 1999). There are reports on occurrence of JMV (JMV-PR) on *J. gossypiifolia* and *J. multifida* from Puerto Rico (USA) (Bird 1957; Brown et al. 1999). JMV was also reported from Cuba on *J. gossypiifolia* (Martinez 2008).

The Jatropha mosaic virus disease (JMVD) has emerged recently and is now widely spread in India. The incidence of JMVD ranged between 13% and 47% in Karnataka state, India, causing significant yield loss (Narayana et al. 2006). The putative *Jatropha mosaic Indian virus* (JMIV) is transmitted through grafting, the dodder *Cuscuta subinclusa* and the whitefly, *Bemisia tabaci*. Phylogenetic analysis of the core coat protein (CP) sequences of JMIV and begomoviruses shows that JMIV groups in a separate cluster close to *Indian* and *Sri Lankan cassava mosaic virus* isolates and shared highest nucleotide identities (90–95%) with them (Narayana et al. 2007). The association of a begomovirus with JMD has been found in north India where it possesses the highest identity level and the closest relationships with *Indian* and *Sri Lankan cassava mosaic virus* isolates (Raj et al. 2008). The begomoviruses causing JMD in the Americas grouped separately from JMIV and shared only 72.8–75.2% nucleotide identity with the core CP and are, thus, distinct.

JMD was identified in *J. curcas* plantations as well as in wild stands in semi arid regions of India (http://www.phytotron.com/jatropha1.htm). Serious disease symptoms were observed on a large number of *J. curcas* plants in various localities of Balrampur District, Uttar Pradesh during the rainy season of the year 2005 (Tewari et al. 2007). The disease was sap transmissible and cleft graft transmissible,

Fig. 10.7 *J. curcas* plant affected with *Jatropha mosaic virus* (Courtesy: Aswatha Narayana et al. 2006)

but not transmitted through seed, dodder and insects (aphids and whiteflies). The disease could not be transmitted to any other plant except *J. curcas*. Attempts made by sap inoculation and grafting to *Nicotiana glutinosa*, *Lycopersicon esculentum*, *Solanum melongena*, *Datura stramonium* and *Carica papaya* were not successful (Tewari et al. 2007).

The disease resembles to some extent with the *tobacco leaf curl virus* (TLCV) (Shanta and Menon 1959). Chlorosis is prominent in case of the disease in *J. curcas*, while not common on plants attacked by TLCV. The virus causing the mosaic disease in *J. curcas* is not transmitted by whiteflies, while TLCV is neither transmitted by whiteflies (Smith 1957) nor by sap (Garga 1960), hence it is considered as a distinct record of mosaic. Metabolic and histopathological alterations of *J. curcas* induced by the Jatropha mosaic begomovirus were studied by HR-MAS NMR spectroscopy and magnetic resonance imaging (Sidhu et al. 2010).

***Symptoms*:** Naturally infected plants showed mosaic, reduced leaf size, leaf distortion, blistering and stunting (Fig. 10.7). Whitefly inoculated plants developed typical symptoms such as veinal netting, chlorotic specks, leaf distortion and stunting of seedlings within 30 days of inoculation (Narayana et al. 2006). The disease was successfully established in healthy plants through grafting using scions from infected cuttings. Graft-inoculated plants produced typical mosaic symptoms within 25 days after graft inoculation. The success rate of disease transmission of JMVD to healthy plants through silverleaf whitefly (*Bemisia tabaci* 'B biotype') was reported to be as high as 40% (Narayana et al. 2006). Although its vector was found in 1994, JMVD has not yet been found in Australia (Jones and Csurhes 2008).

Gao et al. (2010) reported the completion of the nucleotide sequence of the virus. Phylogenetic analysis of the virus genome suggests it is a new strain of *Indian cassava mosaic virus*. The authors suggest that with the genome sequence and the availability of the two infective clones, it may be possible to use double-stranded hairpin RNA or artificial miRNA-mediated RNA interference technology to generate transgenic *J. curcas* lines that would be resistant to this new disease.

Management: The application of ZILLON™ is quite effective against JMVD. Virus free planting material should be propagated to check the disease spread (http://www.phytotron.com/jatropha1.htm). The investigations on epidemiology and development of specific diagnostic techniques have helped to develop suitable management strategies to combat the disease, which is likely to become a serious threat for *J. curcas* cultivation in the country.

African cassava mosaic virus

J. multifida is known to be a host of *African cassava mosaic virus* (ACMV), as a related species of *J. curcas* and, thus, may serve as a reservoir for ACMV. *J. curcas* should not be grown in association with cassava (http://www.jatropha.de/photo-show/index.htm#use) as it is also believed to be capable of transmitting the cassava super-elongation disease (*Sphaceloma manihoticola*) (Achten et al. 2008).

Cucumber mosaic virus

Occurrence of *Cucumber mosaic virus* (CMV) on *J. curcas* in India was reported in Uttar Pradesh, India during 2006 (Raj et al. 2008) where the major symptoms were severe mosaic accompanied by yellow spots. Sap transmission, gel diffusion tests, RT-PCR assays, cloning, sequencing and BLAST analysis indicated a 98–99% identity of the virus responsible for the symptoms with CMV isolates. Sporadic incidence of several other pathogens infecting *J. curcas* plantations was recorded from places to places all over the world (Table 10.1).

Nematode Diseases

Root-Knot Nematode (Meloidogyne javanica; M. incognita)

J. podagrica is registered as a new host for root-knot nematode (*Meloidogyne javanica*) in Brazil (Freire and Mosca 2009; Erum et al. 2005). However, the reaction of *J. curcas* to *M. javanica* was studied in Brazil by Fernandes and Asmus (2007) and found that *J. curcas* was immune to *M. javanica*. Incidence of *M. incognita* has been recorded on *J. podagrica* from Pakistan (Zarina 1996) with maximum population observed during the months of January to April and a rapid decline during summer months.

Table 10.1 Other diseases recorded on *Jatropha* spp.

Pathogen	Host	Country of report	Reference
Amphobotrys ricini	*J. podagrica*	Brazil	Lima et al. (2008)
Botryodiplodia theobromae	*J. podagrica*	China	Fu et al. (2007)
Alternaria sp., *Colletotrichum* sp., *Dothiorella* sp., *Fusarium* sp., *Oidium* sp., and *Xanthomonas* sp.	*J. curcas*	Nicaragua	Padilla and Monterroso (1999)
Cordana musae (leaf spot) and *Chlaropsis thielavioides* (Marginal necrosis)	*J. curcas*	Philippines	Tuan et al. (2009)
Dothiorella gregaria (Stalk rot); *Diplodia* sp. (Root and stem rot)	*J. curcas*	China	Wang et al. (2009)
Elsinoë and *Sphaceloma* spp.	*J. aconitifolia* var. *papaya*	Central and South America	Zeigler and Lozano (1983)
Elsinoë brasiliensis (superelongation disease of cassava)	*J. curcas*	n.a	www.cabi.org
Fusarium solani	*J. glandulifera*	India	Das (1995)
Fusarium spp., *Phytophthora* spp., *Pythium* spp., (Root rot and Damping-off)	*J. curcas*	n.a.	Heller (1996)
Fusarium solani; *Macrophomina phaseolina*; *Pestalotia* sp; *Phoma* sp.	*J. curcas*	Andhra Pradesh, India	Sudhir et al. (2007)
Helminthosporium [*Drechslera*] *tetramera*	*J. curcas*	India	Singh (1983)
Meliola jatrophae (Black mildew)	*Jatropha* sp	n.a	Hosagoudar and Archana (2009)
Nigrospora sphaerica	*J. curcas*	n.a	Kirk (1991)
Passalora ajrekari	*J. curcas*	Ceara state of Brazil	Freire and Mosca (2009)
P. jatrophigena	*Jatropha* sp	Brazil	Braun et al. (2004)
Pseudocercospora jatrophae	*J. curcas*	India	Das and Chattopadhyay (1990)
Powdery mildew and damping-off	*J. curcas* nursery	Zhenkang County	Li et al. (2009)
Five species of fungi (seed quality deterioration)	*J. curcas*	India	Neelu et al. (1996)
Cassava superelongation disease (*Sphaceloma manihoticola*)	*Jatropha*	n.a.	http://www.gardenguides.com
Cassava latent virus-C strain	*J. multifida*	Kenya	Bock et al. (1981)
Bunchy top virus	*J. curcas*	India	http://www.phytotron.com/jatropha1.htm

(continued)

Table 10.1 (continued)

Pathogen	Host	Country of report	Reference
Bacterial diseases			
Angular spot (*Xanthomonas* sp.)	*J. curcas*	Nicaragua during 1993–94	Padilla and Monterroso (1999)
Xanthomonas malvacearum	*J. curcas*	n.a	Hayward and Waterston (1964)
Xanthomonas axonopodis pv. *manihotis*	*J. curcas*	Mexico	CABI UK Centre Annual Report (2009)
Erwinia amylovora (Fire blight)	*J. integerrima*	Lee county, USA	Stephen and Mark (2010) http://lee.ifas.ufl.edu/Hort/, browsed in Dec 2010

Lesion Nematode (Pratylenchus *spp.*)

Pratylenchus roseus was isolated from the soil around the roots of *J. podagrica* from Umerkot, Nawabshah and Karachi, Pakistan (Zarina and Maqbool 1998). *P. roseus* can be separated from all the other species of the genus by the presence of lateral vulval flaps.

Other Nematode Species

Fernandes and Asmus (2007) studied the reaction of *J. curcas* to *Rotylenchulus reniformis* in Brazil and found that *J. curcas* was tolerant to *R. reniformis*. Two nematode species from the family Hemicycliophoridae (Criconematoidea: Nematoda) *viz., Hemicycliophora demani* from *J. gossypiifolia* and *Caloosia exilis* from *J. glandulifera* were reported from Orissa (Ray and Das 1980). A field survey conducted in Northern Samar, Philippines revealed the presence of *Aphelenchoides besseyi* from the soil samples taken from the rhizosphere of *J. curcas* (Tuan et al. 2009).

Insect Pests

A global list of phytophagous insects consisting of 60 species in 21 families and four orders has been compiled in Australia, where *J. curcas* is considered as a weed. At least 15 species from the insect order Heteroptera were reported to affect *J. curcas* in Nicaragua (Manoharan et al. 2006).

Fig. 10.8 Nymph of shield backed bug (*Scutellera nobilis*) attacking *J. curcas* (Courtesy: www.Jatropha.pro)

Scutellarid Pests (Scutellera nobilis, Pachycoris klugii *and* Chrysocoris purpureus)

Scutellera nobilis, Pachycoris klugii and *Chrysocoris purpureus* are the scutellarid bug pests of *J. curcas*. *S. nobilis* is very common in Asia and Africa. Ambika et al. (2007b) conducted studies on the biology, potential damages and management of scutellarid bugs using microbials, plant products and insecticides.

***Scutellera nobilis* (the shield-backed bug or scutellarid bug)** has been considered as a key pest of *J. curcas* plantations in India (Chitra and Dhyani 2006; Sharma 2006), Brazil (Carels 2009) and Nicaragua (*Pachycoris klugii*). Very high to medium incidence of *S. nobillis* on *J. curcas* was also reported in Rajanandgaon District of Chattisgarh (Pankaj 2007; Khande et al. 2008) and Andhra Pradesh (Prabhakar et al. 2008). Another scutellarid bug, *Agonosoma trilineatum*, is also a serious problem on *J. curcas* in India due to its seed-feeding habit (Sharma 2006).

Damage: Damage due to *S. nobilis* is mainly observed during the stage of pod development stage to pod maturity. The pest attacks the fruits by sucking the fluid from the young and premature pods (Fig. 10.8). The pest mainly causes flower fall, fruit abortion and seed malformation (www.Jatropha.de). It also reduces the seed weight. Seed yield losses upto 19% were recorded due to the damage caused by *P. klugii* at a density of 3,500 bugs per hectare (Grimm and Guharay 1998). Feeding by *P. klugii* and *L. zonatus* leads to fruit abortion, hollow and deformed seeds. In Jhansi, *S. nobilis* population was observed to occur at an average of five per bunch, with a maximum of 15 bugs per bunch. The incubation period, nymphal period and adult longevity of *S. nobilis* were reported to be 5.9, 26.9 and 38.8–43.5 days, respectively (Ambika et al. 2007b). The potential damages are positively correlated with the bug population. Two bugs per plant did not cause significant damage, while 20 bugs per plant resulted in significant damages (Sharma and Srivastava 2010).

Fig. 10.9 *C. purpureus* incidence on *J. curcas* (Courtesy: http://jatropha.pro)

Chrysocoris javanus *and* C. purpureus

C. javanus is easily recognized by its bright red colour with black lines across its body. Its length can reach 2 cm. This pest is found in *Ricinus communis* and *J. curcas*. *C. purpureus* (Fig. 10.9) was recorded in North Western provinces, Sikkim, Calcutta, Assam and several parts of South India (Prabhakar et al. 2008), including Pondicherry (Kershaw and Kirkaldy 1908).

Damage: *C. purpureus*, another scutellarid bug pest of *J. curcas* with egg, nymphal and adult longevity of 5.75, 33.49 and 45–51 days, respectively with a fecundity of 25 eggs/female (Ambika et al. 2007b).

Management: *S. nobilis* can be controlled by spraying carbosulfan 25 EC at 1 ml l^{-1} (NOVOD Board 2009). Mechanical removal of *C. javanus* adults during pruning and plant maintenance is recommended.

An egg parasitoid, *Trissolcus* sp. was found to parasitise both the pests, *viz., S. nobilis* and *C. purpureus*. *Beauveria bassiana* was found effective, but neem products were ineffective in controlling the pest. Among the insecticides, carbosulfan (0.025%) was found effective followed by monocrotophos (0.045%) and dichlorvos (0.076%) (Ambika et al. 2007b).

A treatment of two foliar sprays of lambda-cyhalothrin (5 EC at 25 g a.i. ha^{-1}) and imidacloprid (17.8 SL 100 mL ha^{-1}) given at monthly interval followed by carbosulfan (25 EC at 250 g a.i. ha^{-1}) and monocrotophos (36 SL at 500 g a.i. ha^{-1}) was found effective. Spinosad, endosulfan, and entomopathogenous fungi, *viz.,*

Beauveria bassiana and *Metarhizium anisopliae* resulted in moderate reduction in pest population. A maximum reduction of bug population was noticed after the third and seventh day of insecticides and bioagents spray, respectively. Chlorpyriphos (50 EC at 250 g a.i. ha^{-1}) and neem oil (2%) were found less effective and exhibited minimum reduction in bug population as compared to other components of the integrated pest management (Sharma and Srivastava 2010). Control of bugs with *B. bassiana* and *M. anisopliae* resulted in 65% mortality of *P. klugii* and 94% of *L. zonatus*, increasing the yield by 28% (Grimm and Guharay 1998).

Rainbow Shield Bug (Calidea dregii)

This pest is known to pose a threat to *J. curcas* plantations in Guinea-Bissau and has potential to cause much more damage to yield and quality of oil (Nielsen 2010). It has been reported as a minor pest on *J. curcas* in India. It also infests sorghum, maize, rice, okra, sunflower, Noog Abyssinia (*Guizotia abyssinica*), Star Burr (*Acanthospermum hispidum*), *J. podagrica*, and cotton. In Ghana, *C. dregii* Germar infestation on *J. podagrica* as well as oviposition on the flower parts and stems was reported throughout the year (Kaufmann 1966). Although there are number of diferent *Calidea* spp. in Africa, South of Sahara and Arabia, *C. dregii* was only reported on *J. curcas* in Guinea-Bissau and Mozambique (Nielsen 2010). In Mozambique, it occurs in insignificant numbers whereas in Guinea-Bissau, it is a major pest.

Damage: The bug sucks the sap from developing seeds leading to premature seed dropping or incomplete seed development and low mature seed yield. Seed weight from pest affected plants was reduced upto 20% as compared to those from dry areas of Mozambique. In cotton, it leads to fiber staining, which affects the quality although premature boll dropping does not occur (Nielsen 2010). In Tanzania, the development of fewer and smaller seeds due to the pest infestation was also reported on sunflower (Freeman 1939), which demonstrated a large potential of alternative host and increased the threat. It was also found that the infestation increased the *free fatty acid* (FFA) content of the sunflower oil, which results in a drop of transesterification efficiency and biodiesel quality.

Management: It is difficult to kill this pest with insecticide as it feeds and breeds on a wide variety of plants including poisonous ones like *J. curcas* and castor. In addition, their population size increases rapidly with the spreading and devastation rates that one may expect. In Tanzania, early planting of cotton reduced the bug infestations, probably because the boll formation stage coincided with a time when there were other food sources available for the bug. This method cannot be used for perennial plants like *J. curcas*. Pruning can delay or break dormancy and thus influence the flowering time of *J. curcas*. Field trials should be established to assess its feasibility and effectivity. In Ghana (Kaufmann 1966), the bug population was found to be more or less constant. Nymphs eat eggs, thus a higher density of nymphs reduces the number of eggs for the next generation. These bugs are highly mobile and fly around in the area to whatever food source is available. The high number of bugs observed in Guinea-Bissau is probably due to a lack of alternative feed sources in the middle of the dry season.

True Bugs Leptoglossus zonatus (*Dallas*) (*Het.: Coreidae*), Pachycoris klugii *Burmeister* (*Het.: Scutelleridae*), Hypselonotus intermedius *Distant* (*Het.: Coreidae*)

Two species of fruit feeding true bugs, *Leptoglossus zonatus* (Leaf-footed bug) and *Pachycoris klugii* and one flower feeding true bug, *Hypselonotus intermedius*, are known to cause damages to *J. curcas* plantations. In Nicaragua, fruit borers (*L. zonatus, Hyalymenus tarsatus*) and flower borer (*H. intermedius*) has been reported (Grimm and Maes 1997b). The most frequent true bug found in Central America is *P. klugii,* which has been observed throughout the year at densities upto 127 insects per tree or 140,000 per ha of plantation (Wink et al. 2000). *J. curcas*, with its regular fruiting of homogeneous nutritional quality, is as a highly suitable food plant for *L. zonatus* allowing it to maintain its populations throughout the year (Grimm and Somarriba 1999). Among the populations of phytophagous bugs (Heteroptera) monitored by Grimm and Fuhrer (1998) during three growing periods in *J. curcas* plantations of Nicaragua and Cape Verde, *P. klugii* and *L. zonatus* were the two most frequently occurring species.

Damage: The qualitative and quantitative damages caused to the fruits and seeds of *J. curcas* by these true bugs were assessed using field cages (Grimm 1999). All three species caused overall yield reduction of *J. curcas* production. The fruit feeding bugs caused premature fruit abortion and seed malformation. Parameters such as fruit number, seed and seed kernel weight as well as seed length were positively correlated with population density of adult individuals. The oil content of seeds was slightly reduced by the bugs, but protein content remained unchanged. Damage increased with the developmental stage of the larvae. Female *P. klugii* caused less damage than nymphs, while male *P. klugii* caused no significant damage at all. Adult *L. zonatus* of both sexes produced more damages than nymphs. The flower feeding bug *H. intermedius* is a pollinator of *J. curcas* and at high densities, it reduced the number of maturing fruits (Grimm 1999). Grimm and Somarriba (1998) studied the biology and life cycle of the scutellarid bug *P. klugii* in the field and the laboratory in Nicaragua. They found that the species is multivoltine and each female oviposits repeatedly during the rainy season. The adults go hiding during the dry season.

***Management*:** The entomopathogenic fungi, *viz., B. bassiana* and *M. anisopliae* were effective to control populations of scutellarid bugs. Applications of *M. anisopliae* through ultra-low volume droplet of mineral oil at a rate of 1×10^{10} conidia tree^{-1} caused bug mortalities ranging from 65% in *P. klugii* to 94% in *L. zonatus. B. bassiana* increased fruit yield by 28%, and was more effective than malathion or an aqueous extract of ground neem seeds (Grimm and Guharay 1998). The egg parasitoid, *Pseudotelenomus pachycoris* [*Telenomus pachycoris*] was found effective against *P. klugii* in Nicaragua and *Pachycoris torridus* in Sao Paulo, Brazil (Gabriel et al. 1988).

Leaf and Flower Webber Cum Fruit Borer, Salebria (Pempelia) morosalis (*Saalm Uller*)

Salebria (Pempelia) morosalis is reported to only affect few forest species like *Desmodium gangeticum*, *Flemingia* sp., *Uraria lagopides* (Beeson 1941) as well as *J. curcas*, but it does not occur on field crops grown in India. It is a major pest in Jhansi, Uttar Pradesh (Chitra and Dhyani 2006) and an emerging pest in Tamil Nadu (Regupathy and Ayyasamy 2006), Andhra Pradesh (Prabhakar et al. 2008) and Gujarat (Nayak et al. 2008) states of India. *P. morosalis* may likely become a regular pest of *J. curcas* with growing importance due to the extending monoculture of this crop. The occurrence of this pest on *J. cur*cas has also been reported in Kenya (http://betterglobeforestry.com/).

The adult moth is gray with snout like labial palpi in the head and hyaline hind wings. The male is slightly smaller than the female with pointed abdominal tip. The biology of this pest was studied extensively by different workers (Regupathy and Ayyasamy 2006). Ambika et al. (2007a) also studied the biology of leaf webber and the results revealed that egg, larval, pupal, and adult longevity of males and females was 5.83, 28.50, 7.50, 5.67 and 7.25 days, respectively.

Damage: The webber affects the leaves, bark, inflorescence, and fruits during the growth of new flush immediately after rain and flowering stage. It reduces the crop growth and flower formation. The greenish brown/brownish green caterpillars (Fig. 10.10) were observed to feed by digging leaves and remaining in the leaf web (Fig. 10.11). At flowering they bore into peduncle and fruits, which show galleries made of silk and frass. The caterpillar bores into the fruits throwing out faecal matter. The greenish larvae turn pinkish at the time of pupation. It pupates on the fruits. The larvae are seen under a cover of silk, frass or excreta, which extend between flowers or fruits (Regupathy and Ayyasamy 2006).

Management: Tamil Nadu Agricultural University (TNAU) (Paramathma et al. 2004) recommends to spray endosulfan, neem oil (2%), monocrotophos 36 WSC (1.25 ml l^{-1}), profenophos 50 EC (1 ml l^{-1}) or endosulfan (2 ml l^{-1}) (NOVOD Board 2009). Topical application method to assess the toxicity of insecticides to *P. morosalis* was described by Regupathy and Ayyasamy (2006). A dipteran parasite and spider, *Stegodyphus* sp. was reported as a natural controlling agent in Jhansi, India (Chitra and Dhyani 2006). However, the mass culture and release technology is yet to be improved. Among the microbes tested, *Bacillus thuringiensis* was found to be

Fig. 10.10 Leaf webber larvae (*Salebria morosalis*) attacking *J. curcas* leaf (Courtesy: http://jatropha.pro)

Fig. 10.11 Leaf webber damage on *J. curcas* (Courtesy: http://jatropha.pro)

the most effective against leaf webber. The NLC fly ash (125 mg leaf^{-1}) was more effective against leaf webber than paper board fly ash. All the insecticides tested under field conditions, i.e., monocrotophos, profenofos, dichlorvos, chlorpyrifos, quinalphos, endosulfan and methyl parathion [parathion-methyl] were effective in reducing the larval population of leaf webber (Ambika et al. 2007a).

Thrips (Rhipiphorothrips cruentatus, Retithrips syriacus, Selenothrips rubrocinctus)

J. curcas fields in Madurai, Tamil Nadu, India, were severely infested by grapevine thrips (*Rhipiphorothrips cruentatus*) during November 2006 (Shanthi et al. 2007). The incidence was widespread throughout the field, with populations ranging from 200 to 250 nymphs and adults per young leaf. The nymphs were white when they hatched from the eggs, but they soon developed pale red markings. The female thrips were 1.2–1.5 mm long, blackish-brown, with the legs and antennal segments yellow and the fore wings pale with yellowish veins. Male thrips were similar to females, but their pronotum and abdomen were yellow (Shanthi et al. 2007). Damage due to *R. cruentatus* on *J. curcas* was observed in Karnataka, India (Rani and Sridhar 2002). Saturnino et al. (2005) recorded thrips damage on *J. curcas* in Brazil. Thrips were dominant among the insect pests present at the *J. curcas* plantation located in Northern Samar, Philippines (Tuan et al. 2009).

Retithrips syriacus was reported for the first time in Puerto Rico on *J. curcas* leaf during quarantine inspection on 30 April 1993 in San Juan (Medina-Gaud and Franqui 2001). *R. syriacus* incidence was recorded on *J. curcas* in parts of Andhra Pradesh (Prabhakar et al. 2008). In addition to the above two species, *Heliothrips haemorrhoidalis* and *Scirtothrips kenyensis* were reported on *J. curcas* in Kenya (http://betterglobeforestry.com). *Thrips hawaiiensis* and *Scirtothrips dorsalis* were reported to cause damage to *J. curcas* in Andhra Pradesh, India (Raju and Rao 2003).

Damage: Both nymphs and adults of *R. cruentatus* colonize the lower surface of leaves and suck the sap. Younger leaves are preferred by thrips for feeding. Young infested leaves become silvery white initially, later on pale brown and crinkled with roughening of upper surface. Thrips are observed in older leaves with yellow spots on the leaves. In severe cases, shedding of leaves is observed (Shanthi et al. 2007). *R. syriacus* causes leaf mottling, yellowing and browning of leaves and infestation is severe during hot weather condition. This promotes flower dropping and premature shedding of pods.

Management: Research at TNAU, India, revealed that the pest can be controlled by spraying methyl parathion 25 EC (2 ml l^{-1}) or dimethoate 30 EC (2 ml l^{-1}) (NOVOD Board 2009).

Yellow or Golden Flea Beetle (Aphtona *sp.*, Podagrica *spp.*)

The yellow or golden flea beetle (*Aphtona* sp., Halticinae, family Chrysomelidae, Coleoptera) is the main insect pest of *J. curcas* in Africa (Mozambique, Zimbabwe, Tanzania and Kenya). Incidence of *Aphtona* sp., on *J. curcas* plants (few weeks to 1.5 years old) in 6 different planting sites in the East slope of Simba hills, Kenya was noticed in 2006. The beetles were 5 mm long and 2 mm wide, and mostly reddish. These beetles are known to feed on *Euphorbia* species and were introduced in USA for biological control of leafy spurge. Golden flea beetle (*Podagrica* spp.) damage on leaves and shoots of young plants of *J. curcas* was noticed in some areas of Zimbabwe (http://www.jatropha.de/zimbabwe/).

Damage: The larvae and adults feed on young reddish leaves and sometimes on the large green leaves and shoots as well; their larvae also penetrate the roots (Nielsen 2007; Gagnaux 2008). The beetle damages can become severe in young seedlings (not more than 1.5-year-old) and affect the yield in mature plants. Beetle damage was more severe in areas where the vegetation suggested low soil fertility. There are reports that the pest has almost destroyed plantations in Uganda, Mozambique, Mali and Zambia. During 2006, in Mozambique, a mortality rate of 95–100% was experienced in nurseries and in fields with plants upto 3-year-old (Nielsen 2007). The yellow flea beetle (*Aphtona dilutipes*) is reported to cause more severe damage than the golden flea beetle, resulting in upto 100% mortality (Timothy Mahoney, Pers. comm.).

Management: The planting of a trap crop around the *J. curcas* field can'intercept' beetles wherever they try to enter the field. In order to avoid the beetles from leaving the trap crop and moving to the adjacent *J. curcas* field, they should be controlled with an insecticide soon after their arrival. Lambda cyhalothrin has proved to be effective against the beetle. Pesticides containing Chlorpyrifos or Cyphenothrin were also found effective against *Aphtona* spp. Pyrethroids or carbamates (Sevin) are generally effective. For organic farms, neem or insecticidal soap may be applied, but these are less effective. Other insecticides containing pyrethrins (Pyganic) or kaolin clay (Surround) have also sometimes proved to be effective. Insect repellents containing hot pepper or garlic may also provide some control. Commercial formulations of entomopathogenic nematodes may be helpful in reducing flea beetle damage. On soil application, the nematodes attack beetle larvae, reduce root feeding and help to prevent the next cycle of adults. For effectiveness, they should be applied when larvae are present and the soil is wet.

Cutworm (Agrotis ipsilon)

Cutworm is considered as a major pest of *J. curcas* in Philippines (The Philippine Star 2010). Occurrence in Indonesia (http://www.used-cars.co.jp/biotec/jatoropha.pdf) and China (Li et al. 2009) has also been reported.

Damage: *A. ipsilon* attacks seedlings and young plants from the soil surface. The damage is indicated by the cutting of the stem near the soil surface and the plant withering. The assessment of critical pest damages showed that of the five identified insect pests attacking *J. curcas*, cutworms were confirmed as one of the major pests with a threshold of economic loss on yield due to cutworms of ~19%. This level signifies the need for appropriate measures to control this pest. Using the leaf damage index and the critical damage index, researchers established the economic threshold level equivalent to 5–6 larvae/50 plants for cutworms (PCARRD 2009). The researchers also determined that temperature affected the growth and development of cutworms. For instance, a warm temperature of 28.5°C enhanced the reproductive rate of cutworms from egg to larval stage. Likewise, warmer temperatures speed up the hatching of the cutworm eggmass. Under the maximum daily temperature of 28.5°C, it took only a day and a half for the eggs to hatch. Conversely, it took more than a week for the eggs to hatch under the lowest registered temperature of 25°C (PCARRD 2009).

***Management*:** The mechanical control is performed by collecting the larvae around the plant and killing them. Prevention by keeping the field free of weed several weeks before planting will help to reduce the cut worm incidence. Toxic baits such as bran, sawdust, or cassava mixed with insecticides are effective in killing the larvae. The mixture is poured out around the plant, after the attack is noticed. Insecticide can be in liquid form sprayed on the lower part of the stem and the surrounding soil. Granular insecticides can be mixed with soil at the time of ploughing and harrowing. Insecticides, which contain active ingredients such as deltamethrin, thiodicarb, carbofuran, or beta cyfluthrin are effective.

Grasshoppers

Some kinds of grasshoppers (*Valanga nigricornis* and *Locusta migratoria*) may attack *J. curcas* plants anytime (Fig. 10.12). However, in Indonesia, the attack of *L. migratoria* is seldom observed. Generally heavy attack would occur on young plants. Occurrence of grasshopper (*Attractomorpha ranacea*) on *J. curcas* plantations has been reported in South India (Manoharan et al. 2006) and Nicaragua (Grimm and Maes 1997a) with seed yield losses upto 1% (Grimm and Guharay 1998). Giant grasshoppers are known to attack *J. curcas* in Zambia (BAZ 2007). A survey conducted in Philippines revealed that the incidence of leaf chewing grasshoppers is not abundant (Tuan et al. 2009). A grasshopper-like insect, which feeds on *J. curcas* plant was responsible for withering and death of the plant in Swaziland (Friends of the Earth 2009). Surveys conducted in parts of Mexico revealed that grasshoppers and ants are the major insect pests, which are responsible for severe defoliation and death of young plants, hampering the successful establishment of seedlings (Marc 2010). In feeding preference tests in Madras with the grasshopper, *Eyprepocnemis alacris alacris* (Serv.), of the 38 plant species tested, only eight were consumed without reluctance and *J. glandulifera* is one among them (Mralirangan 1978).

Management: Spraying with insecticide is not always successful as the grasshopper attack is periodic and spontaneous. The recommended insecticides are betacyfluthrin, cypermethrin, thiodicarb, MIPC and fipronil, but the application should be careful and wise.

Fig. 10.12 Grasshopper attack on *J. curcas* plant (Courtesy: Jose Ines Bazan-Mota)

Army Worm (Spodoptera litura)

Spodoptera litura was identified as a pest of *J. curcas* during periodical surveys at Barha, near Jabalpur, Madhya Pradesh, in 1992–93 and the incidence reached an extent of 60–70% (Meshram and Joshi 1994). The pest is widely distributed especially in Asia, Pacific and Australia. It has a wide host range covering more than 120 plant kinds, *viz.*, tobacco, corn, paddy, tomato, chilli and legumes including soybean, *J. curcas* and taro. It is identified as a pest of *J. curcas* in India (Dalia Amin 2001).

Damage: Larvae eat the leaves of the young and mature plants and often left the leaves bitted. If the attack is heavy, only veins of the leaves are left, and plant becomes bald (Fig. 10.13).

Management: Natural enemies such as egg parasitoid, *Telenomus spodoptera* (Hymenoptera: Scelionidae), larval parasitoid, *Microplitis manilae* (Hymenoptera: Braconidae), predator from Carabidae, Pathogens, *Nuclear polyhedrosis virus* and *Borrelinavirus litura* were identified against the pest. Mechanical control is done by collecting and killing the egg masses and young larvae. When the larval population is high, insecticide with *Bacillus thuringiensis* or *S. litura*-NPV (S1-NPV) at a concentration of 6×1.0^{11} polyhedral inclusion bodies (PIB ha^{-1}) can be applied. Natural insecticide from neem (*Azadirachta indica*) seeds at a concentration of 4 $g l^{-1}$ of water was also found effective. Proper application of synthetic insecticides such as betacyfluthrin and prothiofos is advised.

Mites

Species of *Jatropha* have been observed to be attacked by mites of different species. Major phytosanitary problems of *J. curcas* in Brazil include attacks by two mite

Fig. 10.13 *Spodoptera litura* larva attacking *J. curcas* leaf (Courtesy: www.fact-foundation.com)

species, the broad mite *Polyphagotarsonemus latus* and the spider mite *Tetranychus bastosi* (Sarmento et al. 2011). Mite damages on *J. curcas* was also reported in Kenya (http://betterglobeforestry.com/).

Broad Mite [*Polyphagotarsonemus latus* (Tarsonemidae)]

The broad mite, *P. latus*, is a polyphagous mite that has been quoted as one of the most important pests of *J. curcas* in Brazil (Carels 2009). *P. latus* was able to complete its life cycle and reproduce on all tested genotypes (Filomena, Bento, Oracília, Gonçalo and Paraguaçú) indicating its damaging potential (Lopes et al. 2010) for *J. curcas*. A strong and permanent attack of the broad mite mostly during the rainy season was also reported on *J. curcas* in Costa Rica (Aguilar et al. 2010). The pest was observed regular on *J. curcas* in Chattisgarh, but sporadic in Tamil Nadu (Regupathy and Ayyasamy 2007) and has been found to be emerging as a major problem in Tamil Nadu, India. The populations of broad mite were the largest during November and the total life cycle of broad mites lasted 6 days; females and males lived for 9 and 7 days, respectively (Kavitha et al. 2007).

A great diversity of mites was found on *J. curcas*, especially predaceous phytoseiids (a species of *Amblyseius* and two species of *Neoseiulus*); *Iphiseiodes zuluagai* Denmark & Muma and *Phytoseiulus macropilis* (Banks), 2 morpho-species of Oribatida, *Tydeus* (Tydeidae) and *P. latus* found. The presence of predatory mites indicated their role in the control of the phytophagous species (Almeida et al. 2010).

Tetranychus sp.

Mites damage the leaf and make the plant weak. Mite is a polyphagous organism that can attack various plants such as cotton, tomato, legumes, citrus, papaya, cassava, peanut and weeds. *J. curcas* with no wax on the flower is more resistant to this pest. In Brazil, *J. gossypiifolia* plant proved to be an important mite reservoir. In all tested samples, two *Tetranychus* species (Acari: Tetranychidae) and one *Neoseiulus* species (Acari: Phytoseiidae) were consistently found, especially on the basal leaves. Leaves of this species are glabrous, differently from leaves of *Jatropha* sp., onto which low levels of a single species of phytophagous mite, *Tetranychus* sp., and a phytoseiid species were found (Almeida et al. 2010). The red spider mite, *T. utricae* is severe in Tamil Nadu, India on plants raised as hedges during warmer months, which may be due to water stress and high temperature (Regupathy and Ayyasamy 2007). The populations of red spider mite were the largest during October and the total life cycle of the spider mite was 6 days (Kavitha et al. 2007).

Damage: Leaves become yellowish and then gets rusted. The shrinking leaf is reddish and then falls on the ground. Mites are commonly found at the lower surface of the leaves and its bites appear as yellow or red spots. Mites can only induce leaf malformation and, ultimately, their fall. These mites spread through the falling leaves blown by the wind or through contact with workers in the garden or estate. In case of broad mite damages (Fig. 10.14), young plants get very thick and sturdy with leathery leaves and salient top veins. Shoots will dry and plants stop growing until a new flush starts with the onset of rain or irrigation. Mite is a typical problem of nursery plants and young plantations. Infected plants sometimes recover, but produce a bunch of flowers in the top.

Management: The basic care of cleaning the plant area by collecting and burning all thrash and infested leaves is important. If the attack is minimal, it is recommended to wait untill pruning time and to cut all the attacked plant parts. Larger areas can be treated with avermectin (1.9 EC at 0.5 ml l–1), dicofol (18.5 EC at 3 ml l–1), or vertimec (1.9 EC at 0.5 ml l–1), or any common miticides. Acaricide or miticide with propargit, dicofol, tetradifon, amitraz and dinobuton as active ingredients can also be used as chemical control agents by spraying upward from below and directly pointed to the mites. Frequency of spraying can be as many times as required. Studies conducted at Maharana Pratap University of Agriculture & Technology, Udaipur revealed that the red mites affect the leaves during the rainy

Fig. 10.14 Damage on *J. curcas* leaf due to broad mite (Courtesy: http://jatropha.pro)

season and are controlled by proparzyte (58% SL, Simba, at 1 ml l^{-1}) (NOVOD Board 2009). Among several chemical assays, abamectin (0.0009%) provided the best control for both broad mites and spider mites (Kavitha et al. 2007).

The natural enemies of mites such as predator from *Phytoseiidae* family, and beetles of *Coccinellidae, Stethorus* sp. attack their eggs and larvae. The suitability of predatory mite species, *viz., Iphiseiodes zuluagai* and *Euseius concordis* in controlling *P. latus* and *T. bastosi* on *J. curcas* was evaluated for the first time by Sarmento et al. (2011) and these authors found that *I. zuluagai* was more efficient than *E. concordis* in reducing populations of *P. latus* and *T. bastosi* under field conditions.

Nettle Caterpillar (Parasa lepida)

Nettle caterpillars (Lepidoptera; family: Limacodidae) attack *J. curcas* periodically. Initially they live in group on a leaf and spread to all parts of the plant as the larvae grow older. The adult female lays its eggs as a mass on the soft part of the plant. The caterpillar is of 1.5–2.5 cm long, green with blue dots lengthwise. The pest moves like snail and produces certain chemical compound, which stings the skin.

Management: Mechanical control is performed by killing young larvae and cocoons by soaking in water or kerosene. In addition to spraying organophosphate insecticides, biocontrol agents such as fungal pathogen (*Cordyceps cocconae*), virus, parasitoid (*Apanteles parasae*) and *Bacillus thuringiensis* are also used to control mites.

Leaf Caterpillar or Castor Semilooper (Achaea janata)

Sporadic occurrence of *A. janata* was recorded on *J. curcas* in selected locations of Tamil Nadu (Regupathy and Ayyasamy 2007). The larvae can eat all the leaves in a short time. Heavy attack will influence the quantity and quality of seeds.

Management: The mechanical collection and burning of old larvae is the most effective mean of combating the pest. Suitable planting distance prevents larvae to migrate from one plant to another. Another way is to throw away the attacked leaves where many young larvae are attached at the lower surface. Until now, there is no variety of *J. curcas* available, which can stand *A. janata*, but there are natural enemies such as *Trichogramma evanescens* (egg parasitoid) and *Microplitis maculipennis* (larval parasitoid). Plant insecticides containing a neem extract can be used as well as a synthetic insecticide such as alfamethrin.

Stem Borer (Ostrinia furnacalis *and* Xyleborus *spp.*)

Damages (Fig. 10.15) by stem borers were mainly noticed on *J. curcas* grown in heavy soils in Tanzania (http://jatropha.pro/). In Indonesia, there are two stem borers, *viz., O. furnacalis* and *Xyleborus* spp. *O. furnacalis* is commonly called as Asian corn borer. Old larvae usually drill stems and cause plant break by the wind. Sometimes there is a hole at the stem basis that indicates the point of entry by a larva.

Management: A good practice is to incinerate all trashes from pruning since the proper maintenance of plants will help to control the population size of this pest. Carbofuran can be used to control the pest.

Termite (Odontotermis *sp.*)

Termite damages occur mainly on laterite red soils, and sandy soils with a very low organic matter content (http://jatropha.pro/). Termites attack plants at their base (Fig. 10.16), causing plant destruction and total loss. Occurrence of natural termite (*Odontotermis* sp.) colonies in West Bengal, India was reported (Chattopadhyay 2009) in an area that causes no harm to plants though in some cases mounds are formed around plant bases, which causes uprooting due to floods of pre-monsoon storm during May-June. The estimated loss is less than 1% so far. Termite incidence on *J. curcas* was also reported in Tanzania (http://www.jatropha.de/photo-show/index.htm#use). However, the termicidal activity of *J. curcas* oil was reported against the Philippine milk termite *Coptotermes vastator* Light (Isoptera: Rhinotermitidae, Acda 2009).

The application of neem infusions prepared by soaking neem leaves in water for 4 days on infested areas is useful against termites as it stops their feeding and finally causes their death by starving. (http://jatropha.pro/).

Fig. 10.15 Stem borer (Courtesy: www.fact-foundation.com) damage on *J. curcas* plant (Courtesy: www.Jatropha.pro)

Fig. 10.16 *J. curcas* plant affected by termites (Source: Jatropha: A Smallholder Bioenergy Crop. The Potential for pro-poor development, Food and Agricultural Organization)

Fig. 10.17 *Nezara viridula* adult sucking on *J. curcas* leaf (Courtesy: www.fact-foundation.com)

Tip Borer Caterpillar (Dichocrosis punctiferalis)

The pest incidence on *J. curcas* was reported in Southeast Asia, Australia and Pacific Islands. In Java, its incidence was found in low land areas upto 1,750 m above the sea level. It attacks usually at the time of flowering initiation. The pest bores the tips of plants and fruits. Females lay their eggs on the soft part of plants. After the eggs hatch into larvae, the attack begins at the tip of young plants, or on seeds of old plants.

Management: The pest incidence can be minimized mechanically by collecting and burning infested shoot tips and seeds. The spraying with monocrotophos and bensultap during flowering or the application of carbofuran prior to flowering and 15–20 days after flowering was found to be useful in minimizing plant infestations.

Stink Bug (Nezara viridula) (*Pentatomidae*)

Stink bug (Fig. 10.17) plays an important role in the tropical region. This pest sometimes attacks *J. curcas* during flowering time, which causes heavy damage to the fruit capsule. This pest is spreading all over the world and is easy to recognize because of its green color. The other host plants are paddy, tomato, legumes, chilli, cotton, potato, soybean and corn. The main damage is not due to the direct suction of plant sap, but by the toxin in its saliva. This toxin may wither the plant and cause the death of its tip and leaves.

Fig. 10.18 *Helicoverpa armigera* larva on *J. curcas* leaf (Courtesy: www.fact-foundation.com)

Management: Eggs and adult insects should be collected and burnt. The pest incidence can be minimized by avoiding alternate host plants in the vicinity. If the pest population is too high, one recommends the application of insecticides such as chlorfluazuron, diflubenzuron, alfamethrin, and lamda cyhalothrin.

Ear Corn Caterpillar (Helicoverpa armigera)

H. armigera is a polyphagous insect that attacks soybean, tomato, chilli, cotton and many other hosts including *J. curcas* (Fig. 10.18).

Management: Applications of chlorphyriphos, lamda sihalothrin, fenvalerate, permethrin or sipermethrin may reduce the population of the pest in the field. Other alternatives are applications of *H. armigera*–NPV virus, eggs of *Trichogramma armigera* parasitoid, and a bioinsecticide powder prepared from *Azadirachta indica* at the concentration of 4 g l^{-1} of water.

Leaf Hopper (Empoasca *sp.*) (*Homoptera*)

Leaf hopper is an important pest of *J. curcas* in tropic and sub-tropic regions. It also attacks tea and other crops. On the field, these hoppers could be found throughout the year, but are very dangerous to the plant on the seed bed. These species were affecting *J. curcas* in industrial cultures in Brazil (Carels 2009).

Damage: Females lay their eggs on the leaf net, close to the leaf ribs at the lower surface. Nymphs and adults suck the plant sap from the lower surface of the leaf and induce it to dry and die. Sometimes, the curly leaf occurred at the tip. The pest attack is influenced by the color of leaf and texture of flowers. *The J. curcas plants with their low* content of leaf carotene and thick flower wax are more tolerant to this pest.

Management: Use of systemic insecticides such as imidacloprid, beta cyfluthrin or carbosulfan on seed bed can minimize the incidence.

Jatropha Leaf Miner (Stomphosistis thraustica) (*Gracillaridae; Lepidoptera*)

The infestation of this insect has been observed by the researchers during 1997–1998 for the first time in Chhattisgarh plains in India (http://www.botanical.com/site/column_poudhia) and later in Karnataka (Rani and Sridhar 2002) and Andhra Pradesh (Arif et al. 2007). The *J. curcas* leaf miner is known as a medicinal insect by the traditional healers of Chhattisgarh plains. The larvae collected just before pupation are considered the best stage for medicine preparation. The newly born larvae are not used. After the collection of larvae, they are dried in the shade, powdered and kept for future use as Galactagogue. The powder is given orally with lukewarm water in order to increase the flow of milk in lactating women (http://www.botanical.com/site/column_poudhia). Leaf mining moth (*Epicephala* sp.: Gracillaridae) is known to attack *J. gossypiifolia* in Australia (Wilson 1997).

Damage: The larvae feed on *Jatropha* species (including *J. curcas* and *J. gossypiifolia*). They mine the leaves of their host plant. Galleries often form several irregular blotch nets per leaf. The moth's larva produces a silvery mine on the upper leaf surface (Fig. 10.19). This moth was first collected in north Queensland in 1989 and has since been observed annually on bellyache bush in late summer. Damage caused by the moth is not sufficient to motivate its control.

Management: The application of *Bacillus thuringiensis* sub sp. *kurstaki* is recommended.

Mealy Bug (Ferrisia virgata)

Mealy bugs are soft body insects belonging to the family Pseudococcidae in the Order Hemiptera. Damages due to the white tailed mealy bug, *Ferrisia virgata*, was recorded from Tamil Nadu, India (Regupathy and Ayyasamy 2007). Cotton mealy bug (*Phenacoccus solenopsis*) incidence was recorded on *J. integerrima* in Multan district of Pakistan (Arif et al. 2009). Mealy bug species, *viz., F. virgata* and *Planococcus* spp. were reported to cause damage to *J. curcas* in Kenya (http://betterglobeforestry.com/).

Damage: Mealy bugs suck the sap from leaves and stems (Fig. 10.20), sometimes from fruits if heavily infested and causes crinkling leaves, dry stems and reduced reproductive parts.

Fig. 10.19 Leaf miner damage of *J. curcas* leaves

Fig. 10.20 Mealybug infestation on a *J. curcas* plant (Courtesy: plantoils.in)

Management: Application of chlorpyrifos or mercaptothion, dimethoate, malathion (50%) is recommended for controlling the infestation (http://jatropha.pro/).

White Grub (Holotrichia consanguinea *and* H. serrata)

White grub is a polyphagous pest whose damages are more severe in sandy and sandy loam soil. Adults are 18–24 mm long and 8–9 mm wide with three segmented thoracic legs and strong mouth parts. The young grubs are translucent, white and 5–6 mm long and beetles emerge out of the soil within 3–4 days after the onset of monsoon. Report of white grub (*H. consanguinea*) occurrence on *J. integerrima* was reported in Punjab, India during 1978 and 1979 (Brar and Sandhu 1982). Both adults and larvae can cause damage to plants. The larvae feed on roots and damage the stems, while grubs feed on fine rootlets, resulting in pale, wilted plants, dying in patches. The caterpillar can cause 20–30% damage to the young plants (http://www.sunplantgroup.com/jat_curcas9.htm).

Management: Repeated ploughing in summer, use of well decomposed organic manure and deep irrigation to restrict the respiration of grubs are some of the good cultural methods of control. The grubs and adults can also be collected and destroyed from host trees around the field. Braconids, dragon flies, *Nuclear polyhedrosis virus* and green muscardine fungus are some of the bioagents against the pest. Cypermethrine (25%) or chloropyriphos (50%) mixed with wood dust are effective against the pest under field conditions.

Giant African Land Snail

The giant land snails are important pests in Africa especially in *J. curcas* seedling nurseries (http://jatropha.pro/). These snails do not like coarse sand around the seedling beds; hence application of coarse sand around the seedling beds keeps them out of the nurseries. The smaller slugs attack the adult plants. They also can be kept away from plants by applying coarse sand around them, however, this method is not very practical on huge plantations. Application of carbamate insecticides can minimize the incidence.

Indian Quarantine Regulations

Reports of incidence of certain other insect pests, *viz.,* white flies, carpenter bee, bugs, weevils, stem and twig girdlers, cushion scale, soft scale etc. on *J. curcas* species through out the world are depicted in Table 10.2.

Table 10.2 Other insect pests recorded on *Jatropha* spp.

Pest	Host	Country of report	Reference
White flies (*Bemisia tabaci*)	*J. gossypiifolia*	Uganda, Kenya	Sseruwagi et al. (2006)
	J. curcas		http://betterglobeforestry.com/
Spiralling whitefly (*Aleurodicus dispersus*)	*J. multifida*	Tirunelveli, Tamil Nadu, India	Babu and David (1999)
Carpenter bee (*Xylocopa fenestra*)	*J. gossypiifolia*	Jodhpur, India	Sharma (1981)
Red pumpkin beetle (*Aulacophora foveicollis*)	*J. curcas*	India	Sharma (2006)
Cotton stainer bug (*Dysdercus suturelius*); moths (*Stomphastis* spp.)	*J. curcas*	Kenya	http://betterglobeforestry.com/
Oxyrachis tarandus Fabricus (Homoptera: Membracidae) Jan-Dec	*J. curcas*	Bihar, India	Ali et al. (2006)
Millepede (*Julus* sp.)	*J. curcas* (loss of seedlings)	—	Heller (1996)
Locust (*Oedaleus senegalensis*)	*J. curcas* (leaves, seedlings)	—	Heller (1996)
Cushion scale (*Pinnaspis strachani*); Wooly aphid (*Ferrisia virgata*)	*J. curcas* (Die back of branches)	—	Van Harten, pers comm (Heller 1996)
Blue bug (*Calidea dregei*)	*J. curcas* (Sucking on fruits)	—	Van Harten, pers comm (Heller 1996)
Flies (*Chrysomya megacephala*), bees (*Apis florea, A. indica* [*Apis cerana indica*] and *Trigona iridipennis*)	*J. curcas*	Andhra Pradesh, India	Raju and Rao (2003)
Phycita sp. (May-June)	*J. curcas*	Bangalore, Karnataka, India	Rani and Sridhar (2002)
Pachycoris torridus	*J. curcas*	Brazil	Soto and Nakano (2002)
Weevil (*Lepropus lateralis*); leaf miners	*J. curcas* nursery	Shuangjiang County	Li et al. (2009)
Nephopteryx larvae	*J. curcas*	Pusa and Mandalay, India	Hampson (1912)
Mylabris pustulata (Thnb.) adults	*J. panduraefolia* flowers	Aligarh, Uttar Pradesh, India	Siddiqui (1983)
White grub, *Holotrichia consanguinea* in 1978 and 1979	*J. integerrima*	Punjab, India	Brar and Sandhu (1982)

(continued)

Table 10.2 (continued)

Pest	Host	Country of report	Reference
Twig girdler (*Oncideres limpida*)	*J. curcas*	India	Prabhakar et al. (2008)
Soft scale (*Megapulvinaria maxima* (Green)	*J. curcas*	Andhra Pradesh	Prabhakar et al. (2008)
Anasa scorbutica (Coreidae)	*J. curcas*	Nicaragua	Grimm (1996)
Hemiptera: *Chelysomidea variabilis*, *Pachycoris torridus* and *Sphyrocoris punctellus*	*J. curcas*	Nicaragua	Grimm and Maes (1997a)
19 species of Coreidae (Heteroptera), 3 species of Rhopalidae and 3 species of Alydidae	*J. curcas*	Nicaragua	Grimm and Maes (1997a)
15 species of Pentatomidae and 1 species of Tessaratomidae	*J. curcas*	Nicaragua	Grimm and Maes (1997c)
Tussock caterpillar *(Orygia postica);* Black hairy caterpillar *(Estigmene lactinea);* Ash weevil *(Myllocerus maculosus);* Grasshopper *(Atractomorpha ranacea);* Neem scale *(Pulvinaria maxima)*; Calotropis leaf hopper bug (*Eurybrachis tomentosa*)	*J. curcas*	India	Manoharan et al. (2006), Regupathy and Ayyasamy (2007)

Three pests have been categorized as quarantine pests of *J. curcas* in India (Plant Quarantine Order 2003). When *J. curcas* plants meant for propagation are imported from USA, care has to be taken that the plants should be free from *Diaprepes abbreviatus* (citrus weevil), *Pseudococcus jackbeardsleyi* (Jack Beardsley mealy bug), and *Armillaria tabescens* (Armillaria root rot): Syn: *Clitocybe tabescens.* Post-entry quarantine growing for a period of 45 days is mandatory. Plants/cuttings meant for propagation from Singapore also should be free from Jack Beardsley mealy bug. All tissue cultured plants from any country should be certified that the tissue cultured plants are obtained from mother stock tested and maintained free from viruses.

Conclusions

J. curcas is susceptible to several pests and diseases. Certain pests with diverse hosts, such as mealy bugs may attack many more plants with the passage of time. Therefore, effective manipulation of weeds and ornamental plants, adopting crop rotation and quarantine measures, etc., will be of high significance, while devising integrated management strategy for such pests.

In the literature, insecticides belonging to different groups have been recommended against some pests; however, main reliance on insecticides may result in resistance, resurgence, environmental hazards and thus leading to discontinuation of their use. In view of the above, focus on alternative control measures is needed. Information on biological parameters of insects and their host preference for feeding and oviposition are very important to develop alternative strategies effective for their control like other important insects.

Systematic research on biotic stress resistance has not been carried out till date in *J. curcas*. Attention to increasing resistance to pests and diseases is needed in the selective breeding of *J. curcas*. A novel interspecific *Jatropha* hybrid "Nandan-4" was developed by the hybridization of *J. gossypiifolia* with *J. curcas* (protected by a patent) and this hybrid is claimed resistant to several pests and diseases (Karanam and Jayakumar 2010) *viz.*, inflorescence & capsule borer (*Pempelia morosalis*), leaf miner (*Neurobathracu rcassi* Busck.), bugs (*Scutellera nobilis* Fabr.) and powdery mildew (*Erysiphe euphorbiae*). There is a definite need for the search of resistant sources in *J. curcas* against other pests also. If neglected, the pests, which are of minor significance, may attain the status of major pests on *J. curcas*.

References

Acda MN (2009) Toxicity, tunneling and feeding behavior of the termite, *Coptotermes vastator*, in sand treated with oil of the physic nut, *Jatropha curcas*. J Insect Sci 9:64

Achten WMJ, Verschot L, Franken YJ, Mathijs E, Singh VP, Aerts R et al (2008) Jatropha biodiesel production and use. Biomass Bioenergy 32:1063–1084. Jatropha: A Smallholder Bioenergy Crop available at www.fao.org/docrep/012

Agriculture Hand Book 165 (1960) United States Department of Agriculture. Washington

Aguilar H, Ureña A, Murillo P (2010) *Polyphagotarsonemus latus*: a conundrum in biodiesel investigation in Costa Rica. Paper presented in the XIII international congress of acarology, Recife-PE, Brazil, 23–7 Aug 2010

Ali MS, Kumar Manoj, Singh R (2006) Host range of *Oxyrhachis tarandus* Fabricus (Homoptera: Membracidae) in woody trees and shrubs of Bihar. Environ Ecol 1:14–16

Almeida EHN, Silva ES, Souza PS (2010) Survey of the fauna of predatory mites on *Jatropha curcas* L. and other species of the same genus in the state of Alagoas, Brazil. Paper presented in the XIII international congress of acarology, Recife-PE, Brazil, 23–7 Aug 2010

Ambika S, Manoharan T, Stanley J, Preetha G (2007a) Biology and management of *Jatropha* shoot webber. Indian J Entomol 69:265–270

Ambika S, Manoharan T, Stanley J, Preetha G (2007b) Scutellarid pests of Jatropha and their Management. Ann Plant Prot Sci 15:370–375

Anitha K, Chakrabarty SK, Sunil N, Prasada Rao RDVJ, Varaprasad KS, Khetarpal RK (2005) Fungi recorded on *Jatropha curcas* L. seed collected in India. Indian J Plant Prot 33:303–304

Arif M, Sharma S, Das SC (2007) Incidence of leaf miner on *Jatropha curcas*—a bio diesel plant in Secunderabad. J Exp Zoolog 10:107

Arif MI, Rafiq M, Ghaffar A (2009) Host plants of cotton mealy bug (*Phenacoccus solenopsis*): a new menace to cotton agroecosystem of Punjab. Intl J Agric Biol 11:163–167

AVA. Agri-food and veterinary authority of Singapore (2010) Pest News, Feb 2010. http://www.ava.gov.sg/AgricultureFisheriesSector/PlantHealthServices

Babu BG, David PMM (1999) New host plant records and host range of the spiralling whitefly, *Aleurodicus dispersus* Russell (Hemiptera: Aleyrodidae). Madras Agric J 86:305–313

BAZ. Biofuel Association of Zambia (2007) *Jatropha curcas*, What do we know? Information to the public. 25 Jan 2007. http://www.thomrobiofuels.com/Baz/Jatropha_curcas
Beeson CFC (1941) The ecology and control of the forest insects of India and the neighbouring countries. Forest Research Institute, Dehra Dun. p 1007
Bird J (1957) A whitefly-transmitted mosaic of *Jatropha gossypiifolia*. Technical paper of the Agric. Exp Stn Puerto Rico 22:1–35
NOVOD Board (2009) 25th annual report 2008–09. Published by NOVOD board, Gurgaon
Bock KR, Guthrie EJ, Figueiredo G (1981) A strain of cassava latent virus occurring in coastal districts of Kenya. Ann Appl Biol 99:151–159
Bocourt PY, López Manes D, López Mesa MO (2005) New host plants for the family Erysiphaceae in Cuba. Fitosanidad 9:73
Brar KS, Sandhu GS (1982) Field biology of the white grub, *Holotrichia consanguinea* Blanchard (Scarabaeidae: Coleoptera) in Punjab. J Soil Biol Ecol 2:32–35
Braun U (1987) A monograph of the Erysiphales (powdery mildews). Beih Nova Hedwigia 89:700
Braun U, Freire F, Das CO (2004) Some cercosporoid hyphomycetes from Brazil—III. Cryptogam Mycol 25:221–244
Brown JK, Idris AM, Torres JI, Bird J (1999) Jatropha mosaic begomovirus variants from weed and cultivated hosts in Puerto Rico. Phytopathology 90:S122
CABI UK Centre Annual Report (2009) Assessing the impact of diseases on the biofuel crop Jatropha. www.cabi.org
Carels N (2009) *Jatropha curcas*: a review. In: Delseny M, Kader JC (eds) Advances in botanical research. Elsevier, Amsterdam, pp 39–86
Chattopadhyay NC (2009) Disease and pest of *Jatropha curcas* in West Bengal, India. http://biofuelexperts.ning.com/forum/topics/disease-and-pest-of-jatropha?com
Chavhan TL, Suryanarayana V, Naik ST (2010) Survey and management of *Pestalotiopsis* leaf blight of *Jatropha* a destructive new disease in Karnataka. Indian Phytopathol 63:110–111
Chitra S, Dhyani SK (2006) Insect pests of *Jatropha curcas* L. and the potential for their management. Curr Sci 91:162–163
Daey Ouwens K, Francis G, Franken YJ, Rijssenbeek W, Riedacker R, Foidl N et al (2007) Position paper on *Jatropha curcas*, state of the art, small and large scale project development. FACT Foundation, Eindhoven
Dalia Amin KH (2001) Final report on global project on development of Jatropha Plant (Global Warming/ Biodiversity/ Antidesertification) XP/GLO/00/024. http://www.jatropha.de/sudan/jatropha-project-proposal.htm
Das N (1995) *Jatropha glandulifera* Roxb.—a new host for *Fusarium solani* (Martius) Sacc. Indian J Mycol Plant Pathol 25:337
Das AK, Chattopadhyay BK (1990) Three new combinations into the genus *Pseudocercospora* Speg. J Mycopathol Res 28:27–32
Debnath M, Verma HN (2008) Effect of phytoprotein treatment on *Jatropha curcas* for wasteland reclamation. Afr J Biotechnol 7:613–616
Dhillon RS, Hooda MS, Jattan M, Chawla V, Bharadwaj M, Goyal SC et al (2009a) Development and molecular characterization of interspecific hybrids of *J. curcas* X *J. integerrima*. Indian J Biotechnol 8:384–390
Dhillon RS, Hooda MS, Pundeer JS, Ahlawat KS, Kumari S (2009b) Development of efficient techniques for clonal multiplication of *Jatropha curcas* L. a potential biodiesel plant. Curr Sci 96:823–827
Erum YI, Musarrat AR, Shahina F (2005) *Jatropha gossypiifolia* (Euphorbiaceae), a new host of *Meloidogyne javanica* in Pakistan. Pakistan J Nematol 23:187–188
Fact Foundation (2010) http://www.fact foundation.com/en/Knowledge_and_Expertise/Jatropha/Jatropha_Pests
Fernandes RS, Asmus GL (2007) Reaction of physic nut (*Jatropha curcas* L.) to *Meloidogyne javanica* and *Rotylenchulus reniformis*. Nematologia Brasil 31:96–99
Freeman P (1939) A contribution to the study of the genus *Calidea* Lapoerte (Hemipt.-Heteropt., Pentatomidae). Trans R Ent Soc Lond 88:139–160

Freire FCO, Mosca JL (2009) Diseases of flowers and ornamental plants in Ceará State, Brazil. Revista Brasileira de Hort Ornamental 15:83–89

Friends of the Earth (2009) Jatropha: wonder crop? Experience from Swaziland. May 16. http://www.foe.co.uk/resource/reports/jatropha_wonder_crop.pdf

Fu G, Huang SL, Wei JG, Yuan GQ, Ren JG, Yan WH et al (2007) First record of *Jatropha podagrica* gummosis caused by *Botryodiplodia theobromae* in China. Australasian Plant Dis Notes 2:75–76

Gabriel D, Calcagnolo G, Tancini RS, Dias Netto N, Petinelli A Jr, Araujo JBM et al (1988) Study of *Pachycoris torridus* (Scopoli, 1772) (Hemiptera: Scutelleridae) and its natural enemy *Pseudotelenomus pachycoris* Lima, 1928 (Hymenoptera; Scelionidae) in *Jatropha* spp. crops. Biologico 54:17–20

Gagnaux PCA (2008) Incidência da entomofauna associada à cultura de Jatrofa (*Jatropha curcas* L) em Moçambique. Thesis, Universidades Eduardo Mondlane, Mozambique

Gao SQ, Qu J, Chua NH, Ye J (2010) A new strain of Indian cassava mosaic virus causes a mosaic disease in the biodiesel crop *Jatropha curcas*. Arch Virol 155:607–612

Garga RP (1960) A leaf distorting virus disease of *Jatropha curcas* Linn. Curr Sci 30:345–346

Goldfarb M, Duarte MEM, Mata MERMC, Nascimento LC do, de Brito NM, Souto FM et al (2010) Incidence of fungus and physiological quality of seeds of *Jatropha curcas* L. after cryogenic storage. Biotemas 23:19–26

Grimm C (1996) Utilization of a life table to quantify damages caused by insects on *Jatropha curcas* (Euphorbiaceae) fruits. Manejo Integrado Plagas 42:23–30

Grimm C (1999) Evaluation of damage to physic nut (*Jatropha curcas*) by true bugs. Entomol Exp Appl 92:127–136

Grimm C, Führer E (1998) Population dynamics of true bugs (Heteroptera) in physic nut (*Jatropha curcas*) plantations in Nicaragua. J Appl Entomol 122:515–521

Grimm C, Guharay F (1998) Control of leaf-footed bug *Leptoglossus zonatus* and shield-backed bug, *Pachycoris klugii* with entomopathogenic fungi. Biocontrol Sci Technol 8:365–376

Grimm C, Maes JM (1997a) Arthropod fauna associated with *Jatropha curcas* in Nicaragua: a synopsis of species, their biology and pest status. In: Gübitz GM, Mittelbach M, Trabi M (eds) Biofuels and industrial products from *Jatropha curcas*. Developed from the symposium "Jatropha 97" Managua, Nicaragua, 23–27 Feb 1997 ISBN 3-7041-0242-3

Grimm C, Maes JM (1997b) Insects associated with physic nut (*Jatropha curcas*) in the Pacific region of Nicaragua. III. Coreoidea (Heteroptera). Rev Nicaragüense de Entomol 42:15–34

Grimm C, Maes JM (1997c) Insects associated with physic nut (*Jatropha curcas*) in the Pacific region of Nicaragua. II. Pentatomidae y Tessaratomidae (Heteroptera). Rev Nicaragüense de Entomol 40:13–28

Grimm PC, Maes JM (1997d) Insects associated with physic nut (*Jatropha curcas* L.) (Euphorbiaceae) in the Pacific region of Nicaragua. I. Scutelleridae (Heteroptera). Rev Nicaragüense de Entomol 39:13–26

Grimm C, Somarriba A (1998) Life cycle and rearing of the shield-backed bug *Pachycoris klugii* in Nicaragua (Heteroptera: Scutelleridae). Entomologia Generalis 22:211–221

Grimm C, Somarriba A (1999) Suitability of physic nut (*Jatropha curcas* L.) as single host plant for the leaf-footed bug *Leptoglossus zonatus* Dallas (Het., Coreidae). J Appl Entomol 123:347–350

Gupta PK, Sharma ND, Singh SR, Singh OP (2007) Fungi associated with medicinal plants of Madhya Pradesh. Ann Plant Prot Sci 15:508–509

Hampson FI (1912) Growth and development of pyralids on various host plant under laboratory conditions. Proc Ent Mtg 4:288

Hayward AC, Waterston JM (1964) *Xanthomonas malvacearum*. [Descriptions of Fungi and Bacteria]. IMI descriptions of fungi and bacteria no. 2 pp. Sheet 12. UK, CAB International

Hegde YR, Chavhan TL (2009/2010) Management of root rot of *Jatropha curcas* in Karnataka. Intl J Plant Prot 2:243–244

Hegde YR, Chavhan T, Patil SJ (2009) *Jatropha curcas*—a new host for *Sclerotium rolfsii*. J Plant Dis Sci 4:230

Heller J (1996) Physic nut. *Jatropha curcas* L. Promoting the conservation and use of underutilized and neglected crops. Institute of Plant Genetics and Crop Plant Research/International Plant Genetic Resources Institute, Gatersleben/Rome. ISBN 92-9043-278-0. p 60

Hosagoudar VB, Archana GR (2009) Host range of meliolaceous fungi in India. J Threat Taxa 1:269–282

http://betterglobeforestry.com/research/pests-and-diseases.html

http://biofuelexperts.ning.com/

http://jatropha.pro/wilt.htm

http://strawberry.ifas.ufl.edu/DiagnosticLab/

http://www.botanical.com/site/column_poudhia

http://www.fact-foundation.com/

http://www.gardenguides.com/

http://www.jatropha.de/photo-show/index

http://www.jatropha.de/sudan/jatropha-project

http://www.jatropha.de/zimbabwe

http://www.k4rd.org/jatropha.htm

http://www.phytotron.com/jatropha1.htm

http://www.rrljorhat.com/

http://www.sunplantgroup.com/jat_curcas9.htm

http://www.used-cars.co.jp/biotec/jatoropha.pdf)

http://www.waterlandasiabio.com

Hudge BV, Datar VV (2009) *In vitro* effect of temperature and humidity against *Alternaria alternata* (Fr.) Keissler causing leaf spot of Jatropha. Ann Plant Physiol 23:129–130

Hudge BV, Datar VV (2010) Study of incidence and severity of leaf spot disease in *Jatropha curcas* L. Intl J Agric Sci 6:355–356

Jones MH, Csurhes S (2008) Pest Plant Risk Assessment: *Physic Nut Jatropha curcas*, Department of Primary Industries and Fisheries, the State of Queensland: 1–28

Kar AK, Das A (1988) New records of fungi from India. Indian Phytopathol 41:505

Karanam KR, Jaya Kumar B (2010) Jatropha interspecific hybrid. United States Patent Application 20100287820. 18 Nov 2010 http://www.freepatentsonline.com/20100287820.pdf

Kaufmann T (1966) Notes on the life history and morphology of *Calidea dregii* (Hemiptera: Pentatomidae: Scutellerini) in Ghana, West Africa. Ann Entomol Soc Am 59:654–659

Kavitha J, Ramaraju K, Baskaran V, Kumar PP (2007) Bioecology and management of spider mites and broad mites occurring on *Jatropha curcas* L. in Tamil Nadu, India. Syst Appl Acarol 12:109–115

Kershaw JCW, Kirkaldy GW (1908) On the metamorphoses of two *Hemiptera heteroptera* from Southern. China Trans Entomol Soc Lond 56:59–62

Khande DM, Aherkar SK, Barkhade UP, Bisane KD (2008) Incidence of *Scutellera nobillis* Fabr. on *Jatropha curcas* in Rajanandgaon District of Chattisgarh. Insect Environ 13:192

Kirk PM (1991) *Nigrospora sphaerica*. [Descriptions of Fungi and Bacteria]. IMI descriptions of fungi and bacteria no. 106 pp. Sheet 1056. UK, CAB International

Kumar S, Sharma S (2010) Studies on the factors affecting pathogenicity of root rot caused by *Rhizoctonia bataticola* in *Jatropha curcas*. Indian For 136:736–741

Latha P, Prakasam V, Kamalakannan A, Gopalakrishnan C, Raguchander T, Paramathma M et al (2009) First report of *Lasiodiplodia theobromae* (Pat.) Griffon & Maubl. causing root and collar rot disease of physic nut (*Jatropha curcas* L.) in India. Australasian Plant Dis Notes 4:19–21

Li Q, Zhao YB, Gao TP, Zhao GY, Xu H (2009) Preliminary Report on Disease and Insect Pests of *Jatropha carcass Nursery*. For Inventory and Plann 01–019

Li YC, Guo QS, Shao QS, Zhang P, Dai XL (2009) Bioassay on herbicidal activity of extracts from *Jatropha curcas*. J Plant Res Environ 18:72–78

Li YC, Luo YP, Lu H, Zhang Y (2010) Infection characteristics of *Jatropha curcas* by powdery mildew in dry-hot valley of Jinshajiang river. Southwest China J Agric Sci 23:1125–1127

Lima BV, Soares DJ, Pereira OL, Barreto RW (2008) Natural infection of *Acalypha hispida* and *Jatropha podagrica* inflorescences by *Amphobotrys ricini* in Brazil. Australasian Plant Dis Notes 3:5–7

Lopes EN, Venzon M, Pallini A, Dias CR, de Oliveira EF, Dias LAS et al (2010) Biological performance of *Polyphagotarsonemus latus* on genotypes of *Jatropha curcas*. Paper presented in the XIII international congress of acarology, Recife-PE, Brazil, 23–27 Aug 2010

Manoharan T, Ambika S, Natarajan N, Senguttuvan K (2006) Emerging pest status of *Jatropha curcas* (L.) in south India. Indian J Agrofor 8:66–79

Marc K (2010) http://biofuelexperts.ning.com/forum/topics/

Martínez Y (2008) Emergence of begomoviruses in Cuba. Revista de Protección Vegetal 23:11–15

Medina-Gaud S, Franqui RA (2001) *Retithrips syriacus* (Mayet), the black vine thrips (Insecta: Thysanoptera: Thripidae) new to Puerto Rico. J Agric Univ Puerto Rico 85:85–89

Meshram PB, Joshi KC (1994) A new report of *Spodoptera litura* (Fab.) Boursin (Lepidoptera: Noctuidae) as a pest of *Jatropha curcas* Linn. Indian For 120:273–274

Mordue JEM (1971) *Pestalotiopsis guepinii*. [Descriptions of Fungi and Bacteria]. IMI descriptions of fungi and bacteria no. 32 pp. Sheet 320. UK, CAB International

Mralirangan MC (1978) Feeding preferences of adults and mandibular morphology in the different instars of *Eyprepocnemis alacris alacris* (Serv.) (Orthoptera: Acrididae). Curr Sci 47:101–104

Naik MK (1993) Unrecorded pathogen on *Jatropha glandulifera* Roxb. from India. Indian J Mycol Plant Pathol 23:332

Narayana ADS, Shankarappa KS, Govindappa MR, Prameela HA, Gururaj Rao MR, Rangaswamy KT et al (2006) Natural occurrence of Jatropha mosaic virus disease in India. Curr Sci 91:584–586

Narayana A DS, Rangaswamy KT, Shankarappa KS, Maruthi MN, Lakshminarayana Reddy CN, Rekha AR et al (2007) Distinct begomoviruses closely related to cassava mosaic viruses cause Indian Jatropha mosaic disease. Intl J Virol 3:1–11. ISSN 1816–4900

Nayak D, Saxena SP, Prajapati VM, Jadeja DB (2008) Seasonal incidence of leaf and flower webber cum fruit borer, *Pempelia morosalis* (Saalam Uller) in *Jatropha curcas* Linn. Insect Environ 14:44–45

Neelu S, Harsh NSK, Alka B (1996) Biodeterioration of *Jatropha curcas* seeds. Ann For 4:52–54

Nielsen F (2007) FN research progress report no 1, Jatropha by FNResearch project: "Jatropha oil for local development in Mozambique" sub-title: "Biofuel for Development and Communal Energy Self-Supply" reporting period: January–July 2007

Nielsen F (2010) Rainbow Shield Bug (*Calidea dregii*). Fact Foundation datasheet

Padilla D, Monterroso D (1999) Preliminary differentiation of diseases in the tempate (*Jatropha curcas*) crop in Nicaragua. Manejo Integrado de Plagas 51:66–69

Pandey A, Shukla AN, Chandra S (2006) *Pestalotiopsis* stem canker of *Jatropha curcas*. Indian For 132:763–766

Pankaj O (2007) Bare facts about poisonous *Jatropha curcas*. eResDocs formal report. http://ecoport.org/ep?SearchType=earticleView&earticleId

Paramathma M, Parthiban KT, Neelakantan KS (2004) *Jatropha curcas*. Forest College & Research Institute, Tamil Nadu Agricultural University, Coimbatore,p 48

PCARRD. Philippine Council for Agriculture, Forestry and Natural Resources Research and Development (2009) http://www.pcarrd.dost.gov.ph/ssentinel/index.php?

Pereira OL, Dutra DC, Dias LAS (2009) *Lasiodiplodia theobromae* is the causal agent of a damaging root and collar rot disease on the biofuel plant *Jatropha curcas* in Brazil. Australasian Plant Dis Notes 4:120–123

Phillips S (1975) A new record of *Pestalotiopsis versicolor* on the leaves of *Jatropha curcas*. Indian Phytopathol 28:546

Plant Quarantine Order (2003) www.plantquarantine.in

Prabhakar M, Prasad YG, Rao GR, Ramakrishna D (2008) Pests of economic importance on *Jatropha curcas* L., a biodiesel plant in Andhra Pradesh, India. Entomology 33:83–86

Raj SK, Kumar S, Snehi SK, Pathre U (2008) First report of *Cucumber mosaic virus* on *Jatropha curcas* in India. Plant Dis 92:171

Raju AJS, Rao SP (2003) Interaction between *Jatropha curcas* L. (Euphorbiaceae) and insects. Insect Environ 9:25

Rani BJ, Sridhar V (2002) Record of insect pests of Jatropha, *Jatropha curcas* Linn.—a medicinal and minor oil seed plant. Insect Environ 8:76–77

Ray S, Das SN (1980) Two new and four known species in the family Hemicycliophoridae (Criconematoidea: Nematoda) from Orissa, India. Indian J Nematol 10:141–147

Reddy JR, Reddy AP (1980) *Erysiphe euphorbiae* sp. on *Jatropha gossypiifolia* L. Curr Microbiol 4:95–97

Regupathy A, Ayyasamy R (2006) Need for generating baseline data for monitoring insecticide resistance in leaf webber cum fruit borer, *Pempelia morosalis* (Saalm Uller), the key pest of biofuel crop, *Jatropha curcas*. Resist Pest Manag Newsl 16:2–5

Regupathy A and Ayyasamy R (2007) Emerging pest scenario in Jatropha curcas plantations and the need for development of ecofriendly pest management. In: Anitha K, Gururaj Katti, Sarath Babu and Varaprasad KS (Eds). Extended Summaries: National Conference on Organic waste utilization and eco-friendly technologies for crop protection. March 15-17, 2007, 253pp. Plant Protection Association of India, NBPGR Regional Station, Rajendranagar, Hyderabad. Pages 67–69

Sarmento RA, Rodrigues DM, Faraji F, Erasmo EA, Lemos F, Teodoro AV et al (2011) Suitability of the predatory mites *Iphiseiodes zuluagai* and *Euseius concordis* in controlling *Polyphagotarsonemus latus* and *Tetranychus bastosi* on *Jatropha curcas* plants in Brazil. Exp Appl Acarol 53(3):203–214

Saturnino HM, Pacheco DD, Kakida J, Tominaga N, Gonçalves NP (2005) Cultivation of *Jatropha curcas* L. Informe Agropecuário 26:44–78

Seier M, Cortat G, Hill L (2009) Preliminary assessment of the rust, *Phakospora jatrophicola*, as a potential biocontrol agent for *Jatropha gossypiifolia*. p 7 in annual report, 2009. http://issuu.com/1.rock/docs/UK-centre_report_2009

Shanta P, Menon KPV (1959) First conference of coconut research workers in India. Thiruvananthapuram, p 1

Shanthi M, Rajavel DS, Baskaran RKM (2007) Severe incidence of *Rhipiphorothrips cruentatus* (Hood) on *Jatropha curcas* Linn. in Madurai, Tamil Nadu. Insect Environ 13:127

Sharma IK (1981) The carpenter bee (*Xylocopa fenestra*) in the Indian Thar Desert. J Bombay Nat Hist Soc 78:408–409

Sharma TK (2006) Insect pests on bio-diesel plant, *Jatropha curcas*. Bionotes 8:103

Sharma S, Kumar K (2009) Root rot of *Jatropha curcas* incited by *Rhizoctonia bataticola* in India. Indian For 135:433–434

Sharma N, Sarraf A (2007) Pest disease management. Expert seminar on *Jatropha curcas* L. agronomy and genetics. FACT Foundation, Wageningen, 26–28 March 2007

Sharma RP, Srivastava CP (2010) Studies on damage potential and integration of some IPM components against scutellerid bug infesting Jatropha in Eastern Uttar Pradesh of India. Intl J Agric Res 5:1116–1123

Sharma S, Kaushik JC, Kaushik N (2001) *Fusarium moniliforme* causing root rot of jatropha. Indian Phytopathol 54:275

Sharma PK, Sangle UR, Ahmad N (2010) A new record of Oidium state of *Erysiphe jatrophe* Doidge from Manipur causing powdery mildew on Jatropha. J Mycol Plant Pathol 40:467

Shukla R, Jamaluddin (2010) Two new leaf spot diseases of *Jatropha curcas* L from central India. J Mycol Plant Pathol 40:479–480

Siddiqui JI (1983) Some new host records for *Mylabris pustulata* Thunbg. Bull Entomol 24:51–52

Sidhu OP, Sanjay A, Uday P, Snehi SK, Raj SK, Roy Raja et al (2010) Metabolic and histopathological alterations of *Jatropha mosaic begomovirus*-infected *Jatropha curcas* L. by HR-MAS NMR spectroscopy and magnetic resonance imaging. Planta 232:85–93

Singh ID (1983) New leaf spot diseases of two medicinal plants. Madras Agric J 70:490

Smith KM (1957) A text book of plant virus diseases, 2nd edn. Churchill, London

Soto SS, Nakano O (2002) Occurrence of *Pachycoris torridus* (Scopoli) (Hemiptera: Scutelleridae) on Barbados cherry plant (*Malpighia glabra* L.) in Brazil. Neotrop Entomol 31:481–482

Sseruwagi P, Maruthi MN, Colvin J, Rey MEC, Brown JK, Legg JP et al (2006) Colonization of non-cassava plant species by cassava whiteflies (*Bemisia tabaci*) in Uganda. Entomologia Experimentalis Appl 119:145–153

Stephen HB, Mark LP (2010) Plant diseases in Lee, Collier, Charlotte and Sarasota Counties, April–June 2010. http://lee.ifas.ufl.edu/Hort/GardenPubsAZ/April-June2010.pdf

Sudhir E, Anitha K, Vinod Kumar, Sunil N, Babu A (2007) Fungal pathogens associated with *Jatropha curcas* germplasm- a potential biopesticide crop. p 68. In: Anitha K, Gururaj Katti, Sarath Babu B, Varaprasad KS (eds) Extended summaries published during the national conference on organic waste utilization and eco-friendly technologies for crop protection during 15–17 March 2007. pp 201–202

TERI. TERI Knowledge centres (2007) Knowledge for rural development. http://www.k4rd.org/index.htm

Tewari JP, Dwivedi HD, Madhavi P, Srivastava SK (2007) Incidence of a mosaic disease in *Jatropha curcas* L. from eastern Uttar Pradesh. Curr Sci 93:1048–1049

The Philippine Star (2010) New pest control for jatropha found. Updated 27 May 2010 12:00 AM http://www.philstar.com/Article.aspx?articleid=578670

Torres-Calzada C, Tapia-Tussell R, Nexticapan-Garcez A, Matin-Mex R, Quijano-Ramayo A, Cortés-Velázquez A et al (2011) First report of *Colletotrichum capsici* causing anthracnose in *Jatropha curcas* in Yucatan, Mexico. New Dis Rep 23:6

Tuan L, Sotto JMEO, Llano A, Punsalan MRO, Ultra VU, Galo EA et al (2009) Common insect pest and diseases of *Jatropha curcas* in the locality of Northern Samar, Philippines. Philippin J Crop Sci 24:28

Wang F, Xiong Z, Xu H, He CZ, Xin PY, Wu K (2009) A Report of 2 Diseases from *Jatropha curcas* L.; For Inventory Plann. 01–022

Watve YG, Diwakar MP, Kadam JJ, Sawant UK, Mundhe VG (2009) Effect of different chemicals, plant extracts and bioagents against *Colletotrichum gloeosporioides* causing leaf spot disease of jatropha. J Plant Dis Sci 4:199–203

Wilson CG (1997) Phytophagous insect fauna of two weeds, *Hyptis suaveolens* (L.) Poit. and *Jatropha gossypiifolia* L., in Australia's Northern Territory. Australian Entomologist 24:55–60

Wink M, Grimm C, Koschmieder C, Sporer F, Bergeot O (2000) Sequestration of phorbol esters by the aposematically coloured bug *Pachycoris klugii* (Heteroptera: Scutelleridae) feeding on *Jatropha curcas* (Euphorbiaceae). Chemoecol 10:179–184

www.cabi.org

www.pestnet.org

Zarina B (1996) Studies on plant parasitic nematodes of ornamental and vegetable plants with special reference to root-knot nematode. Ph.D. thesis, University of Karachi, Karachi. Pakistan Research Repository http://eprints.hec.gov.pk/1397/ deposited on 19 Feb 2007

Zarina B, Maqbool MA (1998) Descriptions and observations on two new and two known species of the genus *Pratylenchus* Filipjev, 1936 (Nematoda: Pratylenchidae) from Pakistan. Pak J Nematol 16:13–24

Zeigler RS, Lozano JC (1983) The relationship of some *Sphaceloma* species pathogenic on cassava and other Euphorbiaceae in Central and South America. Phytopathol 73:293–300

Chapter 11
Phytosanitary Aspects of Jatropha Farming in Brazil

José Carlos Fialho de Resende, Nívio Poubel Gonçalves, Mário Sérgio Carvalho Dias, Carlos Juliano Brant Albuquerque, and Danielle de Lourdes Batista Morais

Introduction

The species *Jatropha curcas* L., commonly known as physic nut, purging nut, *pinhão manso* (in Brazil) is referred to as Jatropha hereafter. It has been studied in many research fields by various international institutions. Jatropha is a promising species that is being now tested as a crop in several countries, justifying the loans and investments that numerous research centers and funding agencies have implemented. Several aspects of its culture have been addressed, notably the plant health since the species was earlier deemed resistant to pests and diseases (Carneiro et al. 2009; Laviola 2011). It is this fame of a rustic plant that has been responsible for the wide acceptance of Jatropha as a crop by farmers in Brazil. Indeed, the latex that oozes from the stem and leaves is caustic and is considered an anti-nutritional factor for various insects. However, under field conditions of high temperature and humidity, fungal diseases and arthropod attacks are common.

J.C.F. de Resende (✉)
Fazenda Experimental de Montes Claros, BR 251, Km 11, CEP 39404-128
Montes Claros, Minas Gerais, Brazil
e-mail: jresende@epamig.br

N.P. Gonçalves
Fazenda Experimental do Gorutuba, MGt 122, Km 150, CEP 39525-000
Nova Porteirinha, Minas Gerais, Brazil

M.S.C. Dias
Fazenda Experimental de Pouso Alegre, Av. Prefeito Tunay, 470 Campus Fátima – Univas, CEP 37550-000 Pouso Alegre, Minas Gerais, Brazil

C.J.B. Albuquerque
Fazenda Experimental de Uberlândia, Caixa Postal 2248, CEP 38402-019
Uberlândia, Minas Gerais, Brazil

D.de.L.B. Morais
Mestranda Unimontes, Bolsista Epamig, Fazenda Experimental do Gorutuba, MGt 122, Km 150, CEP 39525-000, Nova Porteirinha, Minas Gerais, Brazil

N. Carels et al. (eds.), *Jatropha, Challenges for a New Energy Crop: Volume 1: Farming, Economics and Biofuel*, DOI 10.1007/978-1-4614-4806-8_11,

Proper management of pests and diseases should focus on reducing production costs associated to plant protection as well as the environmental costs that are related to agricultural practices. Thus, the correct identification of causal agents is the main objective of pest and disease control programmes. The association of the biology of pest and disease, damage assessment and monitoring methodologies with cultivation practices of Jatropha enables the control of parasite populations in such a way as to cause the least possible impact on environment and human health, thus collaborating and strengthening the sustainability of Jatropha agriculture.

The presence of weeds that emerge in agricultural ecosystems may condition a number of biotic factors acting on a cultivated species, which will affect, not only the productivity, but also the operation of the crop system under production (Silva et al. 2007). In the case of perennial trees, the soil is not renewed every year and is not subject to fallow period when herbicide control should normally be done. In addition, weeds act as hosts for several species of pathogens and arthropods. Together, these factors are responsible for larger losses than when measured apart.

The number of publications in this area has increased in the last few years and, as pointed out by Laviola (2011), pest and disease control is of great importance since it represents around 50% of the productivity of a formalized crop system. Hereafter, we focus on weeds, pests and diseases found in plantations of Jatropha, which can be considered as key problems of this new crop in Brazil.

Harmful Arthropods

Broad mite, two-spotted spider mite, European red mite, thrips, stink bugs, green leafhopper, mealy bugs and termites were recognized as the main organisms that use Jatropha as a food source in Brazil (Saturnino et al. 2005).

Broad mite – Polyphagotarsonemus latus (Banks 1904) (Acari: Prostigmata: Tarsonemidae), a polyphagous, cosmopolitan organism (Gallo et al. 2002) with great economic importance in papaya, beans, cotton, citrus, tomato, mango and grapes, among others, also causes serious damage to Jatropha. In the northern region of Minas Gerais (Brazil), the occurrence of this arachnid has been observed during every months of the year, when new shoots are being produced.

Broad mite occurs preferentially in the young parts of a plant; it is, initially, located in clusters within isolated regions of the field. The symptoms it causes are (1) leaves with a shiny appearance and coriaceous texture (Fig. 11.1), (2) new shoots with short internodes and (3) later, the death of the apical meristem, which causes the appearance of multiple shoots (Fig. 11.2).

When monitoring broad mite threat in a cultivated area, sampling should be made in various locations at random in order to identify broad mite clusters. Mite control must be managed immediately where clusters are found in order to avoid their spread, which would consequently cause an intervention at the level of the whole area. In cotton crop, occurrence of clusters of plants showing leaves with edges turned upwards before tearing is the indicator of mite. The infestation is considered widespread,

Fig. 11.1 Jatropha leaves with symptoms of broad mite

Fig. 11.2 Jatropha plants with multiple shoots resulting from the attack of broad mite (Photo courtesy: Marcos Drumond)

when plants with early symptoms reach 40% of a field, which is the proportion that determines the moment where a control action is needed. Commercial products including profenofos, triazophos and abamectin have acaricide activity however none of these compounds are registered in Brazil for use in Jatropha.

Two-spotted spider mite – Tetranychus urticae (Koch 1836) (Prostigmata: Tetranychidae) is considered of great importance worldwide because of its association with crops of economic importance such as, cotton, strawberry, rose, tomato, bean, soybean, peach tree, castor bean, grape and papaya. Attacks by this arachnid

Fig. 11.3 Jatropha plant with symptoms of European red mite (Photo courtesy: Marcos Drumond)

initially occur in clusters as well, starting on the lower side of leaves. The first signs observed are pinpoint leaf lesions, which gradually fill the entire leaf surface and cause death. Under conditions of high population levels, webs may occur in the lesioned parts and cause great damage to plants.

European red mite – Tetranychus bastosi Tuttle et al. 1977 (Prostigmata: Tetranychidae) has been reported as a pest of *Jatropha* spp. and also of weed amaranth (*Amaranthus viridis* L.), weed black jack (*Bidens pilosa* L.), bam-burral (*Hyptis suaveolens* Poit.), sweet potato (*Ipomoea batatas* (L.) Lam), jiritana (*Ipomoea glabra* Choisy), Bellyache bush (*Jatropha gossypiifolia* L.), mallow (*Malva rotundifolia* L.), blackberry (*Morus nigra* L.) and seedlings of *Manihot pseudoglaziovii* Pax & Hoffmann (Santos et al. 2006). According to Saturnino et al. (2005), attacks of European red mite usually occur in older leaves (Fig. 11.3).

Termites (Isoptera: Termitidae) are insects genuinely Brazilian. They also received popular names, such as white ants, *siriris* or *hallelujahs*. The basic food of termites is live plant materials, recently dead or in various stages of decomposition and also humus. The natural habitat of termites is a forest, but, since termites are currently invading agro ecosystems and damaging crops of economic importance, they become a major problem for humanity. However, literature indicates that only a small proportion of termite species in rural and urban environments should be considered as pests. Many species of termites are important components of the soil fauna in tropical regions because of their role in organic matter fragmentation, playing a key role in the processes of decomposition and nutrient cycling (Menezes et al. 2007).

Termites begin their attacks on roots of both young and adult plants. The attacks initially appear in isolated regions within the area of cultivation (cluster), where plants are found dead or compromised concerning yield capacity and intensely

Fig. 11.4 Final stage of an attack by termites on the root system of Jatropha

producing adventitious roots. Usually, the plant cannot be saved because the root system is already irreversibly damaged when symptoms are clear (Fig. 11.4).

To date there is no product registered in Brazil for pest control of *Jatropha*. A control measure that has been adopted experimentally is the application of termite insecticides used in forest plantations at the level of the plants within a cluster and those in their immediate neighbourhood.

Thrips – Selenothrips rubrocinctus (Giard 1901) (Thysanoptera: Thripidae) is a pest that can cause severe damages to crops, such as Jatropha, grapes, lychee, cashew, mango, avocado, guava, cocoa, rose, annatto, carambola, etc. Females introduce their eggs under the lower surface of leaves and cover them with a secretion that becomes dark when dry. Young thrips appear after 10–12 days. They are yellowish and their two first abdominal segments are red; they also carry a small drop of liquid excrement between the terminal bristles of the abdomen (Fig. 11.5). The duration of the life cycle of this species is about 30 days (Gallo et al. 2002).

The pest *stem borer* of Jatropha – *Sternocoelus (= Coelosternus) notaticeps* Marshall 1925 (Coleoptera: Curculionidae) must, in fact, be considered as *Coelosternus notaticeps*, according to the observations by Bondar (1913) in Bahia (Brazil) reported in Costa Lima (1956). It was found in Tatuí, state of São Paulo (Brazil), causing severe damage to *Jatropha* (Gabriel et al. 1988a).

Adult females lay their eggs in the parenchymal tissue from which larvae hatch; they feed on internal tissues of stems and branches, forming galleries within them. The pupal stage occurs within the tissues and the insect emerges to infect new plants (Ungaro and Regitano Neto 1996).

Damages can be considerable and lead to plant loss when the infestation rate is too high. Plants should be periodically inspected for perforations and residue at the base of stems that could testify attacks by this insect (Fig. 11.6).

Fig. 11.5 Young thrips (*red*) on the abaxial face of a Jatropha leaf (Photo courtesy: Heloisa Mattana Saturnino)

Fig. 11.6 Drilling at the base of the trunk of a Jatropha plant caused by *Sternocoelus notaticeps* Marshall 1925 (Photo courtesy: Dalva Gabriel)

The lack of boron in soil makes plants more susceptible to borers, therefore boron application is recommended for prevention of stem borer. The monitoring of plants, especially during the summer, is also recommended. Stems should be cut and burnt in order to prevent larval development. Chemical control is not advisable because larvae are difficult to reach inside stem tissues (Ungaro and Regitano Neto 1996). Moreover, there are no chemicals registered for Jatropha's stem borer in

Fig. 11.7 Stink bug on Jatropha. A nymph (*white arrow*) sucking a dry fruit (**a**). Variations in drawing and color pattern on bodies of *P. torridus* (**b–g**). An adult female covering its eggs (*white arrows*) (**c** and **d**). Nymphs (*white arrows*) protected by females (**e, f** and **g**)

Brazil. However, the use of aldicarb in experimental treatments for the control of this pest in the state of São Paulo showed promising results (Gabriel et al. 1988a).

Green leafhopper – Empoasca spp. (Hemiptera: Cicadellidae) is one of the most severe pests of Jatropha in northern Minas Gerais. Adults are green, measure about 3 mm and live up to 60 days, on average. Females lay 30–168 eggs (average of 107 eggs per female), each being individually inserted in leaf tissues (Gallo et al. 1988; Quintela 2002). Nymphs and adults are normally located on the lower leaf side, have the same color and move laterally (Quintela 2002). The green leafhoppers complete their life cycle on beans in about 3 weeks (Gallo et al. 1988).

Damages caused by nymphs and adults are physical as a result of the stylus penetration in phloem tissue, which leads to cell disruption, sap granulation and vessel blockage (Ospina 1980). According to Saturnino et al. (2005), the main symptom of attacks by *Empoasca* spp. in Jatropha is the yellowing of leaves, followed by leaf hardening and slight bending down of the leaf blade; occasionally it may also promote flower abortion. After performing leafhopper control, new leaves develop with normal color.

In April 2005 (autumn), population levels of *Empoasca* spp. were high in all areas planted with Jatropha in the North of Minas Gerais and systemic insecticides had to be applied for its control (Saturnino et al. 2005). To date, there is no pesticide registered in Brazil for Jatropha. Since Jatropha is a natural host for leafhoppers, it is important to set up its culture area far enough from its other natural hosts. Insect migration from one crop to the other is to be expected and may cause damages not only because of their own direct action, but also because of their potential for dispersal of the pathogenic agents that these insects may carry.

Another important insect pest is the *stink bug – Pachycoris torridus* Scopoli, 1772 (Hemiptera: Scutelleridae) (Fig. 11.7). Peredo (2002) reported that the species of *Pachycoris* are very similar to each other and there is little difference between the characteristics used for their identification. The similarities between them have promoted confusion concerning their identification. *P. torridus,* commonly known in Brazil as the "bug of *pinhão-bravo*" (Silva et al. 1968), is the best known species of Scutelleridae (Gallo et al. 1988). They are widely distributed in America, being found from the United States (California) to Argentina (Froeschner 1988), occurring more frequently in South America and rarely in Mexico (Peredo 2002).

According to Bondar (1913), this insect appears in Brazil during the summer sucking the sap from leaves and fruits of guava (*Psidium guajava*) and *P. araça.* Saturnino et al. (2005), reported that during the period between June 2004 and the first half of May 2005, this stink bug was found in wild condition on an endemic *Jatropha* sp. from the North and northeast of Minas Gerais. In addition, *P. torridus* was also found on *Oryza sativa* (rice), *Anacardium occidentale* (cashew), *Eucalyptus* spp. (Eucalyptus), *Citrus sinensis* (orange), *Manihot esculenta* (cassava), *Mangifera indica* (mango), *Aleurites fordii* (tung) (Silva et al. 1968), *Malpighia glabra* (acerola) (Sánchez-Soto and Nakano 2002), *Schinus terebinthifolius* (an ornamental species) (Sánchez-Soto et al. 2004) and *Sapium haematospermum,* this last occurrence found in Paraguay (Hussey 1934).

The occurrence of these insects can be verified by examining leaves and fruits by looking for the presence of eggs, nymphs and adults. When found, the insect must be immediately destroyed because they quickly produce large numbers of nymphs and both insect stages may promote fruit alterations because of their sucking activity. Actually, they may cause premature abortion of young fruits and affect endosperm development (Tominaga et al. 2007), cause seed shriveling (Saturnino et al. 2005), affect the oil content and germination potential in mature fruits.

Biological control is one of the best alternatives to reduce the population of the stink bugs. Adults can be parasitized by *Hexacladia smithii* Ashm. (Hymenoptera: Encyrtidae) and by *Trichopoda pilipes* Fabr. (Diptera: Tachinidae) (Costa Lima 1940) and their eggs can be parasitized by *Telenomus (Pseudotelenomus) pachycoris* Costa Lima (1928) (Hymenoptera: Scelionidae) (Costa Lima 1940; Peredo 2002). However, in an experiment carried out in the São Paulo state (Brazil) when *T. pachycoris* (parasitoid) was tentatively used as a control agent, only 27% of the total egg number was parasitized. The low efficiency of the parasitoid is due to the protective activity by stink bug females that leave the nymphs to their own destiny only when they are in their first instar stage, which upset the female parasitoid in its hunting. Thus, the eggs at the periphery of stink bug female spawn are the only one to end up infected (Gabriel et al. 1988a).

In India, Sahai et al. (2011) reported a fruit symptomatology similar to that induced by *P. torridus*, but caused by *Scutellera perplexa* (Hemiptera: Scutelleridae). The nymphal stages of this insect also caused damage on fruits and seeds, such as premature abortion, reduction of fruit weight, seed number and germination potential. Shanker and Dhyani (2006) suggested biological control with *Stegodyphus* sp., *Pseudotelenomus pachycoris* [*Telenomus pachycoris*], *Beauveria bassiana* and *Metarhizium anisopliae*.

Diseases

Saturnino et al. (2005) quoting Viégas (1961) and USDA (1960) report that many pathogens have been found infecting *J. curcas* such as *Cercospora jatrophae-curcas, Helminthosporium tetrâmera, Pestalotiopsis paraguarensis, Pestalotiopsis versicolor, Clitocybe tabescens, Colletotrichum gloesporioides, Elsinoe jatrophae* Bitanc. *Et Jenkins, Fusarium* spp., *Glomerella cingulata, Oidium hevea* Steinm., *Phakopsora jatrophicola* Cumm., *Phytophthora* spp. *Psathyrella subcorticalis* Speg, *Pythium* spp.*, Schizophyllum alneum* (L.) Schoroet. and *Uredo jatrophicola* Arth.

Powdery mildew is an important disease of Jatropha reported in all regions of its cultivation. The causal agent of this disease is a highly evolved obligate fungus that occurs in various parts of the world, in many cultivated species. The characteristic symptom is the formation of whitish, powdery colonies on the surfaces of the aerial parts of living plants (Fig. 11.8), like young branches, flowers, fruits and especially the leaves.

This pathogen usually appears during the dry season, coinciding with the time of natural defoliation of Jatropha. Powdery mildew belongs to family Erysiphaceae. Its imperfect stage corresponds to the genus *Oidium,* a fungus responsible for the disease occurrence in Brazilian conditions. According to Franco and Gabriel (2008), the species that is pathogenic to Jatropha is *Oidium hevea.*

The chief measures used to control powdery mildew are restricted to recommend the use of resistant varieties and chemicals. The application of fungicides is still one of the key control methods of powdery mildew.

The diseases caused by *Colletotrichum* are known as anthracnosis whereas the most commonly associated species is *Colletotrichum gloesporioides. Colletotrichum* spp. are facultative fungal parasites whose control is performed by recommending the use of resistant varieties, fungicide application and disposal of residues. The symptoms are easily noticeable. In the case of Jatropha in the North of Minas Gerais, this fungus has been isolated from necrotic lesions that begin on the leaf edges and move towards the centre of the leaf (Fig. 11.9), which then dries quickly and fall off.

In the state of Ceará (Brazil), anthracnosis of Jatropha occurs almost exclusively during the rainy season and is caused by two *Colletotrichum* species: *C. gloeosporioides* and *C. capsici* (Freire and Parente 2006). Symptoms are similar for both pathogens and the infection starts as small round leaf lesions (0.2–0.5 cm) with light brown color. Spots become dark brown as the disease progresses and in some cases, total leaf necrosis could be observed. Dark-brown color injuries also occur on fruits.

In Nicaragua, Padilla and Monterroso (1999) reported that anthracnosis lesions on Jatropha leaves are necrotic, large and irregularly shaped. Usually, the lesions start on leaf edges, but can eventually start at the center. Anthracnosis due to *C. gloeosporioides* has also been reported in South Korea and confirmed by molecular markers (Jin-Hyeuk et al. 2012).

Alternaria spp. stand out among those causing leaf spots as the one with the largest widespread occurrence in Jatropha. In Nicaragua, the pathogen was found only sporadically, however it has been identified on mature fruits producing pedicel wilting

Fig. 11.8 Symptoms of powdery mildew (*white*) on lower (**a**) upper (**b**) sides of leaves and on fruits (**c**)

and fruit fall (Padilla and Monterroso 1999). The general practice for crop control contaminated by *Alternaria* spp. is recommending the use of resistant varieties, fungicide application and removal of residues.

Fig. 11.9 Anthracnosis necroses on Jatropha leaves

The fungus *Cercospora jatrophae-curcas* is also reported to cause leaf spots on Jatropha. In India, the control of this disease is carried out by spraying Bordeaux mixture at 1% (Swamy and Singh 2006). In the experimental areas of North Minas Technological Center (*Empresa de Pesquisa Agropecuaria de Minas Gerais* – EPAMIG) in Nova Porteirinha the fungi *Cercospora* and *Alternaria* have been isolated from small necrotic lesions of Jatropha leaves (Fig. 11.10). These lesions occur throughout the leaf and may cause total leaf necrosis by convergence in the case of high disease incidence or adverse climatic conditions.

The rust disease was first reported on Jatropha by Viegas (1945) in the state of São Paulo. The causal agent was the fungus *Phakopsora jatrophicola.* According to Franco and Gabriel (2008), this pathogen causes rust disease on leaves (Figs. 11.11a, b) and may lead to plant defoliation. Correia de Sá et al. (2008) claim that rust is one of the main diseases of Jatropha because of the high degree of defoliation it can cause to the plants. These authors observed 100% incidence of rust in 18 month old plants among 50 accessions in the state of Tocantins. However, when severity was

Fig. 11.10 Lesions on Jatropha leaves induced by *Alternaria* spp. and *Cercospora* sp. on upper (**a**) and lower sides of a leaf (**b**)

Fig. 11.11 Rust on the upper side (**a**) and on the lower side (**b**) of Jatropha leaves (Photo courtesy: Jaime Menezes)

assessed in detail, variation in symptom intensity was observed suggesting that the genetic basis for tolerance could be improved by selective breeding.

Roese et al. (2008) reported a rust epidemic in Jatropha plantations in the State of Mato Grosso do Sul between 2006 and 2008. These authors reported that the occurrence of this disease has increased over the years as the total acreage increased. They also quoted a survey conducted by Embrapa in a field of approximately 100 ha, in Dourados (MS), which showed total defoliation caused by the rust in the lower part and partial defoliation in the upper part of Jatropha infected plants. Defoliation occurred right after the reproductive period and affected the fruit and seed production process. Regarding disease control, combinations of flutriafol + thiophanate methyl, pyraclostrobin + epoxiconazole, azoxystrobin + cyproconazole and myclobutanil + azoxystrobin were found effective in controlling the disease in the middle part of plant canopy. Despite the effectiveness of these fungicides, they have not yet been recommended for use in Brazilian commercial plantations because they are not registered at the Ministry of Agriculture, Livestock and Supply for use in Jatropha.

Root diseases caused by soil fungi have promoted the death of Jatropha plants in both experimental areas and commercial plantations in Brazil and worldwide. Fungi of genera *Phytophthora* spp. and *Fusarium* spp. have been isolated from Jatropha with symptoms of collar rot and root rot in northern Minas Gerais.

Fusarium spp. is a cosmopolitan fungus, which has a large number of species known to cause diseases in important agronomic crops. Bedendo (1995) describes that, generally, the symptoms of root rot caused by soil fungi begin with the browning of young roots and progress to the older ones. This gradual darkening starts with a slight brownish color or, in some cases, reddish brown color and then becomes darker as the disease progresses. At the end of the process, necrotic roots become dark-brown or completely black. The browning symptom is accompanied by decomposition process of roots and the fully darkened roots disintegrate when subjected to light finger pressure. The collar rot is characterized by the appearance of stem lesions that can be located just below or above the soil surface. The lesions are usually depressed and also fungal structures may appear on them. In tender stems, the development of lesions can lead to weakening of the lesioned area, which may cause plant overturning. In woody stems, onset of cracking and flaking is observed, which, in addition to local damage, can serve as a gateway to other pathogens. In the field, symptoms occur in clusters, but when there are irrigation furrows, diseased plants appear in the same line due to spread of pathogen structures through water. The first evidence of a disease occurrence caused by a soil fungus appear in plant canopy through symptoms, such as leaf and branch shriveling, leaf yellowing, nutritional deficiency, premature leaf and fruit dropping; this syndrome normally leads to plant death. Young branches of Jatropha dry and break easily upon infection by *Fusarium* spp. and fungal structures can be found over the lesions on branches in the aerial part of plants (Padilla and Monterroso 1999). Although *Fusarium* spp. are being consistently isolated from Jatropha plants with symptoms of stem rot in plantations of North of Minas Gerais, experiments identifying this pathogen and pathogenicity tests on healthy host have not yet been conclusive.

Phytophthora spp. is a fungus that causes root and collar rots in plants in the early stages of development to maturity. Root symptoms induced by these fungi are browning followed by rotting. Franco and Gabriel (2008) described the symptoms of this disease in Jatropha as a manifestation of soft rot where necrotic tissues are darker than normal and exude a liquid with characteristic smell. Symptoms in the canopy are leaf yellowing and wilting that look as a nutritional deficiency. The disease can eventually progress until the branch death. As in the case of *Fusarium* spp., *Phytophthora* spp. are regularly being isolated from Jatropha plants with symptoms of collar rot and root rot in the North of Minas Gerais, but experiments for its identification on healthy host have not been conclusive.

Other soil fungi were also reported in the literature to be pathogenic for *Jatropha*; they are: *Macrophomina phaseolina* and *Rhizoctonia bataticola,* which cause the "collar rot". The characteristic injury symptom of these fungi is a lesion at the base of the main stem that occurs when the soil is waterlogged for long periods or in irrigated monocultures (Swamy and Singh 2006).

Weed Management

The presence of weeds in agricultural ecosystems may condition a number of biotic factors acting on the cultivated species and affecting, not only its productivity, but also its operation (Silva et al. 2007).

The most pernicious weeds for Jatropha are the scandent or climbing ones, such as jitiranas (*Merremia* spp.), morning glory (*Ipomoea* spp.) and bitter melon (*Momordica charantia*), among others. These weeds climb on the trunk and branches causing them to choke, tangling the canopy, providing shade and reducing production or even leading to host death (Saturnino et al. 2005).

An experiment was conducted at EPAMIG in Nova Porteirinha in order to assess the damage caused by weeds in early growth of Jatropha. The predominant weeds of this experimental site were *Merremia* spp., *Ipomoea* spp., *Cenchrus echinatus* and *Brachiaria decumbens.* Reduction of Jatropha green matter became evident 45 days after the coexistence with weeds. After this period, some of the Jatropha plants died due to strangulation and shading caused by weeds (Albuquerque 2012). Shaded plants suffered a drastic reduction in photosynthesis, resulting in a lower growth rate and a less developed root system with lower ability to absorb water and nutrients from soil.

Many weed vine species such as flame vine (*Pyrostegia venusta*), can cause serious injury to perennial species, reducing crop stand due to elevated seedling mortality and also causing trunk deformities. Plants affected by these weeds are almost always the dominant species with the consequence that they do not fully express their genetic potential for biomass production (Pitelli and Marchi 1991). The control of these weeds in early growth of Jatropha is critical for the development of normal plants and to avoid unnecessary costs in replanting.

In a screening of 11 commercial herbicides applied in post-emergence, for selectivity towards weeds and inocuity for Jatropha, haloxyfop-r-metyl, bentazon and

Fig. 11.12 Effect of several herbicides applied in pre-emergence of Jatropha plantlets. *Upper panel*, from left to right: 2.4 D, Imazapyr and Diclosulam (the control is on the *left* of each panel and the treatment on the *right*). *Bottom panel*, from left to right: Imazapyc, Flumetsulam and Trifluralin (the control is on the *left* of each panel and the treatment on the *right*). The word "testemunha" is from Portuguese and means *control* in English

trifluralin were reported affecting the height of Jatropha's shoots (Albuquerque et al. 2008a). All the other products assessed in this experiment (cloransulam-metílico, diclosulam, flumetsulam, 2, 4-D, imazapyr, imazapic, imazethapyr, quinclorac) caused morphological alterations and/or variations in leaf color of seedlings.

In pre-emergence (Fig. 11.12) and post-emergence (Fig. 11.13), trifluralin did not promote any visible toxicity to plants. It did not induce any changes in shoot diameter, shoot dry weight, plant height or stem diameter. Thus, this treatment was concluded to be the closest to an effective control of weeds in real conditions of Jatropha plantations in northern Minas Gerais (Albuquerque et al. 2008b).

In large culture areas, systemic chemicals can be applied between the rows of Jatropha. Among products for weed control, the widely used herbicide known as glyphosate stands out as the most efficient. Glyphosate inhibits the enzyme 5-enolpyruvyl-shikimate 3-phosphate synthase (EPSP) and retards the production of amino acids phenylalanine, tyrosine and tryptophan used for protein synthesis and also for the synthesis of some secondary metabolites, such as vitamins, lignin and hormones. If spray shields are located at the end of spray bars and sprays directed between the rows, the absence of selectivity of this potent systemic herbicide is not important as long as it does not come in contact with Jatropha leaves. The effectiveness of glyphosate depends on the efficiency of a series of processes, such as retention of the herbicide by leaves, adequate penetration, translocation and inhibition of the EPSP active site (Monquero 2003).

Weeds may cause losses in Jatropha cultivation and the lack of herbicides registered for the culture complicates the management of this crop. Environment friendly weed control should be considered for a more sustainable agriculture. A strategy that would be possible in soils without compaction problem, would be the use of desiccants with little residual effect that could be applied before Jatropha planting to give a comparative advantage to Jatropha seedlings. Another interesting strategy

Fig. 11.13 Effect of several herbicides applied in post-emergence of Jatropha plantlets. *Upper panel*, from left to right: 2,4 D (*left*: control, *right*: treatment), Imazapyr (symptoms of leaf and shoot mortality), Bentazon (symptoms of cell mortality on a leaf) and Trifluralin (*left*: control, *right*: treatment). *Medium panel*, from left to right: Quinclorac (symptoms of extended cell mortality on a leaf), Cloransulam-methyl (symptoms of leaf mortality), Diclosulam (symptoms of limited cell mortality on a leaf) and Imazethapyr (symptoms of leaf mortality). *Bottom panel*, from left to right: Imazapyc (symptoms of limited cell mortality on a leaf), Haloxyfop-R-methyl (*left*: control, *right*: treatment), Flumetsulam (induces a reduction of internodes and leaf area giving a rosy shape to the main shoot)

would be the intercropping of legumes between the lines of Jatropha in order to manage weeds and to provide a green manure able to fix nitrogen from air.

To ensure sustainability and competitiveness, Jatropha depends, among other factors, on a technology that meets the standards of quality and productivity reducing environmental impact. This reality brings forth the need for an effective investment in research for the adoption of proper cultural practices.

Conclusions

The management of pests, diseases and weeds is an important part of the technological package to increase productivity levels of Jatropha worldwide. What was once considered a minor problem is now of primer importance in the cultivation of

Jatropha. The diagnosis of major pests and diseases as well as the establishment of appropriate management techniques aiming to optimize cost-benefits will also serve as a guide to other large areas of knowledge, such as genetic improvement and fossil fuel dependency in a feedback process toward Jatropha domestication and industrialization.

Acknowledgements The authors thank the Research Support Foundation of Minas Gerais and Petrobras for financial support between 2007 and 2012.

References

Albuquerque CJB (2012) Controle de Plantas Daninhas na Cultura do Pinhão Manso. Relatório Técnico de Pesquisa Banco do Nordeste / ETENE FUNDECI. EPAMIG, Nova Porteirinha/MG

Albuquerque CJB, Souza IF, Alcântara EM, Saturnino HM, Brant RS (2008a) Aplicação de herbicidas em pós-emergência na cultura do pinhão-manso (*Jatropha curcas* L.). Congresso Brasileiro de Mamona 3. Salvador. Portuguese

Albuquerque CJB, Brant RS, Rocha GR, Jardim RR (2008b) Seletividade de herbicidas para o pinhão-manso. Congresso Brasileiro de Mamona. 3. Salvador. Portuguese

Bedendo IP (1995) Podridões de raiz e colo. In: Bergamin Filho A, Kimati H, Amorim L (eds) Manual de fitopatologia: princípios e conceitos, vol 1. Agronômica Ceres, Portuguese, pp 829–837

Bondar G (1913) Insetos daninhos na agricultura, 2. Bol Agric 14:434–470

Carneiro SMTPG, Ramos ALM, Romano E, Marianowski T, Oliveira JP (2009) Ocorrência de *Phakopsora jatrophicola* em pinhão manso no estado do Paraná. Summa Phytopathol 35(1):73. doi:10.1590/S0100-54052009000100017, Portuguese

Correia de Sá DA, Santos GR, Antunes CG (2008) Incidência e severidade de ferrugem em acessos de pinhão-manso no sul do Estado do Tocantins. Anais do COMBIEN, Uberlândia, Portuguese

Costa Lima A (1940) Insetos do Brasil, 2°. Tomo, capítulo 22, Hemípteros. Série Didática Núm. 3, Escola Nacional de Agronomia, Rio de Janeiro. p 351. Portuguese

Costa Lima A (1956) Insetos do Brasil, 10°. Tomo, capítulo 29, Coleópteros. Série Didática Núm. 12, Escola Nacional de Agronomia, Rio de Janeiro, p 373. Portuguese

Franco DAS, Gabriel D (2008) Aspectos fitossanitários na cultura do pinhão-manso (*Jatropha curcas* L.) para produção de biodiesel. Biológico 70(2):63–64. São Paulo. Portuguese

Freire FCO, Parente GB (2006) As doenças das Jatrophas (*Jatropha curcas* L. e *Jatropha podagrica* Hook.) no Estado do Ceará. Comunicado Técnico. Embrapa, Fortaleza, Portuguese

Froeschner RC (1988) Family Scutelleridae Leach, 1815. The shield bugs. In: Henry TJ, Froeschner RC (eds) Catalog of the Heteroptera or true bugs, of Canada and the Continental United States. EJBrill, New York, pp 684–693, 958

Gabriel D, Calcgnolo G, Tancini RS, Dias Netto N, Petinelli Junior A, Araújo JBM et al (1988a) Estudo com o percevejo *Pachycoris torridus* (Scopoli, 1772) (Hemiptera: Scutelleridade) e seu inimigo natural *Pseudotelenomus pachycoris* Lima, 1928 (Hymenoptera; Scelionidae) em cultura do pinhão paraguaio *Jatropha* spp. Biológico 54(1/6):17–20, Portuguese

Gallo D, Nakano O, Silveira Neto S, Carvalho RPL, Batista GC, Berti Filho E et al (1988) Manual de Entomologia Agrícola, 2nd edn. Agronômica CERES, São Paulo, 649. Portuguese

Gallo D, Nakano O, Silveira Neto S, Carvalho RPL, Batista GC, Berti Filho E et al (2002) Entomologia Agrícola. FEALQ/Biblioteca de Ciências Agrárias Luiz de Queiroz, Piracicaba, 920. Portuguese

Hussey RF (1934) Observations on *Pachycoris torridus* (Scop.), with remarks on parental care in other Hemiptera. Bull Brooklyn Entomol Soc 29(4):133–145

Jin-Hyeuk K, Choi O, Kim J, Kwak Y (2012) First report of Anthracnose disease on *Jatropha curcas* caused by *Colletotrichum gloeosporioides* in Korea. J Phytopathol 160(5):255–257
Laviola BG (2011) Pesquisa. Desenvolvimento e Inovação em Pinhão-manso para produção de Biodiesel. II Congresso Brasileiro de Pesquisa de Pinhão-Manso. Brasília. Portuguese
Menezes EB, Aguiar-Menezes E, Aquino AM, Correia MEF, Souza JH, Mauri R et al (2007) Cupins: taxonomia, biologia, ecologia e sua importância nos sistemas agropecuários, Seropédica. Documentos EMBRAPA Agrobiologia 53 Portuguese
Monquero PA (2003) Dinâmica populacional e mecanismos de tolerância de espécies de plantas daninhas ao herbicida glyphosate. Dissertation, Esalq/USP, Piracicaba (SP). Portuguese
Ospina HFO. (Coord.) (1980) El lorito verde (*Empoasca kraemeri* Ross & Moore) y su control. CIAT¸ Colômbia, p 41. *In: Moreno PR, Nakano O (eds) Activity of buprofezin on the green leafhopper Empoasca kraemeri (Ross & Moore, 1957) (Hemiptera, Cicadellidae) under laboratory conditions. Sci Agric. Piracicaba, 2002;59(3)*
Padilla D, Monterroso D (1999) Diagnostico preliminar de enfermedades del cultivo de tempate (*Jatropha curcas*) en Nicaragua. Manejo Integrado de Plagas. [Internet] Turrialba 51:66–69. Portuguese. Available from: http://web.catie.ac.cr/informacion/rmip/rmip51/padilla.html
Peredo LC (2002) Description, biology, and maternal care of *Pachycoris klugii* (Heteroptera: Scutelleridae). [Internet] Florida Entomologist 85:464–473. Available from: http://www.bioone.org/doi/full/ 10.1653/00154040(2002)085%5B0464%3ADBAMCO%5D2.0.CO%3B2
Pitelli RA, Marchi SR (1991) Interferência das plantas invasoras nas áreas de reflorestamento. Seminário técnico sobre plantas daninhas eo uso de herbicidas em reflorestamento 3. MG, Belo Horizonte, pp 1–11, Portuguese
Quintela ED (2002) Manual de identificação dos insetos e invertebrados: pragas do feijoeiro. Embrapa Arroz e Feijão, Santo Antônio de Goiás, p 52, Documentos. Portuguese
Roese AD, Silva CJ, Goulart ACP, Abrão JS (2008) Ocorrência de Ferrugem no Pinhão-manso, em Mato Grosso do Sul, e efeito de alguns fungicidas no controle da doença. Embrapa, Dourados, (Comunicado Técnico). Portuguese
Sahai K, Srivastava V, Rawat KK (2011) Impact assessment of fruit predation by *Scutellera perplexa* Westwood on the reproductive allocation of Jatropha. Biomass Bioenergy 35:4684–4689
Sánchez-Soto S, Nakano O (2002) Ocorrência de *Pachycoris torridus* (Scopoli) (Hemiptera: Scutelleridade) em Acerola (*Malpighia glabra* L.) no Brasil. Neotrop Entomol 31(3):481–482
Sánchez-Soto S, Milano P, Nakano O (2004) Nova planta hospedeira e novos padrões cromáticos de *Pachycoris torridus* (Scopoli) (Hemiptera: Scutelleridade) no Brasil. Neotrop Entomol 33(1):109–111, Portuguese
Santos HO, Silva-Mann R, Poderoso JCM, Oliveira AS, Carvalho SVA, Boari AJ et al (2006) O ácaro *Tetranychus bastosi* TUTTLE, BAKER & SALES (PROSTIGMATA: TETRANYCHIDAE) infestando germoplasma nativo de *Jatropha* sp., no estado de Sergipe, Brasil. 2° Congresso Brasileiro de Mamona, 2. Aracaju, 4. Portuguese
Saturnino HM, Pacheco DD, Kakida J, Tominaga N, Gonçalves NP (2005) Cultura do pinhão-manso (*Jatropha curcas* L.). Informe Agropecuário 26(26):229, Portuguese
Silva AA, Ferreira FA, Ferreira LR, Santos JB (2007) Métodos de controle de plantas daninhas. In: Silva AA, Silva JF (eds) Tópicos em manejo de plantas daninhas. MG/UFV, Viçosa, pp 63–81, Portuguese
Silva AGA, Gonçalves CR, Galvão DM, Gonçalves AJL, Gomes J, Silva MN et al (1968) Quarto catálogo dos insetos que vivem nas plantas do Brasil. Seus parasitos e predadores. Parte 2, vol 1, Insetos, hospedeiros e inimigos naturais. Ministério da Agricultura, Rio de Janeiro, 622p. Portuguese
Shanker C, Dhyani SK (2006) Insect pests of *Jatropha curcas* L. and the potential for their management. Curr Sci 91(2):162–163
Swamy SL, Singh L (2006) *Jatropha curcas* for biofuel plantations. In: Proceedings of the biodiesel conference towards energy independence – focus on Jatropha, Hyderabad, pp 143–157
Tominaga N, Kakida J, Yasuda EK, Série Agroindústria (2007) Cultivo de pinhão-manso para produção de biodiesel (pragas). Centro de Produções Técnicas, Série Agroindústria. MG, Viçosa, pp 130–133, Portuguese

Ungaro MRG, Regitano Neto A (1996) Considerações sobre pragas e doenças de pinhão-manso no Estado de São Paulo, vol 4, Congresso Brasileiro de Plantas Oleaginosas, Óleos, Gorduras e Biodiesel. MG, Varginha, pp 729–735, Portuguese
Viegas AP (1945) Alguns fungos do Brasil IV: Uredinales. Bragantia, Campinas. 5(1):7–8. Portuguese
Viégas AP (1961) Indice de fungos da América do Sul. Campinas: Instituto Agronômico. p 919
USDA (1960) Index of plant diseases in the United States. Washington. (USDA Crop Research Service. Handbook, 165)

Chapter 12
Phytotechnical Aspects of Jatropha Farming in Brazil

José Carlos Fialho de Resende, José Tadeu Alves da Silva, Fúlvio Rodriguez Simão, Rodrigo Meirelles de Azevedo Pimentel, and Danielle de Lourdes Batista Morais

Introduction

The prospects associated with *Jatropha curcas* L. (hereafter referred to as Jatropha) as a species for biodiesel production have been extensively reported in the literature; however, to date no scientifically valid recommendation for its farming is available for this crop.

Investigations on propagation techniques, crop fertilization, irrigation, plantation spacing, pruning, growth regulators and mechanical harvesting are on-going, and basic trends are already defined. However, official recommendations from government references are necessary to warrant faithful production systems capable of providing acceptable returns on investments to farmers and sustainable oil production.

The *Empresa de Pesquisa Agropecuária de Minas Gerais* (EPAMIG), which can be translated in English as the Agricultural Research Company of Minas Gerais, was incorporated as a public company in 1974. It is the main institution for the implementation of agricultural research in the state of Minas Gerais (Brazil) and functions to provide solutions for agriculture through technological research and development. It also offers specialized services and technical training compatible with customer needs and, more generally, aims to improve the quality of life.

Among the five regional units that Epamig maintains in the state of Minas Gerais, Minas Epamig North has five farms, four in the North of Minas Gerais and one in a

J.C.F. de Resende (✉) • J.T.A. da Silva
Fazenda Experimental de Montes Claros, BR 251, Km 11,
CEP 39404-128, Montes Claros Minas Gerais, Brazil
e-mail: jresende@epamig.br

F.R. Simão • D.d.L.B. Morais
Fazenda Experimental do Gorutuba, MGt 122, Km 150, CEP 39525-000,
Nova Porteirinha, Minas Gerais, Brazil

R.M.d.A. Pimentel
Fazenda Experimental de Caldas, Avenida Santa Cruz, n° 500,
Caixa Postal 33, Bairro Santa Cruz, CEP 37780-000 Caldas, Minas Gerais, Brazil

N. Carels et al. (eds.), *Jatropha, Challenges for a New Energy Crop: Volume 1: Farming, Economics and Biofuel*, DOI 10.1007/978-1-4614-4806-8_12,

region called Vale do Jequitinhonha. Most of Epamig's experiments on Jatropha were carried out on one farm (*Fazenda Experimental Gorutuba*) in North Minas Gerais (Nova Porteirinha). The farm lies at a latitude of 15°03′ South, a longitude of 44°01′ West and at an altitude of 452 m. The predominant soil in this region is a red-yellow *latosol* in flat terrain, but *neosol* is also found. Latosol is a type of soil in an advanced stage of weathering, which is highly evolved as a result of significant changes in its constitutive materials. These soils are deficient in minerals, less resistant to weathering and have low cation exchange capacities. They are generally strongly acidic with low base saturation, well drained and very deep. Neosols include a thin layer of minerals or organic materials and did not experience significant changes due to the low intensity of pedogenic processes. The soils of North of Minas Gerais are typically sandier, with low levels of exchangeable aluminum, phosphorus (P) and organic matter (~0.5%) and are typically of the *dry forest* biome.

Dry forest is characterized by forest vegetation with a predominance of deciduous trees that shed their leaves during the dry season. It is a transition zone between *Caatinga* and *Cerrado*, with characteristics of the *Atlantic Forest* biome because of its diversity in deciduous trees.

Caatinga (from Tupi: caa (kill) + tinga (white) = white forest) is the only exclusively Brazilian biome, which means that most of its biological heritage cannot be found anywhere else on the planet earth. Its name is derived from the whitish landscape presented by the vegetation during the dry season, when most plants shed their leaves and the trunks become dry and whitish. The Caatinga occupies an area of approximately 800,000 km², about 10% of the country, comprising contiguous parts of the states of Maranhão, Piaui, Ceara, Rio Grande do Norte, Paraiba, Pernambuco, Alagoas, Sergipe, Bahia (Northeast Brazil) and the northern part of Minas Gerais. The botanical heritage includes 2,000–3,000 native plant species with xerophytic species in the understory associated with calcareous outcrops.

The Cerrado is the second largest Brazilian biome, extending over an area of 2,045,064 km² in eight states (Central Brazil: Minas Gerais, Goiás, Tocantins, Bahia, Maranhao, Mato Grosso, Mato Grosso do Sul, Piauí and the District Federal). The landscape has a high biodiversity, although less so than the Atlantic Forest. The vegetation is similar to the savanna, with grasses, shrubs and sparse trees. The average annual temperature is 25°C, eventually reaching 40°C in the spring. The minimum registered temperature can reach values close to 10°C or less.

Rainfalls are unevenly distributed between the months of November and March and account for an average of 750 mm annually. The drought normally extends from April to late October and is characterized by a sharp drop in the relative humidity. Strong winds occur between May and August. Even if growth is reduced to minimum rates during the dry season, leaf loss is not observed in 1-year-old plants. The phenomenon of leaf loss during the dry season in northern Minas Gerais is observed only from the second year of the plant. In northern Minas Gerais, Jatropha remains leafy from November to June (8 months, Fig. 12.1a), sheds its leaves during the resting period (winter) from July to September (3 months, Fig. 12.1b) and sprouts again in October (1 month, Fig. 12.1c).

Fig. 12.1 Vegetative cycle of Jatropha in northern Minas Gerais. Vegetative phase (Leafy) from November to June (**a**); without leaves during resting period (winter) from July to September (**b**); and vegetative phase sprouting again in October (**c**)

Depending on regional climatic conditions, the phenology of Jatropha changes according to the time of the year. In the Southeast region of the country, the rainy season is concentrated from October to March. In the northern region of Minas Gerais, especially where Jatropha is planted on a commercial scale for seed production, the rainy season is unevenly distributed in this period (October to March), with peaks between November and December followed by a dry period in January and February and scattered showers in March.

By contrast, in the *Zona da Mata* (a name for a geographic sub-region with specific socio-economic features) in the state of Alagoas in Northeast Brazil, the rainy season is between April and August (70% of total annual rainfall happens in this period), and the dry season occurs from September to February (Souza et al. 2005). Under the conditions of Viçosa, which is included in the *Zona da Mata* of Minas Gerais, Matos (2010) concluded that leaf senescence in Jatropha is due to two factors: a decrease in minimum temperature and an increase in the difference between the maximum and minimum temperatures. Interstingly, leaf senescence was not found to be related to water stress or nitrogen deficiency. The rate of leaf fall increased as the minimum air temperature decreased in the period from January to June.

The leaf fall occurred without any significant relationship to the maximum temperature or soil moisture, which remained near field capacity. The sharp leaf fall in this region occurred when the minimum temperature was below 10°C and the difference between the maximum and minimum temperatures was higher than 20°C.

Corroborating the description of Matos (2010) and Oliveira et al. (2011) from Rio Grande do Sul (southern Brazil), the number of chilling hours directly influences the budding, flowering and fruiting of Jatropha. The climatic differences observed during the experiment were correlated with differences in the phenology of Jatropha over 2 years. In the first year of the experiment, the occurrence of only 160 chilling h guaranteed that fruiting began in November and December, while the incidence of 331 h of cold delayed the onset of fruiting until January of the second year.

After dormancy, the lush production of leaves and inflorescences begins with the first rains that usually occur in mid-October. Thus, the fruit onset of Jatropha in the northern region of Minas Gerais generally begins in November with a main flowering peak in December-January and a secondary one in March-April. In Alagoas, the period from flowering to fruit maturity is 65 days on average (Santos et al. 2010), while it lasts from 43 to 61 days in India (Rao et al. 2008). This period gives a window of fruit harvest in northern Minas Gerais that begins in March (3 months after flowering) and can extend to June-July.

Due to its ability to thrive in a semi-arid climate, Jatropha is well adapted to this region, which lacks good agricultural options in terms of biofuels and agriculture in general. In the early 1980s, some researchers from EPAMIG became interested in Jatropha and started planting trials on various experimental farms in the state of Minas Gerais. They collected data that are stored in the company archives. This information is of significant importance for the continuation of current projects on Jatropha, which were resumed in mid-2004 after having been slowed at the end of the 1980s. Issues relevant to ecological conditions for cultivation, physiological aspects, selective breeding, vegetative propagation, seed quality, soil features, fertilization, weed and irrigation management were carefully addressed. Characterization of harmful arthropods (pests), diseases, harvest, post-harvest and use of co-products of oil extraction for animal feeding are also being pursued.

Under the following, we present the important observations on Jatropha farming that we have made over the past several decades.

Propagation

The longevity of profitable production of Jatropha is estimated to be between 20 and 30 years (Dias et al. 2007). Because Jatropha is a dioecious plant that is generally out-crossing and entomophilous, large variation of pollen spread between individuals is generally observed and may affect the regularity of seed production. In addition to seed propagation, Jatropha can also be propagated asexually by cuttings, grafting or in vitro culture. The rainy season is the best planting period in the field. Plants originating from seeds flower 9 months after their transplantation in the field, whereas plants reproduced through cuttings generally start to flower after 6 months in the field (Saturnino et al. 2005). In general, plants grown from seeds develop a taproot and four lateral roots, and they are economically productive in the fourth year in the field. Plants from cuttings have a less vigorous root system without a taproot, but they exhibit slightly earlier production.

Seed multiplication is recommended because of the better root system (Severino et al. 2006). Indeed, multiplication of Jatropha in Brazil has occurred traditionally by collecting seeds from individual plants growing in hedges, gardens and dwelling neighborhoods (Saturnino et al. 2005). Currently, specialized nurseries are performing this work of seed multiplication on commercial substrates. Seedlings produced in containers of 120 ml are taller, with larger and more abundant leaves than those produced in 50 ml tubes; thus, 120 ml containers are the best for seedling production (Avelar et al. 2005).

Vegetative cloning allows the multiplication of individuals without affecting their genetic structure, which is advantageous to increase the population of elite genotypes (Saturnino et al. 2005). Multiplication through cuttings is most widespread (Lima et al. 2010; Smiderle and Kroets 2008; Vasquez et al. 2010). A possible compromise that has yet to be tested is to carry out Jatropha multiplication *in vitro* by somatic embryogenesis, which should give it a better and uniform rooting system in the field than classical micropropagation by shoot multiplication. Somatic embryogenesis is reported in Jatropha (Cai et al. 2011; Jha et al. 2007; Nunes et al. 2008) and could be a desirable character of a commercial cultivar if the rooting quality is confirmed in the field.

A more pragmatic solution is to produce seedlings from bare roots. This system is among the most cost effective and warrants a better rooting system (Siles et al. 1997). According to this technique, seeds are germinated in high density on a ~50 cm deep sand layer. Seedlings develop a robust pivoting root system and an etiolated aerial part because of the high density. Seedlings are brought to the field in bundles without support, thus reducing costs and improving the planting operations.

The planting of cuttings (clones) is only recommended for replacing plants that are not productive or are attacked by pests or diseases. The stakes for this purpose should be cut from woody branches from plants that have been free of pests and diseases for 1 year. The major limitation for the propagation of cuttings is the large volume of material to be used in commercial fields, as well as the need to know the characteristics of the trees providing the cuttings, emphasizing productivity, health and precocity.

Grafting is an alternative method for propagation of Jatropha. This technique is not used commercially; however, good results have been achieved in experimental fields at EPAMIG by grafting shoots of *J. curcas* onto two wild species, i.e., *Jatropha pohliana* Mull. Arg. and *J. gossypiifolia* L. Cleft (Fig. 12.2). Following this method, a fixation rate of about 90% is to be expected; it is more effective than the simple *English* graft and is the recommended technique for Jatropha (Marques et al. 2007). According to Marques et al. (2008), seedlings of *J. pohliana* obtained from mature seeds without caruncles yield the best rootstocks due to the higher germination rate. Similar results were obtained by grafting *J. curcas* on *J. molissima* Mull. Arg. (Anjos et al. 2007). *J. molissima* is native from Northeast Brazil and therefore is well adapted to the climate and soil conditions of Caatinga (its area of origin), making it an ideal rootstock for Jatropha.

Jatropha's productivity is influenced by spacing. Optimization of the plant arrangement has been investigated by several research institutions in Brazil. Eight spacing schemes (3×1, 3×2, 3×3, 4×1, 4×2, 4×3, 4×4 and 4×5 m) were evaluated for Jatropha planting. Silva et al. (2011) recommended spacings of 4×4 and 4×5 m in the region of Anastasius (Mato Grosso do Sul, Brazil).

The best spacing and planting arrangements for Jatropha are currently under investigation, and these parameters should vary according to the region of cultivation, size of the plant, soil fertility and exploitation system, such as consortium plantation. When intercropped with other cultures, the suggested spacing for Jatropha ranges from 2× 2 m up to 2×8 m (Demartini et al. 2009).

Fig. 12.2 Grafts of *J. curcas* (graft) on *J. pohliana* (rootstock) (Photo courtesy: MS Carvalho Dias)

Fertilizer Application

Jatropha is preferably cultivated in well-structured and deep soil so that the root system can reach the deeper layers and explore the soil as much as possible, ensuring better absorption of water and nutrients, especially when water is scarce. It can grow in poor sandy soils and clay soils of low fertility. However, productivity of the species is better when it is grown in well-drained, deep, and airy soils with medium to high fertility. At EPAMIG, we have observed that the initial development of Jatropha plants grown in clay soils is better than when grown in sandy soil (Fig. 12.3).

Very shallow clay soils with constant humidity, little air and poor drainage should be avoided (Arruda et al. 2004). The root growth of Jatropha has been shown to decrease linearly with the compression rate of the top layer of *quartz-sand dystrophic neosol*, which means that the plant is sensitive to compacted soil (Abreu et al. 2006).

In addition, Jatropha should not be planted in soils where the electrical conductivity (EC) is elevated or where the irrigation water has a high salt content (Vale et al. 2006). We observed a reduction of plantlet height from 19.7 to 13.3 cm using water with EC values of 0.06 dS. m^{-1} and 4.2 dS. m^{-1}, respectively.

Fig. 12.3 Jatropha grown in clay soil (**a**) and sandy soil (**b**) in Northern Minas Gerais

Jatropha adapts to low-fertility soils; however, application of nitrogen (N), phosphorus (P), potassium (K), calcium (Ca) and magnesium (Mg) provides significant increases in production. A favorable level of soil fertility is necessary for the plants to express their yield potential. According to Silva et al. (2009), the macronutrients that most affect (by more than 85%) the production of total dry matter are Ca, Mg and K. The series of the relative importance of macronutrients is Ca > Mg > K > N > P > S. P and S are less important macronutrients because their omissions result in the smallest reductions of dry matter production of the macronutrients. In the greenhouse, Kurihara et al. (2006) observed a highly significant response towards P input, especially in soils with low available P. In research conducted at EPAMIG in Northern Minas Gerais, Moura Neto et al. (2007) found that the relationship (Fig. 12.4) of (1) plantlet height (Fig. 12.4a), (2) stem diameter (Fig. 12.4b), (3) number of leaves (Fig. 12.4c), and (4) weight of roots (Fig. 12.4d), stems (Fig. 12.4e) and leaves (Fig. 12.4f) to P input follows a quadratic function (ax2 + bx + c = 0). Compared to the control without P, these parameters increased under P fertilization by 59% (height), 31% (diameter), 205% (leaf number), 59% (root weight), 87% (stem weight) and 223% (leaf weight). These authors also concluded that P is extremely important for the early development of Jatropha plants, as observed from the effects of increasing doses of P on seedling development (Fig. 12.5).

In sandy soil under controlled conditions, a maximum seed production of 1,538 kg. ha^{-1}. $year^{-1}$ was obtained with 21-month-old plants fertilized with an application of 240 and 400 kg. ha^{-1}. $year^{-1}$ of N and P_2O_5, respectively (Silva et al. 2007a). Maximum seed production (2,137 kg. ha^{-1}. $year^{-1}$) after 33 months, estimated with the equation $= 1{,}390 + 2.24 * N + 2.11 * P_2O_5 - 0.0053 * (P_2O_5)^2$ ($R^2 = 0.70$), was obtained with the application of 240 and 192 kg. ha^{-1}. $year^{-1}$ of N and P_2O_5, respectively (Fig. 12.6). The seed yield increased with age and stabilized between the fifth and sixth year, but increasing doses of N and P_2O_5 did not result in significant effects on the oil content of the seeds (Table 12.1).

As described above, N occupies the fourth place in the ranking of macronutrient importance in Jatropha (Silva et al. 2009), in contrast with castor beans for which N is the most limiting nutrient (Lavres et al. 2005). N is the nutrient that most promotes

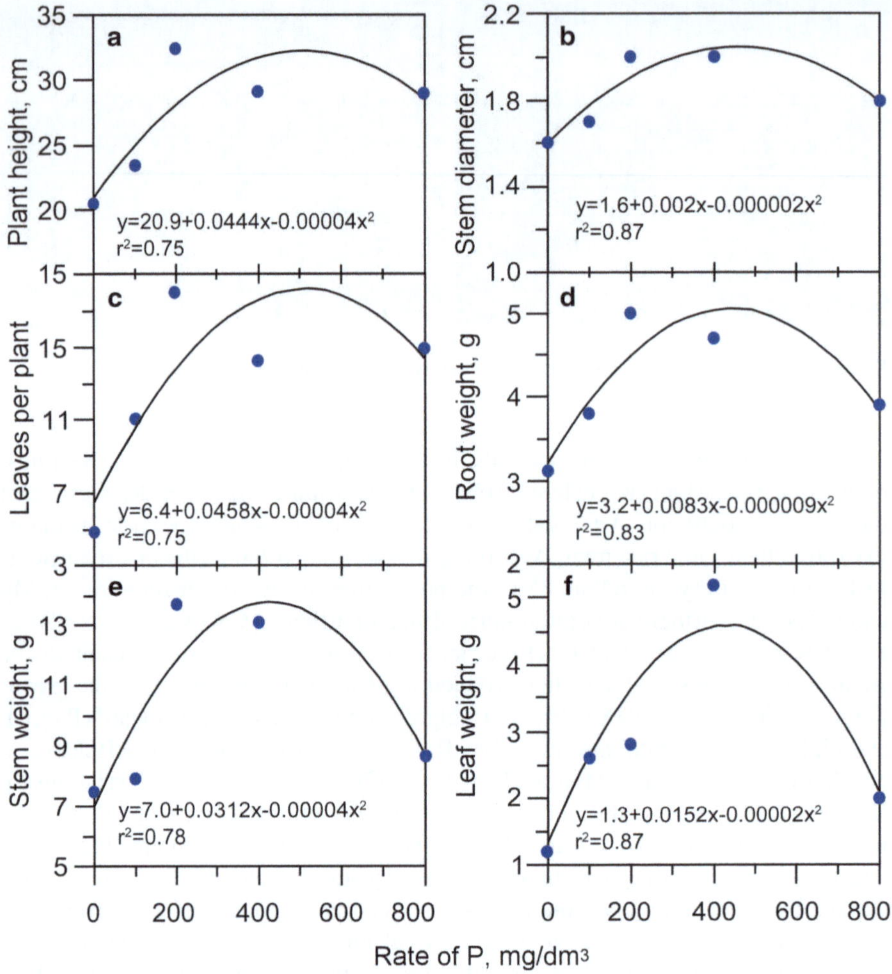

Fig. 12.4 Plant height (**a**), stem diameter (**b**), number of leaves per plant (**c**), root weight (**d**), stem weight (**e**) and leaf weight (**f**) of Jatropha as a function of P input (Moura Neto et al. 2007). The t-test was significant at probability levels of 1% and 5%

plant vegetative development in the presence of water. Inter-cropping with Fabaceae legume species is a way to partially supply the N that is needed for Jatropha development. According to Saturnino et al. (2005), Jatropha is highly productive when grown in areas that receive large amounts of organic manure.

The mass of dry matter produced by Jatropha grown on *yellow-red latosol* with medium texture supplemented with four doses of N and five doses of K was found to increase with N following a quadratic function, but to decrease with K (Silva et al. 2007b), suggesting that Jatropha does not require a large amount of K for its initial development. Thus, the native content of K in soils (87 mg dm^{-3}) is sufficient for the plant demand, as was confirmed by the fact that K application during the first

Fig. 12.5 Response of Jatropha seedlings to increasing P input

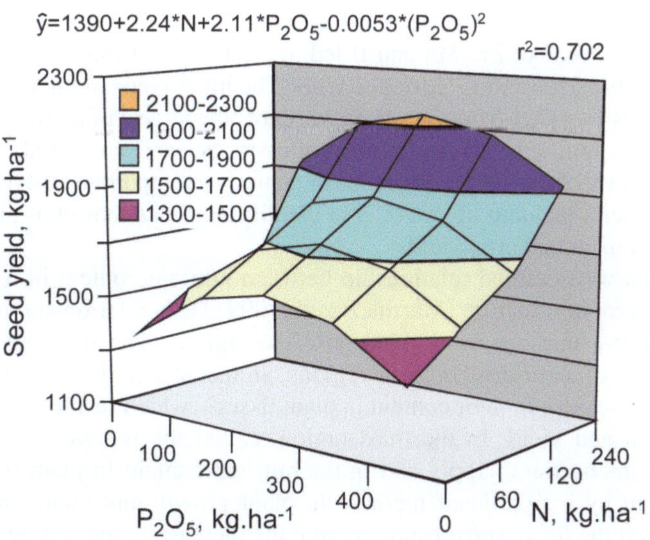

Fig. 12.6 Seed yield of 33-month-old plants as a function of nitrogen and phosphorus application in sandy soils with high availability of phosphorus

Table 12.1 Seed oil content of Jatropha under different doses of N and P_2O_5 application

Rates of N and P_2O_5 (kg ha^{-1} year^{-1})		
N	P_2O_5	Oil content (%)
0	0	34.8
0	100	35.0
0	200	34.0
0	400	35.0
60	0	34.8
60	100	35.4
60	200	34.0
60	400	33.0
120	0	35.2
120	100	34.2
120	200	35.1
120	400	33.1
240	0	35.2
240	100	33.0
240	200	34.7
240	400	32.5
Mean		34.3

5 months do not significantly affect vegetative development (Oliveira et al. 2007). K is expected to be essential at the stages of seed formation and maturation, as K is present in large amounts in mature seeds (CETEC 1983). The absence of K at this stage could be a bottleneck for seed production.

Omission of Fe, Cu, Zn, Mn and B led to reductions of the total dry matter by 84%, 69%, 43%, 31% and 17%, respectively. The importance of micronutrients follows the series Fe > Cu > Mn ~ Zn > B. In Jatropha, the largest micronutrient requirement is for Fe, similar to the requirement of castor beans (Lange et al. 2005).

The diagnosis of the nutritional status of plants is an important tool for the proper use of fertilizers; its main objective is to identify nutrients that limit plant growth, plant development and crop yield.

There is a well-defined relationship between nutrient content in tissues, plant growth and crop production (Martinez et al. 1999). This relationship is characterized by a curve that can be divided into five regions. The first and second are called regions of *disability*. In these regions, an increase in nutrient supply is followed by an increase in their content in plant tissues, which results in an increased plant growth and yield. In the third region, called the region of *adequacy*, an increase in the nutrient supply and in the nutrient content in plant tissues is not accompanied by a significant increase in plant growth and yield. In the fourth region, called the *luxury absorption* region, the increase in the supply of nutrients and their content in tissues is not accompanied by any increase in plant growth or yield, which means that the addition of nutrients does not result in any benefit for

Fig. 12.7 Fifth leaf position on a branch with inflorescence

Table 12.2 Levels of nutrients in the limbo of the fifth leaf collected on a flowering branch in the middle part of a Jatropha plant

N	P	K	Ca	Mg	S	B	Cu	Fe	Mn	Zn
dag kg^{-1}						mg kg^{-1}				
2.93	0.29	2.28	2.11	1.14	0.22	39.7	5.1	139	163	19.2

plant growth or yield, i.e., it is wasted. The fifth region is called the *toxic* region and is characterized by a decline in the plant growth or yield with an increase in the nutrient supply and tissue content.

Studies on Jatropha nutrition are still at an infancy. With the goal of identifying the leaf indicators of nutritional status in Jatropha, we found a significant correlation of N and P in the limbo of the fifth leaf with the doses of N and P applied to the soil; by contrast, only a small correlation was observed between these variables in the leaf petiole. For this correlation to be achieved, the fifth leaf must be (1) counted only among the fully formed leaves, (2) from a branch in the median part of the plant and (3) with an inflorescence (Fig. 12.7). The nutrient levels found in the limbo of the fifth leaf of a highly productive plant are presented in Table 12.2.

Roots exhibit poor development when the soil acidity is below pH 4.5. Limestone should be added at a depth of 20–30 cm approximately 2 months before planting to reduce the acidity of the soil. Correction of the free aluminum content by liming has a positive effect on the development of Jatropha. In a sample of *red clayey latosol*, the addition of 55% limestone to correct for acidity was needed to obtain maximum production (12 g) of the dry mass of Jatropha seedlings. In *sandy neosol*, the maximum dry mass production (24 g) of seedlings was reached by adding 60% limestone

Fig. 12.8 Response of Jatropha to increasing Ca doses corresponding to 0.08 and 1.6 times the lime required to neutralize the pH (Source: Pacheco et al. 2006)

(Kurihara et al. 2006). In typical *ortic sandy neosol*, 55% limestone provided further development of seedlings grown in a greenhouse (Tanure 2006). Several greenhouse surveys have shown that the Jatropha plants develop better when Ca is applied. It is known that Ca stimulates the development of pivoting roots, ensuring better water absorption and formation of secondary roots. These factors are critical for better plant use of soil fertility and for adaptation to stress, particularly under water deficit conditions. Calcium is one of the most important nutrients for root growth because it stimulates water and nutrient uptake by roots. Upon application of combinations of Ca and Mg doses in Jatropha seedlings grown in soil samples with low fertility, Pacheco et al. (2006) found a larger stimulating effect of Ca compared to Mg on seedling development, which indicates that Ca is much more important than Mg to support early seedling growth (Fig. 12.8).

However, the omission of Mg resulted in a decrease in dry matter production statistically equal to that of Ca, showing the importance of Mg as a liming factor of Jatropha development and explaining the benefit of using lime with a higher magnesium content (Silva et al. 2009).

According to the results above, it is possible to suggest preliminary fertilizing recommendations for Jatropha cultivation (Tables 12.3, 12.4 and 12.5) in the semi-arid region of Montes Claros (North of Minas Gerais, Brazil).

Post-planting fertilization is performed in accordance with the projection of the plant canopy. Soil sample analysis must be carried out in the region prior to fertilizer application to calculate the correct dose. Soil analyses must be carried out once a year to assess the evolution of soil fertility. In addition to soil analyses, nutrient analyses of the fifth leaves are also recommended to check which nutrient(s) may be a limiting factor for optimized fruit production. Leaf analysis is an important tool for determining if a fertilization program is in agreement with production simulations.

Table 12.3 Scheme of fertilization at the stage of Jatropha seedling or cutting plantation

Time	Availability of P[1]			Availability of K[1]			Dose of N
	Low	Medium	High	Low	Medium	High	
	P_2O_5			K_2O			
	Kg ha^{-1}						
Planting	150	100	60	20	10	0	0
Days							
30				15	10	5	15
60				15	10	5	15
90				15	10	5	15

[1]Use the criteria for interpretation of each region. On soils low in zinc (Zn), it is recommended to apply 25 g of zinc sulphate per plant in the rainy period

Table 12.4 Jatropha fertilization in the first year of plantation

Time	Availability of P[1]			Availability of K[1]			Dose of N
	Low	Medium	High	Low	Medium	High	
	P_2O_5			K_2O			
	Kg ha^{-1}						
Months[a]							
November	–	–	–	30	20	15	20
January	–	–	–	30	20	15	20
Feb/March	–	–	–	30	20	15	20

[a]Refers to the months of rain, which may vary among regions. On soils low in zinc (Zn), it is recommended to apply 25 g of zinc sulphate per plant in the rainy period

Table 12.5 Jatropha fertilization in the second year of plantation

Time	Availability of P[1]			Availability of K[1]			Dose of N
	Low	Medium	High	Low	Medium	High	
	P_2O_5			K_2O			
	Kg ha^{-1}						
Months[a]							
November	100	60	30	40	30	20	30
January				40	30	20	30
Feb/March				40	30	20	30

[a]Refers to the months of rain, which may vary among regions. On soils low in zinc (Zn), it is recommended to apply 25 g of zinc sulphate per plant in the rainy period

Irrigation

In an eco-physiology study of water and gas exchange, Lima Filho et al. (2007) measured the water potential, gas exchange, stomatal conductance, photosynthesis and transpiration of leaves exposed to sun in 18-month-old plants spaced

2.0 × 2.0 m in irrigated and non-irrigated plots, following a completely randomized design. The water potentials of the irrigated and non-irrigated plants were −0.57 and −0.95 MPa, respectively, at 5:00 am (before sunrise) and −1.4 and −1.7 MPa after sunrise when evapotranspiration is increased. Photosynthesis was much more impaired than evapotranspiration within the levels of water potential and stomatal conductance observed. The parameters of leaf temperature (Tf), photosynthesis, water pressure, stomatal conductance (gs) and transpiration (E) measured at 8:00 am, 10:00 am, 12:00 pm, 2:00 pm and 6:00 pm showed that Jatropha presented the typical behavior of a woody plant under warm climates and a rainy season. The values of gs were higher in the early morning (between 1.0 and 1.5 mol $m^{-2} s^{-1}$) and then fell to values approximately 0.7–0.9 mol $m^{-2} s^{-1}$ between 12:00 am and 2:00 pm depending on the temperature of the largest leaf. The stomata were closed between 12:00 pm and 4:00 pm; however, photosynthesis reached its highest values during this time interval (between 8 and 9 mmol $m^{-2} s^{-1}$), certainly as a result of the higher air evaporative demand (Araújo et al. 2007).

Although the cultivation of Jatropha has been described as drought tolerant to water shortages, we observed a positive effect of an artificial water supply to the crop in periods when it is subjected to water stress. This positive contribution of water to Jatropha development and production has also been observed in regions where rainfalls are higher and evenly distributed. However, successful irrigation of this species requires a rational system together with other necessary agricultural inputs.

Irrigation of Jatropha can be performed using several methods and systems; there is not a system more suitable than the others, but rather advantages and drawbacks for each of the system. Thus, *in situ* experiments can be used to learn about the most appropriate irrigation system. Appropriate irrigation methods for Jatropha are (1) localized irrigation (micro-sprinkling systems and drip), (2) overhead irrigation (central pivot and conventional sprinkling with restrictions) and (3) surface irrigation.

In the localized irrigation methods, water is applied directly to each plant above the root system. In overhead irrigation, water is applied above the plants, resembling natural rain. By contrast, surface irrigation refers to irrigation methods where water is moved from the soil surface to the plants. Linear and central pivot are considered as automated overhead irrigation systems. In linear irrigation, an automated sprinkler moves in a straight line. In central pivot irrigation, automated sprinklers moves in a circle around a central point or pivot.

According to Costa et al. (2008), the root system is an important parameter to be considered for crop irrigation. In addition to providing plant support, it is the main organ responsible for the absorption of water and soil nutrients. As described above, Jatropha plantations derived from seeds exhibit larger vegetative development and fruit production than plantations derived from cuttings, which has been attributed to better root development from effective use of irrigation water. On average, Jatropha cuttings develop five roots, a central and four peripheral roots, indicating good soil use when the appropriate growth conditions are provided.

In farming under irrigation, plant spacing is normally managed according to the plant characteristics. Physical restriction may also occur, depending on the irrigation system used. In experimental areas of irrigated Jatropha, the most widely used spacings have been 4×2 and 5×2 m when mechanical harvesting was applied. The spacing is related to the tree size and irrigation system. For example, "*central pivot*" or "*linear*" systems can be difficult to apply or require pruning. According to Saturnino et al. (2005), Jatropha sheds leaves during the dry season or the cold period. The plant remains dormant until the beginning of next rainy season, and the end of dormancy is marked by new sprouts developing at the tips of branches. However, the periodic leaf loss typical of non-irrigated plantations was not observed under irrigation in Northern Minas Gerais. These results indicate the adaptation potential of Jatropha and an opportunity for further irrigation management strategies. For example, an irrigation pause during certain periods could enable nutrient accumulation, among other benefits.

In northern Minas Gerais, it appears that Jatropha is able to produce and develop under non-irrigated conditions, with only water from the poorly distributed precipitation (rainfall concentrated between the months of November and February) of approximately 1,200 mm. However, preliminary results show positive effects of irrigation including (1) better plant development, (2) increased precocity of production, (3) maximized harvest period, and (4) increased yield. Although tolerant to periods of water shortage, Jatropha needs a proper and constant water supply to achieve its yield potential.

Under irrigation, Jatropha's production starts earlier and is greater. Drummond et al. (1984) reported that in an experimental Jatropha area in Janaúba (northern Minas Gerais), 18-month-old plants under surface irrigation produced 2,500 kg of seeds per hectare at an oil rate of 38% of seed weight. In another region, the seed yield conducted under non-irrigated *yellow-red latosol* at Felixlândia Cerrado (Central region of Minas Gerais) only reached 500 kg ha^{-1}. Thus, although adapted to dry regions and having a thick stem able to store enough water to survive in dry regimes, Jatropha is far less productive under non-irrigated conditions. Jatropha is productive in warm climates with more abundant and regular rainfalls, such as Zona da Mata (Minas Gerais). In irrigated and fertile areas, Jatropha can start producing soon in the second planting year, reaching 2 t ha^{-1} in the third year (unpublished data).

In NNE Minas Agro Florestal Ltda. (Janaúba), flowering initiation took place 7 months after crop planting under *drip irrigation* with a volume of 15 l per plant per week, given in three irrigation events over the course of the week.

MSEA (2008) and Reyadh (1999) reported that 5,000 ha of Jatropha were planted with 3×3 m spacing (1,260 seedlings per hectare) in sandy soils of the desert in the Luxor region (Egypt) under irrigation with effluent water from sewage treatment (EC 1.04 and pH 7.47). The plants did not receive any organic or mineral fertilization other than the nutrients contained naturally in the irrigation water from the effluents of sewage treatment. The production began 18 months after seedling transplantation and reached an average yield of 3–4 kg per plant 2 years after planting. The oldest and largest plants produced between 12 and 18 kg per plant. These reports not only indicated the feasibility of fertirrigation on Jatropha plantations, but also the possibility of using waste water for a productive activity that neither harms human health nor pollutes the environment.

In an area on the experimental farm of EPAMIG in Jaíba (northern Minas Gerais), nitrogen and potassium were successfully applied to a Jatropha plantation using localized irrigation systems (drip and micro-sprinkling).

To determine the yield potential of Jatropha in semiarid conditions, Drummond et al. (2007) compared Jatropha's productivity under dry and irrigated conditions in an experimental field (9°09′ S, 40°22′ W at an altitude of 365.5 m) of Embrapa (ENT-Petrolina, Petrolina, Pernambuco, Brazil). In this region, the average annual rainfall ranges from 400 to 500 mm and is concentrated in February to April. The average temperature is 26.4°C, the average evaporation is 7.4 mm d^{-1}, the average day length is 7.3 h d^{-1} and the annual average relative humidity is 61.8%. Nine rows of 23 plants spaced 2.0×2.0 m with surface irrigation were planted at the beginning of the rainy season. The area was divided into two parts of four rows, separated by a central row. The plants of all rows were grown under non-irrigated conditions until 4 months after planting. After this period, four rows were irrigated each week. Nine months after planting, 63 plants from both of the four dry rows and the four irrigated rows were assessed individually for (1) plant height, (2) stem diameter, (3) branch number at one meter and (4) numbers and weights of fruits and seeds. The results obtained for these parameters showed that Jatropha's performance 4 months after planting was far superior when complemented with irrigation compared to the control plants grown with rain precipitation. The average seed productivity under irrigation was 871 kg ha^{-1}, i.e., which is 3.5 times larger than the control (246 kg ha^{-1}).

In a comparison of vegetative development under dry and irrigated conditions in the region of Vale do Jequitinhonha, plants were spaced 2×2 m, and one irrigated row was separated from the other by a non-irrigated row. Irrigation was performed by dripping. Ninety-six plants, including 48 from dry rows and 48 from irrigated rows, were evaluated after 5 months for plant height and stem diameter. As expected, plants under irrigation exhibited better development in terms of both plant height and stem diameter (Evaristo and Moreira 2008).

An average production of 63.72 and 83.02 g of seeds under non-irrigated and irrigated conditions, respectively, was reported in 9-month-old plants by Coletti et al. (2008). By contrast, Drumond et al. (2008) found that 12-month-old Jatropha planted in a scheme of three rows containing 21 plants spaced 2×2 m produced an average of 50 (330 kg ha^{-1}) and 210 (1,156 kg ha^{-1}) fruits per plant under dry and irrigated (drip) conditions, respectively. More recently, Drumond et al. (2010) reported average seed yields ranging between 2,853 and 3,542 kg ha^{-1} in 12-month-old genotypes under irrigation. However, investigations of Jatropha's productivity are not generally based on field realities. Experiments based on small sample sizes may give unreliable results because of statistical inconstancies. Other important parameters were not considered by these investigators, such as the genotype interaction of this undomesticated species with the edapho-climatic conditions of the environment. However, irrigation and fertirrigation are technologies that have great potential for the cultivation of Jatropha. To be successful, they must be adapted to farming techniques that still need to be optimized for this crop.

Frigo et al. (2008) analyzed the expenses of different energy sources (renewable and non-renewable) of a Jatropha agro-ecosystem under drip irrigation to evaluate its long-term sustainability based on energy balance and use of non-renewable resources. Data from primary (collected in the field through oral reports) and secondary sources (data from bibliographies of the area), as well as manual or mechanical operations such as land cleanup, pruning, rowing, mechanical, digging, seedlings plantations, insecticide application, fungicide manual application, manual weeding, irrigation and harvesting, were used to calculate an energy balance of 2,141.66 MJ ha^{-1}. Thus, for each kg of Jatropha fruit produced (i.e., a gross energy of 12.80 MJ), 4.62 MJ are from non-renewable energy sources, which, in the case of this study, corresponds to fossil fuel sources (fuel, grease and lubricants). Because the energy efficiency is 2.77 for every kg of fruit produced (12.80 MJ), an additional 35.56 MJ of non-renewable energy sources is needed. Finally, the culture efficiency of irrigated Jatropha is 0.36%, meaning that for every 12.80 MJ produced (kg of fruits), 35.67 MJ of fossil fuels are needed as energy input. Thus, due to the heavy use of non-renewable energy in the irrigation process, Jatropha under irrigation is an untenable agro-ecosystem in the long run. Jatropha can be used to convert solar energy into oil with lower energy input than the energy effectively released in the oil after its extraction. However, this study only looked at the first year of cultivation, which is insufficient to draw conclusions about the energy ratio over the 20 years of a Jatropha perennial plantation. Nonetheless, this study may serve as a reference for future analyses needed to identify in which agro-system Jatropha may become sustainable as a member of the energy matrix. There is an imperative need for economic investigations on the financial viability of irrigation of Jatropha before making any recommendation on its use, especially when referring to small- and medium-sized producers who have few resources to invest in such a productive process.

Fruit Harvesting

The fruits of Jatropha are considered mature when they reach a yellow coloration and are generally harvested by hand. The fruits at this stage of development are at the peak of oil accumulation in seeds and get detached easily from stalks. Fructification occurs in bunches, but maturation is not uniform, and flowering occurs continuously as long as there is heat and moisture. Thus, continuous harvesting is needed throughout the maturation period.

An alternative faster and easier method is to shake plants at half their height to allow fruits to fall, which can then be easily collected on a canvas extended on the soil (Saturnino et al. 2005). The drawback of this method is that some fruits do not tear off and others are pulled out when falling on the ground, which also may lead to contamination and the need to collect the fallen fruit.

Farmers and the scientific community believe that the implementation of a mechanical harvesting system for Jatropha would be a critical step to improve crop

Fig. 12.9 Jatropha mechanical harvesting system in Northern Minas Gerais, Brazil (Source: Biojan Ltd). Views are taken from the lateral (**a**), front (**b**) and rear (**c**)

feasibility. Today, equipment used in other crops, such as coffee, is being tested for adaption to Jatropha.

A semi-mechanized system with lateral vibrating fingers is being tested at the Federal University of Viçosa (Minas Gerais). This type of equipment is widely used in the cultivation of coffee. The performance of the prototype has been considered to be excellent when tested on Jatropha (Dias et al. 2007). Ideally, a harvester should cause minimal plant damage and fruit loss. However, tests conducted in the field in Janaúba (Minas Gerais) using this equipment caused extensive damage to branches and plants.

A self-propelled harvester designed for coffee was also tested on Jatropha by Biojan Ltda in Janaúba. This harvester uses a system to drag and shake branches (Fig. 12.9). Fruits are collected by mugs as they fall, led by elevators and dispensed into a cart. According to Biojan, the results are promising because the fruit harvest yield was very good. An evaluation performed shortly after harvesting revealed the occurrence of damage to the trunks, branches, loss of leaves and eventually partial loss of inflorescences. In addition, many immature fruits were collected together with the mature ones. However, despite all the above constraints, the plant recovery was quickly verified, even if the equipment was used throughout the year.

In order for a mechanical harvesting system to be viable, a series of technological advances must be achieved. For example, pruning is a viable alternative to reduce the crown size for mechanical harvesting (Silva et al. 2012). Pruning at 80 cm in the primary branch with thinning below 60 cm and promotion of three secondary branches gave the best results for mechanical fruit harvesting. Dwarf cultivars or those with different architectures should be developed in conjunction with the harvester to allow for better circulation of the equipment over plantation lines. Such dwarf cultivars should present similar flower numbers despite their small size, good/prolific flowering and fructification in discrete periods of the year and greater fruit uniformity within the same cluster. Knowledge of plant physiology for inducing synchronous flowering and maturation under irrigation control, pruning and/or plant regulators are needed. Pruning is a viable alternative to reduce the canopy size for adaptation to mechanical harvesting; however, this method requires human power, which inevitably increases costs. Biophysical analyses of branch, fruit and stem resistance should be carried out to measure the average force needed to collect mature fruits without breaking new inflorescences, productive branches and damaging immature fruits.

Conclusions

Jatropha is an oilseed crop with good potential for biodiesel production; it is hardy, tolerant to drought, widespread through tropical to sub-tropical climates and some temperate regions. However, under inadequate conditions of light, air temperature, relative humidity, rainfall, soil fertility and moisture, this species does not reach the expected productivity. Production costs should be considered based on the specific environment where Jatropha is grown. Damages caused by biotic and abiotic factors also need to be assessed.

Research on the fertilization of Jatropha is still at an early stage in Brazil. To exploit this crop on a large scale, additional information is needed to warrant sustainability of its commercial cultivation.

The information available on the physiology of Jatropha is very meagre. The limited information on the vegetative and reproductive growth in various environments and cropping systems came from observations made in the early stages of plant life without reference to the physiological mechanisms that explain these processes. Investigations on gas exchange and water balance are lacking. Moreover, there is no information on hormonal relationships, nitrogen metabolism, assimilate partitioning, and root physiology, among other things. Molecular biology is in the early stages of applications for understanding the mechanisms of regulation of gene expression in response to environmental stresses.

Irrigation is a technology that holds great potential for cultivation of Jatropha. However, to be successful, it must be adapted to other cultural treatments, performed using good-quality equipment, and maintained periodically to ensure proper performance in the long run. An economic survey is needed to verify the financial viability of irrigation in the cultivation of Jatropha to understand the conditions under which it is recommended.

Because flowering and fruit development is asynchronous in Jatropha, fruit harvesting is one of the main challenges to be overcome for the viability of the culture as an industrial crop. In the present stage of crop development, the process simply needs to be sufficiently efficient to be economically viable, i.e., there must be a positive balance between fruit crop and crop costs; the higher the harvesting frequency, the lower the fruit loss, but the higher the crop cost. On the other hand, if the harvest occurs only once, losses can be so high that they may challenge Jatropha's sustainability. Postharvest steps leading to biodiesel production will be subject to both losses in quantity and quality of oil, which can only be prevented by using several agronomic and industrial technologies. High-quality seeds are needed for storage and industrial processing to ensure good yield and quality of biodiesel.

Acknowledgements The authors thank the Research Support Foundation of Minas Gerais and Petrobras for financial support between 2007 and 2012 as well as Dr. Nicolas Carels for help in editing the manuscript.

References

Abreu HA, Guerra GM, Mesquita DN, Pereira VC, de Assis RL, Silva OA et al (2006) Crescimento aéreo e radicular de pinhão manso sob diferentes níveis de compactação do solo. In: Proceedings of the Congresso da Rede Brasileira de Tecnologia do Biodiesel, vol 1. MCT/ABIPTI, Brasília, pp 144–149, Portuguese

Anjos JB, Drumond MA, Morgado LB (2007) Enxertia de pinhão-bravo com pinhão manso. In: Proceedings of the Congresso Internacional de Bioenergia e Biocombustíveis, Energia de Resultados [CD-ROM]. Teresina, (Embrapa Meio Norte, Documentos 143). 11–15 June 2007. Portuguese

Araújo ECE, Prado CHBA, Veloso MEC, Costa SEM, Freire CL, Novaes P (2007) Curso diário de parâmetros do estado hídrico do pinhão manso (*Jatropha curcas* L.) no período chuvoso do município de Teresina, Piauí. In: Proceedings of the Congresso Internacional de Bioenergia e Biocombustíveis, Energia de Resultados [CD-ROM]. Teresina, (Embrapa Meio Norte, Documentos 143). 11–15 June 2007. Portuguese

Arruda FP, Beltrão NEM, Andrade AP, Pereira WE, Severino LS (2004) Cultivo de pinhão manso (*Jatropha curcas* L.) como alternativa para o semi-árido nordestino. Revista Brasileira de Oleaginosas e Fibrosas 8(1):789–799, Portuguese

Avelar RC, Deperon Júnior MA, Dourado DC, Quintiliano AA, Danfa S, Fraga AC (2005) Produção de mudas de pinhão manso (*Jatropha curcas* L.) em tubetes. In: Proceedings of the 2nd Congresso Brasileiro de Plantas Oleaginosas, Óleos, Gorduras e Biodiesel [CD-ROM], vol 1. MG, Varginha. Biodiesel: combustível ecológico. UFLA, Lavras, pp 298–301, Portuguese

Cai L, Fu L, Ji L (2011) Regeneration of *Jatropha curcas* through efficient somatic embryogenesis and suspension culture. GM Crops Food 2(2):110–117. doi:10.4161/gmcr.2.2.16126

CETEC (1983) Produção de combustíveis líquidos a partir de óleos vegetais: relatório final, vol 2. CETEC, Belo Horizonte, Portuguese

Coletti AJ, Dallacort R, Martins JA, Dalchiavon FC, Silva KD (2008) Produtividade inicial da cultura do pinhão manso em condições irrigadas e de sequeiro, na região de Tangará da Serra-MT. In: Proceedings of the 4th Congresso Interno De Iniciação Científica Da Universidade do Mato Grosso, vol 1. Universidade do Mato Grosso, Cáceres, pp 1–4, Portuguese

Costa EL, Coelho EF, Simão FR, Coelho Filho MA, Oliveira PM (2008) Irrigação da Bananeira. Informe Agropecuário Belo Horizonte 29(245):38–46, Portuguese

Demartini D, Muller MD, Nascimento Junior ER, Fernandes EN (2009) Correlação entre características agronômicas de pinhão manso (*Jatropha curcas* L.) em sistema de consórcio com pastagens. In: Proceedings of the 6th Congresso Brasileiro de Plantas Oleaginosas, Óleos, Gorduras e Biodiesel. Montes Claros. Biodiesel, inovação tecnológica: Revista de resumos. UFLA, Lavras, 6 p. Portuguese

Dias LAS, Leme LP, Laviola BG, Palline A, Pereira OL, Dias DCFS et al (2007) Cultivo de pinhão manso (*Jatropha curcas* L.) para produção de óleo combustível. MG, Viçosa, 40 p. Portuguese

Drummond AO, Purcino AAC, Cunha LHS, Veloso JM (1984) Cultura do pinhão manso. EPAMIG, Belo Horizonte. Não paginado. (EPAMIG. Pesquisando, 131). Portuguese

Drumond MA, Anjos JB, Morgado LB, Paiva LE (2008) Comportamento do pinhão manso no semi-árido brasileiro, resultado do 1° ano. In: Proceedings of the Simpósio Brasileiro de Agroenergia, 2008, Botucatu. Agroenergia e desenvolvimento sustentável. Faculdade de Ciências Agronômicas, UNESP. Available from: http://www.infoteca.cnptia.embrapa.br/bitstream/CPATSA-2009-09/39519/1/OPB1998.pdf. Accessed 30 Jan 2010. Portuguese

Drumond MA, Anjos JB, Paiva LE, Morgado LB, Reis EM (2007) Produção de pinhão manso no semi-árido brasileiro. In: Proceedings of the Congresso Internacional de Bioenergia e Biocombustíveis, Energia de Resultados [CD-ROM]. Teresina, (Embrapa Meio Norte, Documentos 143). 11–15 June 2007. Portuguese

Drumond MA, Santos CAF, Oliveira VR, Martins JC, Anjos JB, Evangelista MRV (2010) Desempenho agronômico de genótipos de pinhão manso no Semiárido pernambucano. Ciência Rural. 2010. Santa Maria. 40(1):44–47. Available from: http://www.scielo.br/scielo.php?script=sci_pdf&pid=S0103-84782010000100008&lng=pt&nrm=iso&tlng=pt. Accessed 30 Mar 2010. Portuguese

Evaristo AB, Moreira TMB (2008) Desenvolvimento de plantas de pinhão manso em regime irrigado e sequeiro na região do Médio Vale do Jequitinhonha. In: Proceedings of the Congresso Brasileiro de Agroenergia & Simpósio Internacional de Biocombustível [CD-ROM]. MG, Uberlândia, 4 p. Portuguese

Frigo MS, Bueno OC, Esperancini MST, Frigo EP, Klar AE (2008) Análise energética do primeiro ano de cultivo do pinhão manso em sistema irrigado por gotejamento. Irriga 13(2):261–271, Portuguese

Jha TB, Mukherjee P, Datta MM (2007) Somatic embryogenesis in *Jatropha curcas* Linn., an important biofuel plant. Plant Biotechnol Rep 1:135–140

Kurihara CH, Roscoe R, Silva WM, Maeda S, Gordin CL, Santos G (2006) Crescimento inicial de pinhão manso sob efeito de calagem e adubação, em solos do Mato Grosso do Sul. In: Proceedings of the 25th Reunião Brasileira de Fertilidade do Solo e Nutrição de Plantas [CD-ROM]. Sociedade Brasileira de Ciência do Solo, Bonito-MT. Portuguese

Lange A, Martines AM, Silva MAC, Sorreano MCM, Cabral CP, Malavolta E (2005) Efeito de deficiência de micronutrientes no estado nutricional da mamoneira cultivar Íris. Pesq Agropec Bras 40:61–67, Portuguese

Lavres J Jr, Boaretto RM, Silva MLS, Correia D, Cabral CP, Malavolta E (2005) Deficiências de macronutrientes no estado nutricional da mamoneira cultivar Íris. Pesq Agropec Bras 40:145–151, Portuguese

Lima Filho JMP, Silva FFS, Lopes AP, Anjos JB, Drumond MA (2007) Comportamento ecofisiológico do pinhão manso (*Jatropha curcas* L.) sob condições semi-áridas. In: Proceedings of the Congresso Internacional de Bioenergia e Biocombustíveis, Energia de Resultados [CD-ROM]. Teresina, (Embrapa Meio Norte, Documentos 143). 11–15 June 2007. Portuguese

Lima RLS, Severino LS, Pereira WE, Lucena AMA, Gheyi HR, Arriel NHC (2010) Comprimento das estacas e parte do ramo para formação de mudas de pinhão-manso. R Bras Eng Agríc Ambiental 14(11):1234–1239, Portuguese

Marques DS, Dias MSC, Saturnino HM, Vitorno Júnior D, Barbosa AP, Barbosa JG (2008) Enxertia de pinhão-manso (*Jatropha curcas* L.) em *Jatropha pohliana* na região Norte de Minas Gerais. Proceedings of the 5th Congresso Brasileiro de Plantas Oleaginosas, Óleos, Gorduras e Biodiesel [CD-ROM]. MG/UFLA, Varginha/Lavras. Portuguese

Marques DS, Saturnino HM, Faria MAV, Santos PG, Morais DLB (2007) Enxertia de pinhão manso (*Jatropha curcas* L.) sobre espécies nativas de *Jatropha* do Norte de Minas Gerais. In: Proceedings of the 4th Congresso Brasileiro de Plantas Oleaginosas, Óleos, Gorduras e Biodiesel. MG/UFLA, Varginha, pp 981–985, Portuguese

Martinez HEP, Carvalho JG, Souza RB. Diagnose foliar (1999) In: Comissão De Fertilidade Do Solo Do Estado De Minas Gerais. Recomendações para o uso de corretivos e fertilizantes em Minas Gerais: 5ª aproximação. Viçosa, pp 143–168, Portuguese

Matos FS (2010) Caracterização fisiológica da senescência foliar em populações de *Jatropha curcas* L. Dissertation, Universidade Federal de Viçosa/MG, Viçosa. Portuguese

Moura Neto A, Silva JTA, Silva IP, Costa EL (2007) Efeito da aplicação de diferentes doses de fósforo no pinhão manso (*Jatropha curcas* L.). In: Proceedings of the 31th Congresso Brasileiro de Ciência do Solo [CD-ROM]. SBCS, Gramado. Portuguese

MSEA – Ministry of State for Environmental Affairs (2008). Environmental affairs agency. The national programme for safe use of treated sewage water for afforestation: planting jatropha in Egypt. Available from: http://www.eeaa.gov.eg/english/main/greencorner.asp. Accessed 11 set. 2008

Nunes CF, Pasqual M, Santos DN, Custódio TN, Araújo AG (2008) Diferentes suplementos no cultivo in vitro de embriões de pinhão-manso. Pesq agropec bras 43(1):9–14, Portuguese

Oliveira EL, Faria MA, Morais AR, Fraga AC, Castro Neto P (2007) Efeito da adubação potássica no crescimento inicial do pinhão manso irrigado por gotejamento. In: Proceedings of the 4th Congresso Brasileiro de Plantas Oleaginosas, Óleos, Gorduras e Biodiesel. MG/UFLA, Varginha/Lavras, p 149. Portuguese

Oliveira RJP, Silva DAS, Lemoes LS, Fonseca ER, Eicholz ED (2011) Estudo da fenologia de pinhão-manso nas safras 2008/09 e 2010/11 em Pelotas-RS. In: Proceedings of the 2nd Congresso Brasileiro de Pesquisas de Pinhão-Manso [Internet]. Brasília, 2011. Available from: http://www.alice.cnptia.embrapa.br/bitstream/doc/920775/1/ESTUDODAFENOLOGIADEPINHAOMANSONASSAFRAS200809E201011EMPELOTASRS.pdf

Pacheco DD, Saturnino HM, Mendes LD, Soares FR, Paula TOM, Prates FBS et al (2006) Produção de massa vegetal e composição mineral de plantas de pinhão manso. In: Proceedings of the 3rd Congresso Brasileiro de Plantas Oleaginosas, Óleos, Gorduras e Biodiesel [CD-ROM]. MG/UFLA, Varginha. Portuguese

Rao GR, Korwar GR, Shanker AK, Ramakrishna YS (2008) Genetic associations, variability and diversity in seed characters, growth, reproductive phenology and yield in *Jatropha curcas* L. accessions. Trees Struct Func 22:697–709

Reyadh M (1999) The cultivation of *Jatropha curcas* in Egypt. In: INTERNATIONAL EXPERT MEETING, 1997, Cairo. Proceedings. Medicinal, culinary and aromatic plants in the Near East. FAO, Cairo, 1999. Available from: http://www.fao.org/documents/show_cdr.asp?url_file=/docrep/x5402e/5402e00.htm. Accessed 30 July 2004

Santos CM, Endres L, Wanderly Filho HCL, Rolim EV, Ferreira VM (2010) Fenologia e crescimento do pinhão-manso cultivado na Zona da Mata do estado de Alagoas, Brasil. Scientia Agraria 11(3):201–209, Portuguese

Saturnino HM, Pacheco DD, Kakida J, Tominaga N, Gonçalves NP (2005) Produção de oleaginosas para o biodiesel. Informe Agropecuário Belo Horizonte-MG 26(229):44–74, Portuguese

Severino LS, Lima RLS, Beltrão NEM (2006) Produção de mudas de pinhão manso. Embrapa Algodão, Campina Grande, (Folder). Portuguese

Siles MC, Montoya A, Vásquez W, Foidl N (1997) Planting *Jatropha curcas* using bare roots. In: Proceedings of the SYMPOSIUM "JATROPHA 97". 1997. Managua, Nicaragua. Biofuels and industrial products from *Jatropha curcas*. University of Technology, Graz, 1997. Available from: http://www.jatropha.de/conferences/abstracts-Jatropha97.htm. Accessed 5 ago. 2004

Silva CJ, Silva YK, Staut LA, Schiavo JA (2011) Produção de pinhão manso em diferentes espaçamentos em Anastácio, MS. In: Proceedings of the 2nd Congresso Brasileiro de Pesquisas de Pinhão Manso [CD-ROM]. II CBPPM, Brasília. Portuguese

Silva EB, Tanure LPP, Santos SR, Resende PS Jr (2009) Sintomas visuais de deficiências nutricionais em pinhão-manso. Pesq Agropec Bras 44(4):392–397

Silva IP, Silva JTA, Moura Neto A, Costa EL (2007a) Resposta do pinhão manso (*Jatropha curcas* L.) a adubação com N e K. In: Proceedings of the 31st Congresso Brasileiro de Ciência do Solo [CD-ROM]. SBCS, Gramado. Portuguese

Silva JTA, Costa EL, Silva IP, Moura Neto A (2007b) Adubação do pinhão manso(*Jatropha curcas* L.) com nitrogênio e fósforo. In: Proceedingsof the 4th Congresso Brasileiro de Plantas Oleaginosas, Óleos, Gorduras e Biodiesel. MG/UFLA, Varginha, p 178. Portuguese

Silva VA, Morais DLB, Kakida J, Ferreira EA, Silva VF (2012) Concentração do ciclo de produção de pinhão-manso por meio de podas de formação ou de produção. Pesq Agropec Bras 47(1):134–137, Portuguese

Smiderle OJ, Kroetz VJ (2008) Produção de mudas de Pinhão-manso propagadas por estaquia. Embrapa Meio Norte. [Comunicado Técnico 22]. Portuguese

Souza JL, Nicácio RM, Moura MAL (2005) Global solar radiation measurements in Maceió Brazil. Renew Energ 30(8):1203–1220

Tanure LPP (2006) Avaliação do crescimento de pinhão manso (*Jatropha curcas* L.) em diferentes níveis de saturação por bases. MG/UFVJM, Diamantina, p 21. (Monograph). Portuguese

Vale LS, Severino LS, Beltrão NEM (2006) Efeito da salinidade da água sobre o pinhão manso. In: Proceedings of the 1st Congresso da Rede Brasileira de Tecnologia de Biodiesel. MCT/ABIPTI, Brasília, 1(1):87–90. Portuguese

Vasquez GH, Lazarini E, Ribeiro TC, Gradela AS, Silva TF, Viana RL (2010) Produção de mudas de pinhão-manso via estaquia. Rev Bras Ol Fibros 14(3):97–105, Portuguese

Chapter 13
Arbuscular Mycorrhizal Fungi for Jatropha Production

Supattra Charoenpakdee, Saisamorn Lumyong, and Bernard Dell

Abbreviations

AMF	Arbuscular mycorrhizal fungi
dS m^{-1}	Deci Siemens per meter
EC	Electrical conductivity
ha	Hectare
K	Potassium
kg	Kilogram
N	Nitrogen
NaCl	Sodium chloride
P	Phosphorus

Introduction

Mutualistic fungi depend on plants for photosynthetic carbon while host plants receive inorganic nutrients and water captured from the soil through fungal hyphae. Most *arbuscular mycorrhizal fungi* (AMF) are obligate symbionts within the monophyletic phylum, Glomeromycota. The AMF form highly branched exchange

S. Charoenpakdee (✉)
Biology Program, Faculty of Science and Technology, Pibulsongkram Rajabhat University, Phitsanulok, 65000 Thailand
e-mail: ansupattra@gmail.com

S. Lumyong
Department of Biology, Faculty of Science, Chiang Mai University, Chiang Mai 50200, Thailand

B. Dell
Sustainable Ecosystems Research Institute, Murdoch University, Western Australia 6150, Australia

N. Carels et al. (eds.), *Jatropha, Challenges for a New Energy Crop: Volume 1: Farming, Economics and Biofuel*, DOI 10.1007/978-1-4614-4806-8_13,

surfaces, known as arbuscules, within the cortical tissue of roots during periods of active root growth. Many AMF also form vesicles, which are lipid-filled structures produced in intercellular spaces. Their function is food storage, but can also serve as reproductive propagules for the fungus in some species. Other structures are auxiliary cells and asexual spores, which are formed in the soil (Fisher and Jayachandran 1999; Sylvia et al. 2005; Trappe 2005; Schüßler et al. 2009). These fungi cannot finish their life cycle or multiply on artificial media without their plant host (Sieverding 1991). The majority of vascular plants develop mycorrhizal associations (Smith and Read 1997; Öpik et al. 2008; Parniske 2008; Siddiqui et al. 2008; Verma 2008) that play important roles in plant fitness including suppression of soil-borne pathogens (Brundrett et al. 1996; van der Heijden and Sanders 2003; Siddiqui et al. 2008), nutrient cycling and soil conservation in natural as well as agricultural ecosystems (Boomsma and Vyn 2008). These fungi are sensitive to some agricultural chemicals; their diversity and abundance are also reduced by fertilizers, especially those containing high amounts of phosphorus (P). Many AMF produce large spores; a fact that facilitates their extraction from soil and observation under microscope (Brundrett et al. 1996; van der Heijden and Sanders 2003; Sylvia et al. 2005; Verma 2008). Hence, field surveys can readily be undertaken since the results are only limited by the capacity to identify AMF using molecular tools. Lastly, inoculation with target mycorrhizal fungi can lead to an improvement in growth and reproduction in many crops (Brundrett et al. 1996; Smith and Read 1997; van der Heijden and Sanders 2003; Öpik et al. 2008; Parniske 2008; Verma 2008) depending on the prevailing edaphic factors.

There is strong interest in bringing marginal lands, which cannot be used for food production, into cultivation for physic nut (*Jatropha curcas* L. hereafter referred to as Jatropha). However, in order for this plant to perform well on these marginal sites, inoculation with symbiotic microorganisms, such as AMF, is likely to be essential due to the loss of soil microflora resulting from past land management and other practices. This report explores the association of AMF with Jatropha including AMF diversity, compatible AMF species and known benefits of AMF for Jatropha growth and development. It draws heavily on work undertaken in northern Thailand, supplemented with studies elsewhere. Knowledge gaps were identified and served as guidelines to investigate future directions for research and management.

The Mycorrhizal Status of Jatropha

Investigations of Jatropha seedlings or cuttings from the field indicate that their roots are generally associated with AMF (see for example Leye et al. 2009; Kamalvanshi et al. 2011; Singh and Jamaluddin 2011). No systematic study has been undertaken on AMF associated with Jatropha in the wild, thus comparisons can only be drawn between Jatropha under cultivation. The analyses by Charoenpakdee et al. (2010a) revealed a strong relationship between Jatropha and AMF in the rhizosphere of Jatropha trees in six provinces of Thailand (Fig. 13.1). Thirty-four AMF morphospecies were identified among which the predominant

Fig. 13.1 Collection sites for AMF associated with Jatropha in six provinces of Thailand. (**a**) CR2: Chiang Rai, (**b**) CM1: Chiang Mai, (**c**) CM2: Chiang Mai, (**d**) CM3: Chiang Mai, (**e**) CM4: Chiang Mai, (**f**) LP1: Lumphun, (**g**) CR1: Chiang Rai, (**h**) KK1: Khon Kean, (**i**) LO1: Loei, (**j**) NK1: Nong Khai

genus, in terms of spore density and species diversity, was *Acaulospora* followed by *Glomus*, *Scutellospora*, *Gigaspora* and *Entrophospora*. Furthermore, seven species dominated the AMF spore populations in many sites (Fig. 13.2). Jatropha roots were heavily colonized in all the field sites sampled (Fig. 13.3). Thus, in both acid

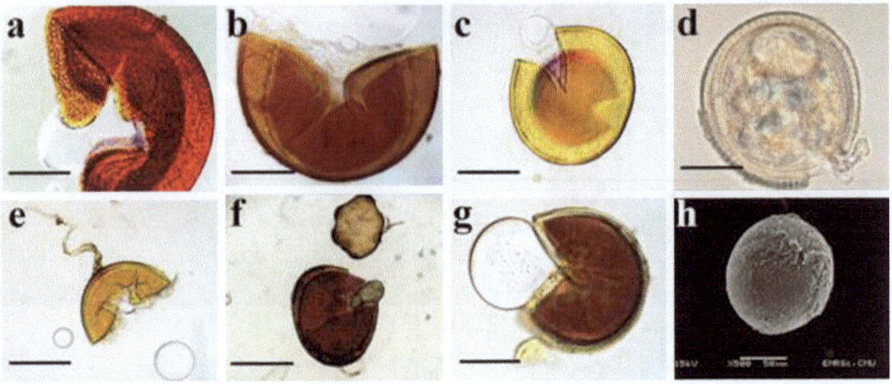

Fig. 13.2 Spore traits of the dominant AMF collected from Jatropha rhizosphere under light (**a–c**, **e–g**) and scanning electron (**h**) microscopes. *Acaulospora foveata* (CMU02) (**a**), *Entrophospora colombiana* (CMU05) (**b** and **h**), *A. dilatata* (CMU09) (**c**), *A. lacunosa* (CMU14) (**d**), *Gigaspora rosea* (CMU29) (**e**), *Gigaspora* sp.01 (CMU28) (**f**), and *A. scrobiculata* (CMU06) (**g**) with Melzer's reagent. Bars: **a**, **b**, **c**, **d**, **g** = 38 μm (40×); e, f = 150 μm (10×); h = 50 μm

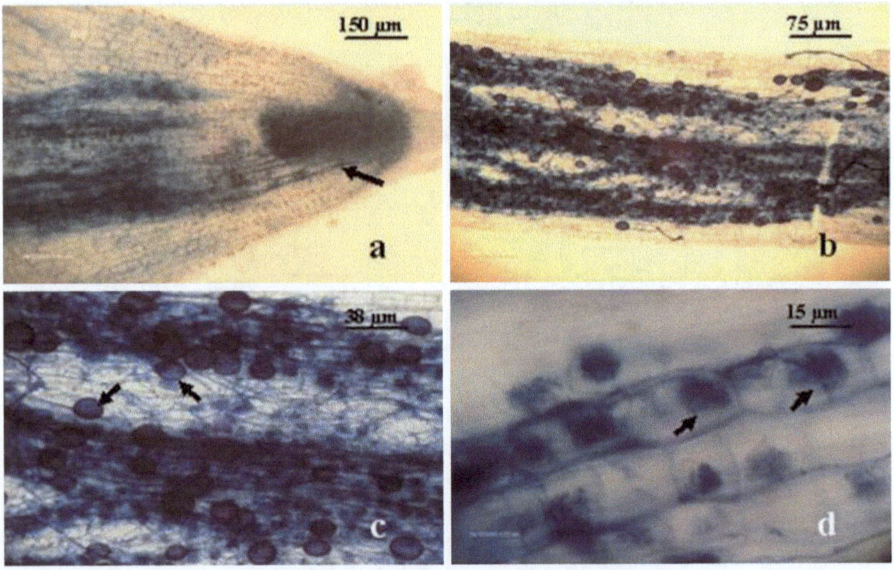

Fig. 13.3 AMF colonization in Jatropha roots. (**a**) AMF structures near the root tip (*arrow*), (**b**) highly infected root, (**c**) vesicles (*arrows*), and (**d**) arbuscules (*arrows*). Bars are 150 μm (**a**), 75 μm (**b**), 38 μm (**c**), 15 μm (**d**), respectively

and alkaline soils in Thailand, Jatropha appears to be readily colonized by AMF under a range of field conditions, including acidic to alkaline soils, soils of low to moderate organic matter content and soils differing greatly in the concentration of available P.

Spore Production of AMF Species Compatible with Jatropha

Jatropha has been introduced into many countries from its natural distribution in the American tropics. The soil microbiota and soil conditions of these new sites can differ markedly from those of its natural habitat. Under these situations, especially where soil microbial diversity has been depleted from soil by disturbance in the past, it may be desirable to introduce compatible beneficial rhizosphere bacteria (Johansson et al. 2004; Desai et al. 2007; Khan 2008; Nehra and Saharan 2011) and fungi. This necessitates the production of inocula for commercial application. In the case of AMF, which cannot complete their life cycle without a host plant (Corkidi et al. 2008), spore inoculum has to be produced in the rhizosphere of compatible host plants (Liu and Wang 2003). A range of techniques are available for producing inoculum including growing plants in sterile sand or in more controlled environments, such as aeroponic and nutrient film systems (Setiadi 2002). Alternatively, when good facilities are available (Pratap and Potty 2011), cultures of transformed roots are preferred for multiplying AMF spores to ensure that only the desirable species are present in the inoculum (Ishii et al. 1997).

Many mycorrhizal fungi have been distributed around the world with the movement of plants prior to the establishment of quarantine services and also because some AMF are compatible with a broad range of hosts. Thus, it is a good practice to first evaluate whether local AMF are effective for Jatropha or not. According to this approach, Jatropha was used as a bait plant for AMF in soils taken from the field in northern and north-eastern Thailand. Although these soils contained over 34 species of AMF, only one species of *Entrophospora* and one species of *Scutellospora* were able to multiply their spores (Fig. 13.4) in the rhizosphere of pot-grown Jatropha, suggesting that the majority of AMF species may be preferentially associated with the roots of weeds and not Jatropha. Each species of AMF was identified on the basis of its spore morphology under microscope. Spore shape, size and ornamentation were used for genus classification. For example, based on light microscopy alone, the AMFs of Fig. 13.4a can be from *Gigaspora* or *Scutellospora* and those of Fig. 13.4d from either *Acaulospora* or *Entrophospora*. Internal structures of spores, such as (1) the *germination shield* can then be used to confirm *Scutellospora* (Fig. 13.4b) and (2) the presence of two *scars* on the outer spore wall in addition to a red to purple reaction of the inner wall layer to Melzer's reagent suggest *Entrophospora* (Fig. 13.4e). Electron microscopy provides additional traits such as spore ornamentation that can be used to confirm some species (Fig. 13.4c, f). Molecular tools, such as the sequence of small (18S) and large subunits (28S) of rDNA genes can be used for final species identification (Charoenpakdee 2009). These two compatible AMFs were then multiplied further using a range of cereals (Fig. 13.5) of which sorghum was the best host plant for AMF sporulation (Charoenpakdee et al. 2010b). Plants such as grasses, which grow fast and have extensive systems of fine hairy roots are ideal for producing AMF spores in sterile sand in pots or in raised beds in the greenhouse. Once compatible species of AMF have been multiplied in sufficient quantity their effectiveness must be

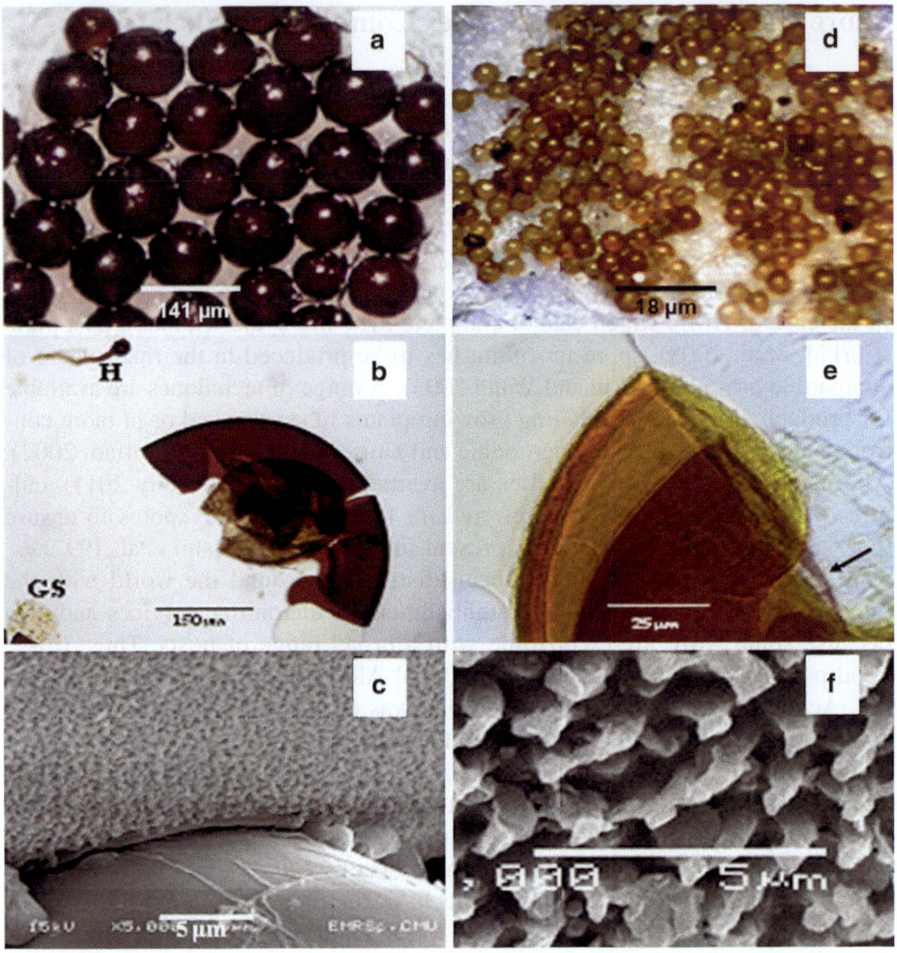

Fig. 13.4 Examples of AMF species trapped with Jatropha viewed under stereomicroscope (**a, d**), compound microscope (**b, e**) and scanning electron microscope (**c, f**). Spores of *Scutellospora* sp. (**a, b, c**) are ~130 µm (**a**). A general view of bulbous subtending hyphae (*H*) and germination shield (*GS*) is given in (**b**). The attachment region of bulbous subtending hyphae is a character that allows species classification (magnification 5,000) (**c**). Spore of *Entrophospora* sp. (**d, e, f**) are ~3 µm (**d**). The spore wall layer appears red to purple in Melzer's reagent (arrow) (**d, e**) and spore ornamentation is evident at magnification 10,000 (**f**). Bars are 141 µm (**a**), 150 µm (**b**), 5 µm (**c**), 18 µm (**d**), 25 µm (**e**) and 5 µm (**f**), respectively

evaluated in field trials before deployment in commercial plantations. Only a few species of AMF have been assessed so far in field trials (e.g., Behera et al. 2010). By contrast, a wider range of AMF species have been evaluated in containers (Feldmann et al. 2008; Leye et al. 2009; Kumar et al. 2010; Ultra 2010) for many food and fiber crops.

Fig. 13.5 Process of AMF spore multiplication in pot culture. Spore inoculum of *Scutellospora* sp. in soil (**a**) and on strips of sterilized paper used as starter inoculum (**b**). Sorghum seedlings are being planted in sterilized soil pre-inoculated with spores (see paper strip) (**c**). The surface of potting mix is kept moist with sterilized coir (host is *Coix lacryma-jobi*) (**d**) and external contamination is controlled by soil cover with a plastic sheet and by pots elevation on bricks (job's tears) (**e**). A case experiment with rice kept on water sheet before harvesting (**f**)

Benefits to be Gained from Symbiosis with AMF

Studies on a wide range of crop species have shown considerable advantages to be gained from roots being associated with compatible AMF. As Jatropha is a relatively new crop for biodiesel production, studies are incomplete. However, it is reasonable to assume that similar benefits as found for many other crops will also apply for this new crop. The following sections describe the main benefits of AMF symbiosis for Jatropha as they are documented in the literature.

Enhancing Nutrient Uptake

AMF play an important role in the mineral nutrient uptake by host plants (Marschner and Dell 1994), which facilitate plant growth and yield (Monzon and Azcon 1996; Hartwig et al. 2002; van der Heijden and Sanders 2003; Davies et al. 2005; Schreiner 2007; Plassard and Dell 2010). AMF produce fine hyphae, which ramify into soil providing a large absorption surface for nutrient uptake. Nutrient capture can take place in sites where root diameter precludes root access especially for elements that are strongly adsorbed to soil surfaces (Brundrett et al. 1996; Smith and Read 1997; van der Heijden and Sanders 2003; Verma 2008). Furthermore, AMF may have access to forms of nutrients that are not readily available to roots including organic forms of P (Jones 1998; van der Heijden and Sanders 2003; Plassard and Dell 2010). AMF may also have a role in the cycling of soil nitrogen (N) (Veresoglou et al. 2011).

Studies undertaken so far with Jatropha indicate that macronutrients are often limiting when this crop is grown on poor soils. Not surprisingly, inoculation with selected AMF can increase the growth of young plants in these soils. In a field trial in India, the application of *Glomus fasciculatum* and *Scutellospora calospora* spores promoted a larger growth of Jatropha compared to the unfertilized control and to plants that were given 10 g NPK fertilizer at establishment (Behera et al. 2010). Similar growth responses to AMF inoculation have been observed on soils with low available P in pot trials. For example, Leye et al. (2009) tested four species of AMF (*Glomus aggregatum, G. fasciculatum, G. interradices,* and *G. mosseae*) and found that growth was stimulated in 4-month old seedlings. However, there was a significant interaction between AMF species and Jatropha cultivars concerning growth stimulation. In the Philippines, Ultra (2010) suggested that AMF should be added in conjunction with 15–30 kg NPK fertilizer/ha to improve nutrient utilization.

In Thailand, growth stimulation of Jatropha's seedlings have been obtained with a range of AMF species (*Acaulospora* sp. no.22, *Entrophospora colombiana*, *Glomus caledonium*, *G. etunicatum*, *Scutellospora* sp. CMU33, a commercial product containing two species of *Glomus* and two species of *Acaulospora*). The values of all parameters referring to plant growth were increased after AMF inoculation (Table 13.1). The increased vigor of inoculated plants can be deduced from their two times larger diameter of their stems (Fig. 13.6, Table 13.1) and their two times larger root system (Charoenpakdee 2009). Despite their preference for specific hosts, AMFs have a wide range of hosts as reported by Zhu et al. (2000). Similarly, application of *G. aggregatum* with organic fertilizer (duck guano) and either phosphate rock or triple superphosphate (TSP) increased fruit production, seed weight and plant height in 2-year old Jatropha cuttings grown in the field in Thailand (Silpachai et al. 2009).

Fig. 13.6 Effect of inoculation with AMF on the growth of Jatropha seedlings for 30 days in 30 cm diameter pots containing 5 kg of sterilized sandy soil; T0=control (no AMF), T1=*Acaulospora* sp., T2=*Entrophospora* sp., T3=*Glomus caledonium*, T4=*G. etunicatum*, T5=*Scutellospora* sp, T6=mixed species (a mixture of 20 spores of each of the above five AMF species) and T7=commercial product (Charoenpakdee 2009)

Table 13.1 Growth of inoculated (T1-T7) and uninoculated (T0) Jatropha seedlings assessed 90 days after transplantation

Treatment	Height (cm)[a]	Stem circumference (cm)[a]	Weight (g)[a]		Colonization (%)[b]
			Fresh	Dry	
T0	11.60±2.29a[c]	3.68±0.56a	16.51±11.10a	5.82±5.40a	0±0.00a
T1	13.35±3.28ab	4.45±0.91ab	28.13±6.11b	10.49±3.90b	73±5.19bc
T2	13.10±1.61ab	4.30±0.24ab	23.79±6.61ab	8.38±2.53ab	75±3.42bc
T3	13.58±1.31ab	4.63±0.75b	25.19±4.83ab	9.21±2.01ab	85±2.64c
T4	13.43±3.33ab	4.75±0.33b	28.68±6.64b	9.13±4.24b	55±5.04b
T5	15.93±1.51b	5.00±0.00b	33.43±2.96b	13.36±2.09b	80±4.51c
T6	14.38±2.22ab	4.93±0.30b	31.85±4.44b	12.74±4.03b	87±2.97c
T7	13.83±2.29ab	4.55±0.64b	30.52±5.44b	11.43±2.57b	77±3.76bc

T0=control, T1=*Acaulospora* sp. no.22, T2=*Entrophospora colombiana*, T3=*Glomus caledonium*, T4=*G. etunicatum*, T5=*Scutellospora* sp., T6=mixed species and T7=commercial product

[a]Values are averages of four replications (mean±standard deviation)

[b]Values are averages of 30 replications (mean±standard deviation)

[c]Letters "a", "b" and "c" stand for the groups of individuals which variance can be considered homogeneous and which mean can be considered significantly different from one another at $P<0.05$ using Duncan's multiple range test. One can see that group "a" appears only for the control, which implies that all the inoculated individuals are significantly different

In Brazil, Jatropha seedlings inoculated with *Gigaspora margarita* or *Glomus clarum* in pots containing 4 kg of sandy soil had higher shoot and root dry matter, plant height, leaf number and total leaf area (Balota et al. 2011). Not surprisingly, in degraded soils AMF can accelerate revegetation including with exotic species, such as Jatropha (Singh and Jamaluddin 2011). The higher success rate of revegetation was demonstrated in India using an inoculum of *G. fasiculatum* and *S. calospora* spore (Behera et al. 2010).

Enhancing Tolerance to Saline Soils

Salt stress can markedly reduce plant growth and can lead to death in sensitive species. In addition to osmotic damage due to the accumulation of salts in plant tissues and to the water deficit in shoots, salinity can also result in mineral nutrient imbalance at the whole plant level. AMF may assist some plants to withstand low levels of salinity stress by enhancing plant biomass and altering host plant physiology (Ruiz-Lozano et al. 1996). Examples of these responses have been demonstrated for food and fiber crops including cotton (Tian et al. 2004), green basil (Enteshari and Hajbagheri 2011), lotus (Sannazzaro et al. 2007), maize (Amerian and Stewart 2001; Feng et al. 2002), tomato (Fritz et al. 2006; Subramanian et al. 2006) and wheat (Daei et al. 2009). Rabie and Almadini (2005) described some of the mechanisms of AMF on ion transport in plants. Mycorrhizal plants have relatively less export of sodium (Na) and chloride (Cl) from the roots to the shoots (Scheloske et al. 2004) and this can have beneficial effects such as reducing Na in leaves as well as increasing membrane stability and concentration of N, P and K (Rinaldelli and Mancuso 1996; Kaya et al. 2009). Moreover, higher rate of K accumulation by mycorrhizal plants under NaCl stress may help in maintaining a large K/N ratio (for biochemical benefits of AMF, see Sannazzaro et al. 2007 and Borde et al. 2011).

Differences between AMF have been observed in their capacity to influence the uptake of Na and Cl (Tian et al. 2004; Daei et al. 2009). Moreover, salinity can have negative consequences on symbiosis by reducing the (1) rate of colonization, (2) rate of spore germination, (3) growth of hyphae in soil and hyphal spread after infection and (4) arbuscule number.

Jatropha is quite sensitive to saline soils. Trial in containers showed the addition of NaCl decreased tap root length and root biomass. However, plants in the same conditions (light, soil pH, soil and air humidity, etc.), but preinoculated with AMF had larger root systems, better leaf water status, less leaf membrane damage (low lipid peroxidation activity), higher solute (proline and sugars) and higher leaf chlorophyll concentrations than uninoculated plants. The authors concluded that inoculation with AMF in the nursery before transplanting in the field is useful for improving the growth of Jatropha in soils with up to 0.5% NaCl (EC of 7.2 dS m^{-1}).

Enhancing Tolerance to Heavy Metals

The plant tolerance to heavy metals such as zinc, chromium, nickel and arsenic is generally improved by mycorrhization (Subramanian et al. 2006; Wu and Xia 2006; Jankong and Visoottiviseth 2008; Turkmen et al. 2008). For example, seedling inoculation with *Glomus mosseae*, *G. intraradices* and *G. etunicatum* resulted in an improved tolerance of *Pityrogramma calomelanos*, *Tagetes erecta* and *Melastoma malabathricum* to arsenic contamination of soil in Thailand

Fig. 13.7 One year old Jatropha seedlings planted in abandoned mine tailings in Mogpog, Marinduque, Philippines. Seedlings were inoculated with a commercial mix of eight AMF species (*Enterphosphora*, *Gigasporea*, *Glomus*) before planting (**a**) or uninoculated for the control (**b**). The inoculated plants had better survival (100%), height (19 cm) and stem diameter (2.1 cm) than the uninoculated plants (70%, 13 cm, 0.8 cm, respectively). Photographs and data are a courtesy of Nelly S. Aggangan (BIOTECH, UPLB, Laguna, Philippines)

(Jankong and Visoottiviseth 2008). Two possible reasons for AMF-mediated arsenate tolerance are: (1) AMF colonization might down-regulate the high-affinity for phosphate or arsenate transport system by enhancing phosphorus uptake, but suppressing arsenic uptake and (2) AMF might increase the efflux of arsenic (as arsenite) from mycorrhizal roots (Jankong and Visoottiviseth 2008). AMF might also reduce plant exposure to metals by their preferential uptake into fungal tissues as well as into extra-cellular glycoproteins in the rhizosphere. Organic amendments in conjunction with AMF could further help to mitigate heavy metal toxicity in soil (Nanda and Abraham 2011).

In the Philippines, Jatropha has been successfully established using AMF on mine tailings contaminated with heavy metals (Fig. 13.7). Inoculation with AMF enhanced survival and growth of seedlings (Lu et al. 2007; Charoenpakdee 2009; Kumar et al. 2010). In general, sequestering of heavy metals in roots of mycorrhized plants reduces heavy metal accumulation in shoots.

Enhancing Tolerance to Water Stress

Mycorrhization strongly affects the growth and tolerance to water deficit of many plants (Borkowska 2002; Kaya et al. 2003; Pinior et al. 2005; Lu et al. 2007). Under unstressed conditions, AMF can (1) increase the relative leaf water content and transpiration rate and (2) decrease stomatal resistance to gas exchanges. Moreover, under continuous drought conditions, mycorrhized plants recovered more quickly their optimal turgor and experienced higher growth rate compared to non-mycorrhized plants (Kaya et al. 2003; Pinior et al. 2005; Lu et al. 2007). The exact mechanisms underlying these physiologic changes are not understood though changes in hormone balances have not been implicated (van Rooyen et al. 2004; Khalvati et al. 2005; Bárzana et al. 2012).

So far, there are no detailed investigations on benefits to be gained from inoculating Jatropha with AMF in water-limited environments. Since Jatropha plantations are being established in some regions with a long dry season where risks of drought are likely to be exacerbated by climate change, it is prudent to undertake research to identify AMF that may enhance survival and oil yield under water-limited conditions. Preliminary observations on rewatering wilted Jatropha seedlings showed that plants inoculated with AMF regained stomatal function more quickly than uninoculated plants (Charoenpakdee, unpublished data).

Enhancing Tolerance to Biotic Stress

The biological control of soil-borne pathogens by AMF has been suggested by many workers because fungi have some ability to produce antimicrobial substances. However, the extent of such pathogen control is likely to be overstated and often pot studies are not confirmed in the field (Sylvia et al. 2005; Li et al. 2007). Biocontrol agents can exert disease suppression by different modes of action, including competition, direct parasitism, antibiosis and the induction of plant resistance mechanisms (Li et al. 2007; Mukerji and Ciancio 2007). Moreover, AMF may enhance host fitness by impairing nematode development. The net effects vary with the environment, plant genotype, nematode species and fungal isolates (Talavera et al. 2001; Gera-Hol and Cook 2005; Oyekanmi et al. 2007). Many changes in plant physiology after AMF infection have been reported including higher chitinase activity in roots (Jothi and Sundarababu 2002).

Application of Mycorrhizal Technology for Jatropha Production

From earlier sections of this report, it is clear that Jatropha can be strongly associated with AMF in the field and that AMF can play an essential role in overall plant fitness, improving the uptake of nutrient and water by the host, protecting plants against toxic

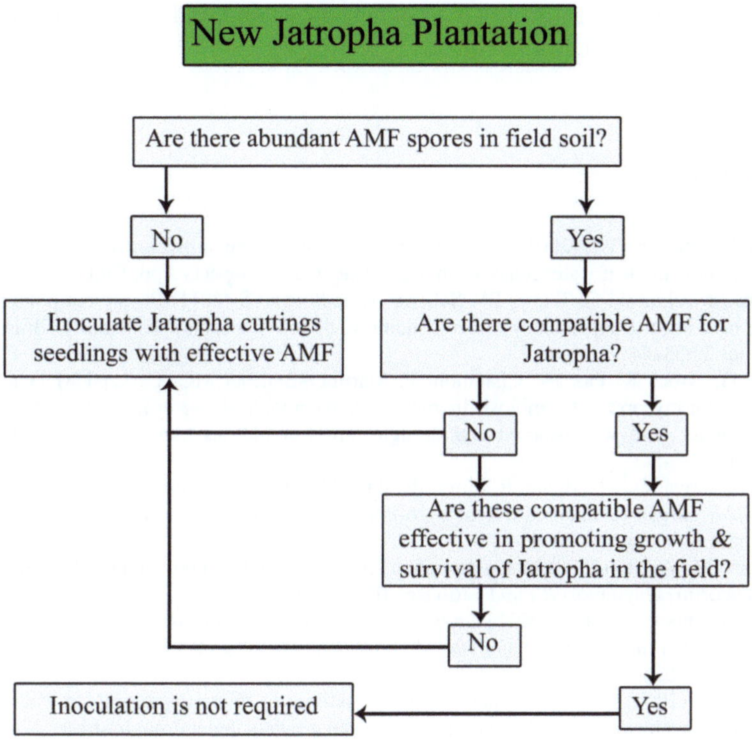

Fig. 13.8 Questions to consider when evaluating the need to inoculate Jatropha with AMF in the context of new commercial plantations

stress, etc. Therefore, managers of Jatropha plantations should request information on the AMF status of the crop being established. The flow chart in Fig. 13.8 shows a hierarchal approach that facilitates decision making as to whether inoculation is desirable or not.

If inoculation is deemed desirable, then it is important to thoroughly test the AMF inoculum and to evaluate its ability to colonize the roots of Jatropha and confer benefit to the crop. Hence, it is most desirable that specific inocula be developed for different site conditions and regions. Investment in the production of these inocula will pay dividends in the future. Managers need to be wary of deploying AMF inoculum that has not been specifically tested and developed for their plantations. It is evident that considerable research and development is still required. However, the task is not difficult since the technology exists and protocols for testing and evaluating AMF are widely available.

Although Jatropha is a robust plant and can grow in marginal environments, oil production for biodiesel will be uneconomic unless a crop with consistent vigor is achieved. Even in soils of larger fertility with less abiotic stress, AMF can be beneficial in nutrient cycling, by increasing the efficiency of fertilizer use and reducing fertilizer loss. Thus Fig. 13.8 should be a component of proper agronomic management in order to obtain optimum yield and economic return (Francis et al. 2005; Ultra 2010).

References

Amerian MR, Stewart WS (2001) Effect of two species of arbuscular mycorrhizal fungi on growth, assimilation and leaf water relations in maize (*Zea mays*). Aspects Appl Biol 63:1–6

Balota EL, Machineski O, Truber PV, Scherer A, de Souza FS (2011) Physic nut plants present high mycorrhizal dependency under conditions of low phosphate availability. Braz J Plant Physiol 23:33–44

Bárzana G, Aroca R, Paz JA, Chaumont F, Martinez-Ballesta MC, Carvajal M et al (2012) Arbuscular mycorrhizal symbiosis increases relative apoplastic water flow in roots of the host plant under both well-watered and drought stress conditions. Ann Bot. doi:10.1093/aob/mcs007

Behera SK, Srivastava P, Tripathi R, Singh JP, Singh N (2010) Evaluation of plant performance of *Jatropha curcas* L. under different agro-practices for optimizing biomass – a case study. Biomass Bioenergy 34:30–41

Boomsma CR, Vyn TJ (2008) Maize drought tolerance: potential improvements through arbuscular mycorrhizal symbiosis? Field Crop Res 108:14–31

Borde M, Dudhane M, Jite P (2011) Growth photosynthetic and antioxidant responses of mycorrhizal and non-mycorrhizal bajra (*Pennisetum glaucum*) crop under salinity stress condition. Crop Prot 30:265–271

Borkowska B (2002) Growth and photosynthetic activity of micropropagated strawberry plants inoculated with endomycorrhizal fungi (AMF) and growing under drought stress. Acta Physiol Plant 24:365–370

Brundrett M, Bougher N, Dell B, Grove T, Malajczuk N (1996) Working with mycorrhizas in forestry and agriculture, vol 32. ACIAR Monograph, Canberra

Charoenpakdee S (2009) Biodiversity and beneficial effect of arbuscular mycorrhizal fungi associated with physic nut (*Jatropha curcas* L.), a potential biofuel plant in Thailand, Ph.D. thesis, Chiang Mai University, Faculty of Science, Thailand

Charoenpakdee S, Phosri C, Dell B, Lumyong S (2010a) The mycorrhizal status of indigenous arbuscular mycorrhizal fungi of physic nut (*Jatropha curcas*) in Thailand. Mycosphere 1:167–181

Charoenpakdee S, Phosri C, Dell B, Choonluechanon S, Lumyong S (2010b) Compatible arbuscular mycorrhizal fungi of *Jatropha curcas* and spore multiplication using cereal crops. Mycosphere 1:195–204

Corkidi L, Evans M, Bohn J (2008) An introduction to propagation of arbuscular mycorrhizal fungi in pot cultures for inoculation of native plant nursery stock. Native Plants J 9:29–38

Daei G, Ardekani MR, Rejali F, Teimuri S, Miransari M (2009) Alleviation of salinity stress on wheat yield, yield components, and nutrient uptake using arbuscular mycorrhizal fungi under field conditions. Plant Physiol 166:617–625

Davies TF, Calderon CM, Huaman Z (2005) Influence of arbuscular mycorrhizae indigenous to Peru and a flavonoid in growth, yield, and leaf elemental concentration of Yungay potatoes. Hort Sci 40:381–385

Desai S, Narayanaiah C, Kumari CK, Reddy MS, Gnanamanickam SS, Rao GR et al (2007) Seed inoculation with *Bacillus* spp. improves seedling vigour in oil-seed plant *Jatropha curcas* L. Biol Fertil Soils 44:229–234

Enteshari S, Hajbagheri S (2011) Effects of mycorrhizal fungi on some physiological characteristics of salt stressed *Ocimun basillicum* L. Iranian J Plant Physiol 1:215–222

Feldmann F, Hutter I, Schneider C (2008) Best production practice of arbuscular mycorrhizal inoculum. Federal Research Centre for Agriculture and Forestry, Messeweg

Feng G, Zhang FS, Li XL, Tian CY, Tang C, Rengel Z (2002) Improved tolerance of maize plants to salt stress by arbuscular mycorrhiza is related to higher accumulation of soluble sugars in roots. Mycorrhiza 12:185–190

Fisher JB, Jayachandran K (1999) Root structure and arbuscular mycorrhizal colonization of the palm *Serenoa repens* under field conditions. Plant Soil 217:229–241

Francis G, Edinger R, Becker K (2005) A concept for simultaneous wasteland reclamation, fuel production, and socio-economic development in degraded areas in India: need, potential and perspectives of *Jatropha* plantations. Nat Res Forum 29:12–24

Fritz M, Jakobsen I, Lyngkjaer MF, Thordal-Christensen H, Kühnemann L (2006) Arbuscular mycorrhiza reduces susceptibility of tomato to *Alternaria solani*. Mycorrhiza 16:413–419

Gera-Hol WH, Cook R (2005) An overview of arbuscular mycorrhizal fungi-nematode interactions. Basic Appl Ecol 6:489–503

Hartwig UA, Wittmann P, Braun R, Hartwig-Räz B, Jansa J, Mozafar A et al (2002) Arbuscular Mycorrhiza infection enhances the growth response of *Lolium perenne* to elevated atmospheric *p*CO2. J Exp Bot 53:1207–1213

Ishii T, Narutaki A, Sawada K, Aikawa J, Matsumoto I, Kadoya K (1997) Growth stimulatory substances for vesicular-arbuscular mycorrhizal fungi in Bahia grass (*Paspalum notatum* Flugge.) roots. Plant Soil 196:301–304

Jankong P, Visoottiviseth P (2008) Effects of arbuscular mycorrhizal inoculation on plants growing on arsenic contaminated soil. Chemosphere 72:1092–1097

Johansson JF, Paul LR, Finlay RD (2004) Mini review microbial interactions in the mycorrhizosphere and their significance for sustainable agriculture. FEMS Micro Ecol 48:1–13

Jones DL (1998) Organic acids in the rhizosphere- a critical review. Plant Soil 205:25–44

Jothi G, Sundarababu R (2002) Peroxidase and chitinase activities in brinjal inoculated with *Meloidogyne incognita* (Kofoid & White) Chitwood and endomycorrhiza. Biol Control 16:161–164

Kamalvanshi M, Kumar A, Jha A, Dhyani SK (2011) Occurrence of arbuscular mycorrhizal fungi in rhizosphere of *Jatropha curcas* L. in arid and semi arid regions of India. Indian J Microbiol. doi:10.1007/s12088-011-0224-0

Kaya C, Higgs D, Kirnak H, Tas I (2003) Mycorrhizal colonization improves fruit yield and water use efficiency in watermelon (*Citrullus lanatus* Thunb.) grown under well-watered and water stressed conditions. Plant Soil 253:287–292

Kaya C, Ashraf M, Sonmez O, Aydemir S, Tuna AL, Cullu MA (2009) The influence of arbuscular mycorrhizal colonization on key growth parameters and fruit yield of pepper plants grown at high salinity. Sci Hortic 121:1–6

Khalvati MA, Hu Y, Mozafar A, Schmidhalter U (2005) Quantification of water uptake by arbuscular mycorrhizal hyphae and its significance for leaf growth, water relations, and gas exchange of barley subjected to drought stress. Plant Biol 7:706–712

Khan AG (2008) Microbial dynamics in the mycorrhizosphere with special reference to arbuscular mycorrhizae. In: Ahmad I, Pichtel J, Hayat S (eds) Plant-bacteria interaction: strategies and techniques to promote plant growth. Wiley-VCH, Weinheim, pp 245–256

Kumar A, Sharma S, Mishra S (2010) Influence of arbuscular mycorrhizal (AM) fungi and salinity on seedling growth, solute accumulation, and mycorrhizal dependency of *Jatropha curcas* L. J Plant Growth Regul 29:297–306

Leye EHM, Ndiaye M, Ndiaye F, Diallo B, Sarr AS, Diouf M et al (2009) Effet de la mycorhization sur la croissance et le développement de *Jatropha curcas* L. Rev Energ Renew 12:269–278

Li B, Ravnskov S, Xie G, Larsen J (2007) Biocontrol of *Pythium* damping-off in cucumber by arbuscular mycorrhiza-associated bacteria from the genus *Parnibacillus*. J Int Organ Biol Control 52:863–875

Liu R, Wang F (2003) Selection of appropriate host plants used in trap culture of arbuscular mycorrhizal fungi. Mycorrhiza 13:123–127

Lu J, Lui M, Mao Y, Shen L (2007) Effect of vesicular-arbuscular mycorrhizae on the drought resistance of wild Jujube (*Zizyphs spinosus* Hu) seedlings. Frontiers Agric China 1:468–471

Marschner H, Dell B (1994) Nutrient uptake in mycorrhizal symbiosis. Plant Soil 159:89–102

Monzon A, Azcon R (1996) Relevance of mycorrhizal fungal origin and host plant genotype to inducing growth and nutrient uptake in Medicago species. Agri Ecosys Environ 60:9–15

Mukerji KG, Ciancio A (2007) Mycorrhizae in the integrated pest and disease. Section 2. In: Mukerji KG, Ciancio A (eds) Management of general concepts in integrated pest and disease management. Springer, Netherlands, pp 245–266

Nanda S, Abraham J (2011) Impact of heavy metals on the rhizosphere microflora of *Jatropha multifida* and their effective remediation. Afr J Biotechnol 10:11948–11955

Nehra V, Saharan BS (2011) Plant growth promoting rhizobacteria: a critical review. Life Sci Med Res 21:1–30

Öpik M, Saks Ü, Kennedy J, Daniell T (2008) Global diversity patterns of arbuscular mycorrhizal fungi-community composition and links with functionality. In: Verma A (ed) Mycorrhiza, 3rd edn. Springer, Germany, pp 274–317

Oyekanmi EO, Coyne DL, Fagade OE, Osonubi O (2007) Improving root-knot nematode management on two soybean genotypes through the application of *Bradyrhizobium japonicum*, *Trichoderma pseudokoningii* and *Glomus mosseae* in full factorial combinations. Crop Prod 26:1006–1012

Parniske M (2008) Arbuscular mycorrhiza: the mother of plant root endosymbioses. Nat Rev Microbiol 6:763–765

Pinior A, Grunewaldt-Stöcker G, Von Alten H, Strasser RJ (2005) Mycorrhizal impact on drought stress tolerance of rose plants probed by chlorophyll a fluorescence, proline content and visual scoring. Mycorrhiza 15:596–605

Plassard C, Dell B (2010) Phosphorus nutrition of mycorrhizal trees. Tree Physiol 30:1129–1139

Pratap CR, Potty VP (2011) Initiation of hairy roots from *Canavalia* sp. using *Agrobacterium rhizogenes* 15834 for the co-cultivation of Arbuscular mycorrhizal fungi, *Glomus microcarpum*. J Agri Tech 7:235–245

Rabie GH, Almadini AM (2005) Role of bioinoculants in development of salt tolerance of *Vicia faba* plants. Afr J Biotechnol 4:210–222

Rinaldelli E, Mancuso S (1996) Response of young mycorrhizal and non-mycorrhizal plants of olive tree (*Olea europaea* L.) to saline conditions. I. Short term electrophysiological and long term vegetative salt effects. Adv Hortic Sci 10:126–134

Ruiz-Lozano JM, Azcon R, Gomez M (1996) Alleviation of salt stress by arbuscular mycorrhizal *Glomus* species in *Lactuca sativa* plants. Physiol Plant 98:767–772

Sannazzaro AI, Echeverría M, Albertó EO, Ruiz OA, Menéndez AB (2007) Modulation of polyamine balance in *Lotus glaber* by salinity and arbuscular mycorrhiza. Plant Physiol Biochem 45:39–46

Scheloske S, Maetz M, Schneider T, Hildenbrandt U, Bothe H, Povh B (2004) Elemental distribution in mycorrhizal and non mycorrhizal roots of the halophyte *Aster tripolium* determined by proton induced X-ray emission. Protoplasma 223:183–189

Schreiner RP (2007) Effects of native and nonnative arbuscular mycorrhizal fungi on growth and nutrient uptake of 'Pinot noir' (*Vitis vinifera* L.) in two soils with contrasting levels of phosphorus. Appl Soil Ecol 36:205–215

Schüßler A, Walker C, Gamper HA (2009) Diversispora celata sp. nov: molecular ecology and phylotaxonomy of an inconspicuous arbuscular mycorrhizal fungus. New Phytol 182:495–506

Setiadi Y (2002) Mycorrhizal inoculum production technique for land rehabilitation. J Manaj Hutan Trop 8:52–64

Siddiqui ZA, Akhtar MS, Futai K (2008) Mycorrhizae: sustainable agriculture and forestry. Springer, Germany

Sieverding E (1991) Vesicular-arbuscular mycorrhizal management in tropical agrosystems. GTZ, Federal Republic of Germany

Silpachai S, Mala T, Phaosang T (2009) Effect of *Glomus aggregatum*, organic and phosphorus fertilizers on the second year growth and yield of physic nut (*Jatropha curcas* L.) cv. India. Kamphaengsean Acad J 7:10–24
Singh AK, Jamaluddin (2011) Status and diversity of arbuscular mycorrhizal fungi and its role in natural regeneration on limestone mined spoils. Biodiversitas 12:107–111
Smith SE, Read DJ (1997) Mycorrhizal symbiosis, 2nd edn. Academic, London
Subramanian KS, Santhanakrishnan P, Balasubramanion P (2006) Responses of field grown tomato plants to arbuscular mycorrhizal fungal colonization under varying intensities of drought stress. Sci Hortic 107:245–253
Sylvia DM, Hartel P, Fuhrmann J, Zuberer D (2005) Principles and applications of soil microbiology. Pearson Prentice Hall, New Jersey
Talavera M, Itou K, Mizukubo T (2001) Reduction of nematode damage by root colonization with arbuscular mycorrhiza (*Glomus* spp.) in tomato-*Meloidogyne incognita* (Tylenchida: Meloidogynidae) and carrot-*Pratylenchus penetrans* (Tylenchida: Pratylenchidae) pathosystems. Appl Ent Zool 36:387–392
Tian CY, Feng G, Li XL, Zhang F (2004) Different effects of arbuscular mycorrhizal fungal isolates from saline or non-saline soil on salinity tolerance of plants. Appl Soil Ecol 26:143–148
Trappe JM (2005) A.B. Frank and mycorrhizae: the challenge to evolutionary and ecologic theory. Mycorrhiza 15:277–281
Turkmen O, Sensoy S, Demir S, Erdinc C (2008) Effects of two different AMF species on growth and nutrient content of pepper seedlings grown under moderate salt stress. Afr J Biotechnol 7:392–396
Ultra VU (2010) Contribution of arbuscular mycorrhiza inoculation on the growth and phosphorus nutrition of jatropha (*Jatropha curcas*) in degraded upland soils of Samar, Philippines. In: 19th world congress of soil science, soil solutions for a changing world, Brisbane, 1–6 Aug 2010, p 67–70
van der Heijden MGA, Sanders IR (2003) Mycorrhizal ecology, ecological studies, 2nd edn. Springer, Germany
van Rooyen M, Valentine A, Archer E (2004) Arbuscular mycorrhizal colonisation modifies the water relations of young transplanted grapevines (*Vitis*). S Afr J Enol Vitic 25:37–42
Veresoglou SD, Sen R, Mamolos AP, Veresoglou DS (2011) Plant species identity and arbuscular mycorrhizal status modulate potential nitrification rates in nitrogen-limited grassland soils. J Ecol 9:1339–1349
Verma A (2008) Mycorrhiza: state of the art, genetics and molecular biology, eco-function, biotechnology, eco-physiology, structure and systematics, 3rd edn. Springer, Berlin
Wu Q, Xia R (2006) Arbuscular mycorrhizal fungi influence growth, osmotic adjustment and photosynthesis of citrus under well-watered and water stress conditions. J Plant Physiol 163:417–425
Zhu YG, Laidlaw AS, Christie P, Hammond MER (2000) The specificity of arbuscular mycorrhizal fungi in perennial ryegrass-white clover pasture. Agri Ecos Environ 77:211–218

Chapter 14
Diversity, Farming Systems, Growth and Productivity of *Jatropha curcas* L. in the Sudano-Sahelian Zone of Senegal, West Africa

Ibrahima Diédhiou, Papa Madiallacké Diédhiou, Khadidiatou Ndir, Roger Bayala, Bassiaka Ouattara, Banna Mbaye, Mohameth Kâne, Djiby Dia, and Idrissa Wade

Introduction

One of the major challenges West African countries are facing and that worsened since the 70s, is the high spatial and temporal variability of quantitative rainfall (IPCC 2007a). This makes agricultural production systems vulnerable and severely limits the opportunities for intensification, thereby constituting the major constraint to achieve the objectives of food self-sufficiency in the region. In this context, climate change poses an additional threat because the forecast is *a priori* an increase of variability in West Africa (IPCC 2007b). In addition, over four million cubic meters of wood are collected annually to meet the energy needs of the populations in Senegal (CSE/MEPN 2005). It is used as firewood, representing 60% of the energy balance of the country and over 80% of total energy consumption of households. Therefore, finding solutions to stabilize farmers' income is a necessity if acceptable living conditions in

I. Diédhiou (✉) • K. Ndir • R. Bayala • B. Mbaye • M. Kâne • I. Wade
Ecole Nationale Supérieure d'Agriculture (ENSA),
Université de Thiès, Thiès, Sénégal
e-mail: ibrahima_diedhiou@hotmail.com

P.M. Diédhiou
UFR des Sciences Agronomiques d'Aquaculture et de Technologies Alimentaires,
Université Gaston Berger de Saint-Louis (UGB), Saint-Louis, Sénégal
e-mail: bilpaco@yahoo.de

B. Ouattara
Centre d'Etudes Régional pour l'Amélioration de l'Adaptation à la
Sécheresse (CERAAS), Thiès, Sénégal
e-mail: obassiaka@yahoo.frD. Dia

Bureau d'Analyses Macro-Économiques, Institut Sénégalais de
Recherches agricoles, Dakar, Sénégal
e-mail: djibydia@gmail.com

N. Carels et al. (eds.), *Jatropha, Challenges for a New Energy Crop: Volume 1: Farming, Economics and Biofuel*, DOI 10.1007/978-1-4614-4806-8_14,

villages are to be maintained and rural exodus reduced. Crop diversification to open up prospects for the promotion of new agricultural industries that can increase farmers' income and thus reduce poverty is increasingly considered one of the preferred solutions. In this regard, some countries of the West African Economic and Monetary Union have taken the option of introducing biofuel crops into their production systems and focusing primarily on *Jatropha curcas* L. This choice is justified by the fact that *J. curcas* provides various products and benefits that cover four main aspects of rural development: the promotion of women (soap production), poverty reduction (sale of seeds and soap), maintenance of soil fertility by controlling erosion (through hedgerows), valuation of oil cake as organic fertilizer and provision of renewable energy for lighting, cooking, windmills and electric generators in order to reduce deforestation. In addition to these socio-economic benefits, *J. curcas* is a species considered to be well adapted to marginal, water scarce and poor soils that are characteristic of the Sahel. The development of *J. curcas* intensive farming in the Sudano-Sahelian region of Africa could however be compromised by inadequate knowledge about its agronomic characteristics. In fact, except for the study of Heller (1996), little work has been devoted to this species in the region.

To overcome this scarcity of reliable data on *J. curcas*, we conducted in Senegal, as part of the project "Interdisciplinary and Participatory Research on the Interactions between Ecosystems, Climate and Societies of West Africa", an exploratory work to contribute to a better understanding of farming systems of this crop, the diversity of its germplasm and its yield potential in the Sudano-Sahelian zone.

Farming Systems of *J. curcas*

The study was conducted in seven villages (Diaglé, Ndiaffé, Diamaguène, Keur Sambaré, Dantakhoune, Keur Bamba and Bambougar) of the Department of Foundiougne (14°08′N and 16°28′W), located in the Senegalese region of Fatick (Fig. 14.1). This site is characterized by a Sudano-Sahelian climate marked by a short rainy season of 4–5 months (June-October) and a long dry season of 7–8 months (November-June). The average monthly temperatures range from minima of 21.9°C in January and maxima of 36.8°C in March (ANAMS 2010). The soils are of tropical ferruginous type leached or not called in Wolof, the local language, "*dior* and *deck*", respectively. Most of these soils are saline (0.5–3 g/l salt) with a relatively high fluorine content (2 mg/l). These saline lands cover large areas and represent a constraint for the development of agriculture (DAT 2001).

The survey that we conducted was on a sample of 24 farms involved in *J. curcas* growing and scattered over seven villages. A questionnaire covering aspects, such as socio-demographic features of farm owners, means of production (labour, agricultural equipment, land) and cropping systems (seed supply, planting patterns, planting density, fertilization, weeding, pesticide treatments, intercropping) was made to collect data and construct a statistical model of the activity.

Fig. 14.1 Location of experimental fields. *Top*: the location of Senegal (*pink*) is given relative to the African continent (*green*). *Bottom*: the location of Foundiougne department (*blue*) is given relative to Senegal (*white*)

It appeared that the motivation of farmers for the cultivation of *J. curcas* is due to a combination of factors. Awareness campaigns were conducted since the middle of the decade 2000–2010 by States, non-governmental organizations (NGOs) and private companies. They focused on the important role that *J. curcas* could play in controlling soil erosion, valorization of marginal land, access to energy and improving farmers' incomes in the West African Sudano-Sahelian zone. Thus, Senegal government started to provide support to install public and private nurseries in order to facilitate farmers' access to seedlings. The fact that *J. curcas* requires only a

Table 14.1 Relative frequency of *J. curcas* according to its type of use

Type of use for *J. curcas*	Frequency (%)
Sale	33.3
Making fences	35.4
Medicines	23.0
Soap	6.3
Wood	2.0

small investment and does not lead to a high workload for farms is a source of motivation for farmers' adoption (Somé 2011). The majority of household chiefs (75%) who have *J. curcas* fields are relatively old (have more than 32 years), illiterate (58.3%) and primarily use family labor, with an average workforce of seven people, which is comparable to the national average (7.97). Those households are very poorly equipped and they combine farming and animal husbandry. Field work is performed manually, but animal traction is also used with bovine traction for 41.7%; horses are used in 20.8% cases while donkeys are used in 54.2% of farms.

There are many reasons for farmers to grow *J. curcas* (Table 14.1). In decreasing order of importance, *J. curcas* is used for:

- The delineation of fields and to protect crops against strayed animals and secure the property over the land used;
- Income that farmers hope to gain from the sale of seeds or oil;
- The medicinal properties, i.e., the latex is used to treat wounds due to its antimicrobial properties and leaves are used as herbal tea to treat respiratory ailments such as colds;
- The making of handmade soap.

Actually, some other factors may play an important role for the adoption of *J. curcas* by farmers. For instance, in Malawi, Mponela et al. (2011) found that age, education level of the household chief, availability of work force for labor and possession of uncultivated land have a positive influence on the allocation of land for *J. curcas* cultivation. In contrast, the ownership of livestock and off-farm incomes discourage farmers to plant this species.

In terms of cropping systems, it appeared that household chiefs are gradually adopting *J. curcas*. Actually, 21.7% of them have started cultivating this crop. The main methods of *J. curcas* planting are hedgerows and intercropping in association with food crops. Hedgerow is an established traditional planting system of *J. curcas* that has spread in many rural areas in Senegal for many years before the recent campaigns for intensive biofuel production. Hedgerows of *J. curcas* are intended primarily to delineate the fields and secondarily to protect crops against livestock strays and to control wind erosion. Planting *J. curcas* in hedgerows is still the largely dominant cropping system in the area of study where it is practiced by 78.3% of farmers surveyed. It is followed by intercropping in association with food crops (13% of farmers) and by monoculture of *J. curcas* that is still marginal (8.7%).

Table 14.2 Relative frequency of food crops used in intercropping with *J. curcas*

Crop	Frequency (%)
Peanut	36
Millet	26
Cowpea	14
Sesame	2
Maize	6
Sorghum	8
Melon	2
Fruit trees	4
Rice	2

Table 14.3 Relative frequency of plantation managements by type

Management	Frequency (%)
Pruning	33.4
Weeding	45.3
Fertilization	4.2
Lack of maintenance	29.2
Phytosanitary treatment	0.0

The association of *J. curcas* with food crops is adopted by farmers because this system allows them to use the same land at the same time for both crops and therefore insure minimum income source, primarily from the food crop and then from *J. curcas*. In such a system, *J. curcas* plants benefit from management provided for food crops (Bazongo 2011). Table 14.2 indicates that food crops most frequently associated with *J. curcas* are, in decreasing order of importance, peanut, pearl millet, sorghum, corn and sesame.

Planting density of *J. curcas* varies from field to field. In hedgerows, the planting density was one plant per meter in a single line (65.2%) or two rows separated by a 50 cm. In monoculture, *J. curcas* is planted in either a 2 × 2 m or 2.5 × 2.5 m spacing (8.6%), which corresponds to a planting density of 1,600 trees/ha. Planting designs in intercropping scenario is 2 m spacing between trees in the row and 10 m spacing between consecutive rows of trees.

In the majority of cases, *J. curcas* is first grown from seeds in nurseries and then transplanted to the field. The establishment of nurseries and field implantation early in the rainy season warrant more vigorous plants. At first glance, this strategy is certainly more costly, but it is the best method for successful plant acclimatization in the field, which is an important safeguard to recover the initial investment. Direct sowing gives poor results mainly due to pest and diseases (Somé 2011).

In terms of crop management, field work in *Jatropha* plots before and after seedling transplantation consists with clearing and weeding (Table 14.3). Fertilization and phytosanitary treatments are rare practices even though our field observations showed that *Jatropha* exhibited symptoms of infestation by pests and diseases in terms of leaf necrosis, defoliation, termites and grasshopper's attacks as well as mite attacks.

Assessment of Seed Traits and Oil Content

On one hand, studying morphological characters of seeds of a natural population is reported to be a useful step towards the assessment of genetic variability in that population (Kaushik et al. 2007). On the other hand, information on morphological attributes of seeds is a necessary step at a very early stage of selective breeding. Therefore, fruits of *J. curcas* were collected from plantations in different agro-climatic zones of the country. Oil characteristics and content were then characterized from seeds of Senegal and foreign accessions introduced from India, Mozambique and Tanzania to assess genetic variability and specificities among traits of *J. curcas*. Thus, seed size and oil content of 30 accessions from Senegal (CE1, CE2, CE3, CE5, CE6, CE8, CE9, CE11, CE12, CE13, CE14, CE15, CE17, CE32, CE33, CE34, CE35, CE38, CE48, CE50, CE51, CE52, CE69, CE77, CE83, CE85, CE89, CE90, CE94 and CE95) and three foreign accessions introduced from Tanzania (CE97), Mozambic (CE98) and India (CE99) (Table 14.4) were assessed by (1) air drying of seeds until constant weight and (2) measure with electronic caliper of parameters, such as maximum length, width and thickness on random aliquots. Seed weight was taken from three replicates of 100 seeds and oil content of kernel (% w/w) was estimated by the soxhlet method using three replicates of 2 g for each sample (Pant et al. 2006).

Seed Traits and Oil Content

Data of seed quantitative traits, oil content and related descriptive statistical analysis are provided in Tables 14.4 and 14.5. It appears that there were significant differences among accessions for seed size. The largest seeds were from Guinean accessions. As an example, the longest seeds were from CE35 while the thickest were from CE90. The smallest seed (width, length and thickness) were observed for CE12 collected in the Sahelian zone. Among the foreign accessions, Indian ones were the largest, close to the Guinean accessions. For ten accessions of *J. curcas* collected in the states of Madhya Pradesh and Maharashtra in central India, Ginwal et al. (2004, 2005) reported seed length from 17.47 to 18.64 mm and seed width from 10.83 to 11.35 mm. Those seeds were 8.47–9.14 mm thick and their oil content ranged from 46.22% to 58.12%. Kaushik et al. (2007) reported seed lengths from 16.00 to 17.63 mm, widths from 10.16 to 10.92 mm and thicknesses from 7.24 to 8.33 mm. In contrast, seeds of accessions from Tanzania and Mozambique were significantly smaller according to all seed size parameters.

Table 14.4 Seed size and oil content variability in *J. curcas* accessions

Accession	Country of origin	Length (mm)	Width (mm)	Thickness (mm)	100-seed weight (g)	Oil content (%)	Climatic zone
CE1	Senegal	17.86	11.17	8.77	64.20	54.83	Guinean
CE2	Senegal	17.60	11.36	8.81	65.30	51.33	Guinean
CE3	Senegal	17.69	11.07	8.44	63.20	46.35	Sahelian
CE5	Senegal	17.85	11.25	8.78	69.30	55.70	Guinean
CE6	Senegal	18.66	11.00	8.86	70.20	51.78	Sahelian
CE8	Senegal	17.58	10.91	8.82	53.45	44.67	Sahelian
CE9	Senegal	17.04	11.01	8.46	43.00	45.20	Sahelian
CE11	Senegal	17.85	11.42	8.68	49.05	42.64	Sahelian
CE12	Senegal	16.30	10.79	8.25	41.65	44.92	Sahelian
CE13	Senegal	17.42	11.14	8.59	53.70	44.66	Sahelian
CE14	Senegal	16.50	10.87	8.38	56.80	51.34	Sahelian
CE15	Senegal	17.20	11.14	8.72	46.65	47.63	Sahelian
CE17	Senegal	18.16	11.03	8.82	59.65	48.74	Guinean
CE32	Senegal	17.14	11.00	8.90	67.50	51.08	Guinean
CE33	Senegal	16.88	11.07	8.66	67.30	55.16	Guinean
CE34	Senegal	18.19	11.29	8.81	68.35	50.94	Guinean
CE35	Senegal	18.69	11.18	8.95	75.40	58.60	Guinean
CE38	Senegal	17.30	11.31	8.57	58.45	50.13	Sahelian
CE48	Senegal	17.91	11.24	8.46	61.50	49.44	Sudanian
CE50	Senegal	17.97	11.10	8.75	61.80	49.61	Sudanian
CE51	Senegal	17.99	11.10	8.87	62.20	51.05	Sudanian
CE52	Senegal	18.66	11.22	8.80	56.40	53.91	Sudanian
CE69	Senegal	17.44	10.94	8.44	51.15	47.46	Sudanian
CE77	Senegal	17.95	11.04	8.45	50.15	48.73	Sudanian
CE83	Senegal	17.69	11.06	8.57	56.30	44.89	Sudanian
CE85	Senegal	17.75	10.97	8.45	39.60	50.20	Sudanian
CE89	Senegal	17.92	11.06	8.55	40.35	49.25	Guinean
CE90	Senegal	18.15	11.36	9.03	64.20	52.00	Guinean
CE94	Senegal	18.50	11.16	8.94	59.30	47.69	Sudanian
CE95	Senegal	17.61	11.01	8.65	50.50	50.40	Sudanian
CE97	Tanzania	16.94	10.69	7.88	60.20	45.64	–
CE98	Mozambic	16.71	10.34	7.85	45.85	36.40	–
CE99	India	18.01	11.06	8.68	61.40	56.20	–
F[a]		13.6[b]	6.63[b]	8.68[b]	21.9[b]	64.0[b]	

[a]F values are result of one-way ANOVA
[b]Statistically significant differences between accessions at $P<0.05$ according to Tukey's test

There were also significant differences between the accessions considering the weight of 100 seeds. Local accessions, such as CE35 from the Guinean zone had the largest values while CE12 from the dry Sahelian zone gave the lowest values. Among foreign accessions, Indian accessions ranked first.

Oil content varied between 42.64% for CE11 and 58.6% for the accession CE35 with significant differences among accessions. Considering foreign accessions, Indian accessions had the highest oil content, followed by those of Tanzania. Thus, the values of seed size and oil content from this study fall into the range reported by Indian authors.

Thus, according to the classification of Kaushik et al. (2007), all the Indian accessions and 10% of African accessions (CE35, CE33 and CE5) could be considered to have a high oil content (> 55% of kernel weight) while 40% of the accessions had moderate oil content (50–55%) and 50% had low oil content (< 50% of kernel weight).

Assessment of Variability of Seed Traits and Oil Content

There is great variability between populations of *J. curcas* in Senegal for seed traits, oil content and yield. A cluster analysis of 30 accessions of *J. curcas* from Senegal based on oil content allowed the identification of three groups of accessions (I, II and III) according to length, width, thickness, weight and oil content of seeds (Table 14.5). Cluster I was composed of 15 accessions including three that originated from the Sahelian zone, six from the Sudanian zone and six others collected in the Guinean zone, i.e., between 5 ° and 20 ° North. This cluster was subdivided into two sub-clusters: Ia, consisting of eight accessions (CE17, CE38, CE48, CE50, CE77, CE85, CE89, CE95) and Ib composed of seven accessions (CE2, CE6, CE14, CE32, CE34, CE51, CE90). Cluster II included ten accessions, mostly collected in the Sahelian zone and also included two sub-clusters: IIa with four accessions (CE3, CE15, CE94, CE69) and IIb with six accessions (CE8, CE9, CE11, CE12, CE13, CE83). Cluster III consisted of five accessions, but had two sub-clusters: IIIa with four accessions (CE1, CE5, CE33, CE52) and IIIb, including a single accession (CE35). Analysis of Table 14.5 shows that there are significant variations between the three groups identified in terms of 100 seed weight and oil content. Group III shows the largest values for all parameters (length, width, thickness of seeds, 100 seed weight and oil content).

Theoretically speaking, the clusters, which are having larger inter-cluster distance and higher mean value, would produce divergent trees (Kaushik et al. 2007; Srivastava et al. 2011). Clusters II and III had the largest inter-cluster distance. Thus, the trees belonging to these clusters could be selected as parents for breeding program. According to this remark, the unique accession of sub-cluster IIIb (CE35) harvested in the locality of Kagnobon in Casamance (Guinean zone) is characterized by a very high weight of 100 seeds (75.4 g) and was exceptionally oil rich with 58.6%. The characteristics of this accession are therefore slightly better than the best performing accessions identified in India

Table 14.5 Composition of Euclidean clusters and cluster mean values obtained by clustering analysis for seed and oil traits in local accessions of *J. curcas*. The parameters used in this analysis were: seed length, seed width, seed thickness, 100 seed weight and oil content

Cluster	Number of accessions	Accession code	Length (mm)	Width (mm)	Thickness (mm)	100 seed weight (g)	Oil content (%)
I	15	CE17, CE38, CE48, CE50, CE77, CE85, CE89, CE95, CE2, CE6, CE14, CE32, CE34, CE51, CE90	17.79 ±0.51	11.12 ±0.16	8.69±0.20	58.44±9.46	50.40±1.07
II	10	CE3, CE15, CE94, CE69, CE8, CE9, CE11, CE12, CE13, CE83	17.47 ±0.57	11.06 ±0.17	8.59±0.21	51.75±6.89	45.61±1.64
III	5	CE1, CE5, CE33, CE52, CE35	17.99 ±0.74	11.18 ±0.07	8.79±0.10	66.52±6.98	55.64±1.78

by Ginwal et al. (2005), Kaushik (2007) and Kaushik et al. (2007). This accession should be taken into consideration for its potential for biodiesel production in Senegal, although the effect of soil types and climatic conditions has yet to be understood. Among the six accessions studied for growth performance, CE96 and CE48 were the more interesting accessions. Their seed yield represented 3.6 times that of CE3, which is the accession with the lowest performance. Therefore, these accessions could be regarded as best adapted to Sudano-Sahelian climate. But, for commercial production of these accessions, knowledge gaps on yield regularity, technical and operational aspects regarding crop farming, fruit harvesting and seed processing have, first, to be fixed. Oil content of seeds and fatty-acid composition are also important information that should also be provided.

The existence of a great phenotypical variability in *J. curcas* was reported in many works in West Africa (Heller 1996; Sanou 2010) and in Asia (Ginwal et al. 2004; Ratree 2004; Kaushik et al. 2007; Srivastava et al. 2011; Wani et al. 2012). This variability may be due to the heterogeneity of growing areas of *J. curcas* whereby the contrasted climatic conditions and soil types should play a major role. Actually, seeds of the different accessions were collected between the isohyets 250 and 1,300 mm. Variations in seed traits, oil content and yield, in relation to plant habitat were already reported for many woody plant species including *J. curcas* (Ginwal et al. 2005; Kaushik et al. 2007), *Azadirachta indica* A. Juss. (Jindal et al. 1999) and *Acacia nilotica* (L.) Willd. (Kundu and Tigerstredt 1997).

Vegetative Growth and Productivity of Local Accessions

The experiment was conducted in Thiès (latitude 14° 42′ N, longitude 16° 57′ W) in Senegal. This site is characterized by a Sudano-Sahelian climate marked by a short rainy season (June-October) and a long dry season (November-May). The maximum and minimum temperatures are 33°C and 23°C in the rainy season and 33°C and 19°C in the dry season. Annual rainfall varies between 600 and 800 mm with a rainfall pattern that exhibits strong variability with a yearly average for the period 1999–2009 of about 480 mm (coefficient of variation of 24%). However, more rainfall was recorded during the 2 years of experimentation with values of 536 mm in 2009 and 645.5 mm in 2010. The soil of Thiès is of sandy loam texture with acidic pH (5.0) and a low fertility (organic carbon and nitrogen contents are less than 5% and 0.3%, respectively). The water holding capacity over a 1 m depth of soil varies from 60 to 100 mm (Cissé 2010).

Senegal can be divided in three major eco-climatic zones that receive different rainfall levels: Sahelian zone, Sudanian zone and Guinean zone (Le Houérou 1989). Compared to Köppen-Geiger's climate classification, these zones correspond

Table 14.6 Sites of collection of Senegalese accessions of *J. curcas* tested at Thiès

Accession code	Origin	Department	Latitude	Longitude	Climatic zone
CE3	Pambal	Tivaouane	14° 58′ N	16° 53′ O	Sahelian
CE38	Noto	Tivaouane	14° 59′ N	17° 01′ O	Sahelian
CE12	Ndialite	Bambey	14° 40′ N	16° 28′ O	Sahelian
CE48	Ndoffane	Ndoffane	13° 55′ N	15° 55′ O	Sudanian
CE34	Sindian	Bignona	12° 57′ N	16°10′ O	Guinean
CE96	Vélingara	Vélingara	13° 07 N	14°08′ O	Guinean

to the following climates: desert climate, steppe climate and equatorial savannah with dry winter climate. The Sahelian zone is between the isohyets 200 and 600 mm, the Sudanian zone receives an annual rainfall between 600 mm and 1,200 mm. The Guinean zone is characterized by an annual rainfall exceeding 1,200 mm.

Six native accessions harvested from Sahelian, Sudanian and Guinean eco-climatic zones of Senegal were planted in a randomized block design with four replications in Thiès (latitude 14° 42′ N, longitude 16° 57′ W). Three accessions (CE3, CE38 and CE12) were collected from Sahelian zone, one accession (CE48) originated from the Sudanian zone and two accessions (CE34 and CE96) were collected from the Guinean zone (Table 14.6).

The comparative performance of these accessions was estimated by measuring plant height, basal diameter, crown diameter, yield and yield components (number of seeds per plant, number of fruits per plant, seed weight per plant, shell weight per tree).

The survey showed that there is no significant difference among the six accessions for growth parameters (Table 14.7). The overall averages were 1.46±0.13 m for plant height, 83.55±6.51 mm for collar diameter and 1.31±0.12 m for canopy diameter. Differences between accessions were significant for yield and its components (Table 14.7). With respect to the yield parameters, the values for the accessions CE48 and CE96, collected in the Guinean and Sudanian zones respectively, were the largest and followed by CE12, while the smallest values were from CE34 (Guinean zone), CE3 and CE38 (Sahelian zone).

The values for growth parameters in a Sudano-Sahelian site under strict rainfed farming conditions (146.5 cm for maximum height, 130.8 cm for crown diameter and 83.5 mm for collar diameter) were similar to those reported in West Africa (Heller 1996; Sanou 2010; Diédhiou 2011) and Asia (Ginwal et al. 2004; Ratree 2004). The average seed yield per hectare obtained in the present study appeared low and varied between 33.36 and 123.22 kg/ha depending on the accession. These results on the performance of *J. curcas* under Sudano-Sahelian climate are in accordance with data

Table 14.7 Changes in growth parameters and yield components in six local accessions 22 months after planting

Accessions	Height (cm)	Collar diameter (mm)	Canopy diameter (cm)	Number of seeds per tree	Fruit weight per tree (g)	Seed weight per tree (g)	Seed yield (kg/ha)	Shell weight per tree (g)
CE38	160.44	85.27	140.02	40.64^{b}	31.44^{b}	20.85^{b}	33.36^{b}	10.59^{c}
CE3	124.42	73.14	112.23	21.28^{b}	18.60^{b}	12.27^{b}	19.63^{b}	6.33^{c}
CE12	153.12	88.18	132.10	82.22^{ab}	71.87^{ab}	44.22^{ab}	70.75^{ab}	27.65^{abc}
CE48	145.89	78.05	121.17	140.28^{a}	116.93^{a}	75.18^{a}	120.29^{a}	41.76^{ab}
CE96	153.63	89.65	132.97	146.38^{a}	120.63^{a}	77.01^{a}	123.22^{a}	43.61^{a}
CE34	141.36	86.98	146.19	48.58^{b}	38.94^{b}	24.46^{b}	39.14^{b}	14.48^{c}

The superscript letters *a*, *b*, *c* and *d* indicate statistical differences between the mean values of different accessions for a given parameter according to Tukey. For each parameter, values in the same column having the same superscript letter are not statistically different at $P<0.05$

from other works in Senegal (Diédhiou 2011) and Burkina Faso (Bazongo 2011). The low yield could be attributed to the short duration of the experiment (2 years) that did not allow plants to reach their maximum ramification, a key yield component. Another reason would be the low soil fertility in the experimentation field. It could be speculated that the soil fertility was not good enough to sustain optimal mineral nutrition for *J. curcas* trees. In fact, a recent work showed that seed yield could be doubled through application of organic manure in a plantation of *J. curcas* (Diédhiou 2011).

Conclusions

This study provides the first scientific data on production practices and actual agronomic performances of *J. curcas* in Senegal. It shows that in Senegal, farmers have started to adopt *J. curcas* as a crop. This finding corroborates the reports by Kane (2010) and Bazongo (2011) showing a gradual breakthrough of this plant in the agricultural production systems in West Africa.

J. curcas is grown in the Sudano-Sahelian West Africa mainly to increase farmers' income, for field protection or to take advantage of its medicinal properties. The dominant cropping systems are hedgerows, association with food crops and monocultures. In the absence of an established farming system for *J. curcas*, care brought to plantation of *J. curcas* is minimal and consist with field clearance and weeding. Fertilization and pest control are almost nonexistent although the pest status of plantations is of concern. The yields of 2 years old plantations that did neither benefit from weeding nor fertilizing were low and far below projections by common reports in the West African sub region. The lack of reliable technical information is the reason for the lack of mastery of farming techniques (planting dates, planting density, nursery management techniques, planting methods) by the farmers. This is currently the major constraint to *J. curcas* industry development (Faye et al. 2011).

The local accessions showed an interesting potential for oil production. Five of them exhibited high oil content at levels similar to the highest values reported for Indian accessions and were introduced in a process of selective breeding. It appeared also that there is a phenotypic variability revealed by the hierarchical analysis in local accessions of *J. curcas* in Senegal that does not reflect the eco-climate diversity of the country. However, other environmental factors including the nature and quality of soils that have not been studied in this work may have influenced the observed phenotypic diversity. In any case, these results are only partial and a molecular analysis is necessary to draw a final conclusion about the relationship between genetic diversity and eco-climatic diversity.

The economic effect of yield variability results in decreasing profitability of *J. curcas* plantations and consequently, negatively affects biofuel industry. Failing to fix this kind of problems would greatly decrease the attractiveness of *J. curcas* industry to potential producers and investors as well. However, plant breeding offers powerful tools to target and select highly productive materials as well as mass propagate elite accessions in view of developing cultivars with optimal yield. It is worth to recall here, as an example, that wheat benefits of 5,000 years domestication while *J. curcas*, at best, benefits 500 years and more realistically ~30 years since its economical value was only seriously recognized in the 1980s. Therefore, breeding for high yielding varieties for a large scale production of *J. curcas* is still a priority if sustainable production of biodiesel is targeted. This option should be coupled with developing appropriate technologies of production in order to optimize the growth and productivity of these accessions.

Acknowledgement The authors thank "Interdisciplinary and Participatory Research on the Interactions between Ecosystems, Climate and Societies of West Africa" (RIPIECSA) for financial support.

References

ANAMS (2010) Base de données de la météorologie du Sénégal; Tableaux climatiques mensuelles de Fatick. Agence Nationale de la Météorologie, Dakar, French

Bazongo P (2011) Introduction du *Jatropha* dans les exploitations agricoles de la zone ouest du Burkina Faso: état des lieux et effet de la plante sur les propriétés des sols et les cultures associées. Master dissertation, Institut du Développement Rural, Université Polytechnique de Bobo Dioulasso, Bobo Dioulasso. French

Cissé C (2010) Etude de la croissance et des réponses à la contrainte hydrique et au stade juvénile de six provenances locales de *Jatropha curcas*. Master dissertation, ENSA, Université de Thiès, Thiès. French

CSE/MEPN (2005) Etat de l'environnement au Sénégal. Ministère de l'Environnement et de la Protection de la Nature, Dakar. French

DAT (2001) Plan régional de développement intégré de Fatick. Direction de l'Aménagement du Territoire, Fatick. French

Diédhiou D (2011) Influence des pratiques paysannes de gestion de la fertilité et d'entretien des plantations sur la croissance et la production de *Jatropha curcas* L.: cas du terroir de Ourour (Bassin Arachidier Sénégalais). Master dissertation, ENSA, Université de Thiès, Thiès. French

Faye A, Fall CS, Dia D, NDour A, Wade I, Diédhiou I (2011) Impact de l'introduction des biocarburants au Sénégal: Évaluation du coût d'opportunité de *Jatropha curcas* L. pour les producteurs dans la zone de Kaolack. RIPIECSA, Cotonou; 18–21 Oct 2011. French

Ginwal HS, Phartyal SS, Rawat PS, Srivastava RL (2005) Seed source variation in morphology, germination and seedling growth of *Jatropha curcas* Linn. in Central India. Silvae Genet 54: 76–80

Ginwal HS, Rawat PS, Srivastava RL (2004) Seed source variation in growth performance and oil yield of *Jatropha curcas* Linn. in Central India. Silvae Genet 53:186–192

Heller J (1996) Physic nut. *Jatropha curcas* L. In: Promoting the conservation and use of underutilized and neglected crops, vol 1. Institute of Plant Genetics and Crop Plant Research/International Plant Genetic Resources Institute, Gatersleben/Italy

IPCC (2007a) The physical science basis – working group I contribution to the fourth assessment report of the intergovernmental panel on climate change. Cambridge University Press, Cambridge, UK

IPCC (2007b) Climate change 2007 – impacts, adaptation and vulnerability – working group II contribution to the fourth assessment report of the intergovernmental panel on climate change. Cambridge University Press, Cambridge, UK

Jindal SK, Vir S, Pancholy A (1999) Variability and associations for seed yield, oil content and tree morphological traits in neem (*Azadirachta indica*). J Trop For Sci 11:320–322

Kane S (2010) Impacts spatiaux et scio-économiques de la culture des biocarburants sur la dynamique des systèmes agraires: Cas de *Jatropha curcas* L. dans la Communauté rurale de Ourour, Département de Guinguénéo. Master dissertation, Université Cheikh Anta Diop, Faculté des Lettres et Sciences Humaines, Dakar. French

Kaushik N (2007) *Jatropha* germplasm characterization for biodiesel production. Energy biosciences strategy for India, New Delhi, 10–11 Sep 2007

Kaushik N, Kumar K, Kumar S, Kaushik N, Roy S (2007) Genetic variability and divergence studies in seed traits and oil content of *Jatropha* (*J. curcas* L.) accessions. Biomass Bioenergy 31:497–502

Kundu SK, Tigerstredt PMA (1997) Geographical variation of seed and seedling traits of neem (*Azadirachta indica* A. Juss.) among ten populations studied in growth chamber. Silvae Genet 46:129–137

Le Houérou HN (1989) The grazing land ecosystems of the African Sahel. Springer, Berlin

Mponela P, Jumbe CBL, Mwase WF (2011) Determinants and extent of land allocation for *Jatropha curcas* L. cultivation among smallholder farmers in Malawi. Biomass Bioenergy 35:2499–2505

Pant KS, Khosla V, Kumar D, Gairola S (2006) Seed oil content variation in *Jatropha curcas* Linn. in different altitudinal ranges and site conditions in H.P. India. Lyonia 11:31–34

Ratree S (2004) A preliminary study on physic nut (*Jatropha curcas* L.) in Thailand. Pak J Biol Sci 7:1620–1623

Sanou F (2010) Production de *Jatropha curcas* et impact de la plante sur les propriétés chimiques du sol: Cas de Bagré (Centre Est du Burkina Faso). Master dissertation, Université Polytechnique de Bobo Dioulasso, Bobo Dioulasso. French

Somé HK (2011) L'introduction de *Jatropha* dans les exploitations agricoles de la zone ouest du Burkina Faso : état des lieux et caractérisation des systèmes de production. Graduation dissertation, Centre Agricole Polyvalent de Matourkou, Bobo Dioulasso. French

Srivastava P, Behera SK, Gupta J, Jamil S, Singh N, Sharma YK (2011) Growth performance, variability in yield traits and oil content of selected accessions of *Jatropha curcas* L. growing in a large scale plantation site. Biomass Bioenergy 35:3936–3942

Wani TA, Kitchlu S, Ram G (2012) Genetic variability studies for morphological and qualitative attributes among *Jatropha curcas* L. accessions grown under subtropical conditions of North India. S Afr J Bot 79:102–105

Part IV
Byproducts

Chapter 15
Assessment of the Potential of *Jatropha curcas* for Energy Production and Other Uses

Siddhartha Proteem Saikia, Animesh Gogoi, Kalpataru Dutta Mudoi, Adrita Goswami, and Debashish Bora

Introduction

Present environmental concerns with climate change led scientists to explore plant based fuels such as biodiesel, because it is renewable, non-toxic, biodegradable and environmentally friendly. In comparison to petroleum diesel, biodiesel decreases the emission of CO_2, sulphur, hydrocarbon, particle matter and smoke during the combustion process (Shu et al. 2007). Furthermore, burning of biodiesel has no net addition to atmospheric CO_2 levels, because it is made from materials produced via photosynthetic carbon fixation. From the economic view point, biodiesel can be produced at low cost on a commercial scale from agricultural and agroforestry crops (Wood 2005). Biodiesel is monoalkyl esters of fatty acids derived from vegetable oils and animal fats that is known as clean and renewable fuel. Biodiesel is usually produced by transesterification of vegetable oils or animal fats with methanol or ethanol (Knothe 2006). The concept of using biofuel in diesel engines has originated more than 100 years ago when Rudolf Diesel tested vegetable oil (i.e., peanut oil) as fuel in his engine. However due to abundant supply of petro-diesel, research and development on vegetable oil were not seriously pursued. It received attention only recently when it was realized that petroleum fuels are dwindling fast and environment friendly renewable substitutes are to be identified (Agarwal and Das 2000). For this purpose many researchers exploited several commercially edible oils as feedstocks for biodiesel. Fortunately, nonedible vegetable oils, mostly produced by oilseed trees and shrubs can provide an alternative to edible oils for production of biodiesel not competing with food resources. The need of nonedible vegetable oils drawn the attention on *Jatropha curcas* (hereafter referred to as Jatropha), which grows in tropical and sub tropical climates across the developing world (Openshaw 2000). The fact that Jatropha

S.P. Saikia (✉) • A. Gogoi • K.D. Mudoi • A. Goswami • D. Bora
Division of Medicinal Aromatic & Economic Plants, North East Institute of Science & Technology (CSIR), Jorhat, Assam 785006, India
e-mail: sp_sidhu@yahoo.com

N. Carels et al. (eds.), *Jatropha, Challenges for a New Energy Crop: Volume 1: Farming, Economics and Biofuel*, DOI 10.1007/978-1-4614-4806-8_15,

may grow and produce on marginal soil not suitable for food crops makes it ideal for use as energy or fuel source. Besides biodiesel production, the seeds of Jatropha provide proteins, which have a descriptive history in both nutrition and therapeutic applications (Mandal and Mandal 2000). From a nutritional perspective, seed storage proteins have always been major players in supplying global protein needs and food energy (Rosegrant et al. 2001; Steinfeld et al. 2006). The larger need for proteins in the livestock sector has accentuated the search for new protein sources that do not conflict with human food security interests. In the current situation, oilseeds able to produce on marginal soils are the potential and preferred choice for protein and other nutrients for livestock, provided they would be free of toxic and anti nutritional factors. The wide adaptability of Jatropha to grow under diverse agro-climatic conditions, marginal lands, thin soil, semi-arid as well as humid conditions theoretically ensure no competition with food crops (Rao et al. 2008). Moreover, the expected large production of biodiesel from Jatropha seeds will result in the availability of high amounts of pressed cake from seeds and kernel meal as by-products, which are rich in proteins of high quality. Unfortunately, the Jatropha's press cake is toxic to animal, but non-toxic varieties exists and research on press cake detoxification is on-going. Thus, in a close future, Jatropha's press cake should be available for animal feeding as well. In addition, the press cake could also be a source for various bioactive molecules having a wide range of activities (Wang et al. 2007). From a pharmaceutical, industrial or agricultural prospective, *Jatropha* spp. are also a rich sources of phytochemicals. Jatropha proteins and peptides have been studied for their role in the plant metabolic activities and defence against predators as well as for their therapeutic and industrial potential. In the present review an attempt has been made to discuss (a) the quality of oil as a biodiesel source, (b) the nutritional quality of seed proteins (specially for animals), (c) the chemistry, biological role and potential application of biologically active cyclopeptides, and (d) other usefulness of *J. curcas* plant parts in our daily life. It is hoped that the state-of-the-art information provided here will stimulate further research and development leading to more intensive, efficient, and sustainable utilization of Jatropha.

Biodiesel

Jatropha, which is believed to be native of Central America (Ratree 2004), belongs to the family Euphorbiaceae. Every part of this plant has its own uses. In the present day context, the major part that is economically important is the seed from which biodiesel can be produced. The utilization of Jatropha oil as a new source of fuel for diesel engine has tremendous scope in contributing to alleviate the pressure on energy resources (Saikia et al. 2009). Jatropha oil can be obtained from seeds by mechanical or solvent extraction with hexane (a non polar solvent) for 6–8 h in a Soxhlet. The extracted oil is then filtered and excess solvent is removed by using a rotary evaporator at 40°C. Finally, the crude oil is stored in a freezer at −2°C for subsequent analysis (Emil et al. 2010). The physicochemical properties include the amount (%) of fatty acid, iodine value, peroxide value, saponification value, viscosity, calorific value, cetane number, flash point, etc. It was found that the values of these properties are significantly different for the crude Jatropha oil compared to the fossil diesel used in diesel engines. This is only because of the presence of higher unsaturated fatty acids like oleic (42.4–48.8%) and linoleic acid (28.8–34.6%) (Emil et al. 2009, 2010). Other major fatty acids of Jatropha oil are the palmitic and stearic fatty acids. As the proportion of oleic and linoleic acids is higher, the Jatropha oil can be classified as oleic-linoleic oil. Physicochemical properties differing from recommended standards may result in poor atomization, coking tendency, carbon deposits and wear. These unwanted features were generally experienced in most of the tests that adversely affects the durability of an engine. It has been reported that the high viscosity and low volatility of vegetable oil (including crude Jatropha oil) are generally observed to be the major drawbacks for their direct injection in diesel engines (Prasad et al. 2000). The high viscosity of vegetable oils cause an increase in smoke levels and the low volatility of the vegetable oils result in oil sticking of the injector or cylinder walls causing deposit formations, which interfere with the combustion process (Sahoo and Das 2009). After several tests with crude vegetable oils for several hours (200 h), it was concluded that as far as power output, thermal efficiency and lubricating oil data are concerned, the 1:3 (v/v) blend of soybean oil and sunflower oil with diesel fuel performed satisfactory (Schlick et al. 1988). However after 200 h of operation, combustion examination of several parts revealed heavy carbon deposits in the combustion chamber, tracers of wear on piston rings as well as on the plunger and injector tips, slight scuffing of the cylinder liner and uneven spray from nozzles (Schlick et al. 1988; Bari et al. 2002). These problems led the investigators to suggest that either different operating conditions or modification of vegetable oils could help in improving the conditions of the engines (Schlick et al. 1988). The transesterification of a vegetable oil to its methyl esters reduces its molecular weight and viscosity and increases its cetane number (Gerhard 1983). The transesterification process has been proven worldwide as an effective mean of biodiesel production and viscosity reduction of vegetable oils (Peterson et al. 1992). Transesterification is the process of reacting triglyceride with alcohol in the presence of a catalyst to produce glycerol and fatty acid esters. Temperature, catalyst type, alcohol to oil proportion and stirring speed have been observed to influence the transesterification process to a greater extent. A comparative study of the physicochemical properties of crude Jatropha oil and its methyl esters with fossil diesel has been carried out to know the actual status of Jatropha oil concerning its biodiesel performance (Singh and Padhi 2009) (Table 15.1). The study show that Jatropha oil can be used as a source of triglycerides for the production of biodiesel by esterification and/or transesterification. Although the biodiesel from refined vegetable oils meets the requirement of a high speed diesel oil, the production of biodiesel from edible oil is currently much more expensive than fossil diesel due to the relatively high cost of edible oil. Thus, there is a need to explore non-edible oils like that of Jatropha for the production of biodiesel as it is easily available in many parts of the world including India and is cheaper compared to edible oils.

Table 15.1 Comparison of fuel properties of Jatropha oil, Jatropha methyl ester and diesel

Property	Unit	Jatropha oil	Jatropha oil methyl ester	Diesel	ASTM D 6751-02	DIN EN 14214
Density at 15°C	Kg m^{-3}	918	880	850	875–900	860–900
Viscosity at 40°C	mm^2 s^{-1}	35.4	4.84	2.60	1.9–6.0	3.5–5.0
Flash point	°C	186	162	70	>130	>120
Pour point	°C	–6	–6	–20	—	—
Water content	%	5	Nil	0.02	<0.03	<0.05
Ash content	%	0.7	Nil	0.01	<0.02	<0.02
Carbon residue	%	0.3	0.025	0.17	—	<0.3
Sulphur content	%	0.02	Nil	—	0.05	—
Acid value	mg KOH g^{-1}	11.0	0.24	0.35	<0.8	<0.50
Iodine value	—	101	104	—	—	—
Saponification value	—	194	190	—	—	—
Calorific value	MJ kg^{-1}	33	37.2	42	—	—
Cetane number	—	23	51.6	46	—	—

Source: Singh and Padhi (2009)
ASTM: American Society for Testing and Materials
DIN: *Deutsches Institut für Normung* (German Institute for Standardization)

Nutritional Source

Both the toxic and non-toxic genotypes of Jatropha exist. Toxic genotypes are prevalent throughout the world while non-toxic genotypes exist only in Mexico (Makkar and Becker 2009). The seeds of *J. curcas* are composed of kernel and shell with an average ratio of 62.2:37.7. The kernel has higher crude protein (22–28%) and oil contents (54–58%) compared to the shell (4–6% crude protein and 0.8–1.4% oil) (Makkar et al. 1998). The seeds also contain antinutrients and toxic factors such as phytate, trypsin inhibitor, lectin, curcin and phorbol esters (PEs). Hence, the use of Jatropha meal and protein isolates prepared from toxic genotypes of Jatropha in animal nutrition is still restricted (Makkar and Becker 2009; Makkar et al. 1997).

Storage Proteins

The storage proteins in seeds are important for germination and also have important nutritional value. The storage proteins extracted from defatted kernel meals of both toxic and non-toxic genotypes revealed that the total protein content is ~89% (Osborne 1924). Glutalins, globulins and albumins shared a major proportion of total protein contributing 56.9%, 27.4% and 10.8%, respectively, whereas prolamins and non extracted proteins were present in minor quantities (0.6% and 4.3%, respectively). Gastric digestions of albumins and globulins by pepsin and pancreatin system had protein digestibilities of 64% and 61%, respectively, whereas a higher digestion rate (95%) was observed for glutalins. The glutelin fraction which forms >50% of proteins leads to low solubility of Jatropha protein (Selje-Assmann et al. 2007), which in turn decreases protein degradation in rumen when compared to soybean proteins. Thus, a high amount of rumen undegradable protein available post-ruminally can be utilized by gastric digestion. The ruminal and gastric protein digestions indicate the potential availability of Jatropha protein in both ruminants and monogastrics. However, the meal should be detoxified before it is incorporated into animal diets (Devappa et al. 2010). It has also been reported that defatted *Jatropha meal* (JM), which was obtained after oil extraction with hexane and defatted *Jatropha kernel meal* (JKM), which was obtained after solvent extraction of kernels (free of shells) have high protein content. The JKM contains higher protein content than JM. Non protein nitrogen represented only 4.7–5.0% of proteins in JKM (Makkar et al. 1998). The rate of digestibility of JKM proteins is high (90%) and the protein fraction has good amino acid composition (Makkar and Becker 2009). The amino acid composition of proteins is often used to define their nutritional quality. The amino acid composition of meals from both toxic and non-toxic genotypes of Jatropha was found similar. The levels of *essential amino acids* (EAA) except lysine were higher than that of the FAO reference and the EAA requirements for chicks and young pigs (Makkar et al. 1997, 1998, 2008; Devappa and Swamylingappa 2008). Similarly, the levels of EAA except isoleucine in Jatropha meal was higher or similar when compared to castor bean meal, and except for lysine, the amino acid profile is comparable with that of soybeans (Makkar et al. 1998). However, in JM and JKM, the presence of antinutritional factors and toxic factors restricts the utilization of these meals in animal nutrition (Makkar and Becker 2009; Devappa and Swamylingappa 2008).

Protein Isolates

Protein isolates are the concentrated forms of plant proteins, generally prepared by solubilizing proteins and removing non protein ingredients. The rate of protein digestibility of such isolates was also approximately 90% (Makkar et al. 1998; Devappa and Swamylingappa 2008). The EAA content (except lysine) of protein isolates was higher than those of FAO/WHO references considering 3–5-year-old children and the amino acid levels in the protein isolates were similar to those in the kernel meal. The calculated values for nutritional indices such as computed protein efficiency ratio (C-PER) (based on the EAA profile and protein digestibility analysis) for JM (1.1), JKM (1.72), protein isolates from JM (1.85) and protein isolates from JKM (2.16) were comparable to or higher than the reported C-PER values for regular animal feed ingredients such as corn meal (1.1), wheat flour (0.8), soy flour (1.3) and quality protein feeds (1.43) (Devappa and Swamylingappa 2008;

Angulo-Bejaranoa et al. 2008). This suggests that Jatropha proteins have good quality and could supplement or replace the conventional protein sources in animal diets.

Animal Nutrition

Most genotypes of Jatropha produce seeds containing several toxic factors such as phytates, trypsin inhibitors, lectin, curcin and phorbol esters (PEs), which makes their seed meal toxic to mice, rats and goats (Goel et al. 2007). The major organs that were affected in these animals were intestine, liver and kidney (Li et al. 2010). PEs were found to be the principal compounds responsible for meal toxicity as shown by feeding fish and mice with their purified fractions (Becker and Makkar 1998; Li et al. 2010). PEs are present in Jatropha meal at levels of 2–4 mg g^{-1} (Makkar and Becker 2009; Makkar et al. 1997).

In the past two decades, several studies have been carried out for the complete detoxification of Jatropha meals. In brief, trypsin inhibitors and lectin were completely deactivated by moist heat (Aderibigbe et al. 1997; Aregheore et al. 2003). However, toxic PEs could not be removed completely due to their stability to heat and chemical degradation (Chivandi et al. 2004; Martinez-Herrera et al. 2006; Devappa and Swamylingappa 2007). Recently detoxification of Jatropha kernel meal and protein isolate had been successfully achieved (Devappa et al. 2010). *The detoxified Jatropha kernel meal* (DJKM) and *detoxified protein isolate* (DPI) prepared from screw pressed cake have been added to fish diet at high level with excellent growth performance and no toxic effects at blood and tissue levels. The DJKM and DPI have high protein content (60% and 90%, respectively) and excellent amino acid composition and these preparations could replace at least 50% of the protein contributed by the high quality fish meal (65% protein) in standard fish diet. It has also been reported that the feeding of Turkey, pigs and broilers with DJKM and DPI resulted in growth response and nutrient utilization comparable to those obtained with concentrates prepared from conventional protein sources without any apparent signs of toxicity (Devappa et al. 2010). On the basis of the results obtained so far on fish and other animal species it has been suggested that the DJKM and DPI are suitable substitutes for fish meal or soybean meal for livestock diets (Makkar and Becker 2009). The meal from non-toxic Jatropha genotypes is free from PEs, but it contains trypsin inhibitors, lectins and phytates at the same levels as the meal from toxic genotypes. The nutritional quality of the non-toxic Jatropha meal, after heat treatment (to inactivate trypsin inhibitor and lectin), evaluated in fish (carp) and rat models, was found to be very high (Makkar and Becker 1999). The meal or the protein isolate obtained from the non-toxic genotype, after heat treatment, could be an excellent protein rich ingredient in feed of ruminant and monogastric animals including fish. Moreover, it has also been reported that along with the storage of proteins and oil, Jatropha seeds are also a source of carbohydrates and other minerals with livable antinutrients levels. The effect of some physical treatments (like soaking, germination and roasting) and some chemical treatments (like $NaHCO_3$, ethanol extraction and NaOH) were successful in inactivating the antinutrients (phytic acid, trypsin inhibitor activity, total phenols and saponins) (Abou-Arab and Abu-Salem 2010). In parts of Mexico (like in Veracruz State), seeds from non-toxic Jatropha genotypes are also consumed by humans, after roasting (Makkar et al. 1998), but the consumption of raw seeds is considered to produce cramps and uneasy feeling in stomach. The protein hydrolysate obtained from ground Jatropha cake have shown to be well solubilised and contained proteins at a rate as high as 71.69%, which is appropriate for further applications in human and animal food (Apiwatanapiwat et al. 2009).

Pharmaceutical Importance

Presently lot of research activities have shown that the cyclopeptides isolated from latex, seeds and roots of many plants possess various biological activities such as cyclooxygenase, acetyl choline esterase and tyrosinase inhibition (Yahara et al. 1989; Morita et al. 1994), immunosuppression (Morita et al. 1997), antimalaria (Baraguey et al. 2000), vasorelaxation (Morita et al. 2005) and cytotoxicity (Mongkolvisut et al. 2006). Jatropha species have been shown to be a rich source of bioactive cyclic peptides, which contain 7–10 residues with a high proportion of hydrophobic amino acids. There is a need to exploit the potential role of cyclic peptides from Jatropha for pharmaceutical applications (Devappa et al. 2010). The cyclic peptides isolated from *J. curcas* (*viz.*, jatrophidin and curcacyclines A & B) only are discussed below, emphasizing their chemistry in view of its potential in agricultural and pharmaceutical sectors.

Jatrophidin

Jatrophidin is an octapeptide isolated from the latex of *J. curcas* (Altei et al. 2008). The latex was partitioned with ethyl acetate, fractionated on Sephadex G15, eluted in solid phase extraction, and purified by HPLC to obtain jatrophidin-I. The amino acid analysis, mass spectroscopy, and 1D/2D nuclear magnetic resonance (NMR) studies demonstrated that jatrophidin-I exists as two conformers of a cyclic structure (Gly-Trp-Leu-Asn-Leu-Leu-Gly-Pro) with the conformational equilibrium of proline residues between *cis* and *trans* forms, indicating that this peptide has more than one conformational state in solution. The isolates of jatrophidin-I had weak antifungal effect against the strains of *Candida albicans*, *C. krusei*, *C. parapsilosis* and *Cryptococcus neoformans* and moderate activity as an acetylcholinesterase inhibitor, when compared with the standard galanthamine.

Curcacyclines A and B

Curcacycline A was isolated from the ethanolic extract of *J. curcas* latex. It is a cyclic octapeptide ($C_{37}H_{66}N_8O_9$; MW 766.97). The amino acid sequence was determined to be cyclo-(Gly^1-Leu^2-Leu^3-Gly^4-Thr^5-Val^6-Leu^7-Leu^8). Curcacycline A displayed a moderate inhibition of (1) the classical pathway activity of human complement and (2) the proliferation of human T-cells (Van den Berg et al. 1995). Curcacycline B is a cyclic nonapeptide ($C_{42}H_{73}N_9O_{10}$; MW 863) isolated from the latex of *J. curcas.* The amino acid sequence was found to be cyclo-(Leu^1-Gly^2-Ser^3-Pro^4-Ile^5-Leu^6-Leu^7-Gly^8-Ile^9). The absolute stereochemistry of amino acids was shown to be *"L"* configuration. It contains mostly hydrophobic residues and one proline, thus differing from cyclic peptides previously isolated from the latex of *Jatropha* spp., which does not contain proline. The structure of curcacycline B was suggested to be a substrate for peptidylprolyl cis-trans isomerase (PPIase), as it has some structural features in common with cyclosporin A (inhibitor of cyclophilins A and B). Curcacycline B enhances PPlase activity by 60% at 30 μM based on an enzymatic experiment involving a human cyclophilin B and R-chymotrypsin rotamase, whereas no modification of cyclophilin B activity was observed in the presence of curcacycline A (Auvin et al. 1997). Curcacycline B from *J. curcas* possesses antimalarial activity (IC_{50} < 10 mM) against *P. falciparum* (Baraguey et al. 2001). Besides the activity of these cyclopeptides, the antibiotic effect of an alcohol extract from *J. curcas* leaves has been observed *in vitro* on *Escherichia coli* and *Staphylococcus aureus* (Zeng et al. 2004). The extract inhibited *E. coli* and *S. aureus* and the activity against *E. coli* was found to be larger than that against *S. aureus.* The poisonous effects of protein fraction, seed oil and ethanol extract from Jatropha seeds were evaluated for insecticidal activity against *Lipaphis erysimi* (Kaltenbach) (Li et al. 2004). The protein fraction did not show any significant effect to *L. erysimi*, while seed oil demonstrated strong contact toxicity. The molluscicidal efficacy of Jatropha seed extracts from Yunnan (China) and Mali (Africa) was compared by Cheng et al. (2001) and these authors did not found any difference of activity between the extracts from these distant countries. Actually, PEs have strong molluscicidal activity (Goel et al. 2007) and the content of PEs in Jatropha seed samples collected from different parts of the world have been of similar order of magnitude.

Other Traditional Uses

Apart from the above mentioned useful properties, Jatropha has many other economic uses. Its seeds and fruits are anthelmenthic, useful in chronic dysentery, thirst, tridosha, urinary discharge, abdominal complaints, biliousness, anaemia fistula and heart diseases (Nasir et al. 1988; Gubitz et al. 1999; Augustus et al. 2002; Akintayo 2004; Franke et al. 2004). It is also applied topically for rheumatism, herpes and pruritus. Jatropha leaves are used in traditional medicine against coughs, as antiseptics after birth, and food for the Tassar silk worms (Gubitz et al. 1999; Kinawy 2010). On the other hand the sap extracted from the leaves is used via external application to treat piles. In addition, the tender twigs of the plant are used for cleaning teeth (Kinawy 2010). The latex is useful for wound healing and other medical uses (Salimon and Abdullah 2008). The deoiled seed cake of Jatropha is suitable as substrate for enzyme production by solid-state fermentation (SSF) and also supported good bacterial growth and enzyme production (Mahanta et al. 2007). Traditionally, various solvent extracts of Jatropha have been taken orally with ripe banana to treat dysentery in adults. The sap from twigs is considered stypic and is used for dressing wounds and ulcers. The bark is rubbed with asafoetida and buttermilk is reportedly taken orally to relieve dyspepsia and diarrhoea. A decoction of bark is used externally for treating rheumatism and leprosy. The decoction of root bark is used to rinse mouth, to relieve toothache and sore throat (Goonasekera et al. 1995; Parotta 2001). The bark is also rich in tannin and produces a purple dye (Openshaw 2000). Moreover Jatropha can be used for erosion control, as living fence, ornamental plant or even as firewood.

Conclusion

Jatropha is a multipurpose plant with many attributes and considerable potential. Its seeds constitute a source of oil, proteins, carbohydrate and minerals with tolerable antinutrient level. The high seed yield of Jatropha oil compared to other vegetable sources is an advantage for selecting this oil to produce cost competitive products. The seeds of Jatropha are nutritionally promising and could allleviate protein malnutrition, which is a major public health problem in the developing world and is still unexplored. The seeds may thus be a good option due to their multipurpose features of Jatropha such as its high level adaptability to environmental factors, applicability of seed oil for biofuel production, and generation of productive value-added co-products.

Acknowledgements SPS thanks Dr. P.G. Rao, Director, North East Institute of Science & Technology, Jorhat for giving his permission to publish this review.

References

Abou-Arab AA, Abu-Salem FM (2010) Nutritional quality of *Jatropha curcas* seeds and effect of some physical and chemical treatments on their anti-nutritional factors. Afr J Food Sci 4:93–103

Aderibigbe AO, Johnson COLE, Makkar HPS, Becker K (1997) Chemical composition and effect of heat on organic matter and nitrogen degradability and some antinutritional components of Jatropha meal. Anim Feed Sci Technol 67:223–243

Agarwal AK, Das LM (2000) Biodiesel development and characterization for use as a fuel. In: compression ignition engine. ASME Gas Turb Power 123:440–447

Akintayo ET (2004) Characteristics and composition of *Parkia biglobbossa* and *Jatropha curcas* oils and cakes. Bioresour Technol 9:307–310

Altei WF, Picchi DG, Barbosa SC, Cilli EM, Giannini MJ, Cardoso-Lopes EM et al (2008) NMR studies, solid phase synthesis and MD/SA simulation as a tool for structural elucidation of new bioactive peptides from the latex of *Jatropha curcas* L. Planta Med 74:1–338

Angulo-Bejaranoa PI, Verdugo-Montoyab NM, Cuevas-Rodrı'gueza EO, Mila'n-Carrillob J, Mora-Escobedod R, Lopez-Valenzuelab JA et al (2008) Tempeh flour from chickpea (*Cicer arietinum* L.) nutritional and physic chemical properties. Food Chem 106:106–112

Apiwatanapiwat W, Vaithanomsat P, Somkliang P, Malapant T (2009) Optimization of protein hydrolysate production process from *Jatropha curcas* cake. World Acad Sci Eng Technol B: Chem Mater Eng 2:161–164

Aregheore EM, Becker K, Makkar HPS (2003) Detoxification of a toxic variety of *Jatropha curcas* using heat and chemical treatments and preliminary nutritional evaluation with rats. S Pac J Nat Sci 21:50–56

Augustus GDPS, Jayabalan M, Seiler GJ (2002) Evaluation and bioinduction of energy components of *Jatropha curcas*. Biomass Bioenergy 23:161–164

Auvin C, Baraguey C, Blond A, Lezenven F, Pousset JL, Bodo B (1997) Curcacycline B a cyclic nonapeptide from *Jatropha curcas* enhancing rotamase activity of cyclophilin. Tetrahedron Lett 38:2845–2848

Baraguey C, Blond A, Correia I, Pousset JL, Bodo B, Auvin-Guette C (2000) Mahafacyclin A, a cyclic heptapeptide from *Jatropha mahafalensis* exhibiting β-bulge conformation. Tetrahedron Lett 41:325–329

Baraguey C, Blond A, Cavelier F, Pousset JL, Bodo B, Auvin-Guette C (2001) Isolation, structure and synthesis of mahafacyclin B, a cyclic heptapeptide from the latex of *Jatropha mahafalensis*. J Chem Soc Perkin Trans 1:2098–2103

Bari S, Yu WC, Lim TH (2002) Performance deterioration and durability issues while running a diesel engine with crude palm oil. J Automob Eng 216:785–792

Becker K, Makkar HPS (1998) Toxic effects of phorbol esters in carp (*Cyprinus carpio* L.). Vet Hum Toxicol 40:82–86

Cheng ZY, Huang SX, Zeng QH, Yang Y, Gao ZQ (2001) Comparison of the indoor molluscidal effects of *Jatropha curcas* L. extract from different places. Chinese J Schisto Cont 13:221–225

Chivandi E, Mtimuni JP, Read JS, Makuza SM (2004) Effect of processing method on the phorbol ester concentration, total phenolics, trypsin inhibitor activity and the proximate composition of the Zimbabwean *Jatropha curcas* provenance: a potential live stock feed. Pak J Biol Sci 7:1001–1005

Devappa RK, Swamylingappa B (2007) Effect of processing methods on the removal of toxic and antinutritional constituents of Jatropha meal: a potential protein source. J Food Sci Technol Nepal 3:88–95

Devappa RK, Swamylingappa B (2008) Biochemical and nutritional evaluation of Jatropha protein isolate prepared by steam injection heating for reduction of toxic and antinutritional factors. J Sci Food Agric 88:911–919

Devappa RK, Makkar HPS, Becker K (2010) Nutritional, biochemical and pharmaceutical potential of proteins and peptides from Jatropha: review. J Agric Food Chem 58:6543–6555

Emil A, Yaako Z, Kumar MNS, Jahim JM, Salimon J (2009) Characteristic and composition of *Jatropha curcas* oil seed from Malaysia and its potential as biodiesel feedstock. Eur J Sci Res 29:396–403

Emil A, Yaakob Z, Kumar MNS, Jahim JM, Salimon J (2010) Comparative evaluation of physicochemical properties of Jatropha seed oil from Malaysia, Indonesia and Thailand. J Am Oil Chem Soc 87:689–695

Franke K, Nasher AK, Schmidt J (2004) Constituents of *Jatropha unicostata*. Biochem Syst Ecol 32:219–220

Gerhard V (1983) Performance of vegetable oils and their monoesters as fuels for diesel engines. Paper no 831358, SAE

Goel G, Makkar HPS, Francis G, Becker K (2007) Phorbol esters: structure, biological activity and toxicity in animals. Int J Toxicol 26:279–288

Goonasekera MM, Gunawardana VK, Jayasena K, Mohammed SG, Balasubramaniam S (1995) Pregnancy terminating effect of *Jatropha curcas* in rats. J Ethnopharmacol 47:117–123

Gubitz GM, Mittelbach M, Trabi M (1999) Exploitation of the tropical oil seed plant *Jatropha curcas* L. Bioresour Technol 67:73–82

Kinawy OSE (2010) Characterization of Egyptian Jatropha oil and is oxidative stability. Energy Sources Part A 32:119–127

Knothe G (2006) Analyzing biodiesel: standards and other methods. J Am Oil Chem Soc 83:823–833

Li J, Yan F, Wu FH, Yue BS, Chen F (2004) Insecticidal activity of extracts from *Jatropha curcas* against *Lipaphis erysimi*. Acta Phytophyl Sin 31:289–293

Li CY, Devappa RK, Liu JX, Makkar HPS, Becker K (2010) Toxicity of *Jatropha curcas* phorbol esters in mice. Food Chem Toxicol 48:620–625

Mahanta N, Gupta A, Khare SK (2007) Production of protease and lipase by solvent tolerant *Pseudomonas aeruginosa* PseA in solid-state fermentation using *Jatropha curcas* seed cake as substrate. Bioresour Technol 99:1729–1735

Makkar HPS, Becker K (1999) Nutritional studies on rats and fish carp (*Cyprinus carpio*) fed diets containing unheated and heated *Jatropha curcas* meal of a non-toxic provenance. Plant Foods Hum Nutr 53:182–192

Makkar HPS, Becker K (2009) *Jatropha curcas*, a promising crop for the generation of biodiesel and value-added coproducts. Eur J Lipid Sci Technol 111:773–787

Makkar HPS, Becker K, Sporer F, Wink M (1997) Studies on nutritive potential and toxic constituents of different provenances of *Jatropha curcas*. J Agric Food Chem 45:3152–3157

Makkar HPS, Aderibigbe AO, Becker K (1998) Comparative evaluation of non-toxic and toxic varieties of *Jatropha curcas* for chemical composition, digestibility, protein degradability and toxic factors. Food Chem 62:207–215

Makkar HPS, Francis G, Becker K (2008) Protein concentrate from *Jatropha curcas* screw-pressed seed cake and toxic and antinutritional factors in protein concentrate. J Sci Food Agric 88:1542–1548

Mandal S, Mandal RK (2000) Seed storage proteins and approaches for improvement of their nutritional quality by genetic engineering. Curr Sci 79:576–589

Martinez-Herrera J, Siddhuraju P, Francis G, Davila-Ortiz G, Becker K (2006) Chemical composition, toxic/antimetabolic constituents, and effects of different treatments on their levels, in four provenances of *Jatropha curcas* L. from Mexico. Food Chem 96:80–89

Mongkolvisut W, Sutthivaiyakit S, Leutbecher H, Mika S, Klaiber I, Moller W et al (2006) Integerrimides A and B, cyclic heptapeptides from the latex of *Jatropha integerrima*. J Nat Prod 69:1435–1441

Morita H, Kayashita T, Kobata H, Gonda A, Takeya K, Itokawa H (1994) Pseudostellarins A-C, new tyrosinase inhibitory cyclic peptides from *Pseudostellaria heterophylla*. Tetrahedron Lett 50:6797–6804

Morita H, Gonda A, Takeya K, Itokawa H, Hirano T, Oka K et al (1997) Solution state conformation of an immunosuppressive cyclic dodecapeptide, cycloleonurinin. Tetrahedron Lett 53:7469–7478

Morita H, Iizuka T, Choo CY, Chan L, Itokawa H, Takeya K, Dichotomins J, Dichotomins K (2005) Vasodilator cyclic peptides from *Stellaria dichotoma*. J Nat Prod 68:1686–1688

Nasir MKA, Memon GM, Valhari MU, Khatri LM (1988) Studies on fixed oil of *Jatropha curcas* seeds. Pak J Sci Ind Res 31:566–568

Openshaw KA (2000) Review of *Jatropha curcas*: an oil plant of unfulfilled promise. Biomass Bioenergy 19:1–15

Osborne TB (1924) The vegetable protein, 2nd edn. Longmans, Green, New York

Parotta JA (2001) Healing plants of peninsular India. CAB International, Wallingford, pp 299–300

Peterson CL, Reece DL, Hammond DL, Cruz R, Thompson JA (1992) Comparison of ethyl and methyl esters of vegetable oils as diesel fuel substitute. In: Proceedings of the alternate energy conference (ASAE 1992), Nashville, TN, pp 99–110

Prasad CMV, Krishna MVSM, Reddy CP, Mohan KR (2000) Performance evaluation of non-edible vegetable oils as substitute fuels in low heat rejection diesel engines. J Inst Mech Eng 214:181–187

Rao VT, Rao PG, Reddy CHK (2008) Experimental investigation of Pongamia, Jatropha and neem methyl esters as biodiesel on C. I. engine. Jordan J Mech Ind Eng 2:117–122

Ratree SA (2004) Preliminary study on physic nut (*Jatropha curcas* L.) in Thailand. Pak J Biol Sci 7:1620–1623

Rosegrant M, Paisner MS, Meijer S (2001) Long-term prospects for agriculture and the resource base. The world bank rural development family: rural development strategy background. World Bank, Washington, DC. Available at http://go.worldbank.org/J1D1O0B6E0

Sahoo PK, Das LM (2009) Process optimization for biodiesel production from Jatropha, Karanja and Polanga oils. Fuel 88:1588–1594

Saikia SP, Bhau BS, Rabha A, Dutta SP, Choudhari RK, Chetia M et al (2009) Study of accession source variation in morpho-physiological parameters and growth performance of *Jatropha curcas* Linn. Curr Sci 96:1631–1636

Salimon J, Abdullah R (2008) Physicochemical properties of Malaysian *Jatropha curcas* seed oil. Sains Malays 37:379–382

Schlick ML, Hanna MA, Schinstock JL (1988) Soybean and sunflower oil performance in diesel engines. Trans ASAE 31:1345–1349

Selje-Assmann N, Makkar HPS, Hoffmann EM, Francis G, Becker K (2007) Quantitative and qualitative analyses of seed storage proteins from toxic and non-toxic varieties of *Jatropha curcas* L. International symposium energy and protein metabolism, European Association Animal Nutrition (Publication No. 124),Vichy, p 625. Available at http://tinyurl.com/37wcfn4

Shu Q, Yang B, Yuan H, Qing S, Zhu G (2007) Synthesis of biodiesel from soybean oil and methanol catalyzed by zeolite beta modified with La^{3+}. Catal Commun 8:2159–2165

Singh RK, Padhi SK (2009) Characterization of Jatropha oil for the preparation of biodiesel. Nat Prod Rad 8:127–132

Steinfeld H, Gerber P, Wassenaar T, Castel V, Rosales M, de Haan C (2006) Livestock's long shadow—environ issues and options. U.N. Food and Agriculture Organization, Rome

Van den Berg AJ, Horsten SF, Kettenes-vanden Bosch JJ, Kroes BH, Beukelman CJ, Leeflang BR et al (1995) Curcacycline A—a novel cyclic octapeptide isolated from the latex of *Jatropha curcas* L. FEBS Lett 358:215–218

Wang MW, Hao X, Chen K (2007) Biological screening of natural products and drug innovation in China. Philos Trans R Soc London B: Biol Sci 362:1093–1105

Wood P (2005) Out of Africa: could Jatropha vegetable oil be Europe's biodiesel feedstock? Refocus 6: 40–44

Yahara S, Shigeyama C, Nohara T, Okuda H, Wakamatsu K, Yasuhara T (1989) Structures of anti-ace and -renin peptides from *Lycii radicis* cortex. Tetrahedron Lett 30:6041–6042

Zeng LH, Yan F, Chen F (2004) In vitro bacteriostasis of *Jatropha curcas* L. extract against chicken *Escherichia coli* and *Staphlococcus aureus*. Chin Poult Sci 8:35–37

Chapter 16
Jatropha curcas Biodiesel, Challenges and Opportunities: Is it a Panacea for Energy Crisis, Ecosystem Service and Rural Livelihoods?

Suhas P. Wani and Girish Chander

Introduction

The current worldwide energy crisis and associated pollution causing global warming are largely due to extensive dependence of energy supply on conventional fossil fuels. Present scenario of depleting world's fossil fuel reserves, increased consumption, rising prices of petroleum products at unprecedented levels, dependence on non-renewable sources and its substantial contribution to environmental pollution and global warming have led researchers, policy makers, environmentalists and industrialists to consider the importance of biofuels such as biodiesel in search of alternatives to existing limitations (Wani et al. 2006). The burgeoning population, fast growing industry and energy hungry new technologies further increase the urgency to look for alternatives to existing limitations. It is estimated by the International Energy Agency (IEA) that increasing population along with increasing incomes in developing countries will contribute a share of 74% to the increase in global primary energy use. India and China will be responsible for 45% of this increase. For India, one driver behind the growing order for energy is the transport sector that currently consumes 27% of total primary oil demand. This share will increase to 47% by 2030 (IEA 2007). Diesel makes up almost 70% of the oil used in Indian road transport. Biodiesel among biofuels is a renewable source of energy, therefore it offers a great potential and mitigate limitation and supplement supplies of fossil fuel and at the same time minimize C emissions. Biofuel and biodiesel, in particular, are derived from biomass and use photo-synthetically fixed C, thus, facilitating recycling of atmospheric CO_2. Meeting the energy requirements by using environment-friendly biofuels will help to decrease the carbon dioxide emission into the atmosphere.

S.P. Wani (✉) • G. Chander
Resilient Dryland Systems, International Crops Research Institute for the Semi Arid Tropics (ICRISAT), Patancheru, Hyderabad 502324, India
e-mail: s.wani@cgiar.org

N. Carels et al. (eds.), *Jatropha, Challenges for a New Energy Crop: Volume 1: Farming, Economics and Biofuel*, DOI 10.1007/978-1-4614-4806-8_16,

Biodiesel from *J. curcas* (hereafter referred to as Jatropha) seed oil is considered as an answer to energy crisis, which can replace depleting fossil fuel and at the same time render environmental services by way of maintaining or reducing CO_2 levels. It also helps in creating other environmental benefits, such as rehabilitation of degraded lands through addition of organic matter through leaf fall, hedge for erosion control and wind break. After extracting oil, the deoiled cake serves as a rich source of organic matter and plant nutrients for crop production. Switching over to biodiesel will have implications in terms of employment generation for various production and market related activities and decentralized energy affecting rural livelihoods. Biodiesel can be a decentralized source of energy through its use in electricity generator, thus addressing the issue of erratic supply in farms with large productivity and use in other domestic or power based livelihood activities. So, we are today facing a question—is Jatropha truly a wonder biodiesel plant, which can be a panacea for issues of depleting fossil fuels, environmental degradation and main streaming of the poor?

Jatropha, the Potential Plant for Biodiesel on Degraded and Marginal Lands

The recent assessments of the Indian government have identified 16% (> 50 m ha) of the geographical area as wasteland (Government of India 2010). Keeping in mind the 221 million poor in India, and the fact that 70% of poor in India are small marginal farmers and landless labourers (Srivastava 2005), it is essential that wasteland development programs are undertaken to generate the needed socioeconomic benefit for poor farmers and labourers. As a plant of the dry regions of Meso-America, Jatropha is suitable to grow on degraded drylands. Distributed by the Portuguese colonial power, Jatropha now grows in many African and South-Asian countries, and is well adapted to semi-arid tropical conditions. Jatropha is drought tolerant and can survive under 200 mm water per year. Its mechanism of leaf fall in dry spells increases the amount of organic matter in the soil, which makes Jatropha not only the one adapted to degraded lands, but the one rehabilitating them. Its dense root system stabilizes soil and arrests erosion. It produces fruits with a high content of non-edible oil that can be used for several purposes and is not browsed by animals. Thus, Jatropha can be a good candidate to rehabilitate the degraded lands mostly in possession of poor and marginal farmers and generate additional income for them. Grass et al. (2008) found that at crude prices above US $ 75 per barrel, Jatropha fuel production on India's wasteland starts to be economically viable.

To develop a strategy that serves both energy and food needs, work has been initiated recently for growing energy plants suitable for the degraded lands. According to a report of the committee on biofuels constituted by Planning Commission, Government of India (2003) out of a large number of oilseed tree species, Jatropha along with *Pongamia pinnata* would best suit the Indian conditions. For raising Jatropha, it is not necessary to sacrifice the land area that is already under cultivation of food and horticultural crops as Jatropha has the inherent ability to thrive on degraded and marginal lands. Jatropha is a fast growing crop, not browsed by cattle and goats with ability to withstand harsh climatic conditions. The National Mission on Biofuels in India has identified about 13.4 million ha for *J. curcas* (and *Pongamia pinnata*) plantations in immediate future and it covers poor, marginal, degraded, fallow, waste and other lands, such as along canals, roads, railway tracks, on farm and property bunds in the arid and semiarid areas. Once success is achieved on potential lands, it should be possible to include lands with low fertility soils, which can be brought under Jatropha plantation in an economically feasible manner to rehabilitate them. By adopting knowledge-based and pro-poor strategy, non-edible oil can be used for biodiesel production. Pro-poor bio-power strategy leads to a win-win-win situation that improves livelihoods, protect environment and allows the energy release from renewable sources.

Coherent National Policy on Biofuels

Biofuels are gaining importance ever since the prices of fossil fuel began skyrocketing due to the reduced supplies and growing concern with environmental pollution. Both developed and developing countries are formulating policies for a mandatory blending of bioethanol and biodiesel (produced from renewable sources) with fossil fuels, resulting in a huge demand for raw materials for producing biofuels.

Realizing the urgency, the Government of India has formulated and approved a National Policy on Biofuel in 2009 along with setting up of an empowered National Biofuel Coordination Committee, headed by the Prime Minister and a Biofuel Steering Committee headed by Cabinet Secretary. Under the approved policy, the country aims to rise blending of biofuels with gasoline and diesel to 20% by the year 2017 (Achten et al. 2010). The policy focuses on indigenous production of biodiesel in waste, degraded and other marginal lands. It incorporates the announcements on Minimum Support Price (MSP) with the provision of periodic revision for biodiesel oilseeds to provide fair price to the growers, which would be based on the actual cost of production of bioethanol. In case of biodiesel, the MSP could be linked to the prevailing retail diesel price. The National Biofuel Policy envisages that biofuels, namely, biodiesel and bioethanol may be brought under the ambit of "Declared Goods" by the Government to ensure unrestricted movement of biofuels within and outside the states. It is also stated in the Policy that no taxes and duty would be levied on biodiesel.

Biodiesel, A Convenient Energy Source with Good Fuel Properties

During the last few decades researchers tried many edible and non-edible oils in compression ignition for different utilities. Due to short supply, India can neither afford usage of edible oils as power source nor can afford to bring in agricultural lands under biofuels. Hence, the cultivation of non-edible vegetable oils on degraded and marginal lands is strategically propagated.

Biodiesel was among the first alternative fuels with bioethanol to really become known to the public. The advantage of biodiesel is that it can be used in existing vehicles with little or no modification required when used as blended. There are energy plants available that will produce a higher yield in kWh per area, but the simplicity of having a fuel that is fully compatible with present fuel and engine technology makes Jatropha biodiesel very attractive.

Jatropha biodiesel has good fuel properties, comparable to or even better than petroleum diesel (Rao et al. 2008). Its cetane number (an indication of its fuel burning efficiency) is 51–52 for biodiesel from Jatropha oil, which is higher than the cetane number of most petroleum diesels. It has 10% built-in oxygen content that helps it to burn completely. The esters of the long-chain fatty acids of biodiesel are excellent lubricants for the fuel injection system. It has a higher flash point than diesel, making it a safer fuel. Other advantages are the almost zero sulphur content and the reduced amount of carbon monoxide, unburned hydrocarbons and particulate matter in the exhaust. But there are a few technical issues that need to be resolved. Biodiesel has a high viscosity at low temperatures, leading to flow problems at these temperatures.

ICRISAT assessed the performance of the vehicles (8 TATA Mobile 207Di, 2 Nissan Diesel and 1 Toyota Qualis) being run since May 2007 on fuel blended (B10) with biodiesel (400 L received from Southern Online Bio Technologies Ltd). The observations and comments by automobile engineer from Farm and Engineering Services at ICRISAT were very encouraging. All vehicles put on fuel mixtures performed normally. There was no starting trouble or pickup problem in any vehicle. There was no abnormal smoke or other specific complaints from the users while driving the vehicle.

Refining Agronomic Practices for Realizing Higher Productivity

Optimizing Fertilizer Practices

Jatropha is well adapted to marginal lands, but it responds to fertilizer application. Nutrient management is one such aspect that may play a pivotal role in economic cultivation in marginal lands. Information available on nutrient requirement of Jatropha is scarce and particularly in wastelands where it is nil. The work to evaluate the effect of nitrogen and phosphorus on growth and productivity of four year-old Jatropha, planted on wastelands at ICRISAT, showed response to fertilizer application. In the on-station study, four-year-old Jatropha plantation recorded 1,290 kg ha^{-1} seed yield in the fertilized control plots (Table 16.1). The seed productivity varied from 1,320 to 1,610 kg ha^{-1} with the application of nitrogen and phosphorus fertilizers. The highest productivity was recorded at 80 kg N and 20 kg P_2O_5 ha^{-1}.

Table 16.1 Effects of nitrogen and phosphorus on seed yield of four-year-old Jatropha planted in wastelands at ICRISAT, India during 2008

Fertilizer/plant	Seed yield (kg ha^{-1})
Control	1,290
N (40) + P_2O_5 (10)	1,320
N (40) + P_2O_5 (20)	1,330
N (80) + P_2O_5 (10)	1,560
N (80) + P_2O_5 (20)	1,610

Source: Wani et al. (2012)

Growth Regulators

Jatropha is a monoecious shrub, where the flowers are unisexual, i.e., male and female flowers are produced in the same inflorescence. Inflorescences produce a central female flower surrounded by a group of male flowers. Generally, there are 1–5 female flowers and 25–90 male flowers per inflorescence with an average male to female flower ratio of 29:1.

Studies were conducted at ICRISAT during 2009 to evaluate the effect of growth regulators like naphthaleneacetic acid (NAA), gibberellic acid (GA), chlorocholine chloride (CCC) and ethrel on the male and flower ratio and yield of Jatropha planted in the year 2004. The results revealed that yields were in general low due to unfavourable weather, but application of either CCC or GA at 90 and 25 mg kg^{-1}, respectively, at the time of flower initiation can improve the flowering characteristics and crop yields.

Mycorrhizae Inoculation

Keeping in view the water and nutrient stress in marginal lands where Jatropha is supposed to grow, the effective absorptive root surface can be increased by inoculating roots with mycorrhizal fungi. Mycorrhizal fungi increases nutrient uptake not only by increasing the absorbing surface area of roots, but also through the release of powerful enzymes into the soil that dissolve hard-to-capture nutrients, such as organic nitrogen, phosphorus, iron and other tightly bound soil nutrients. This extraction process is particularly important in plant nutrition in marginal lands and explains why non-mycorrhizal plants require high levels of fertility to maintain their health. ICRISAT has raised Jatropha following mycorrhizal application to enhance growth and yield of seedlings. One gram of mycorrhizae may be placed below the seed at the time of sowing to enhance growth of the seedlings. Mycorrhized Jatropha seedlings showed higher plant height, stem girth and number of leaves compared to non-mycorrhized, when sampled three months after sowing (Table 16.2).

Table 16.2 Effects of mycorrhizal inoculation on growth of 3-month Jatropha seedlings in nursery at ICRISAT

Treatment	Plant height (cm)	Stem girth (cm)	Number of leaves
Inoculated	47	6.5	16
Non-inoculated	35	5.9	12

Source: Wani et al. (2006)

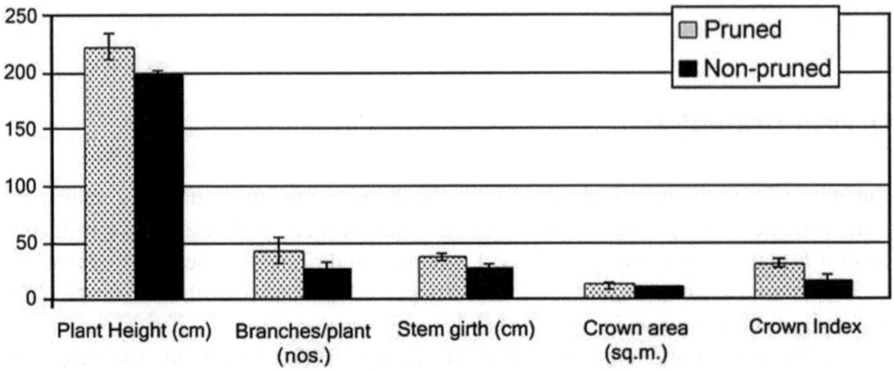

Fig. 16.1 Growth characteristics of Jatropha plants due to summer pruning during the 3rd year of establishment (Source Wani et al. 2009a)

Pruning and Irrigation

Jatropha produces flowers in cymose inflorescences with dichasial cyme pattern in new borne branches and hence, the number of new branchlets determines the number of inflorescences in Jatropha. Therefore, pruning is essential to increase the number of fruiting branchlets in Jatropha, which is carried out by nipping the terminal bud to induce secondary and tertiary branches.

A study was carried out to assess the effects of pruning on canopy characteristics of Jatropha in which half of the plants in the block plantation were pruned at 45 cm in the first year and 75 cm in the second year of growth during the dry season (February–March) when the plants are dormant and the rest half were grown without pruning. During the third year, the top one third of the secondary and tertiary branches of plants were nipped off under the pruning treatment. The effect of pruning was observed on plant height, stem girth at 10 cm above the ground, number of branches, crown area, and volume index during third year (Fig. 16.1). The results showed that pruning significantly ($p<0.05$) influenced plant height (224 cm), stem girth (36 cm), and crown index (29) compared to the non-pruned plants, where the plant height, stem girth and crown index were 197 cm, 27 cm, and 15, respectively. Similarly, the branches per plant were also more numerous (43) in the pruned when compared to the non-pruned (26) plants.

Similarly, Jatropha is well adapted to drought conditions, but shows profuse growth under irrigated conditions compared to no irrigation. As the fruits are borne on new branches in Jatropha, the enhanced growth and number of branches has direct relation to fruit and seed yields.

Water Requirement, Balance and Soil Conservation Under Jatropha cultivation

Soil moisture was monitored in the Jatropha plantation (seedlings planted in November 2004) at ICRISAT from November 2005 using a neutron probe. Weather was monitored at the ICRISAT agrometeorological observatory, Patancheru. Daily reference crop *evapotranspiration* (ETo) was computed following the FAO Penman-Monteith method (1998). Evapotranspiration requirements of Jatropha under ideal soil moisture conditions were estimated based on ETo and crop coefficients measured for different phenophases. Evapotranspiration values under actual field conditions of Jatropha plantation were estimated from soil moisture measurements using the standard water balance equation.

Monthly crop evapotranspiration (Fig. 16.2) values indicate that during April to June, ETo requirements are high due to atmospheric demands as well as the vegetative stage of plantation. However, this is the period in which the actual availability with respect to demand is low. During July to October, soil moisture status is sufficient to satisfy much of the ETo requirements and this period coincides with flowering and fruit set stage. During three years from 2005 to 2008, Jatropha used 75–90% of the rainfall received. Lower relative utilization (%) occurred when the rainfall distribution was erratic, though the rainfall amount was high. In the year 2008, rainfall till 30th

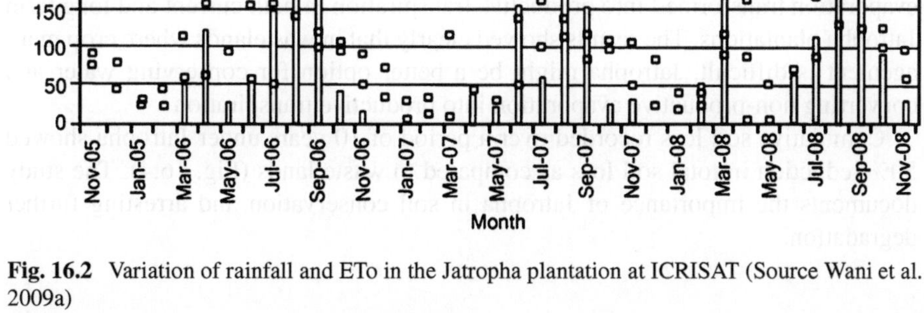

Fig. 16.2 Variation of rainfall and ETo in the Jatropha plantation at ICRISAT (Source Wani et al. 2009a)

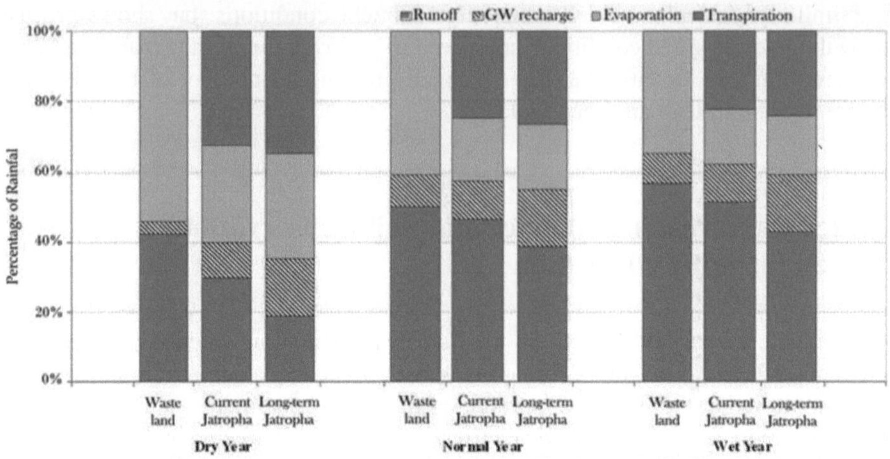

Fig. 16.3 Water balance components of different land management scenarios during dry, normal and wet years (data compiled from 2001 to 2010) (Source Garg et al. 2011)

November was 1,103 mm; however, total rainfall during June and July was only 190 mm compared to the normal 305 mm. The August rainfall was 382 mm compared to the normal 220 mm. There were eight days in the year 2008 with a rainfall of more than 50 mm and long periods of dry spells occurred in June and July. The total ETo use by Jatropha in the year 2008 till November was 820 mm, the highest in the last three years. If the rainfall distribution was good, Jatropha could have used even more water. The study indicated that contrary to the belief that Jatropha needs less water, under favourable soil moisture conditions, Jatropha could use large amounts of water for luxurious growth and high yield.

Combining ArcSWAT and HYDRUS1D modeling tools, water balance components of Jatropha plantation were studied in contrast to wastelands in Velchal, Rangareddy district of Andhra Pradesh, India (Garg et al. 2011). The data showed that under waste lands, 40–60% of the rainfall got lost as runoff, while it was reduced to 20–40% under long term Jatropha plantation (Fig. 16.3). Groundwater recharge doubled in *current* Jatropha and quadrupled in the *long-term* (current Jatropha for long period of time) Jatropha plantation. More than 50% of the non-productive evaporation transformed into productive transpiration in both current and long term Jatropha plantations. The results showed clearly that in wastelands where crop management is difficult, Jatropha might be a better option for conserving water and converting non-productive evaporation into productive transpiration

Cumulative soil loss recorded over a period of 10 years under Jatropha showed 50% reduction in total soil loss as compared to waste lands (Fig. 16.4). The study documents the importance of Jatropha in soil conservation and arresting further degradation.

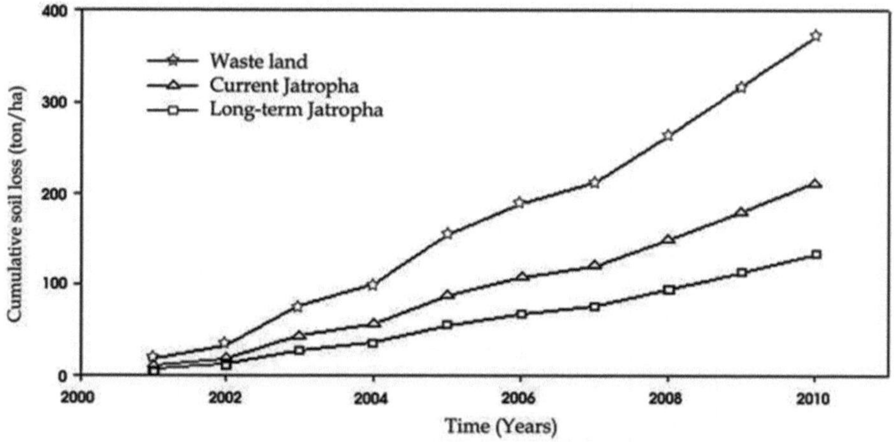

Fig. 16.4 Cumulative soil loss (t ha^{-1}) under different land management conditions (data from year 2001 to 2010) (Source: Garg et al. 2011)

Tapping Germplasm for Potential Productivity

In order to achieve full potential of biodiesel plantations with Jatropha, solution to the problem of low yield needs to be addressed to boost the large scale profitable cultivation. Genotypes influence yield potential, therefore the source of seed assumes greater significance. Genetic diversity in plant species is a gift to humankind as it forms the basis for selection and further improvement. It is desirable to select seed source for multiplication from known plant populations with favorable traits. However, comprehensive work on collection, characterization and evaluation of germplasm for growth, morphology, seed characteristics and yield traits is still in its infancy. The study conducted at ICRISAT (during 2008) on progeny trial evaluation comprising 99 Jatropha accessions collected from different agro-eco regions of India and planted during 2006 showed great variability in plant height (105–274 cm), collar diameter (5.1–18.9 cm), number of branches (9.7–66.3), crown area (0.5–6.3 m^2), volume index (8,056–85,708 cm^3), seed length (11.6–19.3 mm), seed width (9.5–11.7 mm), seed thickness (7.6–11.0 mm), 100-seed weight (38.9–67.1 g) and seed oil content (27.5–40.5). ICJC 06116 recorded the highest 100 seed weight of 67.1 g (Table 16.3). Accession ICJC 06004 recorded the highest plant height (274 cm) and ICJC 06087, the highest volume index (85,708 cm^3) and collar diameter (18.9 cm). The number of branches and crown area were the highest in ICJC 06115 (66.3) and ICJC 06010 (6.3 m^2), respectively. Highest seed length was seen in accession ICJC 06082 (19.3 cm) while accession ICJC 06055 showed the maximum seed width (11.7 cm) and thickness (11.0 cm). The highest oil content for seeds was 40.5% in accession ICJC06019. As a whole, 44% of the accessions have exhibited seed oil content in the range of 35.1–40%.

Table 16.3 Performance of selected *J. curcas* accessions for growth and seed traits at ICRISAT during 2008 (planted 2006)

Genotypes	Plant height (cm)	Collar diameter (cm)	Number of branches	Crown area (m^2)	Volume index (cm^3)	Seed length (mm)	Seed breadth (mm)	Seed thickness (mm)	100 seed weight (g)	Oil content (%)
ICJC 06004	274	16.0	52.1	5.3	76,077	17.9	10.9	9.1	53.9	31.2
ICJC 06010	226	15.0	54.5	6.3	51,563	17.9	10.9	8.6	51.2	37.8
ICJC 06019	228	12.8	41.4	2.4	38,653	16.6	11.4	10.7	52.3	40.5
ICJC 06055	135.9	5.8	14.9	0.8	8,972.8	16.7	11.7	11.0	46.1	34.9
ICJC 06066	168	6.4	23.4	0.8	10,433	16.9	10.6	7.6	54.8	32.1
ICJC 06082	207.5	14.0	32.4	1.5	41,756.3	19.3	10.3	9.3	53.3	35.0
ICJC 06087	233.2	18.9	51.5	3.5	85,708	18.2	11.3	9.0	56.8	37.3
ICJC 06091	202	18.3	51.0	3.4	68,345	17.6	11.5	10.3	64.3	39.1
ICJC 06115	213	14.7	66.3	3.2	48,761	17.3	11.4	10.7	66.7	34.8
ICJC 06116	205	17.6	61.9	3.5	64,976	19.1	11.5	8.9	67.1	29.4
ICJC 06117	201	15.3	62.2	3.7	50,474	17.7	11.6	10.8	58.2	34.7
ICJC 06120	198	10.6	48.5	2.3	23,850	18.7	11.0	8.6	66.6	34.4
Mean (of 99)	204	11.8	37.3	2.0	33,935	17.4	11.1	9.1	52.7	34.7
SEM	14.1	1.3	5.8	0.5	8,288.9	0.2	0.2	0.1	3.4	0.3
CD (5%)	43.6	3.9	17.8	1.3	25,664	0.6	0.5	0.4	11.2	0.8

Source: ICRISAT (2010)

Table 16.4 Marker attributes for AFLP combinations used at ICRISAT

Primer combination	PIC[a]	EMR[b]	MI[c]	RP[d]
E-ACT/M-CTT	0.34	95	32.30	46.82
E-ACA/M-CAA	0.20	88	17.60	23.11
E-ACA/M-CAT	0.25	77	19.25	26.21
E-AGG/M-CAG	0.25	101	25.25	34.81
E-AGG/M-CTA	0.24	119	28.56	39.17
E-ACG/M-CAA	0.24	117	28.08	38.84
E-AGC/M-CTA	0.30	83	24.90	37.53
Minimum	0.20	77	17.60	23.11
Maximum	0.34	119	32.30	46.82
Average	0.26	97	25.13	35.21

Source: Tatikonda et al. (2009)
[a]Polymorphism information content (PIC)
[b]Effective multiplex ratio (EMR)
[c]Marker index (MI)
[d]Resolving power (RP)

The objective for genetic improvement of Jatropha as a crop should aim at a larger proportion of female flowers or pistillate plants, high seed yield with high oil content, early maturity, resistance to pests and diseases, drought tolerance, reduced plant height and high natural ramification of branches. Jatropha is an often cross-pollinated crop and can be improved through mass selection, recurrent selection, mutation breeding, heterosis breeding and interspecific hybridization or biotechnological interventions to bring changes in the desired traits (Divakara et al. 2010).

Assessment of Genetic Variability Using Molecular Markers

Amplified fragment length polymorphism (AFLP) was employed to assess the diversity in the elite germplasm collection of *J. curcas* (Tatikonda et al. 2009). Forty eight accessions collected from six different states of India, were characterised with seven AFLP primer combinations that generated a total of 770 fragments with an average of 110 fragments per primer combination. A total of 680 (88%) fragments showed polymorphism in the germplasm analyzed, of which 59 (8.7%) fragments were unique (accession specific) and 108 (15.9%) fragments were rare (present in less than 10% accessions). In order to assess the discriminatory power of the seven primer combinations used, a variety of marker attributes like *polymorphism information content* (PIC), *marker index* (MI) and *resolving power* (RP) values were calculated (Table 16.4). Although the PIC values ranged from 0.20 (E-ACA/M-CAA) to 0.34 (E-ACT/M-CTT) with an average of 0.26 per primer pair and the MI values were observed in the range of 17.60 (E-ACA/M-CAA) to 32.30 (E-ACT/MCTT) with an average of 25.13 per primer pair, the RP was recognized the real attribute for AFLP to determine the discriminatory power of a primer pair. The RP values for different primer pairs varied from 23.11 (E-ACA/M-CAA) to 46.82 (E-ACT/M-CTT) with an average of 35.21.

Genotyping data obtained for all 680 polymorphic fragments were used to group the accessions analyzed using the UPGMA-phenogram and *principal component analysis* (PCA). Majority of groups obtained in phenogram and PCA contained accessions as per geographical locations. In general, accessions from Andhra Pradesh were found diverse as these were scattered in different groups, whereas accessions from Chhattisgarh showed occurrence of higher number of unique/rare fragments. Molecular diversity estimated in the present study combined with the datasets on other traits will be very useful for selecting the appropriate accessions for plant improvement through conventional as well as molecular breeding approaches.

Mass Multiplication

Seedling quality affects survival, growth and yield of the crop and therefore raising of healthy seedlings is important. Jatropha seedlings can be grown by two methods, namely bare root and container (polythene bag).

In the bare root method, nursery bed is prepared by mixing *farm yard manure* (FYM), soil and sand in equal parts. Soaked seeds are sown at a row spacing of 25 cm and plant to plant spacing of 5 cm. Plants that become ready for transplanting within 6 weeks after germination may be carefully uprooted from nursery beds, wrapped in wet gunny bag and transplanted within 24 h. Before transplanting it should be ensured that enough moisture is available in the pit receiving bare root seedlings.

Seedlings of Jatropha can be raised in poly bags (4″ × 7″, 150 gauge for three to four-month-old seedlings) filled with 2 kg medium comprising equal parts of soil, sand and FYM. Diammonium phosphate (DAP) may be added at 1.0 g per polybag. Good quality seeds having 80% germination should be sown at 1 seed per bag at 2–3 cm depth for getting higher germination success.

The study on evaluation of the effects of propagation techniques in Jatropha in Mali, showed more than 80% survival in plots planted with poly bag seedlings, followed by bare root cuttings as observed after 1 year of plantation. Planting of stem cuttings and direct sowing of seeds in the main field proved less effective and the survival rate was less than 20%.

Benefits of Intercropping in Jatropha During Gestation Period

Because Jatropha takes a minimum of 3–4 years for producing economic yields, the intercrops provide additional income to the farmers during the gestation period. The feasibility of growing drought tolerant field crops as intercrops with Jatropha

Table 16.5 Intercrop yields in strip cropping system with Jatropha at ICRISAT

	Grain yield (t ha^{-1})		
Crop	Year 2005–06	Year 2006–07	Year 2007–08
Sorghum	1.50	–	–
Pearl millet	1.15	1.2	–
Chickpea	1.01	–	–
Sunflower	0.98	–	–
Safflower	0.54	–	–
Soybean	1.06	–	0.51
Pigeonpea	–	0.56	0.62
Mung bean	–	–	0.29

Source: Wani et al. (2009b)

was studied until plantations were three-year-old on Vertisols at the ICRISAT farm, Patancheru, India. Intercrops, such as sorghum, pearl millet, pigeonpea, chickpea, sunflower, safflower, soybean and mung bean were successfully cultivated and evaluated in Jatropha plantations during rainy season and post-rainy seasons. Sorghum, pearl millet, soybean and chickpea yielded more than 1 t ha^{-1} (Table 16.5). The intercrops productivity in terms of grain yield varied from 0.29 tha^{-1} in case of green gram (mung bean) to 1.5 tha^{-1} in case of sorghum.

By-Product Deoiled Cake as Source of Nutrients in Crop Production

Deoiled cake is a by-product of oil extraction that contains all the macro and micronutrients and is an excellent organic fertilizer unlike inorganic fertilizers that supply only few nutrients. Jatropha seeds yield on an average 30% oil and around 70% by-product cake. The cake is suitable for fertilizing plantation or commercial crops. The studies at ICRISAT have indicated that Jatropha cake is relatively rich in nitrogen, phosphorus and sulphur (Table 16.6) compared to leaf and shoot parts.

The contribution of deoiled cake in crop production has been evaluated in maize and soybean crops (Osman et al. 2009; Wani et al. 2006). In an on-farm study by ICRISAT, a replacement of 50% basal N dose in maize through deoiled cake has promoted the maximum yield, which was around 4% more than pure inorganic fertilizers added at a rate of 120:60:40 kg NPK ha^{-1} (Table 16.7). Thus, Jatropha plants grown on degraded lands are useful not only for biofuel, but also for its nutrient rich deoiled cake in order to get higher crop yields with significant cuts on cost and use of chemical fertilizers.

Table 16.6 Chemical composition of deoiled cake from Coimbatore, Tamil Nadu analyzed at ICRISAT, Patancheru, India

Nutrients	Content in doiled cake of Jatropha
Nitrogen (%)	4.91
Phosphorous (%)	0.90
Potassium (%)	1.75
Calcium (%)	0.31
Magnesium (%)	0.68
Zinc (mg kg^{-1})	55
Iron (mg kg^{-1})	772
Copper (mg kg^{-1})	22
Manganese (mg kg^{-1})	85
Boron (mg kg^{-1})	20
Sulphur (mg kg^{-1})	2,433

Source: Wani et al. (2006)

Table 16.7 Effects of Jatropha deoiled cake application on grain yield of maize in Andhra Pradesh, India during 2007

Treatments	Plant height (cm)	Cob length (cm)	Cob diameter (cm)	DMP (kg ha^{-1})	Grain yield (kg ha^{-1})
Absolute control	220	13.44	7.77	11,424	6,640
50% of the basal dressing N (30 kg N ha^{-1}) through deoiled cake	217	17.38	8.04	15,491	9,560
100% of the basal dressing N (60 kg N ha^{-1}) through deoiled cake	228	16.94	8.02	14,193	8,490
100% of the basal dressing N (60 kg N ha^{-1}) through inorganic fertilizer^{-1}	226	16.82	8.12	15,498	9,200
LSD (5%)	NS	1.85	NS	1,240	796

Source: ICRISAT (2008)

C Sequestration and Land Rehabilitation

Carbon sequestration by Jatropha biofuel plantations not only mitigates global warming, but also helps in the reclamation of the degraded lands (Sreedevi et al. 2009; Wani et al. 2006; 2009b). Jatropha plants have a mechanism of drought avoidance by shedding their leaves during moisture stress periods to minimize the water requirement, which proves useful to exploit Jatropha in the rehabilitation of degraded lands. Pruned loppings also add C rich biomass to the degraded soil. In addition to deoiled cake (~500 kg C ha^{-1}), studies have revealed annual addition of around 1,000 kg C ha^{-1} to the soil under Jatropha by means of litter fall and pruned loppings (Table 16.8).

Table 16.8 Total C sequestered through Jatropha plantation as C returned to soil, biodiesel C replacement per year and live plant C

C through Jatropha plantation	Plant part involved	Organic C (kg ha^{-1})
C returned back to soil	Leaf fall	800[a]
	Pruned loppings	150 [a]
	Deoiled cake	495[b]
C replacement in fossil fuel	Jatropha oil	230 [b]
C in live plant	Shoots and roots	5,120

Source: Wani et al. (2012)
[a]Leaf and stem prunings added C every year
[b]Jatropha oil C (Fuel replacement) and deoiled cake added C from fourth year onwards every year

Table 16.9 Microbial population as influenced by Jatropha plantation at Velchal, Andhra Pradesh, India

Microbial parameters	Non-Jatropha plantation soil	Jatropha plantation soil	Coefficient of variation
Bacteria (cfu g^{-1}soil)	8×10^4	1×10^{5a}	54.6
Fungi (cfu g^{-1} soil)	1×10^3	2×10^{3a}	35.6
Actinomycetes (cfu g^{-1} soil)	8×10^2	8×10^{2a}	80.0

Source: Susanna (2009)
[a]Significant at 1%

In addition to C manure through leaf fall, pruned loppings and deoiled cake, the live plants sequester huge C quantities in its above ground and below ground biomass. Studies of Wani et al. (2012) revealed that a plantation at 2×2 m spacing accumulates 4.07 kg $plant^{-1}$ as above ground shoots and 0.81 kg $plant^{-1}$ as below ground roots. Similarly, a plantation at 2×3 m spacing accumulates 5.12 kg $plant^{-1}$ as above ground shoots and 1.02 kg $plant^{-1}$ as below ground roots. This translates to a C sequestration of 6,100 kg ha^{-1} at 2×2 m spacing (2,500 plants) and 5,120 kg ha^{-1} at 2×3 m spacing (1,667 plants). Such huge quantities of C sequestrated in plant biomass is a great ecosystem service we can have and these C credits can be traded, under the Kyoto protocol, with the countries or regions who are not able to manage their C credits (D'silva et al. 2004).

The annual additions of C through added leaves, loppings and decaying roots are enough to give the kick start to microbiological activities and nutrient dynamics. The microbial population is used as an indicator of soil health and thereby land productivity. The studies have revealed a higher level of microbial population complexity under Jatropha plantations (Table 16.9) as compared to non-Jatropha soil.

The falling leaves and loppings are recycle by microbial population, which releases and adds the nutrients to the soil and thereby increases soil fertility. A study from ICRISAT revealed the leaf fall of a three-year-old plant is ~2.6 fold that of a 1 year plant (Table 16.10). The falling leaves in a three-year-old plantation returned around 20 kg each of N and K and around 2 kg of P to the soil.

The cumulative C and nutrients added to soil through Jatropha biomass were documented by Wani et al. (2012). The *in situ* study carried out in a farm at Kothlapur

Table 16.10 Content and amounts of nutrients returned through fallen leaves during 2007 at ICRISAT, Patancheru

			Nutrient recycling by leaf litter					
Age of the plant	Fertilizer dose	Dry leaf (g/plant)	N (%)	P (%)	K (%)	N (kg ha^{-1})	P (kg ha^{-1})	K (kg ha^{-1})
1-year- old	120 g Urea, 170 g SSP, 50 g gypsum	552.5	1.14	0.06	1.1	15.7	0.8	15.2
3-years- old	100 g Urea, 38 g SSP	1,451.1	0.86	0.08	0.95	20.8	2	23

Source: Wani et al. (2009c)

indicated a higher concentration of soil C and available P under the Jatropha plantation as compared to the adjacent uncultivated control grasslands (Fig. 16.5). The C in the surface (0–0.15 m) of soil under four-year-old Jatropha plantation increased by 19% as compared to the adjoining grasslands, which corresponds to about 2,500 kg ha^{-1} additional C sequestered by the plantation in four years. Carbon sequestration in these degraded infertile semi-arid tropical soils serves the dual purpose of reducing the atmospheric CO_2 concentration and increasing the soil organic carbon, which plays a crucial role in soil quality improvement and nutrient availability to plants (Srinivasa Rao et al. 2009). Similarly, available P in surface soil layer under Jatropha increased by as much as fivefold. A positive relationship between soil organic C and available P implies the role of increased organic matter in enhancing P and soil quality as a result of C sequestration (Wani et al. 2003).

Employment Generation and Social Mainstreaming

India is home of 22% of the world poor. Around 221 million poor in India do not have access to a consumption basket; a feature that is considered to define the poverty line. A programme that generates employment is therefore, particularly welcome. The biofuels sector has the potential to successfully rehabilitate the degraded lands and hence to improve the livelihoods of rural people by providing employment and additional sources of income (Wani et al. 2006). Biofuel plantation activity on commercial scale provides employment at village level through plantation, agronomic management, seed collection and through markets for fertilizer, pesticides, fuel and industrial raw material for soap/cosmetics, etc. Developing nations are looking towards biofuels to help reduce their spiraling foreign oil import costs, and to mitigate pollution and global warming. The drylands, often neglected compared to more favorable areas, can contribute importantly to a future powered by biofuels. In India, large tracts of degraded lands not suitable for arable cropping are in the possession of poor farmers. Our challenge and opportunity is to ensure that

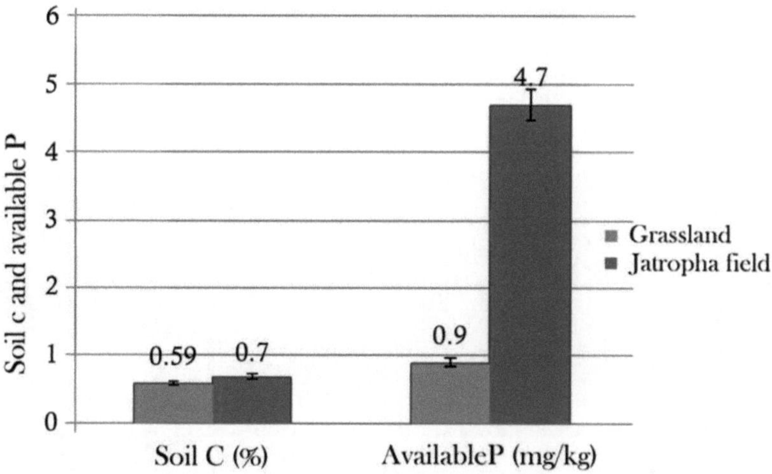

Fig. 16.5 Organic carbon and available P contents in surface (0.0–0.15 m) soil samples under Jatropha plantation in Kothlapur, Andhra Pradesh, India, 2009 (Source: Wani et al. 2011)

poor drylands will not be left behind. The carbon credits gained through switching from fossil fuels to biofuels can also be a source of income to the rural poor as these can be traded with the other developing or developed regions and countries.

A large part of India's population, mostly in rural areas, does not have access to energy services. There is an increasing gap between supply and demand, added with continuous deterioration in quality of power and a low level of access to electricity. Lack of access to affordable energy services among the rural poor seriously affects their chances of benefiting from economic development and improved living standards. Under such circumstances, decentralized power generation using biofuels is the need of the hour. Access to modern decentralized small-scale energy technologies, particularly renewable biofuels, are an important element for effective poverty alleviation policies. A program that develops energy from raw material grown in rural areas will go a long way in providing energy security to the rural people for their domestic use, farm production activities (e.g., irrigation, etc.) and engagement in livelihood activities (e.g., small processing mills) leading to improved incomes, better living and their social main streaming.

Insect Pests and Diseases

Jatropha has been believed to be less susceptible to various insect pests and other diseases merely based on few observations.

Insects, such as the leaf eating beetles, thrips, leaf hoppers, grass hoppers, caterpillars and leaf miner feed on the foliage. Shoot/stem borer and bark eating caterpillar damage the stem. Blue bugs and green stink bug suck on fruits while capsule

Fig. 16.6 (**a**) Shriveled appearance of black rot affected stems, (**b**) black lesions over the stem under the bark (*right*) caused by *Botryosphaeria dothidea,* (**c**) Exudation of reddish brown gummy substance from the late infected stem, (**d**) discoloration of the affected stem, (**e-f**) Initiation of new foliage from affected branches/ground after fungicide spraying in Velchal village plantation, Rangareddy district during July 2009 (Source: Srinivasa Rao et al. 2011)

borer damages the fruits (Wani et al. 2006). The pests may be controlled by spraying endosulfan at 3 ml per litre water or any other pesticide recommended for that particular pest. The galls are formed due to the attack of mites and can be controlled by spraying dicofol at 5 ml of water or wettable sulphur at 3 g per litre of water.

Diseases like root rot, damping off, powdery mildew and leaf spots, cassava mosaic virus are frequently reported when the crop is raised as plantations (Wani et al. 2006; Divakara et al. 2010). Black rot in Jatropha was observed during 2009 and 2010 in plantations in several states in India including Andhra Pradesh, Assam, Chattisgarh and Madhya Pradesh (Srinivasa Rao et al. 2011). Affected plants (Fig. 16.6a–f) showed drying along with shriveling, stem discoloration with sticky reddish brown exudation at the base of plants. Black lesions on the stem under bark and cambium layer were also observed. The symptoms spread to leaves and petioles as 1–3 mm diameter black spots as well as shriveling and gummosis of hard wood, finally leading to death of the infected plant. The causal fungus was identified as *Botryosphaeria dothidea*. Spraying affected plants at early stage of symptoms with bavistin (Carbendazim 50% WP) at the rate of 2 gL^{-1} controlled symptom spread and led to recovery of plants with new leaf growth after the rains.

The studies on regional incidences of insect pests and diseases are necessary for the successful development of an effective package of viable agronomic practices.

Issues Confronting the Upscaling of Biodiesel Production

One of the main problems in getting the biodiesel programme running is the difficulty linked to initiating large-scale cultivation of Jatropha. The success of Jatropha oil for biodiesel production lies in the sustainable and economically viable production of seeds at field level. Farmers do not yet consider Jatropha cultivation remunerative due to the wrong perspective being adopted. Therefore, large-scale cultivation of Jatropha has been put into question and concerns have been raised regarding its success. Actually, the Jatropha cultivation in pilot studies on productive lands has been found to perform badly in terms of net returns when compared with cash crops, which caused serious setback to the program. It should be clear that Jatropha must be grown on lands where cash crops would hardy compete with it in order to warrant its comparative success. Further, diverting arable lands to cultivate such biofuel crops is not a viable proposition, as it will directly affect the food security of a country, which may have further deleterious consequences.

Among other issues challenging upscaling are low yields, lack of high yielding cultivars, high harvesting costs, diseases and pests in block plantations, water balance changes and off-site impacts. Further, in the absence of seed collection and oil extraction infrastructure, it will be difficult to convince entrepreneurs to install transesterification plants. Finally, there is the problem of co-product (glycerol) utilization. The co-product glycerol accounts for about 12% of the biodiesel produced and is of about 88% purity. If alternative means are not quickly found for utilizing glycerol, then its price will plummet due to excess supply.

The farmers are also reluctant in view of:

1. The lack of confidence due to the delay in notifying, publicizing and explaining the government biodiesel policy.
2. The absence of long-term purchase contracts prevents the buy-back arrangements or purchase centers for Jatropha.
3. The lack of availability of certified seeds with higher seed yield and oil content.
4. The lack of incentives and other benefits to farmers.

The government needs to produce measures that seriously address the farmers' concerns. Financial assistance should be given to organizations engaged in developing large-scale training programs for farmers. Subsidized promoting visits of progressive farmers to distant centers of excellence are required to enable them to witness success stories in biodiesel production and propagate their associated know-how.

Acknowledgements The support from National Oilseeds and Vegetable Oils Development, Government of India is duly acknowledged.

References

Achten WMJ, Almeida J, Fobelets V, Bolle E, Mathijs E, Singh VP et al (2010) Life cycle assessment of *Jatropha* biodiesel as transportation fuel in rural India. Appl Energ 87:3652–3660

D'Silva E, Wani SP, Nagnath B (2004) The making of new Powerguda: community empowerment and new technologies transform a problem village in Andhra Pradesh. Global theme on agroecosystems report No 11, ICRISAT, Patancheru, Andhra Pradesh, p 28

Divakara BN, Upadhyaya HD, Wani SP, Gowda CLL (2010) Biology and genetic improvement of *Jatropha curcas* L.: a review. Appl Energy 87:732–742

Garg KK, Karlberg L, Wani SP, Berndes G (2011) *Jatropha* production on watersheds in India: opportunities and trade-offs for soil and water management at the watershed scale. Biofuels Bioprod Biorefining 5:410–430

Government of India (2010) Wasteland Atlas of India, Government of India (GOI), Ministry of Rural Development, Department of Land Resources, New Delhi. Available from www.dolr.nic.in/wasteland_atlas.htm

Grass M, Zeller M, Wani SP, Sreedevi TK (2011) *Jatropha* fuel from India's wastelands: a financial analysis of different *Jatropha* production scenarios linked to possible crude oil price developments. J Fund Renew Energ Appl, Ashdin Publishing

ICRISAT (International Crops Research Institute for the Semi Arid Tropics) (2008) Supporting the Farmers' activities in the value-chain of biofuels, annual project report (October 2007–March 2008), ICRISAT, Patancheru

ICRISAT (International Crops Research Institute for the Semi Arid Tropics) (2010) Harnessing the potential of water-use efficient bio-energy crops for enhancing livelihood opportunities of smallholder farmers in Asia, Africa and Latin America, annual project report (2009–10), ICRISAT, Patancheru

IEA (International Energy Agency) (2007) World energy outlook 2007, China and India insights. International Energy Agency, Paris

Osman M, Wani SP, Balloli SS, Sreedevi TK, Srinivasa Rao Ch, D'Silva E (2009) *Pongamia* seed cake as a valuable source of plant nutrients for sustainable agriculture. Indian J Fert 2(25–6):29–31

Rao YVH, Voleti RS, Hariharan VS, Raju AVS (2008) *Jatropha* oil methyl ester and its blends used as an alternative fuel in diesel engine. Int J Agric Biol Eng 1:32–38

Sreedevi TK, Wani SP, Osman M, Tiwari S (2009) Rehabilitation of degraded lands in watersheds. Proceedings of the comprehensive assessment of watershed programs in India, 25–27 July 2007, ICRISAT, Patancheru, pp 205–220

Srinivasa Rao Ch, Vittal KPR, Venkateswarlu B, Wani SP, Sahrawat KL, Marimuthu S et al (2009) Carbon stocks in different soil types under diverse rainfed production systems in tropical India. Commun Soil Sci Plant Anal 40:2338–2356

Srinivasa Rao Ch, Kumari MP, Wani SP, Marimuthu S (2011) Occurrence of black rot in *Jatropha curcas* L. plantations in India caused by *Botryosphaeria dothidea*. Curr Sci 100(10):1547–1549

Srivastava P (2005) Poverty targeting in India. In: Weiss J (ed) Poverty targeting in Asia. Edward Eigar Publishing Limited, Cheltenham, pp 34–78

Susanna P (2009) Assessing environmental impacts of rehabilitated degraded uplands in watershed with *Jatropha* plantation. M.Sc. dissertation JNTU, Kukatpally, Hyderabad, ICRISAT, Patancheru

Tatikonda L, Wani SP, Kannan S, Beerelli N, Sreedevi TK, Hoisington DA et al (2009) AFLP-based molecular characterization of an elite germplasm collection on *Jatropha curcas* L., a biofuel plant. Plant Sci 76:505–513

Wani SP, Pathak P, Jangawad LS, Eswaran H, Singh P (2003) Improved management of vertisols in the semiarid tropics for increased productivity and soil carbon sequestration. Soil Use Manage 19:217–222

Wani SP, Osman M, D'Silva E, Sreedevi TK (2006) Improved livelihoods and environmental protection through biodiesel plantations in Asia. Asian Biotechnol Dev Rev 8:11–29

Wani SP, Marimuthu S, Sreedevi TK, Srinivasa Rao CH (2009a) Collection, evaluation of germplasm, standardization of agro-techniques and Pilot demonstration for *Jatropha curcas* L. in Rain Shadow Districts of Andhra Pradesh. Annual progress report 2008–09 of RSAD project, ICRISAT, Patancheru

Wani SP, Sreedevi TK, Marimuthu S, Kesava Rao AVR, Vineela C (2009b) Harnessing the potential of *Jatropha* and *Pongamia* plantations for improving livelihoods and rehabilitating degraded lands. In: Proceedings of 6th international Biofuels conference, Winrock International, New Delhi, 4–5 March 2009, pp 256–272

Wani SP, Sreedevi TK, Rockstrom J, Ramakrishna YS (2009c) Rain-fed agriculture—past trend and future prospects. In: Wani SP, Rockstrom J, Oweis T (eds) Rain-fed agriculture: unlocking the potential. Comprehensive assessment of water management in agriculture series. CAB International, Wallingford

Wani SP, Chander G, Sahrawat KL, Srinivasa Rao Ch, Raghavendra G, Susanna P et al (2012) Carbon sequestration and land rehabilitation through *Jatropha curcas* L. plantationin degraded lands. Agric Ecosyst Environ doi:10.1016/j.agee.2012.07.028

Chapter 17
Use of *Jatropha curcas* L. (Non-Toxic Variety) as Traditional Food and Generation of New Products in Mexico

Jorge Martinez Herrera, Cristian Jimenez Martinez, and Norma Guemes Vera

Introduction

Jatropha curcas L. belongs to the family Euphorbiaceae and is considered to have originated from Central America, most probably Mexico. In Mexico, it is found extensively in different regions, namely Sonora, Sinaloa, San Luis Potosi, Guadalajara, Michoacán, Guerrero, Oaxaca, Chiapas, Colima, Tabasco, Yucatán, Quintana Roo, Veracruz, Tamaulipas, Puebla, Hidalgo and Morelos. Some common names for *J. curcas* in Mexico are "piñon", "piñoncillo", "Ashte", "Cak siil", "Chuta", "Chuahuayohuiaxtli", "Cuauyohuatli", "Cuipi", "Que-ca", "Scu-Lu´u", "Tempatl", "Xkakal-che", "Sikil´te", "piñon oil", "Mexican piñon". The edible plants of *J. curcas* grow in several places of Veracruz state, mainly in the north in regions as Castillo de Teayo, Joloapan, Pueblillo, Paso del Correo, Cerro del Carbón, Tlapacoyan, Paso del Progreso, Coyusquihui, Coyutla, Tantoyuca, Tempoal, Chontla, Ilamatlán, Espinal, Papantla, Zozocolco, Misantla, Atzalan y Martínez de la Torre and Puebla state growing in the Sierra nororiental in provenances as Xicotepec, Hueytamalco, Huehuetla, Huitzilan, Xochitlán, Zapotitlán, Zongozotla, San José Acateno, Jonotla, Coatepec, Pahuata and Zoyotla. These regions are called as Totonacapan where the Totonaca culture was developed.

J. curcas occurs between altitudes of 0–1,700 above sea level and in several agro-climatic conditions and plants from the same climatic zone may show morphological differences according to edaphic conditions (Martínez-Herrera et al. 2010).

J.M. Herrera (✉)
Centro De Desarrollo De Productos Bioticos-IPN, Carr. Yautepec-Jojotla km 6. Calle Ceprobi No.8. Col. San Isidro, Yautepec, Morelos, P.O.24.ZIP 62731, Mexico
e-mail: jmartin62@hotmail.com

C.J. Martinez
Escuela Nacional de Ciencias Biologicas-IPN, Dpto Alimentos, DF, Mexico

N.G. Vera
Universidad Autonoma de Hidalgo, Hidalgo, Mexico

N. Carels et al. (eds.), *Jatropha, Challenges for a New Energy Crop: Volume 1: Farming, Economics and Biofuel*, DOI 10.1007/978-1-4614-4806-8_17,

Recently, *J. curcas* has attracted attention of various research organizations, governments, public and international developmental agencies and industries in the tropics and sub-tropics due to its adaptability to semi-arid marginal sites, the possibility of using its oil as a diesel fuel substitute and its role in erosion control. Recently, this species has attracted immense attention in Mexico as well. About four years ago, the Mexican Government initiated a program in which CONAFOR (Consejo Nacional Forestal or National Board of Forestry), MEXICO pays US$ 570 per hectare to a farmer for planting *J. curcas* with a population density of 1,600 plants per hectare (ha). Some states that currently have planted more than 5,000 ha of *J. curcas* have accessions from Michoacan (non-toxic), Veracruz (non-toxic), Chiapas (toxic), Puebla (non-toxic) and San Luis Potosi (toxic) and the plantations would substantially increase in the coming years. The land used for planting *J. curcas* has been the temporarily idle land without any agricultural use or uneven and sloping lands, because the irrigated lands are used for planting staple crops such as corn, beans, sugar cane, sorghum and rice. This year, for the first time, we observed that the leaves, stems and flowers of the non-toxic *J. curcas* were browsed by cows and horses in the dry months in the state of Morelos, Mexico suggesting its non-toxic nature. It is worth noting that toxic genotypes of *J. curcas* have been planted in several countries as a live fence since the leaves and other parts of the plant are not consumed by livestock even in peak summer months when other grass species are sparse. In some places in Mexico, *J. curcas* has been intercropped with beans, peanut, chilli, coffee, chia, medicinal and aromatic plants for the first 2 years, giving additional productive outputs. In addition, *J. curcas* fruits are used as compost or to generate biogas. The seeds from the non-toxic genotype have the added advantage that the press cake can be used as animal feed or as human food (Martínez-Herrera et al. 2010). The fruits of the non-toxic accessions are oval unlike those of the toxic variety which are round to spherical (Fig. 17.1). This difference is important to assist in the determination of a non-toxic variety characterized by the absence of phorbol esters.

Tolerance to Desiccation and Chemical Characterization of *J. curcas* Collected in Totonacapan

Mexican Jatropha nut (*J. curcas*) is a vitally important species, given its potential as an alternative bioenergetic and food source. However, there is little or no information for the conservation of its germplasm. At best, there are reports on collections of genotypes that have high levels of toxicity when eaten. In order to offer alternatives for its conservation and use, in this study the nut (seeds) were evaluated for the degree of tolerance to desiccation and longevity, as well as their proximate chemical analysis and anti-nutritional components from seeds collected in the region of Totonacapan (Puebla and Veracruz, Mexico). To define desiccation tolerance, we used a mixture of seeds from three climatic groups; it was subjected to a

Fig. 17.1 *J. curcas* fruits from Mexico non-toxic (*left*) and toxic (*right*) varieties showing differences in shape and size

12 and 5% desiccation and stored for 3 and 6 months at −18°C. Moreover, seeds with 14, 8 and 5% humidity were stored for 3–6 months in hermetically sealed jars in a cold room at 5°C, in Ecatlan, Puebla (mean annual temperature >18°C, Relative humidity 60%). For the proximate chemical analysis, 30 collections from three different climatic groups were used; the seed flour (without husk/testa) was measured for total raw protein, lipids, ash, composition of fatty acids and anti-nutritional compounds (inhibitors of trypsin, phytates, saponins and phorbol esters). When the humidity content of *J. curcas* seeds was reduced to 5%, the decrease in viability was upto 29%, the germination rate upto 70% and the electrical conductivity upto 10 μS cm^{-1} g^{-1}.

No changes were detected in these variables after 6 months of storage at −18°C. The increase in germination was probably due to a reduction in seed dormancy during the drying process. In the case of seeds stored with 14% humidity in Ecatlan, the decrease in viability was upto 51%, the germination rate upto 52%, and the electrical conductivity upto 28 μS cm^{-1} g^{-1} after 6 months while those stored with 5% humidity were unaffected in any of the variables under different storage conditions. As for the chemical analysis, raw protein varied between 20 and 28% and the fatty acid content ranged between 57% and 69%. Among the saturated fatty acids, the most outstanding ones, based on their concentration were palmitic acid and stearic acids and among the unsaturated acids were oleic and linoleic acids. Most accessions showed values that were undetectable or below the range considered as non-toxic in compounds that inhibit trypsin, phytates, saponins and phorbol esters. No significant effect was detected from the climatic group on the evaluated variables in the year of collection, but significant variation was detected to allow the selection, evaluation and genetic improvement of *J. curcas* to increase the usefulness of this species.

It may be concluded that *J. curcas* seeds can be considered as orthodox. Seed longevity is better maintained at 5% humidity and temperatures below 5°C (Baustista-Ramirez 2010). The evaluated collections showed a higher concentration of proteins and lipids, which represent no risk of toxicity, and can be used as a food both for humans and animals and for use in food industry as well.

Traditional Uses

Based on the empirical knowledge that the seed was used to prepare various dishes and the study of nutritional goodness, it was thought to develop a number of food products from seeds and flour with a beneficial impact on society and particularly in highly marginalized communities where *J. curcas* grows natively.

There are apparently no morphological differences between toxic and non-toxic accessions, except the fruit shape and size, which is why the introduction of toxic genotypes in Totonacapan regions would put at risk an ancestral legacy as well as the loss of biodiversity and the possible poisoning of the residents of these regions by eating toxic seeds mistaken for edible seeds.

The traditional dishes are consumed 1–2 times per week. During holidays, they are consumed on a larger scale to commemorate the day of the dead in Mexico. After roasting the seeds, the farmers add salt, tomato sauce, chilli and prepare a rich snack. The pulacles are bean tamales, wrapped in dried corn husks, the dough is prepared first, followed by the preparation of the baked beans and *J. curcas* seed is ground with onion, zucchini, chayote, added to corn husk, and cooked (Fig. 17.2).

To prepare the pipian, *J. curcas* seeds are roasted, ground and then combined with sesame and pumpkin seeds. To enhance flavour, cut onions and chilli paste is added to the chicken and the preparation is ready for eating.

Besides high quality protein, sulphur amino acids and easy protein digestibility, the promotion of Jatropha can generate additional jobs in the food sector and rescue its Mexican use, which has been replaced by fast food of little nutritional value.

Preparation of Different Foods for Human Consumption

J. curcas *Flour Addition to Wheat Flour Dough and Their Effects on Rheological Properties*

The purpose of this investigation was to study the effects of addition of *J. curcas* flour into wheat flour to elaborate dough with improved protein and better rheology. Dough was prepared by mixing *J. curcas* flour (JF) at rates of 5%, 10%, 15%, 20% or 100% with wheat flour (WF). The protein content of JF was 63.3%. The extensographs showed that increasing the JF percentage from 0% to 20%, the dough became less extensible as indicated by higher ratios of R50/Ex, while the area under the

Fig. 17.2 Traditional Mexican foods from *J. curcas* seeds

curve (i.e., the energy required to break the strength of dough) increased substantially. This indicates that the dough of the blends are still strong and elastic. However, when the amount of JF was raised upto 100%, the dough became very weak, the stability and development time decreased as well as its extensibility and resistance. The adhesiveness decreased particularly in samples prepared with 20% and 100% of JF. The presence of proteins resulted in a decrease in dough firmness and consistency and an increase in its cohesiveness, which favours the production of a high quality product. The best blend for the elaboration of quality bread was obtained with the mixing of 5% JF (Cruz-Villegas et al. 2007).

Evaluation of the Effects of JF on the Quality of Cookies

JF has an important nutritional value for the bread products (baking industry). The aim of the present work was to characterize the dough texture, adhesiveness and extensibility of cookies amended with JF, for which the proximate analysis was done.

The dough was prepared with WF amended with various levels of JF (2.5%, 5%, 7.5%, 10% and 12.5%). Later the rheological analyses (TPA, adhesiveness and extensibility) were performed by using a TA.XT2i texture analyzer (Stable MicroSystems Ltd, Surrey, UK) in compression mode. The chemical composition of the flour was 6.5% protein for WF and 25% protein for JF. The cookies amended with 10% of JF had 24.6% protein content as compared to 6.4% in the regular product, which agreed with other results reported in the literature. The addition of 10% and 12.5% protein content promoted a decrease in firmness and consistency as well as an increase in cohesiveness of dough. Generally, higher amounts of precipitate (30%) did not significantly affect the firmness, consistency or cohesiveness of dough. The adhesiveness increased particularly in samples prepared with 10% and 12.5% of JF. Presence of JF entailed an excellent firmness and consistency of cookies and an increase in their cohesiveness, which favours the production of a high-quality product. Use of JF amended cookies will favor improved protein intake and human nutrition (López et al. 2008).

Sensory and Rheological Properties of Noodles Enrichment with JF

Pasta made from semolina is a good source of complex carbohydrates. Pasta contains relatively high amounts of starch and is low in fat and protein; an alternative to compensate these deficiencies is the use of oilseeds such as *J. curcas*. Therefore, we prepared pasta from semolina enriched with 5%, 10%, 15%, 20% and 25% of JF. The protein content was 13.62% (compared with 11% for commercial pasta) for noodles with 10% of JF and a lysine content of 6.5% (compared with 2.3% for control). Sensory analysis of enriched noodles with 10% of JF was considered as the best. The extensibility of noodles prepared with 10% of JF was 35.3 cm and the R max was 38.9. These results were similar to those of control noodles. The firmness of noodles with 10% JF (0.55 Kf) was similar to control noodles (0.51 Kf). The b* value decreased when percentage of JF was increased. The best treatment was that including 10% JF. (Guemes-Vera et al. 2009).

JF Addition to WF Tortillas and Their Effects on Textural Properties

JF has excellent sulfur amino acid composition and rate of *in vitro* digestibility, which certainly could be an important ingredient in the complementation of different foods and even surpassing protein complementation by beans, corn, soybeans and wheat. The aim of this study was to evaluate the defatted *J. curcas* meal as an ingredient to fortify flour.

Table 17.1 Flour analyses in wheat and *J. curcas*

Sample	Moisture (%)	Protein (%)	Oil (%)	Ash (%)
Wheat	14.84±0.5	10.61±0.1	0.46±0.0	0.30±0.1
J. curcas	7.2%±0.2	31.1±0.1	57.8±0.1	4.7±0.2

The following formulations were used: (1) WF 100%, (2) WF 97.5% and Jatropha meal (HJc) 2.5%, (3) WF 95% and HJc 5%; (4) WF 92.5% and HJc 7.5%; (5) WF 90% and HJc 10%; (6) WF 87.5% and HJc 12.5%; (7) WF 85% and HJc 15%; (8) WF 80% and HJc 20%. These formulations were tested for dough extensibility, dough stickiness, hardness of mass, mass texture profile, determination of protein and fat as per AOAC and sensorial evaluation (AOAC 2005).

The results of proximate analyses of whole flours of wheat and *J. curcas* are shown in Table 17.1. The protein content of WF is similar to that found in other samples of wheat (11.0%) belonging to the durum group; the value of fat was low compared with that reported by Guemes-Vera et al. (2009). The protein content (31%) of HJc was lower than that reported for soybeans, but higher than that of sunflower and cotton. However, HJc oil content (58%) was higher than that reported for soybeans (Villafuerte et al. 2008).

***Extensibility masses*:** Wheat flour (formulation 1) with 660.87 gf was found to be the best followed by formulation 2 (616.7 gf) and formulation 4 (703.84 gf); formulation 8 with 274.6 gf was the worst. Thus, the higher the concentration of HJc in dough, the lower its extensibility.

***Adhesiveness of mass*:** Formulation 2 (21,020 gf) was the closest to the control value (20,247 gf). The adhesion force of a body with a surface is the work necessary to overcome the force of attraction between this body and that surface, which means that the adhesion decrease with the increase of bond strength, i.e., with the HJc proportion.

***Hardness test masses*:** This test showed a value of 2,282.3 gf for the control, which is very close to formulations 2 (2,584.6 gf) and 3 (2,286.9 gf), meaning the force required to achieve a given deformation. It was observed that the dough hardness increased in proportion to its HJc content compared to the control sample.

***Texture profile analysis*:** Formulation 2 (142.13 gf) showed higher hardness when compared with the control (276.39 gf). This can be explained by the fact that the added proteins have an effect on the dough hardness due to their hydration, but without affecting the mass.

Use of *J. curcas* Non-Toxic Press Cake as Animal Feed

***Balanced feed for poultry*:** For this experiment a total of 300 one-day-old chicks were used according to 5 replicates of 20 chicks, which were randomly assigned for a total of 100 chicks per treatment. By mixing corn, soya and *J. curcas* press cake with a commercial diet, we obtained the following treatments: *J. curcas* press cake

Table 17.2 Average weight gain (g) of various treatments in chickens

Week	T1	T2	T3
0	104.26	109.06	103.99
1	237.86	224.95	216.47
2	511.99	400.86	396.91
3	870.37	635.99	646.61
4	1,283.59	869.71	985.76
5	1,721.66	1,167.62	1,243.52
6	2,228.82	1,476.36	1,609.47
7	2,643.64	1,550.45	1,870.43
8	2,923.49	1,662.85	2,125.65

Average of 5 replicates

(T1), corn-soya blend (T2) and *J. curcas* (20%)+(soybean) (T3). The experiment lasted for 2 months Table 17.2. The results show the average body weight gained by chicks according to the three treatments tested. Based on the results, it may be concluded that treatment No.3 is not significantly different from treatment No.1. Therefore, the press cake of *J. curcas* could be used effectively for monogastric and ruminant feeding, which will allow the achievement of the full cake utilization of non-toxic accessions of *J. curcas*. In addition, mixing soybean meal and *J. curcas* cake will ensure partial replacement of soybean meal in animal feed and, thus, allow the decreasing their feeding costs.

Existing Commercial Cultivars and Availability

During the past four years, Mexico has initiated work on the establishment of *J. curcas* plantations in the first instance to obtain seeds and secondly with the purpose of obtaining short-term biodiesel. Some states have chosen the toxic accessions as they are more resistant to pests. However in states where non-toxic accessions are naturally present, the preference is given to them since the seed cake can be utilized as food for animals besides their traditional use in human alimentation. The possibility of using residual products will give added value to the marketing of oil, not excluding of course the use of shell as a biofertilizer. By 2010, the cultivation of this variety in Mexico was about 10,000 hectares, which is expected to double by 2012.

References

AOAC (2005) Official Methods of Analysis Association of Official Analytical Chemists. Washington DC, Chap 4 pp 24–42

Baustista-Ramirez E (2010) Tolerance to desiccation and chemical characterisation of Mexican pine nut (*Jatropha curcas* L.) collected in Totnacapan. Master thesis, Colegio de Postgrauados, Chapingo

Cruz-Villegas R, Martínez-Herrera J, Soto-Simetal S, Perez-Soto E, Hernández-Soto A, Guemes-Vera N (2007) *Jatropha curcas* L. flour addition to wheat flour doughs and their effects on rheological properties. In: AACC international annual meeting, San Antonio, 7–10 Oct 2007

Guemes-Vera N, Flores-Miranda G, Gómez-Montes E, Soto-Simental S, Martinez-Herrera J (2009) Sensory and rheological properties of noodles enrichment with *Jatropha curcas* flours. In: AACC international annual meeting, Baltimore, 13–16 Sept 2009

López L, Martínez-Herrera J, Simental S, Soto E, Chavez J, Fuentes A et al (2008) Evaluation of the effects of *Jatropha curcas* L. flours on the quality of cookies. In: AACC international annual meeting, Honolulu, 21–24 Sept 2008

Martínez-Herrera J, Martínez Ayala AL, Makkar H, Francis G, Becker K (2010) Agroclimatic conditions, chemical and nutritional characterization of different provenances of *Jatropha curcas* L. from Mexico. Eur J Sci Res 39:396–407

Villafuerte I, Martínez-Herrera J, Chávez J, Santamaria M, Fuentes A, Güemes-Vera N (2008) *Jatropha curcas* L. flour addition to wheat flour tortillas and their effects on textural properties. In: AACC international annual meeting, Honolulu, 21–24 Sept 2008

Chapter 18
Jatropha Seeds Oil and Products: Important Properties with Respect to Uses

George Francis

Introduction

Jatropha curcas L. is a shrub or small tree belonging to the family Euphorbiaceae. It is a succulent and is adapted to long periods of droughts and soils low in nutrients. The origin of the plant is thought to be Central and South America. The plant has now become naturalised in most tropical countries of the world. It is being targeted by several countries as a potential source of indigenous bio-energy production. Jatropha is being favoured under the premise that it can thrive on degraded lands that can otherwise not be used for conventional agriculture and requires comparatively less water. Such lands are available in large areas in many tropical countries at a low price.

Even though *J. curcas* has been investigated for quite some time, the kind of comprehensive research and development efforts that would result in the generation of critical information concerning its profitability for the different climatic and edaphic regions have started only recently. Results of such research are trickling in slowly. The present paper attempts to review the existing information on the important quality aspects of *J. curcas* seeds, oil and other by-products such as seed cake relevant to their commercial use.

G. Francis (✉)
Live Energies GmbH, Scharnhauser Strasse 35, Stuttgart D70599, Germany

JATROPOWER Bio-trading Pvt Ltd, 59/28A Bharathi Park 8th Cross Road, Saibaba Colony, Coimbatore 641 011, India
e-mail: george.francis@liveenergies.de

N. Carels et al. (eds.), *Jatropha, Challenges for a New Energy Crop: Volume 1: Farming, Economics and Biofuel*, DOI 10.1007/978-1-4614-4806-8_18,

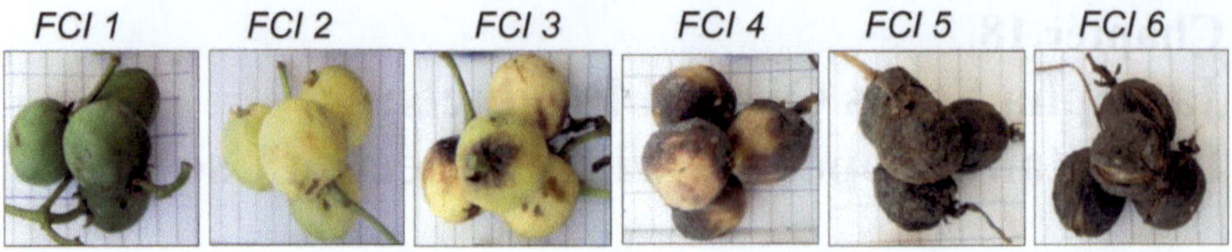

Fig. 18.1 Different stages of maturity in *J. curcas* fruit (After Nuss 2007)

Jatropha Fruits

In practice, at least on small farms, *J. curcas* fruits are picked up from the ground when they fall down. They can also be plucked from trees when the fruits are deep brown or dry. This results in a mix of fruits with different moisture contents and possibly fungal infection. The oil extracted from such mixed collections have a comparatively low quality and have a *free fatty acid* (FFA) content of up to 12% (unpublished data). To maintain high quality of the seeds and oil, *J. curcas* fruits are preferably harvested when they are yellow in colour (FCI 2 and 3 in Fig. 18.1). The plucking force relative to the weight of the fruit is lowest also at this stage. Once harvested, it is best to peel the fruit sheath to expose the black seeds straightaway. The fresh fruit sheaths can be put back as mulch for *J. curcas* plants. The seeds are then to be dried under sunlight on a clean, dust free surface. Under good sun, the moisture content of seeds comes down to below 7% after 3 days.

Alternatively, if there is market demand for the dried husk as fuel, the whole fruits can be dried under sun for 4–5 days and stored as such. The dried fruits can be dehusked before the seeds are to be crushed to extract oil. The dry husk has good burning properties and has an energy content of between 14 and 17 MJ kg^{-1} on a dry matter basis (own unpublished data; Vyas and Singh 2007)

Seeds having less than 7% moisture content can be stored in moisture free conditions for between 3 and 6 months without affecting quality. The time of storage depends on the atmospheric humidity. Dry *J. curcas* seeds can be stored without damage for more than a year if the storage conditions are sterile, cool (around 4°C) and dark.

Evaluation of dried fruits from 17 *J. curcas* provenances showed that the dry fruit weights ranged from 1.5 to 2.8 g and the percentage of seeds between 0.64 and 0.73 (fruits from the second year harvest processed similarly for all provenances, moisture content around 7%, provenances collected from different agro-climatic and geographical zones, cultivated at the Jatropower research farm near Coimbatore; unpublished results).

The FFA content of seed oil extracted from seeds processed as above was almost always less than 1% for all the above 17 provenances (unpublished data) showing their good quality.

Table 18.1 Origin of *J. curcas* seeds and agroclimatic conditions of sites from where they were collected

Origin	Altitude (msl)	Average annual rainfall (mm)	Average temperature (°C)	Climate type
San Jose Acateno, Veracruz	80	1,400	24.0	A(w)
Tenampa, Veracruz	980	920	27.0	A(w1)
Coatzacoalcos, Veracruz "non Toxic"	10	2,500	25.6	Am
Huitzilan, Puebla	900	2,021	18.0	Acf
Xochitlan, Puebla	1,040	1,400	24.0	Acf
Suchiapa, Chiapas	440	1,186	24.4	A(w)
Villaflores, Chiapas	560	1,209	24.3	A(w1)
Cuautla, Morelos	1,300	856	22.6	Aw°
Comalcalco, Tabasco	40	2,675	26.4	Am
Tlaxmalac, Guerrero	940	911	25.0	A(w)
Costa Chica, Guerrero	50	1,200	25.0	A(w1)
Chiapa de Corzo, Chiapas	450	990	26.0	A(w)
Tlapacoyan, Veracruz	430	1,500	18.0	A(w)
Tejabán, Nuevo Urecho, Michoacan	483	853	24.7	A(w)
La Ordeñita, Tepalcatepec, Michoacan	304	621	24.8	A(w1)
San Isidro, Tepalcatepec, Michoacan	402	621	24.8	A(w1)
Corona, Periban, Michoacan	1,420	1,366	28.2	A(w1)
La Cortina, Gabriel Zamora, Michoacan	474	853	26.0	A(w)

A(w) = hot sub-humid region with rains in summer, Am = hot humid climate with abundant rains in summer, Aw° = semi-hot, sub-humid climate with rains in summer, A(w1) = hot sub-humid region with rains in summer, Acf = semi-hot humid climate with rains all year
Source: Herrera et al. (2010)

Table 18.2 Physical characteristics and oil content of seeds of different provenances from Mexico

Origin of the accessions	Average seed weight (g)	Kernel weight (% of seed)	Shell weight (% of seed)	Oil (%)
San Jose Acateno, Veracruz	0.73	68.9	31.1	58.3
Tenampa, Veracruz	0.74	65.9	34.1	57.4
Coatzacoalcos, Veracruz. "non toxic"	0.79	69.7	30.3	52.6
Huitzilan, Puebla	0.68	66.9	33.1	64.5
Xochitlán, Puebla	0.72	67.2	32.8	57.1
Suchiapa, Chiapas	0.83	74.4	25.6	60.4
Villaflores, Chiapas	0.67	73.7	26.3	45.9
Tlaxmalac, Guerrero	0.73	61.7	38.3	57.7
Cuautla, Morelos	0.61	65.9	34.1	58.7
Comalcalco, Tabasco.	0.71	66.5	33.5	56.3
Costa Chica, Guerrero	0.74	62.8	37.2	58.7
Tlapacoyan, Veracruz	0.83	69.5	30.5	56.1
Chiapa de Corzo, Chiapas	0.72	67.2	32.8	55.3
Tejabán, Nuevo Urecho, Michoacán	0.68	64.0	36.0	53.5
La Ordeñita, Tepalcatepec, Michoacán	0.64	63.2	36.8	51.4
Corona, Periban, Michoacán.	0.46	63.8	36.2	51.3
San Isidro, Tepalcatepec, Michoacán	0.67	63.9	36.1	48.9
La Cortina, Gabriel Zamora, Michoacán	0.70	67.1	32.9	51.1

Source: Adapted from Herrera et al. (2010)

Jatropha Seed Composition

The composition of *J. curcas* seeds vary according to provenance, soil characteristics, and climatic conditions on which *J. curcas* plants are grown. The analysis of the physical and chemical characteristics of a collection of *J. curcas* seeds from different climatic conditions in Mexico illustrate this point (data taken from Herrera et al. 2010). Table 18.1 shows the climatic conditions under which the mother plants were grown. Table 18.2 displays the physical characteristics and oil contents of the collected seeds. The variability of the different seed characteristics is obvious from Table 18.2 with seed weights varying from 0.46 to 0.83 g, kernel percentage from 61.7 to 74.4% and kernel oil content from 45.9 to 64.5%. This can be taken to be a representative range for most *J. curcas* provenances. The seed weights of plants growing on very poor soils and under water stress tend to be generally smaller in size. Although quantitative data are lacking, the seeds collected in the rainy season, from July to September are bigger than those collected during the period, November to December (Herrera et al. 2010).

The seed weights of different provenances collected from different agroclimatic and edaphic conditions and grown under similar conditions will also show differences as seen in Table 18.2. Unpublished data from the Jatropower research station showed that seed weight, kernel percentage and kernel oil content varied between 0.41 and 0.72 g, 0.56–0.67% and 42–63%, respectively for 17 provenances with different geographic, climatic and edaphic origins when grown on a common site.

Seed Processing

Dry *J. curcas* seeds are currently crushed using screw presses. The whole seeds with shell are crushed as deshelled kernels alone form a pasty mass that comes through the screw as it is without complete separation of the oil. The seeds are sometimes warmed to about 40°C to make the oil more mobile before crushing. The percentage of oil extracted depends on the oil content of the seeds, but generally the residual oil content of the seed cake after crushing is between 6% and 10% (G. Francis, unpublished data). In practice, 4 kg of dry *J. curcas* seeds generally yield 1 kg of oil and 3 kg of seed cake in small scale screw presses. The oil produced by screw pressing is of sufficiently high quality for conversion to biodiesel.

Until now there has been no attempt to extract the oil through solvent extraction for *J. curcas* seeds. It is economically viable only on a large scale (Adriaans 2006) and such quantities of *J. curcas* seeds are not available currently.

Table 18.3 Important properties of *J. curcas* seed oil

Property	Unit	Value	Limits of DIN V 51605
Density	G cm^{-3}	0.85–0.92	90–93
Kinematic viscosity	cSt at 40°C	34.5	Max. 36
Lower calorific value	MJ kg^{-1}	37.1	Min. 36
Cloud point	°C	4	–
Pour point	°C	1	–
Flash point	°C	256 (186,210)[a]	Min. 220
Acid value	mg KOH g^{-1}	0.48–11	Max. 2
Iodine value	–	95–105	95–125
Sulphur content	mg kg^{-1}	1.6–2000	Max. 10
Oxidative ash content	%	0.02–0.7	Max. 0.01
Oxidation stability	h at 110°C	Up to 16	Min. 6
Cetane number	–	Up to 55	Min. 39
Phorbol esters	mg g^{-1}	3.1–8	–

[a]other literature values; data from own unpublished results, Jindal et al. (2010), Singh and Padhi (2009), Makkar et al. (2009)

Oil Characteristics and Composition

Some of the key properties of screw pressed *J. curcas* oil are presented in Table 18.3. It can be seen that the values of different characteristics vary in a range owing to the differences in seed quality, type of screw press and filter type. Under optimized conditions it is possible to obtain *J. curcas* oil that conforms to the German Pre-Standard for vegetable oils to be used as motor fuel (this standard is the basis for the European standard for vegetable oils as motor fuel that is currently being formulated with some changes and additional requirements) except for ash content. In addition to physical filtering, a chemical adsorbent filtering may be required to reduce the content of elements such as P, Ca and Mg in order to bring down the ash content.

Phorbol esters are toxic substances found in *J. curcas* oil. This is dealt with in detail in a different section below.

The quality of *J. curcas* oil as a feedstock for biodiesel is also determined by the composition of fatty acids. Since standardized *J. curcas* seeds are currently not available there are different compositions (Table 18.4) depending on the origin of the seeds and probably influenced by climatic and environmental conditions under which the mother plants are grown (King et al. 2009).

As biodiesel feedstock, *J. curcas* oil with a higher concentration of oleic acid is desirable as the increase in double bonds reduces the oxidative stability and consequently, the keeping quality of the oil. High oleic acid content also increases the cetane number of the oil. The cetane number is a measure of delay in the combustion of the fuel from ignition and increase in values up to 55 increases the quality of the oil (cross references in King et al. 2009). On the other hand an optimal mix of unsaturated fatty acids is required to ensure higher fluidity and hence, winter suitability for colder regions. Screw pressed *J. curcas* oil, apparently produced from mixed seed lots has been found suitable as a biodiesel feedstock even when the conversion is done in a

Table 18.4 Fatty acid composition of *J. curcas* seeds collected from different sites in Madagascar

Site	Elevation (m)	% fatty acid composition								Calculated CN
		16:0	16:1	18:0	18:1	18:2	18:3	20:0	Others	
Marovoay, Mahajanga	5	14.3	0.7	7.5	51.7	25.3	0.3	0.3	<0.1	55
Ampitolova, Mahajanga	26	14.8	0.8	7.6	48.5	27.8	0.2	0.2	<0.1	55
Kianjavato, Vatovavy Fito Vinany	61	15.5	0.9	6.0	37.4	39.8	0.2	0.2	<0.1	52
Ambohikambana, Moyen Ouest	903	14.9	1.0	5.5	38.1	40.1	0.2	0.2	<0.1	52
Amparaky, Moyen Ouest	1,145	13.4	0.8	6.6	38.6	40.2	0.2	0.2	<0.1	52
Andasy, Soavina	1,084	15.4	1.0	5.4	36.2	41.5	0.2	0.2	<0.1	52
Ankasina, Lac Alaotra	833	15.2	1.0	5.3	35.2	42.8	0.2	0.2	<0.1	51
Trajavona, Ambalavoa	957	15.0	1.0	5.3	34.9	43.3	0.2	0.2	<0.1	51
Fitamatsina, Soavina	1,104	15.0	1.0	5.3	35.0	43.3	0.3	0.2	<0.1	51
Mahavanona, Ambalavoa	964	15.5	1.1	5.3	33.8	43.9	0.2	0.2	<0.1	51

Fatty acids abbreviations; 16:0, palmitic acid; 16:1, palmitoleic acid; 18:0, stearic acid; 18:1, oleic acid; 18:2, linoleic acid; 18:3, linolenic acid; 20:0, arachidic acid. *CN* Cetane number; Table adapted from King et al. (2009)

Table 18.5 Phorbol ester content (expressed in mg $g^{-1}\pm SD$, n=3) of the different fractions obtained during pre-treatment and transesterification of three different *J. curcas* oil samples

	Toxic seed samples		
Parameters	Solvent extracted	Screw pressed	Non-toxic
Crude oil	3.10±0.25	3.77±0.03	ND
Degummed oil	2.48±0.24	3.62±0.19	ND
Acid gums	2.02±0.07	3.35±0.00	ND
Wash water	2.72±0.01	2.08±0.48	ND
Silica-treated oil	2.51±0.33	3.76±0.50	ND
Stripped oil	ND	ND	ND
Fatty acid distillate	ND	ND	ND
Biodiesel	ND	ND	ND
Crude glycerine	ND	ND	ND
Biodiesel wash water	ND	ND	ND

ND not detected; Reproduced from Makkar et al. (2009)

batch processor (own unpublished data). The *J. curcas* oil methyl ester (*J. curcas* biodiesel) produced from such oil has been found to be having the quality requirements prescribed in the European standard EN 14214 (Mandpe et al. 2005).

Phorbol Esters and Toxicity of Jatropha Oil

Phorbol esters are the principal toxic substances present in the toxic variety of *J. curcas* seeds and oil making these inedible. Phorbol esters are toxic when consumed by humans and animals and could also act as carcinogens promoters (Goel et al. 2007). Non-toxic edible varieties of *J. curcas* exist in different regions of Mexico (Herrera et al. 2010). These are often consumed by the local population after cooking (heat destroys other antinutrients in *J. curcas* seeds such as trypsin inhibitors and lectins, e.g., curcin). A comparative analysis of seeds from edible and non-edible varieties revealed that edible seeds lacked phorbol esters (Makkar et al. 1998) giving evidence of their involvement in the toxicity of *J. curcas* seeds.

Phorbol esters are present in concentrations varying from 3.1 to 8 mg g^{-1} in *J. curcas* oil (Makkar et al. 2008). The presence of phorbol esters makes *J. curcas* oil unsuitable for food and feed applications. Even though phorbol esters are not contact poisons, their presence necessitates additional precautions during the processing of *J. curcas* seeds and oils. Care is to be taken that they do not enter the body either through food contamination or through wounds, etc. Makkar et al. (2009) studied the fate of phorbol esters during oil processing and transesterification into biodiesel. The results are reproduced in Table 18.5.

The study found that chemical refining of oil (degumming, neutralization, silica/bleaching, mild deodorization/stripping) as it normally occurs at low temperature does not affect phorbol ester content in the resulting oil. In physical refining (degumming, silica/bleaching, deodorization/stripping at 240–260°C and under vacuum), the deodorization conditions are much more severe and lead, leading to phorbol ester degradation to different levels. Stripping at 200°C did not result in phorbol ester degradation (Haas and Mittelbach 2000), but stripping above 260°C resulted in complete removal of phorbol esters. The biodiesel and glycerol produced through the process outlined in Makkar et al. (2009) were free of phorbol esters; however, these could contain phorbol ester degradation products. At present, no information is available on the nature or toxicity of the degraded products. Research work is currently being conducted on this aspect (Makkar et al. 2009).

Phorbol esters themselves are commercially valuable compounds and find use as a pharmacological tool for the investigation of biochemical processes, such as carcinogenesis and also in many agricultural applications, such as pesticides, molluscicides, insecticides, bacteriocides and fungicides (cross references in Devappa et al. 2010b). *J. curcas* oil could act as a valuable source of this substance. Extraction of a majority of phorbol esters from the oil using methanol did not affect the quality of the oil or biodiesel prepared from it (Devappa et al. 2010a).

Seed Cake Composition, Quality and Potential Uses

Currently, *J. curcas* oil is produced almost exclusively by crushing whole *J. curcas* seeds in small-scale screw presses. The oil produced after suitable filtration procedures are qualitatively adequate for use as pure plant oil fuel or as biodiesel feedstock. *J. curcas* seeds from 17 different provenances grown under similar conditions showed a shell content (%) of between 33.5 and 45.9 and a kernel content of between 55.6 and 66.5 (G. Francis, unpublished data). If an average of 40% shell, 60% kernel and 35% oil content is considered and an oil extraction efficiency of 80% in screw pressing, the seed cake that remains will have an average composition of 56% shell, 35% pressed kernel and 10% residual oil on a dry matter basis. Singh et al. (2008) found 93.8% *total solid* (TS) out of which 92.5% was *volatile solid* (VS) in screw pressed seed cake. The important properties of the seed cake are presented in Table 18.6.

The *J. curcas* seed cake is also relatively rich in important micronutrients, such as Cu, Fe, Mn and Zn (Gaind et al. 2009; Brittaine and Lutaladio 2010). Even if cake from the edible variety of *J. curcas* is considered with non-detectable levels of toxic phorbol esters, this seed cake is not suitable as a feed ingredient for any animal. This is because of the high shell and lignin content that is undigestible for animals. This point is to be especially noted as many publications mistakenly mention the potential of *J. curcas* "seed cake" as an animal feed if the phorbol esters are removed and the cake is heat treated. One could ask what would be the potential of the kernel meal of *J. curcas* seeds where the shell with its lignin content is completely removed. However, such seed meal is unlikely to be produced in the near future on a commercial level as *J. curcas* oil extraction is currently exclusively obtained through screw pressing.

The seed cake is, however, a good organic fertilizer with pesticidal properties. It has been shown to increase yields in pearl millet, maize, cabbage (additional pesti-

Table 18.6 Summarized properties of screw pressed *J. curcas* seed cake on a dry matter basis

Property	Value for screw pressed jatropha seed cake
Organic matter (%)	94.7
Carbon content (%)	55.0
N (%)	3.00–6.50
P (%)	0.65–3.00
K (%)	0.80–1.80
Mg (%)	0.68–1.4
S (%)	0.20–0.35
Lignin content (%)	22–24
Gross Energy (MJ kg^{-1})	19–22
Phorbol esters (mg g^{-1})	0.0–1

Collated from Brittaine and Lutaladio (2010), Gaind et al. (2009), own unpublished data

cidal activity against cutworms), tomatoes, rice and *J. curcas* itself at application levels of 3–10 tons per ha per year (for cross references see Achten et al. 2008, information also from Patolia 2006, personal communication). It degrades slowly and hence the application levels in subsequent years can be reduced after a first application (Patolia 2006). There have been concerns about the persistence and possible accumulation of phorbol esters in the soil on repeated application of seed cake from the non-edible *J. curcas* seeds. This is unfounded as it has been recently shown that phorbol esters are unstable and are degraded in soil. Devappa et al. (2010b) showed that the total phorbol ester concentration caused by application of *J. curcas* seed cake decreased from 0.37 mg g^{-1} in soil to below detection levels in 17 days at 13% moisture levels and 32°C ambient temperature. The degradation rate increased when ambient temperature and soil moisture contents increased. Similar degradation patterns were observed when isolated phorbol esters adsorbed to silica was applied to soil so that the initial level was as high as 2.6 mg g^{-1} of soil.

Because of its high organic content *J. curcas* seed cake is also suitable as biogas feedstock. Singh et al. (2008) reported total biogas production of 348 litres kg^{-1} of total solids from the reactors having *J. curcas* cake at 10% total solids in the substrate after an incubation period of 40 days. The biogas production was 241 litres kg^{-1} of total solids from the reactors having *J. curcas* cake at 15% total solids. Methane in biogas was 66 ± 2% in both cases and microbial cultures were from cow dung slurry (Singh et al. 2008). Previous reports (Staubmann et al. 1997) also showed similar biogas yields for *J. curcas* seed cake. The spent substrate after biogas production can be used as organic manure.

J. curcas Seed Kernel Meal

J. curcas seed kernel meal that could potentially form animal feed ingredients after heat treatment can theoretically be produced from deshelled *J. curcas* kernels from which oil is extracted through solvent extraction. Adriaans (2006) concluded that

Table 18.7 Chemical composition of extracted meal (% dry matter) of the toxic and non-toxic varieties of *J. curcas*

Parameter	Toxic variety	Non-toxic variety
Crude protein	56.4	63.8
Lipid	1.5	1.0
Ash	9.6	9.8
Gross energy (MJ kg^{-1})	18.2	18.0
Neutral detergent fibre	9.0	9.1

Source: Makkar et al. (1998)

solvent extraction using hexane in a continuous process was economical only if over 200 t or more of *J. curcas* seeds are processed per day. He also pointed out the negative environmental impacts of conventional *n*-hexane solvent extraction, such as generation of wastewater, higher specific energy consumption, higher emissions of volatile organic compounds, and possible human health impacts connected to working with hazardous and inflammable chemicals.

Another challenge is the absence of suitable commercial technologies to separate the kernel of *J. curcas* seeds from the shell. This would be necessary for enabling use of the solvent extracted kernel meal as an animal feed ingredient. The separation is difficult because the kernels of many *J. curcas* seeds are attached tightly to the shells even after drying and scraping off the latter results in considerable loss of good kernels (own observations). If suitable separation technology is discovered, it would result in another potentially valuable byproduct namely the shells that could make good fuels for domestic and industrial use with an average energy content of about 20 Mj kg^{-1}. The shells are also suited for making fuel briquettes (Singh et al. 2008).

The future market demand for protein rich animal feed ingredients is expected to stimulate technology development and adoption of the solvent extraction technology when *J. curcas* seeds are available in sufficient quantities.

The chemical composition of *J. curcas* kernel meal after solvent extraction from the edible and non-edible varieties (presented in Table 18.7) is a valuable animal feed ingredients.

In addition to having a high concentration of true proteins, *J. curcas* kernel meal is also having an amino acid composition more or less similar to soybean meal (Makkar et al. 1998). The chemical composition of the edible and non-edible *J. curcas* kernel meal are similar and the only important difference is the presence of phorbol esters in the latter.

The seed kernel meal of edible varieties of *J. curcas* requires only conventional treatments (moist heat treatment to remove trypsin inhibitors and lectins, addition of phytase enzyme to neutralize the phytate content) before use as animal feeds. Feeding trials in common carp with such treated meal from edible varieties has shown that they have high nutritional quality that exceeds that of soybean meal (Nahid Richter, G. Francis and Klaus Becker, unpublished data). Developing plantations with edible varieties of *J. curcas* would therefore be the easiest way of generating seed kernel meal for use as animal feed ingredients.

In the case of toxic varieties, additional treatments are required to remove phorbol esters from the kernel meal. Several treatments have been tried and at least on a laboratory level the *J. curcas* meal can be detoxified (Nahid Richter, G. Francis, Harinder Makkar and Klaus Becker, unpublished data; Kumar et al. 2010). Whether these technologies can viably be integrated into the commercial processing chain of *J. curcas* seeds remains a challenge.

Conclusion

This chapter provides basic information about the harvesting and processing of *J. curcas* seeds that would result in the production of oil of high quality. The important properties of oil, seed cake and seed kernel meal as well as their utilization perspectives are also outlined.

Acknowledgement Support of staff at the Jatropower Laboratory and Research farm, Coimbatore, India is appreciated.

References

Achten WMJ, Verchot L, Franken YJ, Mathijs E, Singh VP, Aerts R et al (2008) *Jatropha* biodiesel production and use. Biomass Bioenergy 32:1063–1084

Adriaans T (2006) Suitability of solvent extraction for *Jatropha curcas*. FACT Foundation, Eindhoven, www.fact-foundation.com/…/FACT_(2006)_Suitability_of_solvent_extraction_for_jatropha_curcas. Accessed 10 Dec 2010

Brittaine R, Lutaladio N (2010) Jatropha: a smallholder bioenergy crop—the potential for pro-poor development, Integrated Crop Management vol 8, FAO, Rome

Devappa RK, Maes J, Makkar HPS, Greyt WD, Becker K (2010a) Quality of biodiesel prepared from phorbol ester extracted *Jatropha curcas* oil. J Am Oil Chem Soc 87:697–704

Devappa RK, Makkar HPS, Becker K (2010b) Biodegradation of *Jatropha curcas* phorbol esters in soil. J Sci Food Agric 90(12):2090–2097

Gaind S, Nain L, Patel VB (2009) Quality evaluation of co-composted wheat straw, poultry droppings and oil seed cakes. Biodegradation 20:307–317

Goel G, Makkar HPS, Francis G, Becker K (2007) Phorbol esters: structure, biological activity, and toxicity in animals. Int J Toxicol 26:279–288

Haas W, Mittelbach M (2000) Detoxification experiments with the seed oil from *Jatropha curcas* L. Ind Crop Prod 12:111–118

Herrera JM, Martinez Ayala AL, Makkar HPS, Francis G, Becker K (2010) Agroclimatic conditions, chemical and nutritional characterization of different provenances of *Jatropha curcas* L. from Mexico. Eur J Sci Res 39:396–407

Jindal S, Bhagwati PN, Narendra SR (2010) Comparative evaluation of combustion, performance, and emissions of jatropha methyl ester and karanj methyl ester in a direct injection diesel engine. Energy Fuel 24:1565–1572

King AJ, He W, Cuevas JA, Freudenberger M, Ramiaramanana DL, Graham IA (2009) Potential of *Jatropha curcas* as a source of renewable oil and animal feed. J Exp Bot 60:2897–2905

Kumar V, Makkar HPS, Becker K (2010) Dietary inclusion of detoxified *Jatropha curcas* kernel meal: effects on growth performance and metabolic efficiency in common carp, *Cyprinus carpio* L. Fish Physiol Biochem 36:1159–1170

Makkar HPS, Aderibigbe AO, Becker K (1998) Comparative evaluation of non-toxic and toxic varieties of *Jatropha curcas* for chemical composition, digestibility, protein degradability and toxic factors. Food Chem 62:207–215

Makkar HPS, Francis G, Becker K (2008) Protein concentrate from *Jatropha curcas* screw-pressed seed cake and toxic and antinutritional factors in protein concentrate. J Sci Food Agric 88:1542–1548

Makkar HPS, Maes J, Greyt WD, Becker K (2009) Removal and degradation of phorbol esters during pre-treatment and transesterification of *Jatropha curcas* oil. J Am Oil Chem Soc 86:173–181

Mandpe S, Kadlaskar S, Degen W, Keppeler S (2005) On road testing of advanced common rail diesel vehicles with biodiesel from the *Jatropha curcas* plants, paper (No. 2005–26–356) presented at International mobility engineering congress and expo 2005, Chennai Trade Centre, Nandambakkam, Chennai organised by SAE India, 23–25 Oct 2005

Nuss P (2007) Investigation of biotechnical conditions of *Jatropha curcas* L. toward gradual harvest mechanisation. Master thesis, Institute of Agricultural Engineering, University of Hohenheim, Stuttgart

Singh RK, Padhi SK (2009) Characterization of jatropha oil for the preparation of biodiesel. Nat Prod Radiance 8:127–132

Singh RN, Vyas DK, Srivastava NSL, Narra M (2008) SPRERI experience on holistic approach to utilize all parts of *Jatropha curcas* fruit for energy. Renew Energy 33:1868–1873

Staubmann R, Foidl G, Foidl N, Gübitz GM, Lafferry RM, Arbizu VMV et al (1997) Biogas production from *Jatropha curcas* press-cake. Appl Biochem Biotechnol 63–65:457–467

Vyas DK, Singh RN (2007) Feasibility study of Jatropha seed husk as an open core gasifier feedstock. Renew Energy 32:512–517

Chapter 19
Value-Addition of Jatropha Cake and Its Utilization as Manure in Jatropha and Other Crops

Arup Ghosh, Jitendra Chikara, and D.R. Chaudhary

Introduction

Jatropha curcas L. (hereafter referred to as Jatropha) holds great potential in future as an oilseed for production of environmental friendly *Jatropha methyl ester* (JME) or biodiesel as it can be grown on wastelands. The Jatropha biodiesel has been demonstrated to be a suitable fuel for unmodified diesel engines (Mandpe et al. 2005; Mukerji 2005; Bacovsky et al. 2006; Fairless 2007; Ghosh et al. 2007a). Byproducts of JME obtained through base catalysis route of transesterification are shell/husk, oil cake, soap, glycerol and potassium sulphate (Ghosh et al. 2010b; patent). For every metric ton of biodiesel produced, roughly thrice the amount of solid Jatopha cake (JC) is produced as one of the main byproducts (Ghosh et al. 2007a). Potential area available for Jatropha plantation in India has been estimated to be about 10 million hectares (ha) or an equivalent of 5.7 million tons of neat biodiesel, assuming on average, a yield of 3.3 tons of seed capsule (i.e. 2 tons of seed) per ha in a year. This generates about 14 million tons of organic JC annually (Ghosh et al. 2007a). Disposal of such enormous amounts of cake generated would be a problem especially since it cannot be used as an animal food in its unprocessed form because of the presence of toxic constituents like phorbol esters (PE), curcin and the presence of other antinutrients and toxins (Martínez-Herrera et al. 2006). Finding a cost effective, environmentally sustainable, long-term solution for utilizing JC is therefore of critical importance and the different ways of addressing this issue are discussed in this chapter.

Jatropha seeds usually contain 28–38% oil, depending upon the genotype and growing conditions. First Jatropha seeds are cooked at 60–70°C for 15–20 min and then pressed in extruder or expeller. Practically, under decentralized model

A. Ghosh (✉) • J. Chikara • D.R. Chaudhary
Discipline of Wasteland Research, Central Salt & Marine Chemicals Research Institute, (Council of Scientific and Industrial Research, New Delhi),
Gijubhai Badheka Marg, Bhavnagar, Gujarat 364002, India
e-mail: arupghosh@csmcri.org

N. Carels et al. (eds.), *Jatropha, Challenges for a New Energy Crop: Volume 1: Farming, Economics and Biofuel*, DOI 10.1007/978-1-4614-4806-8_19,

Fig. 19.1 Mechanical screw press for Jatropha cake production after oil extraction from seeds at a plantation (Chorvadla, Bhavnagar, Gujarat, India)

concept of biodiesel production, i.e., local production for local consumption, wherein, oil extraction from seeds is in close vicinity of plantation (Fig. 19.1) by direct pressing of the entire seeds after removal of the husks, ca., 28–30% (w/w) yield of raw oil from whole seeds has been recorded (Ghosh et al. 2007a). The oily-cake in such cases is left with some residual oil (Reinhardt et al. 2008), usually 5–6% (may be as high as 12% depending upon the type of expeller), which may be further completely de-oiled if solvent extraction is employed subsequent to mechanical pressing, which is more suited under centralized biodiesel production model. However, encouraging decentralized production might well be beneficial to sustainable development if it leads to considerable socio-economic changes (Reinhardt et al. 2007). Thus, seed cake containing residual oil is expected to be produced in larger amount under Indian scenario.

Alternate Uses of Jatropha Cake

JC can be either used as manure or can serve as solid biofuel in power plants due to its high energy value (18.25 MJ kg^{-1}). It can also be used for the production of biogas, while the residual slurry can again serve as a nutrient rich manure. Inherent difficulties make the cake unsuitable for animal nutrition in unprocessed form although research trials have successfully been done towards improving its use as animal food after detoxification. Value addition to obtain non-food related industrial products could be an alternative way of JC utilization. Different alternate uses of cake are discussed as below.

Table 19.1 Elemental composition of Jatropha oil cake* (on dry weight basis) and possible nutrient efflux from soil through cake (assuming 2 $t ha^{-1}$ seed yield annually) disposal outside the plantation

Element	Elemental composition in oil cake	Soil nutrient efflux ($kg\ ha^{-1}$) by oil cake
Carbon (%)	40–48	NA
C/N	10–16	—
Nitrogen (%)	3–4.5	42–63
Phosphorus (%)	0.65–1.2	9.1–16.8
Potassium (%)	0.8–1.4	11.2–19.6
Sulphur (%)	0.2–0.35	2.8–4.9
Copper ($mg\ kg^{-1}$)	18–25	0.03–0.04
Iron ($mg\ kg^{-1}$)	800–1,000	1.12–1.40
Manganese ($mg\ kg^{-1}$)	300–500	0.42–0.7
Zinc ($mg\ kg^{-1}$)	30–50	0.04–0.07

*Cake produced after expelling oil from seed by mechanical screw press
Source: Ghosh et al. (2007a)

Use as Manure in Jatropha and Other Crops

Since India is being already short in arable land *vis-à-vis* burgeoning food demand, it is imperative that biofuel solution should come from feedstocks that does not infringe upon limited land resources. Jatropha suits this criterion by its virtue of being able to be grown on wastelands and is supposed to be cultivated in India only on degraded lands with inherent soil nutrient deficiencies. But, the dilemma is that high Jatropha seed productivity also implies higher efflux of nutrients from the soil (Tablc 19.1). Since the requirement of nutrients of Jatropha appears to be high, it is necessary to provide nutrients in the form of inorganic fertilizers. Recently Mohapatra and Panda (2011) conducted an experiment to optimize the requirements of N:P:K fertilizers for better economic returns from Jatropha.

Much of the nutrients removed by Jatropha seeds are left in the de-oiled cake after oil extraction and if it is disposed off from plantation area, it would result in further deterioration of already degraded land and diminution of the soil fertility as well as productivity, contrary to stated objective of bio-reclamation of wastelands, which would make the production system environmentally unsustainable. Therefore, it is imperative that oil cake – which will be produced to the amount of 1.3–1.5 t/ha- should largely be ploughed back in the field itself (Ghosh et al. 2007a; 2010a).

This could compensate nutrient depletion and at the same time improve the soil physico-chemical properties. Consequently, it would not be of environmental interest to utilize the cake for any other purpose. This conclusion fits well with the concept of "local use of local produce" under the decentralized model of biodiesel production in contrast with the centralized model according to which longer transport distances would be involved to bring back the cake to the plantation and would make it economically and energetically less attractive. Jatropha is demanding in nutrient requirement especially nitrogen if attractive yields are to be realized (Ghosh et al. 2007a). JC is a rich source of micro- and macro-nutrients (Table 19.1), such as

Table 19.2 Seed yield response of Jatropha plants to JC

	Seed yield per plant (g)			
	4 m×3 m		3 m×2 m	
Jatropha cake dose (t ha^{-1})	(2006–2007)	(2007–2008)	(2006–2007)	(2007–2008)
Control	687.2	615.9	447.6	464.8
0.75	776.6	835.3	519.7	546.0
1.50	1,049.9	845.0	627.8	730.0
2.25	1,310.4	987.7	750.2	758.6
3.00	1,519.8	1,820.8	867.1	809.8
S.Em (±)	47.4	83.8	24.3	38.4
LSD (p=0.05)	146.1	258.2	74.9	118.3

NB: The plants received inorganic fertilizers at 45:30:20 N: P_2O_5:K_2O ha^{-1} yr^{-1} prior to the start of the experiment

nitrogen, which makes it a fertilizer with tremendous potential for organic farming and in particular for itself.

Most Jatropha cultivations are likely to take place in poor soils with low water and nutrient capacity under rainfed conditions and high temporal variability in rainfall distribution over the rainy season. Under such conditions, the applied chemical fertilizers may be leached away due to excessive runoff because of excessive rains and proper nutrient supply to Jatropha plants may be hampered at some point during the long Jatropha fruiting season. There is evidence of lower oil content in seeds at the end of the season under similar condition (Ghosh et al. 2010a). Because JC is an organic fertilizer, it releases its essential nutrients in a sustained manner over the growing and fruiting period. The added advantage of de-oiled cakes over similar organic sources of nutrients is its quick mineralization, low C:N ratio that favors quick decomposition, high N content and hence, it can be applied as manure at the crop sowing stage and partially as top dressing as well (Saxena and Singh 1980).

In view of the above considerations, a study was conducted in the third year of plantation (planted in 2003) at CSMCRI Jatropha field station of Mohuda (Orissa), to study the response of Jatropha to its cake used as a fertilizer, in which it was applied to two differently spaced Jatropha plots (4 m×3 m and 3 m×2 m) at five varying doses (0, 0.75, 1.5, 2.25 and 3 t JC ha^{-1}) from June 2005 (2005–06) onwards for 3 years (Ghosh et al. 2007b). The JC application resulted in marked and significant improvement in mean seed yield during the fourth and fifth years of plantation (Table 19.2). The highest seed yield during the fifth year (2007–08) was 1.8 kg $plant^{-1}$ (equivalent to 1.5 t seeds ha^{-1}) and was obtained by application of 3 t ha^{-1} of JC under 4 m×3 m spacing, while the application of a same cake dose in 3 m×2 m spacing only yielded 0.8 kg $plant^{-1}$ (equivalent to 1.4 t seeds ha^{-1}) and the increase in seed yield due to treatments was statistically significant.

In yet another experiment carried out at CSMCRI Jatropha experimental station at Neswad (Bhavnagar, Gujarat), the application of nitrogen (140 kg ha^{-1}) during 2 consecutive years after transplanting through JC fertilization in the second year of

Table 19.3 Effect of substitution of inorganic nitrogen (IN) by Jatropha cake on wheat

Treatment	Seed yield (t ha^{-1})
Control (100% RD of IN[a])	4.34
75% RD dose of IN + 25% N substituted by JC	4.85
50% RD of IN + 50% N substituted by JC	5.36
25% RD of IN + 75% N substituted by JC	5.44
100% RD of N through JC	5.64
S. Em (±)	0.32
CD (p = 0.05)	0.93[b]

NB: [a]RD- Recommended dose of N = 120 kg N ha^{-1}
[b]Highly significant

plantation promoted an increase in seed yield of 35% over the treatments where equivalent amount of nitrogen was applied solely through inorganic fertilizers.

To study the cumulative effect of JC and sulphur fertilization on nutrient content and yield of Jatropha, treatments comprising of 0, 10, 20, 40 kg S ha^{-1} alone and in combination with JC (3.3 and 6.6 t ha^{-1}) applied annually. After 2 years, leaf samples were collected and analyzed for nutrient content (D.R. Chaudhary, unpublished data) and it was found that the N content varied from 1.86 to 2.39%, was non-significantly affected by S applications, and was positively correlated with cake applications. Similarly, N, P (0.20–0.30%) and K (1.70–2.51%) contents were also improved by cake application. The S content in leaves varied from 0.18% to 0.25% and significantly increased at 40 kg S ha^{-1}, which was at par with other sulphur levels applied with cake. Results revealed that the application of cake together with S, influence S uptake by the plant. Seed yield (g $plant^{-1}$) varied from 10.1 to 55.0. The highest seed yield was observed for applications of 20 kg S ha^{-1} with 6.6 t cake ha^{-1}. Oil content was significantly increased from 28.9 (control) to a maximum of 35.11 (40 kg S ha^{-1}).

Jatropha cake was likewise found to be good manure for other crops as well. The effect of the JC produced at CSMCRI's biodiesel pilot plant has been tested on wheat (Table 19.3) at BCKV (Nadia, West Bengal). Significant differences in grain yield was obtained on wheat over controls (where full dose of N was applied by inorganic means) by fertilizing it with Jatropha cake. The largest seed yield (5.64 t ha^{-1}) was obtained in the treatment where entire nitrogen dose was given through JC, which was however not statistically different from both treatments where the mineral fertilizer was substitute with 50% or 75% cake releasing yields of 5.36 and 5.44 t ha^{-1}, respectively. The yield increase over control under treatment having 50% N substitution by JC was 23.5% and appeared to be the optimum substitution rate. The increase in grain yield of wheat by JC application might have been due to increase in plant growth and also perhaps due to availability of nutrients over a sustained period.

In experiments carried out by CSMCRI during 2005–06 where JC was applied to pearl millet, sesame and cotton at 1.5, 1.0 and 2.0 t ha^{-1}, respectively, the

Table 19.4 Effect of Jatropha cake as organic manure on pearl millet, sesame and cotton

Crop	Application (t ha^{-1})	Yield ha^{-1}	
		Control	Treated
Pearl millet	1.5	1.90	2.15
Sesame	1.0	0.56	0.64
Cotton	2.0	1.65	1.90

Table 19.5 Effect of INM using *Jatropha* cake on seed yield of sesame

Treatments	Control	25 kg N ha^{-1}	50 kg N ha^{-1}
Control	475.1e	578.9d	766.7ab
Jatropha cake @ 1 t ha^{-1}	660.8cd	694.0bc	754.5abc
Jatropha cake @ 1.5 t ha^{-1}	748.6abc	813.9a	829.0a
Castor cake @ 1 t ha^{-1}	697.4bc	799.1a	826.3a
FYM @ 5 t ha^{-1}	681.1bc	803.8a	810.0a
Standard error (±)	30.1		
LSD (p=0.05)	85.2		

Means with same letters are not significantly different according to LSD at $p=0.05$

grain (or seed) yield increase by pearl millet, sesame and cotton over controls (without application of cake) was 11.3%, 14.3% and 15.2% in yield, respectively (Table 19.4).

In another experiment of nutrient management (INM) study at CSMCRI during 2006–2007 (unpublished data) where different combinations of mineral N doses and organic N sources (castor cake - CC, Farm yard manure –FYM - and JC) were tested, the seed yields of sesame under JC (1.5 t ha^{-1}), FYM (5 t ha^{-1}) and castor cake (1 t ha^{-1}) applied in combination with mineral nitrogen (25 or 50 kg ha^{-1}) were similar (Table 19.5). The maximum seed yield was obtained by application of 1.5 t ha^{-1} JC and mineral nitrogen at 25 kg ha^{-1}. Larger doses of mineral N (50 kg ha^{-1}) together with Jatropha cake did not result in any yield increase. In the GTZ project in Mali, a trial was carried out with pearl millet where the effect of manure (5 t ha^{-1}), JC (5 t ha^{-1}) and mineral fertilizer (100 kg ammonium phosphate and 50 kg urea ha^{-1}) was investigated. Pearl millet yields per ha were 630 kg for the control, 815 kg for manure, 1,366 kg for press cake and 1,135 kg for mineral fertilizer (Heller 1996).

A similar field experiment was conducted during 2007–08 to find out the effect of de-oiled seed cake of Jatropha and castor (*Ricinus communis* L.) as a source of nutrient alone and in various combinations with mineral sources in spring sunflower (*Helianthus annuus* L.) and maize (*Zea mays* L.) cultivated in sequence (Bodake and Rana 2009). Three organic sources of N (RDN), *viz.*, JC (100%), CC (50%) and FYM (25%) were combined with three doses of mineral N (RDF), *viz.*, 0%, 50% and 75% and applied in sunflower along with a fixed mineral supplement of (80, 26.2 and 30 kg ha^{-1} of N, P and K, respectively). Residual effects of RDN and RDF were evaluated on the succeeding maize. The sunflower seed yield obtained with 25% RDN through JC or CC with 75% RDF was found to be comparable with RDF in the first season. In the second season, replacement of 50% RDN with JC or CC entailed perceptible increase in seed yield over RDF alone. Residual effect of treatments on succeeding maize and system productivity was found to be the most pronounced with 50% RDF + 50% RDN (JC). In the first season, the best *agronomical N use efficiency* (ANUE) was recorded with RDF, closely followed by 75% RDF + 25% RDN (JC or CC). In the second season, best ANUE was obtained with 50% RDF with 50% RDN (JC) closely followed by 50% RDF with 50% RDN (CC). In the second season, integrated use of 50% RDF with 50% RDN through JC recorded the highest ANUE closely followed by 50% RDF with 50% RDN through CC.

Chaturvedi et al. (2009) carried out a field experiment to study the response of garlic to composted de-oiled cake by comparing FYM (12 t ha^{-1}) (control), composted JC (2 t ha^{-1}), composted neem cake (NC) (2 t ha^{-1}), composted JC + tobacco waste (2:1) (2 t ha^{-1}) and urea (120 kg ha^{-1}). The application of 2 t ha^{-1} mixture of composted JC and tobacco waste (in ratio 2:1) resulted in an increase of 22.88% in fresh bulb weight, 29.52% in dried bulb weight per plant and 18.3% in average yield of bulb as compared to control and was comparable to that recorded for plants grown on soil treated with chemical fertilizers. Punia et al. (2010) also reported JC to have better fertilizing effect as compared to other selected cakes (neem, mahua, karanja) on selected vegetable crops (tomato, brinjal, chilli and okra). Thus, based on these studies, it is obvious that the cake can be used as an organic source of nutrients. The organic fertilization through JC enables sustainable agriculture with yield enhancements and is a viable alternative to synthetic inorganic fertilizers.

Effect on Soil and Its Microbial Community

Soil application of JC after oil extraction potentially provides an important disposal method for bioenergy byproduct and it has the added benefit of supplying nutrients to soil, However, JC contains some toxic biochemicals, such as PEs which may inhibit key populations of soil microorganisms. Currently, little information exists on the effect of oilseed cakes on soil microbial ecosystems. A study was carried out by adding different amounts (0.5, 1, and 5% w/w) of Jatropha, camelina, and flax seed cakes and wheat biomass to Weswood loam soils in laboratory to investigate microcosm sensitivity (Wang et al. 2009). Different methods including microbial biomass, quantitative PCR (qPCR) assays, *phospholipid fatty acid profiles* (PLFA), whole soil *fatty acids methyl esters* (FAME) analyses and Biolog EcoPlates were used to investigate community changes over an incubation period of 19 weeks. The total microbial biomass dramatically increased in all treatments with the greatest increase occurring with the oilseed cakes. For the 0.5% and 1% rates of addition, biomass levels returned to the background levels within 2 weeks except for wheat. Biomass remained higher than controls at the 5% application rate until the end of the incubation. Addition of oilseed cakes generally increased the abundance of bacterial and fungal biomarkers, but resulted in lower bacteria to fungi ratios compared to wheat treatments as indicated by PLFA results. The qPCR studies indicated that addition of oilseed cakes resulted in lower ratios of bacteria:fungi indicating community composition changes. Soil amended with Jatropha residues (leaves, fruit shell and deoiled JC) showed that microbial biomass ($FAME_{tot}$) and bacterial and fungal FAME concentrations were significantly higher in soils amended with cake compared to other treatments (Chaudhary et al. 2011). Higher fungal to bacterial FAME ratios were found in residue-amended soils over control, with the highest ratio occurring in fruit shell-amended soil. A significant decrease in the stress indicator (saturated to monounsaturated FAME ratio) was found in cake-amended soil. The fate of PEs in the soil and its bioactivity during the process of degradation was studied by Devappa et al. (2010) and it was concluded that PEs are biodegradable in soils and their degraded products appear to be innocuous. The PEs from seed cake were degraded after 21, 17 and 15 days (at 130 or 230 g kg^{-1} moisture). Increase in temperature and moisture increased the rate of PEs degradation. Using snail (*Physa fontinalis*) bioassay, mortality by PE-amended soil extracts decreased with a decrease in PE concentration in the soil. Similarly, in another study carried out at CSMCRI (unpublished data), it was found that no traces of PE were found in any of the plant parts of groundnut treated with JC in substitution to mineral nitrogen, suggesting that it can be innocuously used as a manure in soil.

Use as a Bio-Pesticide

Apart from its encouraging results for use as organic manure in various crops and itsef, Jatropha has bio-pesticidal properties as well. PEs contained in seeds of Jatropha were reported to be effective as biopesticide against diverse fresh water snails. Snails act as intermediate hosts of schistosomes causing schistosomiasis in many tropical countries (Goel et al. 2007). Aqueous and acetone extracts of neem and Jatropha seed kernel was recommended to control stem borers in upland rice (Amaugo and Emosairue 2003). JC has been proved to possess very good bio-pesticidal properties in control of nematodes and termites. In a study (undertaken in joint collaboration with CSMCRI, Bhavnagar and Anand Agricultural University - AAU, Anand, Gujarat in 2004–2005) to test the efficacy of JC for management of root-knot (*Meloidogyne incognita*) nematodes in tomato, it was found that JC applied at 3 t ha^{-1} to tomato crop could successfully control root knot nematode and was comparable to the chemical treatment (Carbofuran) applied to the control. In addition, the application of 2 t JC ha^{-1} was found to give similar results to NC treatment, which is the reference in organic control of nematodes. Tomato yield significantly increased over control with the increase in dose of cake (Table 19.6).

Studies conducted during 2008–2009 for nematode management in which JC, produced as a byproduct of CSMCRI's biodiesel pilot plant at Bhavnagar, was used as one of the treatments revealed that it works well as an effective nematicide in various crops (Dr. B.A. Patel, personal communication, on the basis of works

Table 19.6 Effect of *Jatropha* cake on root knot index (RKI) in tomato

Treatment	RKI (0–5)*	Tomato Yield (kg t ha^{-1})
Jatropha cake @ 1.0 t ha^{-1}	3.54	27,006
@ 2.0 t ha^{-1}	3.11	29,815
@ 3.0 t ha	2.84	32,716
Neem cake @ 2.0 t ha^{-1}	2.94	39,568
Carbofuran @ 2.0 kg ha^{-1}	2.31	37,037
Control	5	18,271
S. Em. ±	0.24	1,729
CV%	12.6	11.2

*Root Knot index categorized as low to high ranging from 1-5 (1=Free; 5=Maximum nematode infestation)

NB: (a) Organic amendment was incorporated in the soil 10 days prior to seeding

(b) Carbofuran was applied at the time of seeding

(c) Susceptible hybrid (Navin/Rashmi)

carried out by S.K. Patel, A.D. Patel and H.V. Patel at AAU Anand). They concluded that complementation of soil solarization with either NC or mustard cake or JC each at1 kg sq m^{-1} increased growth and development of seedlings by checking root-knot infection (*M. incognita* and *M. javanica*) and thereby produced more number of transplantable seedlings in nursery. No toxic effect of JC was observed on seed germination of tomato. When the treated seedlings were transplanted in nematode infested field, all the cake treatments were significantly superior over carbofuran and control treatments for root-knot management and thereby increased tomato yield. Similarly, control of phytoparasitic nematodes, such as *Meloidogyne spp* in greengram and *Rotylenchulus reniformis* in cowpea through use of NC or JC alone or in combination with the biocontrol agent *Trichoderma viride* was also reported. According to the reports, JC worked well alongside other cakes like NC, CC, etc., in controlling nematodes across various centers under All India Coordinated Research Project on Plant parasitic nematodes (ICAR) on various field crops. NC and JC were recommended for the management of root-knot nematodes infesting bitter gourd in Kerala.

In a laboratory study involving value addition of different cakes for biopesticidal purpose at IIT Delhi, the cold and hot water extracts and methanol extracts of JC provided good control against termites. In field study as well, JC provided better protection of termites over untreated control, although NC was found to be the best. A combination of cakes from Jatropha, Karanj, Mahua and neem provided the best control against termites (Punia et al. 2010).

Use as Animal Food

The cake after oil extraction contains as much as 19–28% protein. The major amino acid in JC was found to be glutamic acid 2.82% and for the others, the series is as follow: arginine 2.19%, aspartic acid 1.74%, leucine 1.24%, serine 0.91%, alanine 0.87%, valine 0.86%, proline 0.82%, glycine 0.81%, phenylalanine 0.78%, lysine

0.75%, threonine 0.71%, isoleucine 0.7%, cysteine 0.51%, histidine 0.43% and tyrosine 0.40% (Waraporn et al. 2009). Despite high protein content, the seed cake as such is not suitable for animal feeding due to the presence of toxins like PEs, hydrocyanic acid, alkaloids, curcin and presence of other antinutrients like tannins, trypsin inhibitors, phytates, saponins and lectins (CRC 1977; Martínez-Herrera et al. 2006; Punia et al. 2010). In addition, the presence of hard outer seed coat (37–42% of seed weight) having high lignin and cellulose makes it further unpalatable. PEs are tetracyclic diterpenoids generally known for their tumor promoting activity and even at very low concentrations, show toxicological manifestations in animals fed diets containing them (Goel et al. 2007). Makkar et al. (1997) reported that trypsin inhibitor activity in the defatted kernels (cake) varied from 18.4 to 27.5 mg of trypsin inhibited g^{-1}. Similarly, a wide variation was observed for saponins (1.8–3.4% as diosgenin equivalent), phytate (6.2–10.1% as phytic acid equivalent), and lectin activity (0.85–6.85 using a latex agglutination test and 51.3–204 using hemagglutination assay) in the cakes. The seed content of PEs varies among different Jatropha cultivars — ranging from undetectable in the Mexican 'non-toxic' varieties (of which the roasted seeds are eaten by humans) to over 6 mg g^{-1} kernel in a toxic variety from India (Francis et al. 2005). Toxicity of Jatropha seeds has been studied extensively in different animal models like goats, sheep, mice, rats, and fish when fed with food containing PEs (Adam 1974; Adam and Magzoub 1975; Makkar and Becker 1999). If JC has to be used for animal nutrition, it can be detoxified and the seeds can be decorticated (i.e., the seed outer cover removed to expose the seed kernels) prior to mechanical and/or solvent extraction. The composition of lignin, cellulose, and hemicellulose in JC after removal of outer seed cover is 0.53%, 6.43%, and 1.94%, respectively, in contrast to 19.46%, 20.3%, and 5.55% present when shells are not removed (Staubmann et al. 2007). The protein rich cake obtained after oil extraction from seed kernel may serve as animal food after detoxification.

Studies conducted at CFTRI, Mysore, revealed that the defatted "*Ghani*" pressed cake, which had 0.072% PEs was reduced to 0.01–0.0079% after treating the samples with 4% NaOH and 2% $Ca(OH)_2$. Autoclaving was the most effective treatment as trypsin inhibitors are heat labile. The roasting, boiling and alkali treatments were most effective in reducing the saponin content by 75.3%, 84.8% and 84.2%, respectively. Alkali soaking and boiling the sample reduced the tannins by 46–50%. The acetone and methanol/water were the most effective solvents in reducing the phytate content by 30% and 34%, respectively. Further works are need to achieve complete detoxification of JC in order to use it in food formulations (Punia et al. 2010), however microbial fermentation is an other route that seems to offer promising results.

Solvent extraction of PEs followed by heat treatment to inactivate lectins in JC was reported to convert the non-toxic cake to a high-quality protein source for livestock (Makkar et al. 1997). Makkar and Becker (1999) explored a non-toxic variety of Jatropha and reported that the heat-treated or untreated cake was suitable for animal feeding. These authors also reported that this variety results in a promising *protein efficiency ratio* (PER) and food conversion ratio with simultaneous decrease in trypsin inhibition and lectin activity when tested in rats and fish carp

(*Cyprinus carpio*). The PER of the diet containing unheated and heated JC was 37% and 86%, respectively, of the casein in rats. The trypsin inhibitor and lectin activities decreased more than 99%, after 45 min of heat treatment. Heat-treated cake of the non-toxic variety of Jatropha was found to be comparable to commercially available soybean cake in nutritional quality for common carp (Goel et al. 2007). On the other hand, heat treatment followed by solvent extraction to remove PEs could result in elimination of most of the antinutrients and toxins from the toxic variety. The cake treated in this manner was found to be innocuous to rats and fish (Makkar et al. 1997).

Use for Biogas Production

Even if detoxification is not an option, the cake may be converted to biogas. When using an anaerobic filter with JC as a substrate (Loading Rate of 13 kg COD $m^{-3}d^{-1}$), 76% of the *chemical oxygen demand* (COD) was degraded and 1 kg degraded COD yielded 355 L of biogas containing 70% methane (Staubmann et al. 1997). An experiment was carried out with *Jatropha* cake supplied by CSMCRI, Bhavnagar to Biogas Research and Extension Centre, Gujarat Vidyapith, Gandhinagar to evaluate biomethanation potential of JC along with buffalo dung (1:4 ratio by wt). Results showed significantly higher (139.20%) biogas production in test (JC + Buffalo dung) over control (buffalo dung only) digesters with methane content of 71.74% during 120 days of experimentation. Nutritive value of effluent slurry of test digester was significantly higher in terms of available nitrogen and potassium, calcium, magnesium and carbonate contents than that of control digesters and could be used as a nutrient rich manure in field. Co-digestion resulted in 92.94% decrease in chemical oxygen demand (Shilpkar et al. 2009). Ali et al. (2010) observed that biogas plant initially charged with pure cattle dung increased the biogas production by upto approximately 25% when gradually replaced with JC (0–100%). Study carried out by Ingenia, Netherlands, reported fairly high (0.81 0.95 m^3/kg dm) biogas production because of the high content of carbon (47.6–48.5%) in JC. They reported that the H_2S concentration in the produced biogas is too low (0.18 mg m^{-3}) to pose a problem for any known application (Visser and Adriaans 2007). Utilization of JC for biogas has also been reported by Singh et al. (2008).

Other Uses

Apart from the above uses, research is being carried out across the world to add value to JC for its gainful utilization. JC, either alone or in combination with potato dextrose agar (PDA) or dextrose has been successfully used as substrate for growth of selected fungi (biocontrol agents). Maximum radial growth of *Paecillomyces variotii* was recorded on JC + dextrose, i.e., 0.36 cm day^{-1} as compared to 0.26 cm day^{-1} obtained under control (PDA) (Punia et al. 2010). As discussed above, the cake after oil extraction contains as much as 19–28% protein which is suitable as a protein source for animal feed after detoxification. Typically, anti-nutrients and toxins could be almost removed by different solvent extractions; however, huge amount of chemical solvents would be discharged as effluents, which need to be treated. Therefore, non-food related products could be an alternative way of JC utilization. Basic hydrolysis of protein in JC using 2.5% NaOH at 50 °C resulted in best degree of hydrolysis (19.93% DH) to obtain most soluble nitrogen. The obtained protein hydrolysate was shown to be well solubilized and contained protein as high as 71.69%, which was appropriate for further applications, such as for light-weight concrete industry (Waraporn et al. 2009). Experimentation on solid-state fermentation of JC showed that, it could be a good source of low cost production of industrial enzymes. The seed cake supported good bacterial growth (*Pseudomonas aeruginosa* PseA strain) and enzyme (protease, 1,818 U g^{-1} of substrate and lipase, 625 U g^{-1} of substrate) production (Mahanta et al. 2008).

Suffice it to say that, besides the obvious implication on Jatropha biodiesel cost, creative applications of by-products could be as important as the application of the biodiesel itself and there is need for intensive research to add value to these products besides identifying new outlets (Ghosh et al. 2007a). The results described above demonstrated viable approaches for utilization of huge amount of JC, which is expected to be released in abundant quantities during biodiesel production.

References

Adam SE (1974) Toxic effects of *Jatropha curcas* in mice. Toxicol 2:67–76

Adam SE, Magzoub M (1975) Toxic effects of *Jatropha curcas* in goats. Toxicol 4:347–354

Ali N, Kurchania AK, Swati B (2010) Bio-methanisation of *Jatropha curcas* defatted waste. J Eng Technol Res 2(3):38–43

Amaugo GO, Emosairue SO (2003) The efficacy of some indigenous medicinal plant extracts for the control of upland rice stem borers in Nigeria. Trop Subtrop Agroecosyst 2:121–127

Bacovsky D, Prankl H, Rathbauer J, Worgetter M (2006) Project title: local and innovative biodiesel. Altener Contract No. 4.1030/C/02-022

Bodake PS, Rana DS (2009) Evaluation of Jatropha (*Jatropha curcas*) and castor (*Ricinus communis*) cake as a source of nutrient and soil amendment in spring sunflower (*Helianthus annuus*)- maize (*Zea mays*) sequence. Indian J Agron 54(3):284–290

Chaturvedi S, Kumar A, Singh B (2009) Utilizing composted Jatropha, neem cake and tobacco waste to sustain garlic yields in indo-gangetic plains. J Phytol 1(6):353–360

Chaudhary DR, Lorenz N, Dick LK, Dick RP (2011) FAME profiling and activity of microbial communities during *Jatropha curcas* L. residue decomposition in semiarid soils. Soil Sci 176:625–633

CRC Critical Review in Toxicology (1977) Higher plant genera and their toxins. Crit Rev Toxicol 7:213–237

Devappa RK, Makkar HPS, Becker K (2010) Biodegradation of *Jatropha curcas* phorbol esters in soil. J Sci Food Agric 90:2090–2097

Fairless D (2007) Biofuel: the little shrub that could—maybe. Nature 449:652–655

Francis G, Edinger R, Becker K (2005) A concept for simultaneous wasteland reclamation, fuel production, and socio-economic development in degraded areas in India: need, potential and perspectives of Jatropha plantations. Nat Resour Forum 29:12–24

Ghosh A, Chaudhary DR, Reddy MP, Rao SN, Chikara J, Pandya JB et al (2007a) Prospects for jatropha methyl ester (biodiesel) in India. Intl J Environ Stud 64:659–674

Ghosh A, Patolia JS, Chaudhary DR, Chikara J, Rao SN, Kumar D et al (2007b) Response of *Jatropha curcas* under different spacing to Jatropha de-oiled cake. In: FACT seminar on *Jatropha curcas* L. agronomy and genetics, FACT Foundation, Wageningen, 26–28 March 2007. Article no.8

Ghosh A, Chikara J, Chaudhary DR, Prakash AR, Boricha G, Zala A (2010a) Paclobutrazol arrests vegetative growth and unveils unexpressed yield potential of *Jatropha curcas*. J Plant Growth Regul 29(3):307–315

Ghosh PK et al (2010b) US patent 7,666,234, 23 Feb 2010, granted to CSIR/CSMCRI

Goel G, Makkar HPS, Francis G, Becker K (2007) Phorbol esters: structure, biological activity, and toxicity in animals. Int J Toxicol 26:279–288

Heller J (1996) Physic nut. *Jatropha curcas* L. Promoting the conservation and use of underutilized and neglected crops. Institute of Plant Genetics and Crop Plant Research/International Plant Genetic Resources Institute, Gatersleben/Rome

Mahanta N, Gupta A, Khare SK (2008) Production of protease and lipase by solvent tolerant *Pseudomonas aeruginosa* PseA in solid-state fermentation using *Jatropha curcas* seed cake as substrate. Bioresour Technol 99(6):1729–1735

Makkar HPS, Becker K, Sporer F, Wink M (1997) Studies on nutritive potential and toxic constituents of different provenances of *Jatropha curcas*. J Agric Food Chem 45:3152–3157

Makkar HPS, Becker K (1999) Nutritional studies on rats and fish carp (*Cyprinus carpio*) fed diets containing unheated and heated *Jatropha curcas* meal of a non-toxic provenance. Plant Foods Human Nutr 53:183–192

Mandpe S, Kadlaskar S, Degen W, Keppeler S (2005) On road testing of advanced common rail diesel vehicles with biodiesel from the *Jatropha curcas* plants, In: Proceedings of SEA INDIA conference, paper No. 2005-26-356

Martı'nez-Herrera J, Siddhuraju P, Francis G, Da'vila-Ortı'z G, Becker K (2006) Chemical composition, toxic/antimetabolic constituents, and effects of different treatments on their levels, in four provenances of *Jatropha curcas* L. from Mexico. Food Chem 96:80–89

Mohapatra S, Panda PK (2011) Effect of fertilizer application on growth and yield of *Jatropha curcas* L. in an Aeric Tropaquept of Eastern India. Not Sci Biol 3(1):95–100

Mukerji R (2005) Biodiesel proved as good as fossil fuel. Deccan Herald (DH Wheels supplement). 21 Sep, p 1

Punia MS, Kureel RS, Kumar V, Pandey A (2010) 5th R&D Report on tree borne oilseeds-2009-10. National oilseeds and vegetable oils development board (NOVOD), Ministry of Agriculture, Govt. of India

Reinhardt GA, Gartner SO, Rettenmaier N, Falkenstein EV (2007) Screening life cycle assessment of Jatropha biodiesel. Institute for energy and environmental research, Heidelberg (IFEU) report7

Reinhardt GA, Ghosh PK, Becker K, Chaudhary DR, Chikara J, Falkenstein EV et al (2008) Basic data for Jatropha production and use. Institute for energy and environmental research, Heidelberg (IFEU), Central salt and marine chemicals research institute (CSMCRI), University of Hohenheim. Heidelberg, Bhavnagar and Hohenheim

Saxena RK, Singh HG (1980) Effect of rates and times of application of castor-cake on potato production on vertisols in Rajasthan. Indian J Agron 25(4):651–654

Shilpkar P, Roal G, Shah M, Shilpkar D (2009) Biomethanation potential of Jatropha (*Jatropha curcas*) cake along with buffalo dung. Afr J Agric Res 4(10):991–995

Singh RN, Vyas DK, Srivastava NSL, Narra M (2008) SPRERI experience on holistic approach to utilize all parts of *Jatropha curcas* fruits for energy. Renew Energy 33(8):1868–1873

Staubmann R, Foidl G, Foidl N, Gübitz GM, Lafferty RM, Arbizu VM, Steiner W (1997) Biogas production from *Jatropha curcas* press-cake. Appl Biochem Biotechnol 63–65:457–467

Staubmann R, Foidl G, Foidl N, Gübitz GM, Lafferty RM, Visser J et al (2007) Anaerobic digestion of *Jatropha curcas* press cake. Project No. 0656.521, Ingenia consultants and engineers of client FACT fuels foundation www.fact-foundation.com/…/Final_report_digestion_**Jatropha**_press_**cake** (Ingenia_0656521-R02) Accessed on 8 Nov 2010

Visser J, Adriaans T (2007) Anaerobic digestion of *Jatropha curcas* press cake. Report produced for FACT, Ingenia Consultants & Engineers, Eindhoven

Wang AS, Gentry TJ, Hons FM, Hu P (2009) Soil microbial communities affected by different oilseed meals and a crop residue. Presented in ASA-CSSA-SSSA annual meetings 1–5 Nov 2009 Pittsburg, PA

Waraporn A, Pilanee V, Phanu S, Taweesiri M (2009) Optimization of protein hydrolysate production process from *Jatropha curcas* cake. World Acad Sci Eng Technol 53:109–112

Chapter 20
Biopesticidal Properties of Seed, Seed Cake and Oil of *Jatropha curcas* L. Against the Polyphagous Lepidopteran Pest *Helicoverpa armigera*

D. Muralidhara Rao, S. Anitha, A. Aravinda, B. Karunakar, and N. Devanna

Introduction

Insect pests are one of the major limiting factors in increasing productivity of many crops. The cost of plant protection in agriculture has been increasing over years (Kumar and Sharma 2008; Jayaraj and Rabindra 1988). Because of increasing problems associated with the use of toxic synthetic insecticides, there is a need for the development of safer alternative crop protectants. Among various options available under the concept of Integrated Pest Management (IPM), bio-pesticides play a key role. Plants produce secondary metabolites that are insecticidal and have diverse modes of action like hormonal, neurological, nutritional or enzymatic effects (Rosenthal and Janzen 1979). These secondary plant metabolites with insecticidal or more generally pesticidal properties are of several types, *viz.*, repellents, antifeedants, phagostimulants and toxins (Philogene 1981). A number of secondary plant metabolites have been extracted from plants.

Biopesticides generally affect only the target pest and closely related organisms, in contrast to broad spectrum, conventional pesticides. Biopesticides are effective in very small quantities, are biodegradable and avoid the pollution problems caused by conventional pesticides (Klocke 2007). When used as a component of IPM program, biopesticides can greatly decrease the use of conventional pesticides, besides maintaining high crop yields. The application of microbial and botanical

D.M. Rao (✉) • S. Anitha • A. Aravinda
Department of Biotechnology, Sri Krishna Devaraya University, Anantapur 515 055, India
e-mail: muralidhararao@yahoo.com

B. Karunakar
Department of Biochemistry, National P.G. College, Nandyal 518 501, India

N. Devanna
Department of Chemistry, Jawaharlal Nehru Technological University, Anantapur 515 055, India

N. Carels et al. (eds.), *Jatropha, Challenges for a New Energy Crop: Volume 1: Farming, Economics and Biofuel*, DOI 10.1007/978-1-4614-4806-8_20,

insecticides is increasing (Dent 1993). Among the plant families studied, Meliaceae, Rutaceae, Asteraceae, Labiateae, Piperaceae and Annonaceae are the most promising families with insecticidal properties (Jacobson 1989; Isman 1995). The search for plant-derived chemicals that have the potential use of crop protectants (insecticides, antifeedants, growth inhibitors) often begins with the screening of plant extracts. The insects under test are fed with plant extracts and monitored for behavior and development. This kind of information is needed to ensure safety to non-target organisms (humans, beneficial insects) (Akhtar and Isman 2004)

Seeds of *Jatropha curcas* (hereafter refered to as Jatropha) are reported to contain 6.6 g H_2O, 18.2 g protein, 38.0 g fat, 33.5 g total carbohydrate, 15.5 g fibre, and 4.5 g ash per 100 g (Duke and Atchley 1983). In addition, Jatropha leaves contain α amyrin, β sitosterol, stigmasterol, and campesterol, 7-keto- β -sitosterol, stigmast-5-ene-3- β, 7- α -diol, and stigmast-5-ene-3 β, 7 β –diol, isovitexin and vitexin.

Rug and Ruppel (2000) found that the oil content of Jatropha seed is 25–30%, the oil is 21% saturated fatty acids and 79% unsaturated fatty acids. However, the amount of oil extracted from seeds and kernels dependents on the method with only ~20% for hand presses and much higher quantity for more sophisticated mechanic presses. The by-product of oil extraction from seeds and kernels is called seed cake. The cake resulting from oil extraction through cottage industry is said to still contain approximately 11% oil (Wink 1993). The seed kernel itself contains about 60% oil that can be converted into biodiesel and used as a substitute for diesel fuel. The seed cake remaining after oil extraction is an excellent source of plant nutrients. However, the presence of high levels of antinutrients prevents its use in animal feeding (Trabi et al. 1997).

Phorbol esters are used to describe the family of naturally occurring compounds that can be referred to as tigliane diterpenes (Evans 1986). Phorbol esters (phorbol-12-myristate 13-acetate) have been identified as the major toxic principle in Jatropha (Makkar and Becker 1997). Adebowale and Adedire (2006) investigated the chemical composition and insecticidal activity of Jatropha seed and evaluated that the oil content of the seed kernels is high (66.4%). Triacylglycerol is the dominant lipid species, while the major triacyglycerol being 1, 2-dioleoyl-3-linoleoyl-rac-glycerol.

Martinez-Herrera et al. (2006) studied Jatropha from four agro-climatic regions of Mexico and found that accessions from all the four regions differed in the contents of crude protein, lipid, fibre contents, starch, total soluble sugars, essential amino acids except lysine, phorbol esters, trypsin inhibitors, saponins and lectins. Makkar et al. (1997) made a comparative evaluation of non-toxic and toxic varieties of *J. curcas* for chemical composition, digestibility, protein degradability and toxic factors. Insecticidal activities of Jatropha oil containing phorbol esters have been reported in *Manduca sexta, Helicoverpa armigera, Aphis gossyii, Pectinophora gossypielaa, Empoasca biguttula, Oncopeltus fasciatus,* etc. (Wink et al. 1997). Insect pests attacking Jatropha and their management potentials were reported by Shanker and Dhyani (2006).

Osoniyi and Onajobi (2003) worked on the humectation action of Jatropha. The latex of Jatropha possesses both procoagulant and anticoagulant activities. Georges et al. (2008) studied the insecticidal activity of eight plant species including *J. curcas* collected from Burkina Faso against *Heliothis virescens* larvae and adult whitefly (*Bemisia tabaci*). The ethyl acetate extract of *S. madagascariensis* was the most active on adult whitefly. Gandhi (1995) studied the larvicidal activity of ethyl acetate, butanol and petroleum ether extracts of Jatropha against the early fourth instar larvae of *Aedes aegypti* L. and *Culex quinquefasciatus*.

The biological effects of these compounds in addition to tumour promotion, bring about a wide range of biochemical and cellular effects, alter cell morphology, serve as lymphocyte mitogens and induce platelet aggregation (Blumberg 1980; 1981). Phorbol esters are defined as "polycyclic compounds in which two hydroxyl groups on neighboring carbon atoms are esterified to fatty acids." Several plant species, such as *Sapium indicum, S. japonicum, Euphorbia frankiana, E. cocrulescence, E. tirucalli, Croton spareiflorus, C. tigilium, C. ciliatoglandulifera, J. curcas, Excoecaria agallocha,* and *Homalanthus nutans* are reported to contain the toxic phorbols (Beutler et al. 1989). Among these species, *J. curcas* has also been reported to possess other potential toxic compounds, such as curcin and hydrocyanic acid (CRC 1977). There are several other plant species that contain different derivatives of phorbols and diterpenes, such as crotonogyne, crytogonone, dimorphocalyx, duvigneaudia, fahrenheitia, maprounea, and plagiostyles (Beutler et al. 1989).

The major toxic principles in Jatropha seeds are phorbol esters (Makkar and Becker 1997). Hirota et al. (1988) named a new phorbol from Jatropha as DHPB (*dihydroxy phenyl dibutyrate*). The effect of phorbol esters was also studied in guinea pigs *in vivo* whereby the topical application of these esters was reported to induce inflammation and epidermal hyperproliferation by inducing DNA synthesis through prostaglandin activation and especially prostaglandin E (Bourin et al. 1982).

Experiments were conducted to assess the pesticidal activity of Jatropha seeds, seed cake and oil under laboratory and field conditions, against the selected Polyphagous pest, Cotton boll worm or *Helicoverpa armigera* (Hubner) (Lepidoptera) attacking musk melon. Biochemical investigations were also carried out with special emphasis on amylase, protease and lactate dehydrogenase.

Solvent Extracts, Biotests and Enzyme assays

Seed (*Oil Technological Research Institute* (OTRI), Anantapur, Andhra Pradesh, India) and cake samples macerated in solvents (methanol, ethanol, petroleum ether and ethyl acetate, benzene, hexane and water) according to 1:3 ratios during 24 h were reduced in a rotary evaporator, to obtain crude extracts and stored in refrigerator at 4 °C (Georges et al. 2008). Larvae of *Helicoverpa armigera* commonly known as cotton boll worm (department of Entomology, Acharya N.G. Ranga Agricultural University, ANGRAU, Anantapur) were cultured on leaves of musk melon (*Cucumis melo*) for feeding. Laboratory and field trials were conducted using 3rd instar larvae since they are responsible for the major crop loss. In laboratory conditions, larvae were fed on fresh leaves pretreated by dipping them in 125, 250 and 500 ppm extract solutions of seed, cake or oil. For field trials, musk melon field plot measuring 460 m^2 located in Katiganikalava village in Anantapur mandal of Anantapur district (Andhra Pradesh, India) were sprayed after infestation with each of the 3 extract solutions. The entire study was conducted according to standard test methods as prescribed by World Health Organisation (1975).

Ten larvae of *H. armigera* (third instar) collected among dead individuals killed extract treatments were then weighed and homogenized in 0.15 M NaCl solution with methanol. After centrifugation (10,000 g for 10 min at 4°C), the supernatant was used to determine the (i) protein concentration of larval extracts by the Bradford method (1976), (ii) amylase activities at 550 nm through the method of Bernfield (1955) as described by Ishaaya and Swirski (1970), (iii) protease activities at 600 nm as described by Snell and Snell (1979), and (iv) lactate dehydrogenase (LDH) activities at 440 nm after 60 min of incubation. All data were statistically consistent at $p = 0.01$ (Sengottayan et al. 2006) using WindowStat version 8, Tukey's Honestly Significance Difference (Tukey's HSD) test and two factorial RBD (Panse and Shukatme 1967).

Pesticide activity of Jatropha extracts

Among the different toxic fractions from oil, the highest average mortality of *H. armigera* larvae under laboratory conditions was registered with 500 ppm of extract obtained with methanol (85.94%), followed by ethanol (68.85%) and petroleum ether (66.14%). The series of extract activity methanol > ethanol > petroleum ether was maintained for all three concentrations tested (Fig. 20.1). Under field conditions, population reduction was observed with methanol and ethanol at all three concentrations tested, but the largest effect was obtained with 500 ppm. At a concentration of 500 ppm, the population reduction was 70.26% with methanol and 67.5% with ethanol (Fig. 20.2).

When different solvent extracts of seeds were tested for average population reduction under laboratory conditions, methanol followed by ethanol and petroleum ether were found to be more effective at 500 ppm. Again, the largest population reduction was obtained with methanol (85.94%) and was not significantly different compared to the methanol extract of oil (Fig. 20.3). When average caterpillar mortality was observed under field conditions, the methanol extract resulted in the significantly largest population reduction corresponding to 47.8%, 61.38% and 82.8% for all three concentrations tested 125 ppm, 250 ppm and 500 ppm, respectively (Fig. 20.4). Similarly to oil, the largest reduction of larval populations was obtained for methanol and ethanol extracts at 500 ppm, i.e., 82.8% and 78.2%, respectively.

Considering cake extracts under laboratory conditions, population reduction was only observed with methanol and it was almost the double at 500 ppm (85.9%) when compared to 250 ppm (46.9%) (Fig. 20.5). under field conditions, 75% and 66% reduction of caterpillar populations were achieved with 500 ppm of methanol and ethanol extracts, respectively (Fig. 20.6). According to our investigations of

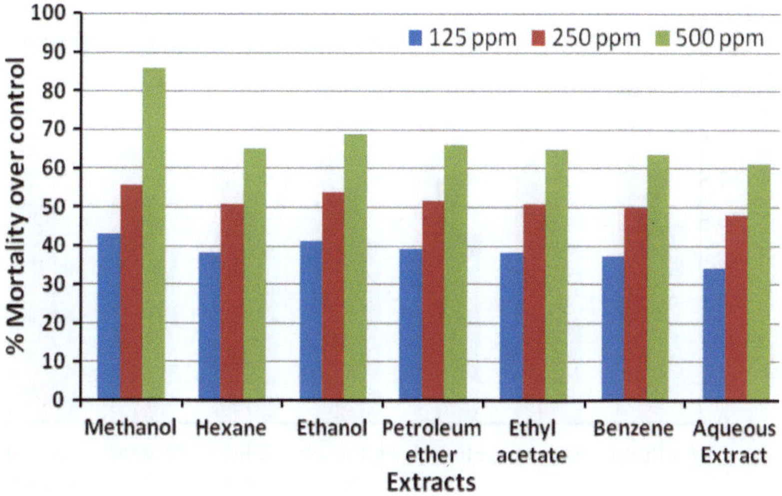

Fig. 20.1 Effect of different solvent extracts of *J. curcas* (oil) on third instar larva of *H. armigera* under laboratory conditions

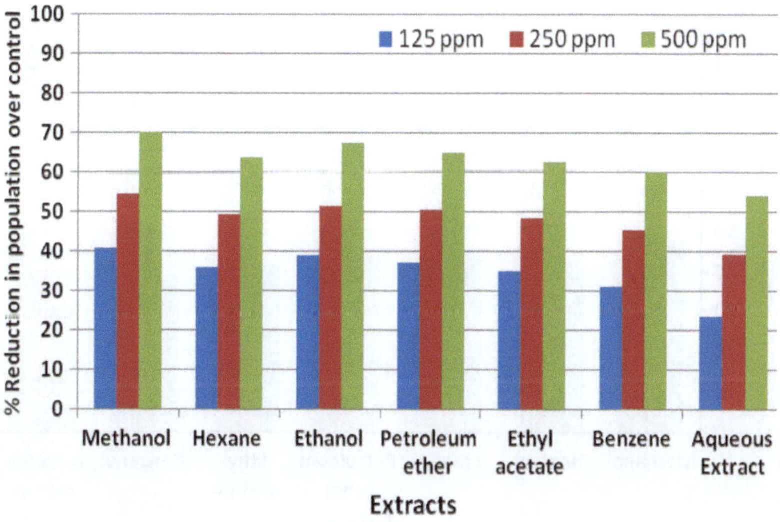

Fig. 20.2 Effect of different solvent extracts of *J. curcas* (oil) on third instar larva of *H.armigera* under field conditions

enzymatic activity in dead caterpillars, methanol extract was found to be the most effective in reducing the amylase activity, i.e., 1.08, 0.74, and 0.67×10^4 μg/mg/min at 125 ppm, 250 ppm and 500 ppm, respectively, compared to control (4.7×10^4 μg/mg/min) (Fig. 20.7).

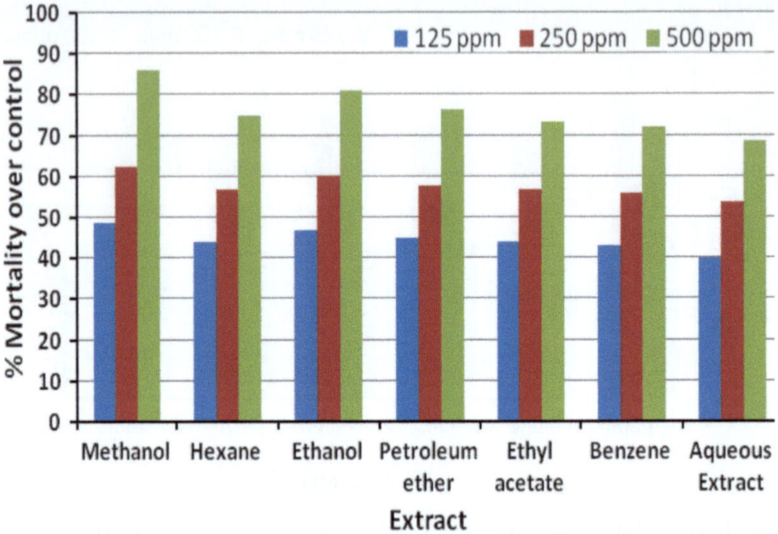

Fig. 20.3 Effect of different solvent extracts of *J. curcas* (seed) on third instar larva of *H. armigera* under laboratory conditions

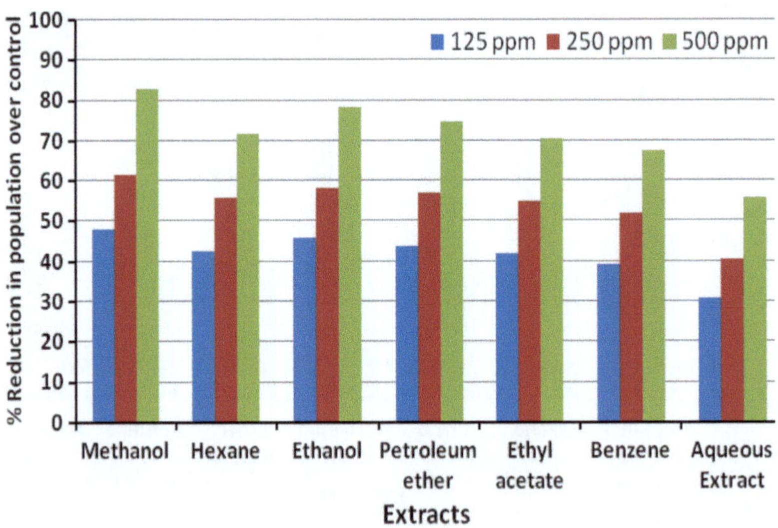

Fig. 20.4 Effect of different solvent extracts of *J. curcas* (seed) on third instar larva of *H.armigera* under field conditions

The dosage effect of oil extracts on protease activity of third instar larvae significantly depended on the solvent type used (Fig. 20.8). The largest reduction of protease activity was found for 500 ppm of methanol extract (1.57×10^4 µg/mg/min), while ethyl acetate, benzene and aqueous extracts did not show significant

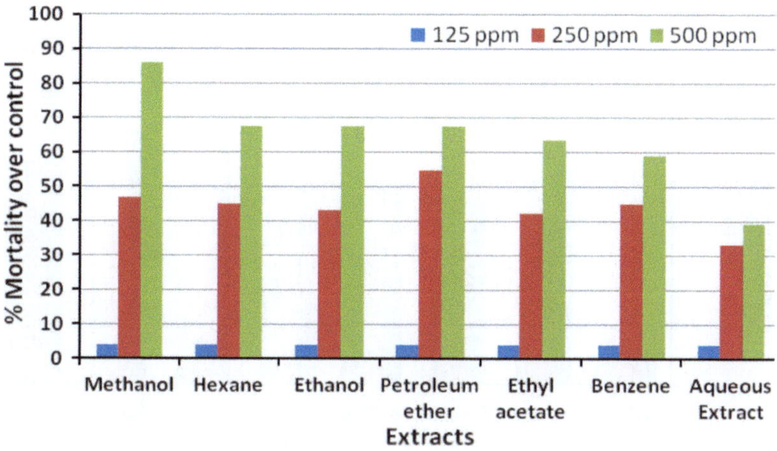

Fig. 20.5 Effect of different solvent extracts of *J. curcas* (meal) on third instar larva of *H. armigera* under laboratory conditions

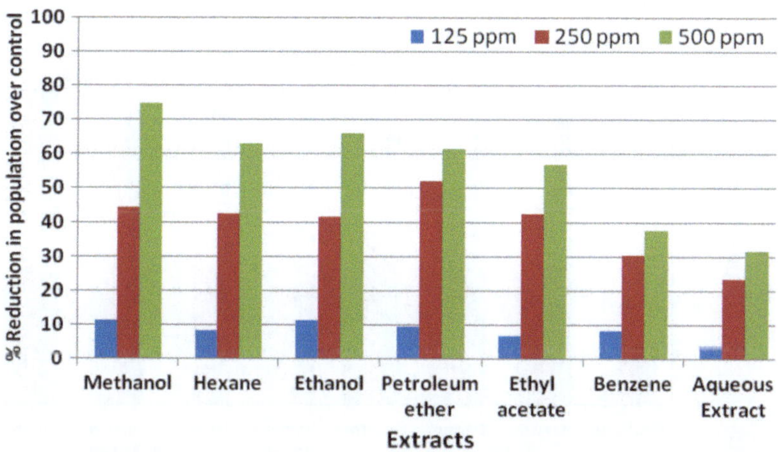

Fig. 20.6 Effect of different solvent extracts of *J. curcas* (meal) on third instar larva of *H.armigera* under field conditions

decrease in protease activity when compared to the control. When LDH activity was investigated in third instar larvae under laboratory conditions, methanol and ethanol extracts at 500 ppm were the most effective (8.42 and 8.48×10^4 µg/mg/min, respectively) (Fig. 20.9).

Amylase activity of third instar larvae treated with seed extracts resulted in similar effects compared to oil extracts with methanol extract at 500 ppm being the most effective treatment (0.57×10^4 µg/mg/min) under lab conditions (Fig. 20.10). Among

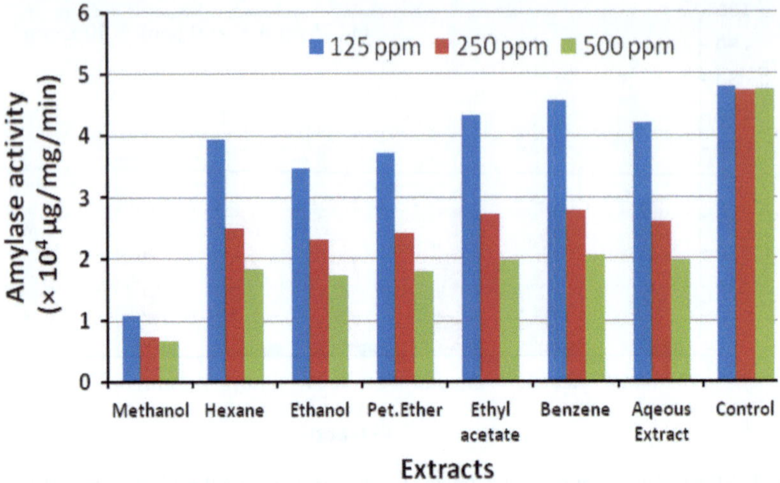

Fig. 20.7 Effect of different solvent extracts of *J. curcas* (oil) on amylase activity of third instar larva of *H. armigera* under laboratory conditions

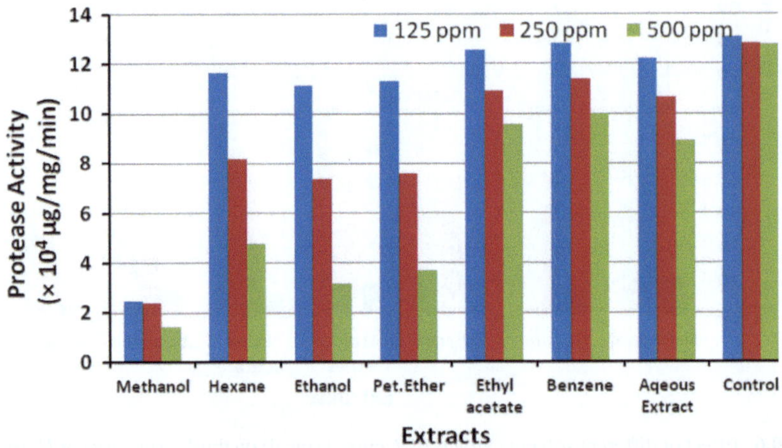

Fig. 20.8 Effect of different solvent extracts of *J. curcas* (oil) on protease activity of third instar larva of *H.armigera* under laboratory conditions

protease activity of seed extract, again methanol was found to be the most effective in decreasing the activity of protease at ~2 μg/mg/min or below (2.11, 2.06 and 1.25×10^4 μg/mg/min) at all concentrations tested (Fig. 20.11). For LDH activity of seed extracts, methanol extract was found to be the most active at all the three concentrations (6.37, 6.44, and 4.10×10^4 μg/mg/min) compared to the other extracts except for ethanol at 500 ppm (Fig. 20.12).

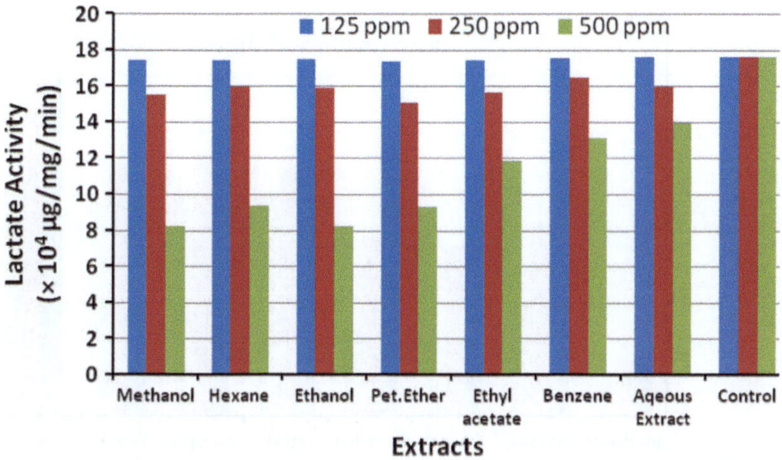

Fig. 20.9 Effect of different solvent extracts of *J. curcas* (oil) on LDH activity of third instar larva of *H. armigera* under laboratory conditions

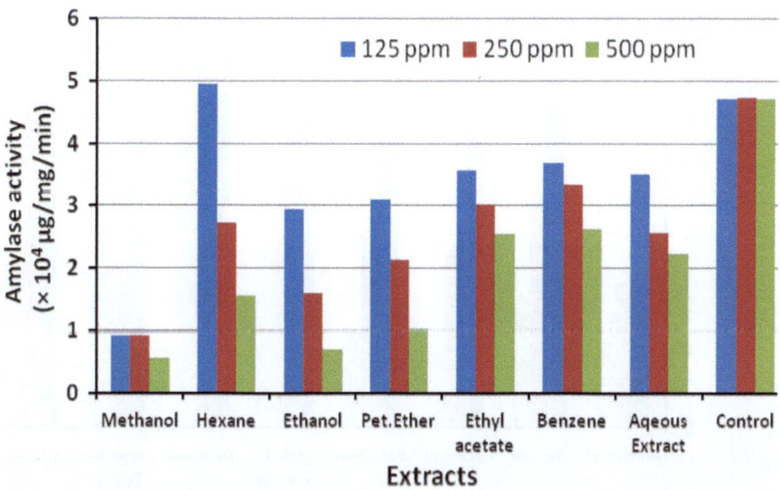

Fig. 20.10 Effect of different solvent extracts of *J. curcas* (seed) on amylase activity of third instar larva of *H.armigera* under field conditions

When amylase activity was tested on third instar larvae under laboratory conditions with different cake extract, the methanol extract was found to be a more effective treatment, as for oil and seeds, in decreasing the amylase activity to 2.16×10^4 µg/mg/min compared to 4.7×10^4 µg/mg/min for the control. It is also interesting to note that the aqueous extract was found to be better than the benzene one (Fig. 20.13). Reduction of protease activity by cake extracts revealed that those of methanol and

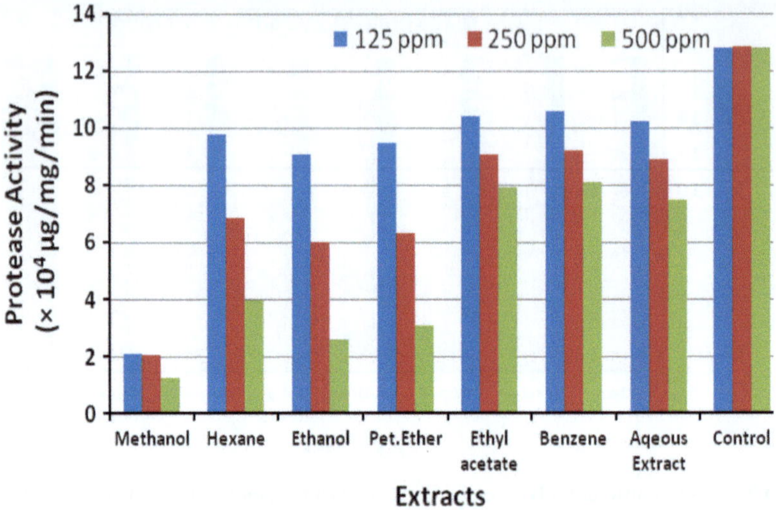

Fig. 20.11 Effect of different solvent extracts of *J. curcas* (seed) on protease activity of third instar larva of *H. armigera* under laboratory conditions

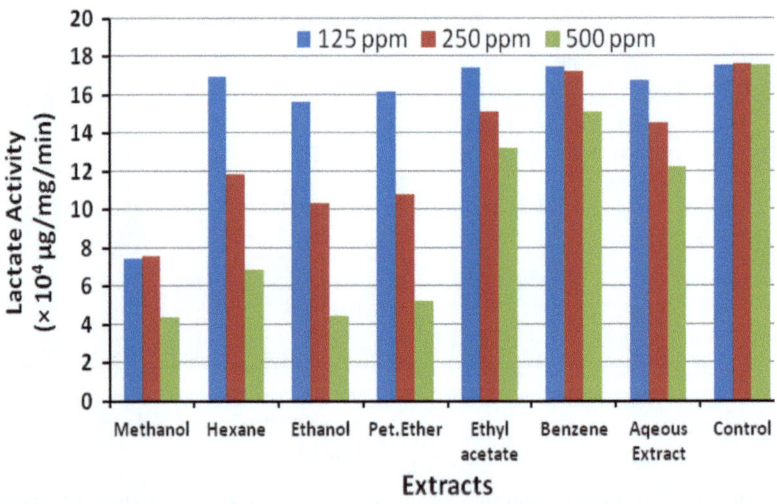

Fig. 20.12 Effect of different solvent extracts of *J. curcas* (seed) on LDH activity of third instar larva of *H.armigera* under field conditions

ethanol at 500 ppm were the most effective in decreasing the activity of this enzyme with 5.96 and 5.98×10^4 μg/mg/min, respectively (Fig. 20.14). When LDH activity was investigated by treating larvae with cake extracts, the methanol and ethanol ones at 500 ppm were found to be the most effective with 2.80 and 2.98×10^4 μg/mg/min, respectively (Fig. 20.15).

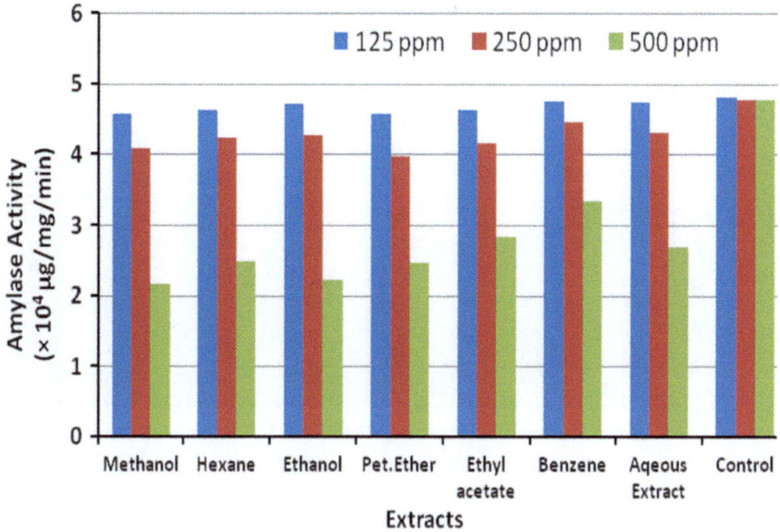

Fig. 20.13 Effect of different solvent extracts of *J. curcas* (meal) on amylase activity of third instar larva of *H. armigera* under laboratory conditions

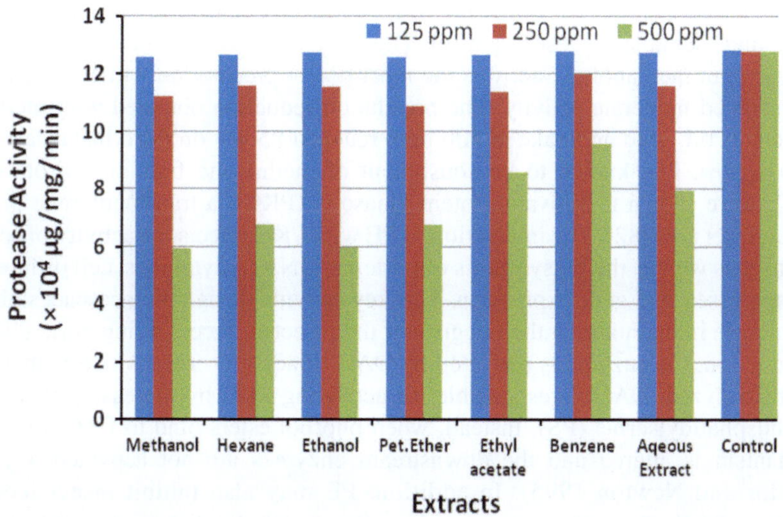

Fig. 20.14 Effect of different solvent extracts of *J. curcas* (meal) on protease activity of third instar larva of *H.armigera* under laboratory conditions

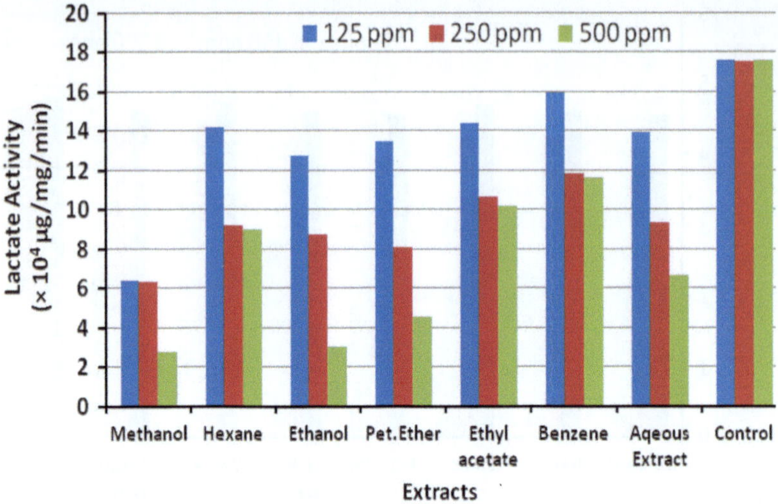

Fig. 20.15 Effect of different solvent extracts of *J. curcas* (meal) on LDH activity of third instar larva of *H. armigera* under laboratory conditions

Conclusion

Jatropha was shown to have insecticidal properties. Among all the extracts tested, methanol, ethanol and petroleum ether extracts showed potent antinutritional and pesticidal activities on *H. armigera*, which indicated that these extracts contained compounds suitable for insect control. These compounds match the PE fraction of oil. 500 ppm methanol extract was the most potent preparation while 125 and 250 ppm showed moderate activity. The population reduction obtained with methanol extracts of oil, seed and cake at 500 ppm reached 85.9% on third instar larvae of *H. armigera*. PEs known to be constituent of methanolic fraction (Hirota et al. 1988), were shown to activate protein kinase C (PKC), a transmembrane protein (Castagna et al. 1982). The interaction of PE with PKC affects the activity of several enzymes as well as the biosynthesis of proteins, DNA, polyamines, cell differentiation processes, and gene expression. This key enzyme of signaling cascades plays a critical role in maintaining the integrity of the insect surface. During normal signal transduction, the enzyme is activated by DAG (diacylglycerol), which is then rapidly hydrolyzed. DAG is responsible for activating PKC by increasing its affinity for phosphatidylserine (PS). Instead, when phorbol esters bind to PKC, the entire mechanism is altered and the downstream enzymes are not activated anymore (Mosior and Newton 1995). In addition, PE may also inhibit insect feeding. Once the level of enzyme activity is lowered, metabolic disturbances takes place thereby leading to insect death (Cox and Willason 1981; Mayzaud and Mayzaud 1985). Aqueous extracts of Jatropha seed and cake contain curcin, a lectin, which is another reason for the decrease in enzyme activities and death of *H. armigera*.

PEs are amphiphilic molecules soluble in a large range of solvent polarities, which may explain the insecticidal activity of the oil (Aravinda 2010). Actually, even if the toxic components of seeds are less soluble in aromatic solvents like benzene, the extracts by this solvent were still moderately toxic. The toxicity of extracts from Jatropha seeds on *Sitophilus zeamaise* was reported by Ohazurike et al. (2003). Thus, solvent extracts of Jatropha are candidates as biopesticides for agricultural pest management.

References

Adebowale KO, Adedire CO (2006) Chemical composition and insecticidal properties of the underutilized *Jatropha curcas* seed oil. Afr J Biotechnol 5:901–906

Akhtar Y, Isman MB (2004) Comparative growth inhibitory and antifeedant effects of plant extracts and pure allelochemicals on four phytophagous insect species. J Appl Entomol 128:32–38

Aravinda A (2010) Investigations on biochemical and biopesticidal properties of *Jatropha* seed, oil and meal. Ph.D thesis. Jawaharlal Nehru Technological University, Anantapur, India

Bernfield P (1955) Amylase alpha and beta. Meth Enzymol 1:149–158

Beutler JA, Ada AB, McCloud TG, Cragg GM (1989) Distribution of phorbol ester bioactivity in the Euphorbiaceae. Phytother Res 3:188–192

Blumberg PM (1980) In vitro studies on the mode of action of the phorbol esters, potent tumor promoters. Part 1. CRC Crit Rev Toxicol 8:153–197

Blumberg PM (1981) In vitro studies on the mode of action of the phorbol esters, potent tumor promoters. Part 2. CRC Crit Rev Toxicol 8:199–234

Bourin MC, Delescluse C, Furstenberger G, Marks F, Schweizer J, Klein-Szanto AJ et al (1982) Effect of phorbol esters on guinea pig skin in vivo. Carcinogenesis 3:671–676

Bradford MM (1976) A rapid and sensitive method for the quantitation of microgram quantities of protein utilizing the principle of protein dye binding. Anal Biochem 72:248–254

Castagna M, Takai Y, Kaibuchi K, Sano K, Kikkawa U, Nishizuka Y (1982) Direct activation of calcium-activated, phospholipid dependent protein kinase by tumor-promoting phorbol esters. J Biol Chem 57:7847–7851

Cox JL, Willason SW (1981) Laminarinase induction in *Calanus pacificus*. Mar Biol Lett 2:307–311

CRC Critical Review in Toxicology (1977) Higher plant genera and their toxins. Crit Rev Toxicol 7:213–237

Dent DR (1993) The use of *Bacillus thuringiensis* as an insecticide. In: Jones G (ed) Exploitation of Mircoorganisms. Chapman and Hall, London, pp 19–44

Duke J, Atchley AA (1983) Proximate analysis. In: Christie BR (ed) The handbook of plant science in agriculture. CRC Press, Boca Raton

Evans FJ (1986) Naturally occurring phorbol esters, vol 7. CRC Press, Boca Raton, pp 213–237

Gandhi VM (1995) Toxicological studies on ratanjyot oil. Food Chem Toxicol 33:39–42

Georges K, Jayaprakasam B, Dalavoy S, Nair G (2008) Pest-managing activities of plant extracts and anthraquinones from *Cassia nigricans* from Burkina Faso. Bioresour Technol 99:2037–2045

Hirota M, Suttajit M, Suguri H, Endo Y, Shudo K, Wongchai V et al (1988) A new tumor promoter from the seed oil of *Jatropha curcas* L., an intramolecular diester of 12-deoxy-16-hydroxyphorbol. Cancer Res 48:5800–5804

Ishaaya I, Swirski E (1970) Invertase and amylase activity in the armoured scales of *Chrysomphalus aonidum* and *Aonidiella aurantii*. J Insect Physiol 16:1599–1606

Isman MB (1995) Leads and prospects for the development of new botanical insecticides. Rev Pest Toxicol 3:1–20

Jacobson M (1989) Botanical insecticides, past, present and future. In: Arnason JT, Philogene BJR, Morand P (eds) Insecticides of plant origin, vol 387, ACS symposium series., pp 1–10

Jayaraj S, Rabindra RJ (1988) Larval extracts and other adjuvants or increased efficacy of nuclear polyhedrosis virus against *Heliothis armigera* larvae. J Biol Control 2:102–105
Klocke JA (2007) Natural plant compounds useful in insect control. Amer Chem Soc Ser 330:396–415
Kumar A, Sharma S (2008) An evaluation of multipurpose oilseed crop for industrial uses (*Jatropha curcas* L.)—A review. Ind Crops Prod 28:1–10
Makkar HPS and Becker K (1997) Studies on nutritive potential and toxic constituents of different provenances of *Jatropha curcas*, J Agric Food Chem 45:3152–3157
Martinez-Herrera J, Siddhuraju P, Francis G, Davila-Ortiz G, Becker K (2006) Chemical composition, toxic/antimetabolic constituents and effects of different treatments on their levels, in four provenances of *Jatropha curcas* L. from Mexico. Food Chem 96:80–89
Mayzaud P, Mayzaud O (1985) The influence of food limitation on the nutritional adaptation of marine zooplankton. Arch Hydrobiol Beih 21:223–233
Mosior M, Newton AC (1995) Mechanism of interaction of protein kinase C with phorbol esters. Reversibility and nature of membrane association. J Biol Chem 270:25526–25533
Ohazurike NC, Onuh MO, Emeribe EO (2003) The use of seed extract of the physic nut (*Jatropha curcas* L.) in the control of maize weevil (*Sitophilus zeamaize* M.) in stored maize grains (*Zea mays* L.). Global J Agri Sci 2:86–88
Osoniyi O, Onajobi F (2003) Coagulant and anticoagulant activities in *Jatropha curcas* latex. J Ethnopharmacol 89:101–105
Panse VG, Shukatme PV (1967) Statistical methods for agricultural workers. ICAR Publication, New Delhi
Philogene BJr (1981) Toxic activities of the plant *Jatropha curcas* against intermediate snail hosts and larvae of schistosomes. Ann Soc Entomol Que 26:177–181
Rosenthal GA, Janzen DH (1979) Herbivores: their interaction with secondary plant metabolites, vol 718. Academic, New York
Rug M, Ruppel A (2000) Toxic activities of the plant *Jatropha curcas* against intermediate snail hosts and larvae of schistosomes. Trop Med Int Health l5:423–430
Sengottayan SN, Kandaswamy K, Chung PG, Murugan K (2006) Effect of neem limonoids on lactate dehydrogenase (LDH) of the rice leaf folder, *Cnaphalocrocis medinalis* (Guenée) (Insecta: Lepidoptera: Pyralidae). Chemosphere 62:1388–1393
Shanker C, Dhyani SK (2006) Insect pests of *Jatropha curcas* L. and the potential for their management. Curr Sci 91:162–163
Snell FD, Snell CT (1979) Calorimetric methods of analysis, 3rd edn. Van Nostrand Company, New York, p 145
Trabi M, Gubitz GM, Steiner W, Foidl N (1997) Toxicity of *Jatropha curcas* seeds. In: Gubitz GM, Mittelbach M, Trabi M (eds) Biofuels and industrial products from *Jatropha curcas*. Dbv-Verlag University of Graz, Graz, pp 173–178
Wink M (1993) Forschungsbericht zum Projekt Nutzung pflanzlicher Öle als Kraftstoffe. Consultant's report prepared for GTZ, Germany
Wink M, Koschmieder C, Sauerwein M, Sporer F (1997) Phorbol esters of *Jatropha curcas* -Biological activities and potential applications. In: Gubitz GM, Mittelbach M, Trabi M (eds) Biofuel and Industrial products from *Jatropha curcas*. Dbv-Verlag University of Graz, Graz, pp 160–166
World Health Organisation (1975) Technical report, resistance of vectors and reservoirs of disease to pesticide. TRS/585

Chapter 21
Phytochemicals in Jatropha Seeds and Potential Agro-Pharmaceutical Applications of *Jatropha curcas* Phorbol Esters

Rakshit K. Devappa, Harinder P.S. Makkar, and Klaus Becker

Introduction

Plant based feedstocks will play an important role in future bio-energy supply. Recently, *Jatropha curcas* L. has been hailed as a promising bio-energy crop (Makkar and Becker 2009). It belongs to Euphorbiaceae family and its seed oil is a good feedstock for production of biodiesel, which meets the European and American Biodiesel Standards (Makkar and Becker 2009; Devappa et al. 2010b). In addition to its seed oil, many co-products can be obtained during the biodiesel production. Some of the important co-products are protein rich seed cake and seed kernel meal, glycerol and biologically active phytochemicals (Makkar et al. 2009a). Seed cake and seed kernel meal are considered to be the major co-products from *J. curcas* based biodiesel industry having potential for use as animal feed. However, the presence of toxic and antinutritional factors limits their efficient utilization (Makkar et al. 1997). In this chapter, antinutritional and toxic compounds in *J. curcas* kernel, seed cake and seed kernel meal are discussed with the aim to understand the main toxic principles and to make the efficient utilization of the seed cake and seed kernel meal as livestock and aquafeeds. In addition, potential utilization of one of the active phytochemicals present in *J. curcas* seeds, *phorbol esters* (PEs), in agricultural and pharmaceutical applications are discussed. *J. curcas* seeds also contain other bioactive compounds: a number of other diterpenes, proteins and peptides. For information on these bioactive moieties readers are referred to Devappa et al. (2010c, f).

R.K. Devappa • K. Becker
Institute for Animal Production in the Tropics and Subtropics, (480b), University of Hohenheim, Stuttgart 70599, Germany

H.P.S. Makkar (✉)
Livestock Production Systems Branch, Animal Production and Health Division, FAO, Rome, Italy
e-mail: Harinder.Makkar@fao.org

N. Carels et al. (eds.), *Jatropha, Challenges for a New Energy Crop: Volume 1: Farming, Economics and Biofuel*, DOI 10.1007/978-1-4614-4806-8_21,

Toxic and Antinutritional Factors Present in Seeds of *J. curcas*

The quality of a feed depends on the availability of nutrients to animals. Any deficiencies, excesses and imbalances of amino acid, vitamins and minerals can limit animal performance and lead to deleterious effects (McDaniel and Freking 2006). The quality of protein supplement is often determined by the quality of amino acids and their bioavailability. Conversely, if plant proteins are to be used, then the role of antinutrients and toxic factors should be considered. The plants produce antinutrients and toxic factors as a self defence against predatory organisms (Ames et al. 1990). The antinutrients are generally defined as substances generated in living systems that by themselves or the intermediary of metabolic products (1) interfere with food or feed utilization; (2) affect the health and reproduction of animals; and (3) produce death or deleterious effects upon high intake (Makkar 1993). Alternatively, they can also be classified as heat labile and heat stable antinutritional compounds. The toxic compounds may also exhibit deleterious effects in animals if inhaled, swallowed or absorbed through the skin. In higher dosages the toxic compounds can often be lethal. The consumption of plant material containing antinutritional or toxic factors by animals produces adverse effects that could vary from decrease in nutrient utilisation, health problems to even death of animals. The information about the structure–activity relationships is important to understand the mechanisms of action of antinutritional or toxic plant compounds and their effects.

The antinutrients and toxic factors present in *J. curcas* seeds are listed in Table 21.1. *J. curcas* has toxic and non-toxic genotypes and the differentiation is solely based on the presence and absence of PEs in them. The non-toxic genotypes, free of PEs were originally available only in Mexico (Makkar and Becker 2009). In addition, recently another non-toxic species of the genus *Jatropha*, i.e., *J. platyphylla* Müll. Arg has been reported from Mexico (Makkar et al. 2011).

Tannins

Tannins are phenolic substances, which upon consumption produce adverse effects such as reduced feed consumption, growth retardation and impaired nutrient absorption (Butler et al. 1986). They contain multiple phenolic hydroxyl groups, which complexes with proteins and to lesser extent with polysaccharides, amino acids and metal ions. The defatted *J. curcas* kernel meal contains very low concentration of total phenols (0.2–0.4%) and tannins (0.02–0.04%). In addition, condensed tannins were not detected in the *J. curcas* kernel meal (Makkar et al. 1998a). However, small amount of tannins were found in bark (outer dark bark: tannins 0.7% and condensed tannins 0.2%; inner green bark: tannins 3.1% and condensed tannins 1.7%; tannins as tannic acid equivalent and condensed tannins as leucocyanidin equivalent) (Makkar and Becker 2009). Similarly, non-toxic *J. platyphylla* kernel meal also has low amounts of tannin (0.17%) (Makkar et al. 2011). Since very low

Table 21.1 Levels of toxic and antinutritional factors in unheated kernel meals of *J. curcas* (toxic and non-toxic genotypes) and *J. platyphylla*

	J. curcas		*J. platyphylla*
Component	Toxic	Non-toxic	Non-toxic
Phorbol esters (mg/g kernel)[a]	2.79	ND	ND
Total phenols (% tannic acid equivalent)	0.36	0.22	0.33
Tannins (% tannic acid equivalent)	0.04	0.02	0.17
Condensed tannins (% leucocyanidin equivalent)	ND	ND	ND
Phytates (% dry matter)	9.40	8.90	8.66
Saponins (% diosgenin equivalent)	2.60	3.40	1.94
Trypsin inhibitor (mg trypsin inhibited per g sample)	21.3	26.5	20.81
Lectin activity (1/mg of meal that produced haemagglutination per ml of assay medium)	51–102	51–102	51–102
Glucosinolates	ND	ND	ND
Cyanogens	ND	ND	ND
Amylase inhibitor	ND	ND	ND
Non-starch polysaccharides (NSP) (% in dry matter)			
Rhamnose	0.2	0.2	0.3
Fucose	0.1	0.1	0.1
Arabinose	2.5	2.7	3.1
Xylose	1.2	1.4	2.0
Mannose	0.3	0.3	0.5
Galactose	1.2	1.2	1.4
Glucose	4.7	4.7	5.7
Glucuronic acid	0.9	0	0
Galacturonic acid	2.6	3	3.0
Total NSP	12.7	13.6	16.0

Source: Makkar and Becker (2009) and Makkar et al. (2011)
[a]As phorbol-12-myristate 13-acetate equivalent. *ND* not detected

levels of tannins are present in the defatted *J. curcas* kernel meal, they cannot be considered as the compounds responsible for causing adverse effects observed on consumption of the meal from toxic genotypes.

Cyanogenic Glycosides, Glucosinolates and Amylase Inhibitors

Cyanogenic glycosides are toxic nitrogenous compounds that on hydrolysis produce hydrogen cyanide (HCN), which is a lethal chemical responsible for halting cellular respiration in aerobic organisms. On the other hand, glucosinolates are sulfur-containing glucosides synthesized by members of the *Brassicaceae* family and generally consist of a sugar entity (β-D-thioglucose), ester bonded to an organic aglycon. These compounds often contribute a bitter or hot taste and may exhibit goitrogenic or antithyroid activity. The glucosinolates are hydrolyzed by enzymes, such as glucosinolase or thioglucosidase into glucose, HSO_4- and one of the following aglycone derivatives: isothiocynates, thiocyanates, nitriles or related compounds, such as oxazolidine-2-thiones, which produce adverse effects. Amylase inhibitors impede the digestive action of α-amylases and proteinases in gut, thereby reducing the starch digestion (Freeman and Beattie 2008). However, cyanogenic glycosides, glucosinolates and amylase inhibitors were not detected in *J. curcas* kernel meal (Makkar et al. 1997, 2011; Makkar and Becker 2009).

Saponins

Saponins are steroid or triterpene glycosides and they can form stable soapy or froth-like formations in aqueous solutions. In plants, saponins may serve as anti-feedants or help in protecting the plant against microbes and fungi. Saponins, due to their bitter taste reduce palatability of plants when present in livestock and aqua-feeds (Sen et al. 1998). In nutritional context, saponins are considered antinutritional compounds, but they are also claimed to have beneficial effects. In principle, saponins act on lipid membranes of the cell and causes haemolysis *in vitro* or when injected intravenously. In general, saponins, as glycosides, have low oral bioavailability, but may be hydrolysed in the intestinal tract and cause systemic toxicity depending on the structure and absorption of the aglycone (European Food Safety Authority 2009). Till now, no individual saponin has been purified from *J. curcas* and tested for its toxicity. The concentration of saponins in toxic and non-toxic genotypes of *J. curcas* was found to be 1.8–2.6 and 3.4% (as diosgenin equivalent) respectively; whereas, in *J. platyphylla* saponin concentration was 1.94% (as diosgenin equivalent). The saponins present in *J. curcas* do not have haemolytic activity (Devappa et al. 2010d; Makkar et al. 1997, 1998b; Makkar and Becker 2009). Low levels of saponins in the toxic genotype and presence of higher level in the non-toxic genotype suggests that *J. curcas* saponins do not elicit any adverse effects.

Trypsin Inhibitors

Protease inhibitors are globular proteins found in many plant derived nutritional ingredients (Norton 1991). They are known to decrease protein digestibility by reducing the activity of pancreatic enzymes (trypsin and chymotrypsin), which are involved in protein digestion (Liener and Kakade 1980; Hertrampf and Piedad-Pascual 2000; Agbo 2008). In soybean, two types of protease inhibitors have been reported. The Kunitz type inhibitor is heat and acid sensitive and the Bowman-Birk type inhibitor is more stable to heat. One molecule of the former inhibits one molecule of either trypsin or chymotrypsin, while one molecule of the latter blocks either two trypsin or chymotrypsin molecules or one trypsin and one chymotrypsin molecule at the same time (Norton 1991). In *J. curcas*, there are no reports on purification and characterization of *trypsin inhibitors* (TI). However, TI activity in the kernel meal of both toxic and non-toxic genotypes of *J. curcas* was found to be similar, ranging from 18.4 to 27.3 TIU (mg trypsin inhibited/g) (Makkar et al. 1997). Makkar and Becker (1999) found that carp (*Cyprinus carpio*) when fed on diets containing *J. curcas* kernel meal of the non-toxic genotype with 24.8 TIU (mg trypsin inhibited/g) and heat-treated (45 min, 121°C, 66% moisture) meal with 1.3 TIU (mg trypsin inhibited/g) had no marked difference in growth performance, indicating that carps were able to tolerate high levels of TI. However, feeding of unheated *J. curcas* kernel meal to monogastric animals, such as poultry, pigs, and fish other than carp may produce adverse effects, since the levels of TI in *J. curcas* kernel meal are similar to that in raw soybean meal (Makkar and Becker 1999). Being proteinaceous in nature, trypsin inhibitor is sensitive to heat and could be easily inactivated by heat treatment. Trypsin inhibitors in *J. curcas* kernel meal could be completely removed by heat treatment (121°C for 30 min) (Aderibigbe et al. 1997; Aregheore et al. 2003; Makkar and Becker 2009). In addition, TIs (such as soybean TI, Kunitz type) have anticancerous properties, such as suppression of ovarian cancer cell invasion by blocking urokinase upregulation. However, no information is available on the biochemical and pharmaceutical properties of TIs from *J, curcas*. In the nutritional context, *J. curcas* TIs could be considered as an antinutrient, which could be inactivated by heat treatment.

Lectin

Lectins are carbohydrate-binding proteins (glycoprotein) and are ubiquitous in nature. They bind reversibly and specifically to carbohydrates and glyco-conjugates, which is responsible for their numerous physiological effects. For example, lectins bind avidly with intestinal glycoproteins on the epithelial surface and interfere with nutrient absorption. Lectins easily evade digestion and then enter in the intestine wherein they possibly bind to the epithelium (Vasconcelos and Oliveira 2004). Some lectins may cause disruption of membrane integrity and initiate a cascade of immune and autoimmune events that ultimately lead to cell death. These lectins also (1) interrupt lipid, carbohydrate and protein metabolism causing atrophy or enlargement of internal organs and (2) amend the biochemical, hormonal and immunological conditions. High consumption of lectins distinctly threatens the growth and health of animals (Vasconcelos and Oliveira 2004). *J. curcas* kernel meal contains lectins at levels of 102 and 51 (inverse of mg meal per ml of assay medium that produced hemagglutination) for toxic and non-toxic genotypes, respectively, the levels are in similar range as present in soybean meal (Makkar et al. 2007). The range of lectin activity observed for both genotype (toxic and non-toxic) meals is almost similar (Table 21.1). *J. curcas* lectin is heat labile and can be inactivated by autoclaving at 121°C for 20 min (Aderibigbe et al. 1997, 2003; Makkar and Becker 2009).

Curcin

Curcin is a toxalbumin, classified as ribosome inactivating proteins (RIPs). Based on their physical properties, RIPs are classified into three groups—Type 1 RIPs, Type 2 RIPs and Type 3 RIPs. Type 1 RIPs are single chained proteins (A chain, ~30 kDa) having N-glycosidation enzymatic activity. The N-glycosidation activity involves the removal of specific adenine corresponding to residue A_{4324} in rat 28S rRNA. These proteins inhibit cell-free protein synthesis *in vitro*, but they are relatively non-toxic to cells and animals. Type 2 RIPs are heterodimeric proteins (~60 kDa) containing A chain (which has similar function as type 1 RIP) and sugar binding B chain, which are joined together by disulfide linkage. The B chain can bind to galactosyl moieties of glycoproteins and or glycolipids present on eukaryotic cell surface, which in turn facilitates retrograde transport of A chain into the cytosol. Whereas, Type 3 RIPs are present as inactive precursors, which need proteolytic processing to become active (Barbieri et al. 1993; Endo and Tsurugi 1988; Lin et al. 2003a).

Curcin purified from the seeds of *J. curcas* is a type 1 RIP (28.2 kDa) with an isoelectric point of 8.54. It exhibits RNA N-glycosidase activity like other type 1 RIPs. Jin-Ping et al. (2005) reported that there are two subfamilies in the curcin gene family (1) present in the endosperm of the seeds and (2) expressed by stress conditions and microbial infestation. In the leaves, a curcin-related RIP (curcin-L) is induced by infection (*Pestalotia funerea* and *Gibberella zeae*), and its expression could be activated by abscisic acid, salicylic acid, polyethylene glycol at temperatures of 4°C and 45°C and by ultraviolet light (Qin et al. 2009). Similarly, Wei et al. (2005) and Huang et al. (2008) have also reported that under stress, drought or fungal infestation, *J. curcas* plants express another protein similar to curcin, called curcin 2 (32 kDa).

The purified curcin isolated from *J. curcas* seeds, exhibited cell-free translation inhibition in the reticulocyte lysate system with an IC_{50} (95% confidence limits) ranging from 0.11 to 0.42 nmol/L (Lin et al. 2003b; 2010). This IC_{50} is higher than those of other RIPs, such as saporin (0.5 nmol/L), luffin A (1 nmol/L) and luffin B (4 nmol/L), and lower than of trichosanthin (0.32 nmol/L) (Barbieri et al. 1993). Recently, Lin et al. (2010) reported that curcin is a glycoprotein (4.9% sugar content) and exhibit haem-agglutinating activity when the concentration is more than 7.8 mg/L. Curcin also exhibited toxicity in fish (*Gambusia*) (94% mortality in 99 h) upon feeding of raw curcin fluid (200 μg) and in mice it exhibited LD_{100} of 1.6 mg/kg body weight (subcutaneous injection after 9 days) (Jiang et al., 2007). Lin et al. (2010) have reported that in mice purified curcin showed oral LD_{50} (semi-lethal dose) of 104.7 ± 29.4 mg/kg body weight and parenteral LD_{50} (semi-lethal dose) of 67.2 ± 10.4 mg/kg body weight. Curcin was also found to inhibit hyphal growth and spore formation in *Pyricularia oryzae* Cavara, *Pestalotia funerea* and *Sclerotinia sclerotiorum*. In addition, crude curcin and purified curcin (curcin I) caused acute toxicity (9.11 and 6.48 mg/mouse) and delayed toxicity (5.83 and 2.21 mg/mouse) in mice (Stripe et al. 1976).

Overall from the above studies it indicates that curcin is toxic upon oral or subcutaneous exposure. Thus, curcin could be considered as a toxic factor in *J. curcas* kernel meal or whole seed cake. In our studies (unpublished), purified curcin from both toxic and non-toxic genotypes of *J. curcas* defatted kernel meal had a similar molecular weight of 28 kDa, demonstrating the presence of curcin in the non-toxic genotype as well. Furthermore, the electrophoresis band pattern (qualitative comparison) of purified curcin, crude curcin fraction prepared from unheated kernel meals (both from toxic and non-toxic genotypes), heat treated kernel meals (both from toxic and non-toxic genotypes) were not similar to each other. The band representing curcin at a region of molecular weight of 28 kDa, was present in unheated crude curcin fractions; whereas, it was absent in the heat treated ones, indicating inactivation of curcin due to heat treatment. However, hitherto, no study has been conducted to compare curcin activity in kernel meals of toxic and non-toxic genotypes of *J. curcas.*

In addition, feeding of the heat treated kernel meals of the non-toxic genotype of *J. curcas* did not show any adverse effects in carp. Furthermore, in the detoxification process (Makkar and Becker 2010) after removal of PEs, the meal is heat treated (121°C for 15 min) to inactivate lectin, curcin and trypsin inhibitors; the feeding of this meal produced excellent results in fish, turkey and pigs; suggesting that curcin is not the main toxin and its adverse effects if any could be removed by heat treatment.

Phorbol Esters

In *J. curcas*, six PEs (factors C_1–C_6) have been characterized (Haas et al. 2002; Haas 2003) and designated as factor C_1, C_2, C_3, epimers C_4, C_5 and C_6, with the molecular formula $C_{44}H_{54}O_8Na$ (mol. wt. 733.37) (Fig. 21.1). These are lipophilic compounds present mainly in oil or kernel. When they are present in oil or kernel generally they are not affected by the heat treatment (Devappa et al. 2010c, d). The concentration of PEs varies from 1–4 mg/g dry matter in defatted kernel meal and from 2–8 mg/g in oil of *J. curcas* (Makkar et al. 1997; Makkar and Becker 2009; Devappa et al. 2010e). In *J. curcas,* the PEs were present in kernels, leaves, stems, flowers, buds, roots, bark (outer brown skin), bark (inner green skin) and wood, but not in latex (Table 21.2) (Makkar and Becker 2009). The PEs are distributed unevenly in the kernel. The distribution was highest (85.7%) in the storage region of endosperm, providing defensive environment for developing embryo during germination. The kernel coat (11.3%), hypocotyl (0.5%) and cotyledon (2.5%) contribute relatively low amount to the total pool of PEs present in the kernel (Devappa et al. 2012c). Other *Jatropha* species such as *J. integerrima* and *J. multifida* contain higher amount of PEs (Table 21.2) when compared with *J. curcas*. Although PEs are lipophilic, they have a strong affinity towards the matrix of kernel meal. Most of the reported studies showed that *Jatropha* PEs exhibit toxicity in a broad range of species, from microorganisms to higher animals (Wink et al. 1997; Devappa et al. 2010d).

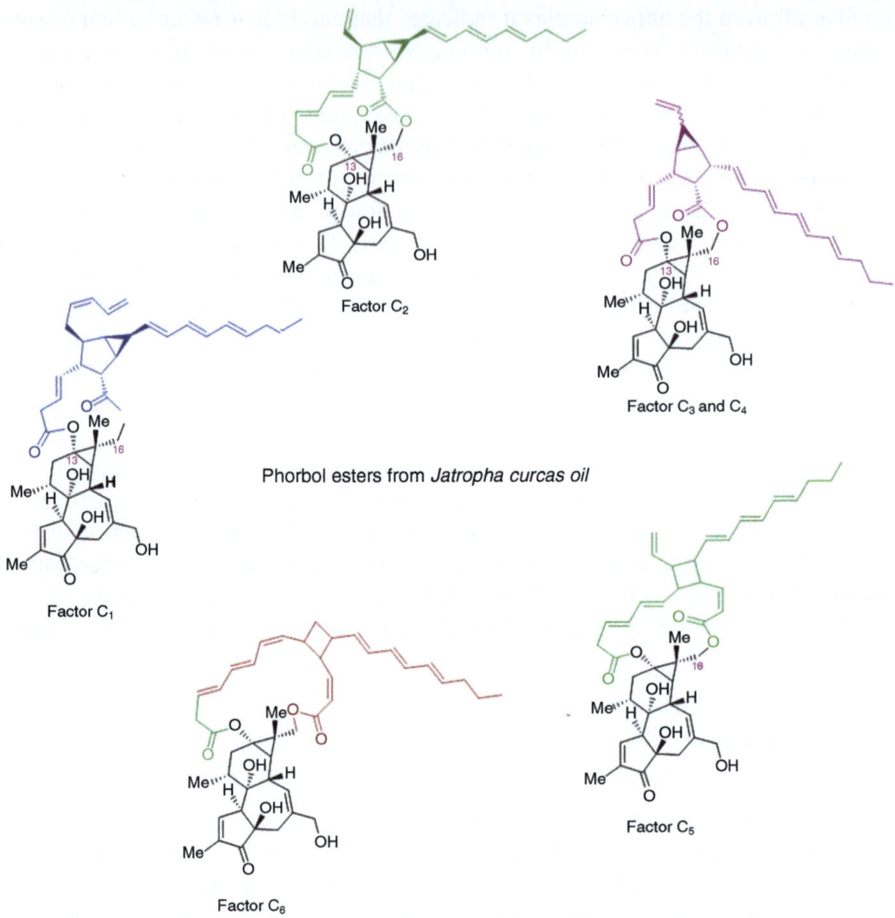

Fig. 21.1 Phorbol esters from *J. curcas* oil (Figure adapted from Devappa (2012) and Haas et al. (2002)

The *Jatropha* PEs are known to be skin irritant (Gandhi et al. 1995). However, their presence in feed material causes severe deleterious effects in animals. In majority of the animal studies (for example, sheep, goat, pig, chicken and fish), the consumption of PE containing feed materials caused severe effects, such as reduced growth, reduced feed intake, marked histological alterations and even mortality (Goel et al. 2007; Adam and Magzoub 1975; El-Badwi and Adam 1992; Devappa et al. 2010d). For example, acetonitrile extract of *J. curcas* (seed or oil) when given to albino rats at an oral dose of 50 mg/kg body mass (single dose) produced mild toxicological, biochemical and histopathological changes. The methanol, petroleum ether and dichloromethane extracts of *J. curcas* fruits caused fetal resorption, indicating pregnancy terminating effect in rats. Most of toxicity studies in higher animals are carried out by force-feeding raw or defatted seed meals, leaves or their

Table 21.2 Phorbol esters (PEs) in different parts of toxic *J. curcas* plants

Parts	PEs (mg/g dry matter)[a]
Kernel[b]	2.00–6.00
Leaves[b]	1.83–2.75
Stems[b]	0.78–0.99
Flower[b]	1.39–1.83
Buds[b]	1.18–2.10
Roots[b]	0.55
Latex[b]	not detected
Bark (outer brown skin)[b]	0.39
Bark (inner green skin)[b]	3.08
Wood[b]	0.09
J. integerrima kernel	7.92
J. multifida kernel	9.09
J. podagrica kernel	4.50
J. glandulifera kernel	2.78

Sources: Makkar and Becker (2009) and Popluechai et al. (2009)
[a]As phorbol-12-myristate 13-acetate (PMA) equivalent
[b]*J. curcas* parts

various organic solvent/aqueous extracts, since the animals do not consume them voluntarily (Devappa et al. 2010d). The studies of Makkar and Becker (1999) unequivocally proved that PEs are the main toxic compounds in *J. curcas* meal. This conclusion is based on the observations that lectin and trypsin inhibitor levels in both toxic and non-toxic genotypes are similar and PEs are absent in the non-toxic genotype. In addition, the purified PEs from *J. curcas* kernel or oil produces the same toxic symptoms in fish as observed on feeding the kernel meal or oil. The tolerance limit of PEs in fish (carp) is as low as 15 ppm (Becker and Makkar 1998). However, there are no studies reporting tolerance limit in higher animals. The PE-containing extracts also showed toxicity against microorganisms, snails, insects, ruminants, mice, rat, chicken and humans (Devappa et al. 2010d). The removal of PEs is necessary for utilization of the seed cake/kernel meal in animal diets. There are many detoxification methods reported which are basically using solvent, microbial and chemical treatment methods. However, most of the methods reported have been unsuccessful in completely detoxifying the kernel meal and protein isolate. Recently, Makkar and Becker (2010) have reported the successful detoxification of *J. curcas* kernel meal and protein isolate. The feeding of detoxified *J. curcas* kernel meal (DJKM) and detoxified protein isolate (DPI) did not affect growth, nutrient utilization and health parameters of fish (carp and rainbow trout—*Oncorhynchus mykiss*). Due to their high protein content of DJKM and DPI (60 and 90%, respectively) and their excellent amino acid composition, they could replace at least 50% of the protein contributed by the fish meal (65% protein) in standard fish diet. Similar results were achieved when these detoxified *J. curcas* products were fed to turkey, pigs and broilers (Makkar et al. 2009a). Results so far obtained on fish and other animal species suggest that DJKM and DPI are ideal substitutes for fish meal or soybean meal for livestock diets (Makkar et al. 2012).

Phytates

Phytic acid (known as inositol hexakisphosphate [IP_6] or phytate when in the salt form) is the principle storage form of phosphorus in most plant seeds. Inositol penta- (IP_5), tetra- (IP_4) and triphosphate (IP_3) are also termed phytates. Generally, phosphate present in phytate molecule is not bio-available for monogastric animals due to lack of digestive enzyme (phytase), while ruminants can make use of phytate-phosphorus due to the presence of phytase in rumen microbes. On consumption, phytates chelate with di and/or trivalent mineral ions such as Ca^{2+}, Mg^{2+}, Zn^{2+}, Cu^{3+}, and Fe^{3+}, resulting in these ions becoming unavailable for consumers (Duffus and Duffus 1991). Thus, these molecules act as an antinutrient for monogastric animals (Liener 1989). Phytates also reduce bioavailability of dietary protein by forming sparingly digestible phytate–protein complexes (Richardson et al. 1985). *J. curcas* kernel meals of both toxic and non-toxic genotypes contain high amount of phytate, ranging from 7.2–10.1% (Makkar et al. 1997, 1998a, b). The kernel meal of, *J. platyphylla* contains 8.6% of phytate (Makkar et al. 2011). Generally, phytates are more soluble in aqueous phase than in organic solvent phase. However, the commonly used processing techniques do not utilize aqueous phase extraction or washing steps to remove phytates from the meal because aqueous treatments also reduce the contents of protein and other nutrients in the processed meal. In addition *Jatropha* phytates are heat stable. Thus, supplementation of phytase becomes inevitable. Recently, Kumar et al. (2011a) reported that feeding of *Jatropha* phytate in the fish (*Nile tilapia*) diet at 1.3% and 5% level exhibited significant reduction in the growth and feed utilization. However, the same diets when supplemented with phytase increased the growth and feed utilization. Since the levels of phytate in the kernel meal are high, efficient utilization of the meal as feed for monogastric animals would require the external supplementation of phytase in the diet. Considering high levels of phytate in the kernel meal and their high stability to heat, *Jatropha* phytate could be considered as an antinutrient.

Non-starch Polysachharides

Non-starch polysaccharides (e.g., cellulose, β-glucans, arabinoxylans, pectins, gums and mucilages) are indigestible by poultry and other monogastric animals and can reduce availability of nutrients within the cells by their encapsulation activity (McDonald et al. 1995; Bedford 2000; Bedford and Schulze 1998). In addition, they can increase digesta viscosity, which, in turn, can reduce enzyme activities in the gastrointestinal track, leading to reduction in nutrient intake and digestibility (Bedford and Schulze 1998). The levels of *non-starch polysaccharides* (NSP) in *J. curcas* kernel meals are almost similar (12.7% and 13.6% for toxic and non-toxic genotypes, respectively). By comparison, the kernel meal of *J. platyphylla*, which is non-toxic, has a higher NSP content of 16% (Makkar et al. 2011). These levels of NSP in *Jatropha* meals are slightly higher than in soybean meal (15.5%) and lower than in other conventional protein-rich feed resources, such as rapeseed meal, cottonseed cake, linseed meal, coconut cake, palm cake and sunflower cake that have 17.8, 16.4, 19.3, 25.0, 36.8, 39.3 and 19.3%, respectively (Makkar et al. 2011; Knudsen 1997). The NSP content in the feed material can be reduced by treating it with NSP-degrading enzymes, such as carbohydrases. However, not all NSPs are responsible for adverse effects on the digestion. Excellent growth performances (Makkar and Becker 2010; Makkar et al. 2011) observed with the diets containing *Jatropha* meal, without the addition of NSP-degrading enzymes; suggest that NSPs do not have any adverse digestive effects up to 50% replacement of fishmeal protein by *Jatropha* kernel meal. Higher levels of incorporation of kernel meal (>75% fishmeal protein replacement) in fish diet, adverse effects on growth, which could be due to the presence of NSPs or incomplete degradation of phytate by the phytase added into the diet (Kumar et al. 2011b). Addition of higher levels of phytase and/or NSP-degrading enzymes could improve our understanding of the role played by *Jatropha* NSPs when present at high concentrations in diets.

Potential Applications of Phorbol Esters

High potency and multiple biological activities of PEs have created fascination and much interest towards these compounds amongst scientists. PEs have long been studied as a tool to understand the molecular mechanisms in cells, especially in the field of cancer research, toxicology, immunology and nutritional sciences (Goel et al. 2007).

Chemistry and Distribution

PEs are widely distributed in plants especially Euphorbiaceae and Thymelaceae families (Beutler et al. 1989; Goel et al. 2007; Haas 2003). Among the many PEs reported, *phorbol 12-myristate-13 acetate* (PMA) (synonym: 12-*O*-tertradeconoylphorbol-13-acetate—TPA) is the most studied one, due to its relatively high potency and commercial availability. PMA has been initially isolated from *Croton tiglium* L. oil. The compound phorbol is a diterpene having a tigliane basic skeleton (Bohm et al. 1935; Silinsky and Searl 2003). Whereas PEs are diversely oxygenated and hydroxylated forms of tigliane skeleton, PEs are highly soluble in organic solvents (e.g., ethanol, methanol, dichloromethane, and dimethyl sulfoxide, among others) and can be easily extracted by solvent extraction or partition techniques (Devappa et al. 2010e). Further, purification is generally carried out using the combination of chromatography, mass spectroscopy and NMR techniques (Haas et al. 2002; Roach et al. 2012). These compounds can also be produced by synthetic methods. More than 50 different types of PEs have been reported (both of natural and synthetic origin) (Haas 2003; Wender et al. 1999, 2008), but very few have been extensively studied with respect to their biological activities.

Phorbol esters exhibit various biological activities, such as tumour promotion, platelet aggregation, apoptosis, cell differentiation, skin irritation and other metabolic effects. Among them, tumour promotion by PEs has been the most studied (Goel et al. 2007; Kinzel et al. 1984; Weinstein et al. 1979). The tumor promoting PEs do not induce tumor when applied alone, but increase the chance of tumor formation on prior application of primary carcinogen (tumor initiators). Majority of cellular activity is mediated through the activation of *Protein kinase C* (PKC), an enzyme which plays an important role in signal transduction pathways and controls cell growth and differentiation (Clemens et al. 1992; Nishizuka 1992). The PEs act as an analogue of *diacyl glycerol* (DAG), a secondary messenger involved in cellular signal transduction pathways. Generally, the biological activities of PEs are structure dependent (Bertolini et al. 2003). For example, the placement of an OH group in ring C makes the phorbol an active (ß form) or inactive type (α form), which results in spatial re-arrangement of D ring and precludes the activation of PKC and other structurally similar PE receptors. The inactive 'α' PEs have similar physicochemical properties, especially lipophilicity, as the active 'ß' phorbols, but are unable to activate PKC due to conformational shifts (Silinsky and Searl 2003).

There are other PEs that are non-tumour promoters (e.g., 12-deoxyphorbol 13-acetate or prostratin, 12-deoxyphorbol 13-propanoate and 12-deoxy phorbol 13-phenylacetate) (Xu et al. 2009) and have at least one of the biological activities of phorbol compounds, such as binding to phorbol receptors (such as PKC), but do not have tumour promoting properties. The PEs are also reported to exhibit non-PKC mediated biological effects, such as neurotransmitter secretion. For detailed information on observed *in vitro* and *in vivo* biological activities, see review articles by Goel et al. (2007) and Silinsky and Searl (2003). The non-PKC enzyme receptors include (a) chimaerins, (b) RasGRP and (c) *Caenorhabditis elegans* Unc-13 and mammalian Munc13s.

The toxicity of any chemical depends on dosage, mode and duration of its exposure. Generally the toxicity increases with the concentration of the toxic compound administered to the test organism. However, at lower non-toxic dosages these chemicals can sometimes exhibit beneficial properties, for example cytotoxic, anti-tumour or anti HIV properties (Goel et al. 2007). Similarly, PEs also exhibit toxicity at higher doses and beneficial biological activities at very low concentrations, acting as a double edged sword. Also not all PEs are toxic. Their activity and potency vary from one type of phorbol ester to another.

The extracts of *J. curcas* plant parts containing PEs are effective in controlling microbes and pests of agricultural interest, suggesting that PEs may have applications as biological control agents (Devappa et al. 2010d). The purified PEs could also be converted or transformed chemically into non-toxic compound, such as prostratin having beneficial activities. Prostratin has been found to be a promising adjuvant in anti-HIV therapy (Wender et al. 2008, 2009). Thus, the beneficial effects of PEs could be exploited depending on the applications.

J. curcas Phorbol Esters and Their Utilization

Jatropha contains at least 6 PEs (Haas et al. 2002) (Fig. 21.1). It should be noted that unless specified the concentration of *J. curcas* PEs is expressed below as equivalent to PMA, a phorbol ester isolated from *Croton tiglium*. During biodiesel production from *J. curcas* oil, many co-products could be obtained. The efficient use of these co-products (such as glycerol, fatty acid distillate and seed cake, among others) would enhance the economic viability of the *J. curcas* biofuel industry. However, the possible presence of PEs in these co-products restricts their efficient utilization. The removal of PEs from the *J. curcas* products would provide considerable opportunities for effective utilization of these products. The presence of PEs in oil also poses occupational health risk. The challenge is to make the co-products and *J. curcas* oil less toxic or toxin free. The best approach would be either to destroy the PEs, select zero phorbol ester genotypes or to extract them carefully as a value added co-product.

Generally, during biodiesel production, *J. curcas* oil is subjected to many treatments (e.g., stripping, degumming and esterification) that cause partial or complete destruction of PEs depending on the treatment conditions (Makkar et al. 2009b). The optimization of suitable methods to best extract the PEs from *J. curcas* oil before subjecting to biodiesel process would make the oil less toxic and more worker-friendly. In addition, PEs could be obtained as a value added product instead of simply allowing them to get destroyed during biodiesel production.

Majority of the PEs are localised in the endosperm portion of the kernel. When kernels are solvent (e.g., hexane) extracted to obtain oil, 70–75% of the total PEs pass into the oil and 25–30% remain in the kernel meal (Makkar et al. 2009b). PEs could be extracted easily from the oil using organic solvents, such as methanol. Devappa et al. (2010e) reported that PEs can be extracted (up to 99.4%) using simple tools, such as magnetic stirrer and ultra turrex with multiple extraction steps (60 min, 4 extraction steps). Extracted PEs were recovered as an enriched fraction (PEEF) (48.4 mg PEs/g), which is 14 fold concentrated compared to PEs in original oil. In addition, the obtained PEEF was highly bioactive when tested in snail bioassay (LC_{100}, 1 μg of PEs/ml). A shortened version of this method is to extract approximately 80% PEs by single extraction step using a magnetic stirrer or an ultra turrex (5 and 2 min, respectively); this simplified method will increase economic viability of the process. In addition, Devappa et al. (2010b, e) reported that the residual oil obtained after PEs extraction has good feedstock quality for biodiesel production. The biodiesel prepared from residual oil met both the European (EN 14214:2008) and American biodiesel standard (ASTM D6751-09) specifications. Overall, the study showed that PEs could be easily extracted by either of the aforesaid two methods with a high yield and the residual oil could be processed to produce high quality biodiesel. Also, the residual oil with a lower PE content is expected to be lesser toxic to the environment and to the workers who have to handle it.

The extracted *J. curcas* PEs present in PEEF could find applications in agriculture and in pharmaceutical industries.

It should be noted that any compound aimed to be used in agricultural applications should fulfil certain basic requirements; for example, extracted/purified compound should remain stable and active during extraction procedures, compounds should have high biological activity and long shelf life, and they should be biodegradable. Devappa et al. (2011) have reported that PEEF has high biological activity when tested in various bioassays and microorganisms. The EC_{50} (48 h) of the PEs present in PEEF was 0.33, 26.48 and 0.95 ppm for snail, brine shrimp and daphnia, respectively. High MIC (minimum inhibitory concentration) values (≥215 ppm) and EC_{50} values (≥58 ppm) were obtained for both the bacterial and fungal species. The shelf life of PEs in PEEF was found to be shorter when stored at room temperature (50% degradation after 132 days) than at 4°C or −80°C (8% and 4% degradation, respectively). During storage, the biological activity of PEEF also decreased with the decrease of PEs concentration in PEEF. PEEF was inactive after 260 days of storage, whereas at 4°C and −80°C, only 27.5% and 32.5% activity was lost after 870 days. The degradation of PEs was due to auto-oxidation, as reflected by changes in fatty acid composition, increase in peroxide value and decrease in free radical scavenging activity of PEEF (Devappa et al. 2009). The supplementation of antioxidants, such as *butylated hydroxyanisole* (BHA), baynox and α-tocopherol was found helpful in protecting PEs against degradation. Furthermore, Devappa et al. (2010a) evaluated PEEF biodegradability in soil. The PEs present in PEEF (2.6 mg/g) in mixture to soil were completely degraded with an increase in temperature and moisture content as well as when they were, first, bounded to silica and, then, mixed to soil after 19, 12, 12 days (at 13% moisture) and after 17, 9, 9 days (at 23% moisture) at room temperature (22–23°C), 32°C and 42°C, respectively. As expected, the decrease in biological activity of PE in soil followed the same pattern as their concentration in soil. The study demonstrated that PEs present in the PEEF are completely biodegradable in soil and the degraded products are innocuous.

The extracted PEs in the form of PEEF has high biological activity, prolonged shelf life at room temperature in the presence of antioxidants (butylated hydroxyl anisole, baynox and α-tocopherol) and complete biodegradability in soil. Considering the above factors, PEEF could be a potential candidate as biocontrol agent against pests. Devappa et al. (2012b) have reported that PEEF exhibited insecticidal activity against *Spodoptera frugiperda* (J. E. Smith), which is a common pest in corn fields and damages maize crop across the tropical/subtropical countries, such as Mexico and Brazil. PEEF exhibited contact toxicity (LC_{50} of 0.83 mg/ml., w/v) against *S. frugiperda*. At higher concentration (0.25 mg/ml, w/v), PEEF reduced food consumption, relative growth rate and food conversion efficiency by 33%, 42% and 38%, respectively. This study showed that PEEF is a promising preparation for controlling the pest larvae and has potential use in agricultural applications as a biocontrol agent. However, further field experiments on the effects of PEEF on *S. frugiperda* are required.

Although *J. curcas* PEs are known to have a wide spectrum biological activity when used as crude form or as a mixture of PEs; individual potency and biological activity of purified *J. curcas* PEs has been little investigated. In addition, similar to other drug candidates if *J. curcas* PEs are to be used in pharma applications the assessment should be made using purified PEs instead of crude PE extracts to avoid interfering results. Roach et al. (2012) reported the purification of *J. curcas* PEs. The factors C_1 and C_2 are purified to homogeneity; factor C_3 was obtained with minute impurities of factors C_1 and C_2 (termed as $C_{3mixture}$) and both factors C_4 and C_5 were obtained as mixture together (termed as (C_4+C_5)). In addition, PE-rich fraction containing all PEs was also obtained in the method described by Roach et al. (2012). It should be noted that in the following discussion, the concentration of PEs are expressed equivalent to factor C_1, a PE purified from *J. curcas* oil. All the *J. curcas* PEs were biologically active when tested in snail and brine shrimp (*Artemia*) bioassays. In the snail bioassay, the order of potency based on EC_{50} was: PE-rich fraction < factor $C_{3mixture}$ < factor C_2 < factor C_1 < factors (C_4+C_5) mixture. In the *Artemia* bioassay, the order of potency based on EC_{50} (ppm, equivalent to *J. curcas* factor C_1) was: factor C_2 < factor $C_{3\,mixture}$ < factor C_1 < factors (C_4+C_5) mixture. In addition, all the *J. curcas* PEs produced platelet aggregation *in vitro* and the order of activity based on ED_{50} was: factor C_2 < factor $C_{3mixture}$ < factor C_1 < factor (C_4+C_5). The study showed that (1) the *J. curcas* PEs obtained (by the method of Roach et al. 2012) are biologically active, (2) PEs differ in their activity amongst themselves and (3) their order of activity depends on the target organism.

In another study, oral (intra gastric) administration of *J. curcas* PEs in the form of PEs rich extract (containing *J. curcas* factors C_1 to C_6 in mixture) was found to be toxic to mice with an LD_{50} of 27.34 mg/kg body mass as PMA equivalent or 0.66 mg/kg body mass as *J. curcas* factor C_1 equivalent. The prominent histopathological symptoms were observed in lung and kidney (Li et al. 2010).

Devappa et al. (2011) reported that *J. curcas* PEs (PEs-rich extract, factors C_1, C_2, $C_{3mixture}$ and C_4+C_5) upon topical application on *reconstituted human epithelium* (RHE) and *human corneal epithelium* (HCE) promoted cellular alterations or disintegration of epithelium layer and also an increased inflammatory response (interleukin-1α and prostaglandin E2 release). Whereas, phosphate buffered saline (experimental blank) did not produce any effect on both RHE and HCE. In the RHE, even the non-toxic oil free of PEs (used as a control; equivalent to the volume used for toxic oil) produced mild cellular and inflammatory response, which was much lower than the response produced by toxic oil. Also in the HCE, non-toxic oil (equivalent to the volume used for toxic oil) produced marked cellular alterations. These results suggest that some factors that cannot be considered as PEs are present in these oils and promote cellular toxicity. However, the presence of PEs in *J. curcas* oil increased the toxicity, both towards the RHE and HCE and the toxicity increased with the PE concentration (Devappa et al. 2011). Similar to PEs from other plant species, *J. curcas* PEs also showed tumour promotion properties when tested *in vitro* using Bhas 42 cells (Devappa et al. 2012a). In the tumour promotion assay, the order of transformed foci/well formation was: PEs-rich extract > factor (C_4 + C_5) > factor $C_{3mixture}$ > factor C_1 > factor C_2. The tumour promotion activity was mediated by the hyperactivation of protein kinase C (PKC). The aforementioned studies demonstrated that *J. curcas* PEs are toxic when administered orally or when applied topically to the skin or eye tissues. Thus, caution is required (e.g., wearing of hand gloves and eye protecting glasses) while using *Jatropha* products and particularly oil from toxic *J. curcas* for various applications.

Wender et al. (2008, 2009) suggested that *J. curcas* PEs could be used as an intermediate feedstock for synthesis of prostratin, a promising adjuvant in anti HIV therapy. Devappa et al. (2012d) in a preliminary study demonstrated that the *J. curcas* PEs in PEEF could be used to synthesize prostratin by converting them first to crotophorbolone and then to prostratin. The methodology followed was that reported by Cairnes et al. (1981) and Wender et al. (2008, 2009). The prostratin synthesized from *J. curcas* PEs had mass and peak retention time similar to the reference prostratin (Sigma, St. Louis), further optimization studies are required to ascertain the synthesis reactions and yield of prostratin synthesized from *J. curcas* PEs (Devappa et al. 2012d).

Apart from high level of biological activity, long shelf life and proper biodegradability, phytochemicals should be available in large quantity with continuous supply and should also be easily extractable. The presence of PEs in high concentrations in oil could make it a novel 'stock' for the agro-pharmaceutical industries. Integration of PEs extraction methods during biodiesel production and their utilization for agro-pharmaceutical applications would increase economic viability and sustainability of the *J. curcas* biodiesel chain in addition to render it more environment and worker friendly.

Acknowledgement Authors are grateful to the Bundesministerium für Bildung und Forschung (BMBF), Berlin, Germany for the financial assistance provided for the research work. The technical assistance of Mr. Vikas Kumar, Mr. Herrmann Baumgartner, Mrs. Beatrix Fischer and Ms. Saskia Pfeffer is also acknowledged.

References

Adam SE, Magzoub M (1975) Toxicity of *Jatropha curcas* for goats. Toxicology 4:347–354

Aderibigbe AO, Johnson COLE, Makkar HPS, Becker K, Foidl N (1997) Chemical composition and effect of heat on organic matter and nitrogen degradability and some anti-nutritional components of *Jatropha* meal. Anim Feed Sci Technol 67:223–243

Agbo NW (2008) Oilseed meals as dietary protein sources for Juvenile *Nile tilapia* (*Oreochromis niloticus* L.). Ph.D. thesis. University of Stirling, Scotland

Ames BN, Profet M, Gold LS (1990) Nature's chemicals and synthetic chemicals: comparative toxicology. Proc Natl Acad Sci USA 87:7782–7786

Aregheore EM, Becker K, Makkar HPS (2003) Detoxification of a toxic variety of *Jatropha curcas* using heat and chemical treatments, and preliminary nutritional evaluation with rats. S Pac J Nat Sci 21:50–56

Barbieri L, Battelli M, Stripe F (1993) Ribosome-inactivating protein from plants. Biochim Biophys Acta 1154:237–282

Becker K, Makkar HPS (1998) Effects of phorbol esters in carp (*Cyprinus carpio* L.). Vet Hum Toxicol 40:82–86

Bedford MR (2000) Exogenous enzymes in monogastric nutrition—their current value and future benefits. Anim Feed Sci Technol 86:1–13

Bedford MR, Schulze H (1998) Exogenous enzymes for pigs and poultry. Nutr Res Rev 11:91–114

Bertolini TM, Giorgione J, Harvey DF, Newton AC (2003) Protein kinase C translocation by modified phorbol esters with functionalized lipophilic regions. J Org Chem 68:5028–5036

Beutler JA, Ada AB, McCloud TG, Cragg GM (1989) Distribution of phorbol ester bioactivity in the Euphorbiaceae. Phytother Res 3:188–192

Bohm R, Flaschentrager B, Lendle L (1935) The activity of substances from Croton oil. Arch Exp Pathol Pharmacol 177:212

Butler L, Rogler J, Mehansho H, Carlson D (1986) Dietary effects of tannins. In: Cody V, Middleton E, Harborne JB (eds) Plant flavonoids in biology and medicine: biochemical, pharmacological, and structure activity relationships. A. R. Liss, New York, pp 141–157

Cairnes DA, Mirvish SS, Wallcave L, Nagel DL, Smith JW (1981) A rapid method for isolating phorbol from croton oil. Cancer Lett 14:85–91

Clemens MJ, Trayner I, Menaya J (1992) The role of protein kinase C isoenzymes in the regulation of cell proliferation and differentiation. J Cell Sci 103:881–887

Devappa RK (2012) Isolation, characterization and potential agro-pharmaceutical application of phorbol esters from Jatropha oil. Ph.D. thesis, University of Hohenheim

Devappa RK, Maes J, Makkar HPS, Greyt WD, Becker K (2009) Isolation of phorbol esters from *Jatropha curcas* oil and quality of produced biodiesel. 2nd International Congress on Biodiesel, The Science and the Technologies, Munich

Devappa RK, Makkar HP, Becker K (2010a) Biodegradation of *Jatropha curcas* phorbol esters in soil. J Sci Food Agric 90:2090–2097

Devappa RK, Makkar HPS, Becker K (2010b) Quality of biodiesel prepared from phorbol ester extracted *Jatropha curcas* oil. J Am Oil Chem Soc 87:697–704

Devappa RK, Makkar HPS, Becker K (2010c) *Jatropha* diterpenes: a review. J Am Oil Chem 88:301–322

Devappa RK, Makkar HPS, Becker K (2010d) *Jatropha* toxicity—a review. J Toxicol Environ Health B Crit Rev 13:476–507

Devappa RK, Makkar HPS, Becker K (2010e) Optimization of conditions for the extraction of phorbol esters from *Jatropha* oil. Biomass Bioenergy 34:1125–1133

Devappa RK, Makkar HPS, Becker K (2010f) Nutritional, biochemical, and pharmaceutical potential of proteins and peptides from Jatropha: review. J Agric Food Chem 58:6543–6555

Devappa RK, Roach J, Makkar HPS, Becker K (2011) In vitro toxicity of *Jatropha curcas* oil phorbol esters. In vitro Biology Meeting, Raleigh

Devappa RK, Makkar HPS, Becker K (2012a) In vitro tumour promotion studies on *Jatropha curcas* phorbol esters. Mutat Res-Gen Tox En (in press)

Devappa RK, Angulo-Escalante MA, Makkar HPS, Becker K (2012b) Potential of using phorbol esters as an insecticide in agricultural applications. Ind Crop Prod 38:50–53

Devappa RK, Makkar HPS, Becker K (2012c) Localisation of antinutrients and qualitative identification of toxic components in *Jatropha curcas* seed. J Sci Food Agric 92:1519–1525

Devappa RK, Malakar CC, Makkar HPS, Becker K (2012d) Pharmaceutical potential of phorbol esters from *Jatropha curcas* oil. J Nat Prod Res (in press)

Duffus CM, Duffus JH (1991) Introduction and overview. In: D'Mello FJP, Duffus CM, Duffus JH (eds) Toxic substances in crop plants. Royal Society of Chemistry, Cambridge, pp 1–21

El-Badwi SM, Adam SE (1992) Toxic effects of low levels of dietary *Jatropha curcas* seed on brown Hisex chicks. Vet Hum Toxicol 34:112–115

Endo Y, Tsurugi K (1988) The RNA N-glycosidase activity of ricin A-chain: the characteristics of the enzymatic activity of ricin A-chain with ribosomes and with rRNA. J Biol Chem 263:8735–8739

European Food Safety Authority (2009) Scientific opinion of the panel on contaminants in the food chain on a request from the European Commission on saponins in *Madhuca longifolia* L. as undesirable substances in animal feed. EFSA J 979:1–36

Freeman BC, Beattie GA (2008) An overview of plant defenses against pathogens and herbivores. Plant Health Instructor. doi:10.1094/PHI-I-2008-0226-01

Gandhi VM, Cherian KM, Mulky MJ (1995) Toxicological studies on ratanjyot oil. Food Chem Toxicol 33:39–42

Goel G, Makkar HPS, Francis G, Becker K (2007) Phorbol esters: structure, biological activity and toxicity in animals. Int J Toxicol 26:279–288

Haas W (2003) Isolation and characterization of the phorbol esters from *Jatropha curcas* seed oil. Ph.D. thesis, Karl-Franzens-Universitat Graz, Germany

Haas W, Sterk H, Mittlebach M (2002) Novel 12-deoxy-16-hydroxyphorbol diesters isolated from the seed oil of *Jatropha curcas*. J Nat Prod 65:1434–1440

Hertrampf JW, Piedad-Pascual F (2000) Handbook on ingredients for aquaculture feeds. Kluwer, Dordrecht

Huang MX, Hou P, Wei EQ, Xu Y, Chen F (2008) A ribosome inactivating protein (curcin 2) induced from *Jatropha curcas* can reduce viral and fungal infection in transgenic tobacco. Plant Growth Regul 54:115–123

Jiang HM, Yang S, Hu DY, Xue W, Song BA (2007) Research progress in pesticidal and medicinal activity of curcin of *Jatropha curcas*. Agrochemicals 46:10–13

Jin-ping Z, Xia-bo Q, Ying X, Fang C (2005) Isolation and analysis on the genomic DNA sequence of members of a curcin gene-family encoding a ribosome-inactivating protein from *Jatropha curcas*. Sichuan Daxue Xuebao (Ziran Kexueban) 42:1042–1046

Kinzel V, Richards J, Goerttler K, Loehrke H, Furstenberger G, Marks F (1984) Interaction of phorbol derivatives with replicating cells. IARC Sci Publ 56:253–264

Knudsen KEB (1997) Carbohydrate and lignin contents of plant materials used in animal feeding. Animal Feed Sci Technol 67:319–338

Kumar V, Makkar HP, Devappa RK, Becker K (2011a) Isolation of phytate from *Jatropha curcas* kernel meal and effects of isolated phytate on growth, digestive physiology and metabolic changes in Nile tilapia (*Oreochromis niloticus* L.). Food Chem Toxicol 49:2144–2156

Kumar V, Makkar HPS, Becker K (2011b) Detoxified *Jatropha curcas* kernel meal as a dietary protein source: growth performance, nutrient utilization and digestive enzymes in common carp (*Cyprinus carpio* L.) fingerlings. Aquacult Nutr 17:313–326

Li CY, Devappa RK, Liu JX, Makkar HPS, Becker K (2010) Toxicity of *Jatropha curcas* phorbol esters in mice. Food Chem Toxicol 48:620–625

Liener IE (1989) Antinutritional factors in legume seeds: state of the art. In: Huisman J, Van der Poel TFB, Liener IE (eds) Recent advances of research in antinutritional factors in legume seeds. Pudoc, Wageningen, pp 6–14

Liener IE, Kakade ML (1980) Protease inhibitors. In: Anonymous (ed) Toxic constituents of plant foodstuffs. Academic, New York, pp 7–71

Lin J, Chen Y, Xu Y, Yan F, Tang L, Chen F (2003a) Cloning and expression of curcin, a ribosome-inactivating protein from the seeds of *Jatropha curcas*. Acta Bot Sin 45:858–863

Lin J, Yan F, Tang L, Chen F (2003b) Antitumor effects of curcin from seeds of *Jatropha curcas*. Acta Pharmacol Sin 24:241–246

Lin J, Zhou X, Wang J, Jiang P, Tang K (2010) Purification and characterization of curcin, a toxic lectin from the seed of *Jatropha curcas*. Prep Biochem Biotechnol 40:107–118

Makkar HPS (1993) Antinutritional factors in foods for livestock. In: Gill M, Owen E, Pollot GE, Lawrence TLJ (eds) Animal production in developing countries. British Society of Animal Production, Edinburgh, pp 69–85

Makkar HPS, Becker K (1999) Nutritional studies on rats and fish carp (*Cyprinus carpio*) fed diets containing unheated and heated *Jatropha curcas* meal of a nontoxic provenance. Plant Foods Hum Nutr 53:182–292

Makkar HPS, Becker K (2009) *Jatropha curcas*, a promising crop for the generation of biodiesel and value-added co-products. Eur J lipid Sci Technol 111:773–787

Makkar HPS, Becker K (2010) Method for detoxifying plant constituents. WO/2010/092143
Makkar HPS, Becker K, Sporer F, Wink M (1997) Studies on nutritive potential and toxic constituents of different provenances of *Jatropha curcas*. J Agric Food Chem 45:3152–3157
Makkar HPS, Aderibigbe AO, Becker K (1998a) Comparative evaluation of nontoxic and toxic varieties of *Jatropha curcas* for chemical composition, digestibility, protein degradability and toxic factors. Food Chem 62:207–215
Makkar HPS, Becker K, Schmook B (1998b) Edible provenances of *Jatropha curcas* from Quintana Roo state of Mexico and effect of roasting on antinutrient and toxic factors in seeds. Plant Foods Hum Nutr 52:31–36
Makkar HPS, Francis G, Becker K (2007) Bioactivity of phytochemicals in some lesser known plants and their effects and potential applications in livestock and aquaculture production systems. Animal 1:1371–1391
Makkar HPS, Kumar V, Karaj S, Kratzeisen M, Tipraqsa P, Muller J, et al (2009a) Sustainable land development and ecosystem conservation through enhancing economic viability of the *Jatropha curcas* based biodiesel production chain using a bio-refinery concept. In: ERSEC International Conference; Beijing
Makkar HPS, Maes J, De Greyt W, Becker K (2009b) Removal and degradation of phorbol esters during pre-treatment and transesterification of *Jatropha curcas* oil. J Amer Oil Chem Soc 86:173–181
Makkar HPS, Kumar V, Oyeleye OO, Akinleye AO, Angulo-Escalante MA, Becker K (2011) *Jatropha platyphylla*, a new nontoxic *Jatropha* species: physical properties and chemical constituents including toxic and antinutritional factors of seeds. Food Chem 125:63–71
Makkar HPS, Kumar V, Becker K (2012) Use of detoxified *Jatropha* kernel meal and protein isolate in diets of farm animals. In: Makkar HPS (ed) Opportunities and challenges in utilization of co-products of the biofuel industry as livestock feed. FAO, Rome
McDaniel J, Freking B (2006) Chap. 9. Goat nutrition, Oklahoma State University. http://meat-goat.okstate.edu/oklahoma-basic-meat-goat-manual-1/Chapter%209%20-%20Nutrition.pdf
McDonald P, Edwards RA, Greenhaugh JF (1995) Animal nutrition, 5th edn. Longman Group, London
Nishizuka Y (1992) Intracellular signaling by hydrolysis of phospholipids and activation of protein kinase C. Science 258:607–614
Norton G (1991) Proteinase inhibitors. In: D'Mello FJP, Duffus CM, Duffus JH (eds) Toxic substances in crop plants. Royal Society of Chemistry, Cambridge, pp 68–106
Popluechai S, Breviario D, Mulpuri S, Makkar HPS, Raorane M, Reddy AR et al (2009) Narrow genetic and apparent phenetic diversity in *Jatropha curcas*: initial success with generating low phorbol ester interspecific hybrids. http://hdl.handle.net/10101/npre.2009.2782.1
Qin X, Zheng X, Shao C, Gao J, Jiang L, Zhu X et al (2009) Stress-induced curcin-L promoter in leaves of *Jatropha curcas* L. and characterization in transgenic tobacco. Planta 230:387–395
Richardson NL, Higgs DA, Beames RM, McBride JR (1985) Influence of dietary calcium, phosphorous, zinc and sodium phytate level on cataract incidence, growth and histopathology in juvenile Chinook salmon *Oncorhynchus tshawytscha*. J Nutr 115:553–567
Roach JS, Devappa RK, Makkar PS, Becker K (2012) Isolation, stability and bioactivity of *Jatropha curcas* phorbol esters. Fitoterapia 83:586–592
Sen S, Makkar HPS, Becker K (1998) Alfalfa saponins and their implication in animal nutrition. J Agric Food Chem 46:131–140
Silinsky EM, Searl TJ (2003) Phorbol esters and neurotransmitter release; more than just protein kinase C? Br J Pharmacol 138:1191–1201
Stripe F, Pession-Brizzi A, Lorenzoni E, Strocchi P, Montanaro L, Sperti S (1976) Studies on the proteins from the seeds of *Croton tiglium* and of *Jatropha curcas*. Biochem J 156:1–6
Vasconcelos IM, Oliveira JTA (2004) Antinutritional properties of plant lectins. Toxicon 44:385–403
Wei Q, Huang MX, Xu Y, Zhang XS, Chen F (2005) Expression of a ribosome inactivating protein (curcin 2) in *Jatropha curcas* is induced by stress. J Biosci 30:351–357

Weinstein IB, Lee LS, Fisher PB, Mufson A, Yamasaki H (1979) Action of phorbol esters in cell culture: mimicry of transformation, altered differentiation, and effects on cell membranes. J Supramol Struct 12:195–208

Wender PA, Kirschberg TA, Williams PD, Bastiaans HMM, Irie K (1999) A new class of simplified phorbol ester analogues: synthesis and binding to PKC and ηPKC-C1B (ηPKC-CRD2). Org Lett 1:1009–1012

Wender PA, Kee JM, Warrington JM (2008) Practical synthesis of prostratin, DPP, and their analogs, adjuvant leads against latent HIV. Science 320:649–652

Wender PA, Warrington JM, Kee J (2009) Process to produce prostratin and structural or functional analogs thereof. US Patent US 2009/0187046

Wink M, Koschmieder C, Sauerweien M, Sporer F (1997) Phorbol esters of *Jatropha curcas*—biological activities and potential applications. In: Gubitz GM, Mittelbach M, Trabi M (eds) Biofuel and industrial products from *Jatropha curcas*. Dbv-Verlag fiir die Technische Universitat Graz, Germany, pp 160–166

Xu R, Zhao W, Jiang C (2009) Ester prodrugs of prostratin and related phorbol compounds. US Patent 2,009,016,358

Chapter 22
Jatropha Pharmacognosy, Phytochemistry and Pharmacology: A Review

Sujatha Samala and Ciddi Veeresham

Introduction

Plant species that can be processed to provide a diesel fuel substitute have captured the interest of scientists in temperate zones. In this category, the following properties of the tropical physic nut (*Jatropha curcas* L., Euphorbiaceae) have won over the interest of various development agencies: as it adapts well to semi-arid marginal sites. Its oil can be processed for use as a diesel fuel substitute and it can be used for erosion control (Verma and Gaur 2009). Although *J. curcas* is of Mexican and Central American origin, it is cultivated in many other Latin American, Asian and African countries as a hedge crop and it was an important export product from the Cape Verde Islands during the first half of this century (Dehgan and Webster 1978).

The genus *Jatropha* contains approximately 170 known species distributed in the tropical and subtropical Africa and America (Heller 1996). There are 12 *Jatropha* species in India, but the research has been confined to nine species only. Among the *Jatropha* species, the most primitive form is *J. curcas* and has the potential to be cultivated for biodiesel and medicinal properties. Distinguishing morphological features of *Jatropha* species are presented in Table 22.1.

J. curcas is believed to have been spread by Portuguese seafarers from its centre of origin in Central America and Mexico *via* Cape Verde and Guinea Bissau to other countries in Africa and Asia. It is now widespread throughout the tropics and subtropics. *J. curcas* oil was used for lighting lamps in olden days. Today, rural communities continue to use it for its medicinal value and for local soap production. In India and in many African countries, *J. curcas* is used as a living hedge to keep out grazing livestock (Brittaine and Lutaladio 2010). It is suitable for the conservation of poorly structured soils and dune stabilization (Katwal and Soni 2003).

S. Samala • C. Veeresham (✉)
University College of Pharmaceutical Sciences, Kakatiya University, Warangal, 506 009, India
e-mail: ciddiveeresham@yahoo.co.in

N. Carels et al. (eds.), *Jatropha, Challenges for a New Energy Crop: Volume 1: Farming, Economics and Biofuel*, DOI 10.1007/978-1-4614-4806-8_22,

Table 22.1 Morphological features, propagation methods and desirable traits of different species of *Jatropha*

Sl no	Species	Native place	Distribution in India	Morphological features	Propagation methods	Oil content (%)	Desirable traits
1	*J. curcas*	Tropical America	In all the states of India	Large shrub, highly branching, cordate–palmately lobed leaves, greenish-yellow flowers, distinct coflorescence, tardily dehiscent fruits with black, ecarunculate seeds	Seed, cutting, grafting, air layering and tissue culture	30–42	High seed yield and oil content
2	*J. gossypiifolia*	Brazil	Commonly found in disturbed soils of all states	Fertile large shrub, profuse branching, cordate leaves, glandular plant parts, dark crimson–purple flowers, violently dehiscent capsules with small brown carunculate seeds	Seed and cuttings	28–30	drought-tolerant and profuse, year round fruiting
3	*J. tanjorensis*	India	Tanjore, Trichy, and Ramnad districts of Tamil Nadu	Sterile shrub, profuse branching, cordate–palmately lobed leaves, margins distinctly serrate, greenish-yellow flowers with crimson-red tinge, no fruit-set	Cuttings	Sterile	Robust and drought hardy
4	*J. multifida*	South America	Ornamental nurseries in all states	Fertile shrub, uniform branching, leaves divided into 5–11 lobes, long petiole and pedunculate, flat-topped cyme, coral-red flowers and fruits are non dehiscent capsules	Seed and by cuttings during spring	32–40	Bigger fruit size and resistant to diseases
5	*J. glandulifera*	India	Black cotton soils of Deccan and Carnatic	Fertile smaller plant, spread and dichotomously branched, narrow leaves with serrated margin, have smooth papery bark, profuse fruiting, but dehisce before maturity	Seed and cuttings	20–27	Profuse fruiting and drought tolerant
6	*J. panduraefolia*	Cuba	Rare in Ornamental in nurseries	Monoecious shrub with slender, graceful branches. Leaves alternate, shallowly cordate at the base, inflorescence terminal cyme, calyx purplish red in colour. Petals twisted in the bud, white hairs inside at the base. Flowering throughout the year, fruits are capsule and purplish green	Seed and cuttings	28	Flowering throughout the year

7	*J. integerrima*	West Indies	Ornamental nurseries in South India	Fertile shrub, sparsely branched, ovate fiddle-shaped leaves, crimson-red flowers, dehiscent capsules, seeds small caruncululate and brown with spots	Cuttings	26–28	Semi-hard woody stem and disease resistant
8	*J. podagrica*	Panama	Ornamental nurseries in Southern and Central India	Fertile, caudiciform shrub, cordate leaves with peltate base, flat-topped corymbose cyme, bright scarlet flowers, fruits dehiscent capsule with big brown ecarunculate seeds	Seed and by division of branches	28–32	Fusarial wilt resistance
9	*J. villosa*	India	Kongan region, Nilgiri, Kanyakumari, and Ramnad districts of Tamil Nadu	Fertile undershrub, shoots rusty—villow, profuse branching, drought-tolerant, evergreen, rhizomatous plant	—	—	Evergreen and rhizomatous plant
10	*J. maheshwarii*	India	Naturally occurs in Southern districts of Tamil Nadu	Fertile evergreen, drought-hardy and rhizomatous plant, leaves long, elliptical and resembles mango leaves	—	—	Drought hardy and rhizomatous plant
11	*J. heynei*	India	Indian Peninsula	Shrub, branching from a tuberous rootstock (weighs about 1 kg). Leaves deeply 3-fid, lobes oblanceolate. Flowers unisexual and small. Both flowers and fruit capsules are green in colour	Seed	—	Tuberous root stock
12	*J. nana*	India	Poona and Mumbai. Endemic to Maharashtra Deccan	Shrub with woody root as thick as finger, stem round, smooth. Leaves 3 lobed/entire with the largest middle lobe. Flowers pedicellate, and few flowered terminal paniculate cymes. Capsule fruit	Seed	—	Woody root system

Source: Rata Krishnan and Paramathma (2009)

Macroscopical Characters

***Root*:** The younger roots of *J. curcas* are grey with longitudinal corrugations on the surface while the older roots are long and anastomosing. Irregularly distributed and well developed lenticels appear globular on the younger parts, but are elongated on the older regions. The adjoining lenticels unite to form a transverse ridge on the root surface and after a certain stage of growth, forms transverse grooves. The transversal section (TS) surface shows light-cream colored wood. *J. curcas* has slightly bitter taste with agreeable odour (Gupta 1985).

***Stem*:** The young stems of *J. curcas* are glandular tomentose. Slightly mature stem has sparsely distributed small, oval lenticels on a smooth surface, which on drying shows fine longitudinal striations. In the older stem, the surface has sparse to dense tubercles. The TS of smooth surface shows a white or creamy-white, soft, spongy wood feebly marked with a few growth rings occupying a major part of the stem. The fresh stem has a characteristic disagreeable odour and taste, but on drying becomes odorless and almost tasteless. In bark portion, the fracture is fibrous, woody, short and splintery (Gupta 1985).

***Leaf*:** *J. curcas* leaves are alternate or sub-opposite, elliptic or oblong-elliptic, obtuse or very often shortly cuspidate, and glabrous. Fully grown leaves are pale dull glaucous green with their base usually round. The leaves turn red before falling. The dried leaf is brittle, having astringent and bitter taste with no distinct odour. The midrib is prominent and pink with 6–10 main nerves prominently raised above the surface on the abaxial side. Petioles are 6–13 mm long (Gupta 1985).

Microscopic Characters

***Young root*:** The young root of *J. curcas* consists of a single layered epiblema with cubical to more or less columnar cells (measure 8–16–21 × 8–12–16–21 μm); of which a number of cells develop into root hairs (Fig. 22.1 3, 2A, B, 3A). A 6–8 layered parenchymatous cortex with cells measuring 12–29–53–70 × 16–25–37–50 μm are present below the epiblema. The endodermis has cells, which are barrel-shaped (measuring 12–16–24–29 × 8–12–16–21 μm) with distinct casparian thickenings on the radial walls (Fig. 22.1 3A). A single layered pericycle below the endodermis encloses a di- to tetrarch stele having usual phloem and xylem elements (Fig. 22.1 2A, B, 3A). A cork cambium arises in the pericycle and gives rise to the usual periderm cells (phellem and phelloderm) (Fig. 22.1 3B). In the younger roots, the phellem cells are rectangular and flat. In lenticels (older stages), two types of cells differentiated from phellem cells, i.e., a flat and compactly arranged layer forming the closing cells and an array of loosely arranged complementary cell toward the outside of lenticels.

***Mature root*:** In TS of mature roots, 11–32 layers of phellem and a wider zone of phelloderm are present in periderm. The phellem cells are thin-walled, subserised and tangentially elongated (measure 45–57–102–156 × 16–36–45–58 μm), and the

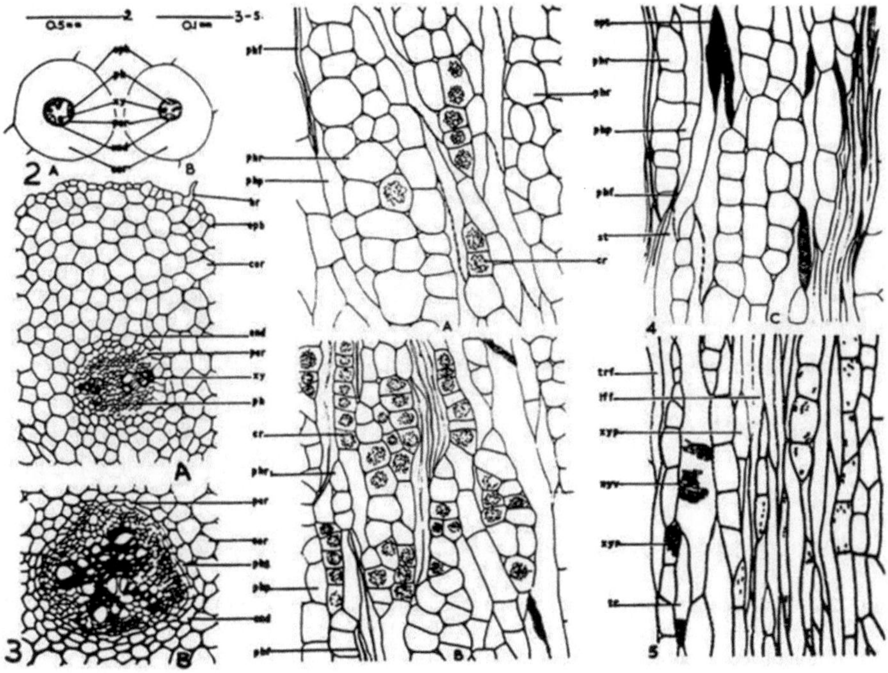

Fig. 22.1 Transverse section of young root (Gupta 1985). (**2A**) Triarch condition of xylem, (**2B**) Diarch condition of xylem, (**3A**) Cellular details of a portion of Fig (**2B**). (**3B**) Details in TS of a slightly mature root showing formation of phellogen in the pericyclic region, (**4**) Mature root in LS through phloem (**A**) outer phloem, (**B**) middle phloem, (**C**) inner phloem, (**5**) Details of mature root in LS through xylem

phelloderm cells vary in shape and size (28–61–102–144 × 20–49–74–107 μm) (Fig. 22.2 6, 8A). The secondary phloem occupies a considerably wider zone and consists of all the usual elements. Some of the phloem elements collapse and get lignified to form ceratenchyma in older barks (Fig. 22.2 8B). The phloem fibres are either solitary or in small groups (Fig. 22.2 6, 8B, C), long (measure 16–20–33–41 × 495–1,402–3,252–3,382 μm), thick-walled and lignified with tapering or truncated ends (Fig. 22.3 10fl–f3). The sieve tubes are usually isodiametric (24–33–37 μm in TS) and have an oblique sieve plate on both, the ends and the longitudinal walls (Figs. 22.1 4 and 22.2 8C). The phloem rays are narrow (uni- and biseriate), broad (3–8 seriate or so) and heterogeneous. The cambium is 1–2 layered (Fig. 22.2 8C). The xylem covers almost three fourth of the root diameter and consists of the usual elements (Figs. 22.2 7 and 22.3 9). The vessels that occur singly or in groups of 2–6 are diffused and porous. The vessel elements vary from cylindrical with outgrowths at the ends that vary from peg like to short wide and drum-shaped during macerations; they measure 56–59–165–198 × 148–338–500–742 μm (Fig. 22.3 10al–a6). Quite distinct perforation rims and bordered type of pits are arranged alternately (Fig. 22.3 10a4). The diffused xylem parenchyma is apotracheal (Fig. 22.3 9A–C). Some of the cells remain unlignified and have simple pits or

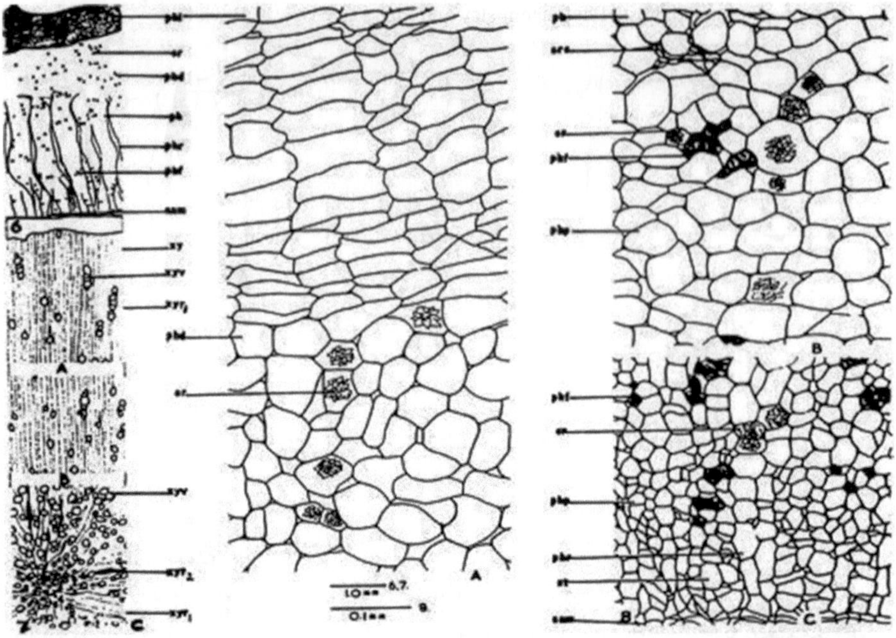

Fig. 22.2 Microscopic characters of *J. curcas* root (Gupta 1985). (**6**) TS of a portion of the mature root-bark, (**7A-C**) TS of a portion of root-wood, (**8**) Cellular details in TS of a portion of Fig. 22.3: (**A**) Portion of the outer bark, (**B**) Middle bark, (**C**) Inner bark

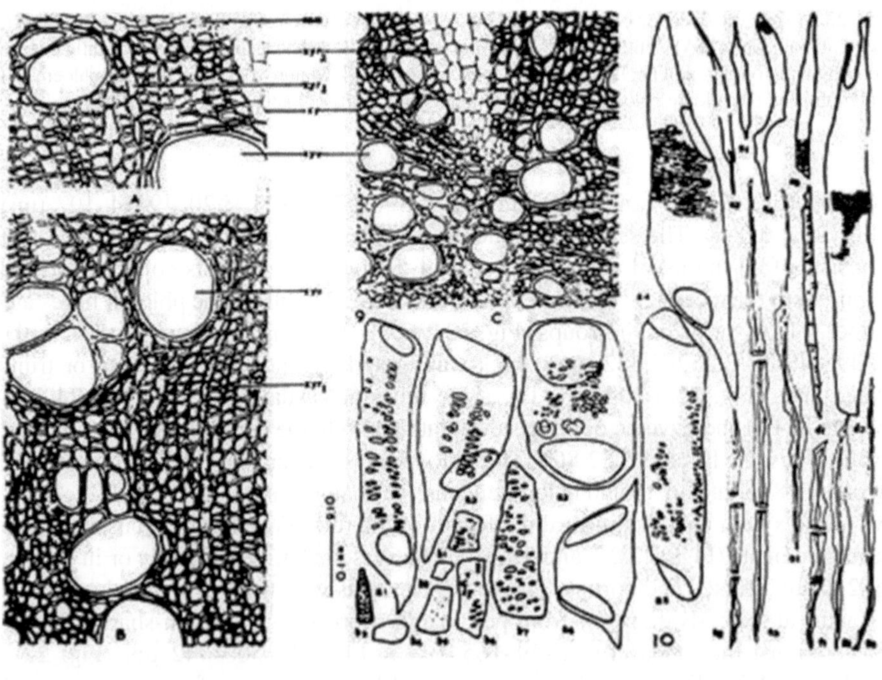

Fig. 22.3 Microscopic characters of root of *J. curcas* (Gupta 1985). (**9**) Cellular details in TS of Fig. 22.2 (**7A**). (**A**) Portion of the outer xylem, (**B**) Middle xylem, (**C**) Inner xylem, (**10**) Isolated elements of mature root: (**al–a6**) Vessel elements, (**bl–b7**) Xylem parenchyma, (**cl–c4**) Tracheids, (**dl–d2**)Tracheid fibres, (**el–e3**) Libriform fibres, (**fl–f3**) Phloem fibres

reticulate thickenings on their walls. The cells adjacent to the vessels are highly lignified and have distinct bordered pits similar to those in vessels (Fig. 22.3 10bl–b7). The tracheids are long (20–33–41–50 × 297–417–570–760 μm) with tapering or branched extremities and bordered pits on the walls (Fig. 22.3 10cl–c4). In wood, libriform xylem fibres are distributed small groups and measure 16–33–41–50 × 594–959–1,270–1,650 μm. These have tapering, truncated or bifurcated ends and appear like phloem fibres in macerations (Fig. 22.3 10el–e3) (Gupta 1985).

***Young stem*:** Young stem of *J. curcas* consists of a single layered epidermis with thin-walled, cubical to tangentially elongated cells (10–16–20–29 × 10–12–16 μm) covered externally with a thin cuticle (Fig. 22.4 11, 12A). The cortex is made up of 6–8 layers of collenchyma (12–19–33–41 × 16–33–45–53 μm) and parenchyma layers. Starch-sheath is absent. Non-articulated laticifers are found all over, but become progressively wider towards the centre (in cortex: 12–16–29–41 μm, phloem: 25–41–47–57 μm and pith: 29–37–47–62 μm) (Fig 22.4 12A–C). The cork cambium, arises in the outermost cortical layer and gives rise to phellem outside and phelloderm inside (Fig. 22.4 13). In older bark, some of the phloem cells collapse to form the ceratenchyma, which may also become lignified.

***Mature stem*:** *J. curcas* bark portion is distinguishable into outer, middle and inner zones in a mature stem with the outer one composed of a few layers of thin-walled, suberised cork cells (12–21–29–41 × 29–41–57–74 μm) arranged in radial rows (Figs. 22.5 17A and 22.6 18A) and the middle bark consisting of somewhat tangentially elongated phelloderm cells, which are comparatively smaller and narrower in the outer few layers. There is no definite demarcation between the phelloderm and the primary cortex (Figs. 22.5 17A and 22.6 18A). However, the laticifers in the latter zone are more frequent (Fig. 22.6 18A). Inner bark is composed of phloem tissue characterized by funnel shaped multiseriate rays, which narrow towards the centre. The phloem, thus, appears arranged in wedge-shaped conical masses (Figs. 22.5 17A and 22.6 18A–C). Phloem fibres occur solitary or in radially arranged groups (Fig 22.6 18B) with tapering or truncated ends (Fig. 22.4 14gl–g4). The sieve tubes are crushed or collapsed at many places in the outer region (Fig. 22.6 18B). The xylem, which occupies more than three fourth parts of the stem diameter, shows a few feebly marked growth rings. Vessels may be arranged radially, singly (solitary) or in groups of 2–12 (Figs. 22.5 17B–D and 22.6 19A–C). Tracheids measure 16–25–33 × 164–471–545–808 μm (Fig. 22.4 14bl–b4), fiber tracheids 16–33–50 × 462–660–842–1,188 μm (Fig. 22.4 14c1–c3) and libriform xylem fibres 16–25–33 × 990–1,320–2,962 μm (Fig. 22.4 14el–e3). The xylem rays are heterogeneous and uni- or bi-seriate. In tangential section these are 1–10–22–45 cells long with individual cells measuring 16–32–48 × 33–48–98–130 μm (Fig. 22.5 16). The pith has polyhedral, oval to almost isodiametric parenchyma (Fig. 22.6 19C) (Gupta 1985). Recently, Idu et al. (2009), have made comparative morphological and anatomical studies in *J. curcas* and *J. tanjorensis* while Nayak and Patel (2010) made pharmacognostic study of leaf and phytochemical screening and evaluation of *J. curcas*.

***Petiole*:** *J. curcas* petiole TS are plano-convex to circular in outline (Fig. 22.7 22) and its epidermis has a moderately thick cuticle on the outside composed of narrow, tangentially elongated cells (Fig. 22.7 23A) with 14–17 layers of collenchyma

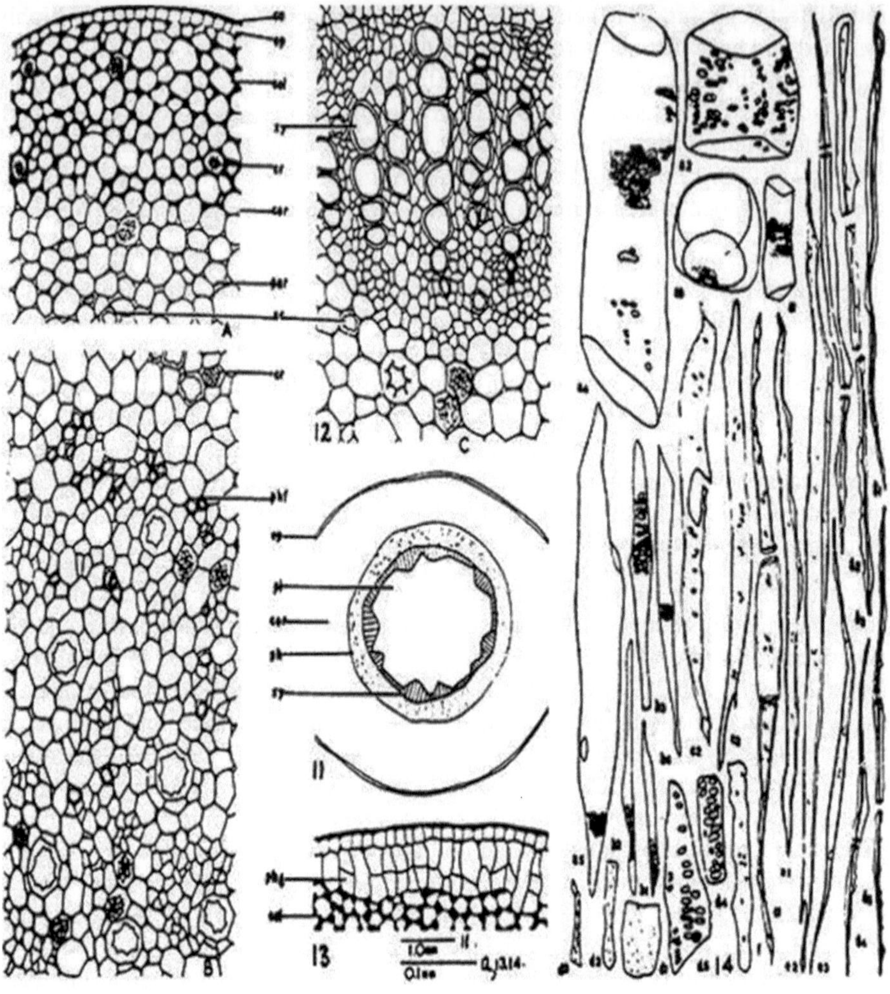

Fig. 22.4 Microscopic characters of the stem of *J. curcas* (Gupta 1985) (**11**) Diagrammatic TS of portion of the young stem. (**12A–C**) Cellular details in TS of portions of (**13**), Details in TS of a portion of slightly mature stem showing formation of phellogen in the outer most cortical layer, (**14**) Isolated elements of the stem: (**al–a5**) Vessel elements, (**bl–b4**) Tracheids, (**cl–c3**) Tracheid-fibres, (**dl–d5**) Xylem parenchyma, (**el–e3**) Libriform fibres, (**f**) Tracheid parenchyma, (**gl–g5**) Phloem fibres

(20–23–49–70 × 37–57–78 μm) and a few layers of parenchymatous cells measuring 20–37–74–111 × 24–66–115 μm. A single layered starch-sheath is present, but no pericycle is discernible (Fig. 22.7 23B). TS in petiole basal ends show a eustele of nine collateral vascular bundles (Fig. 22.7 22), which coalesce to form a large semicircular vascular arc and a small separate bundle on the adaxial side. The phloem

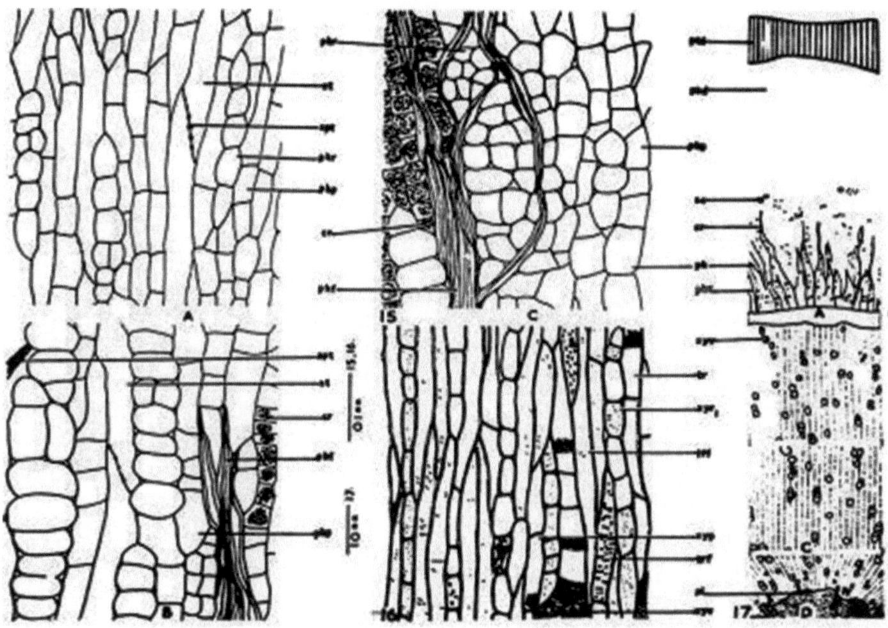

Fig. 22.5 Microscopic characters of the stem of *J. curcas* (Gupta 1985) (**15**) Details of mature stem in LS through phloem: (**A**) a portion of the outer phloem, (**B**) middle phloem, (**C**) inner phloem, (**16**) Details of mature stem in LS through xylem, (**17**) Diagrammatic TS of a portion of the mature stem: (**A**) bark, (**B**) outer xylem, (**C**) middle xylem, (**D**) inner xylem and pith

consists of phloem fibres (diameter 20–23–33–41 μm) and xylem rays arranged radially (Fig. 22.7 23B) whereas the xylem is composed of vessels, tracheids and xylem parenchyma.

***Leaf*:** TS in leaf lamina show its dorsiventral structure as illustrated in Fig. 22.7 20. The epidermal cells on both surfaces are tangentially elongated and have a fairly thick cuticle (Fig. 22.7 21A, B). The upper epidermal cells are larger (20–28–33–41 × 20–33–53–82 μm) than those of the lower surface (20–25–27–33 × 18–29–32–54 μm) and appear polyhedral with comparatively more straight walls than those of the lower epidermal cells (Fig. 22.7 24A, B). Both paracytic and anisocytic stomata (20–25–33 × 33–37–45 μm) are present on adaxial and abaxial sides, but are more frequent on the abaxial side (Fig. 22.7 24A, B). The mesophyll consists of a few layers of spongy parenchyma, a single layered palisade (Fig. 22.7 20A, B) and a few idioblasts containing crystals (Fig. 22.7 21A, B). The mesophyll cells are replaced by thick-walled collenchymatous cells at the margins (8–12–17 × 8–12–19 μm). The primary veins are prominently raised above the surface on the abaxial, but slightly on the adaxial side (Fig. 22.7 20C). Starch sheath is absent and there is only a single crescent-shaped median vascular patch with the xylem facing the upper and the phloem the lower side (Fig. 22.7 20C, 21C) (Gupta 1985).

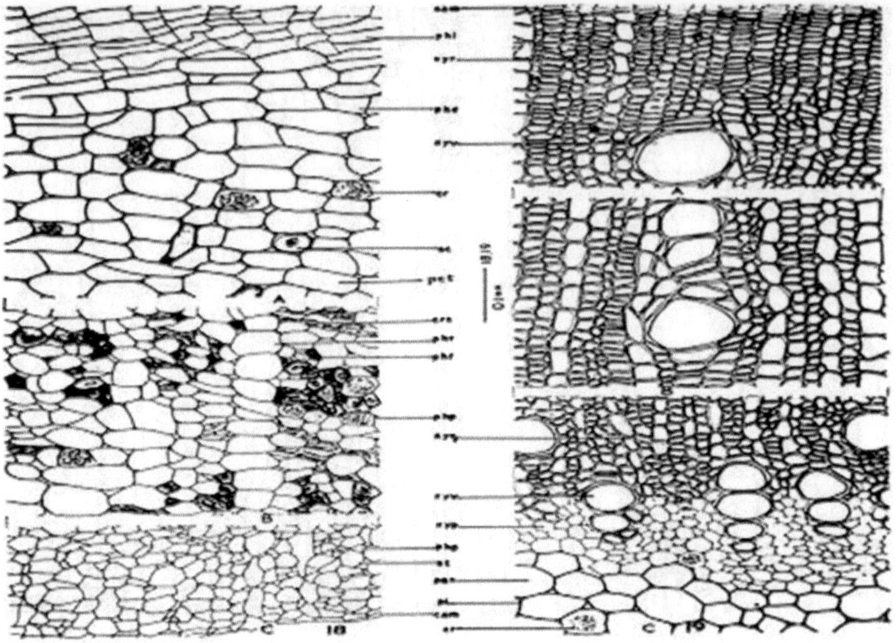

Fig. 22.6 Microscopic characters of the stem of *J. curcas* (Gupta 1985) **(18A–C)** Cellular details in TS of portions of Fig. 22.5 (**17A**). (**19**) Cellular details in TS of portions of Fig. 22.5 (**17B–D**)

Quantitative Microscopy

Quantitative microscopy of leaves of *J. curcas* is shown in Table 22.2. These values are useful for the identification of authentic samples.

Physicochemical Parameters

***Powder characteristics*:** Microscopic examination of leaf powder shows small fragments of lamina with epidermis having anisocytic and paracytic stomata, mesophyll cells, broken parts of tracheary elements and laticifers associated with mesophyll and phloem tissue. The leaf powder is yellowish green, has a slightly astringent and bitter taste with no odour. Whereas the stem powder is grey in colour and has no perceptible odour and taste. On microscopic examination it shows small groups of cork cells and parenchyma, isolated and small groups of broken fibres, tracheids, vessels and broken parts of laticifers filled with dark brown contents. The root powder is light brown, has a slight bitter taste and a pleasant odour. Simple starch grains, calcium-oxalate crystals are found in the root powder and microscopically it is similar to the stem powder. The behaviour of the different powders on treatment with different chemical reagents is shown in Table 22.3. Fluorescence characters of different powders, extractive percentages and ash-values are depicted in Tables 22.4 and 22.5.

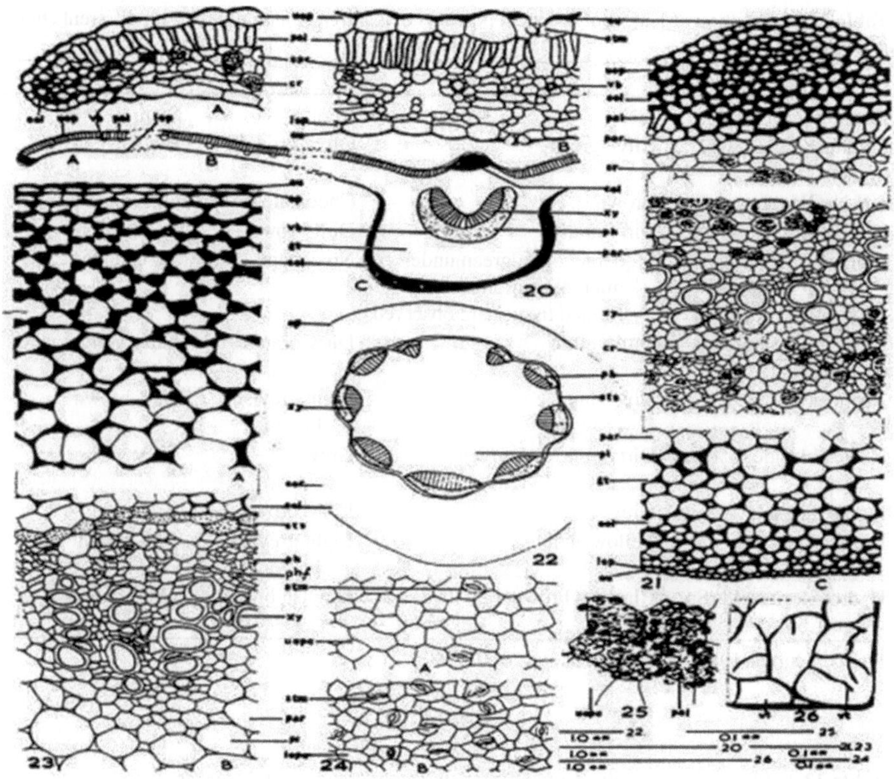

Fig. 22.7 Microscopic characters of the leaf of *J. curcas* (Gupta 1985) **(20)** Diagrammatic TS of a portion of lamina: **(A)** marginal lamina, **(B)** middle lamina, **(C)** mid-rib. **(21)** Cellular details of Fig. **(20)**. **(A)** marginal lamina, **(B)** middle lamina, **(C)** a portion of the mid-rib, **(22)** Diagrammatic TS of a portion of basal end of the petiole. **(23)** Cellular details of a portion of Fig. **(22)**. **(24)** Epidermis of lamina in surface view: **(A)** upper epidermal cells and stomata, **(B)** lower epidermal cells and stomata, **(25)** Palisade cells in surface view, **(26)**, Vein-islets and vein-terminations of lamina in surface view

Table 22.2 Leaf constants of *J. curcas*

Leaf constant	Range	Average
Stomatal number (upper surface)	2–3	2.2
Stomatal number (lower surface)	19–21	20.2
Stomatal index (upper surface)	6.06–8.5	7.076
Stomatal index (lower surface)	29.4–32.3	30.959
Vein-islet number	2–4	3
Vein-termination number	12–42	18.51
Palisade ratio (Base: Middle: Apex)	4.07: 4.1: 4.32	4.07: 4.1: 4.32

Source: Shanti et al. (2010)

Table 22.3 Reaction of leaf, root and stem powders of *J. curcas* on treatment with different chemical reagents

Behaviour of powder			
Reagents	Leaf	Root	Stem
Sulphuric acid (80% soln.)	Olive-green	Olive-brown	Blackish brown
1 N NaOH (aq. soln.)	Brown	Chocolate-brown	Chocolate-brown
Nitric acid (Sp. gr. 1.42)	Dirty orange	Dull-brown	Reddish-brown
Antimony trichloride in $HClO_4$ (saturated soln.)	Moderate olive-green under microscope some red stained tissues are observed	No change	No change
Picric acid (saturated soln.)	Dull-yellow	Dirty orange	Dirty yellow
Acetic acid	No change	Light greyish brown	Buff-brown
Ferric chloride (5% aq. soln,)	Dirty yellowish green	Grey	Grey
Iodine (aq. soln.)	Grey	Light brown	Brown
Lactic acid	Yellowish grey	Light greyish brown	Buff-brown
Hydrochloric acid (Sp. gr. 1.66)	Blackish brown	Greyish brown	Greyish brown

Source: Gupta (1985) and Shanti et al. (2010)

Table 22.4 Fluorescence analysis of leaf, root and stem powders of *J. curcas*

Fluorescence			
Treatment	Leaf	Root	Stem
Drug powder treated with 1 N NaOH (aq. soln.)	Moderate olive-green	Green	Bright green
Drug powder treated with 1 N HCl	Yellowish green	Pale-green	Yellowish green
Drug powder treated with 1 N NaOH in methanol and mounted in nitrocellulose in amylacetate	Dark green	Olive-green	Parrot-green
Drug powder mounted in nitrocellulose in amylacetate	Green	Dark green	Light green
Drug powder treated with 50% H_2SO_4	Green	Dark green	Green
Drug powder as such	Green	White with greenish tinge	Pale-green
Drug powder treated with 1 N NaOH in methanol	Blackish green	Bright yellowish green	Dark green
Drug powder treated with 1 N NaOH (aq. soln.) and mounted in nitrocellulose in amylacetate	Dark olive-green	Dark green	Dirty green
Drug powder treated with 50% HNO_3	Yellowish green	Yellowish green	Yellowish green
Drug powder treated with 1 N HCl and mounted in nitrocellulose in amylacetate	Bright green	Pale-green	Light Yellowish green

Source: Gupta (1985)

Table 22.5 Ash values and extractive percentages of root, stem and leaf of *J. curcas*

Plant-part (%)			
Values	Leaf	Root	Stem
Acid insoluble ash	2.091	1.5	0.51
Total ash	10.55	8.91	8.01
Alcoholic extractive	10.89	4.07	2.752
Water soluble extractive	18.65	9.36	12.08

Source: Gupta (1985)

Phytochemistry of J. curcas

The different extracts of *J. curcas* were screened for the presence of plant bases, glycosides/polysaccharides, reducing sugars, saponins, flavonoids, sterols and tannins. The different phytochemicals found in the extracts were shown in Table 22.6.

Thin layer chromatography: The different solvent systems used for different extractions of *J. curcas* are shown in Table 22.7. In these solvent systems, the phytochemicals in different parts (leaf, stem and root) of the plant are separated clearly.

Phytochemical Composition of Various Parts of J. curcas

Depending on the variety, the decorticated seeds contain 40–60% of oil (Liberalino et al. 1988; Gandhi et al.1995; Sharma et al. 1997; Wink et al. 1997; Makkar et al. 1997; Openshaw 2000). This oil can be used for many purposes such as lighting, as a lubricant, for making soap (Rivera-Lorca and Ku-Vera 1997) and most importantly as biodiesel. *J. curcas* oil contains approximately 24.6% of crude protein, 47.25% of crude fat, and 5.54% of moisture content (Akintayo 2004). The oil fraction of *J. curcas* contains saturated fatty acids mainly palmitic acid (16:0) with 14.1% and stearic acid (18:0) with 6.7%. Unsaturated fatty acids include oleic acid (18:1) with 47.0%, and linoleic acid (18:2) with 31.6%. The chemicals isolated from different parts of *J. curcas* are shown in the Table 22.8 and other chemicals isolated from different *J. curcas* species are presented in Table 22.9.

Medicinal Uses of Jatropha

All parts of *J. curcas* including seeds, leaves and bark, fresh or as a decoction, are used in traditional and folk medicine and for veterinary purposes for a long time (Duke 1988). One of the phytochemical (Curcacycline A) isolated from *J. curcas* showed anti tumour activity (Lin et al. 2003). Molluscicidal, insecticidal and fungicidal activities of phorbol esters from this plant have been demonstrated in laboratory experiments and field trials (Nwosu and Okafor 1995; Solsoloy and Solsoloy 1997). The seed oil can be applied to treat eczema, skin diseases and to soothe

Table 22.6 Phytochemical examination of different extracts of *J. curcas*

Plant part	Observation	Acetone extract	Water extract	Pet ether (60–80) extract	Benzene extract	Chloroform extract	Alcohol extract
Root	Glycosides/polysaccharides	+	+	–	–	–	+
	Flavonoids	–	–	–	–	–	–
	Reducing sugars	+	–	–	–	+	+
	Resins	+	–	–	–	+	–
	Tannins	–	–	–	–	–	–
	Physical appearance	brown	dirty brown	light yellowish green	light yellowish green	light yellowish green	brown
	Average % yield (w/w)	0.414	8.012	0.55	0.184	0.222	8.23
	Sterol	–	–	+	+	–	–
	Bases	–	–	–	–	–	–
	Saponins	–	–	–	–	–	–
Leaf	Glycoside/polysaccharides	+	+	–	–	–	+
	Flavonoids	–	–	–	–	–	–
	Reducing sugars	+	+	–	–	+	+
	Resins	+	–	–	–	+	+
	Tannins	–	–	–	–	–	–
	Physical appearance	pale yellow	dark brown	yellow	yellow	yellow	light brown
	Average % yield (w/w)	0.27	4.968	1.202	0.63	0.184	2.484
	Sterols	–	–	–	–	–	–
	Bases	–	–	–	–	–	–
	Saponins	–	–	–	–	–	–
Stem	Glycoside/polysaccharides	+	+	–	–	–	+
	Flavonoids	–	–	–	–	–	–
	Reducing sugars	+	–	–	–	+	+

Resins	+	–	–	–	+	+
Tannins	–	–	–	–	–	–
Physical appearance	Dirty orange	Reddish brown	Yellowish green	Dark yellowish green	Dirty yellowish green	Brown
Average %yield(w/w)	1.83	15.98	3.324	1.34	0.938	7.4
Sterol	–	–	+	+	+	–
Bases	–	–	–	–	–	–
Saponins	–	–	–	–	–	–

Source: Gupta (1985) and Khafagy et al. (1977)

Table 22.7 Solvent systems used for different extractions of *J. curcas*

Plant extractive	Solvent system
Petroleum ether	Benzene-Chloroform-Acetone (35.5:7.5:5)
Benzene	Benzene-Chloroform-Acetone (2:7:1)
Chloroform	n-butanol-Acetone-Acetic acid-Water (20:25:6:4)
Acetone	n-propanol-Methanol-Acetic acid-Water (20:25:1:5)
Alcoholic	n-propanol-Methanol-water (25:20:5)

Source: Gupta (1985)

Table 22.8 Phytochemicals isolated from different parts of *J. curcas*

Plant parts	Isolated chemicals	Reference
Seeds	Curcin A and B, a lectin phorbol esters, esterases (JEA) and lipase (JEB), mono- and disaccharides like saccharose, raffinose, stachyose, glucose, fructose, galactose are known, but also protein (mostly lectins), and an oil, largely of oleic- and linoleic-acids, curcasin, arachidic-, linoleic-, myristic-, oleic-, palmitic-, and stearic-acids are also reported	Stirpe et al. (1976), Adolf et al. (1984), Makkar et al. (1997), Staubmann et al. (1999b), Chen et al. (2009)
	Triglycerides (by supercritical carbon dioxide extraction)	
Aerial parts	Organic acids (*o* and *p*- coumaric acid, *p*-OH-benzoic acid, protocatechuic acid, resorsilic acid, saponins and tannins	Hemalatha and Radhakrishnaiah (1993)
	Dinorditerpene	Ravindranath et al. (2003)
Latex	Curcacycline A—a cyclic octapeptide	Van den Berg et al. (1995a)
	Curcacycline B—a nonapeptide	Auvin et al. (1997)
	Curcain (a protease)	Nath and Dutta (1991)
Leaves	Cyclic triterpenes stigmasterol, stigmast-5-en-3 β, 7 β -diol, stigmast-5-en-3 β,7 α -diol, cholest-5-en-3 β, 7 β -diol,cholest-5-en-3 fl,7 α -diol, campesterol, β -sitosterol, 7-keto- β -sitosterol as well as the β-D-glucoside of β -sitosterol	Mitra et al. (1970), Khafagy et al. (1977), Hufford and Oguntimein (1987), Staubmann et al. (1999a)
	Flavonoids apigenin, vitexin, isovitexin	
	Leaves also contain the dimer of a triterpene alcohol ($C_{63}H_{117}O_9$) and two flavonoidal glycosides, 5-hydroxypyrrolidin-2-one and pyrimidine-2,4-dione	
Stem bark	β- Amyrin, β-sitosterol and taraxerol from stem- palmarumycins CP1, JC1 and JC2	Mitra et al. (1970), Ravindranath et al. (2004)
Kernel and press cake	Phytates, saponins and a trypsin inhibitor	Makkar and Becker (1997), Wink et al. (1997)
Roots	β - Sitosterol and its β-D-glucoside, marmesin, propacin, the curculathyranes A and B and the curcusones A–D	Naengchomnong et al. (1986, 1994)
	diterpenoids jatrophol and jatropholone A and B, the coumarin tomentin, the coumarino-lignan jatrophin as well as taraxerol	

Source: Kumar and Sharma (2008)

Table 22.9 Phytochemicals isolated from other *Jatropha species*

Species	Plant part	Extract	Isolated compound	Reference
J.gossypiifolia	Stem	Hexane+EtoAC	Cleomiscosin A (Coumarino-lignoid)	Biswanath et al. (2003)
	Stem, Root and Seeds	Petrol	2-Piperonylidene-3-verytryl-3R-γ-butyrolactone (Lignan)	Chatterjee et al. (1981)
	Leaves	Alcoholic	Flavonoids	Sankara Subramanian et al. (1971)
	Stem, Root and seeds	Petrol	Gadain (New Lignan)	Banerji et al. (1984)
	Stem	Petrol	Jatrodien (Lignan)	Biswanath et al. (1996)
	Whole plant	CH_2Cl_2–MeoH (1:1)	Propacin, Jatrophone and its analogues (Diterpenes), Jatropholones A&B and Lignans (Jatrophan, Gadain and Venkatasin)	Biswanath and Venkataiah (2001)
	Whole plant	Petrol	2,3-Bis(hydroxymethyl),6,7 methylene dioxy-1-(3',4'-dimethoxy phenyl)-naphthalene (Arylnaphthalene lignan)	Das and Banerji (1988)
	Whole plant	CH_2Cl_2–MeoH (1:1)	Dinorditerpene	Biswanath and Venkataiah (1999)
	Stem	Hexane	Gossypidien (Lignan)	Biswanath and Anjani (1999)
	Aerial parts	Petrol	Gossypifan (Lignan)	Biswanath and Ratna (1995)
J. multifida	Stem	$CHCl_3$:MeoH	Multifolone and (4E)-jatrogrossidentadione acetate (Diterpenoids)	Biswanath et al. (2010)
	Latex	EtoAC	Multifidin (Cyanoglucoside)	Van den Berg et al. (1995b)
	Latex	H_2O	Multifidol and glucoside (Acylphloroglucinols)	Kosasi et al. (1989a)
	Stem	—	Multifidone (Lathyrane type diterpene)	Biswanath et al. (2009)
	Latex	—	Labaditin (Decapeptide)	Kosasi et al. (1989b)
	Stem	$CHCl_3$–MeoH (1:1)	Multidione (Diterpenoid)	Biswanath et al. (2009)
J. podagrica	Root	Hexane	Japodagrin and japodagrone (Macrocyclic diterpenoids)	Olapeju et al. (2007)
	Latex	—	Podacycline A&B (Peptides)	Van den Berg et al. (1996)
J. gossypiifolia, J. podagrica, J. multifida	Seeds	—	Phorbol derivatives	Adolf et al. (1984)

(continued)

Table 22.9 (continued)

Species	Plant part	Extract	Isolated compound	Reference
J. glandulifera	Roots	MeoH	Jatropholone-A, Fraxetin, and a Coumarino-lignan	Parthasarathy and Pardha Saradhi (1984)
	Root	—	Naphthaquinone esters, 3,3-dimethylacrylyl-shikonin (Pigment)	Ballantine (1969)
J. integerrima	Root	$CHCl_3$	8,9-Seco-rhamnofolane and new rhamnofolane endoperoxide (2-epicaniojane with caniojane and 1,11-bisepicaniojane)	Sutthivaiyakit et al. (2003)
J. macrantha	Stem	MeoH	Catechin, catechin-7-o-β-glucopyranoside and proanthocyanidin β-3 analog (Catechin derivatives)	Benavides et al. (2006)
J. elliptica	Rhizome	Ethanol	Pyridine penta substituted (2,6-dimethyl-4-phenylpyridine-3,5-dicarboxylic acid diethyl ester)	Beatrice et al. (2005)
J. maheshwarii	Stem	Hexane, $CHCl_3$ and MeoH	Friedelin, epi-friedelinol, β-sitosterol, β-sitosterol-3- β-D-glucopyranoside	Viswanathan et al. (2004)
J. pohliana	Latex	EtoAC	Pohlianins A, B & C (Cyclic Peptides)	Guette et al. (1999)
J. grossidentata	Root bark	—	Rhamnofolane derivatives	Jakupovic et al. (1988)
J. dioica	Root	$CHCl_3$	Riolizatrione (Diterpene)	Dominguez et al. (1980)
J. isabelli	Rhizome	Petroleum ether, EtoAC and MeoH	Jatrophone derivative (Jatropholone A&B), Acetyl aleuritolic acid, Cyperenoic acid (Terpenes)	Pertino et al. (2007)
J. weddelliana	Root	Cold hexane + CH_2Cl_2	Jatrowedione (Lathyrane diterpene)	Brum et al. (1998)
J. mahafalensis	Latex	CH_2Cl_2–MeoH (9:1)	Mahafacyclin (Heptapeptide)	Carine Baraguey et al. (2000)

Table 22.10 Uses of different parts of *J. curcas* in medicines

Plant part used	Diseases	Reference
Seeds	To treat arthritis, gout and jaundice	Fojas et al. (1986) and Balaji et al. (2009)
Tender twig/stem	Tooth ache, gum inflammation, gum bleeding, pyorrhoea	Mujumdar and Misar (2004)
Plant sap	Dermatomucosal diseases	Heller (1996)
Plant extract	Allergies, burns, cuts and wounds, inflammation, leprosy, leucoderma, scabies, wound healing and small pox	Nath and Dutta (1997), Villegas et al. (1997), Somashekar et al. (2006)
Water extract of branches	HIV, tumor	Lin et al. (2003)

Source: Heller (1996) and Kaushik and Kumar (2004)

rheumatic pain (Heller 1996). The linoleic acid content in *J. curcas* kernel oil is used for skin care. Goonasekera et al. (1995) showed that various solvent extracts of *J. curcas* have an abortive effect. The oil has a strong purgative action and is also widely used for skin diseases and to soothe pain, such as that caused by rheumatism. The latex itself has been found to be a strong inhibitor of watermelon mosaic virus (Tewari and Shukla 1982). The leaves and latex are used in healing of wounds, refractory ulcers, septic gums, as a styptic in cuts and bruises. A proteolytic enzyme (curcain) has been reported to have wound healing activity in mice (Villegas et al. 1997). Investigation of the coagulant activity of the latex of *J. curcas* showed that whole latex significantly reduced the clotting time of human blood. *J. curcas* root powder in paste form is used for the treatment of inflammation (Mujumdar and Misar 2004). Muanza *et al.* (1995) reported that methanolic extract of *J. curcas* leaves protected the cultured human lymphoblastoid cell against the cytopathic effect of HIV. The methanolic extract of *J. curcas* leaves are reported to have β-blockers, which has potential cardiovascular action in humans. The latex of *J. curcas* contains several alkaloids *viz*., Jatrophine, Jatropham and Curcain with anticancer properties (Thomas et al. 2008). All parts of *J. curcas* show insecticidal properties against insect/pests like cotton bollworm, on pests of pulses, potato and corn (Kaushik and Kumar 2004). Balaji et al. (2009) reported the hepatoprotective activity of methanolic fraction of leaves of *J. curcas* on aflatoxin B1 induced hepatic carcinoma. Jatrophone, a diterpenoid, which was isolated from different *J. curcas* species has been reported to have a reaction with biological thiols (i.e., inhibition of tumor activity), antileukemic activity against P-388 lymphocytic leukemia cytotoxicity against KB cell culture, molluscicidal activity and antinoceptive activity (Santos and Santana 1999; Martini et al. 2000; Taylor et al. 1983). Chaturvedi (2008) investigated the antimicrobial activity of tetramethyl pyrazine, an amide alkaloid isolated from the leaves of *J. curcas* against five human pathogenic bacteria. Bahadur et al. (1997) reported that the extracts of *J. curcas* and *J. gossypiifolia* var. *elegans* showed antifungal activity against two plant pathogenic fungi (*Drechslera spicifera* and *Fusarium oxysporum*). Chopra et al. (1986) reported antibiotic as well as antilepromatic properties of *J. gossypiifolia* and *J. multifida*. Use of various parts of *J. curcas* in the treatment of diseases have been presented in Table 22.10.

Toxicity

Although the seeds are considered to be the most toxic part of the plant, all parts of the *J. curcas* plant contain toxins, such as phorbol esters, curcins and trypsin inhibitors. Varieties commonly growing in Africa and Asia have seeds that are toxic to humans and animals, whereas some varieties found in Mexico and Central America are known to be non-toxic. Just one to three seeds can produce toxic symptoms in humans, mainly those associated with gastro-intestinal irritation. There is acute abdominal pain and a burning sensation in the throat shortly after ingestion of the seeds, followed by nausea, vomiting and diarrhoea (Brittaine and Lutaladio 2010). Various toxins present in the plant are trypsin inhibitors, phytic acid, saponins, curcin, phorbol ester, tannin, hydrocyanic acid, alkaloids (Jatrophine), flavonoids, glycosyl flavonoids, sterols and sterol sapogenins. A non-toxic variety exists in Mexico which is used for human consumption after roasting. It does not contain Phorbol esters (Foidl et al. 1996).

Conclusions

Before exploiting any plant for industrial application, it is imperative to have complete information about its biology, chemistry and all other applications so that the potential of plant could be maximized. The *J. curcas* industry is at a very early stage of development, though there are vigorous efforts to promote it and if successful, it will alter the picture considerably. There are areas in the world where interest in the Jatropha plant is especially strong, such as Central America where it has originated, and Mali, where it is widely grown as a live hedge and a lot of research has been done on biodiesel derived from it. Knowledge of physical properties and their dependence on moisture content of *J. curcas* seed is essential to improve the design of equipment for harvesting, processing and storage of seeds. In the Indian context, development of biodiesel would not only serve to reduce import of fossil diesel, but also in generation of employment opportunities, accelerated rural development and meeting the environmental obligations, such as reduction of green house gases (India can tap the US$ 53 billion global market for carbon trading by promoting biofuel use and production), carbon sequestration, etc. Further, large wastelands could be utilized for the cultivation of *J. curcas* for production of biodiesel. *J. curcas* considered as a wild oilseed of the tropics and subtropics, is now being credited as one of the most promising biofuel crop, ideally suited for growing in the waste lands of the country. This potential biodiesel crop can bring about major economic activity providing income and employment opportunities to the rural communities.

References

Adolf W, Opferkuch HJ, Hecker E (1984) Irritant phorbol derivatives from four *Jatropha* species. Phytochemistry 23:129–132

Akintayo ET (2004) Characteristics and composition of *Parkia biglobbossa* and *Jatropha curcas* oils and cakes. Bioresour Technol 92:307–310

Auvin C, Carine Baraguey C, Blond A, Lezenven F, Pousset JL, Bodo B (1997) Curcacycline B, a cyclic nonapeptide from *Jatropha curcas* enhancing rotamase activity of cyclophilin. Tetrahedron Lett 38:2845–2848

Bahadur B, Reddy SM, Govardhan S, Giridhar P (1997) Anitimicrobial activity in eight species of *Jatropha* L. (Euphorbiaceae). J Indian Bot Soc 77:190–191

Balaji R, Suba V, Rekha N, Deecaraman M (2009) Hepatoprotective activity of methanolic fraction of *Jatropha curcas* on aflatoxin B1 induced hepatic carcinoma. Int J Phys Sci 1:287–296

Ballantine JA (1969) The isolation of two esters of the naphthaquinone alcohol, shikonin, from the shrub *Jatropha glandulifera*. Phytochemistry 8:158–159

Banerji J, Das B, Chatterjee A, Shooler JN (1984) Gadain, A lignan from *Jatropha gossypiifolia*. Phytochemistry 23:2323–2327

Baraguey C, Blond A, Correia I, Pousset JL, Bodo B, Auvin-Guette C (2000) Mahafacyclina A, cyclic heptapeptide from *Jatropha mahafalensis* exhibiting β-bulge conformation. Tetrahedron Lett 41:325–329

Beatrice M, Neuville L, Moreau NJ et al (2005) Multidrug resistance reversal agent from *Jatropha elliptica*. Phytochemistry 66:1804–1811

Benavides A, Montoro P, Bassarello C, Piacente S, Pizza C (2006) Catechin derivatives in *Jatropha macrantha* stems: characterisation and LC/ESI/MS/MS quali-quantitative analysis. J Pharm Biomed Anal 40:639–647

Biswanath D, Anjani G (1999) Gossypidien a lignan from stems of *Jatropha gossypiifolia*. Phytochemistry 51:115–117

Biswanath D, Kashinatham A, Venkataiah B, Srinivas KVNS, Mahender G, Ravinder Reddy M (2003) Cleomiscosin A, a coumarino-lignoid from *Jatropha gossypiifolia*. Biochem Syst Ecol 31:1189–1191

Biswanath D, Padma Rao S, Srinivas KVNS, Ratna D (1996) Jatrodien a lignan from the stems of *Jatropha gossypiifolia*. Phytochemistry 41:985–987

Biswanath D, Ravinder Reddy K, Ravikanth B, Venugopal Raju T, Sridhar B, Usman Khan P et al (2009) Multifidone: a novel cytotoxic lathyrane-type diterpene having an unusual six-membered A ring from *Jatropha multifida*. Bioorg Med Chem Lett 19:77–79

Biswanath D, Ratna D (1995) Gossypifan a lignan from *Jatropha gossypiifolia*. Phytochemistry 40:931–932

Biswanath D, Satya Kumar A, Kumar JN, Venugopal Raju T (2010) A new macrocyclic diterpenoid from *Jatropha multifida*. Nat Prod Res 24:1510–1513

Biswanath D, Venkataiah B (1999) A rare diterpene from *Jatropha gossypiifolia*. Biochem Syst Ecol 27:759–760

Biswanath D, Venkataiah B (2001) A minor coumarino-lignoid from *Jatropha gossypiifolia*. Biochem Syst Ecol 29:213–214

Brittaine R, Lutaladio N (2010) Jatropha: A small holder bioenergy crop. Integr Crop Manag 8:45–60

Brum RL, Honda NK, Mazarin SM, Hess SC, Cavalheiro AJ, Monache FD (1998) Jatrowedione, a lathyrane diterpene from *Jatropha weddelliana*. Phytochemistry 48:1225–1227

Chatterjee A, Biswanath D, Pascard C, Prange T (1981) Crystal structure of a lignan from *Jatropha gossypiifolia*. Phytochemistry 20:2047–2048

Chaturvedi P (2008) The isolation, characterisation and antimicrobial activity of alkaloids of *Jatropha curcas* and *Piper longum*. J Phytol Res 1:53–56

Chen WH, Chen CH, Chang CMJ, Chiu YH, Hsiang D (2009) Supercritical carbon dioxide extraction of triglycerides from *Jatropha curcas* L. seeds. J Supercrit Fluids 51:174–180

Chopra RN, Nayar SL, Chopra IC (1986) Glossary of Indian medical plants. CSIR, New Delhi, p 145

Das B, Banerji J (1988) Arylnaphthalene lignan from *Jatropha gossypiifolia*. Phytochemistry 27:3684–3686

Dehgan B, Webster GL (1978) Three new species of *Jatropha* (Euphorbiaceae) from Western Mexico. Madrono 25:30–39

Dominguez XA, Cano G, Franco R, Villarreal AM, Watson WH, Zarel V (1980) Riolozatrione, a new class of diterpene from *Jatropha dioica* var. *sessiliflora*. Phytochemistry 19:2478

Duke JA (1988) CRC handbook of medicinal herbs. CRC Press, Boca Raton, pp 253–254

Foidl N, Foidl G, Sanchez M, Mittelbach M, Hackel S (1996) *Jatropha curcas* L. as a source for the production of biofuel in Nicaragua. Bioresour Technol 58:77–82

Fojas FR, Garcia LL, Venzon EL, Sison FM, Villanueva BA, Fojas AJ et al (1986) Pharmacological studies on *J. curcas* as a possible source of anti-arrhythmic (β-blocker) agent. The Philipp J Sci 115:317–328

Gandhi VM, Cherian KM, Mulky MJ (1995) Toxicological studies on ratanjyot oil. Food Chem Toxicol 33:39–42

Goonasekera MM, Gunawardana VK, Jayasena K, Mohammed SG, Balasubramaniam S (1995) Pregnancy terminating effect of *Jatropha curcas* in rats. J Ethnopharmacol 47:117–123

Guette CA, Baraguey C, Blond A, Xavier HS, Pousset JL, Bodo B (1999) Pohlianins A, B and C, Cyclic peptides from the latex of *Jatropha pohliana* ssp. Molissima. Tetrahedron 55:11495–11510

Gupta RC (1985) Pharmacognostic studies on 'Dravanti' part-I *Jatropha curcas* Linn. Proc Ind Acad Sci Plant Sci 94:65–82

Heller J (1996) Physic nut. *Jatropha curcas* L. Promoting the conservation and use of underutilised and neglected crops. Institute of Plant Genetics and Crop Plant Research, Gatersleben/ International Plant Genetic Resources Institute, Rome, p 54

Hemalatha A, Radhakrishnaiah M (1993) Chemosystematics of *Jatropha*. J Econ Taxon Bot 17:75–77

Hufford CD, Oguntimein BO (1987) Non-polar constituents of *Jatropha curcas*. Lloydia 41:161–165

Idu M, Timothy O, Onyibe HI, Comor AO (2009) Comparative morphological and anatomical studies on the leaf and stem of some medicinal plants: *Jatropha curcas* L. and *Jatropha tanjorensis* J.L. Ellis and Saroja (Euphorbiaceae). Ethnobot Leafl 13:1232–1239

Jakupovic J, Grenz M, Schmeda-hirschmann G (1988) Rhamnofolane derivatives from *Jatropha grossidentata*. Phytochemistry 27:2997–2998

Katwal RPS, Soni PL (2003) Biofuels: an opportunity for socioeconomic development and cleaner environment. Indian Forester 129:939–949

Kaushik N, Kumar S (2004) *Jatropha curcas* L. Silviculture and uses. Agrobios (India), Jodhpur

Khafagy SM, Mohamed YA, Abdel NA, Mahmoud ZF (1977) Phytochemical study *of Jatropha curcas*. Plant Med 31:274–277

Kosasi S, Van der sluis WG, Labadie RP (1989a) Multifidol and multifidol glucoside from the latex of *Jatropha multifida*. Phytochemistry 28:2439–2441

Kosasi S, Van der sluis WG, Boelens R, Hart LA, Labadie RP (1989b) Labaditin, a novel cyclic decapeptide from the latex of *Jatropha multifida* L. (Euphorbiaceae).Isolation and sequence determination by means of two-dimensional NMR. FEBS Lett 256:91–96

Kumar A, Sharma S (2008) An evaluation of multipurpose oil seed crop for industrial uses (*Jatropha curcas* L.): A review. Ind crop prod 28:1–10

Liberalino AA, Bambirra EA, Moraes-Santos T, Vieira EC (1988) *Jatropha curcas* L. seeds: chemical analysis and toxicity. Arq Biol Technol 31:539–550

Makkar HPS, Becker K, Sporer F, Wink M (1997) Studies on nutritive potential and toxic constituents of different provenances of *Jatropha curcas*. J Agric Food Chem 45:3152–3157

Martini LH, Souza CR, Marques PB, Calixto JB, Yunes RA, Souza DO (2000) Compounds extracted from *Phyllanthus* and *Jatropha elliptica* inhibit the binding of [3 H] Glutamate and [3 H] GMP-PNP in rat cerebral cortex membrane. Neurochem Res 25:211–215

Mitra CR, Bhatnagar SC, Sinha MK (1970) Chemical examination of *Jatropha curcas*. Ind J Chem 8:1047

Muanza DN, Euler KL, Williams L, Newman DJ (1995) Screening for antitumor and anti-HIV activities of nine medicinal plants from Zaire. Intl J Pharmacog 33:98–106

Mujumdar AM, Misar AV (2004) Anti–inflammatory activity of *Jatropha curcas* roots in mice and rats. J Ethnopharmacol 90:11–15

Naengchomnong W, Thebtaranonth Y, Wiriyachitra P, Okamoto KT, Clardy J (1986) Isolation and structure determination of four novel diterpenes of *Jatropha curcas*. Tetrahedron Lett 27:2439–2442

Naengchomnong W, Tarnchompoo B, Thebtaranonth Y (1994) (+)-*Jatropha*, (+)-marmesin, propacin and jatrophin from the roots of *Jatropha curcas* (Euphorbiaceae). J Sci Soc Thai 20:73–83

Nath LK, Dutta SK (1991) Extraction and purification of curcain, a protease from the latex of *Jatropha curcas* L. J Pharm Pharmacol 43:111–114

Nath LK, Dutta SK (1997) Acute toxicity studies and wound healing response of curcain, a proteolytic enzyme extract from the latex of *Jatropha curcas* L. In: Gubitz GM, Mittelbach M, Trabi M (eds) Biofuels and industrial products from *Jatropha curcas*. DBV Graz, Graz, pp 82–86

Nayak BS, Patel KN (2010) Pharmacognostic studies of the *Jatropha curcas* leaves. Intl J PharmTech Res 2:140–143

Nwosu MO, Okafor JL (1995) Preliminary studies of the antifungal activities of some medicinal plants against *Basidiobolus* and some other pathogenic fungi. Mycoses 38:191–195

Olapeju OA, Kayode A, Olusegun E, James BG (2007) Antibacterial diterpenoids from *Jatropha podagrica* Hook. Phytochemistry 68:2420–2425

Openshaw K (2000) A review of *Jatropha curcas*: an oil plant of unfulfilled promise. Biomass Bioenergy 19:1–15

Parthasarathy MR, Pardha Saradhi K (1984) A coumarino-lignan from *Jatropha glandulifera*. Phytochemistry 23:861–869

Pertino M, Schmeda-Hirschmann G, Rodrfguez JA, Theoduloz C (2007) Gastroprotective effect and cytotoxicity of terpenes from the Paraguayan crude drug "yagua rova" (*Jatropha isabelli*). J Ethnopharmacol 111:553–559

Rata Krishnan P, Paramathma M (2009) Potentials and *Jatropha* species wealth of India. Curr Sci 97:1000–1004

Ravindranath N, Ramesh C, Biswanath D (2003) A rare dinorditerpene from *Jatropha curcas*. Biochem Syst Ecol 31:431–432

Ravindranath N, Ravinder Reddy M, Mahender G, Ramu R, Ravi Kumar K, Biswanath D (2004) Deoxypreussomerins from *Jatropha curcas*: are they also plant metabolites? Phytochemistry 65:2387–2390

Rivera-Lorca JA, Ku-Vera JC (1997) Chemical composition of three different varieties of *J. curcas* from Mexico. In: Gubitz GM, Mittelbach M, Trabi M (eds) Biofuels and industrial products from *Jatropha curcas*. DBV Graz, Graz, pp 47–52

Sankara Subramanian S, Nagarajan S, Sulochana N (1971) Flavonoids of the leaves of *Jatropha gossypiifolia*. Phytochemistry 10:1690

Santos AF, Santana AE (1999) Molluscicidal activity of the diterpenoids jatrophone and jatropholones A and B isolated from *Jatropha elliptica* (Pohl) Muell. Arg Phytother Res 13:660–664

Shanti BM, Mukherjee A, Vijayakumar M (2010) Pharmacognostical and phytochemical evaluation of leaves extract of *Jatropha curcas* Linn. Pharmacog J 2:9–14

Sharma GD, Gupta SN, Khabiruddin M (1997) Cultivation of *Jatropha curcas* as a future source of hydrocarbon and other industrial products. In: Gubitz GM, Mittelbach M, Trabi M (eds) Biofuels and industrial products from *Jatropha curcas*. DBV Graz, Graz, pp 19–21

Solsoloy AD, Solsoloy TS (1997) Pesticidal efficacy of formulated *J. curcas* oil on pests of selected field crops. In: Gubitz GM, Mittelbach M, Trabi M (eds) Biofuels and industrial products from *Jatropha curcas*. DBV Graz, Graz, pp 216–226

Somashekar S, Saraswati LU, Alaya LU, Venkata RV (2006) Wound healing activities of bark extract of *Jatropha curcas* Linn in albino rats. Saudi Med J 27:1473–1476

Staubmann R, Schubert-Zsilavecz M, Hiermann A, Kartnig T (1999a) A complex of 5-hydroxypyrrolidin-2-one and pyrimidine-2, 4-dione isolated from *Jatropha curcas*. Phytochemistry 50:337–338

Staubmann R, Ncube I, Gubitz GM, Steiner W, Read JS (1999b) Esterase and lipase activity in *Jatropha curcas* L. seeds. J Biotechnol 75:117–126

Stirpe F, Pession A, Brizzi A, Lorenzoni E, Strochi P, Montanaro L et al (1976) Studies on the proteins from the seeds of *Croton tigilium* and *Jatropha curcas*. Biochem J 156:1–6

Sutthivaiyakit S, Mongkolvisut W, Ponsitipiboon P, Prabpai S, Kongsaeree P, Ruchirawatb S et al (2003) A novel 8, 9-seco-rhamnofolane and a new rhamnofolane endoperoxide from *Jatropha integerrima* roots. Tetrahedron Lett 44:3637–3640

Taylor MD, Smith AB, Furst GT, Gunasekara SP, Bevelle CA, Cordell GA et al (1983) Plant anticancer agents 28. New antileukemic jatrophone derivatives from *Jatropha gossypiifolia*: structural and stereochemical assignment through nuclear magnetic resonance spectroscopy. J Am Chem Soc 105:3177–3183

Tewari JP, Shukla IK (1982) Inhibition of infectivity of two strains of watermelon mosaic virus by latex of some angiosperms. Geobios 9:124–126

Thomas R, Sah NK, Sharma PB (2008) Therapeutic biology of *Jatropha curcas:* a mini review. Curr Pharm Biotechnol 9:315–324

Van den berg AJJ, Horsten SF, van den Kettenes BJJ, Kroes BH, Beukelman CJ, Loeflang B et al (1995a) Curcacycline A: a novel cyclic octapeptide isolated from the latex of *Jatropha curcas* Linn. FEBS Lett 358:215–218

Van den berg AJJ, Horsten SFAJ, Kettenes-Van Den Bosch JJ, Kroes BH, Labadie RP (1995b) Multifidin a cyanoglucoside in the latex of *Jatropha multifida*. Phytochemistry 40:597–598

Van den Berg AJJ, Horsten SFAJ, Bosch JJK, Beukelman CJ, Kroes BH, Leeflang BR et al (1996) Podacycline A and B, two cyclic peptides in the latex of *Jatropha podagrica*. Phytochemistry 42:129–133

Verma KC, Gaur AK (2009) *Jatropha curcas* L.: substitute for conventional energy. World J Agric Sci 5:552–556

Villegas LF, Fernandez ID, Maldonado H, Torres R, Zavaleta A, Vaisberg AJ, Hammond GB (1997) Evaluation of the wound-healing activity of selected traditional medicinal plants from Peru. J Ethnopharmacol 55:193–200

Viswanathan MB, Ramesh N, Ahilan A, Lakshmanaperumalsamy P (2004) Phytoconstituents and antimicrobial activity from the stems of *Jatropha maheshwarii*. Med Chem Res 13:361–368

Wink M, Koschmieder C, Sauerwein M, Sporer F (1997) Phorbol esters of *J. curcas*—biological activities and potential applications. In: Gubitz GM, Mittelbach M, Trabi M (eds) Biofuels and industrial products from *Jatropha curcas*. DBV Graz, Graz, pp 160–166

Chapter 23
Jatropha and Phytoremediation of Metal Contaminated Land

Asha A. Juwarkar, S.K. Yadav, and G.P. Kumar

Introduction

In this age of technology development, the progress of civilization and rapid industrialization have brought the danger of soil pollution. Over the last three decades, there has been an increase in both industrial activities and urbanization. Both aquatic and terrestrial habitats are becoming progressively polluted due to the discharge of pollutants generated from various industries, transportation and fossil fuel burning. Currently the soil, where we grow food, is severely polluted with hazardous chemicals and pathogenic microorganisms, which enter the soil because of unhygienic conditions. Some anthropogenic practices are the major sources responsible for adding a large variety of contaminants to the soil, causing soil pollution. Metals are a major category of globally distributed pollutants, are natural elements extracted from the earth and harnessed for industry and products. Mining can bring much economic prosperity, but large areas of industrial dereliction often result once mining has ceased. This dereliction includes a legacy of abandoned tips and tailings, which contain the waste products of both mining and processing operations. Such materials are often a major source of heavy metal pollution in the local environment owing to dust blow and from the leaching of the products of mineral weathering into nearby water courses. This pollution may have serious detrimental effects upon crops and public health (Smith and Bradshaw 1972). It is now a requirement in most countries that remediation schemes are incorporated at the planning stage of mining and industrial proposals.

A.A. Juwarkar (✉) • S.K. Yadav
Eco-Restoration Division, National Environment Engineering Research Institute (NEERI), Nehru Marg, Nagpur 440020, India
e-mail: aa_juwarkar@neeri.res.in

G.P. Kumar
Plant Nutrition Division, Defence Food Research Laboratory, Mysore, Karnataka, India

N. Carels et al. (eds.), *Jatropha, Challenges for a New Energy Crop: Volume 1: Farming, Economics and Biofuel*, DOI 10.1007/978-1-4614-4806-8_23,

Until now, methods used for their remediation, such as excavation and land fill, thermal treatment, acid leaching and electroreclamation are not suitable for practical applications because of their high cost, low efficiency, large destruction of soil structure and fertility, high dependence on the contaminants of concern, soil properties, site conditions, and so on. Current remediation strategies of heavy metals is primarily based on physico-chemical technologies, which are meant primarily for intensive *in situ* or *ex situ* treatment of relatively highly polluted sites, and thus are not very suitable for the remediation of vast, diffusely polluted areas where pollutants only occur superficially and at relatively low concentrations (Garbisu and Alkorta 2001).

A range of reclamation techniques is available for metalliferous substrates, but complete long term rehabilitation of wasteland (mine wastes/degraded land) can only be achieved through the use of vegetation. Successful revegetation can be a permanent and visually attractive solution and, at the same time, be relatively inexpensive. A vegetation cover can be effective in providing the necessary surface stability to prevent wind-blow of contaminated particulates, and in reducing water pollution by interception of a substantial proportion of incident precipitation. Although revegetation is desirable, metalliferous wastes are very unfavourable environments for plants because of the presence of many growth-limiting factors particularly—residual high levels of heavy metals, macronutrient deficiencies and poor substrate structure. Such features result in most metal wastes being largely devoid of any natural vegetation, even many years after abandonment (Juwarkar et al. 2009). In this context, phytoremediation appears as a valid option since it is best suited for the remediation of these diffusely polluted areas and at much lower costs than other methods (Kumar et al. 1995). The idea of using plants to remove metals from soils came from the discovery that different wild plants, often endemic, naturally accumulate high concentrations of metals in their foliage (Raskin et al. 1994).

Jatropha curcas

J. curcas L. (hereafter referred to as Jatropha) is a bush/small tree native of tropical America. The genus *Jatropha* belongs to tribe Joannesieae in the Euphorbiaceae family and contains approximately 170 known species. Jatropha, can reach a height of 3–5 m, but under favorable conditions it can attain a height of 8 or 10 m. The trees are drought tolerant, deciduous, shedding the leaves in dry season, widely distributed in the wild or semi-cultivated areas in Central and South America, Africa, India and South East Asia (Cano-Asseleih et al. 1989). The first commercial applications of Jatropha were reported from Lisbon, where the oil imported from Cape Verde was used for soap production and lighting. During the last 10 years, Jatropha has gained considerable attention as a potential feedstock of biodiesel and many plantations have been established in tropical and subtropical regions worldwide. Its peculiar features like drought tolerance, rapid growth, easy propagation, higher oil content than other oil crops, small gestation period, adaptation to a wide range of environmental conditions, and the optimum plant size and architecture (that make the seed collection more convenient) make it as a special candidate for further consideration (Abhilash et al. 2010).

Jatropha Plantations on Wastelands in India

Jatropha is regarded as an oil plant with multiple attributes, uses and considerable potential (Openshaw 2000). Importance is given on the plantation of Jatropha on wastelands for environment protection and for fulfilling future energy requirements. Jatropha produces seeds for bio-diesel and it can also restore degraded land environments. Land degradation has become a major threat to world food security with about 2,000 million ha of soil, equivalent to 15% of the Earth's land area. India has vast stretches of degraded land, mostly in areas with adverse agro-climatic conditions, where species of Jatropha can grow easily. Nearly 40% of the land area in India is wasteland. Even 30 million hectares of Jatropha cultivation for biodiesel can completely replace the current usage of fuels in India. Use of 11 million hectares of wasteland for Jatropha cultivation can lead to generation of a minimum 12 million jobs. The Central Salt & Marine Chemical Research Institute (CSMCRI), Bhavnagar, India has successfully cultivated good varieties of Jatropha on marginal land to assess practically realizable seed yields. GB Pant University (India) has planted Jatropha on 140 ha at its farm. The University scientists have selected new high yielding accessions of Jatropha, which have yield potentials of up to 10 t/ha. Efforts of National Oilseed and Vegetable Oil Development Board (NOVOD) include development of quality planting material, improved Jatropha seeds having oil content up to 1.5 times the ordinary seeds. India has about 175 million hectares of wastelands, which remains unutilized. The planning commission of India (2005–2006) has recommended launching of a National Mission on Bio-Diesel with special focus on plantation of Jatropha on wastelands to meet the country's energy demand. The Government of India has identified 400,000 km^2 (98 million acres) of land where Jatropha can be grown, hoping it will replace 20% of India's diesel consumption by 2011 (Biofuel in India, www.en.wikipedia.org).

Potential Uses of Jatropha

In addition to being a source of oil, Jatropha also provides a meal that serves as a highly nutritious and economic protein supplement in animal feed, if the toxins are removed (Becker and Makkar 1998). Jatropha has several advantages over other plant species that can thrive under adverse conditions. It is not eaten by animals and is a vigorous, drought and pest resistant. When planted as a fence, it repels rodents and has phytoprotective action against pests and pathogens providing additional protection to intercropped plants. Cattle have been found to graze in the space between Jatropha rows in large plantations. Various parts of the plant are of medicinal value; its bark contains tannin, the flowers attract bees and thus the plant has a honey production potential. Its wood and fruit can be used for numerous purposes including fuel. It is easy to establish and grows relatively quickly. Nevertheless, Jatropha does mine soil nutrients; it needs mineral and organic fertilizer to increase

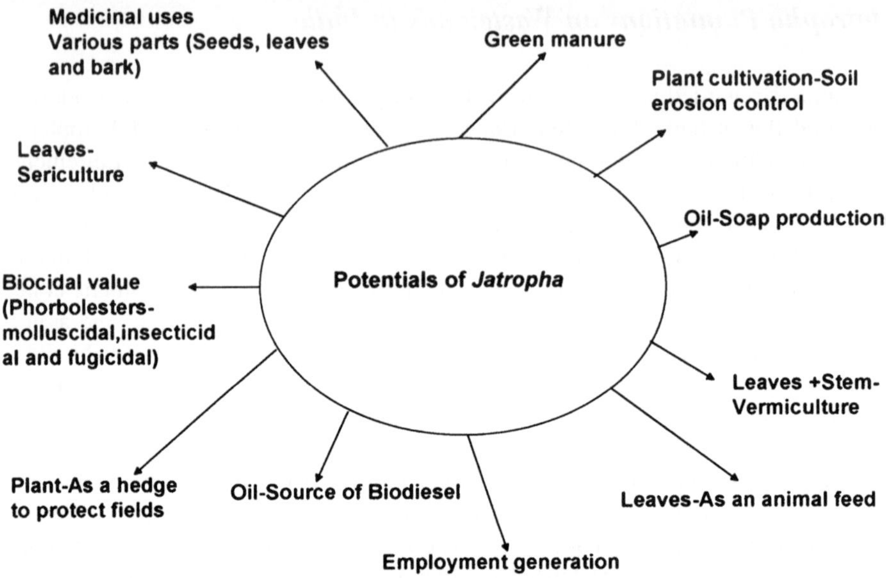

Fig. 23.1 Economic significance of *J. curcas* (Kumar and Sharma (2008)

crop yields. Despite being ecologically friendly, Jatropha is expected to provide income as well as source of alternative energy to rural farmers.

The multiple attributes of Jatropha is presented in Fig. 23.1 and has been extensively reviewed by many workers. The utilization of various parts of Jatropha was reviewed by many scientists and started about one decade ago by Gubitz et al. (1999). In addition, some of the recent review articles provide status and perspectives of Jatropha biodiesel program in various countries [e.g., India (Biswas et al. 2010; Jain and Sharma 2010), UK (Janaun and Ellis 2010), Malaysia (Lim and Teong 2010), China (Ye et al. 2009), Thailand (Siriwardhana et al. 2009)]. Zhou and Thomson (2009), highlighted the current status of biofuel production including the national development targets, strategies, incentives, and policies in Asia's largest producing countries, such as Indonesia, Malaysia, The Philippines, Thailand, China, and India.

Phytoremediation

The term phytoremediation ("phyto" meaning plant and the Latin suffix "remedium" meaning to clean or restore) actually refers to a diverse collection of plant-based technologies that use either naturally occurring or genetically engineered plants for cleaning the contaminated environments (Cunningham et al. 1997). Phytoremediation can be defined as the use of green plants to remove pollutants from the environment or to render them harmless (Berti and Cunningham 2000; Salt et al. 1995). In this respect, plants can be compared to solar driven pumps capable of extracting and concentrating certain elements from their environment (Salt et al. 1995). This technology is being considered as a new highly promising technology for the remediation of polluted sites. The concept of using plants to clean up contaminated environments is not new. About 300 years ago, plants were proposed for use in the treatment of wastewater (Hartman 1975). Phytoremediation technologies that are soil focused are suitable for large areas that have been contaminated with low to moderate levels of contaminants. The primary motivation behind the development of phytoremediation technologies is the potential for low-cost remediation (Ensley 2000). It is also referred to as green technology and can be applied to both organic and inorganic pollutants, present in solid substrates (e.g., soil) and liquid substrates (e.g., water) (Gratao et al. 2005; Salt et al. 1998). Phytoremediation, the use of plants and their associated microbes for environmental cleanup, has gained acceptance in the past 10 years as a cost-effective, noninvasive alternative or complementary technology for engineering based remediation methods. The success of the phytoremediation process depends on many factors (Pulford and Watson 2003):

- Degree of pollutant contamination
- Pollutant bioavailability
- Ability to grow on nutrient-poor soil
- Deep root system
- Fast rate of growth
- Metal-resistance trait

Strategies for Phytoremediation

Phytoremediation can occur through a series of complex interactions between plants, microbes, and soil including accumulation, hyperaccumulation, exclusion, volatilization, and degradation. Plants also stabilize mobile contaminated sediments by forming dense root mats under the surface. In general, phytoextraction and phytovolatization are considered as the main options for the removal of heavy metals and other elemental compounds, whereas phytodegradation and phytostabilisation are applied mostly to organic contaminants (Meagher 2000; Guerinot and Salt 2001). Therefore, phytoremediation can be accomplished by phytoextraction, phytodegradation, phytostabilization, phytovolatization and rhizofiltration. Figure 23.2 revealed the different processes and mechanisms used for phytoremediation (The University of Georgia).

- ***Phytoextraction:*** the use of plants to remove contaminants from soils. Pollutant-accumulating plants are utilized to transport and concentrate contaminants (metal or organic) from the soil into harvestable parts of the roots and aerial parts of the plant; the term is mostly used to refer to metal removal from soils (Kumar et al. 1995).
- ***Phytostabilization:*** the use of plants to reduce the bioavailability of pollutants in the environment. Plants stabilize pollutants in soils by chemically immobilizing contaminants, thus rendering them harmless and reducing the risk of further

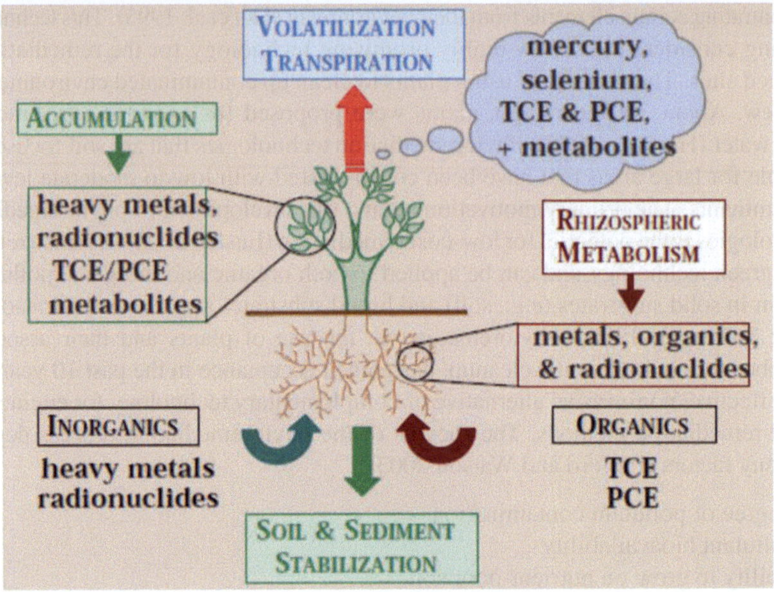

Fig. 23.2 Overview of different processes of phytoremediation

environmental degradation by leaching of pollutants into the ground water or by airborne spread (Prasad and de Oliveira Freitas 2003).

- ***Phytovolatization:*** the use of plants to volatilize pollutants. Plants extract volatile pollutants (e.g., selenium, mercury and arsenic) from the soil and biologically converts them to a gas, which is released via transpiration from the foliage (Ghosh and Singh 2005).
- ***Phytodegradation:*** the use of plants to degrade organic pollutants. Plant roots are utilized to remediate contaminated soils by the breakdown of organic contaminants to simpler molecules, which are stored in the plant tissue (Ghosh and Singh 2005).
- ***Rhizofiltration:*** the approach of using hydroponically cultivated plant roots to remediate contaminated water through absorption, concentration, and precipitation of pollutants. This contaminated water is either collected from a waste site and brought to the plants, or the plants are planted in the contaminated area, where the roots then take up the water and the contaminants dissolved in it (Dushenkov et al. 1995).

Introduction of invasive species for phytoremediation purposes may affect the local biodiversity. Therefore, identification and selection of locally available plant species for phytoremediation research and implementation is one of the challenges that needs to be met and a pre-requisite for successful phytoremediation. Phytoremediation of different types of contaminants requires different general plant characteristics for optimum effectiveness. Careful selection of plant and plant variety is critical, first, to ensure that the plant is appropriate for the climatic and soil conditions at the site, and second, for effectiveness of the phytoremediation of the pollutant at hand. Plant species that are long-term competitors and survivors under adverse changing conditions normally have an advantage. Depending on the climatic and soil conditions, a plant may need resistance or tolerance to diseases, heat, cold, insects, drought, chemicals, and stress to maximize its survival rate. The type, amount and effectiveness of exudates and enzymes produced by a plant's root will vary between species and even within subspecies or varieties of one species. A screening of phytotoxicity and effectiveness of cultivars/varieties might be required on a site-specific basis as an initial step in plant selection. Phytoremediation studies have examined numerous plant species. Terrestrial plant species are more likely to be effective for phytoremediation than aquatic plant species due to their larger root systems.

Jatropha as a Phytoremediator

Scientists all over the world are searching for new plant species suitable to be used in phytoremediation. While selecting a species for phytoremediation several factors have to be taken into account. The species should be fast growing, high biomass producing, with profuse root system, tolerant to adverse environment condition, non edible and economically beneficial (Alkorta et al. 2004). Considering all the options available among non-edible tree, Jatropha is suitable for phytoremediation of contaminated soil/degraded land. It has been identified in India as the most suitable oilseed and has been recommended for plantation on wasteland as it requires minimal inputs for its establishment (Singh 2007; Gunaseelan 2009). Jatropha grows practically all over India under a variety of agro-climatic conditions. Thus, it insures a reasonable production of seeds with very little input.

Jatropha can grow in diverse types of soil conditions (Mangkoedihardjo and Surahmaida 2008). Regarding the potential use of Jatropha for the phytoremediation of coal fly ash, Jamil et al. (2009) suggested that it has the potential of establishing itself on fly ash when provided with basic plant nutrients and also to accumulate heavy metals many folds from fly ash without plant growth diminution. Furthermore, the study performed by Agamuthu et al. (2010) suggested that Jatropha is suitable to be used as a phytoremediator for removing hydrocarbon from contaminated soil.

Studies by various researchers show Jatropha as a potential plant for remediation of soils contaminated with heavy metals. Chehregani and Malayeri (2007) observed that Jatropha has the capability to remove Cd and Pb from soil containing high concentrations of metal contaminated sludge. The roots of Jatropha were suitable for the uptake of heavy metals from soil and sludge, especially Zn. Chromium is also adsorbed effectively by the plant. Thus, Jatropha is very efficient in accumulating heavy metals, causing no damaging effects to the root biomass. Therefore, Jatropha is a suitable plant to use as a phytoremediator to clean heavy metals, in particular As, Cr, Cu, Cd, Hg, Pb and Zn (Chehregani and Malayeri 2007; Juwarkar et al. 2008; Mangkoedihardjo et al. 2008; Ahmadpour et al. 2010). Baker (1981) reported that some plants are able to hasten the transportation of heavy metals from their roots to shoots, which may be related to the rate of metal uptake in the case of Jatropha. The accumulation of heavy metals in Jatropha tends to occur mainly in the roots and to a lesser degree in the stems and leaves. Yadav et al. (2009) reported the bioaccumulation and phyto-translocation of As, Cr and Zn by Jatropha and concluded that the efficiency of heavy metal uptake by plant parts is as follows: shoots < roots. Two possible strategies have been proposed for the use of Jatropha for phytoremediation.

- Jatropha that survive in contaminated soil with minimal uptake of metals into the aerial tissues would be most appropriate for use where distribution of heavy metals to the wider environment or transfer of metals into the food chain is to be avoided.
- Jatropha that accumulates relatively high amounts of metal are desirable if soil remediation is to be achieved by phytoextraction and tree harvesting.

These findings suggest that this plant exhibits the ability for phytoremediation and reclaim of metal contaminated soils. However, plant growth and thus, oil yield can be affected negatively if no appropriate management is applied. The primary conservation benefits to be derived from production of Jatropha concerns soil restoration and management. Establishment of vegetation on heavy metal contaminated lands/soils poses problems because of metal toxicity to plants. The physico-chemical and microbiological properties of heavy metals contaminated soils tend to inhibit plant growth (Sopper 1993). Kumar et al. (2008a) have conducted experiments with respect to survival and growth performance of Jatropha on metalloid and metal contaminated soils with and without organic matter and biofertilizer amendments. They have indicated that Jatropha survived in soils contaminated with very high concentrations of As, Cr and Zn. Organic waste materials and biofertilizer amendments greatly enhanced the height of the plant. Biomass of Jatropha was increased tremendously with the addition of organic waste materials in As, Cr and Zn contaminated soil. Therefore, addition of organic matter and biofertilizer enrich the soil with nutrients and organic matter, which favoured the plant growth on high concentration of metals contaminated soils. Further, the results also showed that Zn enhanced the growth of Jatropha as compared to other metal contaminated soils. The heavy metal accumulation in a plant increased with increasing soil concentrations of heavy metals whereas it decreased when metal contaminated soil is amended with biosludge along with biofertilizers or biosludge alone. It seems that the organic matter present in the biosludge acted as metal chelator thereby reducing the toxicity of metals to plants. The main benefit from Jatropha growth on a metal contaminated soil was reported to its stabilization as metal uptake by biomass was not sufficient for phytoextraction to be significant. However, as stated by Juwarkar et al. (2008), the growth of Jatropha may be stimulated upon soil depletion of the bioavailable metal.

Furthermore, Jatropha is to some extent salt tolerant and has therefore the potential to be grown on saline soil. However, the cultivation on this type of soil can lower its productivity. For instance, Dagar et al. (2006) evaluated Jatropha growth rate at different irrigation treatments with varying levels of salinity. Jatropha showed different biomass (including seeds) growth rates depending on the level of irrigation with saline water. Kumar et al. (2008b) reported that NaCl inhibits the growth of calli and this process is proportional to NaCl levels. This shows that Jatropha can be irrigated with saline water or grown on saline soil, but potentially leading to lower yields and oil content. Both studies show that the dilemma of growing Jatropha on saline soils lies in the potential of using and improving degraded land, and the positive response of Jatropha to (non-saline water) irrigation. Saline stress may also increase on saline lands where water is scarce.

Mechanism of Stress Tolerance in Jatropha

Salt and metals stress in soil are major abiotic stresses especially in arid and semi-arid regions and can severely limit plant growth and yield. Abiotic stress can lead to stomatal closure, which reduces CO_2 availability in leaves and inhibits carbon fixation, exposing chloroplasts to excessive excitation energy, which in turn may increase the generation of *reactive oxygen species* (ROSs) and induce oxidative stress (Parida and Das 2005; Parvaiz and Satyawati 2008). ROSs have the potential to interact with many cellular components, causing significant damage to membranes and other cellular structures. However, an elaborate and highly redundant plant ROS network, composed of antioxidant enzymes and antioxidants, is responsible for maintaining the levels of ROS under tight control. In plant cell, antioxidant enzymes, such as *superoxide dismutase* (SOD), *peroxidase* (POD), *catalase* (CAT), *ascorbate peroxidase* (APX), *glutathione s-transferase* (GST) and *glutathione* (GSH) have been considered as a defensive team, whose combined purpose is to protect cells from oxidative damage (Mittler 2002). The activities of antioxidant enzymes in plants are influenced by the stress conditions where they are growing. These enzymes serves as an antioxidant, directly detoxify metals by conjugating them, forming a non-toxic complex through glutathione-s-transferase catalyzed reaction. Induction of SOD, POD, CAT, APX, GST and GSH provide defenses against metal toxicity and keeps the metabolic activities functional. Yadav et al. (2010) reported that both enzymatic and non-enzymatic antioxidants played significant roles in metal detoxification and accumulation in Jatropha. In addition, increased SOD, POD, CAT and PAL activities in Jatropha suggest tolerance capacity of plants to protect themselves from oxidative damage due to heavy metals and salt (Gao et al. 2008a, b, 2010; Kumar et al. 2008b). Jatropha is of much importance in phytoremediation as it can withstand environmental stress. A novel *betaine aldehyde dehydrogenase* gene (*JcBD1*) helps Jatropha to survive in environmental stress induced by drought, heat, and salt (Zhang et al. 2008). The expression of this novel gene into *E. coli*, results in expression of JcBD1 enzyme that makes it resistance to abiotic stress factors like salt.

Under semi-arid conditions, Jatropha has the potential for reclaiming degraded soils by penetrating soils with its root system, recycling nutrients from deeper soil layers, and providing shadow, thereby reducing risks of soil erosion and desertification (Jongschaap et al. 2010). These studies indicate that Jatropha can successfully be grown in metal, salt and hydrocarbon degraded/contaminated soils and wastelands with suitable amendments to meet the requirements for biodiesel production. Industrial wastes, mostly organic wastes, with no toxicity to plant can be used for cultivation of Jatropha on wasteland. *J. curcas* can also be used to reclaim eroded land and other problematical sites. On degraded lands, planting biofuel crops could help to restore soil fertility and structure if managed appropriately.

Conclusions

Wastelands and degraded lands are characterized by natural physical and biological conditions that are *per se* unfavorable for land-associated human activities. Within this category, land without appreciable vegetative cover or agricultural potential is included. These areas cannot be cultivated under any conditions and, therefore, are not suitable for bioenergy production. Current practice for remediating heavy metal contaminated soils relies heavily on 'dig-and-dump' or encapsulation, neither of which addresses the issue of decontamination of a soil. Phytoremediation, and especially the use of trees, is an emerging and developing technology. Jatropha is a drought resistance species and can easily sustain in all kinds of wastelands. Phytoremediation of polluted soil with non-edible plant like Jatropha offers an environmental friendly and cost effective method for remediating polluted soils. Jatropha produces seeds for biodiesel and it is a multipurpose species with many attributes and considerable potential. Jatropha having a potential for phytoremediation of soil contaminated with heavy metals, salt, and hydrocarbon. It is capable of extracting heavy metals from contaminated soil. The biochemical activity of Jatropha directly indicates resistance offered by plants against stress. Jatropha is a large biomass producing crop with ability to accumulate heavy metals and hence, may be recommended for plantation on heavy metal, salt and oil contaminated land/wasteland when provided with essential plant nutrients. Studies have proven that Jatropha with organic amendments has a potential in reclaiming metal contaminated soil. There is a growing body of knowledge in support of agroforestry for future eco-agriculture, farm diversification and management of heavy metal contaminated sites. Jatropha is suitable for integration in different agroforestry systems, in this sense; Jatropha's domestication can provide a powerful means of eco-friendly socioeconomic management involving income generation, climate change mitigation, soft farming and sustainable development of reclaimed sites.

References

Abhilash PC, Srivastava P, Jamil S, Singh N (2010) Revisited *Jatropha curcas* as an oil plant of multiple benefits: critical research needs and prospects for the future. Environ Sci Pollut Res 18:127–131

Agamuthu P, Abioye OP, Aziz AA (2010) Phytoremediation of soil contaminated with used lubricating oil using *Jatropha curcas*. J Hazard Mater 179:891–894

Ahmadpour P, Nawi AM, Abdu A, Abdul-Hamid H, Singh DK, Hassan A et al (2010) Uptake of heavy metals by *Jatropha curcas* L. planted in soils containing sewage sludge. Am J Appl Sci 7:1291–1299

Alkorta I, Hernandez-Allica J, Becerril JM, Amezaga I, Albizu I, Garbisu C (2004) Recent findings on the phytoremediation of soils contaminated with environmentally toxic heavy metals and metalloids such as zinc, cadmium, lead, and arsenic. Rev Environ Sci Biotechnol 3:71–90

Baker AJM (1981) Accumulation and excluders strategies in the response of plants to heavy metals. J Plant Nutr 3:643–654

Becker K, Makkar HPS (1998) Toxic effects of phorbol esters in carp (*Cyprinus carpio* L.). Vet Hum Toxicol 40:82–86

Berti WR, Cunningham SD (2000) Phytostabilization of metals. In: Raskin I, Ensley BD (eds) Phytoremediation of toxic metals: using plants to clean-up the environment. Wiley, New York, pp 71–88

Biswas PK, Pohit S, Kumar R (2010) Biodiesel from *Jatropha*: can India meet the 20% blending target? Energy Policy 38:1477–1484

Cano-Asseleih LM, Plumbly RA, Hylands PJ (1989) Purification and partial characterization of the hemagglutination from seeds of *Jatropha curcas*. J Food Biochem 13:1–20

Chehregani A, Malayeri BE (2007) Removal of heavy metals by native accumulator plants. Int J Agric Biol 9:462–465

Cunningham SD, Shann JR, Crowley DE, Anderson TA (1997) Phytoremediation of contaminated water and soil. In: Kruger EL, Anderson TA, Coats JR (eds) Phytoremediation of soil and water contaminants, vol 664, ACS Symposium series. American Chemical Society, Washington, DC, pp 2–19

Dagar JC, Tomar OS, Kumar Y, Bhagwan H, Yadav RK, Tyagi NK (2006) Performance of some under-explored crops under saline irrigation in a semiarid climate in Northwest India. Land Degradation Dev 17:285–299

Dushenkov V, Motto H, Raskin I, Kumar NPBA (1995) Rhizofiltration: the use of plants to remove heavy metals from aqueous streams. Environ Sci Technol 30:1239–1245

Ensley BD (2000) Rational for use of phytoremediation. In: Raskin I, Ensley BD (eds) Phytoremediation of toxic metals: using plants to clean- up the environment. Wiley, New York, pp 3–12

Gao S, Ouyang C, Wang S, Xu Y, Tang L, Chen F (2008a) Effects of salt stress on growth, antioxidant enzyme and phenylalanine ammonia-lyase activities in *Jatropha curcas* L. seedlings. Plant Soil Environ 54:374–381

Gao S, Yan R, Cao M, Yang W, Wang S, Chen F (2008b) Effects of copper on growth, antioxidant enzymes and phenylalanine ammonia-lyase activities in *Jatropha curcas* L. seedling. Plant Soil Environ 54:117–122

Gao S, Ou-yang C, Tang L, Zhu J, Xu Y, Wang S et al (2010) Growth and antioxidant responses in *Jatropha curcas* seedlings exposed to mercury toxicity. J Hazard Mater 182:591–597

Garbisu C, Alkorta I (2001) Phytoextraction: a cost-effective plant-based technology for the removal of metals from the environment. Bioresour Technol 77:229–236

Ghosh M, Singh SP (2005) A review on phytoremediation of heavy metals and utilization of its byproducts. Appl Ecol Environ Res 3:1–18

Gratao PL, Prasad MNV, Cardoso PF, Lea PJ, Azevedo RA (2005) Phytoremediation: green technology for the clean up of toxic metals in the environment. Braz J Plant Physiol 17:53–64

Gubitz GM, Mittelbech M, Trabi M (1999) Exploitation of tropical oil seed plant *Jatropha curcas* L. Bioresour Technol 67:73–82

Guerinot ML, Salt DE (2001) Fortified foods and phytoremediation. Two sides of the same coin. Plant Physiol 125:164–167

Gunaseelan VN (2009) Biomass estimates, characteristics, biochemical methane potential, kinetics and energy flow from *Jatropha curcas* on dry lands. Biomass Bioenergy 33:589–596

Hartman WJ Jr (1975) An evaluation of land treatment of municipal wastewater and physical siting of facility installations. Washington DC, US Department of Army

Jain S, Sharma MP (2010) Prospects of biodiesel from *Jatropha* in India: a review. Renew Sust Energy Rev 1:763–771

Jamil S, Abhilash PC, Singha N, Sharma PN (2009) *Jatropha curcas*: a potential crop for phytoremediation of coal fly ash. J Hazard Mater 172:269–275
Janaun J, Ellis N (2010) Perspectives on biodiesel as a sustainable fuel. Renew Sust Energy Rev 14:1312–1320
Jongschaap REE, Corré WJ, Bindraben PS, Brandenburg WA (2010) Claims and Facts on *Jatropha curcas* L.; global *Jatropha curcas* evaluation, breeding and propagation programme. Plant Research International B.V., Wageningen Stichting Het Groene Woudt, Laren. Report 158
Juwarkar AA, Yadav SK, Thawale PR, Kumar P, Singh SK (2008) Effect of biosludge and biofertilizer amendment on growth of *Jatropha curcas* in heavy metal contaminated soils. Environ Monit Assess 145:7–15
Juwarkar AA, Yadav SK, Thawale PR, Kumar P, Singh SK, Chakrabarti T (2009) Developmental strategies for sustainable ecosystem on mine spoil dumps: a case of study. Environ Monit Assess 157:471–481
Kumar PBAN, Dushenkov V, Motto H, Raskin L (1995) Phytoextraction: the use of plants to remove heavy metals from soils. Environ Sci Technol 29:1232–1238
Kumar A, Sharma S (2008) An evaluation of multipurpose oil seed crop for industrial uses (*Jatropha curcas* L.): a review. Industrial Crops and Products, 28:1–10
Kumar GP, Yadav SK, Singh SK, Thawale PR, Juwarkar AA (2008a) Growth of *Jatropha curcas* on heavy metal contaminated soil amended with industrial wastes and *Azotobacter*—a greenhouse study. Bioresour Technol 99:2078–2082
Kumar N, Pamidimarri SDVN, Kaur M, Boricha G, Reddy MP (2008b) Effects of NaCl on growth, ion accumulation, protein, proline contents and antioxidant enzymes activity in callus cultures of *Jatropha curcas*. Biologia 63:378–382
Lim S, Teong LK (2010) Recent trends, opportunities and challenges of biodiesel in Malaysia: an overview. Renew Sust Energy Rev 14:938–954
Mangkoedihardjo S, Surahmaida A (2008) *Jatropha curcas* L. for phytoremediation of lead and cadmium polluted soil. World Appl Sci J 4:519–522
Mangkoedihardjo S, Ratnawati R, Alfianti N (2008) Phytoremediation of hexavalent chromium polluted soil using *Pterocarpus indicus* and *Jatropha curcas* L. World Appl Sci J 4:338–342
Meagher RB (2000) Phytoremediation of toxic elemental and organic pollutants. Curr Opin Plant Biol 3:153–162
Mittler R (2002) Oxidative stress, antioxidants and stress tolerance. Trends Plant Sci 7:405–410
Openshaw K (2000) A review of *Jatropha curcas*: an oil plant of unfulfilled promise. Biomass Bioenergy 19:1–15
Parida AK, Das AB (2005) Salt tolerance and salinity effect on plants: a review. Ecotoxicol Environ Saf 60:324–349
Parvaiz A, Satyawati S (2008) Salt stress and phytobiochemical responses of plants—a review. Plant Soil Environ 54:89–99
Prasad MNV, de Oliveira Freitas HM (2003) Metal hyperaccumulation in plants—biodiversity prospecting for phytoremediation technology. Elec J Biotech 6:285–321
Pulford ID, Watson C (2003) Phytoremediation of heavy metal-contaminated land by trees—a review. Environ Int 29:529–540
Raskin I, Kumar PBAN, Dushenkov S, Salt DE (1994) Bioconcentration of heavy metals by plants. Curr Opin Biotechnol 5:285–290
Salt DE, Blaylock M, Kumar NPBA, Dushenkov V, Ensley D, Chet I et al (1995) Phytoremediation: a novel strategy for the removal of toxic metals from the environment using plants. Biotechnology 13:468–474
Salt DE, Blaylock M, Raskin I (1998) Phytoremediation. Annu Rev Plant Physiol Plant Mol Biol 49:643–668
Singh SK (2007) Global agriculture information network (GAIN), Report IN7047. India Biofuels Annual. pp 5–12
Siriwardhana M, Opathella GKC, Jha MK (2009) Bio-diesel: initiatives, potential and prospects in Thailand: a review. Energy Policy 37:554–559
Smith RAH, Bradshaw AD (1972) Stabilization of toxic mine wastes by use of tolerant plant populations. Trans Instr Mining Metallurg 81:230–237

Sopper WE (1993) Municipal sludge use in land reclamation. Lewis and CRC Press, Berlin
Yadav SK, Juwarkar AA, Kumar GP, Thawale PR, Singh SK, Chakrabarti T (2009) Bioaccumulation and phyto-translocation of arsenic, chromium and zinc by *Jatropha curcas* L: Impact of dairy sludge and biofertilizer. Bioresour Technol 100:4616–4622
Yadav SK, Dhote M, Kumar P, Sharma J, Chakrabarti T, Juwarkar AA (2010) Differential antioxidative enzyme responses of *Jatropha curcas* L. to chromium stress. J Hazard Mater 180:609–615
Ye M, Li C, Francis G, Makkar HPS (2009) Current situation and prospects of *Jatropha curcas* as a multipurpose tree in China. Agroforestry Syst 76:487–497
Zhang FL, Niu B, Wang YC, Chen F, Wang SH, Xu Y et al (2008) A novel betaine aldehyde dehydrogenase gene from *Jatropha curcas*, encoding an enzyme implicated in adaptation to environmental stress. Plant Sci 174:510–518
Zhou A, Thomson E (2009) The development of biofuels in Asia. Appl Energy 86:11–20

Chapter 24
Phorbol Esters and Other Toxic Constituents of *Jatropha curcas* L.

G. Raja Krishna Kumar, V.A. Bapat, and T. Sudhakar Johnson*

Introduction

Jatropha curcas L. (Euphorbiaceae) is a shrub native to South América, which is widely distributed in tropical countries. The plant is stress tolerant, drought resistant, grows in semi-arid and marginal lands. It can be planted in the form of hedges to control erosion and to protect enclosed areas from animals such as goats. *J. curcas* is mainly known as a source of oil, with an oil content of the kernel ranging from 30% to 55% (Parthiban et al. 2009; Varshney and Johnson 2010; Eswaran et al. 2010; Kumar et al. 2011). The oil is mainly used as fuel, in soap manufacturing and as lubricant in wood industry. Different parts of the plant have been traditionally used for various purposes including therapeutic uses. Some of the ethnomedical uses of the extracts of *J. curcas* leaves and roots include use as a remedy for cancer, as an abortifacient, antiseptic, diuretic, purgative and haemostatic (Dalziel 1955). The nut of the plant has also been used traditionally for the treatment of many ailments including burns, convulsions, fever and inflammation. In spite of the myriad of ethnomedical uses for which various parts of the plant are utilized, it is important to note that toxic properties also have been attributed to various plant parts, particu-

G.R.K. Kumar
Plant Metabolic Engineering Group, Reliance Life Sciences Pvt. Ltd., Dhirubhai Ambani Life Sciences Center, R-282, Thane-Belapur Road, Rabale, Navi Mumbai 400701, India

V.A. Bapat
CSIR, Emeritus Scientist, Department of Botany/Biotechnology, Shivaji University, Vidyanagar, Kolhapur 416004, India

T.S. Johnson (✉)
Plant Metabolic Engineering Group, Reliance Life Sciences Pvt. Ltd., Dhirubhai Ambani Life Sciences Center, R-282, Thane-Belapur Road, Rabale, Navi Mumbai 400701, India

*Natural Remedies, Plot No. 5B, Veerasandra Indl. Area, 19th KM Stone, Hosur Road, Electronics City (PO), Bangalore 560100, India
e-mail: thangirala@yahoo.co.uk

N. Carels et al. (eds.), *Jatropha, Challenges for a New Energy Crop: Volume 1: Farming, Economics and Biofuel*, DOI 10.1007/978-1-4614-4806-8_24,

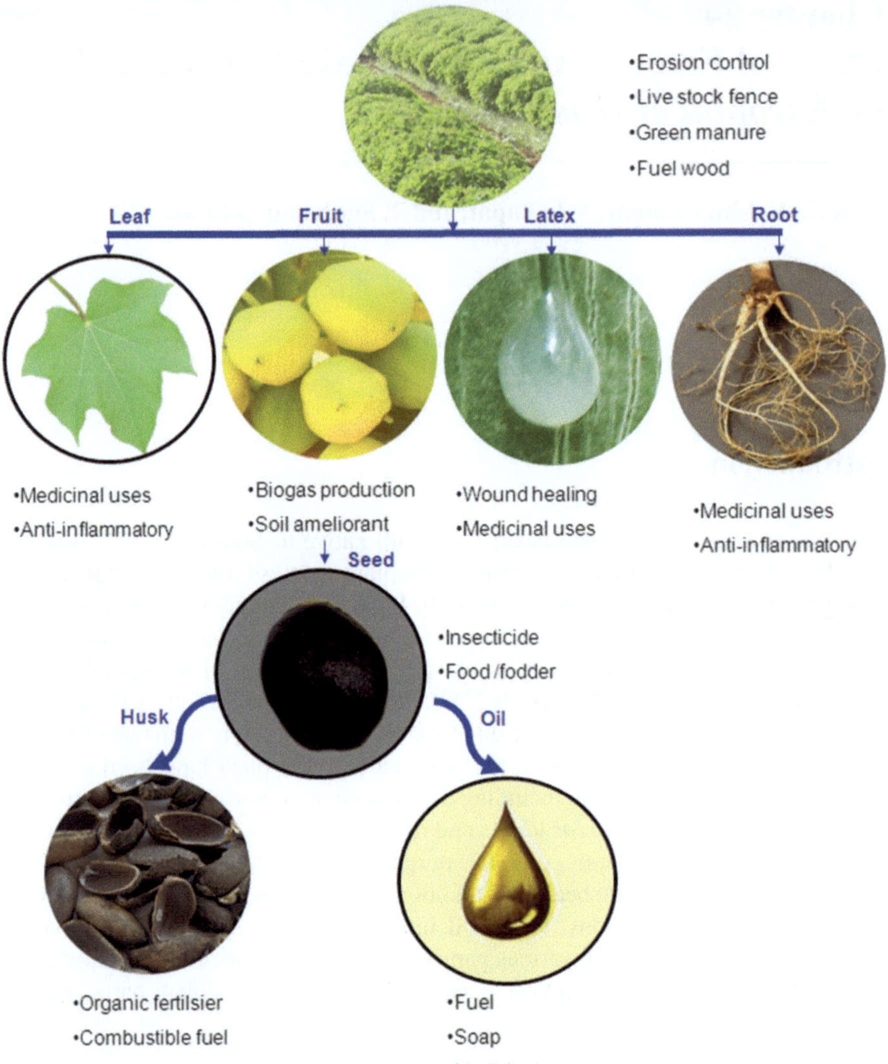

Fig. 24.1 Ethnomedical and multiple uses of *Jatropha curcas*

larly the seeds (Fig. 24.1). Most plant parts of *J. curcas* are reported to be toxic (kernel, leaves, stem, flower, buds, roots, bark, and wood). For example, the seeds contain toxic *phorbol esters* (PE) and antinutritional factors like curcin (a toxalbumin that is highly irritant and produces deleterious effects on blood), trypsin inhibitors, phytates, lectins and other chemicals (Fig. 24.2). The latex is acrid and irritable to skin. Curcain, a protease has been isolated from the latex of *J. curcas.* Among the toxic constituents, PEs are most potent and they exhibit a wide range of biological activities affecting from microorganisms to higher animals. Several studies showed that both the oil and the meal are toxic to humans and animals due to presence of

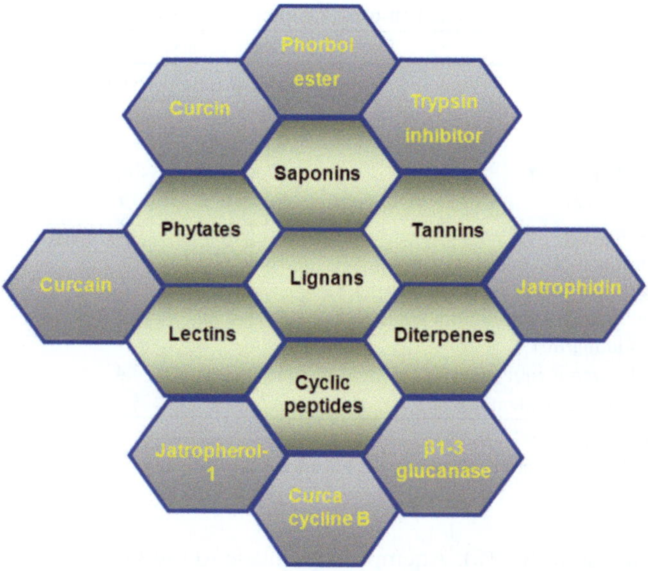

Fig. 24.2 Antinutritional and major toxic components isolated from *Jatropha* species

PEs and curcins. Despite this, the relatively high protein content of *J. curcas* meal can be advantageous since it constitutes rich source of protein, which does not compete with proteins obtained from food crops, such as soybean and wheat. Many chemicals could be isolated from *J. curcas*, and are listed in Table 24.1. The widespread cultivation of toxic varieties of *J. curcas* may increase the frequency of human or animal contact with the plant, seeds or processed products, such as oil and seedcake. Hence, further knowledge on risk factors involved in mechanisms underlying toxicity is needed to make it popular.

The multiple utilities of the crop and interest in *J. curcas* seed oil especially for biodiesel, production has propelled large-scale plantations across Asia, Africa, and South American countries. Knowledge on understanding of antinutritional constituents and developing improved varieties devoid of toxins will enhance its economic benefits. In this paper, we made an attempt to present an account of the toxic constituents present in *Jatropha* and the medicinal properties of *Jatropha* for the benefit of scientific community.

Toxicity of Seeds and Potential of Seed Cake

In the past two decades, *J. curcas* has gained attention due to its oil, which can be used as a feed stock for biodiesel production. Chemical composition of *J. curcas* seeds reveals protein (18%), fat (38%), carbohydrate (17%), fiber (15.5%), ash

Table 24.1 Comparison of toxic and non-toxic varieties of *J. curcas*

Chemical composition	Toxic variety	Non-toxic variety (Mexico)
Crude protein (%)	55.7–61.2	63.8
Lipid (%)	0.8–1.5	1.0
Ash (%)	9.6–10.4	9.8
Gross energy (MJ/kg) (%)	17.8–18.3	18.0
Neutral detergent fiber	8.1–9.0	9.1
Phorbol esters (mg/g)	2.17–2.30	0.11 mg/g
Trypsin inhibitor activity[a] mg/of trypsin inhibited/g of dry matter	18.4–27.3	26.5
Lectin activity[a] (inverse of minimum amount of the samples in mg/ml)	0.85–2.88	1.70
Saponins[a] (as diosgenin equivalent)	1.82–2.0	3.4
Phytates[a] (as phytic equivalent)	7.2–10.1	8.9

Source: Makkar et al. (1997, 1998a, b)
[a]Defatted kernels, Toxic variety (Cape verde, Nicaragua and Ife-Nigeria)

(5.3%) and moisture (6.2%). Attempts were made to use the seed cake as feed for live stock and poultry after detoxification. Seed cake (18.3–25.1%) or meal is the major byproduct after oil extraction. It consists of crude protein (56.4–63.8%), lipid (1–1.5%), ash (9.6–9.8%), gross energy (18–18.2 MJ/kg), neutral detergent fiber (9–9.1) and good balance of essential amino acids, which are higher than in soybean meal. The protein quality of cake or meal is high (Devappa et al. 2010a). Seeds contain kernels and shells with an average ratio of 62.2:37.7. The kernel has higher crude protein (22–28%) and oil (54–58%) contents compared to the shell (4–6% crude protein and 0.8–1.4% oil). The levels of essential amino acids except lysine were higher than those of FAO reference protein (Makkar et al. 1998a).

The seed cake provides a rich source of proteins. The average protein content of the seed cake is 58.1% and has a gross energy content of 18.2 MJ/Kg. In addition to high quality proteins, the cake contains various toxins, co-carcinogens and antinutritional factors. If detoxified, it can provide an excellent animal feed. Seed cake can be valuable as organic fertilizer as it contains more nutrients than both chicken and cattle manure (Francis et al. 2005). The presence of phorbol esters makes the cake also suitable as a biopesticide or insecticide (Francis et al. 2005; Achten et al. 2008). The seed cake can serve as feed for biogas production through anaerobic digestion. Using certain microbes as inoculum, Radhakrishnan (2007) achieved 0.5 m^3 biogas/Kg of solvent extracted kernel cake and 0.6 m^3 biogas/Kg of mechanically de-oiled cake. In mechanical expeller, oil extraction from seeds is an inefficient process. Based on the extraction efficiency, the seed cake will contain 9–12% (v/w) oil. Mahanta et al. (2008) demonstrated a higher value way of use of seed cake that does not involve detoxification, but rather solid state fermentation with *Pseudomonas aeruginosa* to produce enzymes, such as proteases and lipases.

The fact that *J. curcas* seed cake can be used for multiple purposes makes it an important by-product. However, long term effects of seed cake on soils have not been studied.

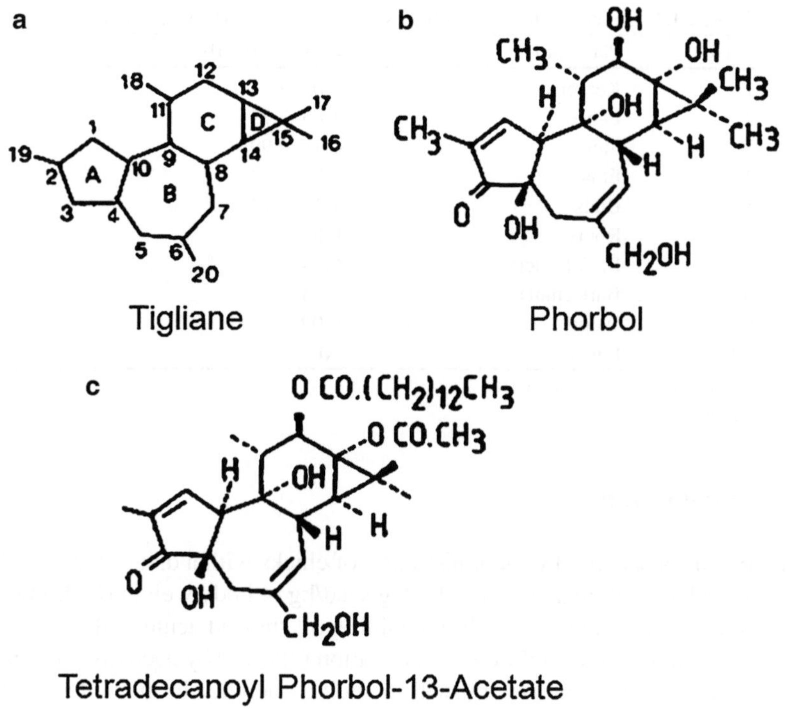

Fig. 24.3 Structure of tigliane and phorbol esters. Source: Evans (1986) and Goel et al. (2007)

Phorbol Esters

The diterpenes are found to be the most potent compounds synthesized by *Jatropha* spp. Among the 20 diterpenes reported from *Jatropha* spp. a group of compounds with tigliane skeleton are called *phorbol esters* (PEs) (Fig. 24.3). The PEs are lipophilic, present mainly in oil and hence, not affected by heat. A comparative analysis of edible and non-edible seed varieties revealed that edible seeds lacked PEs (Makkar et al. 1998a, b). The concentration of PEs in various parts of *J. curcas* is provided in Table 24.2.

The PEs are analogues of diacylglycerol, an activator of protein kinase C (PKC). PKCs act as regulators of many cellular processes. As diacylglycerol has a short biological half-life in the cell, the activation of PKC is transient. Activation of PKC by PEs, however, is much prolonged, which eventually leads to a number of biological processes. Recent study indicated that PEs may contribute to the formation of various cancers by at least two mechanisms. PEs act as co-carcinogens or tumour promoters. They do not cause tumour formation alone, but can lead to the increased risk of tumour formation when there is co-exposure to a chemical carcinogen. The tumour promoting effect of *J. curcas* oil has been demonstrated in mice (Horiuchi et al. 1987). PEs are also known to activate the lytic cycle of the latent Epstein-Barr virus (MacNeil et al. 2003). It has been proposed that exposure to the latex of the plant *Euphorbia tirucalli*, a PE producing plant of the Euphorbiaceae family, is a co-factor for the development of Burkitt's lymphoma in EBV carriers (MacNeil et al. 2003).

Table 24.2 Concentration of phorbol esters in various parts of *J. curcas*

S. No	Plant part	Conc. of phorbol esters (mg/g DW)
1	Kernels	2.0–6.0
2	Leaves	1.83–2.75
3	Stem	0.78–0.99
4	Flowers	1.39–1.83
5	Buds	1.18–2.10
6	Roots	0.55
7	Bark (outer)	0.39
8	Bark (inner)	3.08
9	Wood	0.09
10	Latex	ND

Source: Devappa et al. 2010a
ND not detected

Toxicity Studies of PEs

Mortality was observed after administration of chicks with a diet of 0.5% of *J. curcas* seeds, calves and goats fed with 0.25 g seed/kg of body weight (El-Badwi et al. 1995; Ahmed and Adam 1979). PE containing oil showed acute oral LD_{50} of 6 ml/kg body weight in rats. In addition, skin irritation followed by necrosis was observed in rabbits and rat; dermal toxicity with hemolytic activity in mice. These experiments indicate that PEs could affect the workers who are involved in oil extraction (Carels 2009).

The toxic effect of *J. curcas* extract to mollusks has been demonstrated on *Biomphalaria glabrata, Bulinus globusus* and *Oncomelania hupensis* and also against snails transmitting *Schistosoma mansoni, S. japonicum* and *S. haematobium* (which cause Schistosomiasis) (Liu et al. 1997; Rug and Ruppel 2000). Wink et al. (1997) studied the insecticidal activities of *J. curcas* oil on *Manduca sexta, Helicoverpa armigera, Aphis gossypii, Pectinophora gossypielaa, Empoasca cabiguttula, Callosobruchus chinensis, Sitophilus zeamays, Phthorimaea opercullela, Culex sp., Sesamia calamistis, Busseola fusca, Periplaneta americana, Blatella germanica* and *Oncopeltus fasciatus.* In addition, it also acts against mosquitoes (which cause dengue fever), antibirth activity against house flies, *Monodelphis domestica, Culex quniquefasciatus* (lymphatic filariasis vector).

Recovery and Detoxification

The presence of PEs in seed meal prevents its use as animal feed. Seed meal of *J. curcas* contains crude protein (60–63%) when compared to soybean (45%) and would have high market value if detoxified from PEs and curcin (Makkar et al. 1997, 1998a, b).

There are various physical and chemical methods to destroy the PEs in oil and seed meals. Ahmed and Salimon (2009) tried to refine oil from PEs in Indian,

Malayan and Indonesian species by subjecting to acid degumming, deodorization, bleaching by HCl activated bentonate, neutralization by NaOH and concluded that PE can be reduced significantly by (a) degumming followed by caustic neutralization, (b) bleaching of degummed and neutralized oil and methanol extraction (Haas and Mittelbach 2000; Devappa et al. 2008).

Devappa et al. (2010c) reported different approaches to reduce PEs using various solvents and found that extraction of PEs by different concentrations of methanol at 55°C for 1 h results in reduction of PEs by 98.4%. In addition, mixing methanol: oil ratio of 2:1 (v/w) and stirring on a magnetic stirrer at 60°C for 5 min can effectively reduce PEs from seed oil by 78–80%, which can be applied on a large scale for the production of low PE biodiesel. Solvent extraction of PEs followed by heat treatments to inactivate lectins in *J. curcas* seed meal was reported to convert the non-toxic meal to a high-quality protein source for live stock feeding (Makkar and Becker 1997). Detoxification of seed meal was achieved with a combination of organic solvents and heat/sodium bicarbonate treatments, which resulted in a 48-fold reduction in PE content of seed meal (Martinez-Herrera et al. 2006).

Beneficial Effects of PEs

Some naturally occurring PEs are tumour inhibitors that inhibit human immunodeficiency virus (HIV) replication and possess antileukemic activity. Jatrophone, a macrocyclic diterpenoid isolated from *J. gossypiifolia* showed significant inhibitory activity against cancer cells under *in vitro* and *in vivo* conditions (Kupchan et al. 1970; Goel et al. 2007). Jatrophone also possessed significant antileukemic activity at 27 and 12 mg/kg cytotoxicity against KB cell culture at 0.17 μg/ml. The extracted PEs could be used as biopesticide and insecticide.

Non-toxic Varieties

Although *J. curcas* has been used to extract valuable biodiesel, the toxic substances in seed meal limits its utilization to the fullest extent. Attempts are under way around the world to develop varieties and identifying the existing germplasm that lack PEs. Recently, a non toxic *J. platyphylla* has been reported from Mexico (Makkar et al. 2011). This species is found in warm area (20–29°C) and available on Pacific coast from Sinaloa to Michoacan including the Nayarit and Jalisco states in Mexican deciduous forest. *J. platyphylla* seeds are circular (15.54 ± 1.01 mm) with high oil content (60%), average seed with shell and kernel mass higher than (2–3 fold) the toxic and non-toxic varieties of *J. curcas*. Kernel meal, crude protein and ash content are similar to *J. curcas*.

Makkar et al. (1998b) compared toxic and non-toxic varieties of *J. curcas*. Toxic varieties have PEs ranging from 0.87 to 3.32 mg g^{-1} of kernel while the non-toxic variety from Papantla region, Mexico lacked phorbol esters.

Lacapaxa tribes (Mexico) eat *J. platyphylla* kernel meal (JPKM) after roasting them and are used in preparation of traditional dishes. Makkar et al. (2011) evaluated the antinutritional effect of JPKM with fish (Nile tilapia) bioassay and found that fish remained healthy.

Curcin

Curcin is a toxic glycoprotein or toxalbumin, which belongs to ribosome inactivating protein (RIP) group. RIPs are divided into three types, based on the structure. Type I RIP consists of a single polypeptide chain (~30 kD) and alkaline isoelectric points (pI) of pH 8–10 with or without carbohydrates. Type II RIPs are heterodimeric proteins (~60 kD) consisting of a catalytically active A chain linked to a cell-binding B chain. Type III RIPs are of peculiar type synthesized in an inactive form and will be active after proteolytic processing. RIPs are considered as a defense chemicals. Curcin from *J. curcas* is a type I RIP present in endosperm, fruit and sap and not detected in leaf, root and stem (Fang et al. 2005; Lin et al. 2003). Curcin has 54% homology with ricin A chain (from *Ricinus communis*) and 57% with trichosanthin (*Trichosanthes kirilowii*).

Curcin exhibited hemagglutinating activity, when its concentration was higher than 7.8 mg L^{-1} (Lin et al. 2003). Wei et al. (2005) reported that curcin 2, a 32 kD protein may play a defensive role and is expressed in leaves of seedlings growing under stress conditions like drought, temperature and fungal infections. RIPs are valued for developing antitumor drugs, which selectively target tumor cells (Lin et al. 2003).

Acute toxicity of curcin in mice oral semi-lethal dose LD_{50} was 104.737 ± 29.447 mg kg^{-1}; mice parenteral semi-lethal dose LD_{50} was 67.20 ± 10.445 mg kg^{-1} (Lin et al. 2010). Huang et al. (2008) induced Curcin 2 (RIP) in transgenic tobacco and showed increased tolerance against toacco mosaic virus (TMV) and *Rhizoctonia solani* (a fungal pathogen).

Lin et al. (2003) investigated the effect of curcin on different cell lines and concluded that it inhibits protein synthesis in reticulocyte lysate with an IC_{50} (95% confidence limits) of 0.19 (0.11–0.27) nmol L^{-1}. Mouse myeloma cell line, gastric cancer line and human hepatoma are more sensitive to curcin while Hela cells and normal cells (MRC) were not affected. Further, cysteine containing RIP (curcin) can be used for preparing immunoconjugates, which can act as a chemo therapeutic agent to cure various cancer diseases.

Indian *J. curcas* varieties collected from Kangra and Nasik showed variations in crude protein (23–24.2), lipid (54.8–58.4) neutral detergent fiber (4.5–5.4), ash (3.8–4.4), gross energy (30.4–31.2), trypsin inhibitor activity (24.7–27.5), lectin activity (0.85–6.85), saponins (2.01–2.39) and phytates (8.2–8.6) (Makkar et al. 1997). The toxic and non-toxic varieties of *J. curcas* have good balance of amino acid composition, which is higher than FAO reference protein except lysine.

Other Toxic Constituents

Besides PEs and curcin, *J. curcas* contains several other toxic constituents, such as tannins, saponins, phytates, trypsin inhibitors and lectins (Makkar et al. 1998a, b; Devappa et al. 2010a, b). Chemicals isolated from different parts of *J. curcas* species are presented in Table 24.3 and Fig. 24.2.

Phytates

Phytate is the principal storage form of phosphorus in most plant seeds. Phytates chelate with Ca^{2+}, Mg^{2+} and Fe^{3+} resulting in these ions becoming unavailable in the diet. Non-ruminants, cannot degrade phytate, and hence, presence of phytates in feed reduces the availability of phosphorus. Phytates also form sparingly digestible phytate-protein complexes, thus reducing the availability of dietary protein. *J. curcas* kernel meal from both toxic and non-toxic genotypes contain phytates ranging from 7.2% to 10.1%. Since the levels of phytates in the kernel meal are high, efficient utilization as feed for monogastric animals should require the addition of phytase to feed (Devappa et al. 2010b).

Trypsin Inhibitors

Trypsin inhibitors (TIs) are proteins that inhibit proteolytic enzymes (trypsin and chymotrypsin) during digestion. They are known to decrease protein digestibility. TI activity in the *J. curcas* kernel meal of toxic as well as non-toxic genotypes was found to be ranging from 18.4 to 27.3 mg trypsin inhibited g^{-1} (Makkar et al. 1997). TIs could be inactivated by heat treatment by autoclaving the kernel meal at 121°C for 30 min (Devappa et al. 2010a). It was found that feeding of unheated *J. curcas* kernel meal to poultry, pigs and fish may produce adverse effects.

Lectins

Lectins are carbohydrate binding glycoproteins, which agglutinate red blood cells and affect absorption of nutrients by binding intestinal membrane. When consumed by animals, plant lectins bind to membrane glycosyl groups of the cells lining gastro intestinal tract. They interfere with nutrient digestion and absorption and damage the luminal membrane of the epithelium. When consumed at higher concentrations, lectins threaten the growth and health of animals. The *J. curcas* kernel meal contains lectins at 102 and 51 (inverse of mg meal that produced hemagglutination per ml of assay medium) for

Table 24.3 Chemicals isolated from different parts of *Jatropha* species

Chemical	Chemical type	Plant parts	Species	Biological activity/property	References
Rhamnofolanes	Diterpenes	Root	*J. grossidentata*	Not known	Hirschman et al. (1992)
Curcain	Proteolytic enzyme	Latex	*J. curcas*	Wound healing	Nath and Dutta (1992)
Multifidin	Cyanoglucoside	Latex	*J. multifida*	Not known	Berg et al. (1995)
Choline esterase	Acetylcholine	Leaf and stem	*J. integerrima*	Not known	Gupta and Gupta (1997)
Arabino galacton	Protein	Seeds	*J. curcas*	Inductor of cell differentiation	Zippel et al. (2010)
Jatrowedione	Lathyrane Diterpene	Roots	*J. weddelliana*	Not known	Brum et al. (1998)
Gadain	Lignan	Stem roots Seeds	*J. gossypiifolia*	Not known	Banerji et al. (1984)
Gossypian	Lignan	Aerial parts	*J. gossypiifolia*	Not known	Das and Das (1995)
Japodagrin& japodagrone	Diterpenoids	Roots	*J. podagrica*	Antibacterial activity	Aiyelaagbe et al. (2007)
Multifidone	Lathyrane-diterpene	Stem	*J. multifida*	Not known	Das et al. (2009)
CyclogossineB	Cyclic octapeptide	Latex	*J. gossypiifolia*	Not known	Guette et al. (1997)
Jarophone, Jatropholone A,B, acetyl aleuritolic acid, cyperenoic acid, monoterpenes	terpenes	Root	*J. isabelli*	Not known	Pertino et al. (2007)
Pohlianins A,b, c	Cyclic peptides	Latex	*J. pohliana ssp molissima*	antimalarial	Guette et al. (1999)
Jatrophone		Stem/tuber	*J. elliptica*	—	Calixto and Santana (1987)
Jatrodien	Lignan	Stems	*J. gossypiifolia*	Not known	Das et al. (1996)
Curcacycline A	Cyclic octa peptide	Latex	*J. curcas*	Not known	Berg et al. (1995)
Curcacycline B		Latex	*J. curcas*	Antimalarial activity	Auvin et al. (1997) Baraguey et al. (2001)
Multifidol & multifidol glucoside	Acylphloroglucinol	Latex	*J. multifida*	Not known	Kosasi et al. (1989)
Cleomiscosin A	Coumarino-lignoid		*J. gossypiifolia*	Not known	Das et al. (2003)
Jatropherol-1		Seed	*J. curcas*	Insecticidal activity	Jing (2005), Jing et al. (2005)

Jatrophidin	—	Latex	*J. curcas*	Antibacterial activity	Altei et al. (2008)
Curcasone A–D	—	Root	*J. curcas*	Cytotoxic activity	Chomnong (1990)
Japodic acid	—	Root	*J. podagrica*	Insecticidal activity	Aiyelaagbe and Gloer (2008)
Mahafacyclin	—	Latex	*J. mahafalensis*	Antimalarial activity	Baraguey et al. (2000, 2001)
Chevalierin	—	Latex	*J. chevaliera*	Antimalarial activity	Baraguey et al. (2000, 2001)
β 1,3 Glucanase	—	Seed	*J. curcas*	Antifungal activity	Wei et al. (2005)

toxic and non-toxic genotypes, respectively (Makkar et al. 2007). Like TIs, lectins also could be completely inactivated by autoclaving the kernel meal at 121°C for 30 min.

Curcain

Curcain is a proteolytic enzyme extracted from the latex of *J. curcas* shoots. The curcain was found to have a wound-healing property when tested in mice (Devappa et al. 2010b).

Research on Toxins

Solvent extracts of *J. curcas* showed abortive effect (Goonasekera et al. 1995) in *in vitro* studies conducted in rat. Ethanolic extract of aerial parts of *J. gossypiifolia* when orally administered to rats showed hypotension or reduced the systolic blood pressure in conscious normotensive rats and vasorelaxant activity on rat mesenteric rings precontracted with norepinephrine or $CaCl_2$ (Ca^{2+}) (Abreu et al. 2003). Osoniyi and Onajobi (2003) reported procoagulant activity with low concentration of ethyl acetate fraction of *J. curcas* and anticoagulant activity with butanol extract. Arabinogalactan, a protein isolated from endosperm of *J. curcas* by cold extraction, induced the production of hepatocyte growth factor (HGF) by 2.9-fold, keratinocyte growth factor (KGF) by 1.7 fold, which are key factors of the proliferation control in skin keratinocytes and fibroblasts. In addition, TGFB1 was up regulated by 2.5 fold, which are essential proteins for induction of cellular differentiation in epidermal keratinocytes (Zippel et al. 2010).

In the recent past, scientific investigations have supported few ethnomedicinal treatments. Bhil tribes from Rajasthan are using *J. curcas* root paste to treat local inflammation on humans. Mujumdar and Misar (2004) validated the folkore on rats by using ethanol extract of roots as anti-inflammatory agent. In yet another study, Mujumdar et al. (2000) proved that the methanol extract of *J. curcas* roots has antidiarrhoeal activity in albino mice.

Detoxification

Several attempts were made to nullify the toxic effects of toxins and antinutritional factors present in *J. curcas* seeds and leaves. *J. curcas* seeds in some regions of Mexico were boiled and roasted (Delgado and Parado 1989) and roasting alone (Makkar et al. 1998b) helped to inactivate the toxicity of trypsin inhibitors. Makkar et al. (1998b, 2009) found that after roasting seeds of Mexican variety in hot plate, trypsin inhibitor activity (14.6–28.7 mg trypsin inhibited g^{-1}) was reduced (0.7 mg trypsin inhibited/g to non detectable), and lectin activity was reduced slightly (25.6–52.2 to 6.4 to 12.8 units), but there was no significant reduction in saponins, phytates and phorbol esters. In yet another study, Aregheore et al. (1997) reported that lectin can be completely inactivated by moisture heat treatment at 121 °C for 30 min.

Attempts were also made to degrade the PEs by using an *in vitro* rumen bacterial fermentation system (Makkar and Becker 2010) to use as feed stock. The reports conclude that rumen bacteria were not affected by PEs and did not degrade them. Further, rumen bacteria can degrade significantly tryspin inhibitors, lectins and phytates, which are other antinutrients present in *J. curcas*.

Siddhuraju et al. (2002) attempted to inactivate phorbol esters, phytates, saponins and lectins by giving ionizing radiation treatment to inactivate or remove the aforementioned toxic chemicals.

Martinez-Herrera et al. (2006) used different techniques to detoxify the seed meal from antinutrients and toxins. To inactivate trypsin inhibitors, moist heating at 121°C for 25 min was applied. Irradiation of 10 kHy slightly reduced the phytates while saponins were decreased by ethanol extraction and irradiation. Lectin activity and phorbol ester content (upto 97.9%) were reduced after ethanol extraction followed by 0.07% $NaHCO_3$ treatment. Young leaves of *J. curcas* are perhaps edible if steamed stewed (Ochse 1931; Kumar and Sharma 2008).

Therapeutic Uses of *J. curcas*

In traditional medicine, all plant parts of *J. curcas* have therapeutic uses to cure human and animal ailments.

Leaf

Latin Americans and Carribbeans used leaf extracts or decoction to wash wounds. Tanzanians used leaf extracts to treat skin rashes and oral candidiasis. In Trinidad and Tobago, people used leaves to treat clean sores and in Columbia as an oral medicine to treat venereal disease, in Cameroon for arthritis, in Bahama for heartburn, in Barbadia to cure Marasmus and in Panama for jaundice. Costa Ricans use hot dressing of leaves to cure erysipelas and to cure splenosis. In some places, leaf bath was used for sores, sprains and rashes. In Guatemala, hot leaves are placed on breasts to cure lactagogue. In addition, hot water extracts of leaves are taken orally to induce milk secretion in women after child birth and to treat cough and also used as antiseptic (FAO 2010).

Further, leaf extracts are utilized to treat fever, mouth infection, jaundice, guinea-worms sore, joint rubefacient for rheumatism and paralysis, kidney diuretics, antiaborifacients and pulmonary trouble, to control parasites, malaria and hypertension. Leaf sap was used as febrifuges and to treat hemorrhoids (FAO 2010).

Stem

It was used against pain and stings of bees and wasps, to cure toothache pyorrhea, gum inflammation, to stop bleeding and itching of cuts or scratches and water extract of branches to cure HIV and tumors (FAO 2010).

Root

Root decoction was used as a mouth wash for bleeding gums, toothache, scabies, eczema, ringworms and to treat dysentery and venereal disease like gonorrhea. In Philippines, fisher men use bark as a fish poison. In some regions as folkeric, dried and pulverized root barks are made into poultices and administered orally to treat jaundice and expel worms. Root paste is used to treat local inflammation and antidiarrhoeal activity (FAO 2010).

Latex

Latex was applied to dress sores, inflammed tongs, to cure toothache in Cuba. It was also used to treat burns, hemorrhoids, ringworms and ulcers. Diluted latex in water was used to treat snake poisoning. In Southern Nigeria, latex of *J. gossypiifolia* stem has been used to stop bleeding nose, gums and injured skin (Oduola et al. 2007; FAO 2010).

Seed and Oil

In Mauritius, oil is used to massage ascetic limbs. It has been used as cutaneous, subcutaneous parasitic infection, eczema and skin diseases and to soothe rheumatic pains, tumours and cancers. Further, *J. curcas* soap prepared from oil is used as medicine to treat skin related diseases. Emulsified oil is used to control house flies, weevil pests. Oil extract was found to be effective against cotton bollworm and sorghum stem borers and can be potentially used as fungicide and nematicide. To treat arthritis, gout and jaundice, oil was used internally and externally to stimulate abortion and it is also an ingredient for hair conditioners (Odusote et al. 2002; FAO 2010).

Other Uses

In India, most of agricultural lands are protected by *J. curcas* plantations as a livestock fence because of toxic characters, biodiesel production, and its cost effectiveness.

As Green Manure

De-oiled seed cake, a byproduct of oil extraction has the potential as a fertilizer or biogas production if available in large quantities. It can also be used as a fuel for steam turbines to generate electricity. Being rich in nitrogen, the seed cake is an excellent source of plant nutrients. In a green manure trial with rice in Nepal, the application of 10 t of fresh *J. curcas* biomass resulted in increased yield of many crops (Kumar and Sharma 2008). Another use of *J. curcas* seed cake is as a straight fertilizer. Seed cake can be used as a fuel for steam turbine to generate electricity. Seed cakes are rich in nitrogen and in proteins (50–62%) with all essential amino acids except lysine. Seed cake can be used for low cost enzyme production or simply as plant nutrients and green manure (Kumar and Sharma 2008).

Antibacterial Activity

Aiyelaagbe (2001) showed that roots of *J. multifida* effectively inhibited the growth of gram positive bacteria, such as *Bacillus subtilis* and *Staphylococcus aureus*. *J. gossypiifolia* showed antifungal and antibacterial activity for *Micrococcus luteus, Staphylococcus aureus, Escherichia coli, Pseudomonas aeruginosa, Saccharomyces cerevisiae* and *Aspergillus niger* (Kumar et al. 2006). *J. elliptica* showed resistance against *Staphylococcus aureus* (Marquez et al. 2005). The ethanolic extract of *J. curcas* seed cake exhibited antifungal activities against important fungal phytopathogens *viz., Fusarium oxysporum, Pythium aphanidermatum, Lasiodiplodia theobromae, Curvularia lunata, Fusarium semitectum, Colletotrichum capsici* and *Colletotrichum gloeosporioides* (Donlaporn and Suntornsuk 2010). *J. curcas* extracts were found to inhibit *Colletotrichum musae,* which causes anthracnose disease in bananas (Thangavelu et al. 2004) and leaf extracts have fungicidal activity, which control *Sclerotium* sp., the agent of Azolla disease (Garcia and Lawas 1990).

Bahadur et al. (1997) investigated comparative antimicrobial and antifungal properties using crude extracts derived from mature leaves, stem, seed, root and fresh calli of eight *J. curcas* species, *viz., J. curcas, J. gossypiifolia, J. glandulifera, J. villosa, J. heynei, J. maheshwarii, J. podagrica, J. integerrima* and noted differential response against *Escherichia coli, Bacillus subtilis, B. polymixa, Proteus vulgaris, Fusarium oxysporum, Curvularia lunata.* Fresh callus extract of *J. tanjorensis, J. multifida, J. glandulifera* showed higher antifungal activity. While leaf, stem, and root extracts of *J. gossypiifolia* and *J. tanjorensis,* leaves of *J. multifida* and seeds of *J. integerrima* and *J. curcas* showed antimicrobial activity to various microbes.

Waste Land Reclamation and Agricultural Uses

In agriculture, *J. curcas* helps to control soil erosion, improved rain water infiltration, as a livestock barrier, land demarcation as live fence and green manure. In waste land reclamation, application of *J. curcas* seed cake (3 tha^{-1} containing 3.2% N, 1.2% P_2O_5 and 1.4% K_2O) increased the yield from 93–120% (Ghosh et al. 2007). ICRISAT (India) also reported similar results with a caution note stating that over application may depress yields (Wani et al. 2008). Agamuthu et al. (2010) showed the usefulness of *J. curcas* for phytoremediation to recover the soil contaminated with used lubricating oil from automobiles. In addition, *J. curcas* bioaccumulation potential has been exploited in remediation of soil from heavy metal contamination *viz.,* Al, Fe, Cr, Mn, Ar, Zn, Cd and Pb (Mangkoedihardjo and Surahmaida 2008; Jamil et al. 2009; Santosh et al. 2009). *J. curcas* is prone to pathogens like *Cercospora, Rhizopus oryzae* and insects. Recalcitrant *J. curcas* seeds were soaked in 12% leaf extract of *Clerodendrum aculeatum,* which has increased the agronomic values (Debnath and Verma 2008).

Conclusion

In future, an intensive research on detoxification of toxins, anti nutritional properties and validation of ethnomedicinal properties of *J. curcas* can enhance the feed stock usage, exploration of novel chemicals, medicines, and other value added products to mankind. Furthermore, envolving cell, tissue and molecular biology techniques to develop non toxic varieties with good agronomical traits may help to exploit the potential of the crop. Popularizing *J. curcas* cultivation in waste lands and semi-arid regions will boost the socio economic status of poor people and countries.

References

Abreu IC, Alex Marinho SS, Paes AMA, Freire SMF, Olea RSG, Borges MOR et al (2003) Hypotensive and vasorelaxant effects of ethanolic extract from *Jatropha gossypiifolia* L. in rats. Fitoterapia 74:650–657

Achten WMJ, Verchot L, Franken YJ, Mathijs E, Singh VP, Aerts R et al (2008) *Jatropha* biodiesel production and use. Biomass Bioenergy 32:1063–1084

Agamuthu P, Abioye OP, Abdulaziz A (2010) Phytoremediation of soil contaminated with used lubricating oil using *Jatropha curcas*. J Hazard Mater 179:891–894
Ahmed O, Adam SE (1979) Toxicity of *J. curcas* in sheep and goats. Res Vet Sci 27:89–96
Ahmed WA, Salimon J (2009) Phorbol ester as toxic constituents of tropical *Jatropha curcas* seed oil. Eur J Sci Ind Res 31:429–436
Aiyelaagbe OO (2001) Antibacterial activity of *Jatropha multifida* roots. Fitoterapia 72:544–546
Aiyelaagbe OO, Gloer JB (2008) Japodic acid, a novel aliphatic acid from *Jatropha podagrica* Hook. Rec Nat Prod 2:100–106
Aiyelaagbe OO, Adesogan K, Ekundayo O, Gloer JB (2007) Antibacterial diterpenoids from *Jatropha podagrica* Hook. Phytochemistry 68:2420–2425
Altei WF, Picchi DG, Barbosa SC, Cilli EM, Giannini MJ, Cardoso-Lopes EM et al (2008) NMR studies, solid phase synthesis and MD/SA simulation as a tool for structural elucidation of new bioactive peptides from the latex of *Jatropha curcas* L. Planta Med 74:65
Aregheore EM, Makkar HPS, Becker K (1997) Lectin activity in toxic and nontoxic varieties of *J. curcas* using a latex agglutination test. In: Proceeding of the international conference biofuels and industrial products from *Jatropha curcas*, Nicaragua
Auvin C, Baraguey C, Blond A, Lezenven F, Pousset JL, Bodo B, Curcacycline B (1997) a cyclic nona-peptide from *Jatropha curcas* enhancing rotamase activity of cyclophilin. Tetrahedron Lett 38:2845–2848
Bahadur B, Reddy SM, Goverdhan S, Giridhar P (1997) Antimicrobial activity in eight species of *Jatropha* L. (Euphorbiaceae). J Indian Bot Soc 77:190–191
Banerji J, Das B, Chatterjee A, Shoolery JN (1984) Gadain, a lignan from *Jatropha gossypiifolia*. Phytochemistry 23:2323–2327
Baraguey C, Blond A, Correia I, Pousset JL, Bodo B, Auvin-Guette C, Mahafacyclin A (2000) A cyclic heptapeptide from *Jatropha mahafalensis* exhibiting *β*-bulge conformation. Tetrahedron Lett 41:325–329
Baraguey C, Blond A, Cavelier F, Pousset JL, Bodo B, Auvin-Guette C (2001) Isolation, structure and synthesis of mahafacyclin B, a cyclic heptapeptide from the latex of *Jatropha mahafalensis*. J Chem Soc Perkin Trans 1:2098–3003
Berg VDAJJ, Horsten SFAJ, Kettenes VBJJ, Kroes BH, Labadie RP (1995) Multifidin - a cyanoglucoside in the latex of *Jatropha multifida*. Phytochemistry 40(2):597–598
Brum RL, Hond NK, Mazarin MS, Hess SC, Cavalheiro AJ, Monache FD (1998) Jatrowedione. A lathyrane diterpene from *Jatropha weddelliana*. Phytochemistry 48:1225–1227
Calixto JB, Santana AEG (1987) Pharmacological analysis of the inhibitory effect of Jatrophone, a diterpene isolated from *Jatropha elliptica* on smooth and cardiac muscle. Phytother Res 1:122–126
Carels N (2009) *Jatropha curcas*: a review. Adv Bot Res 50:49–56
Chomnong WN (1990) Investigation of the chemical constituents and cytotoxic activity on *Jatropha curcas*. Newslett Reg Netw Chem Nat Prod. Southeast Asia 14:19–24
Dalziel JM (1955) The useful plants of west-tropical Africa. Crown Agents for Overseas Governments and Administration, London, p 147
Das B, Das R (1995) Gossypifan, a lignan from *Jatropha gossypiifolia*. Phytochemistry 40:931–932
Das B, Rao SP, Srinivas KVNS, Das R (1996) Jatrodien, a lignan from stems of *Jatropha gossypiifolia*. Phytochemistry 41:985–987
Das B, Kashinatham A, Venkataiah B, Srinivas KVNS, Mahender G, Reddy MR, Cleomiscosin A (2003) A coumarino-lignoid from *Jatropha gossypiifolia*. Biochem Sys Ecol 31:1189–1191
Das B, Reddy KR, Ravikanth B, Raju TV, Sridhar B, Khan PU et al (2009) Multifidone: a novel cytotoxic lathyrane-type diterpene having an unusual six-membered a ring from *Jatropha multifida*. Bioorg Med Chem Lett 19:77–79
Debnath M, Verma HN (2008) Effect of phytoprotein treatment on *Jatropha curcas* for wasteland reclamation. Afr J Biotechnol 7:613–616
Delgado MJ, Parado TE (1989) Potential multipurpose agroforestry crops identified for the Mexican Tropics. In: Wickens GE, Haq N, Day P (eds) New crops for food and industry. Chapman and Hall, London

Devappa RK, Darukeshwara J, RathinaRaj K, Narasimhamurthy K, Saibaba P, Bhagya S (2008) Toxicity studies of detoxified *Jatropha* meal (*Jatropha curcas*) in rats. Food Chem Toxicol 46:3621–3625

Devappa RK, Makkar HPS, Becker K (2010a) *Jatropha* toxicity—a review. J Toxicol Environ Health Part B 13:476–477

Devappa RK, Makkar HPS, Becker K (2010b) Nutritional, biochemical, and pharmaceutical potential of proteins and peptides from *Jatropha*: review. J Agric Food Chem 58:6543–6545

Devappa RK, Makkar HPS, Becker K (2010c) Optimization of conditions for the extraction of phorbol esters from *Jatropha* oil. Biomass Bioenergy 34:1125–1133

Donlaporn S, Suntornsuk W (2010) Antifungal activities of ethanolic extract from *Jatropha curcas* seed cake. J Microbiol Biotechnol 20:319–324

El-Badwi SM, Adam SE, Hapke HJ (1995) Comparative toxicity of *Ricinus communis* and *Jatropha curcas* in brown Hisex chicks. Dtsch Tierarztl Wochenschr 102:75–77

Eswaran N, Parameswaran S, Sathram B, Anantharaman B, Kumar GRK, Johnson TS (2010) Yeast functional screen to identify genetic determinants capable of conferring abiotic stress tolerance in *Jatropha curcas*. BMC Biotechnol 10:23

Evans FJ (1986) Environmental hazards of diterpene esters from plants. In: Evans FJ (ed) Naturally occurring phorbol esters. CRC press, Boca Raton, FL, pp 171–215

Fang R, Shenghua W, Xia GZ, Liang G, Qin W, Ling T et al (2005) Identification of curcin by western blot in calli generated from explants of *Jatropha curcas* L. J Sich Uni (Nat Sci Edn) 42:206–209

FAO report (2010) *Jatropha*: a smallholder bioenergy crop. In: Brittaine R, Lutaladio N (eds) The potential for pro-poor development, vol 8. Food and Agriculture Organization of the UN, Rome, pp 1–96

Francis G, Edinger R, Becker K (2005) A concept for simultaneous wasteland reclamation, fuel production, and socio-economic development in degraded areas in India: need, potential and perspectives of *Jatropha* plantations. Nat Resour Forum 29:12–14

Garcia RP, Lawas P (1990) Potential plant extracts for the control of Azolla fungal pathogens. Phillipp Agri 73:343–348

Ghosh A, Patolia JS, Chaudhary DR, Chikara J, Rao SN, Kumar D et al (2007) Response of *Jatropha curcas* under different spacing to *Jatropha* de-oiled cake. FACT Foundation, The Netherlands. Available from http://www.fact-fuels.org. Accessed March 2012

Goel G, Makkar HPS, Francis G, Becker K (2007) Phorbol esters: structure, biological activity and toxicity in animals. Int J Toxicol 26:279–288

Goonasekera MM, Gunawardana VK, Jayasena K, Mohammed SG, Balasubramaniam S (1995) Pregnancy terminating effect of *Jatropha curcas* in rats. J Ethnopharmacol 47:117–123

Guette AC, Baraguey C, Blond A, Pousset J, Bodo B, Cyclogossine B (1997) A cyclic octapeptide from *Jatropha gossypiifolia*. J Nat Prod 60:1155–1157

Guette AC, Baraguey C, Blond A, Xavier HS, Pousset JL, Bodo B (1999) Pohlianins A, B and C, cyclic peptides from the latex of *Jatropha pohliana* ssp. *Molissima*. Tetrahedron 55:11495–11510

Gupta A, Gupta R (1997) A survey of plants for presence of cholinesterase activity. Phytochemistry 46:827–831

Haas W, Mittelbach M (2000) Detoxification experiments with the seed oil from *Jatropha curcas* L. Ind Crops Prod 12:111–118

Hirschmann SG, Tsichritzis F, Jakupovic J (1992) Diterpenes and a lignan from *Jatropha grossidentata*. Phytochemistry 31:1731–1735

Horiuchi T, Fujiki H, Hirota M, Suttajit M, Suganuma M, Yoshioka A et al (1987) Presence of tumor promoters in the seed oil of *Jatropha curcas* L. from Thailand. Jpn J Cancer Res 78:223–226

Huang MX, Hou P, Wei Q, Xu Y, Chen F (2008) A ribosome-inactivating protein (Curcin 2) induced from *Jatropha curcas* can reduce viral and fungal infection in transgenic tobacco. Plant Growth Regul 54:115–123

Jamil S, Abhilash PC, Singh N, Sharma P (2009) *Jatropha curcas*: A potential crop for phytoremediation of coal fly ash. J Hazard Mater 172:269–275

Jing L (2005) Study of insecticidal active components in *Jatropha curcas* L. seed on its extraction, isolation and toxicity action mechanism. Ph.D. dissertation, Sichuan University. http://www.fabiao.net/thread-1513285-1-1.html

Jing L, Fang Y, Ying X, Wenxing H, Meng X, Syed MN, Fang C (2005) Toxic impact of ingested Jatropherol-I on selected enzymatic activities and the ultra structure of midgut cells in silkworm, *Bombyx mori* L. J Appl Entomol 129:98–104

Kosasi S, Van Der Sluis WG, Labadie RP (1989) Multifidol and multifidol glucoside from the latex of *Jatropha multifida*. Phytochemistry 28:2439–2441

Kumar A, Sharma S (2008) An evaluation of multipurpose oil seed crop for industrial uses (*Jatropha curcas*): a review. Ind Crops Prod 28:1–10

Kumar PV, Chauhan NS, Padh H, Rajani M (2006) Search for antibacterial and antifungal agents from selected Indian medicinal plants. J Ethnobiol Pharmacol 107:182–188

Kumar GRK, Eswaran N, Johnson TS (2011) Isolation of high-quality RNA from various tissues of *Jatropha curcas* for downstream applications. Anal Biochem 413:63–65

Kupchan SM, Sigel CW, Matz MJ (1970) Jatrophone, a novel macrocyclic diterpeniod tumor inhibitor from *Jatropha gossypiifolia*. J Am Chem Soc 92:4476–4477

Lin J, Yan F, Tang L, Chen F (2003) Antitumor effects of curcin from seeds of *Jatropha curcas*. Acta Pharmacol Sin 24:241–246

Lin J, Zhou X, Wang J, Jiang P, Tang K (2010) Purification and characterization of curcin, a toxic lectin from the seed of *Jatropha curcas*. Prep Biochem Biotechnol 40:107–108

Liu SY, Sporer F, Wink M, Jourdane J, Henning R, Li YL et al (1997) Anthraquinones in *Rheum palmatum* and *Rumex dentatus* (Polygonaceae and phorbol esters) from *Jatropha curcas* (Euphorbiaceae) with molluscicidal activity against the schistosomias vector snails *Oncomelania, Biomphalaria* and *Bulinus*. Trop Med Int Health 2:179–188

MacNeil A, Sumba OP, Lutzke ML, Moormann A, Rochford R (2003) Activation of the Epstein–Barr virus lytic cycle by the latex of the plant *Euphorbia tirucalli*. Br J Cancer 88:1566–1569

Mahanta N, Gupta A, Khare SK (2008) Production of protease and lipase by solvent tolerant *Pseudomonas aeruginosa* PseA in solid-state fermentation using *Jatropha curcas* seed cake as substrate. Bioresour Technol 99:1729–1735

Makkar HPS, Becker K (1997) Potential of *Jatropha curcas* seed meal asa protein supplement to livestock feed, constraints to its utilization and possible strategies to overcome constraints to its utilization. In Proceedings Jatropha 1997: International symposium on Biofuels and Industrial Products from *Jatropha curcas* and other tropical oil seed plants, February 23–27, Managua, Mexico.

Makkar HPS, Becker K (2010) Are phorbol esters degraded by rumen microbes? J Sci Food Agric 90:1562–1565

Makkar HPS, Becker K, Sporer F, Wink M (1997) Studies on nutritive potential and toxic constituents of different provenances of *Jatropha curcas*. J Agric Food Chem 45:3152–3157

Makkar HPS, Aderibigbe AO, Becker K (1998a) Comparative evaluation of non-toxic and toxic varieties of *Jatropha curcas* for chemical composition, digestibility, protein degradability and toxic factors. Food Chem 62:207–215

Makkar HPS, Becker K, Schmook B (1998b) Edible provenances of *Jatropha curcas* from Quintana Roo state of Mexico and effect of roasting on antinutrient and toxic factors in seeds. Plant Foods Hum Nutr 52:31–36

Makkar HPS, Francis G, Becker K (2007) Bioactivity of phytochemicals in some lesser known plants and their effects and potential applications in livestock and aquaculture production systems. Animal 1:1371–1391

Makkar HPS, Jeroen M, Becker K (2009) Removal and degradation of phorbol esters during pre-treatment and transesterification of *Jatropha curcas* oil. J Am Oil Chem Soc 86:173–181

Makkar HPS, Kumar V, Oyeleye OO, Akinleye AO, Angulo-Escalante MA, Becker K (2011) *Jatropha platyphylla*, a new non-toxic *Jatropha* species: physical properties and chemical constituents including toxic and antinutritional factors of seeds. Food Chem 125:63–71

Mangkoedihardjo S, Surahmaida (2008) *Jatropha curcas* L. for phytoremediation of lead and cadmium polluted soil. World Appl Sci J 4:519–522

Marquez B, Neuville L, Moreau NJ, Genet JP, Santos AFD, Andrade MCCD et al (2005) Multidrug resistance reversal agent from *Jatropha elliptica*. Phytochemistry 66:1804–1811
Martinez-Herrera J, Siddhuraju P, Francis G, Da'vila-Ortız G, Becker K (2006) Chemical composition, toxic/antimetabolic constituents, and effects of different treatments on their levels, in four provenances of *Jatropha curcas* L. from Mexico. Food Chem 96:80–89
Mujumdar AM, Misar AV (2004) Local anti inflammatory activity of *Jatropha curcas* roots in mice. J Ethnopharmacol 90:11–15
Mujumdar AM, Upadye AS, Misar AV (2000) Studies on antidiarrhoeal activity of *Jatropha curcas* root extract in albino mice. J Ethnopharmacol 70:183–187
Nath LK, Dutta SK (1992) Wound healing response of the proteolytic enzyme curcain. Ind J Pharmacol 24:114–115
Ochse JJ (1931) Vegetables of the Dutch East Indies. A. Ashernand Co/Hacquebard, Amsterdam, reprinted 1980
Oduola T, Popoola GB, Avwioro OG, Oduola TA, Ademosun AA, Lawal MO (2007) Use of *Jatropha gossypiifolia* stem latex as a hemostatic agent: how safe is it? J Med Plant Res 1:14–17
Odusote OM, Abioye AO, Rotibi MO (2002) *Jatropha curcas* seed oil Linn (Euphorbiaceae) contraceptive activity and on oral formulation. Nig Qt J Hosp Med 12:44–47
Osoniyi O, Onajobi F (2003) Coagulant and anticoagulant activities in *Jatropha curcas* latex. J Ethnopharmacol 89:101–105
Parthiban KT, Kumar RS, Thiyagarajan P, Subbulakshmi V, Vennila S, Rao MG (2009) Hybrid progenies in *Jatropha*—a new development. Curr Sci 96:815–823
Pertino M, Schmeda-Hirschmann G, Rodrıguez JA, Theoduloz C (2007) Gastroprotective effect and cytotoxicity of terpenes from the Paraguayan crude drug "yagua rova" (*Jatropha isabelli.*). J Ethnopharmacol 111:553–559
Radhakrishnan P (2007) Contribution of de-oiled cakes in carbon sequestration and as a source of energy, in Indian agriculture—need for a policy initiative. In: Proceedings of the 4th international biofuels conference, New Delhi, 1–2 Feb 2007. Winrock International India, New Delhi
Rug M, Ruppel A (2000) Toxic activities of the plant *Jatropha curcas* against intermediate snail hosts and larvae of schistosomes. Trop Med Int Health 5:423–430
Santosh KV, Juwarkar AA, Kumar GP, Thawale PR, Singh SK, Chakrabarti T (2009) Bioaccumulation and phyto-translocation of arsenic, chromium and zinc by *Jatropha curcas* L.: impact of dairy sludge and biofertilizer. Bioresour Technol 100:4616–4622
Siddhuraju P, Makkar HPS, Becker K (2002) The effect of ionising radiation on antinutritional factors and the nutritional value of plant materials with reference to human and animal food. Food Chem 78:187–195
Thangavelu R, Sundararaju P, Sathiamoorthy S (2004) Management of anthracnose disease of banana caused by *Colletotrichum musae* using plant extracts. J Hort Sci Biotechnol 79:664–668
Varshney A, Johnson TS (2010) Efficient plant regeneration from immature embryo cultures of *Jatropha curcas*, a biodiesel plant. Plant Biotechnol Rep 4:139–148
Wani S, Sreedevi TK, Marimuthu S (2008) Pro-poor biodiesel initiative for rehabilitating degraded drylands. In: International consultation on pro-poor Jatropha development, IFAD, Rome, 10–11 April 2008. Available from http://www.ifad.org/events/jatropha/. Accessed March 2012
Wei Q, Liao Y, Chen Y, Wang SH, Xu Y, Tang L et al (2005) Isolation, characterization and antifungal activity of β- 1,3-glucanase from seeds of *Jatropha curcas*. S Afr J Bot 71:95–99
Wink M, Koschmieder C, Sauerwein M, Sporer F (1997) Phorbol esters of *J. curcas:* biological activities and potential applications. In: Gübitz GM, Mittelbach M, Trabi M (eds) Biofuel and industrial products from *Jatropha curcas*. Dbv-Verlag Univ. Graz, Graz, pp 160–166
Zippel J, Wells T, Hensel A (2010) Arabinogalactan protein from *Jatropha curcas* L. seeds as TGFβ1-mediated inductor of keratinocyte *in vitro* differentiation and stimulation of GM CSF, HGF, KGF and in organotypic skin equivalents. Fitoterapia 81:772–778

Part V
Biofuel

Chapter 25
Biodiesel Production from *Jatropha curcas* Oil

P.P. Chakrabarti and R.B.N. Prasad

Introduction

Renewable fuels are gaining more importance as a replacement to non-renewable petroleum based fuels. The major reasons for the development of various renewable resources for fuel production are to achieve the security of energy required for transportation, to produce environmentally benign fuels, to use in the engines already developed for various types of vehicles, reduce health hazards and to provide the user an economic advantage. Biodiesel is one of the most important renewable fuels that falls into this category that do not require any modification of the existing engines and also exhibit significant advantages over petroleum derived diesel fuel (Jayed et al. 2009). The concept of using food as a fuel dates back to 1895 when Rudolf Diesel developed the first diesel engine to run on food oil. Diesel demonstrated his engine at the World Exhibition in Paris in 1900 using peanut oil as fuel.

Biodiesel is the name widely used for alkyl esters of fatty acids made from vegetable oil or animal fats. Biodiesel prepared from any vegetable oils exhibits similar kinematic viscosity, better fuel properties like higher combustion efficiency and cetane number (Demirbas 1998) in addition to excellent biodegradability (Mudge and Pereira 1999; Speidel et al. 2000) compared to petro diesel. As biodiesel generally does not contain any sulfur and aromatic molecules, the level of toxic emissions is also lower compared to that of diesel fuel (Harrington 1986; USEPA Report 2002). In fact, pollutants like unburnt hydrocarbons, particulate matters and carbon monoxides are found to be less in the biodiesel exhaust emissions. The NO_X emission of biodiesel is found to be little higher compared to diesel. Most of the biodiesel samples prepared from several vegetable oils can be used in the engine without addition of any lubricant (Jain and Sharma 2010a; Canakci et al. 2006; Knothe et al.

P.P. Chakrabarti • R.B.N. Prasad (✉)
Centre for Lipid Research, Indian Institute of Chemical Technology, Hyderabad 500007, India
e-mail: rbnprasad@iict.res.in

N. Carels et al. (eds.), *Jatropha, Challenges for a New Energy Crop: Volume 1: Farming, Economics and Biofuel*, DOI 10.1007/978-1-4614-4806-8_25,

2006; Lopez et al. 2009). The production of biodiesel is less complicated as it does not require any underground drilling and complicated plants like petroleum refining (Juan et al. 2011). As biodiesel can be produced locally, it would have more flexibility as far as fuel security is concerned and will not have complicated tax systems as it prevails for petroleum products (Jain and Sharma 2010a).

However, biodiesel has certain disadvantages compared to petroleum derived diesel fuel. It has lower calorific value, low volatility and biodiesel is found to be slightly corrosive against copper and brass. On the other hand, biodiesel prepared from saturated-rich vegetable oils, exhibits higher pour and cloud points and cannot be used at extreme cold atmosphere. If biodiesel is produced with proper care and used judiciously, it can be used as the most viable alternative to non renewable fuels like diesel.

The biodiesel business worldwide is a complex phenomenon and depends on many issues that cannot be controlled so easily. Because of the use of food crops as biodiesel feedstock, the price of oils sky-rocketed to a very high level and as a result, the biodiesel price also goes up making this economically unviable as an alternative fuel. Large scale mono-crop energy plantations are required to cater to the feed stock for biodiesel production. The major issues today are the policies adopted by local governmental agencies for sustainability of both food and non-food agricultural commodities. In the changed scenario, where people are advocating for second generation biofuels, Jatropha has once again got some impetus as it does not have any direct conflict with any food crop.

Common Feedstocks for Biodiesel Production

Biodiesel can be prepared from various types of raw materials like vegetable oils, animal fats, fish oils, algal oils, used frying oils and even from vegetable oil processing by-products like soap stock and fatty acid distillates. However, biodiesel is predominantly prepared worldwide either from refined or raw vegetable oils, mostly employed for edible purpose (Patil and Deng 2009). The type of vegetable oil used for biodiesel production is the parameter that has the greatest contribution on demographic condition, climatic consequences and finally on the basis of economic feasibility. Majority of the large scale biodiesel production units in USA depend on soybean, sunflower and corn oils as the biodiesel feedstock, whereas in Europe and Canada, rapeseed/canola and sunflower are the main feed-stocks for biodiesel production (Ma and Hanna 1999). *Used frying oils* (UFO) are also used commercially as the feedstock for biodiesel production in Europe and other countries (Kulkarni and Dalai 2006; Gui et al. 2008; Leung and Guo 2006). In south-east Asia, particularly in Malaysia, Indonesia and neighboring countries, palm oil is used as the raw material for biodiesel preparation (Ma and Hanna 1999). Another important raw material in some countries for biodiesel preparation is animal fat, predominantly the tallow fats and lard (Ma et al. 1998). In some other Asian countries including India and also the African countries, who are net importers of vegetable oils, it is very difficult to use edible oils for production of biodiesel. However, India, due to its tropical climatic conditions and vast terrain region has oil from more than 100 species of oilseed trees. These oils can be targeted as the raw material for biodiesel production as they are cheaper compared to other edible oils and there exists huge scope for rural development by utilizing these crops that can be grown in forests and in rain shadow areas. India, hence is projecting non edible oils like *Jatropha curcas* (Jatropha) and *Pongamia pinnata* (karanja) as the main feedstock for the preparation of biodiesel (Jain and Sharma 2010a). In India, a total of 10 lakh (in India, 1 lakh = 100,000) hectares (ha) of land are under Jatropha cultivation whereas Mozambique (3 lakh ha), Indonesia (2 lakh ha), Malawi (55,000 ha), Malaysia (22,000 ha), Brazil (20,000 ha), Cambodia (20,000 ha), Madagascar (17,000 ha), South Africa (15,000 ha) are also cultivating Jatropha in a big way (Comprehensive Jatropha Report 2010, Biozio). Other countries like China, Tanzania, Nicaragua has also shown interest on producing biodiesel from Jatropha oil and planted Jatropha trees (Lu et al. 2009; Eijck and Romjin 2008; Foidl et al. 1996). Countries like Burkina Faso, Ghana, Swaziland, Zambia also followed similar approach. However, the exact extent of cultivation, crop yield and use for production of biodiesel in these countries are yet to be published in the peer reviewed literature. Later, *Pongamia pinnata* (karanja) was also chosen as the raw material for biodiesel production in India. States like Chhattisgarh, Rajasthan, Jharkhand, West Bengal, Andhra Pradesh, and Karnataka have already initiated organized cultivation of both crops. A more practical assessment on yield, production cost, technology options for production of biodiesel having qualities that can match international specifications, economies, and impact on environment can only be done after few years of evaluation.

Apart from all these important raw materials, biodiesel has been prepared from various other oilseed crops and animal fats. Vegetable oils like rice bran oil (Bak et al. 1996; Ju and Vali 2005), castor oil (Conceicao et al. 2007) cottonseed oil (Kose et al. 2002), groundnut oil (Ahmad et al. 2009), linseed oil (Sendzikiene et al. 2005), karanja oil (Karmee and Chadha 2005) rubber seed oil (Satyanarayana and Muraleedharan 2010), mahua oil (Kapilan and Reddy 2008), low value lipids from soap stock (Wang et al. 2007), fish oils (Lin and Li 2009) were also evaluated for their suitability as biodiesel feedstock. However, in this chapter, Jatropha oil is compared with only the major feedstocks for biodiesel production for better evaluation of the oil as biodiesel raw material.

The issue of choosing the raw material for biodiesel preparation is a little complex. Apart from the economic standpoint, issues like food versus fuel, environmental aspects, and regulatory restrictions may control the biodiesel market in a particular region and the feedstock is chosen accordingly. In this chapter, emphasis is given only on the technical aspects rather than other issues. The suitability of a particular raw material may be evaluated by comparing the physico-chemical properties of the particular oil with that of the commonly used cleaner raw materials like sunflower, soybean, rapeseed, etc. (Tables 25.1 and 25.2). The major disadvantages of the non edible oils are the presence of higher amount of *free fatty acids* (FFA) and also higher amounts of gums and other unsaponifiable components. These

Table 25.1 Physico-chemical properties of major biodiesel raw materials

Type of raw materials	Acid value	Iodine value	Sap value	Unsaponi-fiable matter (%)	'P' content (ppm)	Density (g cm^{-3})	Kinematic viscosity cst at 40°C	Flash point (°C)	Heating value (MJ Kg^{-1})	Cetane number	References
Soybean oil	0.2–1.0	114–138.5	188.5–201.6	0.5–1.6	1244	0.91	32.9	254–324	39.6	37.9	Leung et al. (2010), Hammond et al. (2005), Srivastava and Prasad (2000)
Rapeseed oil	0.5–1.8	97–108	168–181	0.5–1.2	Upto 1155	0.91	35.1	278–282	39.7	37.6	Leung et al. (2010), Przybylski et al. (2005), Srivastava and Prasad (2000)
Sunflower oil	0.1–0.2	118–141	182–194	0.5–1.5	165–396	0.92	32.6	316	39.6	37.1	Leung et al. (2010), Grompone (2005), Srivastava and Prasad (2000)
Canola oil	0.4–1.2	110–126	188–192	0.5–1.2	Upto 825	0.914–0.917	38.2	275–290	—	—	Leung et al. (2010), Przybylski et al. (2005)
Corn oil	0.05 (After RBD)	127–133	187–193	1.3–2.3	10–80	0.91	34.9	332–338	39.5	39.5	Leung et al. (2010), Moreau (2005), Feng et al. (2002), Srivastava and Prasad (2000)
Cottonseed oil	0.5–0.6	98–118	189–198	<1.5	200–300	0.91	**18.2**	293.3	39.5	41.8	Leung et al. (2010), O'Brien et al. (2005), Srivastava and Prasad (2000)
Palm oil[a]	0.1	50.6–55.1	190.1–201.7	0.15–0.99	4	0.92	39.6	267	—	42.0	Leung et al. (2010), Basiron (2005), Srivastava and Prasad (2000)

Tallow	—	35–48	193–202	—	—	0.92	—	—	40.05	—	Leung et al. (2010), Ma and Hanna (1999)
Used oils	2.5	—	—	—	—	0.90	44.7	—	—	—	Leung et al. (2010)
Karanj oil	5.06	86.5	187	2.6	—	0.91	27.8	205	34.0	—	Leung et al. (2010)
Jatropha oil	0.29–28	95.2–106.6	190.1–192.4	0.79–1.08	290	0.92	29.4	240	38.5	45.0	Leung et al. (2010), Gubitz et al. (1999), Foidl et al. (1996), Singh and Padhi (2009)

[a]Refined, bleached and deodorized

Table 25.2 Fatty acid composition of major biodiesel raw materials

Oil	Myristic	Palmitic	Palmitoleic	Stearic	Oleic	Linoleic	Linolenic	Arachidic	Eicosenoic	Behenic	Lignoceric	References
Soybean	0.1	10.8	0.2	3.9	23.9	52.1	7.8	0.3	0.1	0.2	0.3	Dubois et al. (2007)
Rapeseed	0.1	5.1	0.2	1.7	60.1	21.5	9.9	0.6	1.4	0.3	0.2	Dubois et al. (2007)
Sunflower	0.1	6.4	0.1	4.5	22.1	65.6	0.5	0.3	0.2	0.8	0.2	Dubois et al. (2007)
Canola[a]	0.1	3.6	0.2	1.5	61.6	21.7	9.6	0.6	1.4	0.3	0.2	Przybylski et al. (2005)
Corn	—	12.3	0.1	1.9	27.7	56.1	1.0	0.4	0.3	0.1	0.1	Dubois et al. (2007)
Cottonseed	0.8	24.2	0.7	2.3	17.4	53.2	0.2	0.2	0.1	0.1	0.1	Dubois et al. 2007
Palm[b]	1.1	43.8	0.2	4.4	39.1	10.2	0.3	0.3	0.1	0.1	0.1	Dubois et al. (2007)
Tallow[c]	2.7–4.8	20.9–28.9	—	7–26.5	30.4–48.0	0.6–1.8	0.3–0.7	0.9	0.3–1.7	0.1	—	Haas (2005)
Karanja	—	10.6	—	6.8	49.0	19.0	—	4.1	1.4	5.3	2.4	Bringi and Mukherjee (1987)
Jatropha	0.1	15.3	—	6.6	40.1	35.9	0.2	0.2	—	—	—	Dubois et al. (2007)

[a]Also contains 22:1–0.1%

[b]Also contains 8:0–0.1%, 10:0–0.1%, 12:0–0.4%

[c]Also contains 14:1–0.7–0.8%, 17:0–1.0% and trans fatty acids 11.0–14.6%

oils require multiple steps of reactions like degumming, esterification and transesterification for the production of biodiesel and hence increase the cost of production (Leung et al. 2010). These raw materials may result in production of biodiesel with lower ester contents and may not be able to meet the stringent specification for biodiesels laid by different standard authorities unless proper methodologies are employed for the production of biodiesel. Animal fats and some other vegetable oils containing higher amounts of saturated fatty acids exist as solid form at room temperature and may cause problems during the production and usage of biodiesel (Patil and Deng 2009; Sahoo and Das 2009). Hence, physico-chemical properties and fatty acid composition of the raw materials have to be analyzed initially to evaluate the suitability of the particular raw material as biodiesel feedstock. Table 25.1 describes detailed physico-chemical properties of major oils being used as raw materials for biodiesel production along with Jatropha oil. The data indicates that non edible oils like karanja and Jatropha contain higher content of FFAs and these oils require two step processes for the preparation of biodiesel. The fatty acid composition of major biodiesel raw materials is given in Table 25.2. The fatty acid profile of Jatropha oil shows that it has a very balanced composition for biodiesel as it is rich in unsaturation with about 20–25% of saturated fatty acids.

Extraction Methods for Oil Production from Jatropha Seeds

Jatropha seeds look somewhat similar to castor seed, and are dark brown to black in colour. The kernel constitutes 66–68% of the total seed. The oil content in the kernel varies from 46 to 58% (Bringi 1987). In another report, the proximate analysis of seeds showed that it contained around 24.6% crude protein, 47.25% crude oil and 5.54% of moisture (Akintayo 2004). Recent reports are showing that the oil contents of different varieties of Jatropha seeds in India are in the range of 30–37% by weight (Singh and Padhi 2009; Jain and Sharma 2010a; Sunil et al. 2008).

Ripe fruits of Jatropha can be easily collected; a modified apple picker may also be used to collect fruits in a cotton bag (Henning 2003). Before dehulling or decorticating, the seeds are to be dried to get better quality of oil. The whole seeds also can be expelled for oil. However, the oil yield becomes lower compared to that of dehulled/decorticated seeds if seeds are expelled directly. The seeds should be stored in a well ventilated store room after drying. Manual dehulling of Jatropha seed is also possible, however, this is time consuming and mechanical dehuller/decorticators are available for Jatropha seeds and no specific problem were reported concerning the decorticating process (Bringi 1987). After decortication, seed and hulls are separated by winnowing or sieving. Some processors add 20–25% whole seed along with the dehulled seeds for ease of operation during expelling for the extraction of oil from the seeds.

Generally, vegetable oils are extracted either by expelling using different types of screw presses or by solvent extraction using hexane. If the oil content of seeds or oil bearing materials is less than 30%, oil must be recover through solvent extraction. Jatropha oil can be expelled from seeds with the expellers used for other seedoils without major changes. Double crushing in expellers produces 28.5% oil and if similar procedure is followed for whole seed, the oil yield is around 25% (Bringi 1987). In African countries, hand presses are used for expelling of oil and by using Bielenberg Ram type of manual expeller 60–65% oil present in the seeds could be extracted, whereas 75–80% oil could be extracted by Sundhara type mechanical expellers (Henning 2003). Some researchers have slightly modified the existing expellers used for other oils and developed one mechanical expeller comprising three components: dehuller that removes the hard outer shell of dried seeds, the crusher that makes dehulled seeds into finer pieces and the oil expeller that presses the oil out of the crushed seeds (Iwayan 2008). Monforts and Reiners types of expellers are used in Nicaragua (Foidl et al. 1996). The process is very simple to scaled up for large capacities and operators can be trained very easily. Well known Sayari and Komet expellers are also used for extraction of Jatropha oils.

The extraction of oil by using screw press or expellers leaves 10–15% of residual oil in the expelled cake. Oils are to be recovered from the expelled cake to have better economy of the process. The residual oil is generally extracted by using hexane. Since this oil would be used for non-edible purposes, one may think of using hexane extraction directly from the decorticated seed materials as well. In a recent report, various aspects of solvent extraction of Jatropha seeds were discussed (Sayyar et al. 2009). It was observed that, the best results as far as maximum extractability is concerned was obtained with hexane at around 68°C with coarse particle size of 0.5–0.75 mm, 8 h of extraction and 6:1 solvent : seed material ratio. After solvent extraction, less than 1% of residual oil is left in the deoiled cake.

Jatropha oil contains several antinutritional and toxic components. Proper measures are to be taken to avoid any possible contamination before using the same expeller or solvent extraction plant for other oilseeds particularly for edible oils. It is advisable to avoid such practices.

In recent years, researchers have tried ultrasonication, enzymatic extraction (Shah et al. 2005), and supercritical carbon dioxide extraction (Min et al. 2010) to recover Jatropha oil from its seeds. However, these are still in research phase and only the conventional methods of expelling and solvent extraction are being currently used for extraction of Jatropha oil.

Toxicity Studies of Jatropha Oil

It is reported that Jatropha seed is having some compounds that contravene the Chemical Weapons Convention of Geneva (Parawira 2010). Scientific studies revealed that Jatropha seeds are highly toxic for human beings, common animals and birds. Because of its toxicity, Jatropha is known as 'black vomit nut', 'purge nut', physic nut', 'pinoncillo', 'poison nut tree', 'graveyard tree', 'hell oil', 'oil infernal' etc. (Akintayo 2004; Gubitz et al. 1999; Makkar et al. 1998; Staubamann et al. 1999). Feeding studies showed that the seed and oil were toxic to mice (Adams 1974) and rats (Liberalino et al. 1988) and also to larger animals like goats, sheep and calves (Ahmed and Adam 1979a, b). In a more recent report, the poisoning of 11 children by Jatropha has been revealed in Mauritius (Rai and Lakhanpal 2008). In their study, Liberalino et al. (1988) reported that rats were killed within 2–3 days when fed with raw or cooked seeds and after 6–8 days when fed with raw or cooked oil or defatted seed meal. Rats could survive 14–16 days after feeding with roasted and cooked seeds. Histopathological examinations showed gastrointestinal inflammation, heart and kidney disorders and damages, haemorrhages and necrosis of livers, etc. in all the birds or animals fed with Jatropha seed or oil (Adam 1974; Adam and Magzoub 1975; Ahmed and Adam 1979a; Abdu-Aguye et al. 1986; Liberalino et al. 1988). It has been reported that rats had diarrhoea and inflammation in the intestine when fed with an Indian variety of Jatropha oil and LD_{50} has been found to be 6 ml per kg body weight (Gandhi et al. 1995). In the medical literature severe toxicoses were reported; the symptoms being similar to that of organophosphate pesticide intoxication (Koltin et al. 2006).

Treatment procedure for Jatropha poisoning are also reported in the literature. Generally, consumption of 3–5 seeds results in severe nausea, pain in abdomen, gastric irritation and vomiting and diarrhoea (Gubitz et al. 1999). In some cases, patients were dehydrated and then given intravenous fluid replacement (Joubert et al. 1984; Abdu-Aguye et al. 1986; Mampane et al. 1987). In some places, antidotes of Jatropha poisoning are available. In Mexico, Jatropha poisoning is treated by using annatto seeds (*Bixa orellana* L.) as an antidote (Rivera-Lorca and Ku-Vera 1997). In South Africa, a watery extract of *Peltophorum africanum* is used as antidote (Mampane et al. 1987).

The toxicity of seeds was attributed to a hemagglutinin called curcin (Felke 1913). This toxic protein was found to inhibit protein synthesis in *in vitro* studies. However, the toxicity was found to be much lower compared to ricin or abrin (Stripe et al. 1976). Though, presence of some lectins were also referred for the toxicity, it was found that similar amounts of lectins were present in non-toxic Mexico varieties of Jatropha and hence concluded that lectins are not the major toxic principle present in Jatropha seeds (Makkar et al. 1998; Parawira 2010). The toxicity is primarily due to the presence of some toxalbumin, cyanic acids and toxic phorbol esters (Nath and Dutta 1991; Ito et al. 1983). Some more antinutritional components identified were phytates, saponins and a trypsin inhibitor (Wink et al. 1997). Various methanolic extracts of seeds had shown contraceptive and abortive action on rats (Mameesh et al. 1963; Goonasekera et al. 1995). The seed kernels of Jatropha were found to have between 0.03% and 3.4% of phorbol esters that can be present in oil also (Horiuchi et al. 1987; Hirota et al. 1988; Sauerwein et al. 1993; Wink et al. 1997). These compounds activate protein kinase C (PKC), which is an enzyme that plays an important role in signal transduction and development of most cells and tissues (Azzi et al. 1992). Prolonged interaction between phorbol esters and PKC would be the cause of tumorigenesis (Rotenberg et al. 1991).

However, some varieties of Jatropha seeds from Mexico are reported to be used for feeding birds, animals and for preparing some traditional dishes without any antinutritional effects. White winged dove is reported to have Jatropha seeds as a feed (Rivera-Lorca and Ku-Vera 1997). Chickens and pigs also consume Jatropha seeds in Mexico (Gubitz et al. 1999). Some reports showed that mice had normal weight gain when fed with another Mexican variety of Jatropha seed (Panigrahi et al. 1984). It was reported that in Mexico, boiled or roasted seeds are sometimes used for preparing some traditional delicacies (Schmook and Serralta-Peraza 1997).

Extraction of oil from huge quantity of Jatropha seeds will also generate double quantity of deoiled cake. This nitrogen rich by-product of Jatropha oil biodiesel industry cannot be used as animal/poultry feed if it is not detoxified properly. Efforts were made to detoxify oil and deoiled Jatropha cake (Haas and Mittelbach 2000). Usual methods of oil refining like degumming, bleaching, deodorization were employed. Other treatments like $NaHCO_3$ treatment, extraction with alcohols were also tried. However, even the best detoxification method reported could remove only half of the phorbol esters and since the initial contents were very high, the toxicity level could not be reduced to safe limits (Martinez et al. 2006). Moreover, detoxification was performed only on laboratory scale and the process is highly complicated. This is very difficult to perform at larger scale (Parawira 2010). The economy of Jatropha biodiesel industry could have changed dramatically, if deoiled cake could be detoxified properly and be used as animal/poultry feed (Mahanta et al. 2008). Therefore, detoxification of Jatropha deoiled cake seems to be difficult and additional investigations are still need.

Physico-chemical Properties of Jatropha Oil

The physico-chemical properties like free fatty acid content, iodine value, unsaponifiable matter, triglyceride content, phosphorus content, etc. are very important for the evaluation of a given oil as biodiesel feedstock and for the definition of pre-treatment procedures as well as the economy of the whole process. Table 25.3 describes the physico-chemical properties of Jatropha seed oils from different countries and shows that in most case characteristics are similar except for free fatty acid content, which depends on the quality and handling of seeds.

The oils from several Jatropha sources were also characterized for their physico-chemical properties. The oil from one Indian variety of Jatropha was found to have 918 kg m^{-3} density at 15°C, flash point of 176°C, pour point of −6°C, carbon residue of 0.3% and calorific value of 33 MJ kg^{-1} (Singh and Padhi 2009). Jatropha oil from Thailand was found with 0.52% FFA, a 0.52%, kinematic viscosity at 40°C of 34.84 mm^{-2} S^{-1} and density at 15°C of 918.6 kg m^{-3} (Nakpong and Wootthikanokkhan 2010). In one report, the flash point of the Jatropha oil was found to be 240°C and the phosphorus, calcium, magnesium and iron contents were determined to be 290, 56, 103 and 2.4 ppm, respectively (Gubitz et al. 1999). The data available in the literature is still not sufficient to make a generalized comment. Perhaps, a more

Table 25.3 Physico-chemical properties of different varieties of Jatropha oil samples

Property	Carboverde[a]	Nicaragua[b]	Nigeria[c]	India[d]	Myanmar[e]
Moisture (%)	0.07	—	—	—	0.2
Free fatty acid (%)	0.29–0.40	0.60–1.27	1.66–1.86	10.2	22.6
Iodine value	95.2	106.6	104.5–105.9	98	100.1
Saponification value	192.4	190.1	197.5–200.3	—	208.3
Unsaponifiable matter (%)	1.08	0.79	0.7–0.9	0.8	—
Viscosity (cSt)	38.8	37.0	17.1	35.4	—

[a]Foidl et al. (1996)
[b]Kywe and Oo (2009)
[c]Gandhi et al. (1995)
[d]Akintayo (2004)
[e]Singh and Padhi (2009)

detailed characterization of the oil will be possible with a larger number of analyses.

Fatty Acid composition of Jatropha oil: One of the most important characteristics of the oil, which defines its suitability as biodiesel feedstock, is the fatty acid composition. If more saturated acids are present in the oil, the biodiesel prepared from it would have problem of pour point as at lower temperature it would cease to flow. On the other hand, if it has more unsaturation, it may have lower oxidative stability. This may result in rancidity and may affect its fuel properties. The fatty acid composition of different sources of Jatropha oils showed that most of them have more or less similar pattern of fatty acid composition. Table 25.4 shows the fatty acid composition of representative oil sources from different countries.

Pre-treatment Methods for Biodiesel Preparation

Vegetable oils contain FFAs and phospholipids that may cause problems during biodiesel preparation. Before the transesterification reaction for the conversion of vegetable oils to biodiesel, necessary pre-treatment steps like degumming followed by neutralization or esterification have to be performed to overcome these problems. For clean oils like sunflower, soybean, rapeseed, etc. simple degumming process is sufficient before proceeding for transesterification. On the contrary all the non-edible oils, animal fat and used cooking oils require very elaborate pre-treatment methods like degumming and esterification before transesterification if stringent international specifications of biodiesel are to be met. Jatropha oil also has such impurities and thorough pre-treatment methodologies are to be developed for various Jatropha oil samples depending on their impurity profiles. The basic pre-treatment methodologies generally followed are described here.

The first pre-treatment step is the degumming. Most of the vegetable oils contain gums and mucilages. Gums are nothing but phospholipids and they act as surface

Table 25.4 Fatty acid composition (wt%) of different varieties of Jatropha oils

Fatty Acid	Carboverde[a]	Nicaragua[b]	Nigeria[c]	India 1[d]	India 2[e]	Thailand[f]
Capric	0.1	0.1	—	—	—	—
Myristic	0.1	0.1	—	0.1	1.4	—
Palmitic	15.1	13.6	19.5	15.3	15.6	13.77
Palmitoleic	0.9	0.8	—	—	—	—
Stearic	7.1	7.4	6.8	6.6	9.7	6.77
Oleic	44.7	34.3	41.3	40.1	41.0	41.68
Linoleic	31.4	43.2	31.4	35.9	32.1	35.55
Linolenic	0.2	0.2	—	0.2	—	—
Arachidic	0.2	0.3	—	0.2	—	—
Behenic	0.2	—	—		—	—

[a]Foidl et al. 1996
[b]Kywe and Oo (2009)
[c]Gandhi et al. (1995)
[d]Bringi (1987)
[e]Akintayo (2004)
[f]Singh and Padhi (2009)

active agents. If these gums are present in the oil in large quantities, they may create problems during biodiesel preparation as they form an emulsion that prevent layer separation of glycerol and methyl esters. This pre-treatment step allow the removal of phospholipids, metal ions and other impurities from oil. These contaminents are critical since the quality of the degummed oil is generally judged by its phosphorous and metal contents. If not removed effectively, these impurities may eventually interfere with the subsequent processing steps. Phospholipids present in oils are broadly classified as hydratable and non-hydratable (Hvolby 1971). While hydratable types are removed from the oil by simple water degumming step (Carlson and Scott 1991; Zhang et al. 1994), non-hydratables require acid treatment. Calcium and magnesium ions bind to phosphatidic acid and phosphatidyl ethanolamine very strongly at pH 7. It was found that the binding of bivalent ions is 1,000 times stronger than the binding of monovalent ions, such as sodium and potassium ions at pH 6 and pH 7 (Hvolby 1971). For this reason, under the conditions of water degumming, such compounds do not become hydratable. The addition of a sufficiently strong acid liberates free phosphatidic acid and phosphatidyl ethanolamine, and thus converts them to hydratable compounds. Bivalent metal ions bind with the added acids like citric or phosphoric acids and can be removed with water. However, these acids must be thoroughly mixed to achieve the desired result. Acids like acetic acid and acetic anhydride are also capable of dissociating phosphatidic acid and phosphatidyl ethanolamine from bivalent metal ions. It is of utmost importance that metal complexes formed with added acid must not be oil-soluble; otherwise they would remain in the oil.

Commonly used degumming processes are Alcon process (Kock 1978), dry degumming, acid degumming (Hvolby 1971), super degumming (Ringers and Segers 1976; Segers 1982) uni-degumming (Van de sande and Segers 1989) special degumming (Carlson 1985), total degumming (Dijkstra and Van Opstal 1985) and enzymatic degumming (Chakrabarti et al. 2009). Amongst all these degumming techniques, water degumming and acid degumming processes are followed by most of the industries. These techniques are generally modified according to particular constraints. Jatropha oil samples were found to contain up to 300 ppm of phosphorus. According to BIS and ASTM specifications, biodiesel should have less than 10 ppm of phosphorus. It is, therefore, left to the processor to choose a degumming technique for getting the final product having the required specifications.

In neutralization method, primarily the FFAs present in the oil are neutralized using caustic alkali. This method is of utmost importance as apart from removal of FFAs, it removes almost all the impurities (except few colouring bodies and odorous compounds) present in the oil. The neutralization is done by using aqueous caustic solution. This forms sodium salts of fatty acids called soap. In the subsequent steps of bleaching and deodorization for the removal of colouring bodies and odorous compounds, respectively, oils and fats are heated to a high temperature. In general, biodiesel industry does not require bleaching and deodorization unless the quality of the oil is very bad. Therefore, most of the impurities should be removed in the neutralization step in case the free fatty content is less than 2% in the oil. As far as biodiesel industries are concerned, a judicious decision is to be taken for whether to go for neutralization or not. The acid number for the biodiesel should not exceed beyond 0.5 as per the standard specifications. For oils having FFA above 3%, esterification has to be done for converting the FFAs to methyl esters and then the whole mixture is treated for transesterification reaction. When the FFA content is less than a specified value, biodiesel manufacturers may apply neutralization to reach a specific quality level. Since some Jatropha oils have higher amounts of FFAs, it becomes the prerogative of the manufacturers to decide whether to use the neutralization step or not.

Very few studies are available for Jatropha oil in this direction. Yong et al. (2009) reported the degumming process for Jatropha oil by treating the oil with phosphoric acid (0.2% of the oil mass) along with water (3 times the phospholipids present in oil) at 80°C, for 30 min to obtain degummed oil before proceeding for the production of biodiesel. In another study, Lu et al. (2009) reported that phospholipids were eliminated during pre-esterification of Jatropha oil and a separate degumming operation was unnecessary.

Preparation of Biodiesel from Jatropha Oil

Vegetable oils can be used as energy source in four different ways through the direct use of vegetable oil or as a product of micro-emulsion, thermal cracking and transesterification. Vegetable oils have several limitations for direct use in diesel engines as the high viscosity would damage the engine by causing coking and trumpet formation (Agarwal and Agarwal 2007). The fuel obtained from micro-emulsion and thermal cracking methods lead to incomplete combustion due to a low cetane number and energy content (Leung et al. 2010). The most traditional methodology for the preparation of biodiesel from vegetable oils is transesterification. Transesterification was conducted as early as 1853. One of the first uses of biodiesel (transesterified vegetable oil) was for powering heavy vehicles in South Africa before World War II. Transesterification is the reaction of oil with an alcohol to form biodiesel and a by-product, glycerol. It is in principle the action of one alcohol displacing another from an ester. As the reaction is reversible, an excess of alcohol is usually used to force the equilibrium to the product side. The stoichiometry for the reaction is 3:1 alcohol to oil; however in practice this is usually increased to 6:1 to increase the product yield. A catalyst is usually used to speed up the reaction and may be a base, an acid or an enzyme. The alkalis that are generally used include sodium hydroxide, potassium hydroxide, carbonates and corresponding sodium and potassium alkoxides, such as sodium methoxide, ethoxide, propoxide and butoxide. Sodium hydroxide is the most common alkali used due to economical reasons and availability. Alkali-catalyzed reactions are used more often commercially than acid ones because their are faster.

Technology for the preparation of biodiesel from low FFA and high quality oils is less difficult since it only involves the step of transesterification (Fig. 25.1). In principle, the chemistry of transesterification is very simple and only the efficient conversion rate of oil to fatty acid methyl esters is very important. In the current commercial processes based on crude feedstock, an excess alkali is added to remove all FFAs (in case FFA is less than 2–3%. The main challenge of the biodiesel technology is handling oils like Jatropha with high content of FFAs. For alkali-catalyzed transesterification, water and FFA are not favorable to the reaction; so anhydrous triglycerides and alcohol are necessary to minimize the production of soap. Soap production decreases the amount of esters and the separation of glycerol and esters becomes difficult.

The typical product mixture of a transesterification reaction contains fatty acid esters, triglycerides, monoglycerides, diglycerides, glycerol, alcohol and catalyst in varying concentrations. The primary goal is the removal of the esters from the mixture, maintaining low costs and ensuring a high purity of the product. Glycerol in its pure form is seen to be a secondary product of the reaction as it can be sold to various industries. The removal and resale of glycerol is essential to keep the cost of production competitive. The remaining mixture contains by-products that should have minimal contaminants if the conversion rate is high and alcohol that would be distilled off.

There are some concerns over hydrolysis inhibiting the conversion in transesterification. Alternative separation techniques have not yet been adequately developed. Washing and settling are well established for commercial batch processes for clean oils like sunflower, rapeseed, etc. but may have several problems for high FFA containing oils like Jatropha oil. As commented above, the refining of biodiesel products from oil with high FFA content can be technically difficult and can substantially increase the production costs. Research into the equipment required for a low cost online method to analyze the extent of conversion

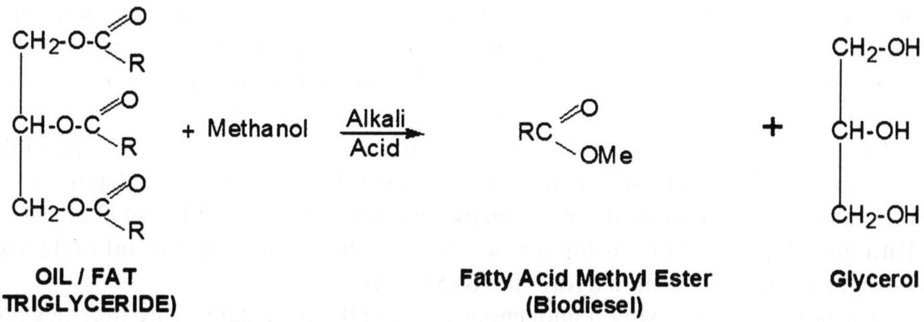

Fig. 25.1 Transesterification of vegetable oils

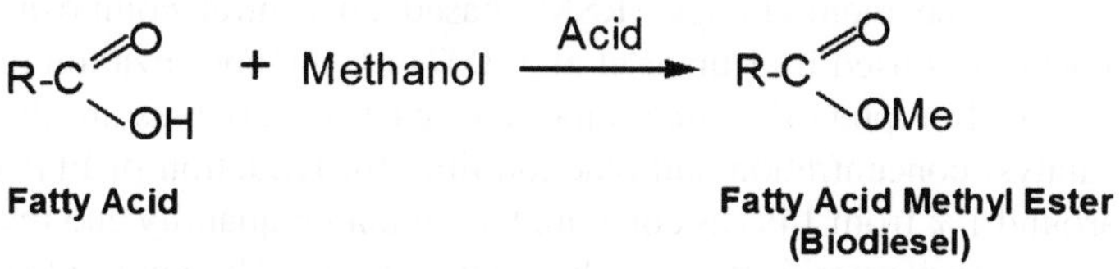

Fig. 25.2 Esterification of fatty acids present in Jatropha oils

as the purity of biodiesel must be invested in conformity to international standards.

In case of high FFA oils, initially the fatty acid has to be esterified using acid catalyst like sulphuric acid followed by neutralization of the catalyst (Fig. 25.2).

The major problem associated to acid catalysts is the formation of salts during neutralization and also the conversion rate that is not very high compared to that of transesterification. The inefficient conversion rates of acid catalyst may result in the formation of biodiesel without matching the International Standards. Problems with the generated water during esterification reaction have also not yet been fully addressed.

Several methodologies have been reported for the preparation of biodiesel from Jatropha oil in which the first step was esterification to reduce the FFA content in the feedstock by esterification route employing sulphuric as a catalyst in order to reduce the fatty acid content to less than 1% (Lu et al. 2009; Syam et al. 2009) followed by a transesterification step envolving potassium hydroxide (Lin and Fang 2008; Berchmans et al. 2010a, b; Syam et al. 2009; Lu et al. 2009), sodium hydroxide (Berchmans and Shizuko 2008; Lin and Fang 2008; Boonmee et al. 2010; Tapnes et al. 2008; Chitra et al. 2005; Jain and Sharma 2010b, c; Nakpong and Wootthikanokkhan 2010) and sodium methoxide (Lin and Fang 2008; Yong et al. 2009; Guo et al. 2008) for the conversion of triglyceride (oil) into biodiesel at a rate higher than 97%. The data generated during these studies indicated that the optimized conditions vary in the following limits for esterification: molar ratio of oil to methanol, 6–12%; catalyst concentration, 0.5–1 wt% H_2SO_4; reaction time,

30 min; reaction temperature, 60–65°C; transesterification: molar ratio of methanol to oil at 6:1; alkaline catalyst loading, 1–1.3%; reaction temperature, 60–65°C; and reaction time of 20–60 min. Large scale production of biodiesel from 25 kg of Jatropha oil has resulted in 24 kg of biodiesel as compared to laboratory scale (Chitra et al. 2005). The literature indicates that a better conversion rate is possible by implementing transesterification with a lower molar ratio of methanol to oil (4:1) and a lower amount of potassium hydroxide (0.55% w/w) (Tiwari et al. 2007; Patil and Deng 2009) by using a lower molar ratio of methanol to oil (4:1) and lower amount of potassium hydroxide (0.55% w/w).

The advantages of using homogenous acid or alkaline catalysts are that they are cheaper with a high catalytic activity for the production of high quality biodiesel in a short time (Helwani et al. 2009; Kulkarni and Dalai 2006). By contrast, their drawback is the generation of huge amounts of effluents.

Response surface methodology (RSM) based on central composite rotatable design (CCRD) was used by Kumar et al. (2007) for the optimization of biodiesel production from Jatropha oil. Three important reaction variables are the methanol quantity, catalyst concentration and reaction time for reduction of FFA content of the oil to around 1% from 14% as compared to methanol quantity and reaction time for carrying out transesterification of the pretreated oil. The study also generated quadratic polynomial equations for predicting acid value and transesterification using RSM. Boonmee et al. (2010) also reported a full quadratic model for the preparation of biodiesel from Jatropha oil by the program using response surface methodology (RSM) with an R2 and adjusted R2 of 97% and 94%, respectively. The optimum conditions for transesterification were: methanol-to-oil molar ratio of 6:1, 1% wt/wt sodium hydroxide and 90 min reaction time to obtain a yield of biodiesel up to 99.9%. The resulting Jatropha biodiesel properties satisfied both the ASTMD 6751 and EN 14214 biodiesel standards.

In another study, potassium carbonate was used as a catalyst for the transesterification reaction of Jatropha oil (Baroi et al. 2009). From this study it was observed that the transesterification of Jatropha oil to biodiesel appeared to be completed within 15 min when a 5 wt% (based on the wt. of the oil) potassium carbonate, 6:1 methanol to oil molar ratio, 60°C or a 4 wt% potassium carbonate, 9:1 methanol to oil molar ratio and 60°C reaction temperature were used.

The main problem using homogeneous catalysts like alkali or acids is that they are used in large quantities, which cause various problems, such as corrosion, catalyst separation, acid waste, environmental pollution. Therefore, there is a need to replace the homogeneous catalysts with solid catalysts, which are easy to handle and environmental friendly. The by-product glycerol remains diluted with water using homogenous catalysts and pose a problem when delivered to the environment where it becomes a waste product. With a solid catalyst, glycerol will be almost free of water and its purification is very simple. In addition, the pellet form of solid catalyst can be used in fixed-bed reaction processes and it can be very easily recycled several times.

The esterification of carboxylic acids by solid acid catalysts is important considering that low-cost lipid feedstock contains high concentrations of FFAs. Therefore, it is expected that a good solid acid catalyst must be able to carry out simultaneously both esterification and transesterification. Since esterification and transesterification share a common molecular pathway, evidence about catalyst reactivity for esterification also provides evidence about transesterification and *vice versa.*

Solid acid catalysts were employed by few researchers for both esterification and transesterification methods. SiO_2-HF was reported to be excellent heterogeneous catalyst to reduce the FFA content of Jatropha oil from 7.9% to 0.3% in 180 min before transesterification employing methanolic sodium hydroxide (Corro et al. 2010). Lu et al. (2009) reported calcining metatitanic acid as solid acid catalyst (4 wt% of oil) and the conversion of FFAs was higher than 97% at 90°C in 2 h at a 20:1 molar ratio of methanol to FFAs. Liu and Kong (2009) reported the transesterification of Jatropha oil with higher FFAs, using SO_4^{2-}/TiO_2-Al_2O_3 as a solid acid catalyst instead of traditional liquid acid and alkali catalysts. The experimental results indicated that the solid acid catalyst had a higher reactivity and stability on the transesterification of Jatropha oil at 130°C; methanol/oil molar ratio, 15:1; solid acid catalyst dosage, 4%; mixing speed 480 rpm; co-solvent hexane/oil wt. ratio 1:4; reaction time 3 h with a purity of 97.6%. The fuel properties of Jatropha biodiesel met the national BD100 standard. Solid acid catalyst Na/SiO_2 was also employed for the preparation of biodiesel from Jatropha oil under ultrasonication conditions (Kumar et al. 2010). The optimal conditions for biodiesel production are the molar ratio oil to methanol 1:9; catalyst concentration, 3 wt% of oil and 15 min reaction time under atmospheric pressure. Solid catalyst and ultrasonication reduced the reaction time compared to the conventional batch processes with a biodiesel yield of 98.5% in optimal conditions of: molar ratio of oil to methanol 1:9, catalyst concentration, 3 wt% of oil and 15 min reaction time. Vyas et al. (2009) reported the use of potassium nitrate (KNO_3) on alumina (Al_2O_3) as solid base catalyst for transesterification of Jatropha oil to biodiesel however, the catalyst can be reused only for 3 times. Super solid base of calcium oxide was also used as heterogeneous catalyst for biodiesel production from Jatropha oil. Commercial calcium oxide was immersed in ammonium carbonate solution to increase the base strength and calcined at 900°C for 1.5 h and the catalyst resulted in biodiesel yields of 93%. The catalyst could be reused at least 3 times without significant loss of catalytic activity. Solid acid catalyst prepared from crude glycerol obtained from biodiesel process was found to be one of the best esterification catalysts for the conversion of free fatty acid present in any vegetable oils (Prabhavathi Devi et al. 2007).

The disadvantages caused by chemical catalysts can be overcome by using biocatalysts particularly lipases for the preparation of biodiesel by employing greener approach (Haas and Foglia 2005). Biocatalysts are usually lipases, however conditions need to be well controlled to maintain the activity of the catalyst. They are stable, do not require co-enzymes and will often tolerate organic solvents. Enzyme catalyzed reaction is more efficient, highly selective, involves less energy consumption, produces less waste (Akoh et al. 2007) and it is recyclable as enzymes can be immobilized onto a support (Robles-Medina et al. 2009). Glycerol recovery in enzymatic process is easier as it would produce high grade glycerol as compared to chemical processes. Recent patents and articles have shown that reaction yields and times are still unfavourable compared to base-catalyzed transesterification for commercial application. Several lipases were employed either in free form or in immobilized state for this purpose. Immobilized *Candida antarctica, C. rugosa. Thermomyces lanuginosus, Rhizopus oryzae, Pseudomonas fluorescens, P. cepacia* were some of the lipases exploited for the preparation of biodiesel from different vegetable oils and fatty acids. As lipases can simultaneously esterify and transesterify FFA and triacylglycerols, low quality oils like high FFA oils, acid oils, restaurant greases, oils isolated from spent bleaching earth were also attempted to convert oil into biodiesel in reasonably good yields. The cost of lipase production is still the main bottleneck to commercialize the lipase-catalyzed process for low value product like biodiesel. However, there is a lot of scope for biotechnological revolution in this area.

Shah et al. (2004) reported the preparation of biodiesel from Jatropha oil employing *Chromobacterium viscosum, C. rugosa,* porcine pancreas lipases with an oil to alcohol (methanol and ethanol) ratio of 1:4, lipase concentration of 10 wt% of oil at 40°C for 8 h. Reaction was carried out with and without immobilization of lipase on celite and with and without water. The biodiesel yield varied from 62% to 92% depending on the reaction conditions. Three lipases from *Candida antarctica, Thermomyces lanuginosus* and *Rhizomucor meihei* were employed for the preparation of biodiesel from Jatropha oil using co-solvent (25% pentanol and 75% isooctane) to avoid lipase deactivation caused by methanol and glycerol. Conversions were up to 98% at 45°C with 7.5% w/w of *C. antarctica* lipase in 24 h (Su and Wei 2008). Kumari et al. (2009) used the *Enterobacter aurogenes* lipase immobilized on activated silica with ethanolamine for transesterification of Jatropha oil with only 68% yield, in spite of using t-butanol as a solvent to eliminate lipase deactivation.

The short-chain alcohols, such as methanol and ethanol are commonly used as acyl acceptors for the biodiesel production. However, the use of an excess alcohols would lead to enzyme inactivation. In addition, glycerol would block the active sites of the enzyme resulting in low enzyme activity. Stepwise addition of methanol may overcome the problem of lipase deactivation in solvent-free system in biodiesel production. Jatropha oil was converted to ethyl esters using *P. cepacia* lipase immobilized on celite via stepwise addition of ethanol to obtain more than 91% conversion (Shah et al. 2004). In another study, Jatropha oil was transesterified under a solvent-free system in a stepwise addition of methanol using Novozyme 435 and *R. oryzae* lipases with *R. oryzae* lipase resulting in 5% higher conversions of biodiesel (Tamalampudi et al. 2008). Novel acyl acceptors, such as methyl acetate, ethyl acetate and propan-2-ol were reported recently for the interesterification of various oils into biodiesel. Ethyl acetate was employed as the acyl acceptor for the production of the biodiesel from crude oils of Jatropha, karanja and sunflower using Novozyme 435 and the yields of ethyl esters were 91.3%, 90.0% and 92.7%, respectively (Modi et al. 2006). With ethyl acetate, the relative lipase activity could be maintained over 12 cycles while it did not pass over the 6th cycle whith ethanol used as an acyl acceptor. Su and Wei (2008) studied *in situ* extraction and transesterification of Jatropha oil from kernels using methyl and ethyl acetates with 86% and 87% yields, respectively. Modi et al. (2007) also reported propan-2-ol as an acyl acceptor for the Novozyme 435 catalyzed transesterification reaction for the production of biodiesel from crude Jatropha, karanja and sunflower oils with good yields. Lipase reusability could be maintained over 12 cycles with propan-2-ol as the acyl acceptor.

Tang et al. (2007) reported the transesterification of the crude Jatropha oil catalyzed by micro-sodium hydroxide in supercritical and subcritical methanol for enhancing the transesterification activity. Another study (Gan et al. 2010) reported the transesterification of Jatropha oil at subcritical methanol conditions with p–toluene sulfonic acid as a catalyst for the preparation of biodiesel and the optimum conditions were found to be: reaction temperature, 170°C; molar ratio of methanol to oil 40: 1; mass ratio of catalyst to oil 0.75% and reaction time of 30 min. The production of biodiesel from raw rice bran oil (RBO) in supercritical methanol was studied by adopting response surface method to examine the effect of molar ratio of methanol to oil, reaction temperature and reaction time on the conversion of biodiesel and the optimum reaction condition could be performed at the temperature of 295.9°C within 40.5 min with 25.1 molar ratio of methanol to oil (Han et al. 2009). Transesterification reaction without using any catalyst requires a high temperature above the critical temperature of alcohol and this is called as supercritical method. Rathore and Madras (2007) investigated the possibility of using the supercritical method on methanol and ethanol for biodiesel production from crude Jatropha oil and obtained conversion rates of upto 95% at 400°C under 20 MPa for both methyl and ethyl esters. In another study, supercritical reaction was carried out at milder conditions of 320°C under 8.4 MPa with a molar ratio of 1:43 of Jatropha oil to methanol and the yield of methyl ester was 100% (Hawash et al. 2009). FFA content of the Jatropha oil used by Rathore and Madras (2007) was about 10% whereas it was 2% in the oil used by Hawash et al. (2009). The water released during supercritical reaction is proportional to FFA content while the biodiesel yield is inversely proportional. Supercritical reactive extraction was employed for the simultaneous extraction and production of biodiesel from Jatropha seeds by Lim et al. (2010) in a relatively short total operating time (45–80 min). Particle size of seeds (0.5–2.0 mm), reaction temperature (200–300°C) and pressure were investigated; results were found to be encouraging with reaction conditions of 300°C, 240 MPa, a methanol to solid ratio of 10 ml g^{-1} and a hexane to seed ratio of 2.5 ml g^{-1} of Jatropha oil was also extracted using supercritical carbon dioxide and then subjected to subcritical hydrolysis (Chen et al. 2010). The hydrolyzed fatty acid was further reacted with methanol under supercritical esterification to obtain 99% of biodiesel from Jatropha oil in 15 min under 11 MPa at 290°C with 33% v/v of hydrolyzed oil to methanol. In another study, Ilham and Saka (2010) developed a two step process to produce biodiesel from Jatropha oil employing dimethyl carbonate instead of methanol to obtain 97% biodiesel in 15 min at 300°C under 9 MPa. During this process, glyoxal as by-product is produced instead of glycerol and the method is not affected by FFA content.

Li et al. (2006) reported a method featuring the addition of 0.02–0.3% of rosemary antioxidant to Jatropha oil at a wt. ratio of 1: (2–3), heating to 90–130°C, maintaining for 20–80 min to completely dissolve rosemary antioxidant in the oil. The method greatly inhibited the oxidation of Jatropha oil, thus prolonged its storage period. In another method, propyl gallate, pyrogallol, synergistic agents like trisodium citrate, dodecyl glucoside, and/or metal ion complexing agent EDTA disodium were added to Jatropha oil or blend diesel (obtained by mixing Jatropha oil and fossil diesel at a volume ratio of 5–99%), heating to 35–45°C, and maintaining for 30–90 min under stirring (Gao et al. 2010). The synergistic agent used in this article can play the role of antioxidant without generating free radicals or precipitates.

Engine tests with Jatropha oil were carried out in Thailand and showed satisfactory performance. A 50 h continuous test was conducted using transesterified Jatropha oil, diesel fuel and their blends in two diesel engines with small pre-combustion chambers (Gubitz et al. 1999; Recep et al. 2000). Biodiesel can be blended in various proportions with fossil diesel to create a biodiesel blend or can be used in its pure form. It can be used in compression ignition engines with very little or no engine modifications because it has properties similar to petroleum diesel (Banapurmath et al. 2008; Devanesan et al. 2007). Arumugham (2009) reported the exhaust gas emission characteristics of diesel blends prepared with Jatropha biodiesel (0%, 10%, 20%, 30%, 40% and 50%) employing a stationary diesel engine without any engine modifications. A single cylinder four stroke diesel engine was tested at various loads with the blended fuel at the speed of 1,500 rpm. Exhaust gas emissions were measured using an AVL 5 gas analyzer and a smoke meter. The results indicated that the brake thermal efficiency of diesel is higher at all loads followed by blends of Jatropha oil and diesel. The specific fuel consumption was found to be even lower than the conventional diesel for blends up to B20. The brake thermal efficiency for B10 and B20 were also closer to diesel and the CO_2 emissions were found to be lower than diesel. There was a marginal increase in the smoke opacity and NOx. The author claimed that the increase in opacity can be effectively managed by engine optimization.

Effective utilization of by-products like deoiled-cake, glycerol (Demirel et al. 2007; Zhou et al. 2008), phytochemicals and fine chemicals would certainly help the economics of biodiesel.

Physico-chemical Characterization of Jatropha Biodiesel

The physico-chemical properties of Jatropha biodiesel are well comparable with that of fossil diesel and that is the reason why Jatropha oil is being projected as an excellent source of biodiesel from non edible oils. Table 25.5 describes the physico-chemical properties of Jatropha biodiesel along with the biodiesel prepared from common feedstocks. The data revealed that Jatropha biodiesel is almost matching with biodiesel from other feedstocks. The properties of Jatropha biodiesel conforms to all the international specifications of biodiesel. However, proper care has to be taken during the production of biodiesel for obtaining specified acid value, when the raw Jatropha oil contains higher amounts of FFAs. Iodine value and pour points of Jatropha biodiesel are very attractive for its exploitation in colder regions. The cetane number of Jatropha biodiesel is very high compared to other oilseed sources.

Table 25.5 Comparison of biodiesel properties prepared from different raw materials

Oil	Acid Value (mg KOH g-1)	Iodine Value	Kinematic Viscosity (cst at 40°C)	Density (g/cm³)	Cetane Number	Heating Value (MJ/kg)	Flash Point (°C)	Cloud Point (°C)	Pour Point (°C)	References
Soybean	0.15	138.7	4.08	0.885	52	40	176	0	–4	Singh and Singh (2009), Liu et al. (2008), Leung et al. (2010), Ivanoiu and Rusnac (2009)
Rapeseed	0.25–0.45	132	4.3–5.83	0.88–0.888	49–50	45	163	–3	–6	Rashid and Anwar (2008), Ibiari et al. (2010)
Sunflower	0.24	142.7	4.9	0.88	49	45.3	183	4	–5	Winayanuwattikun et al. (2008), Rashid et al. (2008)
Canola	0.48	103.8	3.53	0.88–0.9	56	45	136	–3	–4	Winayanuwattikun et al. (2008), Leung et al. (2010), Patil and Deng (2009)
Corn	0.45	120.3	3.39	0.88–0.89	58–59	45				Winayanuwattikun et al. (2008), Leung et al. (2010), Patil and Deng (2009)
Palm	0.1–0.2	60.07	4.9–5.6	0.88	58.3–62	34	164	12	15	Singh and Singh (2009), Leung et al. (2010), Hameed et al. (2009)
Tallow	0.65	126	—	0.856	59	—	150	19	16	Leung et al. (2010), Bhatti et al. (2008)
Used frying oil	0.15–0.25	123–130	4	0.864–0.900	—	—	160–170	—	—	Leung and Guo (2006), Leung et al. (2010), Feleizardo et al. (2006)
Karanja	0.62	92	5.43	0.883	60–61	42	150	22	—	Sahoo and Das (2009), Patil and Deng (2009), Leung et al. (2010), Gopinath et al. (2009), Karmee and Chadha (2005)
Jatropha	0.26	108.4	4.78	0.864	61–63	39.4	160–162	0	–6	Winayanuwattikun et al. (2008), Leung et al. (2010), Patil and Deng (2009), El Diwani et al. (2009)
ASTM Specification 6751-11a	0.5	—	1.9–6.0	—	47 minimum	—	130°C minimum	To report	—	ASTM Specifications

Conclusions

Jatropha is one of the best renewable oilseed for biodiesel production due to its fatty acid composition and low content of unsaponifiable matter. However, the major drawback of this oil is its higher FFA content. Hence, a lot of care has to be taken during the harvesting and handling of Jatropha seeds. If Jatropha oil contains less than 2% FFAs, alkali-catalyzed transesterification is the most suitable process for the preparation of biodiesel. In case, FFA content is larger than 2%, a two step process consisting of esterification followed by transesterification is an appropriate choice for biodiesel production. Active research is going on to replace homogenous catalysts with heterogeneous catalysts (either chemical or enzymatic) for a greener process and a success in this direction may change the scenario of biodiesel industry. The specifications for biodiesel are designed to ensure that consumers will not experience operational problems from the fuel use. Hence, efforts should be made to optimize the process parameters of biodiesel to meet stringent global specifications. Quality fuel would provide the consumer with improved air quality and enhanced operability. Poor quality fuel would create operability problems and increased maintenance activity. Hence, it is really necessary to look for newer approaches for the production of biodiesel particularly for feedstocks with heterogeneous quality. In the present scenario, biodiesel industry must give more importance to by-product utilization from biodiesel to make the industry sustainable.

References

Abdu-Aguye I, Sannusi A, Alafiya-Tayo RA, Bhusnurmath SR (1986) Acute toxicity studies with *Jatropha curcas* L. Hum Exp Toxicol 5:269–274

Adam SEI (1974) Toxic effects of *Jatropha curcas* L. in mice. Toxicology 2:67–76

Adam SEI, Magzoub M (1975) Toxicity of *Jatropha curcas* L. for goats. Toxicology 4:388–389

Agarwal D, Agarwal AK (2007) Performance and emissions characteristics of Jatropha oil (preheated and blends) in a direct injection compression ignition engine. Appl Therm Eng 27:2314–2323

Ahmad M, Rashid S, Khan MA, Zafar M, Sultana S, Gulzar S (2009) Optimization of base catalyzed transesterification of peanut oil biodiesel. Afr J Biotechnol 8:441–446

Ahmed OMM, Adam SEI (1979a) Effect of *Jatropha curcas* L. on calves. Vet Pathol 16:476–482

Ahmed OMM, Adam SEI (1979b) Toxicity of *Jatropha curcas* in sheep and goats. Res Vet Sci 27:89–96

Akintayo ET (2004) Characteristics and composition of *Parkia biglobbossa* and *Jatropha carcus* oils and cake. Bioresour Technol 92:307–310

Akoh CC, Chang SW, Lee GC, Shaw JF (2007) Enzymatic approach to biodiesel production. J Agric Food Chem 55:8995–9005

Arumugham V (2009) Performance and emission characteristics of CI engine fuelled with non edible vegetable oil and diesel blends. J Environ Res Dev 4:212–221

Azzi A, Boscoboinik D, Hensey C (1992) The protein kinase C family. Eur J Biochem 208:547–557

Bak YC, Choi JH, Kim SB, Kang DW (1996) Production of biodiesel fuels by transesterification of rice bran oil. Korean J Chem Eng 13:242–245

Banapurmath NR, Tewari PG, Hosmath RS (2008) Performance and emission characteristics of a DI compression ignition engine operated on Honge, Jatropha and sesame oil methyl esters. Renew Energy 33:1982–1988

Baroi C, Yanful EK, Bergougnou MA (2009) Biodiesel production from *Jatropha curcas* oil using potassium carbonate as an unsupported catalyst. Can Int Chem Reactor Eng 7:72

Basiron Y (2005) Palm oil. In: Shahidi F (ed) Bailey's industrial oil and fat products, vol 2, 6th edn. Wiley, New Jersey, pp 333–429

Berchmans HJ, Shizuko H (2008) Biodiesel production from crude *Jatropha curcas* seed oil with a high content of free fatty acids. Bioresour Technol 99:1716–1721

Berchmans HJ, Morshita K, Takarada T (2010a) Kinetic study of hydroxide-catalyzed methanolysis of *Jatropha curcas*-waste food oil mixture for biodiesel production. Fuel. doi:10.1016/j.fuel.2010.01.017

Berchmans HJ, Morshita K, Takarada T (2010b) Kinetic study of methanolysis of *Jatropha curcas*-waste food oil mixture. J Chem Eng Jpn 43:661–670

Bhatti HN, Hanif MA, Qasim M, Ataur R (2008) Biodiesel production from waste tallow. Fuel 87:2961–2966

Boonmee K, Chuntranuluck S, Punsuvon V, Silayoi P (2010) Optimization of biodiesel production from Jatropha oil (*Jatropha curcas*) using response surface methodology. Kasetsart J Nat Sci 44:290–299

Bringi NV (ed) (1987) Non-traditional oilseeds and oils of India. Oxford & IBH, New Delhi, p 226

Bringi NV, Mukherjee SK (1987). In: Bringi NV (ed) Nontraditional oilseeds and oils of India, Oxford & IBH, New Delhi, 1987, p 147

Canakci M, Erdill A, Arcaklioglu E (2006) Performance and exhaust emissions of a biodiesel engine. Appl Energy 83:594–605

Carlson KF (1985) Degumming and neutralization methods. Paper presented at the 76th annual meeting of American oil chemists' society, Honolulu

Carlson KF, Scott JD (1991) Recent developments and trends: processing of oilseeds, fats and oils. Inform 2(12):1034–1060

Chakrabarti PP, Rao BVSK, Roy SK, Devi BLAP, Rani KNP, Vandana V, Kalyani C, Gadam K, Kale V, Prasad RBN (2009) Process for the pretreatment of vegetable oils for physical refining, US Patent No. 7494676

Chen C, Chen W, Ming CJ, Lai S, Tu C (2010) Biodiesel production from supercritical carbon dioxide extracted Jatropha oil using subcritical hydrolysis and supercritical methylation. J Supercrit Fluid 52:228–234

Chitra P, Venkatachalam P, Sampathrajan A (2005) Optimization of experimental procedure for biodiesel production from alkaline catalyzed transesterification of *Jatropha curcas* oil. Energ Sust Dev 9:13–18

Comprehensive Jatropha Report (2010) available at www.biozio.com

Conceicao MM, Candeia RA, Silva FC, Bezerra AF, Fernandes VJ Jr, Souza AG (2007) Thermoanalytical characterization of castor oil biodiesel. Renew Sust Energ Rev 11:964–975

Corro G, Tellez N, Ayala E, Mainez-Ayala A (2010) Two step biodiesel production from *Jatropha curcas* crude oil using SiO_2.HF solid catalyst for FFA esterification step. Fuel 89:2815–2821

Demirbas A (1998) Fuel properties and calculation of higher heating values of vegetable oils. Fuel 77:1117–1120

Demirel S, Lehnert K, Lucas M, Claus P (2007) Use of renewables for the production of chemicals: glycerol oxidation over carbon supported gold catalysts. Appl Catal B Environ 70:637–643

Devanesan MG, Viruthagiri T, Sugumar N (2007) Transesterification of *Jatropha* oil using immobilised *Pseudomonas fluoroscens*. Afr J Biotechnol 6:2497–2501

Dijkstra AJ, Van Opstal M (1985) Process for producing degummed vegetable oils and gums of high phosphatydic acid contents. US Patent 4,698,185

Dubois V, Breton S, Linder M, Fanni J, Parmentier M (2007) Fatty acid profiles of 80 vegetable oils with regard to their nutritional potential. Eur J Lipid Sci Technol 109:710–732

Eijck JV, Romjin H (2008) Prospect of *Jatropha* Biofuel in Tanzania: an analysis with strategic niche management. Energy Policy 36:311–325

El Diwani G, Attia NK, Hawash SI (2009) Development and evaluation of biodiesel fuel and by-products from Jatropha oil. Int J Environ Sci Tech 6:219–224

Feleizardo P, Correia MJN, Raposo I, Mendes JF, Berkemeier R, Bordado JM (2006) Production of biodiesel from waste frying oils. Waste Manag 26:487–494

Felke J (1913) Aber die giftstoffe der samen von *Jatropha curcas*. Land Vers Stat 82:427–463

Feng F, Myers DJ, Evengelista MPH, Miller KA, Johnson LA, Singh SK (2002) Quality of corn oil obtained by sequential extraction processing. Cereal Chem 79:707–709

Foidl N, Foidl G, Sanchez M, Mittelbach M, Hackel S (1996) *Jatropha curcas* L. as a source for the production of biofuel in Nicaragua. Bioresour Technol 58:77–82

Gan MY, Chen Q, Pan D, Wei S (2010) Study on the preparation of biodiesel oil from *Jatropha curcas* oil in subcritical methanol. Guangzhou Huagong 38:70–72

Gandhi VM, Cherian KM, Mulky MJ (1995) Toxicological studies on Ratanjyot oil. Food Chem Toxicol 33:39–42

Gao C, Ye Y, Zeng G, Li C, Ou L (2010) Method for improving oxidation stability of *Jatropha curcas* oil and mixture of *Jatropha curcas oil* and diesel fuel. Faming Zhuanli Shenqing 11 pp, Chinese Patent Application No. 2009-10094981

Goonasekera MM, Gunawardana VK, Jayasena K, Mohammad SG, Balasubramaniam S (1995) Pregnancy terminating effect of *Jatropha curcas* in rats. J Ethnopharmacol 47:117–123

Gopinath A, Puhan S, Nagarajan G (2009) Relating the cetane number of biodiesel fuels to their fatty acid composition: a critical study. J Automobile Eng 223:565–583

Grompone MA (2005) Sunflower oil. In: Shahidi F (ed) Bailey's industrial oil and fat products, vol 2, 6th edn. Wiley, Hoboken, p 163

Gubitz GM, Mitelbach M, Trabi M (1999) Exploitation of the tropical oilseed plant *Jatropha curcas* L. Bioresour Technol 67:73–82

Gui MM, Lee KT, Bhatia S (2008) Feasibility of edible oil vs non edible oils vs waste edible oils as biodiesel feedstock. Energy 33:1646–1653

Guo JB, Yang G, Peng QT, Ma JL (2008) Study on biodiesel preparation with *Jatropha curcas* oil as raw material. Kezaisheng Nengyuan 26:27–29

Haas MJ (2005) Animal fats. In: Shahidi F (ed) Bailey's industrial oil and fat products, vol 1, 6th edn. Wiley, Hoboken, p 163

Haas MJ, Foglia TA (2005) Alternate feedstocks and technologies for biodiesel production. In: Knothe G, Gerpen J, Krahl J (eds) The biodiesel handbook. AOCS Press, Champaign, pp 42–61

Haas W, Mittelbach M (2000) Detoxification experiments with the seed oil from *Jatropha curcas* L. Ind Crops Prod 12:111–118

Hameed BH, Lai LF, Chin LH (2009) Production of biodiesel from palm oil (*Elaeis guineensis)* using heterogeneous catalyst: an optimized process. Fuel Process Technol 90:606–610

Hammond EG, Johnson LA, SU C, Wang T, White PJ (2005) Soybean oil. In: Shahidi F (ed) Bailey's industrial oil and fat products, vol 2, 6th edn. Wiley, Hoboken, pp 577–653

Han X, Huang X, Ma X (2009) Production of biodiesel from rice bran oil in supercritical methanol. Taiyangneng Xuebao 30:677–681

Harrington KJ (1986) Chemical and physical properties of vegetable oil esters and their effect on diesel fuel performance. Biomass 9:1–7

Hawash S, Kamal N, Zaher F, Kenawi O, El Diwani G (2009) Biodiesel fuel from Jatropha oil via non-catalytic supercritical methanol transesterification. Fuel 88:579–582

Helwani Z, Orthman MR, Aziz N, Kim J, Fernando WJN (2009) Solid heterogeneous catalyst for transesterification of triglycerides with methanol: a review. Appl Catal A Gen 363:1–10

Henning R (2003) Jatropha Booklet, Bagani GbR, Weissenberg, Germany, pp 10–15

Hirota M, Suttajit M, Suguri H, Endo Y, Shudo K, Wongchai V (1988) A new tumor promoter from the seed oil of *Jatropha curcas* L. an intramolecular diester of 12-deoxy-16-hydroxy phorbol. Cancer Res 8:5800–5804

Horiuchi T, Fujiki H, Hirota M, Suttajit M, Suganuma M (1987) Presence of tumor promoters from the seed oil of *Jatropha curcas* L. from Thailand. Jpn J Cancer Res (Gann) 78:223–226

Hvolby A (1971) Removal of non-hydratable phospholipids from soybean oil. J Am Oil Chem Soc 48:503–509

Ibiari NN, El-Enin SAA, Attia NK, El-Diwani G (2010) Ultrasonic comparative assessment for biodiesel production from rapeseed. J Am Sci 6:937–943

Ilham Z, Saka S (2010) Two-step supercritical dimethyl carbonate methods for biodiesel production from *Jatropha curcas* oil. Bioresour Technol 101:2735–2740

Ito Y, Yanase S, Tokuda H, Krishishita M, Ohigashi H, Hirata M et al (1983) Epstein barr virus activation by tung oil. Extracts of *Aleurites fordii* and its diterpene ester 12-o-hexadecanoyl-16-hydroxphorbol-13-acetate. Cancer Lett 18:87–95

Ivanoiu IA, Rusnac LM (2009) Influence of reaction time and the catalyst amount on the properties of biodiesel from palm oil in comparison with biodiesel from soybean oil. Chem Bull "POLITECHNICA" Univ (Timisoara) 54:89–92

Iwayan RP (2008) Jatropha oil extractor equipment, ESD in TVET 2010 conference proceedings, held in Manila, Philippines, pp 238–243

Jain S, Sharma MP (2010a) Prospects of biodiesel from Jatropha in India: a review. Renew Sust Energ Rev 14:763–771

Jain S, Sharma MP (2010b) Kinetics of acid base catalyzed transesterification of *Jatropha curcas* oil. Bioresour Technol 101:7701–7706

Jain S, Sharma MP (2010c) Biodiesel production from Jatropha oil. Renew Sust Energ Rev 14:3140–3147

Jayed MH, Masjuki HH, Saidur R, Kalam MA, Jahirul MI (2009) Environmental aspects and challenges of oilseed produced biodiesel in Southeast Asia. Renew Sust Energy Rev 13:2452–2462

Joubert PH, Brown MM, Hay IT, Sebata DB (1984) Acute Poisoning with *Jatropha curcas* (purging nut tree) in children. S Afr Med J 65:729–730

Ju YH, Vali SR (2005) Rice bran oil as a potential resource for biodiesel: a review. J Sci Ind Res 64:866–882

Juan JC, Kartika DA, Wu TY, Hin TYY (2011) Biodiesel production from Jatropha oil by catalytic and non-catalytic approaches: an overview. Bioresour Technol 102:452–460

Kapilan N, Reddy RP (2008) Evaluation of methyl esters of mahua oil (*Madhuca indica*) as diesel fuel. J Am Oil Chem Soc 85:185–188

Karmee SK, Chadha A (2005) Preparation of biodiesel from crude oil of *Pongamia pinnata*. Bioresour Technol 96:1425–1429

Knothe G, Sharp CA, Ryan TW (2006) Exhaust emission of biodiesel, petrodiesel, neat methyl esters and alkanes in a new technology engine. Energy Fuels 20:403–408

Kock M (1978) German Patent DE 2,722,245

Kose O, Tutter M, Aksoy HA (2002) Immobilized *Candida antarctica* lipase-catalyzed alcoholysis of cotton seed oil in a solvent-free medium. Bioresour Technol 83:125–129

Kotlin D, Uziel Y, Schneidermann D, Kotzki S, Wolach B, Fainmesser P (2006) A case of *Jatropha multifida* poisoning resembling organophosphate intoxication. Clin Toxicol (Phila) 44:337–338

Kulkarni MG, Dalai AK (2006) Waste cooking oil—an economical source of biodiesel. Ind Eng Chem Resour 45:2901–2913

Kumar TA, Kumar A, Raheman H (2007) Biodiesel production from Jatropha oil (*J. curcas* L.) with high free fatty acids: an optimized process. Biomass Bioenergy 31:569–575

Kumar D, Kumar G, Poonam, Singh CP (2010) Ultrasonic-assisted transesterification of *Jatropha curcas* oil using solid catalyst, Na/SiO_2. Ultrason Sonochem 17:839–844

Kumari A, Mahapatra P, Garlapati VK, Banerjee R (2009) Enzymatic transesterification of Jatropha oil. Biotechnol Biofuels 2. doi:10.1186/1754-6834-2-1

Kywe TT, Oo MM (2009) Production of biodiesel from Jatropha oil (*Jatropha curcas*) in pilot plant. World Acad Sci Eng Technol 50:477–483

Leung DYC, Guo Y (2006) Transesterification of neat and used frying oils: optimization of biodiesel production. Fuel Process Technol 87:883–890

Leung DYC, Wu X, Leung MKH (2010) A review on biodiesel production using catalyzed transesterification. Appl Energy 87:1083–1095

Li C, Gao Y, Ou L, Hu Z (2006) Preparation of diesel fuel by blending jatropha oil with petroleum distillates, Faming Zhuanli Shenqing Gongkai Shuomingshu, 6 pp, Chinese Patent Application No. 2006–10010811

Liberalino AAA, Bambira EA, Moraes-Santos T, Viera EC (1988) *Jatropha curcas* L. seeds: chemical analysis and toxicity. Arquivos de Biologiae Technologia 31:539–550

Lim S, Hoong SS, Teong LK, Bhatia S (2010) Supercritical fluid reactive extraction of *Jatropha curcas* L. seeds with methanol: a novel biodiesel production method. Bioresour Technol 101:7169–7172

Lin J, Fang L (2008) Production of biodiesel from *Jatropha curcas* oil and its economic benefits. Huagong Jinzhan 27:1977–1981

Lin CY, Li RJ (2009) Fuel properties of biodiesel produced from the crude fish oil from the soapstock of marine fish. Fuel Process Technol 90:130–136

Liu J, Kong QY (2009) Preparation of biodiesel from *Jatropha curcas* L. oil by solid acid catalyst. Changsha Ligong Daxue Xuebao, Ziran Kexueban 6:92–96

Liu X, He H, Wang Y, Zhu S, Piao X (2008) Transesterification of soybean oil to biodiesel using CaO as solid base catalyst. Fuel 87:265–273

Lopez JM, Gomez A, Aparicio F, Javier Sanchez F (2009) Comparison of GHG emission from diesel, biodiesel and natural gas refuse trucks of the city of Madrid. Appl Energy 86:610–615

Lu H, Liu Y, Zhou H, Yang Y, Chen M, Liang B (2009) Production of biodiesel from *Jatropha curcas* L. oil. Comput Chem Eng 33:1091–1096

Ma F, Hanna MA (1999) Biodiesl production: a review. Bioresour Technol 70:1–15

Ma F, Clements LD, Hanna MA (1998) Biodiesel fuel from animal fat. Ancillary studies on transesterification from beef tallow. Ind Eng Chem Resour 37:3768–3771

Mahanta N, Gupta A, Khare SK (2008) Production of protease and lipase by solvent tolerant *Pseudomonus aeruginosa* PseA in solid state fermentation using *Jatropha curcas* seed cake as substrate. Bioresour Technol 99:1729–1735

Makkar HPS, Aderibigbe AO, Becker K (1998) Comparative evaluation of non-toxic and toxic varieties of *Jatropha curcas* for chemical composition, digestibility, protein degradability and toxic factors. Food Chem 62:207–215. *Pongamia pinnata* (karanja) oil for production of biodiesel. J Sci Ind Res 63:913–918

Mameesh MS, El-Hakim LM, Hassan A (1963) Reproductive failure in female rats fed the fruit of seeds of *Jatropha curcas*. Plant Med 11:98–102

Mampane KJ, Joubert PH, Hay IT (1987) *Jatropha curcas*: use as a traditional Tswana medicine and its role as a cause of acute poisoning. Phytother Res 1:50–51

Martinez JH, Siddhuraju P, Francis G, Davila GO, Becker K (2006) Chemical composition, toxic/antimetabolic constituents and effects of different treatments on their levels in four provenances of *Jatropha curcas* L. from Mexico. Food Chem 96:80–89

Min J, Li S, Hao J, Liu N (2010) Supercritical CO_2 extraction of Jatropha oil and solubility correlation. J Chem Eng Data 55:3755–3758

Modi MK, Reddy JRC, Rao BVSK, Prasad RBN (2006) Lipase-mediated transformation of vegetable oils into biodiesel using propan-2-ol as acyl acceptor. Biotechnol Lett 28:637–640

Modi MK, Reddy JRC, Rao BVSK, Prasad RBN (2007) Lipase-mediated conversion of vegetable oils into biodiesel using ethyl acetate as acyl acceptor. Bioresour Technol 98:1260–1264

Moreau RA (2005) Corn oil. In: Shahidi F (ed) Bailey's industrial oil and fat products, vol 2, 6th edn. Wiley, Hoboken, p 158

Mudge SM, Pereira G (1999) Stimulating the biodegradation of crude oil with biodiesel preliminary results. Spill Sci Technol Bull 5:353–355

Nakpong P, Wootthikanokkhan S (2010) Optimization of biodiesel production from *Jatropha curcas* oil via alkali catalyzed methanolysis. J Sust Energy Environ 1:105–109

Nath LK, Dutta SK (1991) Extraction and purification of curcin, a protease from the latex of *Jatropha Curcas* Linn. J Pharm Pharmacol 43:111–114

O'Brien RD, Jones LA, King CC, Wakelyn PJ, Wan PJ (2005) Cottonseed oil. In: Shahidi F (ed) Bailey's industrial oil and fat products, vol 2, 6th edn. Wiley, Hoboken, pp 173–279

Panigrahi S, Francis BJ, Cano LM, Burbage MB (1984) Toxicity of *Jatropha curcas* seeds from Mexico to rats and mice. Nutr Rep Int 29:1089–1098

Parawira W (2010) Biodiesel production from *Jatropha curcas*: a review. Sci Res Essays 5:1796–1808

Patil PD, Deng S (2009) Optimization of biodiesel production from edible and non-edible vegetable oils. Fuel 88:1302–1306

Prabhavathi Devi BLA, Gangadhar KN, Sai Prasad PS, Jagannadh B, Prasad RBN (2007) Glycerol-based carbon catalyst for the preparation of biodiesel. ChemSusChem 2:617–620

Przybylski R, Mag T, Eskin NAM, Mcdonald BE (2005) Corn oil. In: Shahidi F (ed) Bailey's industrial oil and fat products, vol 2, 6th edn. Wiley, Hoboken, pp 61–121

Rai DK, Lakhanpal P (2008) *Jatropha curcas* L. Poisoning in paediatric patients in Mauritius. Int J Paediatr Neonatal 8:1–6

Rashid U, Anwar F (2008) Production of biodiesel through optimized alkaline catalyzed transesterification of rapeseed oil. Fuel 87:265–273

Rashid U, Anwar F, Moser BR, Ashraf S (2008) Production of sunflower oil methyl ester by optimized alkali-catalyzed methanolysis. Biomass Bioenergy 32:1202–1205

Rathore V, Madras G (2007) Synthesis of biodiesel from edible and non-edible oils in supercritical alcohols and enzymatic synthesis in supercritical carbon dioxides. Fuel 86:2650–2659

Recep A, Selim C, Huseyin SY (2000) The potential of using vegetable oil fuels as fuel for diesel engines. Int J Energy Convers Manage 42:529–538

USEPA Report (2002) A comprehensive analysis of biodiesel impacts on exhaust emission. Draft technical report, USEPA

Ringers HJ, Segers JC (1976) German Patent 2,609,705

Rivera-Lorca JA, KU-Vera JC (1997) Rumen digestion of raw, roasted and boiled seeds of *Jatropha curcas* from Chiapas, Mexico. In: Gubitz GM, Mittelbach M, Trabi M (eds) Biofuels and industrial products from *Jatropha curcas* L. DBV Graz, Graz, pp 167–72

Robles-Medina A, Gonzales-Moreno PA, Esteban-Cerdan L, Molina-Grima E (2009) Biocatalysis: towards ever greener biodiesel production. Biotechnol Adv 27:398–408

Rotenberg SA, Calogeropoulou T, Jawrosky JS, Weinstein IB, Rideout D (1991) A self-assembling protein kinase C inhibitor. Proc Natl Acad Sci USA 88:2490–2494

Sahoo PK, Das LM (2009) Process optimization for biodiesel production from Jatropha, karanja and polonga oils. Fuel 88:1588–1594

Satyanarayana M, Muraleedharan C (2010) Methyl ester production from rubberseed oil using two-step pretreatment process. Int J Green Energ 7:84–90

Sauerwein M, Sporer F, Wink M (1993) Insect-toxicity of phorbol esters from *Jatropha curcas* seed oil. Plant Med Suppl 59:686

Sayyar S, Abidin ZZ, Younus R, Muhammad A (2009) Extraction of oil from Jatropha seeds—optimization and kinetics. Am J Appl Sci 6:1390–1395

Schmook B, Serralta-Peraza L (1997) *J. curcas*. Distribution and uses in the Yucatán Peninsula of Mexico. In: Gubitz GM, Mittelbach M, Trabi M (eds) Biofuels and industrial products from *Jatropha curcas* L. DVB Graz, Graz, pp 53–57

Segers JC (1982) Superdegumming, a new degumming process and its effect on the effluent problems of edible oil refining. Fette Seifen Anstrichm 88:543–546

Sendzikiene E, Kareviciene VM, Janulis P (2005) Oxidation stability of biodiesel fuel produced from fatty wastes. Polish J Environ Stud 14:335–336

Shah S, Sharma S, Gupta MN (2004) Biodiesel preparation by lipase-catalyzed transesterification of *Jatropha* oil. Energy Fuel 18:154–159

Shah S, Sharma A, Gupta MN (2005) Extraction of oil *Jatropha curcas* L. seed kernels by combination of ultrasonication and aqueous enzymatic oil extraction. Bioresour Technol 96:121–123

Singh RK, Padhi SK (2009) Characterization of Jatropha oil for the preparation of biodiesel. Nat Prod Radiance 8:127–132

Singh SP, Singh D (2009) Biodiesel production through the use of different sources and characterization of oils and their esters as the substitute of diesel: a review. Renew Sust Energ Rev 14:200–216

Speidel HK, Lightner RL, Ahmed I (2000) Biodegradability of new engineered fuels compared to conventional petroleum fuels and alternative fuels in current use. Appl Biochem Biotechnol 84–86:879–897

Srivastava A, Prasad R (2000) Triglyceride-based diesel fuel. Renew Sust Energ Rev 4:111–133

Staubamann R, Ncube I, Gubitz GM, Steiner W, Read JS (1999) Esterase and lipase activity in *Jatropha curcas* L. seeds. J Biotechnol 75:117–126

Stripe F, Pession Brizzi A, Lorezoni E, Strosshi P, Montanaro L, Sperti S (1976) Studies on the proteins from the seeds of *Crotontiglium* and of *Jatropha curcas* L. Biochem J 156:1–6

Su E, Wei D (2008) Improvement in lipase-catalysed methanolysis of triacylglycerols for biodiesel production using solvent engineering method. J Mol Catal B Enzyme 55:118–125

Sunil N, Varaprasad KS, Sivaraj N, Suresh Kumar T, Abraham B, Prasad RBN (2008) Assessing *Jatropha curcas* L. germplasm in-situ—a case study. Biomass Bioenergy 32:198–202

Syam AM, Yunus R, Ghazi TIM, Yaw TCS (2009) Methanolysis of *Jatropha* oil in the presence of potassium hydroxide catalyst. J Appl Sci 9:3161–3165

Tamalampudi S, Talukder MR, Hama S, Numata T, Kondo A, Fukuda H (2008) Enzymatic production of *Jatropha* oil: a comparative study of immobilized-whole cell and commercial lipases as a biocatalyst. Biochem Eng J 39:185–189

Tang Z, Wang L, Yang J (2007) Transesterification of the crude *Jatropha curcas* L. oil catalyzed by micro-NaOH in supercritical and subcritical methanol. Eur J Lipid Sci Technol 109:585–590

Tapnes NCO, Aranda DAG, Carneiro JWM, Antunes OAC (2008) Transesterification of *Jatropha curcas* oil glycerides: theoretical and experimental studies on biodiesel reaction. Fuel 87:2286–2295

Tiwari AK, Kumar A, Rehman H (2007) Biodiesel production from Jatropha oil with high free fatty acids: an optimized process. Biomass Bioenergy 31:569–575

Van de Sande RKLM, Segers JC (1989) Methods of refining of glyceride oils. European patent EP 0,348,004

Vyas AP, Subrahmanyam N, Patel PA (2009) Production of biodiesel through esterification of Jatropha oil using KNO_3/Al_2O_3 solid catalyst. Fuel 88:625–628

Wang ZM, Lee JS, Park JY, Wu CZ, Yuan ZH (2007) Novel biodiesel production technology from soybean soapstock. Korean J Chem Eng 24:1027–1030

Winayanuwattikun P, Kaewpiboon C, Piriyakananon K, Tangtong S, Thakernkernkrit W, Chulalaksananukul W et al (2008) Potential plant oil feedstock for lipase-catalyzed biodiesel production in Thailand. Biomass Bioenergy 32:1279–1286

Wink M, Koschmieder C, Sauerwein M, Sporer F (1997) Phorbol esters of *J. curcas*—biological activities and potential applications. In: Gubitz GM, Mittelbach M, Trabi M (eds) Biofuels and industrial products from *Jatropha curcas*. DVB Graz, Graz, pp 160–166

Yong KD, Gou K, Ming J, Zhang P, Lu Y (2009) Testing study on pretreatment of Jatropha oil and ester-exchange for biodiesel. Kezaisheng Nengyuan 27:28–32

Zhang F, Koseoglu SS, Rhee KC (1994) Effects of expander process on phospholipids in soybean oil. J Am Oil Chem Soc 71:1145–1148

Zhou ChH, Beltramini JN, Fan YX, Lu GQ (2008) Chemoselective catalytic conversion of glycerol as a biorenewable source to valuable commodity chemicals. Chem Soc Rev 37:527–549

Chapter 26
Performance, Emission and Combustion Characteristics of Preheated and Blended Jatropha Oil

Avinash Kumar Agarwal and Atul Dhar

Introduction

Fatty acids from vegetable oils can be described as similar in molecular structure, chain length and carbon to hydrogen ratio (C:H) *vis-a-vis* alkanes from conventional diesel (Srivastava and Prasad 2000; Crookes et al. 1997). However, they differ from the later because of presence of oxygen in their molecular structure. Vegetable oils have higher kinematic viscosity, density, and lower cetane number, lower stoichiometric mixture ratio and specific enthalpy of combustion compared to mineral diesel (Crookes et al. 1997). Vegetable oils can be used directly or blended with mineral diesel in order to be operated in compression ignition engines (Agarwal 2007). Even Rudolf diesel, inventor of CI engine, emphasized the use of vegetable oils to run CI engine during the 1900 world exhibition in Paris and demonstrated his invention by using peanut oil as a fuel (Bryant 1976). Various vegetable oils and their blends with mineral diesel have been tested worldwide, e.g., Jatropha oil (Agarwal and Agarwal 2007; Agarwal and Dhar 2009, 2010; Pramanik 2003; Gangwar and Agarwal 2008; Forson et al. 2004; Kumar et al. 2003), Karanja oil (Bajpai et al. 2009; Mahanta et al. 2006; Agarwal and Rajamanoharan 2009), rubber seed oil (Ramadhas et al. 2005), palm oil (Almeida et al. 2002) etc. Higher concentrations of vegetable oils have been used by preheating the fuel before injection into the cylinder (Agarwal and Agarwal 2007; Agarwal and Dhar 2010; Almeida et al. 2002; Bari et al. 2002; Nwafor 2003, 2004). Life cycle impact assessment study for comparing the effect of *straight vegetable oil* (SVO) and biodiesel on the environment has indicated SVO route as more environment friendly with emphasis on its localized production and utilization (Fore et al. 2011; Esteban et al. 2011). With this

A.K. Agarwal (✉) • A. Dhar
Engine Research Laboratory, Department of Mechanical Engineering,
Indian Institute of Technology, Kanpur 208016, India
e-mail: akag@iitk.ac.in; atuldhar@iitk.ac.in

N. Carels et al. (eds.), *Jatropha, Challenges for a New Energy Crop: Volume 1: Farming, Economics and Biofuel*, DOI 10.1007/978-1-4614-4806-8_26,

Table 26.1 Physico-chemical properties of mineral diesel and Jatropha oil (Agarwal and Dhar 2009).

Property	Diesel	Jatropha
Density (kg/m^3)	833.7	921.8
Kinematic viscosity @ 40°C (cSt)	2.71	34.33
Calorific value (MJ/kg)	43.06	41.85
Flash point (°C)	48	180
Carbon residue %, (w/w)	0.08	0.74
Ash content %, (w/w)	0.014	0.036
Carbon %, (w/w)	83.12	76.56
Hydrogen %, (w/w)	14.72	13.19
Nitrogen %, (w/w)	0.45	0.34
Copper corrosion grade	1a	1a

background, this chapter describes detailed performance, emission and combustion characteristics of preheated and unheated Jatropha oil blends in unmodified compression ignition engines.

Utilization of Jatropha oil with Blending

Viscosity of straight Jatropha oil at 40°C is 34.33 cSt, which is very high in comparison to conventional CI engine fuels (Table 26.1). So, straight Jatropha oil needs to be modified to bring its combustion properties closer to those of mineral diesel. The fuel modifications are mainly aimed at reducing the viscosity to eliminate problems related to its flow and atomization. Four techniques are proposed to reduce the viscosity of vegetable oils namely pre-heating/pyrolysis, dilution/blending, micro-emulsion and transesterification (Agarwal 2007). Viscosity of Jatropha oil decreases after blending it with mineral diesel. The viscosity of 20% blend of Jatropha oil with mineral diesel (J20) was reported to be 4.19 cSt (Agarwal and Agarwal 2007; Agarwal and Dhar 2009), which is within the prescribed ASTM limit for CI engine fuels (6 cSt at 40°C). This section describes the performance, emission and combustion characteristics of 10%, 20%, 50% and 100% Jatropha oil blends in a single cylinder direct injection compression ignition engine operating at a constant speed of 1,500 rpm without any engine hardware modification *vis-à-vis* mineral diesel.

Performance Characteristics

The performance characteristics of Jatropha oil blends compared to mineral diesel (Fig. 26.1) are evaluated using standard equipments and parameters (1) the *brake*

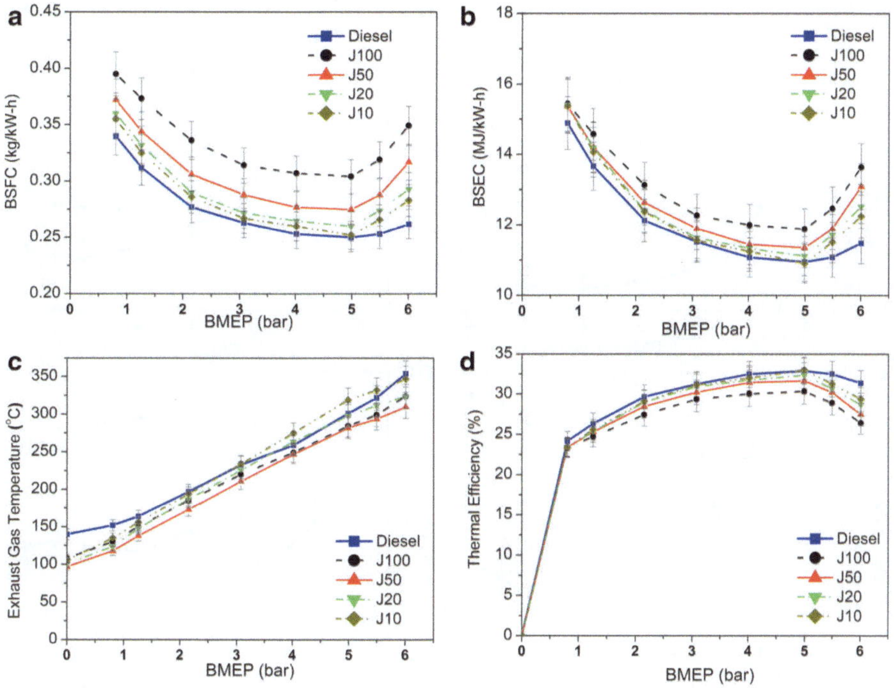

Fig. 26.1 Performance characteristics of Jatropha oil blends (**a**) BSFC, (**b**) BSEC, (**c**) exhaust gas temperature and (**d**) thermal efficiency (Agarwal and Dhar 2009)

mean effective pressure (BMEP) is a quantity related to the operation of a piston engine and is a valuable measure of an engine's capacity to do work independent of its size; (2) the *brake specific energy consumption* (BSEC) is the energy consumed per unit power produced by the engine; (3) the *brake specific fuel consumption* (BSFC) is a measure of the mass of fuel consumption for producing unit power and (4) the *thermal efficiency* measures the efficiency of the engine, i.e., the relation between the heat-content of a fuel that is consumed and the mechanical work output produced or in other words, the ratio of *brake power output to fuel energy input*.

BSFC was found to increase with higher proportion of Jatropha oil in the blend compared to diesel in the entire load range. Calorific value of Jatropha oil is relatively lower compared to that of diesel, therefore increasing proportion of Jatropha oil in blend decreases the calorific value of the blend, resulting in increased BSFC. Thermal efficiency of Jatropha blends is lower than that of mineral diesel. However, thermal efficiency of blends up to J20 is very close to mineral diesel. Oxygen present in the fuel molecules improves the reference of combustion characteristics, but higher viscosity and poor volatility of vegetable oils lead to poor atomization and inferior combustion characteristics. Therefore, thermal efficiency was found to be relatively lower for higher blends concentrations compared to that of mineral diesel. The exhaust gas temperature with blends having higher percentage of Jatropha oil was higher than mineral diesel. Exhaust gas temperature for lower blends was comparable to mineral diesel.

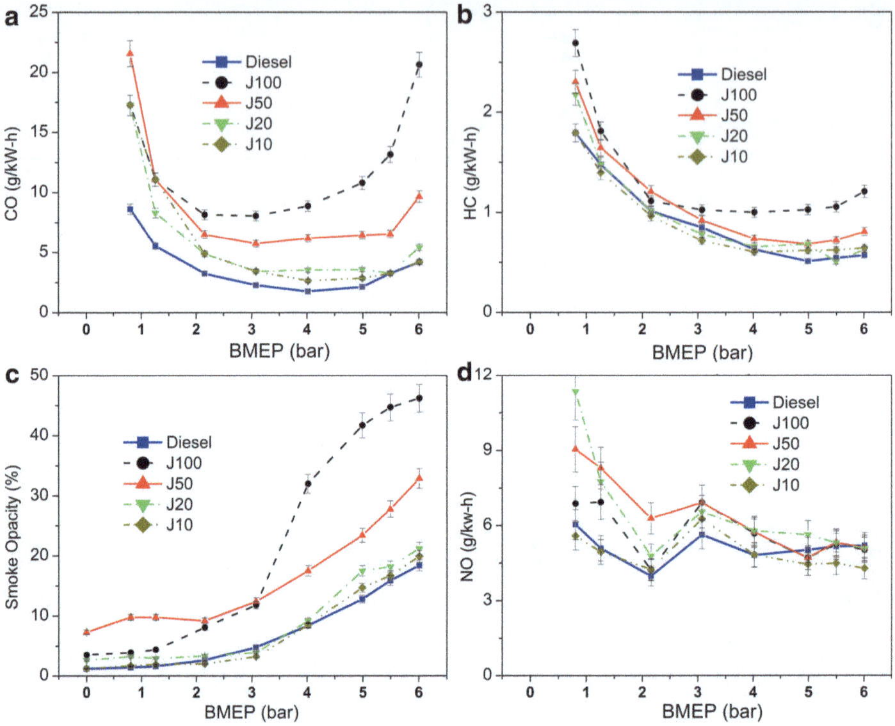

Fig. 26.2 Emission characteristics of Jatropha oil blends (**a**) carbon mono oxide emissions, (**b**) hydrocarbon emissions (**c**) smoke opacity and (**d**) nitrogen oxide emissions (Agarwal and Dhar 2009)

Emission Characteristics

According to the emission characteristics of Jatropha oil blends (Fig. 26.2), the brake specific carbon monoxide (CO) emission first decreases then increases with increasing engine load for all fuels (Fig. 26.2a). Higher the engine load, richer is the fuel-air mixture to be burned therefore more CO is produced due to relative lack of oxygen. For lower Jatropha oil blends (up to J20), CO emissions for Jatropha oil are close to mineral diesel, but higher blends exhibit significant increase in brake specific CO emissions at higher BMEP (above 3 bar). Jatropha oil blends exhibit higher *hydrocarbon* (HC) emissions compared to diesel (Fig. 26.2b). It can be observed that HC emissions increase with increasing proportion of Jatropha oil in blends. The smoke opacity increases with increase in Jatropha oil concentration in blends particularly at higher loads (Fig. 26.2c). Higher smoke opacity may be due to poor atomization of Jatropha oil. Bulky fuel molecules and higher viscosity of Jatropha oil results in poor atomization of fuel blends. Nitrogen oxide (NO) emission was the lowest for J10 and NO emissions for other Jatropha oil blends were higher than mineral diesel (Fig. 26.2d). High exhaust gas temperature and oxygen in the exhaust are responsible for higher NO formation with increase of Jatropha oil concentration in blends.

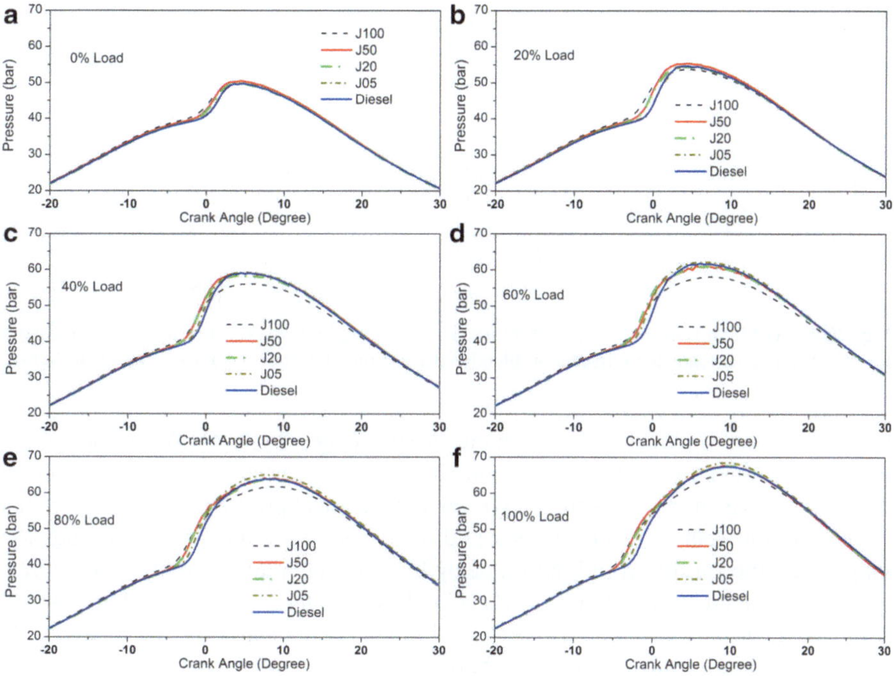

Fig. 26.3 Variation of in-cylinder pressure of Jatropha oil blends at (**a**) 0%, (**b**) 20%, (**c**) 40%, (**d**) 60%, (**e**) 80% and (**f**) 100% rated load (Agarwal and Dhar 2009)

Combustion Characteristics

The variations of in-cylinder pressure with crank angle for mineral diesel, 5%, 20%, 50% blends and 100% Jatropha oil at different engine operating conditions are shown in Fig. 26.3. The *crank angle* is measured in 'degree' and is used to report the piston position during the combustion process. Four strokes of engine operation are completed in −360° to 360° rotation of crank shaft. At −360°, piston is at its highest point, known as the *top dead center* (TDC) and suction stroke is completed from −360° to −180°. Inducted air is compressed during −180° to 0° rotation of crank shaft when piston moves from *bottom dead center* (BDC) to TDC. 0 to 180° rotation of crank shaft denotes expansion stroke, when piston moves from TDC to BDC. 180° to 360° rotation represents exhaust stroke when piston moves from BDC to TDC. At low engine loads, cylinder pressure trends are almost similar for different Jatropha oil blends. Jatropha oil blends are showing progressively earlier pressure rise with respect to mineral diesel for higher engine loads suggesting lower ignition delay for Jatropha oil blends. For low engine loads, vegetable oil blends show higher peak pressure, but at higher engine loads, diesel and low concentration blends give higher peak pressure compared to J100. For 80% (Fig. 26.3e) and 100% (Fig. 26.3f) load, 5% Jatropha blend shows slightly higher peak pressure compared to mineral diesel. At all engine loads, combustion starts earlier for Jatropha oil blends than mineral diesel and the rate of pressure rise is slower for vegetable oil

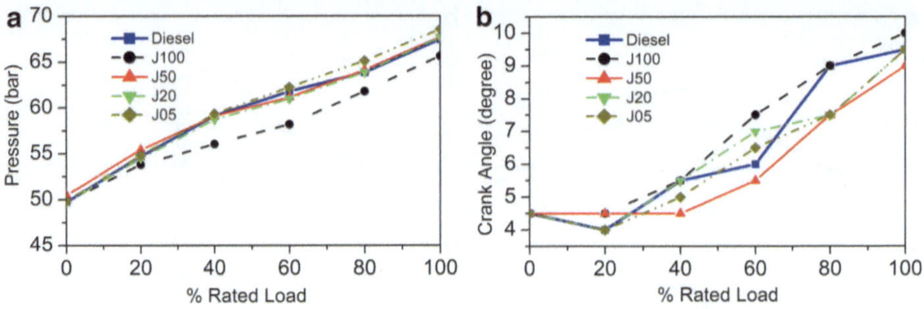

Fig. 26.4 (**a**) Maximum in-cylinder pressure (**b**) Crank angle of maximum in-cylinder pressure for different engine loads for Jatropha oil blends *vis-à-vis* mineral diesel (Agarwal and Dhar 2009)

blends because of slower burning characteristics (Figs. 26.3c–f). When the engine load is increased, the combustion point shifts to start earlier for all fuels. Ignition delay represents the time taken in physical and chemical pre-flame reactions. In this study, ignition delay was not measured directly; however, the start of combustion may reflect the variations in ignition delay because fuel pump and injector settings were kept identical for all test fuels. Combustion starts earlier for Jatropha oil (Fig. 26.3) partially owing to a shorter ignition delay and partially due to advanced injection timing because of a higher bulk modulus, i.e., lower compressibility and higher density of Jatropha oil. In spite of the higher viscosity and lower volatility of the Jatropha oil, its ignition delay seems to be lower than that of mineral diesel. This may possibly be because a complex and rapid pre-flame chemical reactions, which take place at high temperatures. Because of the high in-cylinder temperature existing during fuel injection, Jatropha oil may undergo thermal cracking resulting in lighter compounds that could ignite earlier. The combustion of Jatropha blends however seems to be slower essentially because of bulkier and complex fuel molecules of vegetable oils, which essentially take longer time for releasing the heat therefore leading to slower rate of heat release. This can be confirmed by the *rate of heat release* (ROHR) diagrams.

The maximum in-cylinder pressure varies according to load and blends (Fig. 26.4a). At higher engine loads, the peak pressure for mineral diesel is comparable to Jatropha oil blends except J100. In case of different Jatropha oil blends, the difference in peak pressure is not significant. For Jatropha oil blends, mineral diesel's combustion delay and volatility of Jatropha oil causes comparable peak pressures. The location of this peak pressure (Fig. 26.4b) is also comparable for all Jatropha oil blends with that of mineral diesel and this is within a narrow band of 3 crank angle degrees for all fuel blends under investigation. Maximum in-cylinder pressure is attained within 4–10 crank angle degrees after TDC for all fuel blends under different load conditions. At very low engine loads (particularly idling and 20% rated load), combustion starts later for mineral diesel compared to Jatropha oil blends because of the longer ignition delay. As evident from the pressure and crank angle diagrams at low-load condition for mineral diesel (Figs. 26.3a, b), combustion starts near TDC at 1,500 rpm. As a result, the peak cylinder pressure attains a

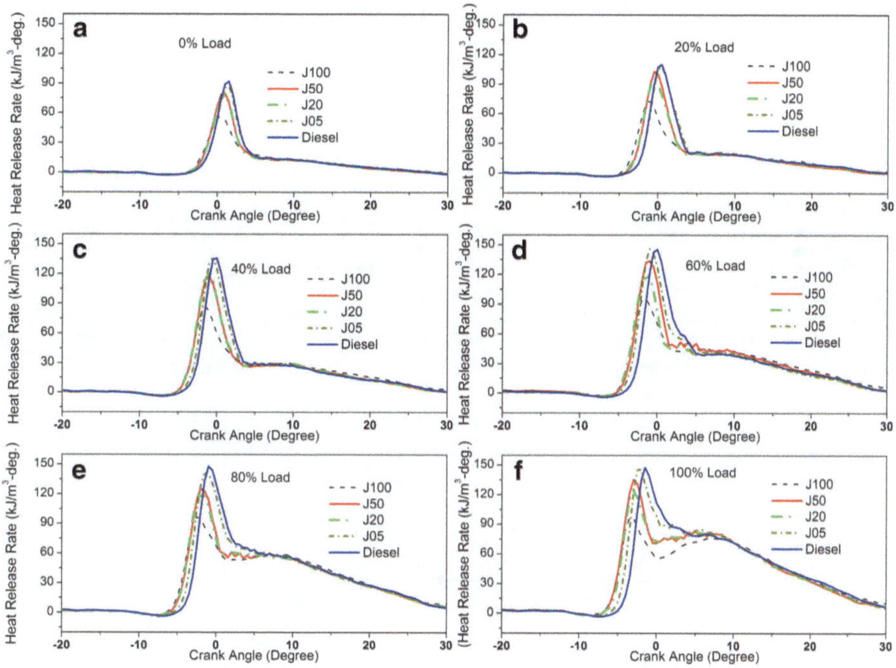

Fig. 26.5 Variation in heat release rate of Jatropha oil blends at (**a**) 0%, (**b**) 20%, (**c**) 40%, (**d**) 60%, (**e**) 80% and (**f**) 100% rated load (Agarwal and Dhar 2009)

lower value as it is further away from the TDC in the expansion stroke at low engine loads (Fig. 26.4b). Shorter ignition delay of Jatropha oil tends to advance the position of peak pressure while its low volatility and high viscosity retards the position of peak pressure. Since these two effects act in opposite direction, there is no directional trend of position of peak pressure with the variation in concentration of Jatropha oil in fuel blends.

The heat release rate of Jatropha oil blends also varies according to engine operating conditions (Fig. 26.5). All Jatropha blends experience identical combustion stages similar to mineral diesel, such as ignition delay, premixed combustion, mixing controlled combustion or diffusion combustion, and late combustion. Start of injection timing of engine was 26° BTDC. Because of the fuel vaporization in the beginning of the ignition delay, a negative heat release is observed around crank angle −7° and, after initiation of combustion, heat release becomes positive. After the ignition delay, premixed fuel-air mixture burns rapidly, followed by diffusion combustion with burn rate controlled by fuel-air mixing process. It can be observed that combustion starts earlier for Jatropha blends under all engine operating conditions. The premixed combustion heat release is always higher for mineral diesel owing to higher volatility and better mixing of diesel with air. Another reason may be the longer ignition delay of mineral diesel, which leads to a larger fuel accumulation in the combustion chamber at the time of the premixed combustion, leading to a higher rate of heat release. One can notice that at higher engine loads, the mixing

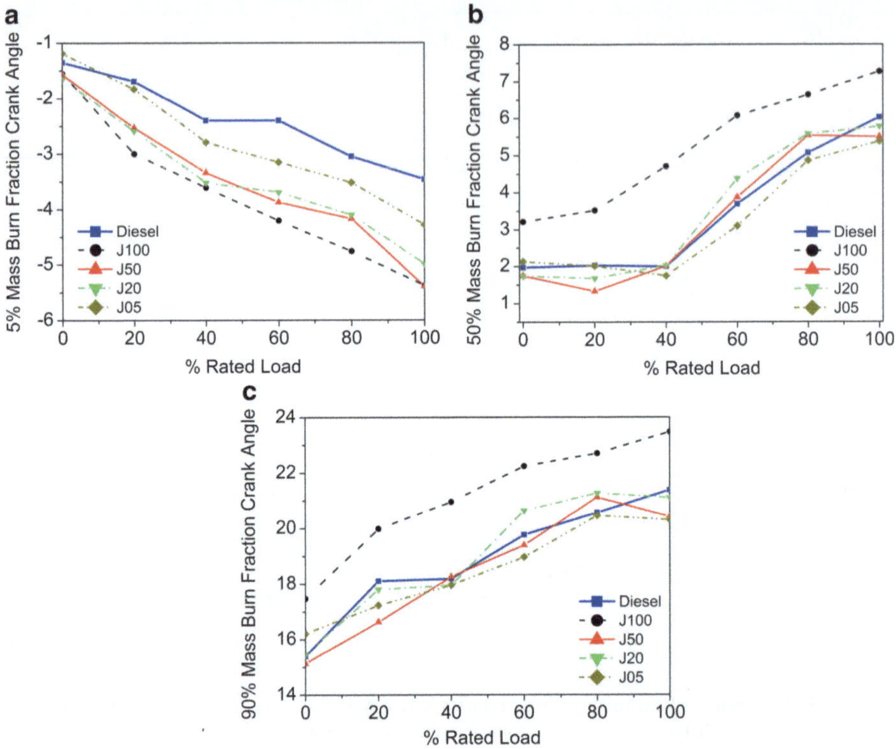

Fig. 26.6 Crank angle for (**a**) 5%, (**b**) 50% and (**c**) 90% mass fraction burned for unheated Jatropha oil and blends *vis-à-vis* mineral diesel (Agarwal and Dhar 2009)

controlled combustion is dominant (as observed by a second peak) for Jatropha oil blends (Figs. 26.5e, f). This is possibly due to longer combustion duration of larger fuel molecules in vegetable oils.

Considering crank angle for 5% mass fraction burned (Fig. 26.6a), one may observe that the first 5% fuel corresponding to Jatropha oil blends burn earlier than mineral diesel and it burns successively earlier for an increasing proportion of Jatropha oil in blends. This is due to the earlier start of combustion for Jatropha oil blends, as suggested above, i.e., progressively lower ignition delay for increasing blends of Jatropha oil compared to mineral diesel. As the engine load is increased, this deviation increases because, at higher loads, the start of combustion crank angle decreases. Fig. 26.6b shows that the crank angle degree for 50% mass fraction burned at different engine load conditions remains almost same for all blends and diesel except for J100, where slower burning of Jatropha is observed. Slower burning of J100 may be due to higher flash point and higher viscosity of Jatropha oil, which hinders atomization and vaporization of J100. Fig. 26.6c shows the crank angle degree for 90% mass fraction burned at different engine load conditions. The time of 90% mass fraction burning is comparable for Jatropha blends except J100 again because blends containing mineral diesel probably accelerate the combustion process. More fuel mass is required in case of Jatropha oil blends because of lower

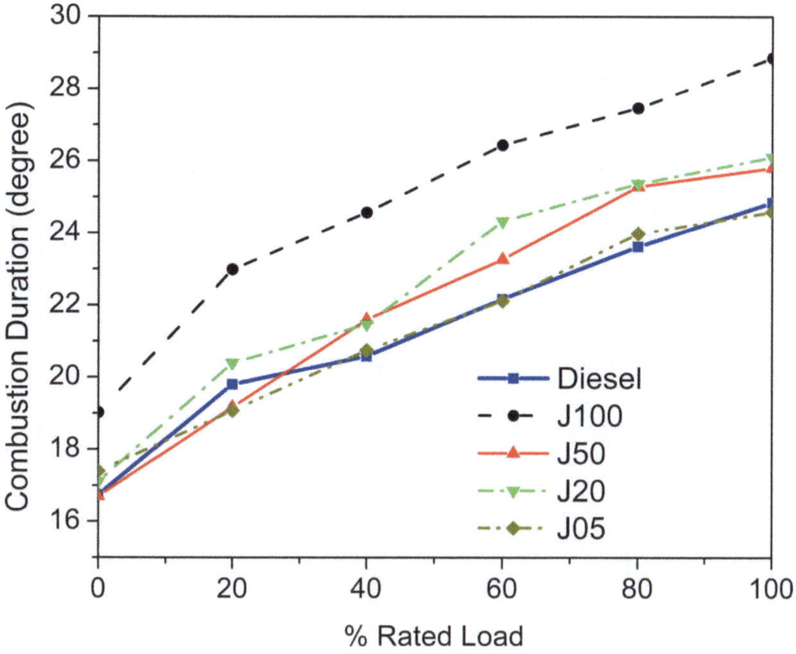

Fig. 26.7 Combustion duration for unheated Jatropha oil and blends *vis-à-vis* mineral diesel (Agarwal and Dhar 2009)

calorific value of these blends *vis-à-vis* mineral diesel. These factors lead to longer combustion duration for Jatropha oil blends compared to mineral diesel.

The variation in combustion duration for different blends at different engine loads is shown in Fig. 26.7. The *Crank angle degree* (CAD) from 5% mass burn to 90% mass burn has been taken as the reference of combustion duration for comparing different fuels. Combustion duration increases with increase in engine load owing to increase in the quantity of fuel injected. Combustion duration was observed to be longer for Jatropha oil blends than mineral diesel. The increase in combustion duration correlates well with increase in proportion of Jatropha oil in blends, which confirms slower combustion characteristics of Jatropha oil.

Utilization of Jatropha oil with Preheating

For investigating the performance, emission and combustion characteristics of preheated Jatropha oil, a shell and tube type heat exchanger was used to preheat the oil with waste heat of exhaust gases. Heat exchanger consists of an inner pipe and an outer shell. Fins were brazed to the inner pipe to increase the heat transfer area between vegetable oil and exhaust gases. One supply pipe connection is provided on each side of the heat exchanger for inlet and outlet of the vegetable oil. A thermocouple was provided in the heat exchanger to measure the temperature of the heated vegetable oil, close to the exit point. The temperature of Jatropha oil was

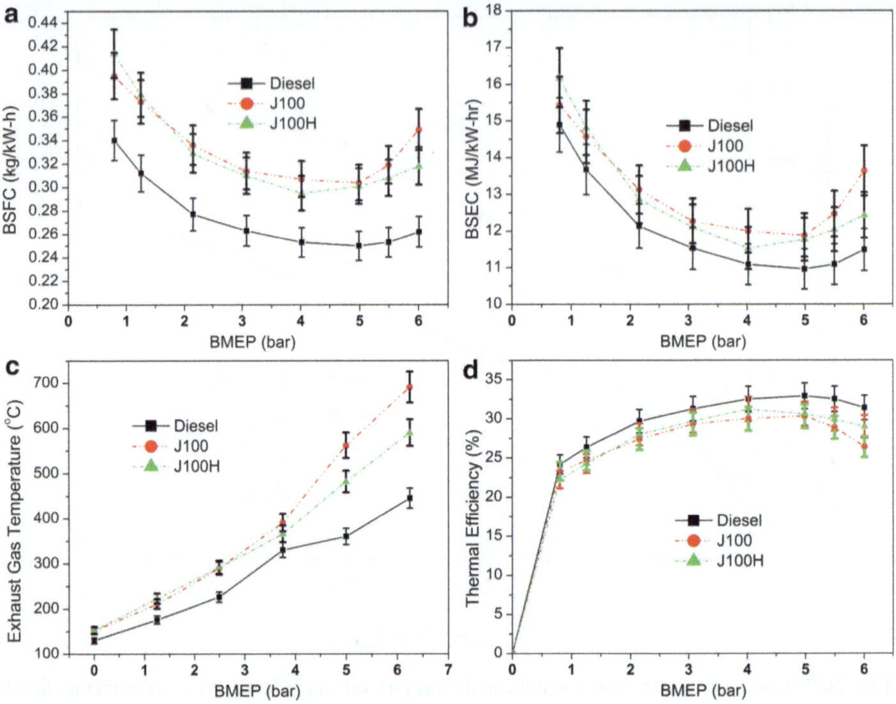

Fig. 26.8 Performance characteristics of preheated (J100H) and unheated Jatropha (J100) oil blends (**a**) BSFC, (**b**) BSEC, (**c**) exhaust gas temperature and (**d**) thermal efficiency (Agarwal and Dhar 2010)

maintained at 100°C at the exit of the heat exchanger. The flow of exhaust gas toward the heat exchanger was regulated by a bypass valve, which in turn regulated the temperature of oil at heat exchanger exit. All the performance, combustion and emission tests were carried out at optimum fuel injection pressure (200 bars) for minimum BSFC, highest thermal efficiency and lowest smoke opacity for this engine.

Performance Characteristics

In this section, we compare the performance characteristics of preheated Jatropha oil with unheated Jatropha oil and mineral diesel (Fig. 26.8). BSFC of preheated Jatropha oil was found to decrease with preheating in the entire load range compared to unheated Jatropha oil (Fig. 26.8a). But, BSFC (Fig. 26.8a) and BSEC (Fig. 26.8b) of preheated Jatropha oil is higher than mineral diesel. Lower calorific value of Jatropha oil and poor atomization due to high viscosity are responsible for higher BSFC of Jatropha oil. The exhaust gas temperature increases with preheating

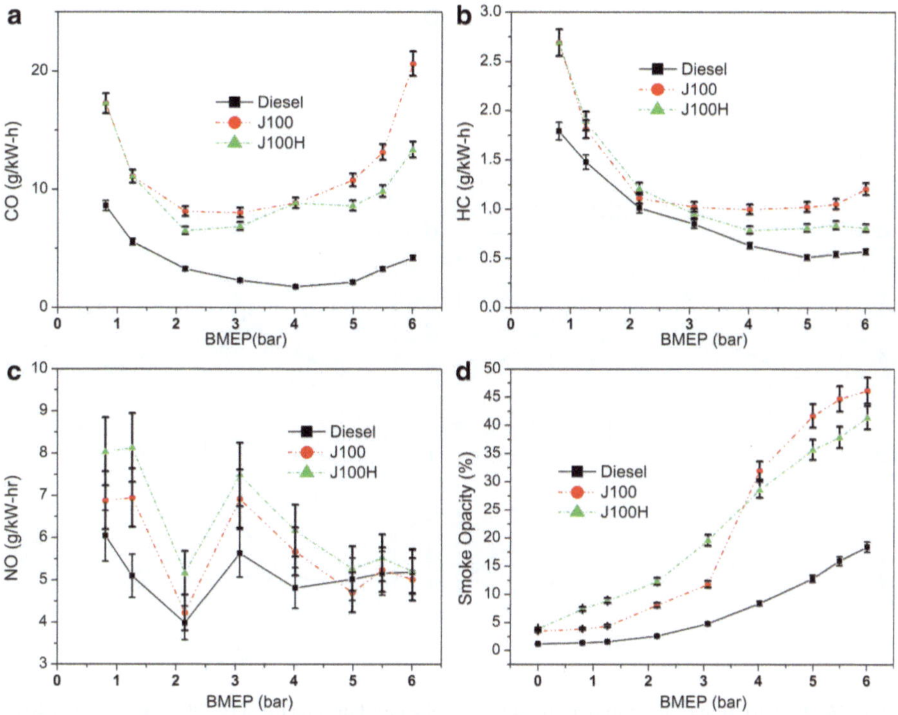

Fig. 26.9 Emission characteristics of preheated and unheated Jatropha oil blends (**a**) carbon mono oxide, (**b**) hydrocarbons, (**c**) nitrogen oxide and (**d**) smoke emissions (Agarwal and Dhar 2010)

of Jatropha oil (Fig. 26.8c). After preheating the vegetable oil, viscosity reduces hence superior and atomization into smaller droplets results in better air-fuel mixing as well as improved thermal efficiency (Fig. 26.8d).

Emission Characteristics

Carbon monoxide (CO) emissions were observed to increase with increasing engine load (Fig. 26.9a). At higher loads, richer fuel to air mixture is burned thus more CO is produced due to relative lack of oxygen. CO emissions for both heated and unheated Jatropha oil were higher than mineral diesel. Heating the oil results in reduction of CO emissions (Fig. 26.9a). Both heated and unheated Jatropha oils show higher HC emissions compared to mineral diesel. Preheating Jatropha oil results in reduction of HC emissions (Fig. 26.9b). Higher exhaust gas temperature and oxygen concentration results in an increased NO emission for preheated Jatropha oil (Fig. 26.9c). The smoke opacity for unheated as well as preheated

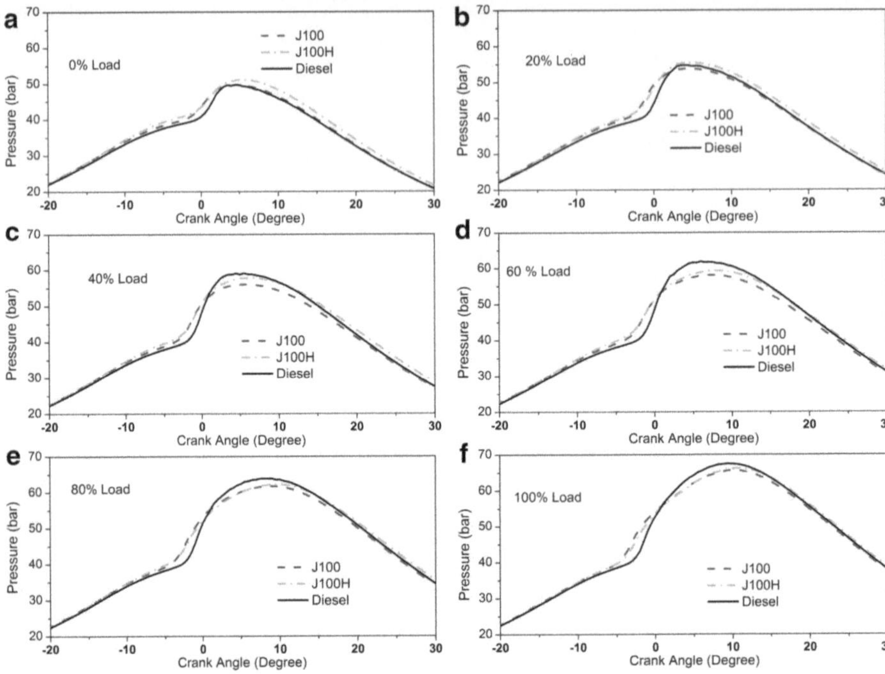

Fig. 26.10 Variation of in-cylinder pressure of preheated Jatropha oil at (**a**) 0%, (**b**) 20%, (**c**) 40%, (**d**) 60%, (**e**) 80% and (**f**) 100% rated load (Agarwal and Dhar 2010)

Jatropha oils was higher than that of mineral diesel, but was lower for preheated Jatropha oil compared to unheated oil (Fig. 26.9d).

Combustion Characteristics

The variations of in-cylinder pressure with crank angle for heated and unheated Jatropha oil at different engine operating conditions are shown at Figs. 26.10a–f. From this figure, it can be noticed that at low engine loads, pressure trends are almost overlapping for the three fuels. Heated and unheated Jatropha oil show earlier pressure rise *vis-à-vis* mineral diesel. Jatropha oil shows higher peak pressure at lower engine load (Figs. 26.10a, b) and relatively lower peak in-cylinder pressure for higher engine load in comparison to mineral diesel (Figs. 26.10c–f). Peak pressure of heated Jatropha oil is always higher than unheated Jatropha oil at all engine load conditions suggesting that the heat release is actually affected in a positive manner by preheating Jatropha oil because preheating reduces viscosity, which resulting in better air-fuel mixing. At all engine loads, combustion starts earlier for Jatropha oil (heated and unheated) *vis-à-vis* mineral diesel. Ignition delay for all fuels decreases

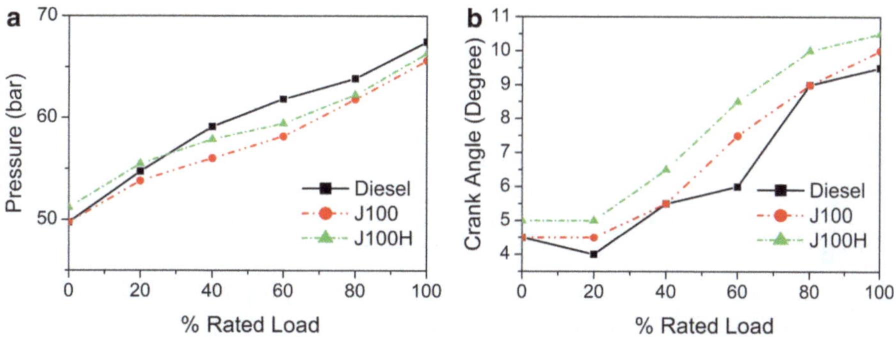

Fig. 26.11 Variation of (**a**) maximum in-cylinder pressure and (**b**) crank angle position of maximum in-cylinder pressure with engine load (Agarwal and Dhar 2010)

as the engine load increases because the temperature inside the cylinder increases due to heat transfer in different modes at high engine loads, which reduces the physical delay part of the ignition delay. The start of combustion reflects variation in ignition delay because fuel pump and injector settings were not changed for any fuel. Combustion starts earlier for heated Jatropha oil (Fig. 26.10a–f) due to reduction in fuel viscosity. In spite of slightly higher viscosity and lower volatility of the unheated Jatropha oil, its ignition delay seems to be lower than that of mineral diesel perhaps because complex and rapid pre-flame chemical reactions taking place at higher temperatures. As a result of the high in-cylinder temperature existing during fuel injection, Jatropha oil may undergo thermal cracking; as a result, lighter fractions are produced, which ignite earlier, resulting in a shorter ignition delay. Comparison with unheated oil shows that preheating of Jatropha oil enhances this effect further.

The peak pressure for mineral diesel is similar to that of heated and unheated Jatropha oils (Fig. 26.11a) for all loads. The crank angle, at which the peak cylinder pressure is attained for all fuels at different engine operating conditions, indicates that as load increases, peak in-pressure shifts away from TDC (Fig. 26.11b). Preheating of Jatropha oil leads to delay in achieving peak pressure and also peak pressure is attained later in the expansion stroke.

The heat release rate for heated and unheated Jatropha oil was compared to mineral diesel at different engine operating conditions (Fig. 26.12). Because of the vaporization of the fuel accumulated at the beginning of ignition delay, a negative heat release is observed; afterwards combustion is initiated and heat release becomes positive. Jatropha oils exhibit identical combustion stages as mineral diesel. After the ignition delay, premixed fuel-air mixture burns rapidly, followed by combustion diffusion, where the burn rate is controlled by fuel-air mixing. It can be observed that combustion starts earlier for unheated and heated Jatropha oil under all engine operating conditions *vis-à-vis* mineral diesel. The premixed combustion heat release is higher for mineral diesel in spite of a later start of combustion, owing to higher volatility and better mixing of mineral diesel with air. Another reason for larger premixed combustion heat release may be longer ignition delay of mineral

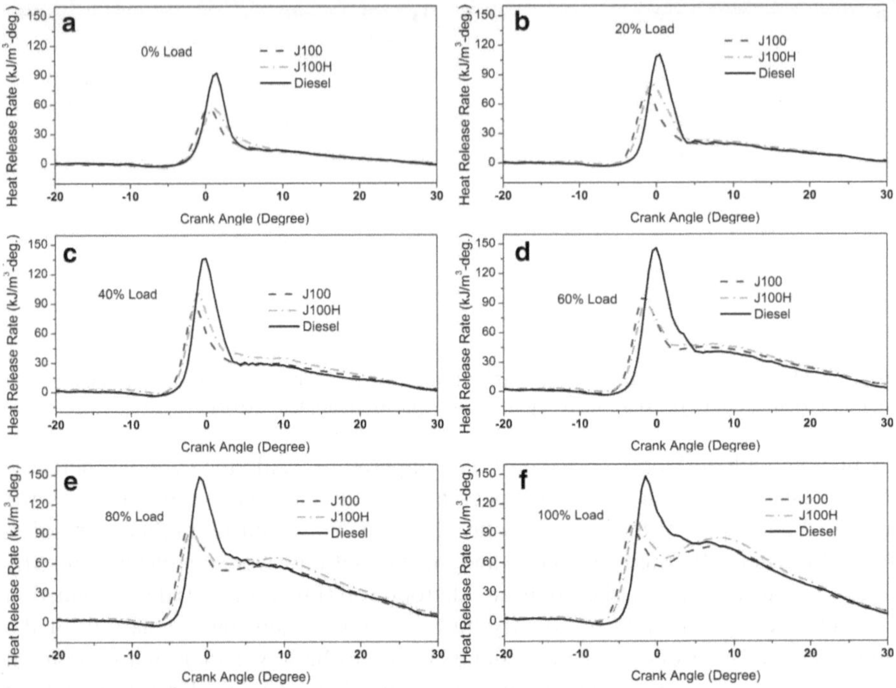

Fig. 26.12 Variation of heat release in preheated Jatropha oil combustion at (**a**) 0%, (**b**) 20%, (**c**) 40%, (**d**) 60%, (**e**) 80% and (**f**) 100% rated load (Agarwal and Dhar 2010)

diesel, which leads to a larger amount of fuel accumulation in the combustion chamber at the time of premixed combustion stage, leading to a higher rate of heat release. One can clearly observe that preheating of Jatropha oil actually increases the rate of heat release, which is quite a desirable effect expected from preheating of vegetable oils.

The analysis of crank angle of 5% mass fraction burned shows a faster combustion for heated Jatropha oil (Fig. 26.13a). Even unheated Jatropha oil takes lesser time than mineral diesel for 5% mass burning. This is due to an earlier start of combustion for Jatropha oil, as suggested above, because of volatile compounds present in Jatropha due to preheating of fuel. In relation to the crank angle degree for 50% mass fraction burned at different engine operating conditions, both unheated and heated Jatropha oils take more time for 50% burning than mineral diesel (Fig. 26.13b). The crank angle degree for 90% mass fraction burned at different engine load conditions (Fig. 26.13c) show similar trend as that of 50% mass fraction burn. More fuel mass is required in case of Jatropha oil because of its low calorific value compared to mineral diesel and this may possibly be one of the major factor contributing to longer combustion duration observed for Jatropha oil. The variation in combustion duration of heated and unheated Jatropha oil *vis-à-vis* mineral diesel at different engine loads can be expressed as the crank angle duration from 5% to 90% mass

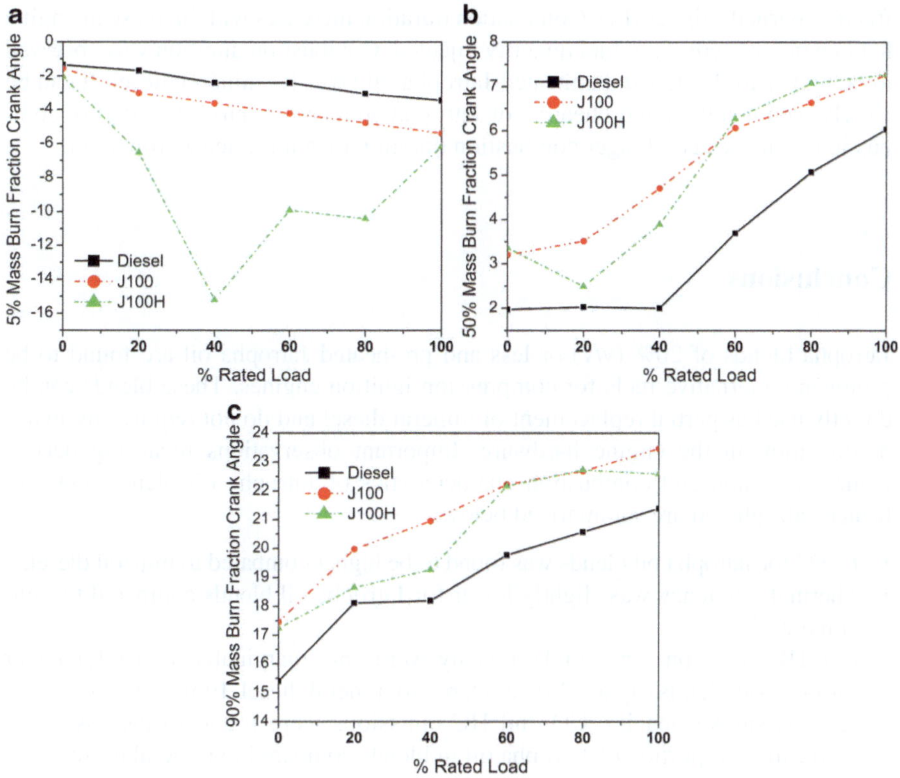

Fig. 26.13 Crank angle for (**a**) 5%, (**b**) 50% and (**c**) 90% mass fraction burned for preheated and unheated Jatropha oil *vis-à-vis* mineral diesel (Agarwal and Dhar 2010)

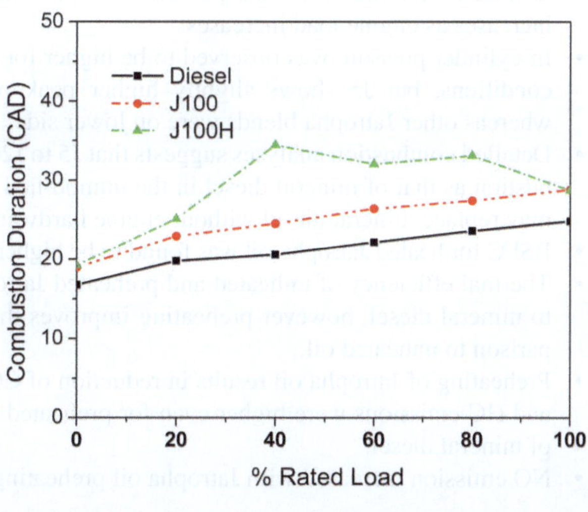

Fig. 26.14 Combustion duration for preheated and unheated Jatropha oil *vis-à-vis* mineral diesel (Agarwal and Dhar 2010)

fraction burned (Fig. 26.14). Combustion duration increases with increase in engine load owing to increase in fuel quantity injected. Combustion duration was observed to be higher for heated and unheated Jatropha oil *vis-a-vis* mineral diesel. It can be clearly seen that though preheating of Jatropha oil results in an earlier start of combustion, it also causes longer combustion duration than unheated Jatropha oil.

Conclusions

Jatropha blends of 20% (v/v) or less and pre-heated Jatropha oil are found to be promising alternative fuels for compression ignition engines. These blends can be directly used as partial replacement of mineral diesel and do not require any major modification in the engine hardware. Important observations regarding performance, emission and combustion characteristics of Jatropha oil blends and pre-heated Jatropha oil are summarized below:

- BSFC for Jatropha oil blends was found to be higher compared to mineral diesel.
- Thermal efficiency was slightly lower for Jatropha oil blends compared to mineral diesel.
- CO, HC emissions and smoke opacity were only marginally higher for lower Jatropha oil blends (up to J20) compared to mineral diesel. Emission parameters such as smoke opacity, CO and HC emissions were found to increase with increasing proportion of Jatropha oil in blends compared to mineral diesel.
- NO emission was lowest for 10% Jatropha blend and it was even lower than mineral diesel. Other blends exhibit relatively higher NO emission in comparison to mineral diesel.
- Combustion phases are very similar for Jatropha oil blends (lower) and mineral diesel.
- Combustion duration for Jatropha oil blends is longer than mineral diesel and it increases as engine load increases.
- In-cylinder pressure was observed to be higher for mineral diesel under all load conditions, but J5 shows slightly higher peak pressure than mineral diesel whereas other Jatropha blends were on lower side.
- Detailed combustion analyses suggests that J5 to J20 gave exactly identical combustion as that of mineral diesel in the unmodified engine and that these blends may replace mineral diesel without engine hardware modification.
- BSFC for heated Jatropha oil was found to be higher compared to mineral diesel.
- Thermal efficiency of unheated and preheated Jatropha oil was lower compared to mineral diesel, however preheating improves the thermal efficiency in comparison to unheated oil.
- Preheating of Jatropha oil results in reduction of CO and HC emissions, but CO and HC emissions were higher even for preheated Jatropha oil compared to hat of mineral diesel.
- NO emission increased with Jatropha oil preheating.

- Heated Jatropha oil shows shorter combustion delay however slower heat release rate than unheated Jatropha oil.
- Combustion duration for both heated and unheated Jatropha oils is higher than that of mineral diesel, but pre heating was found to increase the combustion duration of Jatropha oil.
- Maximum in-cylinder pressure was obtained for mineral diesel at higher load, but at lower load, higher peak in-cylinder pressure was obtained with heated Jatropha oil.
- Detailed combustion analyses suggests that heated Jatropha oil gives identical combustion to mineral diesel. Hence, waste heat of exhaust gas can be effectively utilized to preheat the Jatropha oil for improving its combustion performance in a diesel engine.

Acknowledgement Generous financial support from Technology Systems Group, Department of Science and Technology, Government of India for carrying out this research is gratefully acknowledged. Authors are also thankful to ASTM international and SAE International for permitting the reproduction of data from the copyrighted sources.

References

Agarwal AK (2007) Biofuels (alcohols and biodiesel) applications as fuels for internal combustion engines. Prog Energy Combustion Sci 33:233–271

Agarwal D, Agarwal AK (2007) Performance and emissions characteristics of Jatropha oil (preheated and blends) in a direct injection compression ignition engine. Appl Thermal Eng 27:2314–2323

Agarwal AK, Dhar A (2009) Performance, emission and combustion characteristics of Jatropha oil blends in a direct injection engine. SAE 2009-01-0947, SAE Special Publication SP-2239

Agarwal AK, Dhar A (2010) Experimental investigation of preheated Jatropha oil fuelled direct injection compression ignition engine: Part-I, performance, emission and combustion characteristics. J ASTM Int 7:1–13

Agarwal AK, Rajamanoharan K (2009) Experimental investigations of performance and emissions of Karanja oil and its blends in a single cylinder agricultural diesel engine. Appl Energy 86:106–112

Almeida SCAD, Belchior CR, Nascimento MVG, Vieira LDSR, Fleury G (2002) Performance of a diesel generator fuelled with palm oil. Fuel 81:2097–2102

Bajpai S, Sahoo PK, Das LM (2009) Feasibility of blending Karanja vegetable oil in petro-diesel and utilization in a direct injection diesel engine. Fuel 88:705–711

Bari S, Lim TH, Yu CW (2002) Effects of preheating of crude palm oil (CPO) on injection system, performance and emission of a diesel engine. Renew Energy 27:339–351

Bryant L (1976) The development of the diesel engine. Technol Cult 17:432–446

Crookes RJ, Kiannejad F, Nazha MAA (1997) Systematic assessment of combustion characteristics of bio-fuels and emulsions with water for use as diesel engine fuels. Energy Convers Manag 38:1785–1795

Esteban B, Baquero G, Puig R, Riba JR, Rius A (2011) Is it environmentally advantageous to use vegetable oil directly as biofuel instead or converting it to biodiesel? Biomass Bioenergy 35:1317–1328

Fore SR, Lazarus W, Porter P, Jordan N (2011) Economics of small-scale on-farm use of canola and soybean for biodiesel and straight vegetable oil biofuels. Biomass Bioenergy 35:193–202

Forson FK, Oduro EK, Donkoh EH (2004) Performance of Jatropha oil blends in a diesel engine. Renew Energy 29:1135–1145
Gangwar HK, Agarwal AK (2008) Combustion characteristics of Jatropha oil blends in a transportation engine. SAE Paper No. 2008-01-1383
Kumar MS, Ramesh A, Nagalingam B (2003) An experimental comparison of methods to use methanol and Jatropha oil in a compression ignition engine. Biomass Bioenergy 25:309–318
Mahanta P, Mishra SC, Kushwah YS (2006) An experimental study of *Pongamia pinnata* L. oil as a diesel substitute. J Power Energy Proc I Mech E: Part A 220:803–808
Nwafor OMI (2003) The effect of elevated fuel inlet temperature on performance of diesel engine running on neat vegetable oil at constant speed conditions. Renew Energy 28:171–181
Nwafor OMI (2004) Emission characteristics of diesel engine running on vegetable oil with elevated fuel inlet temperature. Biomass Bioenergy 27:507–511
Pramanik K (2003) Properties and use of *Jatropha curcas* oil and diesel fuel blends in compression ignition engine. Renew Energy 28:239–248
Ramadhas AS, Jayaraj S, Muraleedharan C (2005) Characterization and effect of using rubber seed oil as fuel in the compression ignition engines. Renew Energy 30:795–803
Srivastava A, Prasad R (2000) Triglycerides-based diesel fuels. Renew Sust Energy Rev 4:111–133

Chapter 27
Jatropha Oil Transesterification and Byproducts

Rosenira Serpa da Cruz, Ivon Pinheiro Lôbo, José Faustino Souza de Carvalho Filho, Rafael Costa Amaral, and Felipe Oliveira Souza

Biodiesel

The first energy source used by humanity was biomass, which has been in use for thousands of years. Biomass energy has been gradually substituted after the discovery of huge quantities of cheap fossil fuels, which proved to be more convenient for daily use and transport purposes (Thornley 2006). The use of biomass energy on a large scale has been considered a promising option, which could contribute to sustainable development in the environmental, social and economic sectors (Trevisani et al. 2007; Serrano-Ruiz et al. 2010).

In terms of environmental characteristics, biomass energy does not contribute significantly to global warming because of its almost completely closed carbon cycle. In other words, the greater part of *carbon dioxide* (CO_2) released by biomass combustion is reabsorbed on agricultural areas where it is incorporated into new biomass. In social matters, the demand of workforce by cultivation of biomass for energy purposes could be an excellent opportunity for work diversification and income creation in rural areas and become an efficient instrument to avoid rural depopulation. From an economic point of view, the biomass use for energy production means reduction of the dependency on importation of crude oil. The price of crude oil has been rising continuously forcing the society as well as industry and trade sectors to seek for the diversification of energy sources (Serrano-Ruiz et al. 2010).

Biodiesel is an example of exploration of biomass energy and appeared as a substitute for mineral diesel at a moment of large oscillations of crude oil price on the world market. The search for a substitute of fossil diesel is valid because of its strategic importance for the world economy. The fact that large part of the world diesel production is used for transportation and agricultural sectors has motivated many

R.S. da Cruz (✉) • I.P. Lôbo • J.F.S. de Carvalho Filho • R.C. Amaral • F.O. Souza
Departamento de Ciências Exatas e Tecnológicas, Universidade Estadual de Santa Cruz Road, Ilhéus/Itabuna km 16 CEP 45 662-900, Ilhéus, Bahia, Brazil
e-mail: roserpa@uesc.br

N. Carels et al. (eds.), *Jatropha, Challenges for a New Energy Crop: Volume 1: Farming, Economics and Biofuel*, DOI 10.1007/978-1-4614-4806-8_27,

countries to search for alternative solutions to achieve auto-sufficiency in matter of energy.

Biodiesel is compatible with conventional diesel fuel and both fuels can be blended in different proportions. Biodiesel brings some benefits compared to conventional diesel fuel mainly because it is renewable (Atadashi et al. 2010). Emission of *carbon monoxide* (CO), *sulfur oxides* (SO_x) and *particulate matters* (PM) can be lower when biodiesel is blended to diesel. The *polycyclic aromatic hydrocarbons* (PAH) rate of diesel is around 35% while biodiesel does not emit PAHs. Therefore, PAH emissions are reduced by 2.7%, 6.3% and 17.2% for B2 (blend of 2% of biodiesel and 98% diesel), B5 (blend of 5% biodiesel and 95% conventional diesel fuel) and B20 (blend of 20% biodiesel and 80% conventional diesel fuel), respectively, compared to pure diesel (Corrêa and Arbilla 2006). The reduction of PAH emissions by diesel-biodiesel blends is important since PAHs are carcinogenic to humans.

Jatropha curcas as a Feedstock for Biodiesel Production

Various oilseed species have been selected worldwide as feedstock for biodiesel production. Soybean, canola, corn, sunflower and palm oil are the most used oilseeds for biodiesel production. Other oilseeds, such as coconut and *J. curcas* (hereafter referred to as Jatropha) have been studied as potential feedstock for biodiesel production. The choice for a specific oilseed is based on considerations related to climate and soil conditions, productivity and oil quality. Another important aspect to be considered is the non-competition with food production of the non-edible oils, such as Jatropha, used frying oils, and animal tallow, in biodiesel production. Since plant oils account for about 70% to 95% of total biodiesel production costs, the use of non-edible oils and fats, such as Jatropha and waste cooking oils can significantly reduce production costs (Kiros et al. 2011; Aroua et al. 2011).

Considering the oil productivity, Jatropha is one of the most efficient oil seeds, only surpassed by microlagae, coconut, and oil palm (Hoekman et al. 2012). Jatropha oil is rich in mono-unsaturated fatty acids, which lead to very good biodiesel qualities (Vaknin et al. 2011; Wang et al. 2011). Neither Jatropha oil nor the press cake of its seeds are fit for human or animal consumption because of the presence of toxic components. In addition, Jatropha may be grown on marginal land not suitable for food crops and therefore does not compete with food production (Rodríguez et al. 2011). The press cake resulting from oil extraction has a high protein content and therefore can be used as organic fertilizer or as an ingredient of animal feeds after elimination of toxic compounds (Vaknin et al. 2011). There is a need for further investigations on selective breeding, agroclimatic zoning, farming and harvesting technologies that could promote the expansion of Jatropha as an economically viable and efficient feedstock for biodiesel production without prejudice to food security.

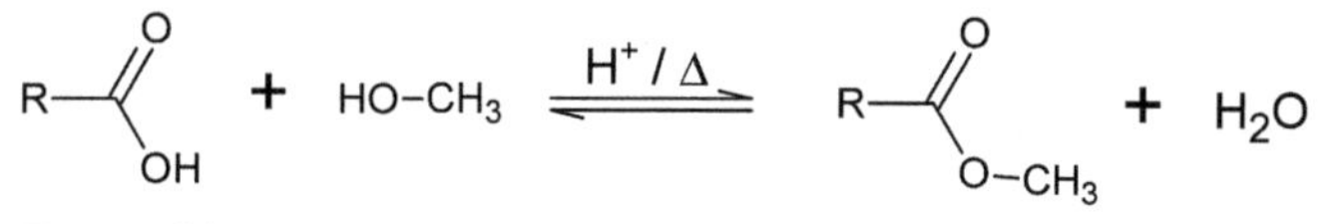

Fig. 27.1 Esterification reaction: The carbon chain of the fatty acid is represented by R

Biodiesel production is conventionally obtained by triglycerides transesterification in presence of a short chained alcohol and an alkaline or acid catalyst. Jatropha oil is prone to deterioration caused by improper handling and storage. Improper handling and exposure of the oil to atmospheric air and sunlight promotes the increase of *free fatty acids* (FFA) concentrations, which in turn make the transesterification through alkaline reaction impossible. Actually, FFAs deactivate the catalyst and lead to saponification, i.e., the production of larger quantities of soap, which hamper the separation of the glycerin and reduce the yield of methyl esters.

Many oil pretreatments have been proposed to reduce the adverse effects of FFAs on biodiesel production, including steam distillation, extraction by alcohol and esterification by acid catalysts. However, the esterification of FFAs with methanol in the presence of acidic catalysts is the most commonly applied method because of the simplicity of the process. In the first step, oil FFAs are converted into esters (esterification, Fig. 27.1) in the presence of an acid catalyst. In a second step, the remaining triacylglycerides are transesterified via conventional alkaline catalysis (Berchmans and Hirata 2008; Koh and Ghazi 2011).

Another important requirement of the oil is a low water content because the presence of water can induce hydrolysis of monoesters as well as of alkoxides (catalyst) and reduce reaction yield (Schuchardt et al. 1998). Other factors of the neat Jatropha oil, such as higher acidity, higher humidity and gums (phospholipids) may affect the biodiesel quality by deactivating the alkaline catalyst with consequent reduction of ester yields and biodiesel quality. Degumming with phosphoric acid minimizes the problems related to saponification and humidity and reduces significantly the amount of phospholipids in the oil (Syam et al. 2009; Berchmans and Hirata 2008).

Composition of Alkyl Esters and Biodiesel Quality

Biodiesel quality is strictly related to the efficiency of the production process, the efficiency of storage and also to feedstock quality. Therefore, standards have been established (Table 27.1) to ensure the (1) quality of the emissions caused by combustion, (2) performance and integrity of motors and (3) security during transport and handling of biodiesel. These standards also allow the monitoring of possible product degradation during storage.

Table 27.1 Quality Standards for biodiesel

Feature	Unit	EN 14214	ASTM D6751
Specific mass[a]	kg/m3	860–900 at 15°C	—
Kinematic Viscosity at 40°C[a]	mm^2/s	3.5–5.0	1.9–6.0
Water and sediments, max.[a]	% volume	—	0.05
Flash point, min.[a]	°C	120	130
Distillation; 90% vol. recuperated, max.[a]	°C	—	360
Carbon residue, max.[a]	%m	In 10% of distillation residues 0.3	In 100% of the sample 0.05
Sulfated ash, max.[a]	%m	0.02	—
Total sulfur, max.[a]	mg kg^{-1}	10	15
Copper strip corrosion 3 h a 50°C, max.[a]	—	1	3
Cetane number[a]	—	51 (min.)	47 (min.)
Cold filter plugging point, max.[a]	°C	by region	—
Pour point - PP[a]	°C	by region	—
Cloud point - CP[a]	°C	—	Register
Sodium and potassium, max.	mg kg^{-1}	5	—
Calcium and magnesium, max.	mg kg^{-1}	5	—
Phosphorus, max.	mg kg^{-1}	10	10
Total contamination, max.	mg kg^{-1}	24	—
Ester content, min.	%m	96.5	—
Acid Number, max.	KOH mg g^{-1}	0.5	0.5
Free glycerol, max.	%m	0.02	0.02
Total glycerol, max.	%m	0.25	0.24
Monoglycerides	%m	0.8 (max)	—
Diglycerides	%m	0.2 (max)	—
Triglycerides	%m	0.2 (max)	—
Methanol or Ethanol, max.	%m	0.2	—
Iodine index	g I_2 per 100 g	120 (max)	—
Oxidation Stability at 110°C, min	H	6	—
Water, max.	mg kg^{-1}	500	500
Linoleic acid	%m	12 max	—
Methyl esters with more than 4 unsaturations	%m	1 max	—

[a]Typical parameters for mineral diesel standards

The quality parameters are used as a system to score possible feedstock contaminants, such as phosphorus, sulfur, calcium and magnesium, besides the efficiency of the production process through the monitoring of viscosity, free glycerol content, FFA, soap, residual alcohol, catalyst residues, and water. During storage, biodiesel humidity may be absorbed and oxidizing processes may occur, which result in alterations of water content, viscosity, acidity as well as peroxide accumulation. The confidence interval of all these modifications is also patronized by quality standards. At the moment, the American quality standards of the *American Society of Testing and Materials* (ASTM D6751) and the standards of the *European Committee of Normalization* (Comité Européen de Normalisation—CEN), (EN 14214) appear to be the best known and are used as references for other standards. Among the

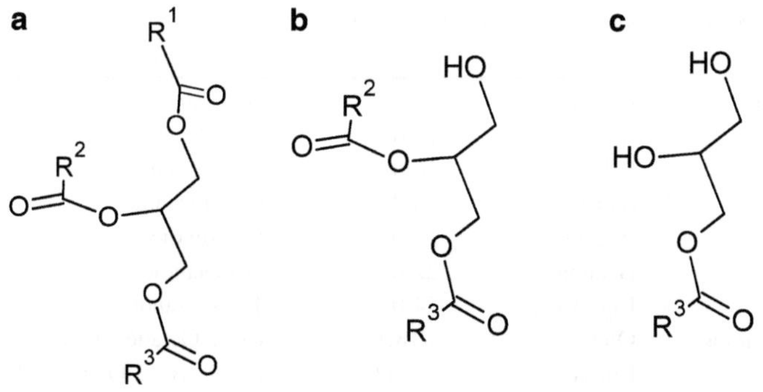

Fig. 27.2 Molecular structures of acylglycerols: (**a**) triacylglycerol, (**b**) diacylglycerol, (**c**) mono acylglycerol

parameters specified in those standards appear the ones derived from the standardization of mineral diesel and those originating from the analyses of plant oils that are commonly applied in the oleo-chemical industry. These parameters intend to evaluate biodiesel quality in terms of (1) possible feedstock contamination, (2) the process efficiency, (3) molecular structure of alkylic esters and (4) evolution of biodiesel quality during storage.

The characterization of feedstock in relation to fatty acid content is very important because of its influence on some biodiesel properties, such as kinematic viscosity, specific mass, cold filter plugging point, oxidation stability and combustion efficiency (Lôbo et al. 2009). Fatty acids are present in oils and fats in form of acylglycerols (Fig. 27.2), which are esters of fatty acids with glycerin. The triacylglycerols (Fig. 27.2a) make up 98% (w/w) of fats and oils (Ong et al. 2011). Fatty acids can be encountered in very small proportions in form of FFA as mono- (Fig. 27.2c) and diacylglycerols (Fig. 27.2b).

In Fig. 27.2, the parameters R_1, R_2, and R_3 represent the carbon chains of fatty acids. The most common fatty acids have between 12 and 22 carbon atoms, counted from the carboxyl carbon (Table 27.2).

After the transesterification reaction, biodiesel contains different quantities of alkyl esters, whose proportions should correspond to the ones present in their respective precursors, i.e., the feedstock fatty acids. The molecular structures of esters in biodiesel can vary according to the size of carbon chains as well as the quantity and position of unsaturations. Table 27.3 shows the composition of different plant oils, in relation to fatty acid content. Fatty acid chains with 18 carbons are the most frequent, whereas C16:0 is predominant in palm oil and beef tallow. Other plant oils may present a totally different composition, such as babassu oil, where short chained fatty acids are predominant, and olive oil, which mainly consists of C18:1.

The percentage of unsaturated fatty acids is an important parameter for the identification of different feedstocks. Olive oil, Jatropha oil, and rapeseed oil differ from other unsaturated oils by their high content of mono-unsaturated fatty acids. However, olive oil is edible and therefore cannot be considered as a feedstock for

Table 27.2 Most frequent fatty acids in oils and fats

	Fatty acid	Structure	Formal name
Saturated acids	Lauric	12:0	Dodecanoic
	Myristic	14:0	Tetradecanoic
	Palmitic	16:0	Hexadecanoic
	Stearic	18:0	Octadecanoic
	Arachidic	20:0	Eicosenoic
	Behenic	22:0	Docosanoic
	Lignoceric	24:0	Tetracosanoic
Unsaturated acids	Oleic	C18:1	*Cis*-9-Octadecenoic
	Linoleic	C18:2	*Cis*-9-*cis*-12-Octadecadienoic
	Linolenic	C18:3	*Cis*-9-*cis*-12-*cis*-15-Octadecatrienoic
	Gadoleic	C20:1	11-Eicosenoic
	Erucic	C22:1	*Cis*-13-Docosenoic

biodiesel. This oil has been included in Table 27.3 only for the purpose of comparison. Polyunsaturated fatty acids contribute to the formation of carbon deposits in the burning chamber of motors because of their tendency to polymerize and hence, mono-unsaturated fatty acids are preferable. It is important to note here that the proportions of fatty acids in oil can vary according to different climatic and soil conditions where the same oilseed species is grown (Nzikou et al. 2009; Rodríguez et al. 2011). Biodiesel with different physical and chemical properties are produced from different oilseeds according to the fatty acid composition of their oil (Table 27.4).

Kinematic Viscosity and Specific Mass

The formation of carbon deposits in the combustion chamber is also associated to the kinematic viscosity of biodiesel. High fuel viscosity causes combustion heterogeneity due to lower atomization efficiency, which leads to such residue deposition. Biodiesel has a significant higher viscosity compared to fossil diesel. The kinematic viscosity is a property brought about by the molecular structure of the alkyl esters that compose a biodiesel; it has the tendency to increase with the length of carbon chains and to decrease with the rate of unsaturation. The addition of a second double bound reduces significantly the viscosity of a fatty acid, whereas the position of the unsaturations has little effect (Knothe and Steidley 2005). Some other factors can also contribute to the change of biodiesel viscosity. The presence of soap residues as well as native glycerides (mono-, di-, and triglycerides) that did not react during the (trans) esterification process cause an increase in viscosity. Products resulting from oxidative degradation also contribute to viscosity increase. Therefore, triglyceride content and viscosity are important parameters for the monitoring of production process and storage of biodiesel. The viscosity of Jatropha biodiesel is similar to that of biodiesels from other predominately unsaturated oils (Table 27.4) and is within the recommendations of North American and European standards. In contrast,

Table 27.3 Percentage of the principal fatty acids in different oils and fats[a]

Oil	C8:0	C10:0	C12:0	C14:0	C16:0	C16:1	C18:0	C18:1	C18:2	C18:3	Saturated acid	Unsaturated acid
Corn	—	—	—	—	6.0	—	2.0	44.0	48.0	—	8.0	92.0
Cottonseed	—	—	—	—	28.3	—	0.9	13.3	57.5	—	29.2	70.8
Palm	0.8	0.5	0.3	1.1	42.5	0.2	4.2	41.3	9.5	0.3	49.4	51.3
Rapeseed	—	—	—	—	3.5	0.1	0.9	54.1	22.3		4.4	76.5
Soybean	—	—	—	—	11.9	0.3	4.1	23.2	54.2	6.3	16.0	84.0
Sunflower	—	—	—	—	6.4	0.1	2.9	17.7	72.9	—	9.3	90.7
Tallow	—	—	—	—	29.0	—	24.5	44.5	—	—	53.5	44.5
Olive	—	—	—	—	11.6	1.0	3.1	75.0	7.8	0.6	14.7	84.4
Babassu	5.0	6.0	44.0	17.0	8.0	—	4.5	14.0	2.0	—	84.5	16.0
J. curcas	—	—	—	—	14.2	0.7	7.0	44.7	32.8	0.2	21.2	78.4

[a]References: Balat (2011), Ramos et al. (2009), Lima et al. (2007), Hoekman et al. (2012).

Table 27.4 Properties of biodiesel of different feedstocks. U.S. and European standardization[a]

Property	Olive	Rapeseed	*J. curcas*	Palm	Soybean	Sunflower	Corn	Babassu	Tallow	EM 14214	ASTM D6751-08
Kinematic viscosity 40°C (mm^2/s)	4.5	4.2–4.6	4.3–4.8	4.5–5.7	4.2–4.6	4.2–4.6	4.2	3.6	4.7–5.1	3.5–5.0	1.9–6.0
Flash point (°C)	178	170	147–172	163–183	111–178	175–183	170–171	127	124	101 (min.)	93 (min.)
Cetane number	57	51–60	52–59	52–79	49–51	49–51	53–56	63	59	51 (min.)	47 (min.)
Oxidative stability 110°C (h)	3.3	2.0–6.3	3.7–8.0	4.0	1.3–1.7	0.8	1.2	>20	0.44–1.2	6 (min.)	3 (min.)
Cold filter plug point (CFPP, °C)	–6	10–12	8	10–9	(–4)–(–5)	(–3)–(2)	(–12)–(–3)	–4	(13)	Country specific	—
CP (°C)	—	—	6	15	(–2)–1	1	—	4	—	—	Report
Iodine value (g I_2/100 g)	84	109–116	62–110	51–57	125–128	129–132	101–101	—	52–66	120 (34ab.)	—
Densidade 15°C, kg/m^3	—	879–888	862–884	873–888	879–885	860–878	883	875	878–867	860–900	—

[a]References: Mittelbach and Schober (2003), Lebedevas and Vaicekauskas (2006), Becker and Makkar (2008), Ramos et al. (2009), Singh and Singh (2010), Nakpong and Wootthikanokkhan (2010), Parawira (2010), Hayyan et al. (2010), Wang et al. (2011), Jain and Sharma (2011), Balat (2011), Ong et al. (2011), Santos et al. (2011), Hoekman et al. (2012)

the viscosity of palm oil and tallow biodiesels is higher than the limits recommended by these standards. Babassu biodiesel, which is characterized by a high degree of saturation, has a low viscosity because of its high content of short chained esters. The specific mass of biodiesel is also directly connected to the structure of its molecules. The longer the carbon chains of alkyl-esters, the larger the specific mass. On the other hand, specific mass decreases as the rate of unsaturation in molecules increases. The presence of impurities, such as alcohol or triglycerides in biodiesel, can also affect specific mass.

The biodiesel specific mass also affects its volumetric energy content. Compared to mineral diesel, biodiesel contains 11% oxygen in esters of fatty acids, which brings a decrease of about 10% calorific energy, but on the other hand, this difference is reduced to about 5% due to the higher density of biodiesel (Hoekman et al. 2012).

Ignition Quality

The ignition quality of biodiesel is mainly determined by the cetane number of alkyl esters. Similar to the octane number, the cetane number indicates the ignition delay of a fuel in a motor engine. The higher the cetane number, the shorter is the ignition time. The cetane number increases with the length and saturation of unbranched carbon chains (Vaknin et al. 2011). Therefore, a tendency towards higher cetane numbers is observed in palm oil biodiesel, whereas in babassu biodiesel, this value is relatively high due to its saturated character. The cetane number in Jatropha biodiesel does not differ significantly from other biodiesel fuels. In the scale of cetanes, hexadecane (cetane) is defined as standard with a value of 100, whereas for 2,2,4,4,6,8,8,—heptamethylnonane, this value is defined as 15. Cetane tests are conducted in four-stroke cycle engines with single cylinder and ignition by compression that is designed to test fuel samples for diesel motors. Compared to mineral diesel, biodiesel has larger cetane numbers.

Crystallization Behavior

Fatty acid composition of oils and fats is of great importance for use of biodiesel in regions with cold climates. Actually, biodiesel can become solid at low temperatures, cause interruptions of fuel flow, clogging of fuel filters and hamper engine starting. The tendency of a biodiesel to solidify increases with the size of its fatty acid chains and their level of saturation. This tendency can be measured by three laboratory tests:

- *Cloud point* (CP) that determines the fuel temperature at which first crystals start to form (ASTM D2500);
- *Cold-filter plugging point* (CFPP) that determines the fuel temperature at which a fuel looses its ability to be filtered. The testing methods are EM 116, which are similar to the American low temperature flow test (LTFT, ASTM D 4539);

- *Pour point* (PP) that determines the temperature where a fuel looses its fluidity (EM 3016).

Considering a fuel sample, CP is always associated to a higher temperature than CFPP and PP, which makes it a more restrictive parameter. The tendency of a fuel to solidify at low temperatures decreases with increasing Polyunsaturation, as shown in Table 27.3. Jatropha biodiesel has a CP of 6°C. In B20, the CP is significantly reduced (Singh and Singh 2010). Even though the polyunsaturated character favors the reduction of CP and CFPP, unsaturation also turns biodiesel more susceptible to degradation by oxidization. Jatropha and rapeseed biodiesels are more stable compared to other biodiesel fuels with higher levels of polyunsaturation.

Oxidation Stability

The oxidation stability is directly related to the degree of unsaturation of alkyl esters and also to double bond position in a carbon chain. The molecule susceptibility to oxidative degradation increases with the number of unsaturations. Therefore, alkyl oleate is more stable than alkyl linoleate, which in turn is more stable than linoleate. High temperatures and exposition to air have a high influence on biodiesel stability. Exposition to light and traces of metals also promote faster oxidative degradation (Santos et al. 2011). The initial phase of this process is known as peroxidation and consists of three steps:

- *Initiation*—formation of free radicals by removing one hydrogen from a carbon-hydrogen-bond in a carbon chain;
- *Propagation*—formation of peroxide radicals by reaction of free radicals with oxygen. The free peroxide radicals attack the carbon chains of esters and form hydroperoxides as well as new free radicals, which give continuity to the propagation;
- *Termination*—combinations of radicals that form undissolvable compounds with higher molar masses.

The undissolvable compounds that result from oxidative degradation cause formation of deposits in fuel injection system of engines. Natural antioxidants in plant oils promote a higher stability to oxidation (e.g., tocopherols), however, antioxidants can be lost during oil refining by thermal degradation or during the biodiesel production process. Usually, artificial antioxidants, such as pyrogallol (PY), tert-butylhydroquinone (TBHQ), butylated hydroxyanisole (BHA) and propyl gallate (PrG) are added to increase the oxidative stability of biodiesel (Santos et al. 2011; Maia et al. 2011).

Viscosity, peroxide index and, above all, the Rancimat induction period are parameters that can be used to monitor the oxidative degradation of biodiesel during storage. The Rancimat method is a test in agreement with standard EM 14112 for the determination of biodiesel oxidative stability with a minimal induction period of

6 h. During the test, a stream of air is passed through a sample of alkyl ester (biodiesel), which is maintained in a reaction vessel at 110°C. During the process, volatile organic compounds are formed, such as organic acids with low molecular mass. These compounds are transported by the air stream to another vessel containing distilled water where the organic acids are detected by increasing the conductivity of the system. The time span until the detection of the first volatiles organic acids is referred to as induction period.

Transesterification of Oils and Fats for Biodiesel Production

The use of plant oils as fuel in cars dates from the development of the diesel engine by Rudolf Diesel in 1898, which was initially designed to run on peanut oil. For economic as well as for technical reasons, plant oils have been substituted by petrol diesel (Demirbas 2003). Suggestion: During more than half of the last century, low prices and abundant offer of crude oil derivatives had a decisive influence on the choice of fossil diesel as automotive fuel.

After prolonged use of plant oil as fuel in diesel engines, carbon and gum deposits start to form in the combustion chamber, caused by oxidation and polymerization reactions. These problems are associated to higher viscosity, low volatility, and polyunsaturation of plant oils (Leung et al. 2006).

Different strategies have been investigated to solve these problems, such as blending of plant oil with diesel, micro-emulsions of oil with alcohols, cracking, and transesterification. The transesterification of oils and fats showed the most promising results and is at the moment considered the technological state of the art for industrial biodiesel production (Helwani et al. 2009).

The reaction between an ester and an alcohol, in which the alkoxide group of the ester is exchanged with the one from the alcohol, is known as transesterification. The esters used in biodiesel production are *triacylglycerols* (TAG) (the major components of animal or plant oils and fats) induced to react with a short chained alcohol. This kind of reaction happens by nucleophilic substitution, which is typical for carboxylic derivatives and can be catalyzed by acids as well as by bases. The transesterification of TAGs results in a mixture of alkyl esters with linear chains called biodiesel (Fig. 27.3) with glycerin as co-product.

Because biodiesel (alkyl esters) has simpler molecular structures than its precursors, the TAGs, its viscosity is lower. Thus, their atomization a combustion chamber is more efficient and the combustion is significantly more homogeneous and complete, causing less residue deposits in the motor than it is with the combustion of TAGs.

Short chain alcohols, such as methanol and ethanol are usually used for the transesterification of oils and fats. On the commercial scale, methanol is preferred because it is more reactive, which results in lower reaction temperatures, shorter reaction time and requires lower alcohol quantities (Jain and Sharma 2010). Ethanol, on the other hand, is considerably less toxic, cheaper, renewable and produces biodiesel fuels with higher cetane numbers and better lubricity. The great drawback of

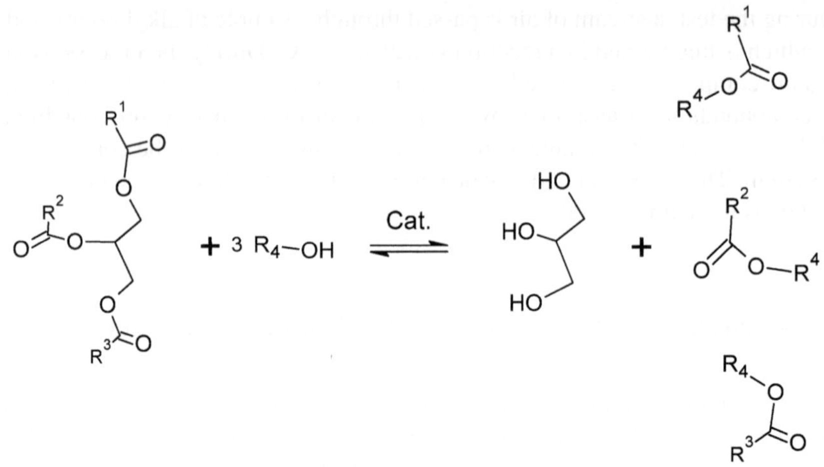

Triacylglycerol Alcohol Glycerol Fatty acid alkyl esters (biodiesel)

Fig. 27.3 Transesterification reaction of trialkyl glycerids

ethanol lies in its lower reactivity, which leads to higher reaction costs (e.g., higher reaction temperature, larger alcohol to oil ratio and longer reaction time). Another drawback of ethanol lies in the fact that this alcohol causes a higher dispersion of glycerin in biodiesel, which hampers its separation by decantation.

The molar stoichiometric alcohol to oil ratio for the transesterification is three moles of alcohol per one mole of triglyceride. This reaction is reversible, therefore it is recommended to work with an excess of alcohol and/or continuous removal of glycerin in order to dislocate the equilibrium of the reaction and to obtain major yields. For methanol, the molar ratio commonly used is 6:1 while for ethanol ratios of 9:1 up to 12:1 are necessary (Bondioli and Bella 2005; Sharma et al. 2008; Leung et al. 2010).

Simply speaking, the production of biodiesel can be described as a process in three steps: (1) the production step, where the mixture of the components and transesterification takes place; (2) the separation step, where the alcohol and glycerin co-product is separated from the produced biodiesel; and (3) the purification step, where impurities, such as residual glycerin, soaps, catalyst and residual alcohol are removed from biodiesel.

After the separation step, the biodiesel fuel can be submitted to a second transesterification step with the objective to obtain higher yields. The alcohol excess can be recovered in the separation step and can be reused for subsequent transesterifications. The crude glycerin can be used for biologic processes (Robra et al. 2010; Khanal and Nitayavardhana 2011) and, depending on its quality, can also be purified and be used for the production of high value products, such as cosmetics, explosives, plastificants, polymers, etc. (Rivaldi et al. 2007; Mota et al. 2009) as described at the end of this chapter.

The transesterification reaction of oils or fats is realized in the presence of acid, alkaline or enzymatic catalysts, which can be in homogeneous or heterogeneous form in relation to the reaction environment.

Among the homogeneous alkaline catalysts, the most frequently used in industrial scale are the sodium or potassium alkoxides. This kind of catalyst is very reactive and results in yields higher than 98% in the transesterification reaction, however, this catalyst is sensitive to the water present in oils and fats. The alkoxides can be produced by the reaction of sodium or potassium hydroxides with the alcohol used for transesterification (Bronsted-Lowry acid). This strategy brings about lower costs and satisfactory yields. However, the quantity of hydroxide has to be monitored closely because the water resulting from the neutralization reaction between the alcohol and the hydroxide promotes saponification (Leung et al. 2010).

The homogeneous catalysts have two drawbacks: they cannot be reused and they impose an additional washing step in order to remove the catalyst from biodiesel together with the residual glycerin. The necessary washing steps and treatment of their effluents bring about additional costs to the process. Much research has been realized on the development of reusable heterogeneous catalysts suitable for the transesterification of oils and fats. Besides the simplification of the purification steps of biodiesel and glycerin as well as the reduction of effluents, the major benefit of this technology is that it allows the process of transesterification to be continuous.

The following sections will discuss some processes for biodiesel production focusing on Jatropha oil as a feedstock.

Processes for Biodiesel Production

Among the different processes for biodiesel production in industrial, pilot or laboratory scale, we will discuss separately the processes under (1) conventional homogeneous catalysts, (2) heterogeneous catalysts, (3) *in situ* transesterification, (4) super or subcritical conditions, and (5) boosted by microwaves or ultrasound.

Homogeneous Catalysts

As already indicated, the transesterification by alkaline homogeneous catalysis has become the most commonly applied process for industrial biodiesel production because of its high yields. The principal catalysts used for the conversion of Jatropha oils and fats are sodium or potassium hydroxides and alkoxides.

Different authors have evaluated the use of homogeneous basic catalysts for biodiesel production from Jatropha oil (Jain and Sharma 2010; Berchmans et al. 2010; Sahoo and Das 2009). With sodium alkoxide or even with KOH, yields higher than 97% can be obtained, with reaction times close to 30 min (Syam et al. 2009). Yields of over 98% esters were achieved in the transesterification of pre-esterified

Jatropha oil using KOH 1.3% (w/w) as catalyst and a molar ratio of methanol:oil of 6:1, during 20 min at 64°C (Lu et al. 2009).

The acid homogeneous catalysis (Bronsted—Lowry acids) is little applied compared to the alkaline homogeneous process because it demands longer reaction times and temperatures close to the boiling point of acids due to slower alcoholysis kinetics, which makes the whole process more costly. However, unlike the alkaline catalysis, the acid catalysis can be advantageous in the transesterification of oils and fats with a high FFA content, transforming them into alkyl esters by esterification in a preliminary step (Fig. 27.1).

As a matter of fact, the acid catalysis is amply applied in biodiesel production where low-cost feedstocks with higher acidity are generally used. These feedstocks can be waste cooking oils, animal fats and some other oils with higher acidity like, for example, Jatropha oil. Among the mostly used catalysts are the Bronsted acids HCl, H_2SO_4 and H_3PO_4.

The mechanism of action of the Bronsted catalysts (which contain ionisable hydrogene) in transesterification consists in the protonation of the carbonyl group of the esters in the triglycerides forming a more reactive carbocation susceptible to the neutrophilic attack of the alcohol. The acid catalysis has to be conducted in an anhydrous medium because the acids catalyze the hydrolysis reaction in which fatty acids are formed (see Fig. 27.1). As a consequence, the presence of water can provoke the hydrolysis of triglycerides resulting in the competitive formation of fatty acids and reduced ester yield.

Technological alternatives that involve an acid catalyst, such as the hydroesterification with hydrolysis of TAG followed by the esterification of FFA and simultaneous conduction of transesterification and esterification, are currently being evaluated (Shuit et al. 2010a). These processes allow the use of any fatty raw material independently of its acidity and water content (Freire et al. 2010).

Heterogeneous Catalysts

A technological alternative for transesterification and esterification reactions in biodiesel production is the use of heterogeneous acid or alkaline catalysts. Among the main benefits of these solids are the possibility of recovery and reuse of the catalyst and a significant reduction in steps of product purification. The technical and environmental benefits compared to homogeneously catalyzed processes led to intense research activities for stable solids able to efficiently promote esterification and transesterification reactions, which at the moment pose one of the principal technological challenges of biodiesel industry. The literature offers different reviews about this topic including different classes of solids like oxides and inorganic salts, organic acids and bases, layered double hydroxides, ion exchange resins, zeolites, heteropolyacids, ionic liquids and biocatalysts, among others (Ng et al. 2010; Ramos et al. 2011; Sharma et al. 2011; Singh Chouhan and Sarma 2011; Kiros et al. 2011).

Species like zeolites ($Na_{21.9}$ $K_{7.5}$ $Ti_{16.5}$ $Si_{77.5}$ O_{208}), hydrotalcites ($Mg_6Al_2(OH_{16})$ $CO_3.4H_2O$) and metallic oxides (CaO, MgO, SrO) were extensively applied in biodiesel synthesis reactions (Lam et al. 2010). Among the quoted classes, calcium and magnesium oxides most attracted the attention of researchers due to their pronounced alkaline force, low solubility in methanol and simple synthesis. In scientific literature, countless papers can be found about the use of these oxides in various proportions and conditions like, for example, the very interesting work of Huaping et al. (2006), where calcium oxide (CaO) was used as a catalyst in the transesterification of Jatropha oil. The reaction was conducted at 70°C for 2.5 h with a catalyst concentration of 1.5% (w/w) and a molar ratio for alcohol:oil of 9:1. Besides these conditions, the authors boosted the alkaline force of the catalyst by a treatment with ammonia carbonate $(NH_4)_2CO_3$ followed by calcination at 900°C. According to the authors, an increase of 26.5% of the alkaline force of CaO was observed, resulting in a oil to biodiesel conversion of approximately 93%. The catalyst was reused in another three cycles, maintaining a conversion rate of 92%.

Another interesting work was developed by Vyas et al. (2009) where a potassium nitrate supported by aluminum (KNO_3/Al_2O_3) was synthesized for the methanolic transesterification of Jatropha oil. The reaction was conducted with an alcohol:oil ratio of 12:1 and a concentration of catalyst of 6% (w/w) for 6 h at 70°C. According to the authors, the conversion in the first cycle reached 87%. Reuse of the catalyst for two more cycles was possible after calcination at 500°C for 4 h, but a reduction of the activity of 9% and 72%, respectively, was observed in relation to the initial value.

Though the alkaline homogeneous and heterogeneous catalysts have shown a higher efficiency, research is being focused on the development of acidic solids capable of simultaneous catalysis of esterification and transesterification reactions. This way, feedstocks with high FFA contents like Jatropha and castor oils could be processed without neutralizing steps, with the consequence of eliminating problems, such as reactor corrosion and difficult regeneration of the catalyst. Among the principal classes of acid solids used in the simultaneous transesterification and esterification, the emphasis is on metallic oxides (ZrO_2, TiO_2, SnO_2 and ZnO), zeolites with acid sites ($Na_2Al_2Si_3O_{10}.2H_2O$), ion exchange resins (polystyrene sulfonic acid), sulfated or sulfonated inert solids (SO_4, SO_3H) and heteropoliacids ($H_3PW_{12}O_{40}.6H_2O$) (Lam et al. 2010).

Endalew et al. (2011) proposed the development of acid and basic heterogeneous catalysts for the production of biodiesel samples from Jatropha oil with a high index of FFA (9% w/w). The study was divided in two parts. In the first, pure calcium oxide with high acidity and doped with lithium was used for oils with strong acidity. In the second part of the work, scientists investigated the possibility of simultaneous oil esterification and transesterification. For this end, the catalysts were mixed with ferrous sulfate—$Fe_2(SO_4)_3$. The conditions for the oil methanolysis were: alcohol:oil ratio of 6:1, catalyst concentration of 5% (w/w), reaction temperature of 60°C, reaction time of 3 h and constant agitation of 300 rpm. According to the authors, the alkaline catalyst was inefficient since the conversion of oil to biodiesel was insignificant and a large amount of soap was formed. On the other hand, simultaneous reactions with a mixture of acid and alkaline solids were efficient with conversion rates of 93.4% for CaO:$Fe_2(SO_4)_3$ and 96% for Li-CaO:$Fe_2(SO_4)_3$.

In Situ *Transesterification*

In general, the conventional methods for the production of biodiesel from plant oils start with oil extraction. The extraction can be done by mechanical (seeds pressing) or chemical extraction, usually by hexane or chloroform, followed by a purification step (deparaffinization, degumming, dehydration, deacidification, dephosphorization, etc.). Further on, the oil is submitted to an esterification or transesterification process depending principally on the FFA content of the oil. The pretreatment steps of oil are responsible for over 70% of the total production costs when refined oils are used as feedstock (Helwani et al. 2009).

The *in situ* transesterification, also known as *reactive extraction*, is attractive because the preliminary steps of the oil purification are not necessary. In this reaction, extraction and esterification/transesterification steps occur simultaneously in only one process. In this process, the alcohol acts as an extraction solvent and transesterification agent. In some cases, the solvent or a mixture of solvents is added to the reaction in small quantities, acting as co-solvents, with the objective to improve reaction yield. The *in situ* reaction reduces the expenses of the extracting solvent and eliminates the costs of oil pretreatment, therefore the process becomes very simple, faster and less expensive. Several examples in the literature report reactive extraction of oil from soybean, algae, castor bean, sunflower, etc.

For the reactive extraction, the feedstock, usually an oilseed, has to be treated before it can be submitted to the reaction process. The seeds are ground and sieved in order to obtain particles of uniform size generally in the range of millimeters. The principal objective of the procedure is to increase the contact surface between feedstock and alcohol in order to improve extraction efficiency. An evaluation of the influence of particle size of grounded Jatropha seeds was conducted with particle sizes varying from 0.355 to 1 mm. According to these studies, particle sizes between 0.355 and 1 mm showed little variation of extraction efficiency (reaching 60%). The particles smaller than 0.355 mm showed better extraction efficiencies with yields up to 90% of oil effectively extracted (Shuit et al. 2010b). Following grinding, the seeds are dried at 70°C to avoid interference with humidity.

In reactive extraction, homogeneous alkaline catalysts are usually used because of their larger benefit compared to acid catalysts. However, when a large amount of FFAs are present in the sample, the use of an alkaline catalyst results in soap formation and hampers phase separation. For this reason, the acid catalysis is a better alternative for reactive extraction of oils with high FFA index, such as that of Jatropha. Table 27.5 reports some works on *in situ* transesterification of Jatropha oil.

Transesterification with Supercritical and Subcritical Alcohols

The transesterification reaction of oils and fats can be done with higher alkyl ester yields and in a shorter reaction time in absence of catalysts, but using supercritical alcohols. In this process, the reaction medium is kept under high temperatures and

Table 27.5 Results of some works concerning the transesterification *in situ* of Jatropha oil seeds.

Alcohol	Quantity of alcohol (ml) per gram of seed	Temperature (°C)	Catalyst/Quantity (%w/w)	Yield/Time (%/min)	Reference
Methanol	10.5	60	H_2SO_4/21.80	98.10/600	Shuit et al. (2010b)
Ethanol	8.5	65	H_2SO_4/23.75	80.93/1080	Shuit et al. (2010c)
Ethanol	7	30	CH_3ONa/2	99.98/120	Ginting et al. (2012)
Methanol	10	65	NaOH/0.1 mol L^{-1}	98.00/60	Kaul et al. (2010)
Methanol	7.5	60°	H_2SO_4/15	99.80/1440 (24 h)	Shuit et al. (2010a)

high pressures favoring the solubilization of reagents. Although reaction speed increases with temperature, the upper temperature limit is 350°C to avoid alkyl esters and glycerol decomposition. High medium pressure also increases yield since higher molar density facilitates the contact among molecules.

Besides the fact that the reaction can be conducted without catalysts, the transesterification in supercritical conditions has other advantages compared to conventional transesterification, such as short reaction times, simplified separation and purification steps, significant reduction of process effluents, no water interference with reaction yield and FFA conversion into alkyl esters. The last two benefits allow the recommendation of supercritical transesterification for the production of biodiesel from cheap residual oils and fats.

Maintaining the reaction conditions with high temperatures and pressures requires a higher energy demand and special equipment, which make the process expensive (Patil et al. 2009). Another drawback of transesterification in supercritical conditions is the necessity of a high molar ratio of alcohol to oil (>40:1) to obtain satisfactory yields (Pinnarati and Savage 2008).

In this context, research on co-solvents focuses on the increase of alcohol solubilization in the reaction medium together with reaction conditions at lower temperature and pressure (Sawangkeaw et al. 2007). The co-solvents used for this purpose are hexane, tetrahydrofurane, propane and carbon dioxide, the latter being preferred because of its low cost and extremely low toxicity.

Different research works on transesterification of Jatropha oil in supercritical conditions have been realized. The great majority of these investigations conducted reactions with methanol as the transesterification agent. An evolution of this strategy is the application of supercritical fluid to reactive extraction of oil from Jatropha seeds, with simultaneous transesterification (Lim et al. 2010; Lim and Lee 2011; Niza et al. 2011).

The transesterification reaction in subcritical conditions follows the same principles as the supercritical reaction except that temperatures and pressures are always kept below the critical point (Chen et al. 2010). Keeping the system in subcritical conditions brings about technical, economical and environmental benefits. Unfortunately, subcritical fluids are of little efficiency to promote transesterification reactions, compared to reactions conducted in supercritical conditions (Chen et al. 2010; Ilham and Saka 2010).

Transesterification Boosted by Microwave Radiation

The use of microwave radiation is very common in chemistry laboratories in order to promote and accelerate digestion of samples, synthesis of organic compounds, desorption and extraction of substances. Microwave radiation provokes constant changes of orientation of polar molecules and oscillation of electric fields as well as a higher probability of shocks between molecules due to increased Brownian motion, which both are possible causes for the increase of the medium temperature and subsequent reaction speed up. Another explanation could be that under these conditions the formation of loaded species in transition state is favored (Souza and Miranda 2011).

When applied in biodiesel production, the microwave radiation promotes higher transesterification yields in a shorter time. However, the application of microwave radiation does not turn the use of catalysts obsolete. Catalysts under microwaves may be acidic, alkaline, or enzymatic, in homogeneous or heterogeneous media (Yuan et al. 2009; Barnard et al. 2007; Da Ros et al. 2011; Liao and Chung 2011). Besides a considerable reduction in reaction time, the use of microwaves brings about benefits, such as reduced energy consumption and simpler process of biodiesel purification. Microwaves can also be applied to promote the esterification of FFA, which enables the use of residual oils for biodiesel production without previous treatment (Kim et al. 2011).

Microwaves allow alkyl ester yields larger than 97% from Jatropha oil with as much as 3.1% FFAs, without previous treatment and in a few minutes of transesterification. Following the conventional process, 90 min esterification of FFAs and 60 min alkaline transesterification were necessary to reach the same biodiesel yield. Both results were obtained with potassium hydroxide as a catalyst in the proportion of 1% in relation to the mass of the oil and a molar alcohol:oil ratio of 7.5:1 (El Sherbiny et al. 2010). However, attention to the reaction kinetics has to be given when using microwaves, because after the time still necessary to obtain the maximum biodiesel yield, a reduction of the yield can be observed. This phenomenon seems to be due to equilibrium dislocation of the transesterification reaction or to possible alkyl ester degradation (El Sherbiny et al. 2010). Even though results are in favor of transesterification boosted by microwaves, additional investigations are still necessary for its application at industrial scale.

Transesterification Boosted by Ultrasound

Ultrasound is usually applied by sonochemistry in various processes like food processing, stabilization of emulsions, homogenization, atomization, etc. The application of ultrasound in the transesterification reaction intends to promote the mass transfer between alcoholic and oily phases by increasing the contact between these reagents and the catalyst in order to obtain an increased reaction yield. The dispersion of the alcohol in the oil occurs due to the formation and collapse of microbubbles within the liquid (Mahamuni and Adewuyi 2009). This phenomenon is called *cavitation* and can be induced by sound waves, which cause zones with high pressure and temperatures where the necessary energy for the transesterification is provided. Regarding the process, ultrasound transesterification occurs with higher efficiency, in shorter time and with lower energy consumption than conventional transesterification (Mahamuni and Adewuyi 2009).

The application of ultrasound in the transesterification of Jatropha oil has been reported by different authors. Table 27.6 summarizes the principal information (reaction conditions, reagents, catalysts, etc.) and results available on ultrasound transesterification.

Glycerin: A Useful Co-product from Biodiesel Production

Glycerin can be obtained from microbial fermentation, petrochemical feedstocks, saponification in soap production and transesterification for biodiesel production, where it is the major co-product. Theoretically, every 100 L of biodiesel yields 10 L of glycerin (0.3 kg glycerin for each 1 gal biodiesel), a colorless, odorless, viscous and sweet tasting hygroscopic nontoxic liquid (Singhabhandhu and Tezuka 2010).

The separation of biodiesel and glycerin is carried out after transesterification. The difference in density between biodiesel (light) and glycerol (heavy) is sufficient for the application of techniques, such as gravitational settling or centrifugation. The upper ester phase contains the main product—methyl esters (biodiesel). After the glycerin has been removed, this phase still can show traces of alcohol, catalyst, glycerin and water. Beside these products, traces of unreacted glycerides (mono-, di- and tri-acylglycerides) can also be found. With the objective to attend the international specification for commercialization, the use of a purification step is mandatory. Several different separation and purification techniques of biodiesel have been investigated (Sdrula 2010; Aroua et al. 2011; Caramão et al. 2012).

The lower glycerin phase typically consists of glycerol (C_3H_5OH, propane-1,2,3-triol) and varying proportions of impurities. These impurities include methyl esters, water, soapstock, alcohol, inorganic salts (catalyst residue), unreacted mono-, di-, triglycerides and a variety of other *matter organic non-glycerol* (MONG) in varying quantities, which depend on the feedstock and chemical process used (Robra et al. 2010; Hájek and Skopal 2010; Astals et al. 2011). The methanol typically is removed from the glycerol-rich phase and reused, leaving what is known as a crude glycerin. The biodiesel industry generates millions of tons of residual crude glycerin each year and the amount is increasing rapidly along with the growth of biodiesel production.

In order to obtain a commercially valued product, the crude glycerin has to undergo an expensive refining process to comply with the technical standards required by the consumer industry. As a consequence, biodiesel producers refine the crude glycerin through filtration and/or centrifugation, chemical additions and fractional vacuum distillation, to produce various commercial grades (Pagliaro and Rossi 2008). It is important to note that if glycerin is destined to food, cosmetic or drug industry, further treatments, such as bleaching, deodorizing and ion exchange are needed to remove all contaminants (FFA and their salts) to meet the requirements of the *United States Pharmacopeia* (USP) and the *Food Chemicals Codex* (FCC).

Table 27.6 Jatropha Biodiesel production using transesterification of oil boosted by ultrasound

Frequency/Power	Alcohol	Molar ratio Alcohol:oil	Temperature (°C)	Quantity of catalyst for the mass oil (% w/w)	Yield/Time (%/min)	Reference
30 kHz/—	methanol	9:1	50	NaOH/1	93.0/30	Vyas et al. (2011)
24 kHz/120 W	methanol/ethanol (mixture)	3/3:1	50–70	KOH/0.75	98.0/7	Kumar et al. (2011a)
20 kHz/1,500 W	methanol	9:1	60	K/Al_2O_3/15	97.0/60	Larpkiattaworn et al. (2010)
24 kHz/100 W	methanol	9:1	Autogenous	$NaSiO_2$/3	98.5/15	Kumar et al. (2010)
24 kHz/100 W	methanol	4:1	Autogenous	Enzima (Chromobacterium viscosum)/5	84.5/30	Kumar et al. (2011b)

There are different separation technologies that can be used for the ultra-purification of glycerin, such as vacuum distillation, supercritical fluid extraction followed by appropriate fractionation steps and *membrane separation technology* (MST). Sdrula (2010) conducted the ultra-purification of biodiesel glycerin by applying MST (a combination of nanofiltration, microfiltration and electrodialysis), which resulted in a purity >99.5%, thus proving the efficiency of this method.

Hájek and Skopal (2010) carried out treatments of the glycerol phase to obtain glycerol with a purity of 86% (w/w) (without distillation) also containing a mixture of fatty acids and alkyl esters in a ratio of 1:1 or only a mixture of fatty acids. According to the treatment, the successive removal of alkaline substances and esters, production of fatty acids by saponification of the remaining esters and subsequent neutralization of alkaline substances results in fatty acids. The authors claim that the treatment is simple and environmentally friendly because no special chemicals or equipment are required and all products can be recycled.

With the growth of biodiesel production, the increasing quantities of crude glycerin have become an abundant and cheap resource with a remarkable impact on glycerin market. Since glycerin is an attractive carbon source for the production of fuels and chemicals, its recovery and further processing has attracted the attention of scientific community in recent years. Many researchers are seeking to develop new technologies and/or optimize technologies already established in order to create benefits from this glycerin surplus (Galan et al. 2009; Mota et al. 2009; Behr et al. 2008; Pagliaro et al. 2007). Moreover, its utilization can contribute to the economic viability of biodiesel production (Romero et al. 2010; Singhabhandhu and Tezuka 2010).

Theodoropoulos et al. (2011) studied the concept of integrated biorefineries and examined alternative schemes for the co-production of biodiesel and succinic acid, which can be used as a building block for a number of commodities and specialty chemicals. Considering the different uses of crude glycerin from biodiesel, four different biorefinery schemes can be distinguished: (1) the disposal of crude glycerin as a waste, (2) the purification (by distillation) of crude glycerin to at least 80%, (3) the purification of glycerin to at least 95% and (4) the production of succinic acid from glycerine through fermentation. Theodoropoulos et al. (2011) concluded that the co-production of succinic acid can increase the overall profit of a biorefinery by 60% for a 20 years plant lifetime. These results indicate the importance of glycerol when it is utilized as a key renewable building block for the production of commodity chemicals.

Numerous processes have been reported for the conversion of glycerol to higher value added chemicals and fuels. For more details, few excellent reviews (Behr et al. 2008; Pagliaro et al. 2007; Corma et al. 2007; Jérôme and Barrault 2011) and books (Krause et al. 2007; Jacobs et al. 2007; Pagliaro and Rossi 2008) are found in the literature. Figure 27.4 summarizes some of the routes for the transformation of glycerin and its main products and applications.

Besides noble applications that require different degrees of purification (Fig. 27.4), crude glycerin can be applied as feedstock in biological processes where a wide variety of high-value products, such as enzymes (phytase), propylene glycols (1,3-propanediol) (1,3-PD), organic acids (propionic and succinic acids) and biodegradable thermoplastic polyesters (poly(3-hydroxybutyrate)–P-3HB) can

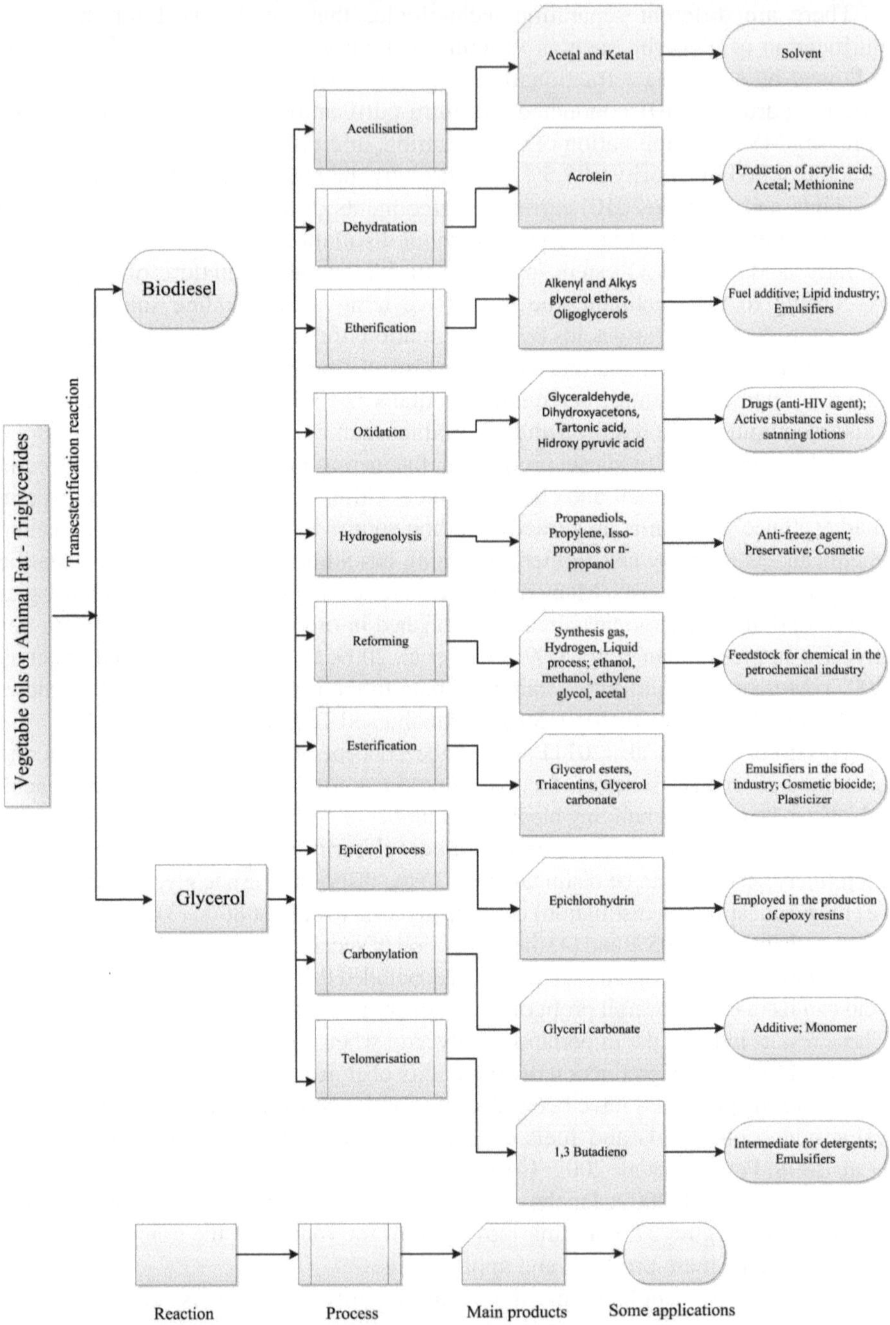

Fig. 27.4 Routes for the transformation of glycerin as well as its main products and applications (Ciriminna et al. 2006; Suppes et al. 2006; Jacobs et al. 2007; Pagliaro et al. 2007; Krause et al. 2007; Corma et al. 2007; Behr at al. 2008; Fajula et al. 2008; Mota et al. 2009; Galan et al. 2009; Lee et al. 2010; Özgür and Uysal 2011)

be obtained. Fungal fermentation offers an alternative option for crude glycerol utilization and can be easily integrated into biodiesel plants due to its minimal requirement of additional operations/processes in existing plants. The integration of fungal technology into biodiesel production could provide an opportunity to produce an inexpensive food-grade fungal protein for animal feed and offers an alternative option for glycerol utilization (Khanal and Nitayavardhana 2011).

Crude glycerin also appears to be a suitable carbon source for anaerobic microbiological processes. One alternative solution for the valorization of crude glycerin from biodiesel production is the anaerobic treatment of crude glycerin in biodigesters for biogas production, a renewable and versatile energy (Robra et al. 2010; Castrillón et al. 2011).

References

Aroua MK, Atadashi IM, Abdul Aziz AR (2011) Biodiesel separation and purification: a review. Renew Energy 36:437–443

Astals S, Ariso M, Galí A, Mata-Alvarez J (2011) Co-digestion of pig manure and glycerine: experimental and modelling study. J Environ Manag 92:1091–1096

Atadashi IM, Aroua MK, Abdul Aziz A (2010) High quality biodiesel and its diesel engine application: a review. Renew Sustain Energy Rev 14:1999–2008

Balat M (2011) Potential alternatives to edible oils for biodiesel production—a review of current work. Energy Convers Manag 52:1479–1492

Barnard TM, Leadbeater NE, Boucher MB, Stencel LM, Wilhite BA (2007) Continuous-flow preparation of biodiesel using microwave heating. Energy Fuel 21:1777–1781

Becker K, Makkar HPS (2008) *Jatropha curcas*: a potential source for tomorrow´s oil and biodiesel. Lipid Technol 20:104–107

Behr A, Eilting J, Irawadi K, Leschinski J, Lindner F (2008) Improved utilization of renewable resources: new important derivates of glycerol. Green Chem 10:13–30

Berchmans HJ, Hirata S (2008) Biodiesel production from crude *Jatropha curcas* L. seed oil with a high content of free fatty acids. Bioresour Technol 99:1716–1721

Berchmans HJ, Moroshita K, Takarada T (2010) Kinetic study of hydroxide-catalyzed methanolysis of *Jatropha curcas* waste food oil mixture for biodiesel production. Fuel. doi:10.1016/j.fuel.2010.01.01

Bondioli P, Bella LD (2005) An alternative spectrophotometric method for the determination of free glycerol in biodiesel. Eur J Lipid Sci Technol 107:153–157

Caramão EB, Manique MC, Faccin CS, Onorevoli B, Benvenutti EV (2012) Rice husk ash as an adsorbent for purifying biodiesel from waste frying oil. Fuel 92:56–61

Castrillón L, Fernández-Nava Y, Ormaechea P, Marañón E (2011) Optimization of biogas production from cattle manure by pre-treatment with ultrasound and co-digestion with crude glycerin. Bioresour Technol 102:7845–7849

Chen CH, Chen WH, Chang CMJ, Setsua I, Tu CH, Shieh CJ (2010) Subcritical hydrolysis and supercritical methylation of supercritical carbon dioxide extraction of Jatropha oil. Sep Purif Technol 74:7–13

Ciriminna R, Palmisano G, Della Pina C, Rossi M, Pagliaro M (2006) One-pot electrocatalytic oxidation of glycerol to DHA. Tetrahedron Lett 47:6993–6995

Corma A, Iborra S, Velty A (2007) Chemical routes for the transformation of biomass into chemicals. Chem Rev 107:2411–2502

Corrêa SM, Arbilla G (2006) Aromatic hydrocarbons emissions in diesel and biodiesel exhaust. Atmospheric Environ 40:6821–6826

Da Ros PCM, de Castro HF, Carvalho AKF, Soares CMF, de Moraes FF, Zanin GM (2011) Microwave-assisted enzymatic synthesis of beef tallow biodiesel. J Ind Microbiol Biotechnol 39:529–536

Demirbas A (2003) Biodiesel fuels from vegetable oils via catalytic and non-catalytic supercritical alcohol transesterifications and other methods: a survey. Energy Convers Manag 44:2093–2109

El Sherbiny SA, Refaat AA, El Sheltawy ST (2010) Production of biodiesel using the microwave technique. J Adv Res 1:309–314

Endalew AK, Kiros Y, Zanzi R (2011) Heterogeneous catalysis for biodiesel production from *Jatropha curcas* oil (JCO). Energy 36:2693–2700

Fajula F, Tanchoux N, Pariente S (2008) Etherification of glycerol with ethanol over solid acid catalysts. Green Chem 11:1256–1261

Freire DMG, Aranda DAG, Oliveira EC, Sousa JS (2010) Application of lipase from the physic nut (*Jatropha curcas* L.) to a new hybrid (enzyme/chemical) hydroesterification process for biodiesel production. J Mol Catal B: Enzym 65:133–137

Galan MI, Bonet J, Sire R, Reneaume JM, Plesu AE (2009) From residual to useful oil: revalorization of glycerine from the biodiesel synthesis. Bioresour Technol 100:3775–3778

Ginting MSA, Azizan MT, Yusup S (2012) Alkaline *in situ* ethanolysis of *Jatropha curcas*. Fuel 93:82–85

Hájek M, Skopal F (2010) Treatment of glycerol phase formed by biodiesel production. Bioresour Technol 101:3242–3245

Hayyan A, Alam MZ, Mirghani MES, Kabbashi NA, Hakimi NINSM, Siran YM, Tahiruddin S (2010) Sludge palm oil as a renewable raw material for biodiesel production by two-step processes. Bioresour Technol 101:7804–7811

Helwani Z, Othman MR, Aziz N, Fernando WJN, Kim J (2009) Technologies for production of biodiesel focusing on green catalytic techniques: a review. Fuel Process Technol 90:1502–1514

Hoekman SK, Broch A, Robbins C, Ceniceros E, Natarajan M (2012) Review of biodiesel composition, properties, and specifications. Renew Sust Energy Rev 16:143–169

Huaping Z, Zongbin W, Yuanxiong C, Ping Z, Shijie D, Xiaohua L (2006) Preparation of biodiesel catalyzed by solid super base of calcium oxide and its refining process. China J Catalysis 27:391–396

Ilham Z, Saka S (2010) Two-step supercritical dimethyl carbonate method for biodiesel production from *Jatropha curcas* oil. Bioresour Technol 101:2735–2740

Jacobs P, DHondt E, Sels B (2007) Catalytic transformation of glycerol. In: Centi G, van Santen RA (eds) Catalysis for renewables: from feedstock to energy production. Wiley–VCH Verlag GmbH & Co. KGaA, Weinheim, pp 223–255. ISBN 978-3-527-31788-2

Jain S, Sharma MP (2010) Biodiesel production from *Jatropha curcas* oil. Renew Sust Energy Rev 14:3140–3147

Jain S, Sharma MP (2011) Correlation development for the effect of metal contaminants on the thermal stability of *Jatropha curcas* biodiesel. Energy Fuel 25:1276–1283

Jérôme F, Barrault J (2011) Use of hybrid organic-siliceous catalysts for the selective conversion of glycerol. Eur J Lipid Sci Technol 113:118–134

Kaul S, Porwal J, Garg MO (2010) Parametric study of Jatropha seeds for biodiesel production by reactive extraction. Am Oil Chem Soc 87:903–908

Khanal SK, Nitayavardhana S (2011) Biodiesel-derived crude glycerol bioconversion to animal feed: a sustainable option for a biodiesel refinery. Bioresour Technol 102:5808–5814

Kim D, Choi J, Seol SK, Há YC, Vijayan M, Jung S et al (2011) Microwave-accelerated energy-efficient esterification of free fatty acid with a heterogeneous catalyst. Bioresour Technol 102:3639–3641

Kiros Y, Endalew AK, Zanzi R (2011) Inorganic heterogeneous catalysts for biodiesel production from vegetable oils. Biomass Bioenergy 35:3787–3809

Knothe G, Steidley KR (2005) Kinematic viscosity of biodiesel fuel components and related compounds. Influence of compound structure and comparison to petrodiesel fuel components. Fuel 84:1059–1065

Koh MY, Ghazi TIM (2011) A review of biodiesel production from *Jatropha curcas* L. oil. Renew Sustain Energy Rev 15:2240–2251

Krause AOI, Karinen RS, Viinikainen TSB (2007) Conversion of glycerol into traffic fuels. In: Centi G, van Santen RA (eds) Catalysis for renewables: from feedstock to energy production. Wiley–VCH Verlag GmbH & Co. KGaA, Weinheim, pp 209–222. ISBN 978-3-527-31788-2

Kumar D, Kumar G, Poonam, Singh CP (2010) Ultrasonic-assisted transesterification of *Jatropha curcas* oil using solid catalyst, Na/SiO_2. Ultrason Sonochem 17:839–844

Kumar D, Johari R, Kumar PG (2011a) Fast, easy ethanomethanolysis of *Jatropha curcas* oil for biodiesel production due to the better solubility of oil with ethanol in reaction mixture and assisted by ultrasonication. Ultrason Sonochem 19(4):816–822

Kumar G, Kumar D, Poonam, Johari R, Singh CP (2011b) Enzymatic transesterification of *Jatropha curcas* oil assisted by ultrasonication. Ultrason Sonochem 18:923–927

Lam MK, Lee KT, Mohamed AR (2010) Homogeneous, heterogeneous and enzymatic catalysis for transesterification of high free fatty acid oil (waste cooking oil) to biodiesel: a review. Biotechnol Adv 28:500–518

Larpkiattaworn S, Jeerapan C, Tongpan R, Tongon S (2010) Ultrasonic on transesterification reaction for biodiesel production. 7th Biomass Asia Workshop, Jakarta, 29–1 Nov–Dec 2010

Lebedevas S, Vaicekauskas A (2006) Use of waste fats of animal and vegetable origin for the production of biodiesel fuel: quality, motor properties, and emissions of harmful components. Energy Fuel 20:2274–2280

Lee JG, Lee SH, Son YI, Choi YC, Yoon SJ (2010) Gasification of biodiesel by-product with air or oxygen to make syngas. Bioresour Technol 101:1227–1232

Leung DYC, Koo BCP, Guo Y (2006) Degradation of biodiesel under different storage conditions. Bioresour Technol 97:250–256

Leung DYC, Wu X, Leung MKH (2010) A review on biodiesel production using catalyzed transesterification. Appl Energy 87:1083–1095

Liao C, Chung T (2011) Analysis of parameters and interaction between parameters of the microwave-assisted continuous transesterification process of Jatropha oil using response surface methodology. Chem Eng Res Des 89:2575–2581

Lim S, Lee KT (2011) Effects of solid pre-treatment towards optimizing supercritical methanol extraction and transesterification of *Jatropha curcas* L. seeds for the production of biodiesel. Sep Purif Technol 81:363–370

Lim S, Hoong SS, Teong LK, Bhatia S (2010) Supercritical fluid reactive extraction of *Jatropha curcas* L. seeds with methanol: a novel biodiesel production method. Bioresour Technol 101:7169–7172

Lima JRO, Silva RB, Silva CCM, Santos LSS, Santos Junior JR, Moura EM et al (2007) Biodiesel de babaçu (*Orbignya* sp.) obtido por via etanólica. Quim Nova 30:600–603, Portuguese

Lôbo IP, Ferreira SLC, da Cruz RS (2009) Biodiesel: parâmetros de qualidade e métodos analíticos. Quim Nova 32:1596–1608, Portuguese

Lu H, Liu Y, Zhou H, Yang Y, Chen M, Liang B (2009) Production of biodiesel from *Jatropha curcas* L. oil. Comput Chem Eng 33:1091–1096

Mahamuni NN, Adewuyi YG (2009) Optimization of the synthesis of biodiesel via ultrasound-enhanced base-catalyzed transesterification of soybean oil using a multifrequency ultrasonic reactor. Energy Fuel 23:2757–2766

Maia ECR, Borsato D, Moreira I, Spacino KR, Rodrigues PRP, Gallina AL (2011) Study of the biodiesel B100 oxidative stability in mixture with antioxidants. Fuel Process Technol 92:1750–1755

Mittelbach M, Schober S (2003) The influence of antioxidants on the oxidation stability of biodiesel. JAOCS 80:817–823

Mota CJA, da Silva CXA, Gonçalves VLC (2009) Glycerochemistry: new products and processes from glycerin of biodiesel production. Quim Nova 32:639–648

Nakpong P, Wootthikanokkhan S (2010) Optimization of biodiesel production from *Jatropha curcas* L. oil via alkali-catalyzed methanolysis. J Sustain Energy Environ 1:105–109

Ng KYS, Salley SO, Kim M, Mohan S, DiMaggio C, Yan S (2010) Advancements in heterogeneous catalysis for biodiesel synthesis. Top Catal 53:721–736

Niza NM, Tan KT, Ahmad Z, Lee KT (2011) Comparison and optimisation of biodiesel production from *Jatropha curcas* oil using supercritical methyl acetate and methanol. Chem Pap 65:721–729

Nzikou JM, Matos L, Mbemba F, Ndangui CB, Pambou-Tobi NPG, Kimbonguila A et al (2009) Characteristics and composition of *Jatropha curcas* oils, variety Congo-Brazzaville. Res J Appl Sci Eng Technol 1:154–159

Ong HC, Mahlia TMI, Masjuki HH, Norhasyima RS (2011) Comparison of palm oil, *Jatropha curcas* and *Calophyllum inophyllum* for biodiesel: a review. Renew Sustain Energ Rev 15:3501–3515

Özgür DO, Uysal BZ (2011) Hydrogen production by aqueous phase catalytic reforming of glycerine. Biomass Bioenergy 35:822–826

Pagliaro M, Rossi M (2008) The future of glycerol: new uses of a versatile raw material. The Royal Society of Chemistry, Cambridge, UK. ISBN 978-0-85404-124-4

Pagliaro M, Ciriminna R, Kimura H, Rossi M, Della Pina C (2007) From glycerol to value-added products. Angew Chem Int Ed 46:4434–4440

Parawira W (2010) Biodiesel production from *Jatropha curcas*: a review. Sci Res Essays 5:1796–1808

Patil PD, Gude VG, Deng S (2009) Transesterification of *Camelina sativa* oil using supercritical and subcritical methanol with cosolvents. Energy Fuel 24:746–751

Pinnarati T, Savage PE (2008) Assessment of noncatalytic biodiesel synthesis using supercritical reaction conditions. Ind Eng Chem Res 47:6801–6808

Ramos MJ, Fernández CM, Casas A, Rodríguez L, Pérez A (2009) Influence of fatty acid composition of raw materials on biodiesel properties. Bioresour Technol 100:261–268

Ramos LP, Wypych F, Silva FR, Cordeiro CS (2011) Heterogeneous catalysts for biodiesel production. Quim Nova 34:477–486

Rivaldi JD, Sarrouh BF, Fiorilo R, da Silva SS (2007) Glicerol de biodiesel: Estratégias biotecnológicas para o aproveitamento do glicerol gerado da produção de biodiesel. Biotecnologia Cien Desenvolv 37:44–51, Portuguese

Robra S, da Cruz RS, de Oliveira AM, Almeida Neto JA, Santos JV (2010) Generation of biogas using crude glycerin from biodiesel production as a supplement to cattle slurry. Biomass Bioenergy 34:1330–1335

Rodríguez RP, Perez LG, Alfonso M, Duarte M, Caro R, Galle J et al (2011) Characterization of *Jatropha curcas* oils and their derived fatty acid ethyl esters obtained from two different plantations. Biomass Bioenergy 35:4092–4098

Romero AA, Campelo JM, Clancy J, Datta B, Lovett JC, Luque R (2010) Biodiesel as feasible petrol fuel replacement: multidisciplinary overview. Energy Environ Sci 3:1706–1721

Sahoo PK, Das LM (2009) Process optimization for biodiesel production from Jatropha, Karanja and Polanga oils. Fuel 88:1588–1594

Santos NA, Damasceno SS, Araújo PHM, Marques VC, Rosenhaim R, Fernandes VJ et al (2011) Caffeic acid: an efficient antioxidant for soybean biodiesel contaminated with metals. Energy Fuel 25:4190–4194

Sawangkeaw R, Bunyakiat K, Ngamprasertsith S (2007) Effect of co-solvents on production of biodiesel via transesterification in supercritical methanol. Green Chem 9:679–685

Schuchardt U, Sercheli R, Vargas RM (1998) Transesterification of vegetable oils: a review. J Braz Chem Soc 9:199–210

Sdrula M (2010) A study using classical or membrane separation in the biodiesel process. Desalination 250:1070–1072

Serrano-Ruiz JC, West RM, Dumesic JA (2010) Catalytic conversion of renewable biomass resources to fuels and chemicals. Ann Rev Chem Biomol Eng 1:79–100

Sharma YC, Singh B, Upadhyay SN (2008) Advancements in development and characterization of biodiesel: a review. Fuel 87:2355–2373

Sharma TC, Singh B, Korstad J (2011) Latest developments on application of heterogeneous basic catalysts for an efficient and eco friendly synthesis of biodiesel: a review. Fuel 90:1309–1324

Shuit SH, Lee KT, Kamaruddin AH, Yusup S (2010a) Reactive extraction of *Jatropha curcas* L. seed for production of biodiesel: process optimization study. Environ Sci Technol 44:4361–4367

Shuit SH, Lee KT, Kamaruddin AH, Yusup S (2010b) Reactive extraction and *in situ* esterification of *Jatropha curcas* L. seeds for the production of biodiesel. Fuel 89:527–530

Shuit SH, Lee KT, Kamaruddin AH, Yusup S (2010c) Reactive extraction for production of biodiesel from *Jatropha curcas* L. seed using ethanol as alcohol source. In: Proceedings Venice 2010, third international symposium on energy from biomass and waste, Venice, 8–11 Nov 2010

Singh Chouhan AP, Sarma AK (2011) Modern heterogeneous catalysts for biodiesel production: a comprehensive review. Renew Sustain Energy Rev 15:4378–4399

Singh SP, Singh D (2010) Biodiesel production through the use of different sources and characterization of oils and their esters as the substitute of diesel: a review. Renew Sustain Energy Rev 14:200–216

Singhabhandhu A, Tezuka T (2010) A perspective on incorporation of glycerin purification process in biodiesel plants using waste cooking oil as feedstock. Energy 35:2493–2504

Souza R, Miranda LSM (2011) Irradiação de micro-ondas aplicada à síntese orgânica: uma história de sucesso no Brasil. Quim Nova 34:497–506, Portuguese

Suppes GJ, Chiu CW, Dasari MA, Sutterlin WR (2006) Dehydration of glycerol to acetol via catalytic reactive distillation. AiChE J 52:3453–3458

Syam AM, Yunus R, Ghazi TIM (2009) Methanolysis of Jatropha oil in the presence of potassium hydroxide catalyst. J Appl Sci 9:3161–3165

Theodoropoulos C, Webb C, Binns M, Vlysidis A (2011) A techno-economic analysis of biodiesel biorefineries: assessment of integrated designs for the co-production of fuels and chemicals. Energy 36:4671–4683

Thornley P (2006) Increasing biomass based power generation in the UK. Energy Policy 34:2087–2099

Trevisani L, Fabbri M, Negrini F, Ribani PL (2007) Advanced energy recovery systems from liquid hydrogen. Energy Convers Manag 48:146–154

Vaknin Y, Ghanim M, Samra S, Dvash L, Hendelsman E, Eisikowitch D et al (2011) Predicting *Jatropha curcas* seed-oil content, oil composition and protein content using near-infrared spectroscopy—a quick and non-destructive method. Ind Crops Prod 34:1029–1034

Vyas AP, Subrahmanyam N, Patel PA (2009) Production of biodiesel through transesterification of Jatropha oil using KNO_3/Al_2O_3 solid catalyst. Fuel 88:625–628

Vyas AP, Verma JL, Subrahmanyam N (2011) Effects of molar ratio, alkali catalyst concentration and temperature on transesterification of Jatropha oil with methanol under ultrasonic irradiation. Adv Chem Eng Sci 1:45–50

Wang R, Hanna MA, Zhou W, Bhadury PS, Chen Q, Song B et al (2011) Production and selected fuel properties of biodiesel from promising non-edible oils: *Euphorbia lathyris* L., *Sapium sebiferum* L. and *Jatropha curcas* L. Bioresour Technol 102:1194–1199

Yuan H, Yang BL, Zhu GL (2009) Synthesis of biodiesel using microwave absorption catalysts. Energy Fuels 23:548–552

Chapter 28
The *In Situ* Biodiesel Production and Its Applicability to Jatropha

Keat Teong Lee and Steven Lim

Introduction

Jatropha is a genus of the family of euphorbiaceae and consists of over 170 of different plants, shrubs and trees species. Most of them are succulent plants with water-retaining features that make them highly adaptable to semi-dry climates or soil conditions.

Recently, *Jatropha curcas* L (JCL) has successfully attracted considerable interest from the mass media and scientific community due to its suitability and advantages as an alternative feedstock for biodiesel production. A research report carried out by Goldman Sachs Global Investment Research quoted JCL as one of the most promising candidates for biodiesel production (Currie 2007). Firstly originated from Central America, JCL was introduced into India by Portuguese traders during 1600 for its medicinal application. Nowadays, JCL is mainly cultivated in tropical and subtropical regions, such as Mexico, India and Africa. JCL trees are used traditionally for live fencing, erosion control and soil improvement. The oil is extracted from seeds for manufacture of soaps, candles, lubricants and dyes. The leaves are known to contain several medicinal nutrients. JCL was still a relatively wild and unexplored species until the publication of its promising potential as biodiesel feedstock. In 2010, a team of researchers from Kazusa DNA Research Institute (Japan) successfully sequenced the whole genome of JCL, which in turn spurred more fundamental research to improve its properties and applications (Sato et al. 2010). Table 28.1 lists the distribution of *Jatropha* plantations and development around the world.

K.T. Lee (✉) • S. Lim
School of Chemical Engineering, Universiti Sains Malaysia, Engineering Campus, Seri Ampangan, 14300, Nibong Tebal, Pulau Pinang, Malaysia
e-mail: chktlee@eng.usm.my

N. Carels et al. (eds.), *Jatropha, Challenges for a New Energy Crop: Volume 1: Farming, Economics and Biofuel*, DOI 10.1007/978-1-4614-4806-8_28,

Table 28.1 Global development of *J. curcas* plantation in 2008

	Countries							
	Commercial activity		Political support (legislation/government program)			Scales (millions ha)		
Regions	Strong (>5,000 ha)	Starting (100–5,000 ha)	In place	In progress	Banned	Current (2008)	Projected (2015)	Number of projects
Asia	China, India, Indonesia, Bangladesh, Thailand, Myanmar, Vietnam, Laos	Cambodia, Malaysia, Philippines, Taiwan, Sri Lanka	China, India, Indonesia, Philippines, Thailand, Myanmar,	Laos, Cambodia, Malaysia	Australia	0.796	9.2	104
Africa	Ghana, Madagascar, Tanzania, Zambia, Mozambique, Swaziland	Cameroon, Ethiopia, Kenya, Nigeria, Mali, Malawi, Senegal, Uganda, Namibia, Zimbabwe	Senegal, Mali, Nigeria, Ethiopia, Zimbabwe	Ghana, Cameroon, Uganda, Kenya, Tanzania, Angola, Namibia, Zambia, Madagascar	South Africa	0.119	2.0	97
Latin America	Brazil, Mexico	Argentina, Belize, Costa Rica, Peru, Ecuador, Guatemala, Honduras, Dominican Republic, Colombia	Mexico	Colombia, Brazil, Peru, Argentina, Honduras, Ecuador, Bolivia		0.021	1.6	41
Total	16	24	12	19	2	0.936	12.8	242

Source: The Global Exchange for Social Investment (GEXSI 2008)

Table 28.2 Comparison of properties between *J. curcas* biofuels and mineral diesel.

Parameters	Mineral diesel	*Jatropha* oil	*Jatropha* biodiesel
Density kg/l (15/40°C)	0.84–0.85	0.86–0.93	0.86–0.88
Cloud point (°C)	−14.0	2.0	4.0
Flash point (°C)	80	110–240	170–192
Pour point (°C)	3	−3	4.2
Cetane number	47.8	51.0	57–62
Sulphur content (%)	1.0–1.2	0–0.13	negligible
Kinematic viscosity (cSt at 40°C)	4–8	36.9	4.84–5.65
Net calorific value (MJ/l)	42–46	38.2	38.45–41.0
FFA content (%)	—	5.0–8.5	< 1.0

When reaching maturity, the colour of JCL fruits successively changes from green to yellow and, finally, brown. JCL seeds are separated from the fruit pericarp (shell) and undergo either mechanical pressing (small scale) or solvent extraction (large scale) to obtain lipids. Brownish seed are husk plus kernel and peeling off the husk gives the whitish seed kernel. Seed kernel contains primarily oil and protein while protected by a white thin layer of seed coat made up of fibers. Most of the oil is stored in the seed kernel and by separating the seed husk from the seeds, higher oil extraction yield in the range of 45–60% can be obtained especially through solvent extraction. If seeds are not dehusked, the average rate of oil extraction reported from literature normally ranges between 30% and 40% (Pramanik 2002).

Characteristics of extracted JCL oil can differ greatly according to different environments, cultivation conditions and genetics. Table 28.2 shows the comparison of the main characteristics between fossil diesel, JCL oil and biodiesel (Achten et al. 2008). Investigations have also shown that the processing and storage of JCL oil can affect its quality. In order to produce high quality fuel in the form of biodiesel, certain criteria must be fulfilled. In general, high quality biodiesel fuel from JCL must possess characteristics as below:

- Low content of phosphorus, sulfur, ash and moisture
- Low content of contaminants from fuel processing such as methanol and glycerin
- High oxidative stability to prevent fuel degradation
- Low acidity value
- Low viscosity
- High heating value
- Low pour point (lowest temperature at which the fuel stops to flow)
- Low cloud point (lowest temperature at which dissolved solids are precipitated as second phase)
- Low flash point (lowest temperature for the fuel to ignite and sustain combustion)
- High cetane number (shorter ignition delay during the combustion)

Table 28.3 Comparison between *J. curcas* and first generation biofuel crops for biofuels production

Advantages	*J. curcas*	First generation biofuel crops
Land requirement	Can grow in degraded and wasted land	Require huge amount of arable land
Cultivation	Less fertilizers requirement Less rainfall	Require a lot of fertilizers and water to grow
Pests and disease	Higher resistance Not eaten by animals due to toxicity	Susceptible to attacks by pests and plant diseases
Food	No conflict with food source	Competing with food sources
By-product value	High by-product usage and value	Moderate by-product usage and value
Other usage	Soil protection barrier to protect against soil erosion	Biomass for combustion
Oil content	High oil content and easy for extraction	Moderate oil content
Future development (R&D)	Has great potential for improvement and optimization	Most crops has been optimized and room for improvement is limited

Crude JCL oil is relatively viscous and slow-drying. It is odourless and colourless when fresh, but becomes yellowish after standing. *Free fatty acid* (FFA) content of crude JCL oil can vary from as low as 2–3% to 10–14% (w/w), depending on the extraction conditions. FFA content must be as low as possible to avoid acidic transesterification and improve biodiesel storage stability. There are four main types of fatty acids in JCL oil, i.e., palmitic, stearic, oleic and linoleic acids. Oleic and linoleic acids are unsaturated fatty acids with double bonds and usually constitute over 50% of the composition of JCL oil. Although fatty acid unsaturation improves the fuel's flow properties at cold temperature, it favors its oxidation during long term storage; a factor which must be taken into consideration as well. Combustion of JCL biodiesel in internal combustion engine has been proven to lead to reduction in several critical atmospheric pollutants, such as *hydrocarbons* (HC), *carbon dioxide* (CO_2), *carbon monoxide* (CO), sulphates, *polycyclic aromatic hydrocarbons* (PAH) and *particulate matter* (PM) (Pandey et al. 2011). Gas emission of CO_2 equivalent from usage of JCL biodiesel is found to be about 78% lower than the usage of corresponding fossil diesel and thus may contribute significantly towards carbon reduction. However, its higher oxygenated content will also give rise to the increase emission of *nitrogen oxides* (NO_x) compared to fossil diesel. By-products from JCL after biodiesel production are seed cake, fruit shell and seed husk. Its seed cake is known to be rich in protein content (58.1% compared to soybean 48.0%) and may be useful for livestock feed or fertilizers after detoxification (Aregheore et al. 1998). Meanwhile, fruit shell and seed husk can be used for direct combustion to provide energy as had been demonstrated by trials in a gasification plant (Achten et al. 2008). Further advantages of using JCL as feedstock for biodiesel production compared to other conventional first generation biofuel crops (such as soybean, rapeseed and sunflower) are listed in Table 28.3.

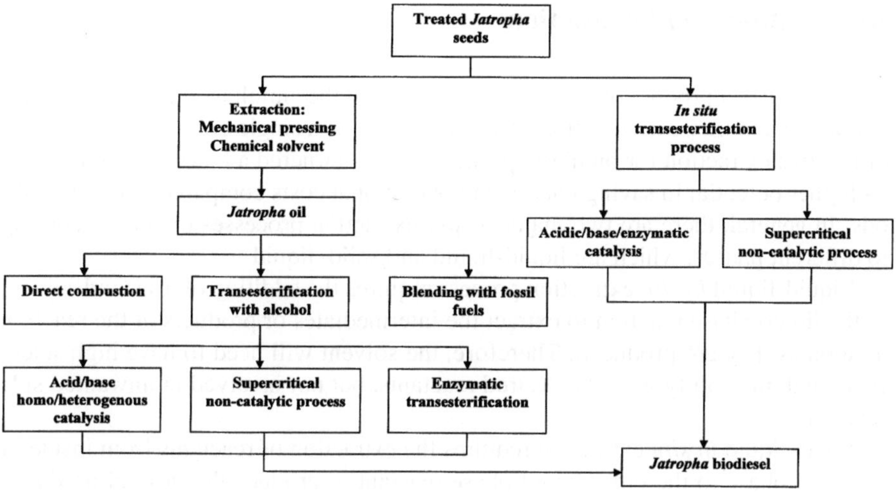

Fig. 28.1 Reaction pathway for the conversion of *J. curcas* seeds into biodiesel

There are several pathways available to convert JCL into biofuels as depicted in Fig. 28.1. Several of these processes have been commercially established while a few others are still in intensive research phase. Conventionally, JCL seeds will first undergo a process of oil extraction to obtain the oil and other lipids. There are many oil extraction processes available, which can be grouped as either mechanical or chemical. Mechanical methods, such as pressing and expulsion, are simple and easy to be carried out individually (Karaj and Muller 2011), but are only suitable for small scale process due to lower yield. Advanced chemical extraction processes, such as aqueous oil extraction (Shah et al. 2005), enzyme-assisted three phase partitioning (Shah et al. 2004) and supercritical CO_2 extraction (Min et al. 2010) can provide higher yield, but are more suitable for large industrial scale. Afterwards, it can be subjected to direct combustion for energy or blending with fossil fuels. Kumar et al. (2003) found that direct JCL oil injection in diesel engines lead to lower brake thermal efficiency, higher energy consumption and higher exhaust gas temperature than fossil diesel. Meanwhile, blending of JCL oil with methanol was also found to be unfavourable due to increase in ignition delay and higher emissions of HC and CO (Kumar et al. 2003). Consequently, transesterification of JCL oil into biodiesel with the aid of short-chain alcohol is highly preferred from a technical point of view. Apart from the more popular and established catalytic transesterification, more advanced transesterification techniques are being investigated including supercritical non-catalytic transesterification and also enzymatic transesterification. Recently, the search for a low cost process of biodiesel production has promoted the study of *in situ* transesterification, which combines the extraction and transesterification process in a single step.

In Situ Biodiesel Production

In situ biodiesel production is generally referred to the simultaneous process where extraction and transesterification proceed together in a single unit operation. *In situ* or reactive extraction for biodiesel production has garnered a lot of attention due to its higher potential in saving energy and operational costs compared to other methods. In general, there are two major *in situ* extraction processes divided according to different phases, which are liquid-liquid and solid–liquid.

Liquid-liquid *in situ* extraction process requires the addition of a second solvent to the liquid phase reaction to extract the intermediates or products of the reaction as soon as they are produced. Therefore, the solvent will need to have high selectivity and miscibility with the desired reactants, but not involved in unwanted side reactions.

Solid–liquid *in situ* extraction requires the extraction of reactants from inside of the solid phase to the other liquid phase reactant to produce the desired products. Solid crops biomass, such as oil seeds, trunks and leaves contain a lot of desirable extracts including lipids, HC, fatty acids and specialty chemicals. After being extracted, they can then be further processed and refined into higher value products. In the context of biodiesel production, solid–liquid *in situ* transesterification enabled the direct contact between the oil-bearing solid materials (seeds) with the transesterification reagent (usually a short-chain alcohol). The alcohols will also act simultaneously as a oil extraction agent. This will eliminate the step of oil extraction process, which usually consumes a lot of energy and time. Furthermore, potential loss of oil yield can be minimized by reducing the number of industrial operations. The main drawback for *in situ* transesterification is normally due to the chemical properties of the transesterification reagent. Conventionally in biodiesel production, short-chain alcohols, such as methanol and ethanol are highly preferred as transesterification reagents due to their low cost and high reactivity. Unfortunately, they have low miscibility with triglycerides or fatty acids, which resulted in low extraction efficiency. However, this can be circumvented with the help of appropriate catalysts (acid or base), co-solvents or supercritical conditions to increase the kinetic of the extraction process substantially. Most of the *in situ* transesterification processes for biodiesel production are summarized in Table 28.4.

The first trials of *in situ* transesterification go back to 1963 when it was employed in analytical studies to detect fatty acid compositions of materials from different organisms (Abel et al. 1963). When analytical conditions, such as water content, catalyst, heating and non-polar solvents are appropriate, *in situ* transesterification can provide the same qualitative and quantitative results compared to conventional methods with less experimental variations. Later on, Carrapiso et al. (2000) in their analytical study of fatty acids profile in Iberian pig subcutaneous adipose tissue discovered that *in situ* transesterification gave higher fatty acid concentration for all types of fatty acids compared to the conventional method of lipid extraction followed by a transesterification step. The study showed that higher amount of lipids could be obtained by *in situ* transesterification and that oxidation of reactive unsaturated fatty

Table 28.4 Comparison of different *in situ* transesterification methods applied by different authors for the production of biodiesel

Year	Process	Materials	Objectives	Operating conditions	Results	Literature
1963	Acid catalytic reaction	Bacteria Boron trichloride Methanol	Developing an effective analytical method for lipids from bacteria for gas chromatography analysis	0.5 g of bacteria 10 ml of methanol 1 g of BCl_3 12 min	Usage of *in situ* transesterification in microorganism classification through lipids profile offer higher speed and simplicity	Abel et al. (1963)
1993	Acid catalytic reaction	Rice bran H_2SO_4 Methanol Ethanol	Comparing the effect of different reacting solvents to *in situ* transesterification of rice bran	50 g rice bran 200 ml solvent 5 ml concentrated H_2SO_4 1–4 h	Different transesterification solvents give rise to different esters compositions and purity	Ozgul-Yucel and Turkay (1993)
1996	Acid catalytic reaction	Soybean Methanol Ethanol H_2SO_4 n-propanol n-butanol	Comparing the effect of different reacting solvents to *in situ* transesterification of soybean	50 g of soybean 150 ml of solvent 6 ml concentrated H_2SO_4 65°C temperature 3 h	The yield and purity of the products are highly dependent on the extraction process	Kildiran et al. (1996)
2003	Acid catalytic reaction	Rice bran H_2SO_4 Ethanol n-propanol iso-propanol	Comparing the effect of different reacting solvents to *in situ* transesterification of rice bran	50 g rice bran 200 ml solvent 5 ml H_2SO_4 1 h	Solubility of neutral oil in different monohydroxy alcohols will affect the quality and yield of monoester content	Ozgul-Yucel and Turkay (2003)
2004	Base catalytic reaction	Soybean Methanol NaOH	Investigating the potential of using base catalytic *in situ* transesterification for soybean biodiesel production	5 g of soybean 7.5–30 ml methanol 0.05–0.5 N NaOH 2.5–10 h	Alkaline catalyst achieved higher extraction of oil from soybean and transesterification	Haas et al. (2004)
2006	Base catalytic reaction	Soybean Methanol NaOH	Investigating the effect of moisture removal from soybean for *in situ* transesterification	5 g of soybean 10–20 ml methanol 0.07–0.13 N NaOH 4–16 h	Removal of moisture from soybean seeds can reduce the usage of methanol by 60% and catalyst by 56%	Haas and Scott (2007)

(continued)

Table 28.4 (continued)

Year	Process	Materials	Objectives	Operating conditions	Results	Literature
2007	Base catalytic reaction	DDGS MBM Methanol NaOH	Determining the feasibility of alternative feedstock (DDGS and MBM) for biodiesel production using *in situ* transesterification	5 g of feedstock 12–24 ml methanol 0.2–0.5 N NaOH 0.5–4.5 h	Both feedstocks were found to be suitable for biodiesel production after subjected to certain pre-treatment to maximize the extraction and transesterification process	Haas et al. (2007)
2008	Base catalytic reaction	Cottonseeds NaOH Petroleum ether Methanol	Investigating the application of *in situ* transesterification for biodiesel production using cottonseeds	25 g milled cottonseeds 100–200 ml methanol 0.05–0.10 mol/L NaOH 30–65°C 1–5 h	High yield of biodiesel was successfully obtained from cottonseeds with the help of co-solvent petroleum ether	Qian et al. (2008)
2009	Acid and base catalytic reaction with carbon dioxide-expanded methanol	Soybean CO_2 NaOH H_2SO_4 Methanol	Exploring the possibility of using gas-expanded system for *in situ* transesterification	22.5 g of soybean 27–54 ml sodium methoxide (30% in alcohol) 0–3.6 N H_2SO_4 7.38 MPa pressure 25°C, 57°C and 121°C 0.5–10 h	CO_2-expanded methanol was effective in lowering the solvent requirements (by one-third) and increasing reaction time (by 2.5 fold) without affecting biodiesel yield and quality for *in situ* transesterification of soybean	Wyatt and Haas (2009)
2011	Base catalytic heterog-enous reaction	Microalgae (*Nanno-chloropsis* sp.) Mg-Zr solid base catalyst Methanol Methylene dichloride	Studying the role of heterogeneous catalyst for i*n situ* transesterification of microalgae	1.0 g of dry microalgae 22.5–60.0 ml of solvent (Methanol/methylene dichloride = 2:1 v/v) 2–10 wt% of Mg-Zr 65°C temperature 4 h	Higher biodiesel yield was achieved compare to conventional biodiesel production method and the catalyst can be separated easily	Li et al. (2011)

acids after extraction was also lower. Considering the application of *in situ* transesterification to biodiesel production, Harrington and D'Arcy-Evans (1985) were among the first to successfully produce biodiesel from homogenized sunflower seeds by *in situ* acid-catalyzed transesterification. According to their method, 20 g of sunflower seeds were macerated in 150 ml of methanol with 6 ml of concentrated sulphuric acid (H_2SO_4) and heated under reflux condition in a reaction flask. *In situ* transesterification process was investigated in order to decrease the lipid losses due to imperfect separation of hull-kernel and reduce the usage of highly inflammable solvent (hexane). Yield improvement of over 20% with similar product properties was being reported compared to conventional synthesizing method (pre-extraction before transesterification) due to improved interaction between solvent (methanol) and lipids with the aid of acidic catalyst. Acid labile lipids with lower hexane miscibility were found to react *in situ* and contributed to higher biodiesel yield. Following Harrington's study, Siler-Marinkovic and Tomasevic (1998) investigated further the properties of operating parameters corresponding to the model of *in situ* transesterification of sunflower seeds. Employing similar setup as in Harrington's experiment, Siler-Marinkovic and Tomasevic analyzed the effect of methanol to oil molar ratio, amount of catalyst, reaction temperature and time of transesterification process. The results of their investigations showed that the molar ratio of alcohol to oil was the dominant factor. Molar ratios greater than 200:1 should generally be applied to achieve a good conversion. Room temperature (30°C) was found to be adequate for the process although higher reaction temperature (up to 65°C) could decrease the reaction time from 4 h to slightly more than 1 h. In optimum conditions with the weight of H_2SO_4 equal to the weight of oil in the seeds, a 98% (w/w) transesterification rate could be achieved. The properties of the esters produced by *in situ* transesterification were found to be essentially the same as those obtained by the conventional process although the cloud point was slightly lower. In another experiment, Zeng et al. (2009) pre-treated sunflower seeds with sodium hydroxide (NaOH), methanol and diethoxymethane (DEM), a co-solvent. They found that the three molar ratios (methanol/oil, DEM/oil and catalyst/oil) were the primary factors affecting biodiesel yield. The molar ratio of methanol required for optimum biodiesel yield was successfully reduced from above 200:1 to 101:1 upon DEM addition. A 97.7% biodiesel yield containing 0.74% FFA was obtained in 13 min with a DEM/oil molar ratio of 57.85:1 under room temperature and agitation speed of 150 rpm. The study concluded that DEM-assisted *in situ* transesterification of sunflower seeds could promote faster reaction with lower amount of catalyst (NaOH) and methanol. Thus, DEM would play the roles of both extraction solvent and catalyst.

Rice bran is readily available and cheaper than many refined vegetable oils, which makes it attractive for biodiesel production. Thus, Ozgul-Yucel and Turkay (1993) investigated the *in situ* transesterification of rice bran with sulphuric acid as a catalyst with respect to the comparative effect of methanol and ethanol. About 50 g of rice bran was mixed with 200 ml methanol or ethanol and refluxed with 5 ml concentrated H_2SO_4 under magnetic stirring for 1–4 h. They concluded that the relative composition and amount of esters were dependant on the choice of the solvent.

On one hand, the amount of methyl esters obtained by *in situ* transesterification with methanol was directly proportional to the amount of fatty acid content in rice bran. On the other hand, *in situ* transesterification with ethanol gave less pure ethyl esters due to higher miscibility of oil components contained in rice bran. Further investigations were also carried out by the same authors on rice bran with different monohydroxy alcohols (ethanol, n-propanol, iso-propanol and n-butanol) in order to study the effects of FFA, moisture and alcohol chain size on *in situ* transesterification (Ozgul-Yucel and Turkay 2003). With the same experimental setup as that of the previous work, these authors found that more FFA and less neutral oil (total oil content without FFA, which includes partial glycerides and unsaponifiable matter) will dissolve in the alcohol phase by *in situ* transesterification than by conventional solvent extraction with ethanol. This might be due to the dependency on solubility of each component of oil in the alcohol for *in situ* esterification and alcoholysis. For feedstock with high FFA content such as rice bran (up to 80% and above), total ester (biodiesel) yield depended strongly on solubility and conversion of FFAs in the alcohol phase. Increasing solubility of other oil components (neutral oil) in the alcohol phase is correlated with decreasing of FFA solubility and lower biodiesel yield. The experimental work showed that the solubility of neutral oil in alcohol increased with increasing molecular weight of alcohol, i.e., a decreasing solubility of FFAs in alcohols of increasing chain length. Consequently, the authors concluded that methanol was the optimal solvent for *in situ* transesterification of rice bran for biodiesel production. In contrast, *in situ* transesterification with soybean led to different results than rice bran (Kildiran et al. 1996). In this system, methanol alcoholysis gave the lowest esters yield (22%) compared to higher chain alcohols, such as ethanol and propanol using H_2SO_4 as catalyst. This showed that methanol was a poor extraction solvent for soybean and since the reaction rate depends heavily on the extraction phase, the biodiesel quality and quantity will be affected. Soybean oil typically has low FFA content and therefore is largely unaffected by the decreasing FFA solubility in the alcohol phase as for rice bran. Due to the low transesterification yield with acidic catalyst, Haas et al. (2004) tested *in situ* transesterification on soybean using alkaline catalyst, NaOH at normal room temperature with a 226:1 molar ratio of methanol to oil and a 1.6:1 molar ratio of catalyst to oil. On one hand, they found a 95% of lipids extraction in soybean after 8 h of maceration and as a consequence, 5% of lipids remained in the soybean residue after completion of the reaction. On the other hand, the methyl ester conversion rate of extracted lipids was 84% and this brings the total biodiesel conversion rate from soybean seed homogenate to be at 80% (w/w). This conversion rate was almost four times that of acid-catalyzed *in situ* transesterification. One of the reasons proposed for such a difference of lipid extraction rate was that alkaline alcohols could destroy intracellular compartmentalization in soybean and thus promote faster solubilisation and transesterification. However, both studies still required a large amount of solvent for the process. Thus, Haas and Scott (2007) tested the dehydration effect of soybean seed homogenate before *in situ* transesterification. Soybean flakes were mixed with NaOH dissolved in methanol in screw-capped bottles at room temperature. The bottles were shook continuously by orbital shaking to keep the flakes well suspended. A 100%

transesterification was achieved within 10 h with 12 ml of 0.1 N NaOH in methanol. On the other hand, the soybean preparation with native moisture required 30 ml of 0.09 N NaOH in methanol to yield 97% biodiesel. The authors believed that most phospholipids in soybean flakes were converted to esters, which in turn led to a slight increase of maximum biodiesel yield.

In another study, Wyatt and Haas (2009) investigated CO_2 to reduce the amount of solvents required for *in situ* transesterification of soybean. Supercritical CO_2 had been actively explored in the past to replace hexane as the extraction solvent. However, direct use of supercritical CO_2 for *in situ* transesterification was deemed impractical due to its apolar nature, which decreases the solubility of many polar reagents and catalysts. Furthermore, its supercritical conditions of operation require large pressure (>100 bar) and reactor volume. In order to negate these effects, CO_2 was dissolved in traditional liquid solvent to form a gas–liquid mixture capable of both increasing the solubility of other reagents and promote faster reaction. The experiment was carried out in a stainless steel reactor of 300 ml. CO_2 was introduced into the reactor either through manual bleeding through gas valve or with a syringe pump. In alkaline-catalyzed reaction, the introduction of CO_2 into the system actually decreased the biodiesel yield from 91.0% to 62.2% with a majority of un-reacted lipids remaining in soybean flakes. The reason for lower biodiesel yield in CO_2-solvent mix was the formation of carbonic acid derivative between CO_2 and sodium methoxide (MeONa), which increased acidity of the system and neutralized the alkaline catalyst. By contrast, acidic-catalyzed *in situ* transesterification was able to take advantage of the CO_2 added to the system by maintaining biodiesel yield at 88% even though the amount of MeONa was lowered from 54 ml to 36 ml. Increase in reaction speed by as much as 2.5 fold was also observed upon CO_2 addition. The authors also suggested that CO_2 increment results in larger liquid phase by increasing solubility of non-polar lipids, which in turn resulted in higher transesterification rate. Moreover, because of its inherent acidity when dissolved in liquid phase, CO_2 acts as an acidic catalyst. Haas et al. (2007) tested alkaline catalyzed transesterification on two other potential biodiesel feedstocks: *distillers dried grains with soluble* (DDGS, distillers grains are co-products produced from the fermentation of grains for alcohol) and *meat and bone meal* (MBM, a product of the rendering industry). They concluded that certain solid pre-treatment, such as grounding and drying was essential to improve *in situ* transesterification from lipid-bearing materials. By using *central composite response surface* (CCRS) to optimize the process, DDGS at 6.9% moisture content required 24 ml of methanol for complete lipid transesterification (99% yield). Removal of 70% of the moisture successfully reduced the methanol volume to 14 ml. In addition, preliminary neutralization of feedstocks acidity also prevented saponification upon alkaline catalyze (NaOH) and, thus, allowed the use of lower NaOH amount for effective transesterification. *In situ* transesterification using homogenous alkaline catalyst was also performed on cottonseed by Qian et al. (2008). In this study, cottonseed homogenate was incubated with methanol and petroleum ether as co-solvent. Under optimum conditions, approximately 98% conversion of oil to biodiesel could be obtained with 0.1 mol/l NaOH and 135:1 methanol to oil molar ratio within 3 h. Recycling of methanol was also investigated by mixing recovered methanol with fresh methanol in the subsequent experimental run. The optimum recycling ratio was found to be 0.2. Beyond this value, the biodiesel yield was discovered to drop substantially due to impurities. *In situ* transesterification also helped to reduce the amount of free and total gossypol levels in cottonseed meal far below the level recommended by *Food and Agricultural Organization* (FAO) standard. The seed homogenate resulting from *in situ* transesterification was, thus, suitable as animal protein feed source, i.e., a co-product with significant economical potential. Li et al. (2011) tried heterogeneous solid alkaline catalyst for *in situ* transesterification of microalgae. Solid catalyst can be separated easily from the residue and thus contributes to process simplicity and waste streams reduction. Mg-Zr is a solid alkaline catalyst that can be prepared in the laboratory and mixed with dried microalgae powder (*Nannochloropsis* sp.) in an amended Soxhlet extractor with reflux and magnetic stirrer. The reactor was heated to 65°C with an oil bath and took 4 h for complete reaction. A mixture of methanol and methylene dichloride in a volume ratio of 2:1 was used as the extraction and transesterification solvent. At optimum biodiesel conversion, the transesterification yield was only 28% (w/w) of the total lipid content. This value (28%) seems to be low for an optimum yield, but total lipids of microalgae contain significant amount of non-fatty acid lipids, which are not convertible to methyl esters. The transesterification conditions were 45 ml of solvent and 10% (w/w) catalyst while for the conventional two-step extraction-transesterification process, only 22.2% (w/w) maximum biodiesel yield could be obtained.

Process Applicability to *Jatropha*

Even though *in situ* transesterification is very promising, not all oilseed feedstocks are suitable for this process. Since extraction and transesterification processes proceed simultaneously, the solid oil-bearing material must have adequate surface area for the oil to be extracted easily by the reactants. This can be manually achieved by certain solid pre-treatment, such as grounding and sieving (Lim and Lee 2011). Although solid pre-treatments will incur additional cost, adequate application can almost double the extraction yield and biodiesel content, which in turn can compensate for the initial investment depending on the scale and duration. In this context, JCL seeds can undergo solid pre-treatment processes relatively easily due to its moderate sturdiness and good porosity. Furthermore, it also has relatively high oil content and uniform oil distribution, which make it suitable for simultaneous extraction and transesterification process. Composition of JCL oil is also favorable for *in situ* transesterification since it does not contain too many impurities for side-reactions. On the other hand, JCL seedcake can be easily separated by simple filtration, which simplifies downstream separation process. However, JCL oil contains high rate of FFA (>1.0 wt%) and moisture, which is detrimental to biodiesel yield by conventional NaOH transesterification due to soap formation. Hence, *in situ* biodiesel production is expected to improve the cost-effectiveness of JCL biodiesel. Currently, two reaction pathways have been investigated for *in situ* transesterification processes, which are catalytic and non-catalytic. Catalytic *in situ* transesterification normally involves homogenous alkaline or acidic catalyst, such as NaOH and H_2SO_4, respectively. Enzymatic *in situ* transesterification is not normally carried out due to enzyme deactivation by alcohol and glycerol. However, it is becoming feasible through substitution of short-chain alcohols by short-chain alkyl acetates as alternative acyl acceptors in the transesterification process (Su et al. 2007). The *in situ* transesterification of JCL seeds is summarized in Table 28.5.

Alkaline Catalytic Reaction

The most common alkaline catalyst used in transesterification is NaOH. For alkaline catalytic *in situ* transesterification, Kaul et al. (2010) successfully produced up to 98% biodiesel yield with NaOH as catalyst and methanol as reagent. Hailegiorgis et al. (2011) investigated the effect of adding *Cetyltrimethylammonium bromide* (CTMAB), a *phase transfer catalyst* (PTC) for *in situ* transesterification of JCL seeds with ethanol and NaOH. CTMAB was mixed with ethanol and added to the JCL seed homogenate in the reactor. Almost 99.5% of biodiesel yield was obtained compared to 89.2% without the use of PTC. Ginting et al. (2012) studied up to three different alkaline catalysts (NaOH, potassium hydroxide and *sodium methoxide*—MeONa) for the same *in situ* ethanolysis transesterification process and reported 99.98% of biodiesel yield with MeONa as the best catalyst.

Kaul et al. (2010) also studied the effect of particle size of seed homogenates in the ranges <0.85 mm, 0.85–2.46 mm and >2.46 mm. They found that JCL seed homogenate in the range of >2.46 mm gave the most consistent results compared to the other two. This was not in agreement with the other two studies as both of them sieved their seeds homogenate to particle size between 0.3 and 0.5 mm. The differences may be due to the effect of moisture content in the seeds since Kaul et al. (2010) did not dry their homogenate before transesterification. In addition, the seed shell has also been included to the homogenate, which was not the case in the other two studies. Kaul et al. (2010) also concluded that 9.85 ml/g (methanol/seed) was the best ratio compared to 3.28 ml/g and 6.57 ml/g that were also investigated. By contrast, Hailegiorgis et al. (2011) obtained 9 ml/g (ethanol to seed) as the optimum ratio and 7.5 ml/g when PTC was being added; Ginting et al. (2012) fixed the ratio of ethanol to seeds at 7 ml/g according to their particular conditions. Hailegiorgis et al. (2011) proposed that PTC could form soluble complexes, which would increase the reaction rate between ethanol and triglycerides in different immiscible phases.

From the three catalyst concentrations studied, Kaul et al. (2010) observed that the highest biodiesel yield was obtained at 3.94% (w/w). Lower concentration of catalyst was found to be insufficient to catalyze the reaction due to high FFA content. On the other hand, Hailegiorgis et al. (2011) successfully reduced the amount of NaOH required for optimal reaction by introducing PTC. With 1 molar ratio of

Table 28.5 Comparison of different *in situ* transesterification methods applied for the production of biodiesel using JCL seeds

Year	Solvent (molar ratio solvent: seed)	Catalyst	Miscellaneous	Reaction time (hours)	Operating temperature and pressure	Biodiesel yield (%)	Literature
2007	Methyl acetate Ethyl acetate (7.5 ml/g)	30% w/w enzyme (lipase)	5 g of feed Stirrer : 180 rpm	36	50°C Room pressure	86.1 87.2	Su et al. (2007)
2009	Methanol 7.5 ml/g n-hexane 0.75 ml/g	H_2SO_4 15 wt%	20 g of feed Size:<0.355 mm	24	60°C Room pressure	99.8	Shuit et al. (2010a)
2010	Methanol 9.85 ml/g	NaOH 0.1 M	14 g of feed Seed size: >2.46 mm Mechanical stirring	0.5	65°C Room pressure	98	Kaul et al. (2010)
2010	Methanol 10.5 ml/g	H_2SO_4 21.8 wt%	20 g of feed Size:<0.355 mm	10	60°C Room pressure	98.1	Shuit et al. (2010b)
2010	Methanol 10.0 ml/g n-hexane 2.5 ml/g	Supercritical condition	20 g of feed Size:<1.0 mm	1	300°C 24 MPa	103.5	Lim et al. (2010)
2011	Methanol (1.5:1) molar ratio n-hexane 2.5:1 ml/g	Self-germinated lipase	4 g of feed Stirrer: 180 rpm	8	35°C Room pressure	90.2	Gu et al. (2011)
2011	Ethanol 9.0 ml/g	NaOH 0.675wt% CTMAB/NaOH 1:1 molar ratio	20 g of feed Size: 0.3–0.5 mm Stirrer: 400 rpm	2.5	30°C Room pressure	99.5	Hailegiorgis et al. (2011)
2011	Ethanol 7.0 ml/g	NaOH 2.5 wt% KOH 2.5 wt% MeONa 2.0 wt%	20 g of feed Size: <0.5 mm Stirrer: 600	2	30°C Room pressure	99.98	Ginting et al. (2012)
2011	Methanol 13.3 ml/g n-hexane 3.3 ml/g	Supercritical condition	15 g of feed Size:<1.0 mm	1	300°C 24 MPa	128.78	Lim and Lee (2011)

PCT to NaOH being added, CTMAB managed to reduce NaOH usage by 33.3%, i.e., from 1.013% (w/w) to 0.675% (w/w). Ethanolysis usually required higher amount of catalyst compared to methanol due to lower miscibility of lipids in ethanol. Ginting et al. (2012) noted that MeONa at 2.0% (w/w) provided the highest biodiesel yield compared to NaOH and KOH. The absence of hydroxide group in MeONa apparently reduces the rate of saponification, which would otherwise reduce the total biodiesel yield. Ginting et al. (2012) found that biodiesel yield remains almost constant in the temperature range studied (30–70°C). Thus, they concluded that the temperature factor was not important. In contrast, Hailegiorgis et al. (2011) found that 50°C reaction temperature was the optimum for biodiesel production. However, the optimum temperature dropped to 30°C with the addition of CTMAB. Both studies of *in situ* ethanolysis transesterification stated around 2 h for complete reaction. By contrast, *in situ* methanolysis transesterification is faster with 30 min for complete reaction (Kaul et al. 2010). These results prove that shorter chain alcohol is more reactive than longer chain for *in situ* transesterification.

Acid Catalytic Reaction

In an experiment of *in situ* transesterification of JCL seed homogenate catalyzed with H_2SO_4, Shuit et al. (2010a) investigated the role of (1) n-hexane used as co-solvent to enhance methanolysis, (2) homogenate particle size and (3) reaction time on biodiesel yield. They found that particle size smaller than 0.355 mm were more favorable both in terms of extraction efficiency and biodiesel yield. The higher biodiesel yield at small particle size was attributed to the fact that smaller particles had less mass transfer resistance and larger surface area for reaction. The drawback of this method was the long reaction time (24 h) required to reach 99.8% biodiesel yield. In order to reduce the reaction time, the authors conducted an optimization study using CCRS based on the same experimental setup, i.e., reaction temperature (30–60°C), reaction time (1–24 h), methanol to seed ratio (5–20) and catalyst loading (5–30% w/w), but without n-hexane. According to this optimization, catalyst loading was found to have the most significant effect on biodiesel yield while methanol to seed ratio affected the least. The low interaction of methanol amount with biodiesel yield was attributed to the fact that a large volume of methanol was already used with the consequence that the range of variation explored would not affect much the biodiesel yield. Interaction effect was also detected from the catalyst loading and the reaction temperature. When the catalyst loading was low (<11.35% w/w), increase in temperature did not have any significant effect on the biodiesel yield. However, at higher catalyst loading (23.75% w/w), an increase of reaction temperature from 38°C to 53°C can increase the biodiesel yield by almost 50%. This proved that the *in situ* transesterification of JCL seed homogenate was a catalyst-dependent reaction and that the reaction time could be reduced to 10 h with a rate of biodiesel as high as 98.1%.

Enzymatic Reaction

The enzymatic *in situ* transesterification of JCL seed homogenate has been carried out with Novozym435 (a lipase from *Candida antarctica*) (Su et al. 2007). In order to avoid enzyme deactivation, methyl and ethyl acetate were used as the *in situ* solvents. In this system, glycerol is no longer produced, but replaced by triacetylglycerol or triacetin. Triacetin could potentially improve biodiesel properties, such as cloud point and pour point. In comparison with the alcohol based conventional two-step transesterification, the *in situ* transesterification of JCL showed improvement by as much as 39% biodiesel yield for methyl acetate and 9% for ethyl acetate. The authors pointed out that the improvements might be due to the poisoning effect of short-chain alcohol on the enzyme activity. Moreover, n-hexane extracts higher amount of impurities (phospholipids) from the seed homogenate compared to alkyl acetate and these impurities may also affect the enzyme activity. Additional investigation of enzymatic kinetic of *in situ* transesterification showed that biodiesel yield increased exponentially during the first 12–16 h. After that period, the rate of reaction decreased slowly until reaching almost zero after 36 h. This decreasing rate efficiency may be due to the depletion of fatty acids and the increase of triacetylglycerol. An optimum point for the effect of solvent to seed ratio was found at 7.5. Below that value, the solvent was found to be insufficient to perform the reaction effectively. Meanwhile, exceeding that value would cause oil dilution and thus result in lower biodiesel yield. The water content in the seed homogenate was another important parameter since enzymes requires water for their biological activity. However, too much water would result in hydrolytic reactions, which would compete with the transesterification reaction. The optimum water content for JCL seed homogenate was found to be 4.62%. The highest biodiesel yield was found to be 86.1% and 87.2% for methyl and ethyl acetate, respectively. In this case, the difference between methyl acetate and ethyl acetate was almost negligible at the same optimum conditions.

One of the main drawbacks to use enzymes as catalysts is their prohibitive cost that makes them economically unattractive. In order to overcome the cost factor, Gu et al. (2011) took advantage of the plant lipase found in germinating JCL seeds to carry out the *in situ* transesterification with methanol and n-hexane. Plant lipase had the advantage of being low cost and easy to purify. In this context, lipase derived from germinating JCL seeds could replace the role of catalyst and thus reduce the cost in chemicals and separation process. Gu et al. (2011) showed that the lipase activity was found to be at its highest level 4 days after germination and that the oil content still remained at similar level. N-hexane was chosen as the co-solvent on the basis that it is an effective solvent for lipid extraction and also promoted the activity of lipase. The optimal ratio of n-hexane to germinating seeds was found to be 2.5 ml/g; higher ratio value would dilute the oil and methanol interface. Since excessive methanol levels could inhibit the lipase activity, the proportion of methanol to germinated seeds was carefully adjusted at the optimum value of 1.5:1. The reaction temperature was kept at 35°C in order to prevent enzyme denaturation. The optimal water content of seeds was 2.9% allowing a rate of 87.2% biodiesel to be reached. While this technology is promising, more studies are required concerning seed germination and cleaning for commercial production.

Supercritical Reaction

Supercritical transesterification for biodiesel production has been thoroughly investigated for some time. Fluid in a supercritical phase can be considered as an intermediate state between liquid and gas. Several distinctive characteristics, such as low viscosity, high diffusion coefficients, variation of density and dielectric constant as a function of pressure were attributed to this special state. Consequently, supercritical fluids (SCF) are an excellent extraction solvents as well as chemical reaction reagents. In supercritical condition, lipids are completely miscible with reagents, such as methanol and ethanol, which enable the reaction to proceed at a high rate without the need for a catalyst. Moreover, non-catalytic supercritical fluids had also been proven to be superior for biodiesel production in terms of reaction time, product separation, *fatty acid methyl esters* (FAME) yield and process simplicity compared to conventional biodiesel processing. In addition, supercritical transesterification is normally insensitive to the water and FFA content. However, this process requires high temperature and pressure, which make the process risky and costly. A higher alcohol to oil ratio is also needed compared to conventional catalytic transesterification in order to obtain high FAME yield.

Lim et al. (2010) chose supercritical methanol for *in situ* transesterification of JCL seed homogenate. Supercritical methanol has milder supercritical condition (240°C, 8.1 MPa) and low boiling point (65°C) that allows easier products separation. N-hexane was also used as co-solvent to facilitate lipid extraction from seed homogenate. The effects of homogenate particle size, reaction temperature and pressure were investigated. Homogenate particle size <1.0 mm was found to provide the largest biodiesel yield. Although extraction efficiency was found to be high (65%) even at low temperature and pressure, biodiesel yield was low and only increased exponentially once supercritical condition of methanol was attained. Despite n-hexane to be an efficient extraction solvent at low temperature, it is only when supercritical methanol is formed that it can react with the extracted lipids to form biodiesel. The highest biodiesel yield was recorded at 103.5%, which exceeded the maximum theoretical yield from n-hexane extraction. This proved that supercritical condition had the potential to extract even more lipids from the JCL seeds when compared to conventional n-hexane solvent extraction (defined as 100%). The larger extraction rate by supercritical methanol was attributed to the high pressure exerted during supercritical condition, which expelled lipids trapped deep inside the core of the homogenized seeds. Lim and Lee (2011) found that seed homogenate pre-treatment, such as drying temperature, drying time, sieving and deshelling were important factors to ensure optimum biodiesel yield. However, since most of these pre-treatments are energy-intensive, their implementation is a compromise between cost and return. The authors suggested that decorticated seeds without sieving and drying at 75°C were sufficient to obtain a FAME yield as high as 128.8% using the same reference value as the study outlined above. From most of the comparison with conventional two-step biodiesel production method, supercritical *in situ* transesterification clearly exhibited higher biodiesel yield efficiency.

Conclusion

In situ transesterification process has shown promising future as a viable method to produce biodiesel. By eliminating the energy-intensive phase of solvent or mechanical extraction and their respective preparation stages, it might even be feasible for the people in rural area or under-developed countries to produce biodiesel on their own. However, there are still several hurdles that biofuel research needs to overcome in order to make it attractive for large-scale commercial biodiesel production. First of all, chemical reagent input still need to be reduced as currently the process requires a large amount of solvents, such as methanol or ethanol. Secondly, *in situ* transesterification mechanism is still not being fully understood and well established. Therefore, more research is needed to explore this mechanism, improve the reaction efficiency and find solvents that best fit the reaction. Last but not the least, by-products of the process should also be developed in order to improve the economic viability. If not toxic, the leftover from the homogenate of *in situ* transesterification could be utilised as either animal feed or subjected to further biomass processing and produce additional energy.

Acknowledgments The authors would like to acknowledge Universiti Sains Malaysia for the financial support given (Research University Grant No: 814062) and USM Vice-Chancellors Award of a student scholarship to Steven Lim.

References

Abel K, de Schmertzing H, Peterson JI (1963) Classification of microorganism by analysis of chemical composition 1. Feasibility of utilizing gas chromatography. J Bacteriol 85: 1039–1044

Achten WMJ, Verchot L, Franken YJ, Mathijs E, Singh VP, Aerts R et al (2008) *Jatropha* biodiesel production and use. Biomass Bioenergy 32:1063–1084

Aregheore EM, Makkar HPS, Becker K (1998) Assessment of lectin activity in a toxic and a non-toxic variety of *Jatropha curcas* using latex agglutination and haemagglutination methods and inactivation of lectin by heat treatments. J Sci Food Agric 77:349–352

Carrapiso AI, Timón ML, Petrón MJ, Tejeda JF, García C (2000) In situ transesterification of fatty acids from Iberian pig subcutaneous adipose tissue. Meat Sci 56:159–164

Currie J (2007) Food, feed and fuels: an outlook on the agriculture, livestock and biofuel markets. The Goldman Sachs Group. Available from: http://www.gceholdings.com/pdf/GoldmanReport FoodFeedFuel.pdf

Ginting MSA, Azizan MT, Yusup S (2012) Alkaline in situ ethanolysis of *Jatropha curcas*. Fuel 93:82–85. doi:10.1016/j.fuel.2011.08.062

Gu H, Jiang Y, Zhou L, Gao J (2011) Reactive extraction and in situ self-catalyzed methanolysis of germinated oilseed for biodiesel production. Energy Environ Sci 4:1337–1344

Haas JH, Scott KM (2007) Moisture removal substantially improves the efficiency of in situ biodiesel production from soybeans. J Am Oil Chem Soc 84:197–204

Haas MJ, Scott KM, Marmer WN, Foglia TA (2004) In situ alkaline transesterification: an effective method for the production of fatty acid esters from vegetable oils. J Am Oil Chem Soc 81:83–89

Haas MJ, Scott KM, Foglia TA, Marmer WN (2007) The general applicability of in situ transesterification for the production of fatty acid esters from a variety of feedstocks. J Am Oil Chem Soc 84:963–970

Hailegiorgis SM, Mahadzir S, Subbarao D (2011) Enhanced in situ ethanolysis of *Jatropha curcas* L. in the presence of cetyltrimethylammonium bromide as a phase transfer catalyst. Renew Energy 36:2502–2507

Harrington KJ, D'Arcy-Evans C (1985) Transesterification in situ of sunflower seed oil. Ind Eng Chem Prod Res Dev 24:314–318

Karaj S, Muller J (2011) Optimizing mechanical oil extraction from *Jatropha curcas* L. seeds with respect to press capacity, oil recovery and energy efficiency. Ind Crops Prod 34:1010–1016

Kaul S, Porwal J, Garg MO (2010) Parametric study of Jatropha seeds for biodiesel production by reactive extraction. J Am Oil Chem Soc 87:903–908

Kildiran G, Ozgul-Yucel S, Turkay S (1996) In situ alcoholysis of soybean oil. J Am Oil Chem Soc 73:225–232

Kumar MS, Ramesh A, Nagalingam B (2003) An experimental comparison of methods to use methanol and *Jatropha* oil in a compression ignition engine. Biomass Bioenergy 25:309–318

Li Y, Lian S, Tong D, Song R, Yang W, Fan Y et al (2011) One-step production of biodiesel from *Nannochloropsis* sp. on solid base Mg–Zr catalyst. Appl Energy 88:3313–3317

Lim S, Lee KT (2011) Effects of solid pre-treatment towards optimizing supercritical methanol extraction and transesterification of *Jatropha curcas* L. seeds for the production of biodiesel. Sep Purif Technol 81:363–370

Lim S, Hoong SS, Teong LK, Bhatia S (2010) Supercritical fluid reactive extraction of *Jatropha curcas* L. seeds with methanol: a novel biodiesel production method. Bioresour Technol 101:7169–7172

Min J, Li S, Hao J, Liu N (2010) Supercritical CO_2 extraction of *Jatropha* oil and solubility correlation. J Chem Eng Data 55:3755–3758

Ozgul-Yucel S, Turkay S (1993) In situ esterification of rice bran oil with methanol and ethanol. J Am Oil Chem Soc 70:145–147

Ozgul-Yucel S, Turkay S (2003) Fatty acid monoalkylesters from rice bran oil by in situ esterification. J Am Oil Chem Soc 80:81–84

Pandey KK, Pragya N, Sahoo PK (2011) Life cycle assessment of small-scale high-input *Jatropha* biodiesel production in India. Appl Energy 88:4831–4839

Pramanik K (2002) Properties and use of *Jatropha curcas* oil and diesel fuel blends in compression ignition engine. Renew Energy 28:239–248

Qian J, Wang F, Liu S, Yun Z (2008) In situ alkaline transesterification of cottonseed oil for production of biodiesel and nontoxic cottonseed meal. Bioresour Technol 99:9009–9012

Sato S, Hirakawa H, Isobe S, Fukai E, Watanabe A, Kato M et al (2010) Sequence analysis of the genome of an oil-bearing tree, *Jatropha curcas* L. DNA Res 18:65–76

Shah S, Sharma A, Gupta MN (2004) Extraction of oil from *Jatropha curcas* L. seed kernels by enzyme assisted three phase partitioning. Ind Crops Prod 20:275–279

Shah S, Sharma A, Gupta MN (2005) Extraction of oil from *Jatropha curcas* L. seed kernels by combination of ultrasonication and aqueous enzymatic oil extraction. Bioresour Technol 96:121–123

Shuit SH, Lee KT, Kamaruddin AH (2010a) Reactive extraction and in situ esterification of *Jatropha curcas* L. seeds for the production of biodiesel. Fuel 89:527–530

Shuit SH, Lee KT, Kamaruddin AH, Yusup S (2010b) Reactive Extraction of *Jatropha curcas* L. seed for Production of Biodiesel: Process Optimization Study. Environ Sci Technol 44:4361–4367

Siler-Marinkovic S, Tomasevic A (1998) Transesterification of sunflower oil in situ. Fuel 77: 1389–1391

Su EZ, Xu WQ, Gao KL, Zheng Y, Wei DZ (2007) Lipase-catalyzed in situ reactive extraction of oilseeds with short-chained alkyl acetates for fatty acid esters production. J Mol Catal B: Enzym 48:28–32

The Global Exchange for Social Investment (GEXSI 2008) Global Market Study on Jatropha and our project in Madagascar. JatrophaWorld 2008 Miami

Wyatt VT, Haas MJ (2009) Production of fatty acid methyl esters via the in situ transesterification of soybean oil in carbon dioxide-expanded methanol. J Am Oil Chem Soc 86:1009–1016

Zeng J, Wang X, Zhao B, Sun J, Wang Y (2009) Rapid in situ transesterification of sunflower oil. Ind Chem Eng Res 48:850–856

Chapter 29
Combustion of *Jatropha curcas* Oil, Methyl Esters and Blends with Diesel or Ethanol in a CI Engine

N.R. Banapurmath, V.S. Yaliwal, Y.H. Basavarajappa, S.S. Hiremath, S.V. Khandal, R.S. Hosmath, N.M. Girish, A.V. Tumbal, and P.G. Tewari

Introduction

Vegetable oils are a very promising alternative to diesel oil since they are renewable and have similar properties. Many investigators have studied the use of vegetable oils in diesel engines. Vegetable oils offer almost the same power output with slightly lower thermal efficiency when used in diesel engine (Srivastava and Prasad 2000; Vellguth 1983; Demirbas 2005; Jajoo and Keoti 1997; Recep et al. 2000; Pramanik 2003; Nabi et al. 2006). Reduction of engine emissions is a major research aspect in engine development with the increasing concern on environmental protection and the stringent exhaust gas regulation. Vegetable oils are a mixture of straight chain organic compounds that are mainly triglycerides with a number of branched chains of different lengths. Research towards the use of vegetable oils as fuels for

N.R. Banapurmath (✉) • R.S. Hosmath • N.M. Girish • A.V. Tumbal • P.G. Tewari
Department of Mechanical Engineering, B.V.B. College of Engineering and Technology, Hubli, 580031, India
e-mail: nr_banapurmath@rediffmail.com; nr_banapurmath@bvb.edu

V.S. Yaliwal
Department of Mechanical Engineering, S.D.M. College of Engineering and Technology, Dharwad, 580002, India

Y.H. Basavarajappa
Department of Mechanical Engineering, Tontadarya College of Engineering, Gadag, India

S.S. Hiremath
Department of Mechanical Engineering, Basaveshwar Engineering College, Bagalkot, 580002, India

S.V. Khandal
Department of Mechanical Engineering, KLEIT, Chikodi, 580002, India

N. Carels et al. (eds.), *Jatropha, Challenges for a New Energy Crop: Volume 1: Farming, Economics and Biofuel*, DOI 10.1007/978-1-4614-4806-8_29,

compression ignition (CI) engine has yielded encouraging results (Agarwal and Agarwal 2007; Zhang and Wang 2007; Radhwan et al. 2007; Scholl and Sorenson; 1993; Demirbas 2005; Carmen et al. 2007; Nwafor 2000). The oil from *Jatropha curcas* L. (Ratanjyot, hereafter referred to as Jatropha) is extracted from its seeds. The use of neat vegetable oils causes some problems to CI engines when used for long time. These problems are due to high viscosity, low volatility and polyunsaturated character of neat vegetable oils (Srivastava and Prasad 2000; Vellguth 1983). Some of the common problems of vegetable oils in diesel engines in the long run are coking, trumpet formation on the injectors, carbon deposits, oil ring sticking, thickening and gelling of lubricating oil as a result of contamination by the vegetable oils (Vellguth 1983). Different methods, such as pre-heating, blending, ultrasonically assisted methanol transesterification and supercritical methanol transesterification (Demirbas 2005; Carmen et al. 2007; Nwafor 2000; Barsic and Humke 1981) are being used to reduce the viscosity and make vegetable oil suitable for engine applications. In the present investigation, biodiesel of Jatropha, i.e., *methyl esters of Jatropha's oil* is referred to JOME. Biodiesel properties were determined and their combustion and emission characteristics were studied on a four-stroke single-cylinder direct-injection CI engine to check their feasibility as CI engine fuels.

Characterization of Jatropha Oil and JOME

Jatropha is a large shrub and belongs to the family Euphorbiaceae occurring almost throughout India and tropics. It has a long productive period of around 50 years and could yield economic returns annually.

Properties of Jatropha Oil, JOME and Blends with Ethanol or Diesel

The properties of diesel, neat Jatropha oil, JOME, a blend of 80% diesel with 20% JOME (B20) and a blend of 80% JOME with 20% ethanol (BE20) have been determined and summarized in Table 29.1. To prepare BE20, a stirrer was mounted inside the fuel tank in order to prevent separation of ethanol and JOME. The fuel blend was prepared just before starting the experiment to ensure that fuel mixture remains homogeneous during the experiment (Banapurmath and Tewari 2010). This procedure was sufficient since JOME does not contain aromatics, the ambient temperature was higher than 30°C and the ethanol used in this experiment was dry. In real life applications, an additive should be added to the mixture to stabilize ethanol and JOME phases. Akzo Nobel Surface Chemistry and Lubrizol Corporation developed and produced a low cost additive, which makes it possible to blend ethanol with diesel (the so called Dieshol) to get a stable and clear fuel (see Lü et al. 2004).

Table 29.1 Properties of diesel, Jatropha oil, JOME, B20 and BE20

Property	Diesel	Jatropha oil	JOME	B20	BE20
Density, kg m^{-3}	840	917	870	855	862
Specific gravity	0.840	0.917	0.870	0.872	0.862
Kinematic viscosity at 40°C, centistokes	3.5	44.5	5.65	3.99	4.4
Flash point, °C	56	280	170	75	140
Calorific value, Kj kg^{-1}	43,000	35,600	38,450	39,800	35,834
Cetane number	40–48	40	44	—	—

Availability and Economic Value of Oil

India has rich and abundant resources of both edible and non-edible oilseeds. The production of methyl/ethyl esters from edible oils is much more expensive than that of diesel fuels. This is due to the relatively high costs of vegetable oils (about 4 times the cost of diesel in India). Therefore, there is a need to search and explore alternative feedstocks for the production of biodiesels. Non-edible oil from source of Jatropha is easily available in many parts of India and is cheap compared to edible oils. Jatropha is a drought-tolerant, perennial plant with capability to grow on marginal soils requiring very little irrigation and grows in all types of soils. At present, the production of Jatropha seeds is about 0.8 kg m^{-2} per year. The oil content of Jatropha seed ranges from 30% to 40% (w/w) and from 45% to 60% when considering it on kernel basis. Fresh Jatropha oil is slow drying, odorless and colorless; it turns yellow with aging. The tree is occasionally seen on roadsides in India; Jatropha tree stands tall with some trees being over 5–10 ft high. Jatropha can grow on almost all soil types ranging from stony to sandy or to clayey, but prefers good aeration (Achten et al. 2008; Meher et al. 2004; Vivek Gupta 2004; Ghadge and Hifjur 2005). The price of crude petroleum and the cost of transporting diesel through long distances to remote markets play a key role in evaluating the economical feasibility of biodiesels. The cost of producing methyl/ethyl esters from edible oils is at present much more expensive than fossil diesel fuel. The cost of biodiesel can be reduced if non-edible oils are used instead of edible oils (Agarwal 2006). Barnwal and Sharma (2005) have presented the economic feasibility of different vegetable oils including edible and non-edible oils.

Experimental Heat-Release Rate Estimation

The heat-release rate of the fuel causes a variation of gas pressure and temperature within the engine cylinder. It strongly affects the fuel economy, power output and engine emissions (David et al. 1975; Brunt and Platts 1999; Hayes et al. 1986; Aman 1985; Rocco 1993; Homsy and Arvind 1997; Heywood 1988; Banapurmath et al.

2005). In-cylinder pressure and *top dead center* (TDC) signals were acquired through a high-speed digital data acquisition system connected to a computer. The data from 100 consecutive cycles were recorded and processed with homemade software to obtain combustion parameters. The heat release rate was calculated by first law analysis of the pressure crank angle data. A program was developed to obtain the ensemble averaged pressure crank angle data of 100 cycles. Heat release analysis is done within the framework of the first law of thermodynamics. For a single-zone model, cylinder contents were assumed to be homogeneous. The net heat-release rate using single-zone heat-release model is given below (Heywood 1988):

$$\frac{\partial Q_n}{\partial \theta} = (\frac{\partial Q_g}{\partial \theta}) + (\frac{\partial Q_w}{\partial \theta}) \tag{29.1}$$

$$\frac{\partial Q_n}{\partial \theta} = (\frac{\gamma}{\gamma - 1})\, p(\frac{\partial v}{\partial \theta}) + (\frac{1}{\gamma - 1}) v(\frac{\partial p}{\partial \theta}) + (\frac{\partial Q_w}{\partial \theta}) \tag{29.2}$$

Where,

γ- is specific heat ratio,
p- is the pressure at a given crank angle,
v is the volume of the cylinder at a given crank angle,
$\partial Q_n/\partial\theta$ is the net heat release rate,
$\partial Q_w/\partial\theta$ is the heat transfer rate to the cylinder wall,
$\partial Q_g/\partial\theta$ is the gross heat release rate.

Experimental Test Rig

Experiments were conducted on a four-stroke single cylinder direct-injection water-cooled CI engine as shown in Fig. 29.1 and whose specifications are given in Table 29.2. The engine was always operated at a rated speed of 1,500 rpm. It was operated on Jatropha oil and JOME at optimum injection timings (19° BTDC) and injection pressures (220 bars). For determining the optimum conditions, tests were conducted initially at three injection timings of 19°, 23°, 27° BTDC and injection pressures were varied from 200 to 300 bar. The fuel flow rate was measured on a volumetric basis using a burette and stopwatch. The engine and dynamometer were coupled. The *Engine Soft* software was used to get the outputs. Engine Soft serves most of the engine testing application needs including monitoring, reporting, data entry and data logging. The software evaluates power, efficiencies, fuel consumption, heat release and allows the plotting of pressure vs. crank angle diagrams. It is configurable as per engine setup. This Engine Soft helps to get various graphs at different operating conditions. The emission characteristics were measured by using Hartridge Smokemeter (London, UK) and equipment from AVL (New Delhi, India) during the steady state operation. The tests were conducted with neat Jatropha oil, JOME, B20 (diesel 80% + JOME 20%) and BE20 (JOME 80% + Ethanol 20%).

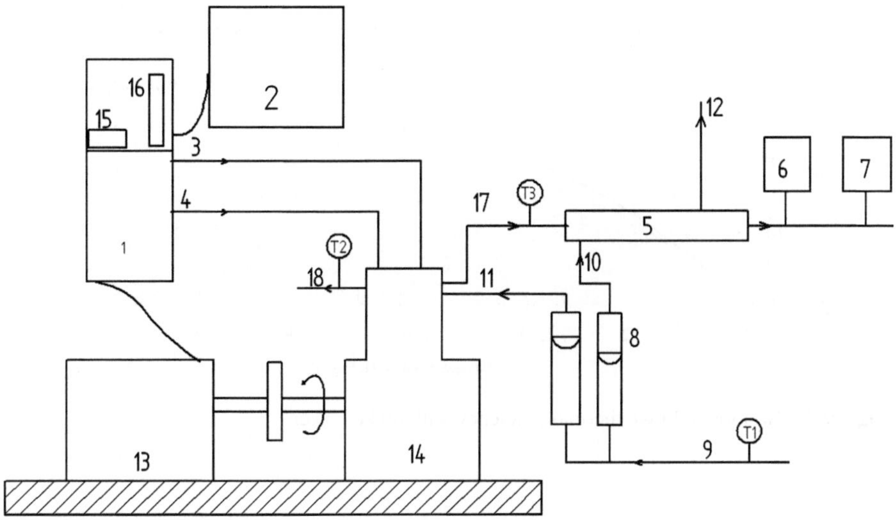

Fig. 29.1 Experimental set up of CI engine. (*1*) Control Panel, (*2*) Computer system, (*3*) Diesel flow line, (*4*) Air flow line, (*5*) Calorimeter, (*6*) Exhaust gas analyzer, (*7*) Smoke meter, (*8*) Rota meter, (*9*, *10*) Calorimeter inlet water temperature, (*11*) Inlet water temperature, (*12*) Calorimeter outlet water temperature, (*13*) Dynamometer, (*14*) CI engine, (*15*) Speed measurement, (*16*) Burette for fuel measurement, (*17*) Exhaust gas outlet, (*18*) Outlet water temperature, (*T1*) Inlet water temperature, (*T2*) Outlet water temperature, (*T3*) Exhaust gas temperature

Table 29.2 Specification of CI engine

Particulars	Specification
Make and type	Kirloskar, TV1, Single cylinder, four stroke CI engine.
Software used	Engine Soft
Brake power	5.2 kW
Bore (D) and Stroke (L)	87.5 and 110 mm
Compression ratio	17.5:1
Dynamometer	Eddy current, 7.5 kW at 1,500–3,000 rpm Model AG-10

Performance Characteristics

The effect of brake power on *brake thermal efficiency* (the dimensionless performance measure of a device that uses thermal energy) for diesel, Jatropha oil, JOME, B20 and BE 20 is shown in Fig. 29.2. There is a steady increase in efficiency as the engine load increases. The brake thermal efficiency is always lowest with the neat Jatropha oil compared to JOME. This is due to poor mixture formation obtained with Jatropha oil as a result of its lower volatility, higher viscosity and density

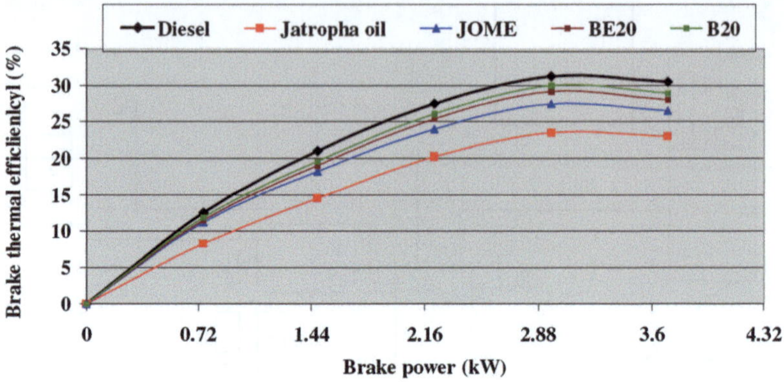

Fig. 29.2 Variation of brake thermal efficiency with brake power

compared to JOME. Blending diesel with JOME (B20) also resulted in improved overall properties of JOME and B20 resulted in performance closer to diesel engine operation with higher BTE compared to JOME. This might be attributed to improved spray pattern and reduced viscosity of B20 blend compared to pure JOME. Similarly, blending JOME with ethanol (BE20) results in higher BTE compared to JOME. This could be also due to reduced viscosity and better spray pattern of BE20 blend. The brake thermal efficiency obtained with Jatropha oil, JOME, BE20 and B20 were found to be 23.5%, 27.5%, 29.2% and 30.01% compared to 31.25% for diesel at 80% engine load, respectively.

Emission Characteristics

The values of smoke opacity recorded for Jatropha oil, JOME, BE20 and B20 were found to be 76, 67, 65 and 62 HSU, respectively, compared to 56 *Hartridge smoke unit* (HSU) for diesel in the engine operated at 80% load (Fig. 29.3). The smoke opacity for Jatropha oil was higher in comparison to JOME. The heavier molecular structure and higher viscosity of Jatropha oil results in poor atomization leading to higher smoke emission. Smoke opacity was observed to be higher for Jatropha oil and JOME compared to diesel. JOME performance is only improved further by blending with diesel (B20) or ethanol (BE20). However the calorific value of these blends still remains lower than that of fossil diesel. This could be attributed to the improved combustion of B20 probably because of its better mixing with oxygen as well as its better atomization and spray pattern due to its reduced viscosity. Similarly, BE20 also shows lower smoke opacity compared to JOME supposedly for the same reasons just outlined for B20. Another reason might be that significant portion of the injected fuel BE20 burns in the premixed mode, which results in lower smoke.

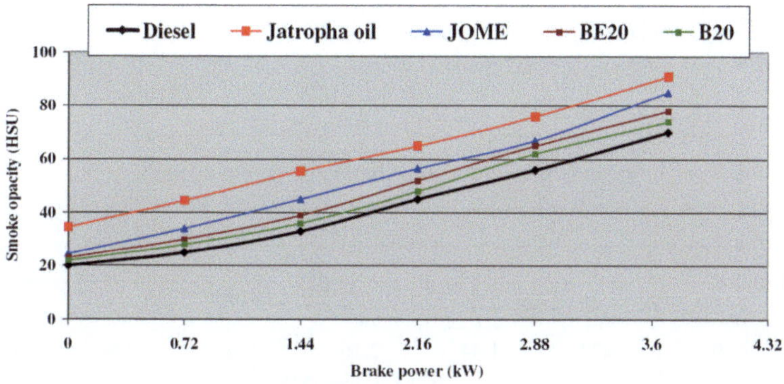

Fig. 29.3 Variation of smoke opacity with brake power

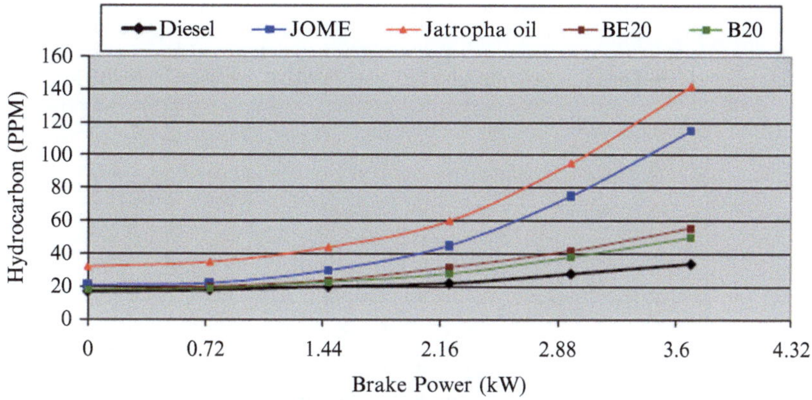

Fig. 29.4 Variation of hydrocarbon with brake power

The effects of brake power on *hydrocarbon* (HC) and *carbon monoxide* (CO) emissions for diesel, Jatropha oil, JOME, B20 and BE20 are shown in Figs. 29.4 and 29.5. HC emissions with neat Jatropha oil and JOME were found to be higher compared to the standard diesel oil, followed by BE20 and B20 fuel blends. Again, it is the relatively poor atomization and lower volatility of neat Jatropha, JOME and blends compared to diesel that is responsible for this trend. The HC emission levels were found to be 75, 95, 42 and 38 PPM for Jatropha oil, JOME, BE20 and B20, respectively, compared to 28 PPM for diesel in the engine operating at 80% load. However, B20 and BE20 resulted in nearly same HC emissions as of diesel at lower engine load operation, most probably because of the improved combustion associated with the better spray pattern and reduced viscosity of these blends. However, the HC emission levels increased at higher load because of free fatty acids (FFA) that can be still in significant proportion in JOME compared to biodiesel from other plant sources (Banapurmath et al. 2005).

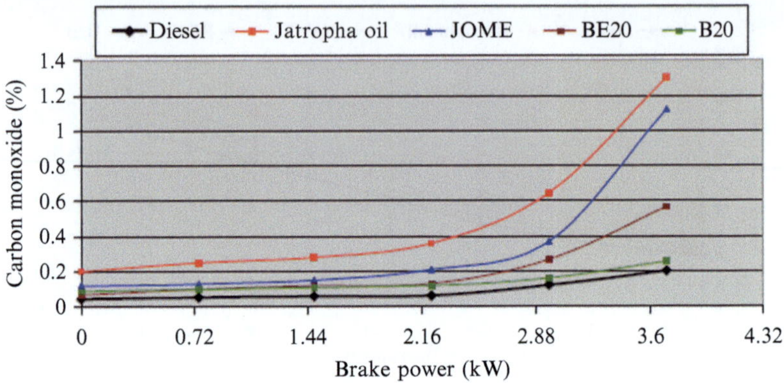

Fig. 29.5 Variation of carbon monoxide with brake power

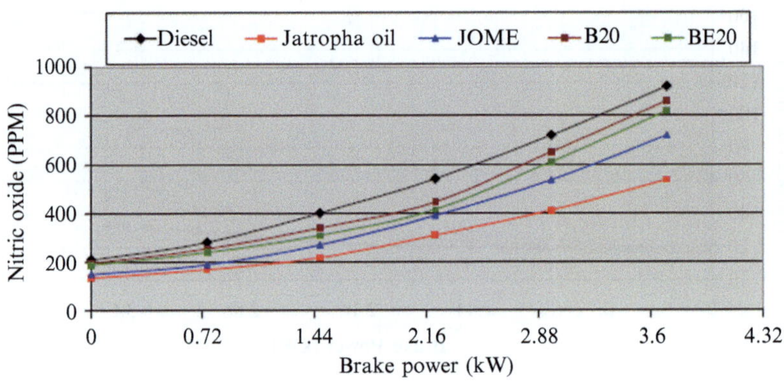

Fig. 29.6 Variation of nitric oxide with brake power

As expected, CO emission (Fig. 29.5) also showed a trend similar to that observed for HC. Carbon monoxide values were 0.645%, 0.37%, 0.265% and 0.16% for Jatropha oil, JOME, B20 and BE20, respectively, compared to 0.12% with diesel in operation at 80% engine load.

The effect of brake power on *nitrogen oxides* (NOx) emissions is shown in Fig. 29.6. NOx also contribute to greenhouse effect and should be avoided as much as possible. It has been observed that NOx emissions were higher for diesel followed by B20 and BE20 compared to neat Jatropha oil and JOME. Heat release rates of Jatropha oil and JOME and their blends were lower during premixed combustion phase, which will lead to lower peak temperatures. Nitrogen oxides formation strongly depends on peak temperature and oxygen availability, which explains the observed phenomenon. The NOx emissions obtained in the engine at 80% load

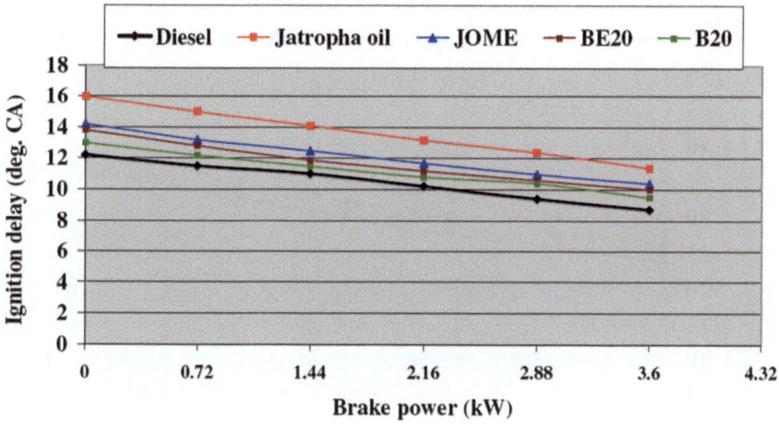

Fig. 29.7 Variation of ignition delay with brake power

with Jatropha oil, JOME, B20 and BE20 were 412, 535, 650 and 610 PPM, respectively, compared to 720 PPM for diesel. The NOx emissions for BE20 were observed to be lower than for diesel and B20 because of higher latent heat vaporization of ethanol slightly lowering the combustion temperature.

Combustion Characteristics

The effect of brake power on ignition delay is shown in Fig. 29.7. The ignition delay is calculated in relation to the injection timing. Values of the ignition delay were 12.4°, 11°, 10.6° and 10.4° of *crank angle* (deg. CA) for Jatropha oil, JOME, BE20 and B20, respectively, compared to 9.4 deg. CA with diesel in the engine at 80% load. Jatropha oil and JOME showed longer ignition delays compared to diesel. As expected, B20, BE20 blends showed shorter ignition delay than JOME and Jatropha oil, with B20 the closest to diesel. B20 and BE20 has shorter ignition delay compared to Jatropha oil and JOME because of their reduced viscosity, density and increased volatility that facilitate a better mixing of air and fuel inside the engine cylinder.

The combustion duration shown in Fig. 29.8 was calculated based on the duration between the start of combustion and 90% cumulative heat release. The combustion duration increased with an increase in the power output for all fuels because of the corresponding increase in the quantity of fuel injected. The combustion duration for the engine operating at 80% load was 42, 37.8, 35 and 33.6 deg. CA, for Jatropha oil, JOME, BE20 and B20 blends, respectively, compared to 32.4 deg. CA for diesel. The highest combustion duration was observed with Jatropha oil followed by JOME due to longer diffusion combustion phase. However, Fig. 29.8 shows that the

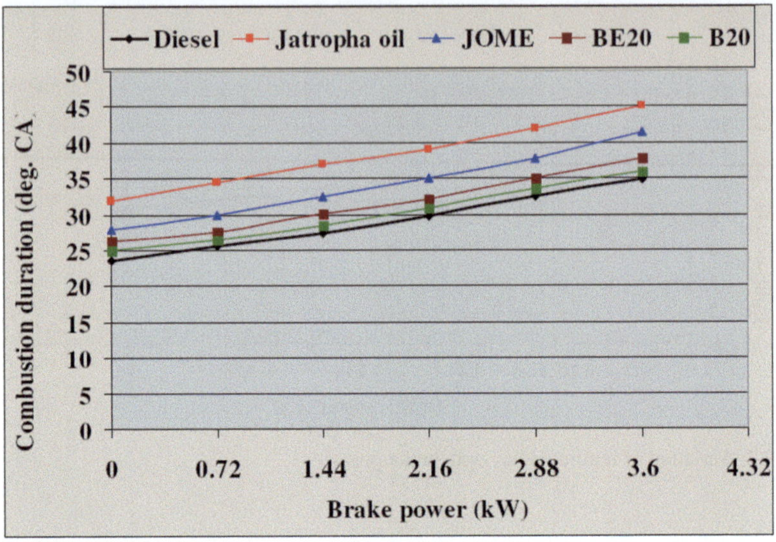

Fig. 29.8 Variation of combustion duration with brake power

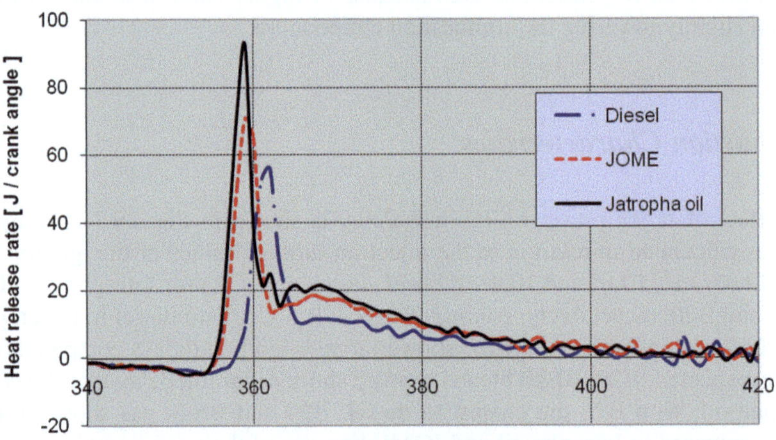

Fig. 29.9 Heat release rate for the tested fuels

combustion duration was reduced with B20 and BE20 blends compared to Jatropha oil and JOME because of a better mixing of air and fuel with the consequence of higher rate of heat release.

All fuels experience a rapid premixed burning (main peak) followed by a diffusion (shoulder) of combustion that is typical for naturally aspirated engines (Fig. 29.9). After the ignition delay period, the premixed fuel air mixture burns

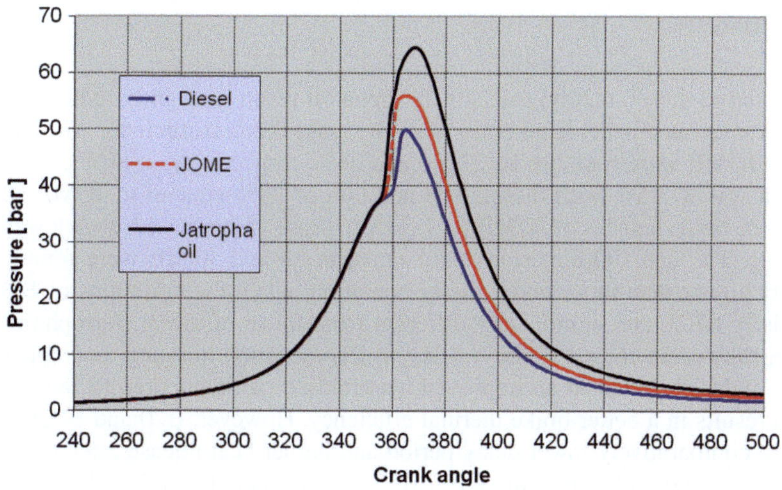

Fig. 29.10 Variation of peak pressure with crank angle

rapidly releasing heat at a very fast rate, followed by a diffusion of combustion where the burning rate is controlled by the availability of fuel–air mixture. The phase corresponding to premix burning is associated with a high rate of heat release that occurs quickly after the largest piston course when diesel is used as a fuel. Thus, under diesel operation, the peak combustion associated to higher peak pressure is transformed in mechanical work to push the piston from *top dead center* (TDC) back to *bottom dead center* (BDC). With neat oil and JOME, a larger part of the energy released by premixed burning is released before TDC and, thus, not efficiently transformed into work. This could be the reason for the higher thermal efficiency with diesel. However, JOME shows improvement in rate of heat release compared to neat Jatropha oil. The rate of heat release for Jatropha oil and JOME is found to be higher during the phase of combustion diffusion compared to diesel. The higher combustion rates during the diffusion stage of Jatropha oil lead to higher exhaust temperatures and lower thermal efficiency.

The variation of peak pressure with crank angle ($p - \theta$ diagram) is shown in Fig. 29.10. The observed values of peak pressure for Jatropha, JOME and diesel were 68.2, 58.5 and 50 bars, respectively, at full engine load. In a compression ignition engine, peak pressure depends on the combustion rate in initial stages, which in turn is influenced by the amount of fuel taking part in the uncontrolled combustion phase. The premixed or uncontrolled combustion phase is governed by the ignition delay period and by the mixture preparation during the delay period. Higher viscosity and lower volatility of the Jatropha oil and JOME that leads to poor atomization and mixture preparation with air during the ignition delay period are the reasons for this observed trend.

Conclusion

Compared to diesel, methyl esters of Jatropha oil result in slightly reduced performances associated with higher emissions. The brake thermal efficiency with Jatropha oil and JOME were found to be 27.5% and 26%, respectively, at 80% engine load when it was 31.25% with diesel. The addition of 20% ethanol to JOME (BE20) brings the performances of JOME very close to those of diesel with even lower NOx emission. HC and CO emissions with Jatropha oil and JOME were found to be slightly higher than those under diesel operation; again the emissions of B20 and especially BE20 are significantly different from those of diesel. Jatropha oil and JOME showed increased ignition delay and combustion duration as compared to diesel. JOME resulted in an improved heat-release rate compared to Jatropha oil, which results in a better brake thermal efficiency. However, B20 and BE20 blends showed comparatively lower delay period and higher heat release rates very much comparable to diesel, especially when BE20 is considered. On the whole it is seen that operation of the engine was smooth on JOME and blends. The neat Jatropha oil and JOME tested results in a slightly reduced thermal efficiency and increased smoke, HC and CO levels. Thus, the existing CI engine can be operated on JOME and blends without any major engine modification.

References

Achten WMJ, Verchot L, Franken YJ, Mathijis E, Singh VP, Aerts R et al (2008) Jatropha biodiesel production and use. Biomass Bioenergy 32:1063–1084

Agarwal AK (2006) Biofuels (alcohols and biodiesel) applications as fuels for internal combustion engines. Int J Prog Energy Combust Sci 33:233–271

Agarwal D, Agarwal AK (2007) Performance and emission characteristics of a *Jatropha* oil (preheated and blends) in a direct injection compression ignition engine. Int J Appl Therm Eng 27:2314–2323

Aman CA (1985) Cylinder pressure measurement and its use in engine research. Paper no. 852067, Society of Automotive Engineers, USA

Banapurmath NR, Tewari PG (2010) Performance, combustion, and emissions characteristics of a single-cylinder compression ignition engine operated on ethanol–biodiesel blended fuels. Proc I MechE Part A: J Power Energ 224:533–543

Banapurmath NR, Tewari PG, Basavarajappa YH, Yaliwal VS (2005) Performance of Honge (*Pongamia pinnata*) oil blends in a diesel engine. In: XIX National conference on innovations in civil engineering (XIXNCICEC), Annamalai University, Chidambaram, 21–23 Dec 2005

Barnwal BK, Sharma MP (2005) Prospects of biodiesel production from vegetable oils in India. Int J Renew Sust Energ Rev 9:363–378

Barsic NJ, Humke AC (1981) Performance and emission characteristics of a naturally aspirated diesel engine with vegetable oil fuels. Paper no. 810262, Society of Automotive Engineers, USA, pp 95–109

Brunt MFJ, Platts K (1999) Calculation of heat release rate in direct injection diesel engines. Paper no. 1999-01-0187, Society of Automotive Engineers, USA

Carmen S, Vinatoru M, Maeda Y (2007) Aspects of ultrasonically assisted transesterification of various vegetable oils with methanol. Ultrason Sonochem 14:380–386

David LR, Krieger RB, Lienesh JH (1975) Measurement and analysis of engine pressure data. Paper no. 750026, Society of Automotive Engineers, USA
Demirbas A (2005) Biodiesel production from vegetable oils via catalytic and non-catalytic supercritical methanol transesterification methods. Int J Prog Energ Combust Sci 31:466–487
Ghadge SV, Hifjur R (2005) Biodiesel production from Mahua (*Madhuca indica*) oil having high free fatty acids. Biomass Bioenergy 28:601–605
Hayes TK, Savage LD, Soreson SC (1986) Cylinder pressure data acquisition and heat release analysis on a personal computer. Paper no. 860029, Society of Automotive Engineers, USA
Heywood JB (1988) Internal combustion engine fundamentals. McGraw-Hill, New York
Homsy SC, Arvind A (1997) An experimental heat release rate analysis of a diesel engine operating under steady state conditions. Paper no. 970889, Society of Automotive Engineers, USA
Jajoo BN, Keoti RS (1997) Evaluation of vegetable oils as supplementary fuels for diesel engines. IN: Proceedings of the XV national conference on I.C. engines and combustion, Anna University, Chennai
Lü XCL, Yang JG, Zhang WG, Huang Z (2004) Effect of cetane number improver on heat release rate and emissions of high speed diesel engine fueled with ethanol–diesel blend fuel. Fuel 83:2013–2020
Meher LC, Naik SN, Das LM (2004) Methanolysis of *Pongamia pinnata* (Karanja) oil for production of biodiesel. J Sci Ind Res 63:913–918
Nabi MdN, Akhter MdS, Mhia Md, Zaglul S (2006) Improvement of engine emissions with conventional diesel–biodiesel blends. J Bioresour Technol 97:372–378
Nwafor OMI (2000) Effect of advanced injection timing on the performance of rapeseed oil in diesel engines. Int J Renew Energy 21:433–444
Pramanik K (2003) Properties and use of *Jatropha curcas* oil and diesel fuel blends in compression ignition engine. Int J Renew Energy 28:239–248
Radhwan MS, Ismail MA, Elfeky SMS, Abu-Elyazeed MOS (2007) Jojoba methyl ester as a diesel fuel substitute: preparation and characterization. Int J Appl Therm Eng 27:314–322
Recep A, Selim C, Huseyin SY (2000) The potential of using vegetable oil fuels as fuel for diesel engines. Int J Energ Convers Manag 42:529–538
Rocco VDI (1993) Diesel engine in cylinder pressure data analysis under TDC setting error. Paper no. 930595, Society of Automotive Engineers, USA
Scholl KW, Sorenson SC (1993) Combustion analysis of soyabean oil methyl ester in a direct injection diesel engine. Paper no. 930934, Society of Automotive Engineers, USA
Srivastava A, Prasad R (2000) Triglycerides-based diesel fuels. Renew Sust Energ Rev 4:111–133
Vellguth G (1983) Performance of vegetable oils and their monoesters as fuels for diesel engines. SAE paper no. 831358, Society of Automotive Engineers, USA
Vivek Gupta AK (2004) Biodiesel production from Karanja oil. J Sci Ind Res 63:39–47
Zhang H, Wang J (2007) Oil from biomass corncob tar as a fuel. Int J Energy Convers Manag 48:1751–1757

Chapter 30
Potential of Jatropha as an Energy Crop

Sébastien Bonnet and Shabbir H. Gheewala

Introduction

Biofuels have gained recognition worldwide as major substitutes to fossil fuels particularly in the transport sector. As a result of energy security issues, oil price volatility and environmental concerns, many nations have investigated and set policies promoting the domestic production and conversion of biomass feedstock into biofuels. It is anticipated that by 2050, biofuels could provide 27% of total transportation fuel and contribute notably to the replacement of diesel, kerosene and jet fuel (IEA 2011).

Because of national initiatives for biofuel promotion, bioethanol and biodiesel have seen their production increase by 23% and 43% over the period 2001–2007. At present, biofuels (100 billion liters) provide 3% of road transport fuel (on an energy basis) worldwide. In Brazil, a pioneer in the development of biofuel industry, biofuels meet 21% of the road transport demand while they meet just about 4% in the US and 3% in EU countries with however a target of 10% to be achieved by 2020. In India, an emerging giant along with China in the world economy, a 20% share of biofuel in the transport sector is targeted by 2017 (Silalertruksa 2011; IEA 2011).

The development of biofuels is supposed to provide new economic opportunities and drive countries towards improved living and environmental standards. The use of biofuels from locally produced feedstocks is expected to reduce nations' dependency on imported fossil energy and contribute to climate change mitigation via

S. Bonnet • S.H. Gheewala (✉)
The Joint Graduate School of Energy and Environment,
King Mongkut's University of Technology Thonburi, Bangkok, Thailand

Center for Energy Technology and Environment, Ministry of Education,
Bangkok, Thailand
e-mail: shabbir_g@jgsee.kmutt.ac.th

N. Carels et al. (eds.), *Jatropha, Challenges for a New Energy Crop: Volume 1: Farming, Economics and Biofuel*, DOI 10.1007/978-1-4614-4806-8_30,

reduced global warming. In developing countries largely depending on agriculture, biofuels are expected to contribute to the revitalization of the agricultural sector and to boost rural economy. This is notably the case for several important biofuel producers in Asia including, China, India, Indonesia, Malaysia, the Philippines and Thailand.

However, it is also important to recognize that the increased use and trade of biofuels need to be sustainably framed to avoid other potential impacts including (1) issues of land use and land use change, (2) food versus fuel competition, (3) increase of other environmental problems, such as water stress, (4) marginalization of small farm holders and (5) cost issues affecting their competitiveness with fossil fuel.

Feedstocks for Biodiesel Production

Biodiesel is one of the renewable fuels that can be used in the transport sector as an alternative to fossil diesel. Biodiesel can be produced from vegetable oils extracted from oilseed crops the most promising of which are oil palm, rapeseed, sunflower and *Jatropha curcas* L. (referred to as Jatropha, hereafter). It can also be produced from animal fats and waste oils (Sukkasi et al. 2010; IEA 2011). These oils and fats can be converted to biodiesel via different methods, the most common being the chemical process referred to as transesterification requiring vegetable oil (or fat), ethanol (or methanol) and a catalyst (usually sodium or potassium hydroxide or methoxide). Biodiesel can be used in regular diesel engines without any modification for its most common use in the form of B1 (1% biodiesel with 99% diesel), B2 or B20. Biodiesel contains about 90% of the energy content of conventional diesel, but has a higher cetane number and provides better lubrication, which tends to increase the overall fuel economy (USAID 2009; Uriarte 2010).

There are regional differences in the types of feedstocks grown worldwide for oil production based on suitability of local conditions including weather and soil characteristics among others. For instance, soybean is the largest feedstock grown in the US for oil production, while rapeseed is largely grown in Europe, Germany being the largest producer and oil palm is a major feedstock in tropical countries, Malaysia and Indonesia being the top producers. Fruit and oil yield of feedstocks are important parameters for successful commercialization of biodiesel. Table 30.1 gives

Table 30.1 Yields of some important oilseed feedstocks

Feedstock	Yield of oil (kg/ha)	Yield of oil (liters/ha)
Soybean	375	446
Rapeseed	1,000	1,190
Palm oil	5,000	5,950
Jatropha	1,590	1,892

Based on Uriarte (2010)

average values of these parameters for important biodiesel feedstocks that may however vary according to climatic conditions, soil type and agricultural inputs, such as water and fertilizers. It is oil palm that provides the largest fruit and oil yield, which is the reason for its extensive cultivation in Malaysia and Indonesia as well as in Thailand even if at a much lower scale. Jatropha also proves to be a potential energy crop given its fruit and oil yield are larger than rapeseed or soybean.

Jatropha, an Alternative Feedstock for Biodiesel Production

Characteristics of Jatropha

Jatropha belongs to the Euphorbiaceae family. It is a woody drought resistant bush or small tree (3–10 m high), which can grow on marginal land requiring only small amount of water, fertilizer and pesticides to grow. Jatropha originates from South America and is now widely found in tropical and sub-tropical regions in Africa and Asia. Areas for Jatropha cultivation are located in a belt ranging from 35° S to 30° N. However, this belt represents only an approximation of areas of potential Jatropha cultivation (Li et al. 2010). Thus, although Jatropha can grow in a wide range of conditions, a better crop performance is observed for cultivation under semi-arid conditions with average temperatures in the range of 20°C to 28°C and rainfall between 250 and 3,000 mm. Well drained sandy soil and gravely soils are also best suited for Jatropha cultivation (Openshaw 2000; Achten et al. 2008).

As an undomesticated plant, Jatropha seed yields are reported to be in the range of 0.4–12 t $ha^{-1}year^{-1}$ (Openshaw 2000; Achten et al. 2008; Li et al. 2010). However, such figures are still much debated as it depends on the extrapolation that has been made from annual yields of individual trees. Jatropha yields are affected by a wide range of parameters including site characteristics (rainfall, soil type and fertility), genetics, plant age, cultivation practices (propagation method), spacing between trees, pruning, fertilization, irrigation, etc. Seed yields are reported to range between 0.2 kg and 2 kg per tree. In semi-arid areas, dry seed production is in the range of 2–3 t $ha^{-1}year^{-1}$. However, for sites with good soil and water supply (average annual rainfall of 900–1,200 mm) or with optimal management practices, dry seed yields of 5 t $ha^{-1}year^{-1}$ can be achieved (Achten et al. 2008). These estimations are somewhat consistent with other studies which estimate a dry seed yield in the range of 0 (desert) to 7.62 t $ha^{-1}year^{-1}$ (tropical rainforest) or 1.5–7.8 t $ha^{-1}year^{-1}$ (Jongschaap et al. 2007; Li et al. 2010).

Although Jatropha could grow in variable conditions including on poor soil and with limited water resources (only rainfall), it is quite clear that Jatropha yields are also much influenced by the conditions under which the trees are cultivated. But, before promoting Jatropha for biodiesel production as for any other energy crop, it is imperative that a full chain analysis be considered as a preliminary step to address the energy gain or loss of renewable energy production. This is exemplified in the next section using Thailand as a case study.

Energy Potential of Jatropha

Biodiesel in Thailand

Thailand is an agricultural country that heavily relies on fossil energy importation. Hence, the country has been intensively promoting the use of renewable energy including from biomass over the decade of 2000–2010. The introduction of liquid biofuels in the transportation sector is one of the options pursued by the Thai government. Thailand is in fact the only Asian economy to date to embrace biofuels in the main consumer markets (Sukkasi et al. 2010), with clear policy targets. With regards to biodiesel, according to the 15 years renewable energy development plan (2008–2022), national targets of production by years 2011, 2016 and 2022 have been set to 3, 3.64 and 4.5 million liters/day, respectively (DEDE 2008). B2 is available nationwide and B5 has already been launched in the market as a voluntary program with more than 3,400 service stations offering this blend (Sukkasi et al. 2010). According to policy targets for biodiesel, B10 is expected to be available for nationwide use from 2012 (EPPO 2009).

Although biodiesel is mainly produced from palm oil, Jatropha is also being considered by the Thai government which has encouraged research and development before setting commercial planting. Already several private initiatives for Jatropha cultivation have been implemented and several small scale production units were launched at community level for local production and use of crude Jatropha oil. The oil price of Jatropha is 3–10 times larger than that of diesel at a commercial scale jeopardizing its competitiveness in Thailand. However, although Jatropha is generally considered as a perennial plantation (economic life up to 35 years, Uriarte 2010), annual plantation is also an option that may offer benefits since it is a fast growing tree that could yield fruit from the first year (Foidl et al. 1996). Aside from the oil, co-products including wood from the tree and peel from the seed could be used for energy while seed cake, a co-product from the seed oil extraction process, could be used either as an energy source or as organic fertilizer. This option may help to improve the energy balance of Jatropha and therefore lead to economic and environmental (notably GHG) benefits. Such benefits can be assessed at an initial stage via evaluation of the energy balance since fossil energy contribute to a large extent to the costs and GHG emissions associated to biomass systems (Nguyen et al. 2007).

Net Energy Balance and Net Energy Ratio

The *Net Energy Balance* (NEB) is a method that can be used to evaluate the energy performance of a particular system. NEB is simply the difference between energy output and input. For fossil energy, the NEB is negative as expressed by the second law of thermodynamics, which dictates that when converting one form of energy to another more useful form some energy will be lost. However, for biofuels it may

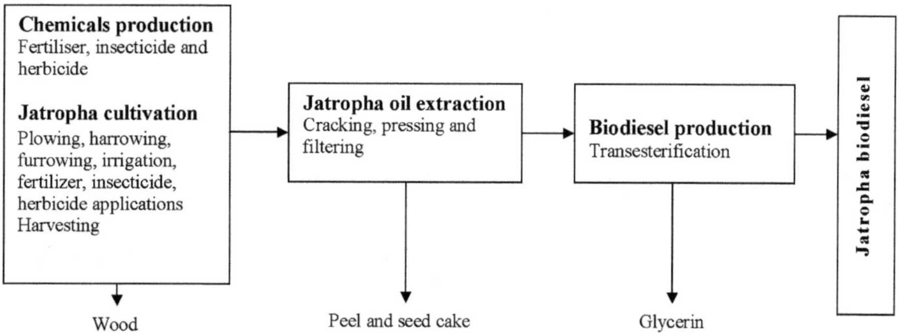

Fig. 30.1 Jatropha biodiesel production system

become positive since the solar energy consumed by green biomass during photosynthesis to build its organic matter is considered energetically free (Shapouri et al. 2006). NEB is therefore a useful indicator to evaluate the energy efficiency of biofuels. Another related indicator is the *net energy ratio* (NER), which is simply the ratio of energy output to energy input and which if above 1 indicates a net energy gain.

Jatropha Biodiesel Production Systems

As stated earlier, two plantation systems for Jatropha can be considered, which include the perennial plantation system for which Jatropha is cultivated over a 20 years period and the annual plantation where the tree is harvested every or few year(s). For energy assessment, Jatropha plantation, oil extraction process and biodiesel production unit as well as transportation at all stages need to be considered. The overall system of Jatropha biodiesel production is depicted in Fig. 30.1.

In terms of energy inputs, one includes: (1) fertilizer production and field application, (2) Jatropha cultivation and fruit harvest and (3) biodiesel production. In terms of energy outputs, one includes: (1) biodiesel from Jatropha oil and (2) all co-products including wood, peel, seed cake and glycerin. The factors used for the energy assessment are reported in Table 30.2. Although data are relevant for Thailand, they are also consistent with information available from the international literature. More details about the inventory for Jatropha plantations as well as information related to transportation can be found in Prueksakorn and Gheewala (2008).

Perennial Plantation

For perennial plantations, as depicted in Fig. 30.1, land preparation includes ploughing, harrowing and furrowing and is done only in the first year of plantation. The crop density is an important parameter since along with rainfall and soil nutrient

Table 30.2 Energy factors for Jatropha biodiesel production

Product	Unit (MJ/kg)	Product	Unit (MJ/kg)
Fertilizer production:		Diesel use:	
– Nitrogen (N)	87.9	– Fuel energy per kg of diesel	43.1 (36.4 MJ/L)
– Phosphorus (P)	26.4	– Energy for producing diesel	9.6 (8.1 MJ/L)
– Potassium (K)	10.5	Electricity use: – For present electricity mix in Thailand	36 MJ electricity from 100 MJ primary energy
Herbicides production:		JCL biodiesel	37.3–39.65
– Glyphosate	452.5	Crude glycerin	25.6
– Paraquat	458.4	Wood (air dry)	16.54–15.5
Methanol (MeOH)	38.08	Peel (air dry)	11.1–13.07
Seed cake (as fertilizer)	6.22	Seed cake (as fuel stock)	18.81–25.1

Note: (a) specific gravity of diesel = 0.845 kg/L, (b) specific gravity of JME = 0.88 kg/L
Based on Prueksakorn et al. (2010)

level it affects fruit yield. For the case of Thailand, the crop density is in the range of 1,100–3,300 trees per ha depending on the site. Since water supply is important for good fruit yield, irrigation is generally applied, except in areas where rainfall exceeds 2,500 mm per year. Fertilizers are applied in the field to ensure good nutrient supply, i.e., 625 kg of NPK (15:15:15) per year, on average. Herbicides are also used along with manual removal and animal grazing. On average, 3 L glyphosphate (48% w/v) per ha per year are used during the rainy season and 2 L paraquat (27.6% w/v) per ha are used during the dry season. Insecticides are usually not used.

Ratoon Plantation

The study of the potential benefits of a ratoon plantation (planting each five years) of Jatropha is rather new and therefore only one pilot study is available in Thailand at Kasetsart University. The crop density for such a plantation is 10,000 trees per ha. Land preparation is performed once every 5 years. Water is supplied via irrigation every 15 days. N, P, K fertilizers, are applied at an annual rate of 277, 223 and 405 kg per ha, respectively. As for perennial plantation, glyphosate (48% w/v) is also applied for ratoon plantation of Jatropha, but at a rate of 25 L per ha per year. The trees are cut every year and regeneration starts from the tap root.

Jatropha Oil Extraction and Biodiesel Production

Jatropha seeds are dried and seeds separated from the peels using a cracking machine working at a rate of 100–120 seeds per hour. The oil is extracted from the seeds using a five hp screw press (12.2–20 L oil per h) and processed with a two hp filtrating machine (150–170 L oil per h). An 80 L batch reactor (4 kWh energy requirement per batch) is used for the transesterification of Jatropha oil using methanol as the reactant and sodium hydroxide as the catalyst.

Product Outputs

For perennial plantation in Thailand, the yield of dry fruit is increasing during the first 6 years to reach 12–13 t per ha on an annual basis. It then remains stable for the remaining 14 years. The seed yield is 70% that of dry fruit and the oil is 23% that of seeds. The amount of dry wood collected during the first year is 9 t per ha, on average (from reducing tree density in some plantations from 1 × 1 m to 2 × 2 m) while for the last year it amounts to 4 t per ha. For ratoon plantations, a constant yield of 7 t of fruit per ha and 24.5 t of wood per ha are obtained each year. Independently of the plantation system, the conversion efficiency for biodiesel is 95% and the ratio of biodiesel to glycerin is 3.5:1.

Energy Assessment Results of Jatropha Biodiesel

Considering the system of perennial plantation for biodiesel production, investigations from Thailand show that with reasonable supply of water (irrigation) and nutrients (fertilizers), the NER is about 1.4. From an energy prospective, this indicates that Jatropha cultivation solely for the purpose of biodiesel production is an option of moderate interest.

When looking at the potential of including co-products in the assessment of energy balance, assuming that a market exists for such products, the NER of perennial plantation rises to 6 with a NEB of 4,720 GJ per ha of plantation over 20 years. For ratoon-annual plantation on the other hand a NER of 7.5 is achieved with a NEB of 9,860 GJ per ha over 20 years. Thus, although both plantation systems show a very positive net energy balance, the value for ratoon-annual plantations is twice that of perennial plantations. This is indicative of the potential benefits that ratoon plantations of Jatropha may provide when co-products are being valorized for energy purposes.

Looking further into the details of the energy assessment when co-products are valorized, it is observed, as shown in Fig. 30.2, that the highest contributor to energy cost for perennial plantations is from the agricultural management (38%) and biodiesel production (30%). Fertilizers and methanol production are contributing the largest life cycle energy consumption with 30% and 36%, respectively. For ratoon plantation, the agricultural management is by far the largest contributor to energy cost with 68%, fertilizer production dominating the life cycle energy cost with 53%. An important strategy to enhance the energy performance of both plantation systems would therefore be to optimize fertilizer application via good agricultural practices, which would also lead to potential economic and environmental benefits.

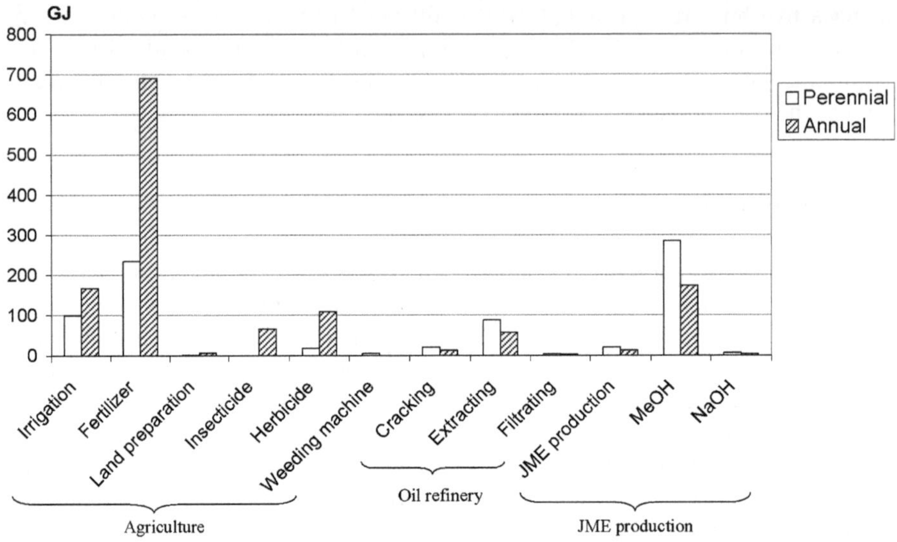

Fig. 30.2 Breakdown of energy inputs for ratoon-annual and perennial plantation systems

This is particularly true for ratoon plantations since energy consumption for fertilizer production is three times that for perennial plantations, i.e., 700 GJ versus 180 GJ (per ha over 20 years).

With regard to total energy output, Fig. 30.3 shows that for perennial plantation it is estimated to be about 6,000 GJ per ha over 20 years, half of which is from the seed cake. Jatropha biodiesel comes next contributing only half the energy output of seed cake or 1,500 GJ (per ha over 20 years). Other co-products including, peel, crude glycerin and wood contribute altogether another 1,500 GJ (per ha over 20 years). In this energy assessment, although seed cake is considered as a fuel source, it is important to recognize that it could also be used as organic fertilizer. It would then enable to substitute chemical fertilizers, which are rather energy intensive to produce as observed from the energy cost assessment detailed above. However, IFEU (2009) reported that estimates based on the use of seed cake as fertilizer showed reduced energy benefits compared to its use as an energy source. This is still to be balanced with potential environmental benefits that may arise from the use of seed cake as fertilizer as well as economic advantages of using seed cake as either fuel or fertilizer.

As stated earlier, a major interest in promoting ratoon plantation of Jatropha lies in the enhanced energy benefits that may arise from the use of wood each year (from cutting the trees) as a source of energy and the economic benefits it may provide to compensate for the relatively low economic benefits from biodiesel. The energy

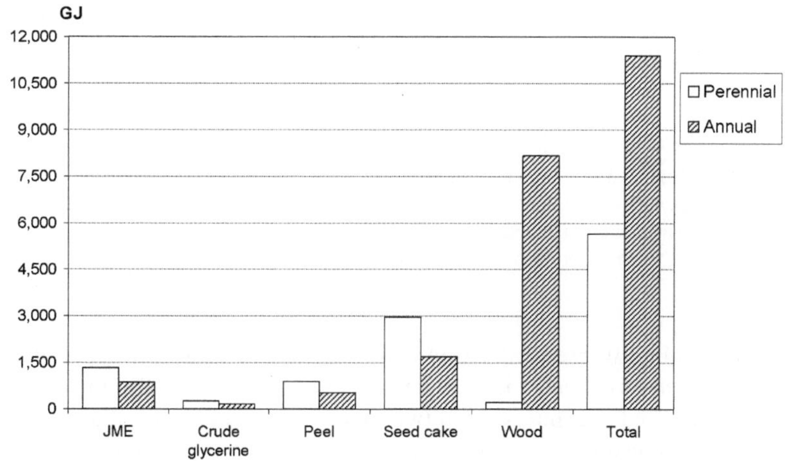

Fig. 30.3 Energy outputs for ratoon-annual and perennial plantation systems

assessment reveals that the total energy output from such a plantation system is about 11,500 GJ per ha over 20 years. This is twice the output achieved from a perennial plantation with wood contributing about 70% of the life cycle energy output. So although ratoon plantations are characterized by lower fruit yields and therefore lower biodiesel, glycerin, peel and seed cake than perennial plantations, wood enables by far to outweigh such reduced energy benefits. However, it is important to note that such assessment is based on a pilot study from Thailand. Hence its applicability under real conditions remains to be confirmed.

The energy assessment of Jatropha plantation illustrated for the case of Thailand demonstrates that Jatropha holds potential as an energy crop for as long as the focus is not only on large scale production of biodiesel. Rather, optimization of the use of all co-products generated along the life cycle of Jatropha production for oil and biodiesel should be considered. This is also confirmed by other studies highlighting that Jatropha biodiesel is unlikely the most profitable option for Jatropha plantations, but rather a sub-optimal option (Openshaw 2000; Achten et al. 2008).

Comparison with Other Energy Crops

For comparative purposes, focusing on NER and NEB, two other feedstocks with potential for biodiesel production are considered here, i.e., oil palm (main feedstock for commercial production of biodiesel in Thailand) and coconut. For both feedstocks, the phases considered in the energy assessment include, farming, oil

Table 30.3 NER and NEB of potential energy feedstocks for biodiesel production

Feedstock	NER	NEB[a] (GJ/ha)
Coconut Methyl Ester	4.16[b]; 4.36[c]	67
Palm Methyl Ester	2.42[b]; 3.58[c]	101
Jatropha Methyl Ester (perennial)	1.42[b]; 6.03[c]	236

Based on Pleanjai and Gheewala (2009)
[a]NEB values with co-products included in the energy assessment
[b]NER values when co-products are not included in the energy assessment
[c]NER values when co-products are included in the energy assessment

extraction, refining, transesterification and intermediate transport. NER and NEB assessment for all three feedstocks on an annual basis (Table 30.3) indicate that the NER of oil palm and coconut is about two to three time larger than that of Jatropha. However, when considering co-products, based on the assumptions that adequate infrastructure for transport and utilization of all the biomass are available, Jatropha shows larger potential energy benefits than the other two feedstocks. This confirms the potential assets of Jatropha as an energy crop provided co-product utilization is promoted.

Key Considerations for Jatropha Promotion as an Energy Crop

In many developing countries worldwide, Jatropha has attracted attention over the past decade as a potential crop for large scale production of biodiesel. This expectation is partly due to the reported ability of the crop to thrive in poor conditions including marginal land. Such characteristics were anticipated to provide attractive energy, economic and environmental benefits. However, this vision failed to recognize the implication such conditions may have on the feedstock productivity and therefore on seed and oil yields.

To maximize the benefits of Jatropha for biodiesel production, it is essential to optimize the yield of the crop. In addition, other applications including the utilization (1) of seedcake for animal feed, fertilizer or fuel production and (2) of peel and wood as a source of energy are also essential. This can be achieved via proper agricultural practices including cultivation on adequate land with appropriate inputs of water and nutrients among others. However, this is to be done ensuring that no additional environmental risks are generated including (1) loss of biodiversity, (2) competition for water resources and fertilizers and (3) degradation of soil of surrounding ecosystems. In this regard, Jatropha cultivation should not result in deforestation for instance as it may negate the benefits due to the damage thereof. Also it should not be encroaching on private properties, protected areas or even marginal land, which some communities may rely on as it may disrupt their livelihoods.

Jatropha is a potential energy crop showing positive NER and NEB results and therefore possible environmental and economic gains. This is partly why several developing countries are assessing large scale cultivation of the crop for biodiesel production. In China, the Yunnan province is witnessing the construction of refineries with annual capacity of 100,000 t biodiesel for which commercial Jatropha plantations are also expected to be developed. In Lao PDR, there is also a plan to plant 2 million ha of biofuels crops by 2020, Jatropha being one of the feedstock considered with already 50,000 ha planted in 2007. India and Vietnam are conducting small-scale experiments to optimize the yields of crop and oil. However, large scale development plans require appropriate policies, infrastructure and management plans, which are still lacking in developing countries (Sukkasi et al. 2010).

It appears therefore that Jatropha is of particular relevance for community based cultivation and use. At such scale, with minimum requirements for infrastructure, the crop may contribute to provide farmers with additional sources of energy; first from the oil because of its properties similar to that of diesel that make it directly usable in agricultural machineries, such as the commonly found one piston engines that are easy to maintain and repair, but also from the co-products of oil processing including peel, wood and seedcake. This can contribute to diversify the sources of energy that are available to small communities and render them accessible and affordable. In Lao PDR for instance, a small holder Jatropha program was reported to enhance the welfare of farmers by raising their net income by 75% (Sukkasi et al. 2010). In Thailand, in 2004, the department of agricultural extension considered to promote small scale cultivation of Jatropha in unused areas including fences around farms, temples, road-side, canals and swamps. The purpose of this plan was to promote the production and self use of Jatropha oil as fuel for agricultural engines based on the sufficiency economy's concept initiated by His Majesty the King of Thailand. From an initial 17 sites, the scheme expanded within 2 years to 257 sites with 12,850 families involved and benefiting from the initiative (Ladawan Na Ayudhaya 2009).

In light of the discussion above, promotion of energy crops should not focus only on the end-product (biofuel), but also on how to maximize the overall benefits that can be obtained from the crop so as to contribute to sustainable development (Gheewala 2011). The crops should be promoted so as to contribute to economic progress, social development and environmental protection with particular emphasis on (1) human welfare, (2) poverty reduction, (3) access to affordable and diversified energy sources, (4) environmental protection with particular focus on reduction of GHG emissions, (5) climate change mitigation and adaptation and (6) biodiversity preservation. Small-scale cultivation of energy crops particularly for countries that are in the process of development and rich in biomass resources, e.g., China, India, Lao PDR or Thailand, may contribute to bringing socio-economic benefits to the communities concerned by providing them with the basic needs for energy and contribute with good planning to sustainable development (Sukkasi et al. 2010). The development of cooperatives is one possible option for small scale production of Jatropha. Such type of organization would enable to provide farmers with safer welfare and strengthen their negotiation power. Under this system, farmers could grow their area of crop energy and own a part of the capital of the cooperative that would process biodiesel and eventually own agriculture equipment that would remain for the benefit of all cooperators.

Acknowledgments The authors would like to acknowledge Dr. Nicolas Carels, Dr. Savitri Garivait, Ms. Narumon Ladawan Na Ayudhaya and Mr. Kritana Prueksakorn for their kind inputs. The funding from the Joint Graduate School of Energy and Environment is also acknowledged.

References

Achten WMJ, Verchot L, Franken YJ, Mathijs E, Singh VP, Aerts R et al (2008) Jatropha biodiesel production and use. Biomass Bioenergy 32:1063–1084

DEDE (2008) The 15 years renewable energy development plan (2008–2022). Department of Alternative Energy Development and Efficiency, Ministry of Energy, Thailand. Available from www.dede.go.th

EPPO (2009) Enery strategy. Energy Policy and Planning Office, Ministry of Energy, Thailand. Available from www.eppo.go.th

Foidl N, Foidl G, Sanchez M, Mittelbach M, Hackel S (1996) *Jatropha curcas* L. as a source for the production of biofuel in Nicaragua. Bioresour Technol 58:77–82

Gheewala SH (2011) Life cycle assessment (LCA) to evaluate environmental impacts of bioenergy projects. J Sust Energy Environ 2:35–38

IEA (2011) Technology roadmap – biofuels for transport. International Energy Agency, Paris. Available from www.iea.org

IFEU (2009) Screening life cycle assessment of Jatropha Bodiesel. Available from www.ifeu.de/landwirtschaft/pdf/jatropha_report_111207.pdf

Jongschaap REE, Corré WJ, Bindraban PS, Brandenburg WA (2007) Claims and facts on *J. curcas* L.: Global *Jatropha curcas* evaluation, breeding and propagation programme. Plant Research International, Wageningen, The Netherlands

Ladawan Na Ayudhaya N (2009) Potential of *Jatropha curcas* derived biodiesel for rice farmers. Master thesis, The Joint Graduate School of Energy and Environment, Bangkok

Li Z, Lin B-I, Zhao X, Sagisaka M, Shibazaki R (2010) System approach for evaluating the potential yield and plantation of *Jatropha curcas* L. on a global scale. Environ Sci Technol 44:2204–2209

Nguyen TLT, Gheewala SH, Garivait S (2007) Full chain energy analysis of fuel ethanol from cassava in Thailand. Environ Sci Technol 41:4135–4142

Openshaw K (2000) A review of *Jatropha curcas*: an oil plant of unfulfilled promise. Biomass Bioenergy 19:1–15

Pleanjai S, Gheewala SH (2009) Full chain energy analysis of biodiesel production from palm oil in Thailand. Appl Energy 86:S209–S214

Prueksakorn K, Gheewala SH (2008) Full chain energy analysis of biodiesel from *Jatropha curcas* L. in Thailand. Environ Sci Technol 42:3388–3393

Prueksakorn K, Gheewala SH, Malakul P, Bonnet S (2010) Energy analysis of Jatropha plantation systems for biodiesel production in Thailand. Energy Sustain Dev 14:1–5

Shapouri H, Wang M, Duffield JA (2006) Net energy balancing and fuel-cycle analysis. In: Dewulf J, Van Langenhove H (eds) Renewables-based technology – sustainability assessment. Wiley, Chichester

Silalertruksa T (2011) Sustainability assessment of biofuels for transport in Thailand. Ph.D. thesis, The Joint Graduate School of Energy and Environment, Thailand

Sukkasi S, Chollacoop N, Ellis W, Grimley S, Jai-In S (2010) Challenges and considerations for planning toward sustainable biodiesel development in developing countries: lessons from the Greater Mekong Subregion. Renew Sustain Energ Rev 14:3100–3107

Uriarte FA (2010) Biofuels from plant oils. ASEAN Foundation, Jakarta

USAID (2009) Biofuels in Asia – an analysis of sustainability options. USAID – Asia report prepared under the ECO-Asia clean development and climate program. Contract No. EPP-1-100-03-00013-00. Task Order 9

Index

N. Carels et al. (eds.), *Jatropha, Challenges for a New Energy Crop: Volume 1: Farming, Economics and Biofuel*, DOI 10.1007/978-1-4614-4806-8,

Q

R

S

MIX
Papier aus verantwortungsvollen Quellen
Paper from responsible sources
FSC® C105338

If you have any concerns about our products,
you can contact us on
ProductSafety@springernature.com

In case Publisher is established outside the EU,
the EU authorized representative is:
Springer Nature Customer Service Center GmbH
Europaplatz 3, 69115 Heidelberg, Germany

Printed by Libri Plureos GmbH
in Hamburg, Germany